Conversion Factors (continued)

Specific Energy	1 Btu/lbm	= 2.326 kJ/kg	1 kJ/kg	= 0.4299 Btu/lbm
	1 Btu/lbmol	= 2.326 kJ/kgmol	1 kJ/kgmol	= 0.4299 Btu/lbmol
	1 cal/gmol	= 1.8 Btu/lbmol		

Specific Entropy, Specific Heat, Gas Constant

1 Btu/lbm $\cdot R \equiv 4.1868$ kJ/kg $\cdot$ K 1 kJ/kg $\cdot$ K $= 0.238846$ Btu/lbm $\cdot$ R

1 Btu/lbmol $\cdot$ R $\equiv 4.1868$ kJ/kgmol $\cdot$ K

1 cal/gmol $\cdot$ K $\equiv 1$ Btu/lbmol $\cdot$ R

Specific Volume	1 ft^3/lbm	= 0.062428 m^3/kg	1 m^3/kg	= 16.018 ft^3/lbm

* See also E. A. Mechtly, *The International System of Units*, National Aeronautics and Space Administration

Physical Constants*

Avogadro's number	$N_A = 6.0222 \times 10^{26}$/kg mol
Boltzmann constant	$k = 1.3806 \times 10^{-23}$ J/K
Universal gas constant	$\bar{R} = 8314.3$ J/kg mol-K
	$= 1545.3$ ft-lbf/lb mol-°R
	$= 1.986$ Btu/lb mol-°R
	$= 1.986$ cal/g mol-K
	$= 82.057$ cm^3-atm/g mol-K
Speed of light	$c = 2.9979 \times 10^8$ m/s
Planck constant	$h = 6.6262 \times 10^{-34}$ J-s
Stefan–Boltzmann constant	$\sigma = 5.6696 \times 10^{-8}$ W/m^2-(K)^4
Faraday constant	$F = 9.6487 \times 10^7$ C/kg mol
Electron charge	$e = 1.6022 \times 10^{-19}$ C
Electron rest mass	$m_e = 9.1096 \times 10^{-31}$ kg
Gravitational constant	$G = 6.6732 \times 10^{-11}$ N-m^2/kg^2

* SOURCE: E. A. Mechtly, *The International System of Units*, National Aeronautics and Space Administration Report No. SP-7012, 1969.

ENGINEERING THERMODYNAMICS

Fundamentals and Applications

ENGINEERING THERMODYNAMICS

Fundamentals and Applications

Second Edition

Francis F. Huang

Department of Mechanical Engineering
San Jose State University

Macmillan Publishing Company
New York

Collier Macmillan Publishers
London

Copyright © 1988, Macmillan Publishing Company, a division of Macmillan, Inc.

Printed in the United States of America.

Earlier edition copyright © 1976 by Francis F. Huang

Macmillan Publishing Company
866 Third Avenue, New York, New York 10022

Collier Macmillan Canada, Inc.

Library of Congress Cataloging in Publication Data

Huang, Francis F.
   Engineering thermodynamics: fundamentals and applications
   Francis F. Huang. –– 2nd ed.
      p.   cm.
      Includes index.
      ISBN 0-02-357391-0
      1. Thermodynamics.   I. Title.
   TJ265.H78 1988
   621.402'1––dc 19

                                                          87-16832
                                                          CIP

Printing: 1 2 3 4 5 6 7 8   Year: 8 9 0 1 2 3 4 5 6 7

Preface to the Second Edition

In the intervening years since the first edition of this book appear in 1976, I have benefited enormously from the many suggestions, comments, and encouragement that I have received from my peers, my colleagues, and our students. The book has been extensively rewritten, essentially based on their input. The purpose of the book, however, remains substantially unchanged.

This new edition, as with the first edition, is written as a text for the first course in thermodynamics taken by students in all branches of engineering. The philosophy behind the design of this book may be summarized as follows:

1. Let the readers look at the "big picture" of the subject matter right from the beginning. It is my belief and experience that students can increase their comprehension and retention of information much more effectively if they can seen how the parts fit into the whole.
2. Develop the basic equations (energy accounting equation, entropy accounting equation, conservation of mass equation, and the property equation) in a concise and clear manner before the entire package is being put to use. This will enhance the awareness of the students that the package contains essentially all the necessary tools in engineering thermodynamics.
3. Make use of real-world applications to illustrate the power of engineering thermodynamics and to give the students repeated exposure to the general methodology of problem solving.
4. Place a strong emphasis on the rudiments of design, and frequently remind the students that a real-world design decision always involves trade-offs. Indeed, life itself is full of compromises.

The major changes that have been made in this new edition are given below.

1. A comprehensive introductory chapter is used to accomplish the following:

    (a) Present the basic concepts and definitions in a coherent manner.

    (b) Expose the reader to several different engineering systems to illustrate the usefulness of engineering thermodynamics and to acquaint the reader with the different kinds of hardware used in the design of these systems.

    (c) Show the reader that the basic structure of engineering thermodynamics is essentially relatively simple.

    (d) Introduce the reader to the general methodology of problem solving.

2. A chapter on availability, irreversibility, and availability analysis has been added. Exergy, the concept central to second-law analysis, is introduced in an intuitively acceptable manner. The exergy accounting equation, which is simply the combined first and second law of thermodynamics, is formulated in a manner parallel to that for the energy accounting equation and the entropy accounting equation.

3. The SI system of units has been adopted. But engineers in English-speaking countries must be familiar with both SI and English units for many years to come. Consequently, example problems are almost evenly divided in the use of the two systems of units.

4. The number of problems at the end of each chapter has been increased significantly. In addition, there is also greater variety and complexity in many of these problems.

5. The sections on gas turbine power plant design have been expanded to reflect the versatility of this prime mover in contemporary applications.

6. A fairly comprehensive section on cogeneration systems has been added to reflect the importance of this concept in energy conservation.

7. Many problems for which the use of a computer is either required or recommended have been added. These problems have been selected primarily on the basis that they serve to enhance the reader's understanding of engineering thermodynamics.

8. Thermodynamic data in SI units have been added.

    The sum of these changes and additions should make this new edition a more effective and up-to-date learning text for our students. It is my sincere hope that the reader, after using this text, will have acquired not only a working knowledge of engineering thermodynamics, but also the motivation for continued formal or self-education in this important and fascinating subject.

San Jose, California   **Francis F. Huang**

Acknowledgments for the Second Edition

I am indebted to a great many individuals who have either directly or indirectly made an important contribution to the preparation of this edition. Only a few will be mentioned by name here but all have a special place in my heart.

I first wish to thank Dr. Robert F. Clothier, my colleague at San Jose State University for almost thirty years, for his sustaining support and encouragement over the years. I will always treasure the stimulating discussions we have on pedagogical techniques in general and on the substance of thermodynamics in particular.

Next I wish to thank all the users of the first edition for their suggestions and critical comments. As I have mentioned in the preface to this edition, this book has been extensively rewritten essentially based on their input.

I also wish to thank the following reviewers who have read the entire manuscript: Professor Robert F. Clothier of San Jose State University, Professor Leonard M. Goldman of the University of Rochester, Professor John Hester of California State University, Sacramento, Professor John Sarris of the University of New Haven, and Professor Bedru Yimer of the University of Kansas. Their perceptive comments are gratefully acknowledged.

For an opportunity to attend the 1981 Gordon Research Conference on Thermodynamic Analysis, I am grateful to Professor K. S. Spiegler and Professor E. P. Gyftopoulos. It was a most exhilarating experience for me to be able to interact with and to learn from so many outstanding engineers and scientists on the topic of second-law analysis.

For their support for a 1985 General Electric Faculty Fellowship, I wish to thank my colleagues and Dr. Jay D. Pinson, Dean of Engineering, at San Jose State University. This timely award made it possible for me to translate my interest in second-law analysis into some tangible results.

For his assistance in the selection and program preparation of computer problems, I am indebted to Tim Naumowicz, my research assistant. I greatly appreciate his perspective on the use of computers in the teaching and learning of engineering thermodynamics from a young engineer's point of view.

Finally, I owe a special debt of gratitude to my wife, Yuen, for her love, patience and understanding. Together with our children, Raymond, Stanley and Helen, she provided the invaluable support, good company and diversions that made the writing of this edition a most pleasant adventure.

Francis F. Huang

Preface to
the First Edition

This book is written as a text for the first course in thermodynamics taken by students in all branches of engineering. It is based on an approach that I have developed in a period of about ten years. It represents my attempt to teach the science of thermodynamics spiced with a strong engineering flavor. It differs from the usual thermodynamics texts essentially in two aspects.

First, it is my belief that students will have a better understanding of the power and beauty of thermodynamics if they can see from the beginning how the parts fit into the total picture. To this end, I have in [Chapters 1–6], using the postulate approach and the macroscopic point of view on matter, formulated the entire structure of engineering thermodynamics. This entire structure, including the energy equation, the entropy equation, and the property equation, is then repeatedly applied in the rest to the book to processes, devices, and systems involving different substances. With this approach, I hope to enhance the student's awareness of the fact that effective utilization of our energy and other natural resources in engineering design must always involve two of the most important laws of nature, namely, the first law and the second law of thermodynamics.

This book differs also from most texts on thermodynamics in the amount of emphasis it places on the rudiments of engineering design. I believe that it is never too early to introduce engineering students to the idea of design, which may be defined as a decision-making process requiring broad, comprehensive knowledge (technical and otherwise), practical experience, and, frequently, creative imagination. I believe that engineering students should be reminded frequently that a real-world design decision always is arrived at only after compromises (trade-offs) have been made.

My presentation of the fundamentals is dictated by an attempt to relate important concepts as much as possible to what the students already know or can intuitively accept. It is sufficiently general and rigorous. I believe that the coverage of material is comprehensive enough so that this book should be suitable not only for students taking thermodynamics as a terminal course, but also for those contemplating further study of this fascinating subject.

In deducing the second law of thermodynamics, I have taken advantage of the idea of Hatsopoulos and Keenan by defining the concept of entropy in terms of the concept of unavailable energy of a system. By giving entropy this physical interpretation right from the beginning, it is hoped that much of the mystery surrounding the concept of entropy is dispersed.

This book contains enough material for a one-year course. However, it has been structured in such a manner that the instructor has a great deal of flexibility in making suitable selections to meet his or her course objectives. Since I have summarized the entire structure of engineering thermodynamics in Chapter 7, one might want to rush through the first seven chapters so that one can get to the applications as soon as possible, and then use the applications as the apparatus to reinforce the student's understanding of the basic principles.

It was my original plan to experiment more with my approach before putting it into print. However, knowing my own limitations, I have decided to put this approach forward now so that it might attract the attention of persons who are more capable of improving upon it. Paraphrasing a Chinese saying, I hope that by throwing out a brick I can attract a piece of jade.

F. F. H.

Acknowledgments to the First Edition

It is impossible to list all the persons and events that have contributed to the creation of this book. I do wish to acknowledge with gratitude the following.

First, I wish to acknowledge my indebtedness to all authors on whose work I have drawn. I have benefited enormously from studying many outstanding books on thermodynamics. In particular, the books by H. B. Callen, J. H. Keenan, G. N. Hatsopoulos, M. Tribus, M. W. Zemansky, and E. A. Guggenheim have exterted a great influence in molding my understanding of the subject.

For their critical comments and helpful suggestions, I am grateful to Dr. Robert F. Clothier of San Jose State University, Dr. David Mage of Environmental Protection Agency, Glennon Maples of Auburn University, Arthur R. Foster of Northeastern University, John F. Bridge of Ohio State University, and Clifford A. Wojan of the Polytechnic Institute of New York. I am particularly grateful to Dr. Clothier, as many of my ideas were developed with his encouragement and support during the years when he was the Chairman of the Department of Mechanical Engineering at San Jose State University.

Two years of Science Faculty Fellowship from the National Science Foundation contributed heavily to my intellectual growth. A sabbatical leave from San Jose State University gave me the opportunity to obtain valuable input from the academic community and the industrial world.

I wish to express my appreciation to Dr. H. D. Baker, Professor Emeritus of Columbia University, from whom I received guidance in research and from whom I developed my appreciation of the importance of experimental as well as analytical studies of thermodynamics.

I also wish to thank Mrs. Lois Bowman and Mrs. Gertrude Brewster for their patience and skill in editing and typing the manuscript; the many students on whom the material was tried over the years; Dr. Philip M. Blair of San Jose State University for being my "sounding board" during the years we shared an office; and my secretary, Mary C. Bowers, for her assistance in taking care of the many small but necessary details.

Finally, I wish to thank my wife, Yuen. Without her patience and encouragement, the completion of this book would not have been possible. To her this book is dedicated.

Contents

6.   *Some Consequences of the Second Law of Thermodynamics with an Introduction to Multicomponent Systems*   **213**

13. *Thermodynamics of a Simple Compressible Substance* **635**

List of Symbols

A	Area	F_f	Force due to sidewall friction
a	Acceleration	$\mathscr{F}$	Generalized force
a, A	Specific Helmholtz function and total Helmholt function	g	Acceleration due to gravity
AF	Air-fuel ratio	g_c	Constant that relates force, mass, length, and time in Newton's second law of motion
B, C, D	Second, third, and fourth virial coefficients in the density series of the virial equation of state	g, G	Specific Gibbs function and total Gibbs function
B', C', D'	Second, third, and fourth virial coefficients in the pressure series of the virial equation of state	G_p	Gibbs function of products of chemical reaction
B_f	Exergy content in fuel	G_R	Gibbs function of reactants
B_p	Exergy content in process heat	ΔG_r	Gibbs function change for a complete unit reaction
C	Number of components	h, H	Specific enthalpy and total enthalpy
C	Arbitrary constant	H_p	Total enthalpy of products of chemical reaction
c	Velocity of sound		
c_p	Constant-pressure specific heat		
c_v	Constant-volume specific heat	H_R	Total enthalpy of reactants
e, E	Specific energy and total energy	ΔH_r	Enthalpy change for a complete unit reaction
E_{AV}	Available energy	$\mathscr{H}$	External magnetic field
E_{UA}	Unavailable energy	HHV	Higher heating value of a fuel
E_{kin}	Kinetic energy		
E_{pot}	Potential energy	I	Irreversibility
$\mathscr{E}$	Electromotive force	i	Electric current
f	Fugacity	k	Boltzmann constant
f	Fanning friction factor	k	Ratio of specific heats, c_p/c_v
F	Degree of freedom$F$		
F	Force		
F_p	Force due to fluid pressure		

K_p	Equilibrium constant for ideal gas reaction	$\dot{Q}$	Rate of heat transfer
L	Length	Q_p	Process heat
lbf	Pound force	Q_{rev}	Heat transfer for a reversible process
lbm	Pound mass	Q_{irr}	Heat transfer for an irreversible process
lb mol	Pound mole		
LHV	Lower heating value of a fuel	R	Gas constant for a particular gas, $\bar{R}/M$
lw, LW	Lost work per unit mass and total lost work	$\bar{R}$	Universal gas constant
LW_0	Thermodynamic lost work	R_{pH}	Power-to-heat ratio
m	Mass	r_c	Cut-off ratio
$\dot{m}$	Mass rate of flow	r_p	Pressure ratio
M	Mach number	r_v	Compression or expansion ratio
M	Molecular weight		
mep	Mean effective pressure	$(r_p)_{opt}$	Optimum pressure ratio
n or N	Number of moles	s, S	Specific entropy and total entropy
n	Polytropic exponent		
p	Pressure	S_{gen}	Entropy generation
P	Number of phases	S_p	Total entropy of products of chemical reaction
p_C	Pressure at critical point		
p_R	Reduced pressure, p/p_C	S_R	Total entropy of reactants
p_r	Relative pressure as used in the gas tables	ΔS_r	Entropy change for a complete unit reaction
p_i	Partial pressure of the ith component in a mixture	T	Absolute temperature
p_a	Partial pressure of dry air in an air–water–vapor mixture	T_C	Temperature at critical point
		T_R	Reduced temperature, T/T_C
p_v	Partial pressure of water vapor in an air–water–vapor mixture	T^A	Temperature of system A
		t_d	Dry-bulb temperature
		t_w	Wet-bulb temperature
		u, U	Specific and total internal energy
p_g	Saturation pressure of water vapor corresponding to temperature of an air–water–vapor mixture	v, V	Specific volume and total volume
		V_i	Partial volume of ith component in a mixture
		v_C	Specific volume at critical point
p_{gw}	Saturation pressure of water vapor corresponding to wet-bulb temperature of an air–water–vapor mixture	v_r	Relative volume as used in the gas tables
		v_R	Reduced volume, v/v_C
		$v_{r'}$	Pseudo-reduced volume
		$\bar{V}$	Velocity
q, Q	Heat transfer per unit mass and total heat transfer	w, W	Work transfer per unit mass and total work transfer

$\dot{W}$	Rate of work transfer or power	η_N	Nozzle efficiency
W_{ad}	Work transfer for an adiabatic process	η_{reg}	Regenerator effectiveness
		θ	Empirical ideal-gas temperature
W_{rev}	Work transfer for a reversible process	κ	Isothermal compressibility
W_{irr}	Work transfer for an irreversible process	μ	Chemical potential
x	Mole fraction	μ	Magnetization
x	Quality of a two-phase mixture	μ_{JT}	Joule–Thomson coefficient
X	Generalized displacement	ν	Stoichiometric coefficient
y	Moisture content of a two-phase mixture	π	Osmotic pressure
		ρ	Densityρ
Y	Young's isothermal modulus of elasticity	σ	Stefan-Boltzmann constant
Y	Liquid yield	τ	Tension
z	Compressibility factor	τ	Time
z_C	Compressibility at the critical point	ϕ	Function of temperature for an ideal gas,
Z	Elevation		
Z	Quantity of electric charge		$$\phi \equiv \int_{T_0}^{T} c_p \frac{dT}{T}$$

Greek Letters

			as defined in the gas tables
α	Coefficient of linear expansion	ϕ	Relative humidity
α	Isentropic compressibility	ϕ, Φ	Specific and total availability of a closed system (non-flow exergy)
β	Coefficient of expansion		
β_R	Coefficient of performance of a refrigerator		
		ψ, Ψ	Specific and total flow availability of a steady-state steady-flow process (flow exergy)
β_{HP}	Coefficient of performance of a heat pump		
		Ω	Thermodynamic probability
ε	Reaction variable	ω	Specific humidity
ε_f	Exergy factor of fuel		
ε_p	Exergy factor of process heat	**Subscripts**	
η_f	Fuel utilization efficiency	cv	Control volume
η_{II}	Second-law efficiency	f	Formation
η_{th}	Thermal efficiency	f	Property of saturated liquid
η_c	Adiabatic compressor efficiency	fg	Difference in property for saturated vapor and saturated liquid
η_p	Adiabatic pump efficiency		
η_T	Adiabatic turbine efficiency		
		g	Property of saturated vapor
η_D	Diffuser efficiency		

i	Property of ith component in a multicomponent system	$\bar{d}$	Infinitesimal change in a path function
i	Property of saturated solid	$\equiv$	Identity symbol, used when the equation defines a quantity
if	Difference in property for saturated liquid and saturated solid	Δ	Finite change in a property, Δ = final − initial
ig	Difference in property for saturated vapor and saturated solid	$\left(\dfrac{\partial x}{\partial y}\right)_z$	Partial derivative of x with respect to y, keeping z constant
0	Stagnation property		
0	Property at reference state		

Superscripts

$^{-}$	Bar over symbol denotes property on a molal basis	$\oint$	Integration around a cycle
*	Ideal gas state	Q_H	Heat transfer between system and high-temperature body
*	Property at the throat of a nozzle	Q_L	Heat transfer between system and low-temperature body
0	Property at standard-state condition	Q_{12}	Amount of heat transfer corresponding to a change from state 1 to state 2
HR	Heat reservoir		
WR	Work reservoir		

Special Notations

d	Infinitesimal change in a property	W_{12}	Amount of work transfer corresponding to a change from state 1 to state 2

To the Student

You are about to begin the study of one of the most general and fascinating subjects in science—thermodynamics, the science that deals with energy and its relation to matter. Thermodynamics is most general because every technological system involves the utilization of energy and matter. In fact, since engineering thermodynamics plays such a vital role in the design of so many processes, devices, and systems for use in the home as well as in industry, it is no exaggeration to say that it touches our daily life. Engineering thermodynamics is also very important in our search for solutions to problems in connection with energy crises, shortages of fresh water, air pollution, and garbage disposal in the cities. In short, engineering thermodynamics is most relevant in our search for a better quality of life.

Thermodynamics is most fascinating because so many varied results have been derived from so little. Consequently, it is a subject that has attracted a variety of minds, from engineers to biologists to mathematicians. The entire structure of thermodynamic theory is essentially very simple. In fact, it has been pointed out by a panel of distinguished contemporary thermodynamicists that the towering structure of literally hundreds of mathematical relations in thermodynamics has at its foundations just three basic equations— energy, entropy, and properties.

This book has two main parts. In Chapters 1–6, the entire structure of thermodynamics, including the development of the three basic equations, is presented. In Chapters 7–16, the entire structure of thermodynamics is then applied to a wide range of engineering problems.

Although the structure of thermodynamics is simple, it is also quite subtle. Many of the ideas and concepts employed in constructing the theory are rather abstract and formal. As a consequence, the material contained in Chapters 1–6 certainly will not be fully understood at once. But it will take on real meaning and significance as it is being applied over and over during the study of Chapters 7–16. By gaining confidence based on understanding the entire structure of thermodynamics, you will have at your disposal a powerful tool for use in the design of engineering processes, devices, and systems.

Francis F. Huang

ENGINEERING THERMODYNAMICS

Fundamentals and Applications

PART I: Fundamentals

CHAPTER 1

Basic Concepts and Definitions

Thermodynamics, like other physical sciences, is based on observations of nature. It is a discipline that involves the formulation of a number of primitive and intuitive concepts derived from common experience. As such, it employs a number of terms, such as energy, equilibrium, property, system, process, work, and heat, which are used in everyday language but have very precise definitions in thermodynamics. Although all these terms may appear somewhat abstract and formal at the early stage of discussion, they will begin to assume reality and significance as we start applying them in the solution of practical problems.

Since understanding and retention of new or unfamiliar material would be increased if we first look at the "big picture" of the subject matter, we have presented in this chapter a snapshot of the entire structure of engineering thermodynamics, showing how the parts fit into the whole. We will see that this "package" will be used repeatedly in later chapters. Careful study of this chapter is therefore essential in order to apply the methodology of engineering thermodynamics effectively.

1.1 What Is Engineering?

Engineering is a profession. More important, it is generally considered to be a creative profession. In fact, much of the history of civilization may be recognized as the history of engineering. Consequently, it is difficult to define engineering in a few words. One definition, set forth by the Engineers' Council for Professional Development (now the Accreditation Board for Engineering and Technology), is as follows:

Engineering is the profession in which a knowledge of mathematics and natural sciences gained by study, experience and practice is applied

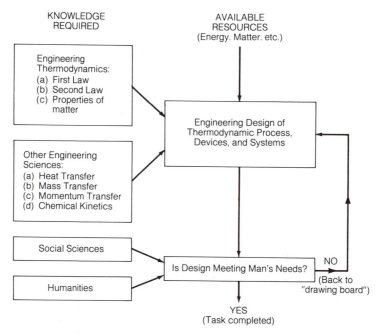

Figure 1.1 Flowchart on Application of Engineering Thermodynamics

with judgment to develop ways to utilize, economically, the materials and forces of nature for the benefit of mankind.[1]

Other definitions of engineering have been given from time to time, but they are all basically the same. In less sophisticated language, engineering may simply be considered to consist of human activities involving effective utilization of our resources to meet human needs.

In order to come up with a solution that is scientifically sound for a given engineering problem, we require knowledge of engineering sciences. However, we must keep in mind that a scientifically sound solution is not necessarily beneficial to humankind. To judge whether or not our solution is good for humankind, we require knowledge of the social sciences and humanities (Fig. 1.1). Because of the nature of this book, our discussions will be devoted almost exclusively to the scientific aspects of engineering. On certain occasions, we shall comment on the social, political, and economic aspects of engineering.

[1] *42nd Annual Report.* Engineers' Council for Professional Development, New York, 1974.

1.2 What Is Thermodynamics?

Historically, the science of thermodynamics was developed to provide a better understanding of devices, known as heat engines, that absorb heat from a high-temperature source and produce useful work. It is not surprising, then, to find the following definition of thermodynamics appearing in many earlier books on thermodynamics: "Thermodynamics is the science that deals with relations between heat and work."

We have come a long way, of course. Thermodynamics is now exceedingly general in its applicability. It is not only important in engineering, it is also important in physics, chemistry, and the life sciences. Consequently, many definitions of thermodynamics are available in the literature. It would be instructive for us to ponder a few of these.

According to Hatsopoulos and Keenan,[2] "thermodynamics is the science of states and changes in state of physical systems and the interaction between systems which may accompany changes in state."

According to Callen,[3] "thermodynamics is the study of the macroscopic consequence of myriads of atomic coordinates which, by virtue of the statistical averaging, do not appear explicitly in a macroscopic description of a system."

According to Epstein,[4] "thermodynamics deals with systems which, in addition to mechanical and electromagnetic parameters, are described by a specifically thermal one, namely, the temperature or some equivalent of it. Thermodynamics is essentially a science about the conditions of equilibrium of systems and about the processes which can go on in states little different from the state of equilibrium."

According to Kestin,[5] "the science of thermodynamics is a branch of physics. It describes natural processes in which changes in temperature play an important part. Such processes involve the transformation of energy from one form into another. Consequently, thermodynamics deals with the laws which govern such transformation of energy."

---

[2] G. N. Hatsopoulos and J. H. Keenan, *Principles of General Thermodynamics*, John Wiley & Sons, Inc., New York, 1965.

[3] H. B. Callen, *Thermodynamics*, John Wiley & Sons, Inc., New York, 1960.

[4] P. S. Epstein, *Textbook of Thermodynamics*, John Wiley & Sons, Inc., New York, 1937.

[5] Joseph Kestin, *A Course in Thermodynamics*, Blaisdell Publishing Co., Waltham, Mass., 1966.

According to Van Wylen and Sonntag,[6] "one very excellent definition of thermodynamics is that it is the science of energy and entropy."

In this book we consider *thermodynamics* simply as the science that deals with energy, matter, and the laws governing their interactions.

We observe that each definition contains seemingly familiar terms. Actually, all these terms must in turn be defined before we can fully appreciate the significance of each definition. There are, in fact, different means that we can employ to develop the structure of thermodynamics, depending on the way we define some of the key words in the definitions above. For example, thermodynamics may be developed in a highly sophisticated manner by making use of information theory,[7] or it may be treated in an exceedingly pragmatic way. The approach to thermodynamics in this book is dictated by an attempt to relate important concepts as much as possible to what the reader already knows or can intuitively accept. It is essentially a classical one.

The foundation of thermodynamics lies in our observation and correct generalization of the behavior of the real world. The importance of thermodynamics lies in the fact that its logical structure has made it possible for us to predict the behavior of the real world. Einstein was so impressed with the theory of classical thermodynamics that he had this to say:

> A theory is the more impressive the greater the simplicity of its premises is, the more different kinds of things it relates, and the more extended is its area of applicability. Therefore, the deep impression which classical thermodynamics made upon me. It is the only physical theory of universal content concerning which I am convinced that, within the framework of the applicability of its basic concepts, it will never be overthrown.[8]

We should mention, however, that thermodynamics, like other sciences, did not progress tidily. Its logical structure, often taken for granted now, is really the result of the work of many great men of science. Unfortunately, the scope of this book does not permit us to go into the fascinating history of thermodynamics. But some important dates, events, and names in the historical development of thermodynamics are given in Table 1.1.

[6] G. J. Van Wylen and R. E. Sonntag, *Fundamentals of Classical Thermodynamics*, John Wiley & Sons, Inc., New York, 1973.

[7] Myron Tribus, *Thermostatics and Thermodynamics*, Van Nostrand Reinhold Company, Inc., New York, 1961.

[8] P. A. Schlipp, ed., *Albert Einstein: Philosopher–Scientist*, Vol. I, Harper & Row, Publishers, Inc., New York, 1959, p. 33.

Table 1.1 Important Dates, Events, and Names in the Historical Development of Thermodynamics[a]

Date	Event
1798	Count Rumford (Benjamin Thompson) began the quantitative study of the conversion of work into heat by means of his famous cannon-boring experiments.
1799	Sir Humphry Davy studied the conversion of work into heat by means of his ice-rubbing experiments.
1824	Sadi Carnot published his famous thesis "Reflections on the Motive Power of Fire," which includes the new concept of cycle and the principle that the reversible cyclic engine operating between two heat reservoirs depends only on the temperatures of the reservoirs and not on the working substance.
1842	Mayer postulated the principle of conservation of energy.
1847	Helmholtz formulated the principle of conservation of energy, independent of Mayer.
1843–1848	James Prescott Joule laid the experimental foundation of the first law of thermodynamics by performing experiments to establish the equivalence of work and heat. We now honor this great scientist by using J to denote the mechanical equivalent of heat.
1848	Lord Kelvin (William Thomson) defined an absolute temperature scale based on the Carnot cycle.
1850	Rudolf J. Clausius was probably the first to see that there were two basic principles: the first and second laws of thermodynamics. He also introduced the concept of U, which we now call the internal energy.
1865	Clausius stated the first and second laws of thermodynamics in two lines: 1. The energy of the universe is constant. 2. The entropy of the universe tends toward a maximum.
1875	Josiah Willard Gibbs published his monumental work "On the Equilibrium of Heterogeneous Substances," which extends thermodynamics in a general form to heterogeneous systems and chemical reactions. This work includes the important concept of chemical potential.
1897	Max Planck stated the second law of thermodynamics in the following form: "It is impossible to construct an engine which, working in a complete cycle, will produce no effect other than the raising of a weight and the cooling of a heat reservoir."
1909	Carathéodory published his structure of thermodynamics on a new axiomatic basis, which is entirely mathematical in form.

[a] See also E. Mendoza, "A Sketch for a History of Early Thermodynamics," *Physics Today*, Vol. 14, No. 2 (1961), p. 32

1.3 Definition of Engineering Thermodynamics

Having introduced the words engineering and thermodynamics as separate entities, we may now define *engineering thermodynamics* in the following manner:

> Engineering thermodynamics is the subject that deals with the study of the science of thermodynamics and the usefulness of this science in the engineering design of processes, devices, and systems involving the effective utilization of energy and matter for the benefit of mankind.

In this definition are several key words, the meaning of which will become clear in the material to follow. The term *engineering design*, however, deserves some comment at this point. Engineering design may be defined as a decision-making process requiring broad, comprehensive knowledge (technical and otherwise), practical experience, and frequently, creative imagination. There is usually more than one possible design for a given engineering problem. But a feasible design, besides being scientifically sound, must also be compatible with the requirements of reliability, cost, space, and standard of environmental quality. Indeed, the final selection of a design is usually based on a great deal of analysis and synthesis, with consideration of many relevant factors, some of which may be entirely nontechnical in nature.

1.4 Applications of Engineering Thermodynamics

Since engineering thermodynamics plays a vital role in the design of so many devices, processes, and systems for use in the home as well as in industry, it is no exaggeration to say that it touches our daily life. Engineering thermodynamics is in action when we start our car in the morning and when we "raid" the refrigerator at night. Engineering thermodynamics is in action inside the space suit of an astronaut. Engineering thermodynamics is in action in an air-separation plant that produces oxygen for use in the steel-manufacturing industry and for use in sustaining life in the hospital. Engineering thermodynamics will be in action in our search for solutions to problems in connection with energy crises, shortages of fresh water, and garbage disposal in the cities. Engineering thermodynamics is most relevant in our search for a better quality of life. It is indeed a fascinating subject that challenges our imagination and creative ability.

To illustrate the usefulness of engineering thermodynamics, we provide in this section a brief description of several engineering thermodynamic systems. Since many students have had limited exposure to much of the equipment (hardware) and many of the processes used in the creation of such systems, the description of such equipment and processes should make the study of them more relevant and meaningful as they are encountered in the future.

Simple Steam Power System

An industrialized society consumes a large quantity of energy, much of which is in the form of mechanical or electrical energy. It is appropriate to begin our discussion on applications with the description of a simple steam power system of the type shown schematically in Fig. 1.2. The objective of this system is to convert thermal energy (heat) to mechanical or electrical energy, which is much more convenient to use. In Fig. 1.2 water enters the steam generator (a heat exchanger) as high-pressure liquid and leaves as high-pressure and high-temperature steam after receiving thermal energy from the heat source. The steam then expands in the turbine (a work-producing device) to a low-pressure and low-temperature condition, producing useful work in the process. The low-pressure and low-temperature steam leaving the turbine is then condensed in a condenser (a heat exchanger) by giving up thermal energy to a heat sink. The pressure of the condensate is increased in a pump (a work-absorbing device) to the pressure at the inlet of the steam generator so that the entire process may begin all over again.

The fluid we pump around in the system is called the working substance of the system. In the case of a steam power plant, the working substance is water. Depending on the available heat source, the

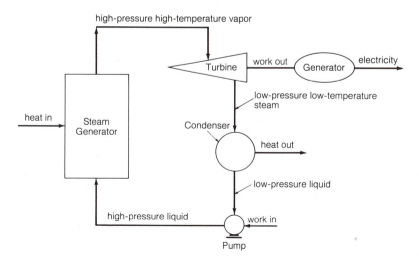

Figure 1.2 Simple Steam Power Plant

working substance could also be ammonia, an organic fluid, or a substance that has not yet been developed. But the basic principles governing the design are the same.

The thermal energy added to the working fluid may come from the burning of a fossil fuel (coal, oil, natural gas) or from a nuclear reaction. It could also come from the sun in the form of thermal radiation. It could even come from the ocean, where an adequate temperature gradient may exist between the surface and a point way below the surface, in which case we would have a system commonly known as OTEC (ocean thermal energy conversion).

We will discover that a large portion of the thermal energy absorbed by the working substance in the steam generator must be rejected in the condenser. This is why large modern steam power plants are usually located near an ocean, lake, or river where a large quantity of cooling water is available. If cooling water is limited, heat may be rejected to the atmospheric air through the use of a cooling tower (also a form of heat exchanger), which is more expensive than a condenser. The design of the system and the equipment involved are all influenced by the heat source, heat sink, and the working substance to be selected.

Gas-Turbine Cogeneration System

Cogeneration is an old engineering concept involving the production of both electrical energy and thermal energy (steam or process heat) in one operation, thereby utilizing fuel more effectively than if the desired products were produced separately. This concept has received a great deal of attention in recent years because of the need to conserve energy. Since the heart of a cogeneration system is a prime mover with waste heat at a usable temperature, it is not surprising that the requirements of cogeneration may be met in many ways. A cogeneration system making use of a gas-turbine power plant as the prime mover is shown in Fig. 1.3.

In Fig. 1.3, the pressure of the atmospheric air is increased in the compressor (a work-absorbing device). The compressed air enters the combustor (a furnace) and reacts with the injected fuel to produce a high-temperature gas (products of combustion). The hot gas then expands in the gas turbine (a work-producing device) to essentially atmospheric pressure, producing useful work, part of which is used to drive the compressor. The gas leaving the gas turbine is still quite hot (around 500°C). It is therefore a good heat source for the production of steam in the heat recovery steam generator (a heat exchanger).

An important design parameter for any cogeneration system is the electrical-to-thermal energy ratio, which is also known as the power-

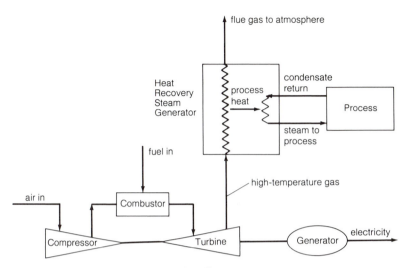

Figure 1.3 Gas-Turbine Cogeneration System

to-heat ratio. Since the economic value of electrical energy is three or four times that of thermal energy, the cost-effectiveness of a cogeneration system is directly related to its power-to-heat ratio. For a given amount of process heat needed, the gas-turbine cogeneration system is capable of producing significantly more power than many other systems. Consequently, combustion gas turbines have been widely used in cogeneration applications for many years.

Mechanical Refrigeration System

Refrigeration may be called the production of "cold." It is simply the process of removing unwanted heat from a cold body so that the cold body will remain cold. For many centuries natural ice and snow provided the only means of refrigeration. Now, different types of refrigeration systems have been developed for a wide range of applications. Refrigeration has provided air conditioning for our homes and offices. Refrigeration has provided the cold storage facilities we need for the preservation of food and other materials. Refrigeration at extremely low temperature has also contributed to significant advances in medical and scientific research. Figure 1.4 is a mechanical refrigeration system involving the use of mechanical components.

In Fig. 1.4, high-pressure liquid enters a throttling valve (a pressure-reducing device) and leaves as a cold mixture of vapor and liquid. This cold mixture, being at a temperature lower than that of the cold body, will absorb the unwanted heat in the evaporator (a heat exchanger), thereby producing refrigeration, which is the objective of the system. The low-pressure and low-temperature vapor

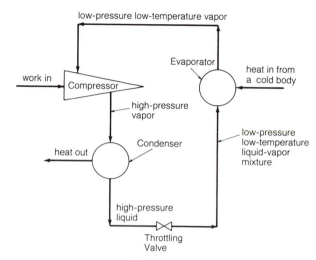

Figure 1.4 Mechanical Refrigeration System

leaving the evaporator is compressed to a higher pressure in the compressor (a work-absorbing device). The high-pressure vapor leaving the compressor is then condensed in the condenser (a heat exchanger) by giving up heat to another body, such as the atmosphere. The high-pressure liquid leaving the condenser is the liquid at the inlet of the throttling valve.

The working substance circulating in a refrigeration system is called the refrigerant. Many refrigerants are available; which is best to use depends on the application. Thus the design and size of equipment involved will vary, but the principle of operation is the same.

Heat Pump System

If we look at a refrigeration system as a whole, we could have considered it as a device to "pump" heat from a low-temperature region to another region at a higher temperature at the expense of work input. We call the system a refrigerator if our objective is to remove heat from a low-temperature region. If our objective is to supply heat to the high-temperature region, we would call the system a heat pump. Consequently, the same principles govern the design and operation of both refrigerators and heat pumps.

Just as we have different types of refrigeration systems, we also have different types of heat pump systems. A heat pump system involving the use of mechanical components is shown in Fig. 1.5. It is known as an air-to-air heat pump because we use the outdoor air to provide the heat input in the evaporator and make use of the room air

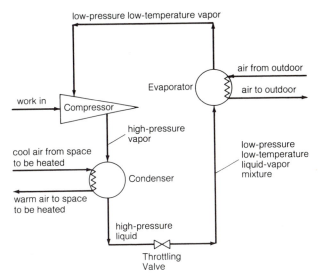

Figure 1.5 Air-to-Air Mechanical Heat Pump System

as the carrier of heat to the space to be heated. This heat pump system is the most common type in use today. We will observe that the equipment involved in this system is identical to that used for the mechanical refrigeration system. This is why a properly designed system could provide air conditioning in the summer and space heating in the winter.

Solar Heating System

Energy for space heating may be supplied in a variety of ways. In a heat pump system, it is obtained at the expense of electrical energy (work) which is a valuable commodity. On the other hand, solar energy is cost free, pollution free, and renewable. However, it is diffuse and intermittent, requiring rather costly solar collectors and storage system. Whether solar energy is economical or not depends on many factors, such as the cost of conventional energy, location, advances in solar technology, and government policy.

In Fig. 1.6 is shown a solar heating system with water storage. It has two water circuits and one air circuit. The main circulation pump forces the water in one circuit through the solar collectors (heat exchangers) to receive the solar energy that will be stored in the water tank. When space heating is needed, the fan (a compressor with very small pressure differential across it) will drive the air through one side of the heat exchanger in which it is heated by the hot water coming from the water storage tank through the use of another pump.

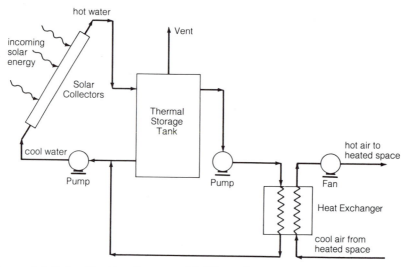

Figure 1.6 Solar Heating System with Water Storage

Reverse-Osmosis Water Desalination System

We need a large quantity of fresh water for drinking, irrigation, and industrial purposes. In many parts of the world, this need is adequately met by nature with rainfall. Unfortunately, serious water shortages exist in some parts of the world, resulting in much human suffering. Fortunately, we have developed different methods to produce fresh water from seawater or brackish water. One promising desalination scheme making use of the principle of reverse osmosis (also known as hyperfiltration) is shown in Fig. 1.7.

The heart of this scheme is the semipermeable membrane inside the reverse-osmosis module that will permit only fresh water to go through. Since energy is needed to separate seawater into fresh water and brine, the pressure of seawater entering the reverse-osmosis module must be greater than the osmotic pressure of the seawater

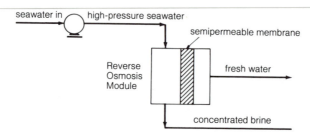

Figure 1.7 Reverse-Osmosis Module

involved. This required pressure is different for different membranes. A commercial unit operates with a pressure of about 55 atm at the inlet. The great appeal of this system is that the only energy needed is the work input at the pump, which turns out to be relatively small. However, reverse-osmosis modules are relatively expensive.

Gas-Liquefaction System

We may liquefy a gas (such as nitrogen) if we first cool it at high pressure and then drop its pressure across a throttling valve. A gas-liquefaction scheme making use of this phenomenon is shown in Fig. 1.8. The high-pressure low-temperature fluid enters the throttling valve as a gas and leaves as a very cold liquid–vapor mixture at essentially atmospheric pressure. The liquid is separated out as the product. The cold vapor is used to cool the high-pressure gas in the heat exchanger. The feed and the recirculation vapor are compressed to the needed high pressure in the compressor, which is usually water jacketed so that the temperature of the gas leaving the compressor is about the same as the temperature of the gas entering the compressor.

Gas-Separation System

We may separate a gaseous mixture (such as air) into its constituents by taking essentially the following steps:

1. Produce a very cold liquid–vapor mixture through the use of a gas-liquefaction system.

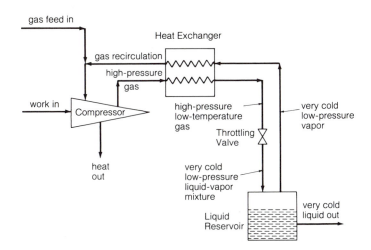

Figure 1.8 Simple Gas-Liquefaction System

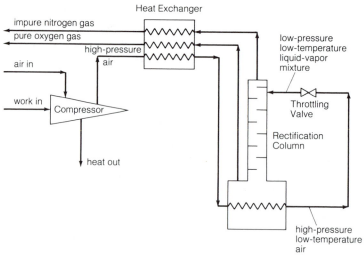

Figure 1.9 Simple Air-Separation System

2. Feed the very cold liquid–vapor mixture into a vessel known as a rectification column (a distillation device) to carry out the separation process.

An air-separation system employing the foregoing principles is shown in Fig. 1.9. As the very cold liquid–vapor mixture cascading down the top of the rectification column, a series of evaporations and condensations will occur. Because nitrogen boils at a lower temperature than does oxygen, more and more nitrogen vapor is formed and more and more oxygen is condensed. Although pure oxygen may be obtained with this scheme, the nitrogen product will have a trace of oxygen because of the thermodynamic behavior of this binary (two-component) mixture.

Fuel Cells

In a fossil-fuel-burning power plant, the conversion of the chemical energy in the fuel to the end product of electrical energy is a multi-step process: from chemical energy to thermal energy to mechanical energy to electrical energy. It would be desirable to convert chemical energy into electrical energy directly. The fuel cell is an electrochemical device that makes this direct energy conversion process possible.

Shown schematically in Fig. 1.10 is a simple hydrogen–oxygen fuel cell, which consists of essentially three parts: the anode, the cathode,

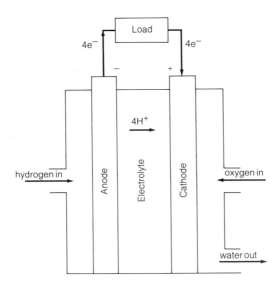

Figure 1.10 Hydrogen–Oxygen Fuel Cell

and the electrolyte. The anode and the cathode are separated by the electrolyte, which could be a base (such as KOH, potassium hydroxide) or an ion-exchange membrane such as those in the fuel cells that provided the power supply for the *Gemini* spacecraft.

The fuel (hydrogen) enters at the anode side and the oxidizer (oxygen) enters at the cathode side. At the anode, hydrogen will be ionized according to the following reaction:

$$2H_2 \rightarrow 4H^+ + 4e^-$$

The electrons so produced flow through an external load resistance to do useful work. With an ion-exchange membrane as the electrolyte, the ions will flow to the cathode side. At the cathode, the ions, the electrons, and the oxygen will combine to form water according to the following reaction:

$$4H^+ + 4e^- + O_2 \rightarrow 2H_2O$$

The water produced in this fuel cell is a by-product of the system which may be used for life-support purposes, as in the case of the fuel cells carried on board the *Gemini* spacecraft.

The attractiveness of a fuel cell is that it can convert considerably more of the chemical energy in the fuel to electrical energy than can the most efficient fossil-fuel-burning power plants. But the hydrogen–oxygen fuel cells are just too expensive for terrestrial applications. However, effort is being made toward the development of a fuel cell that will operate with a hydrocarbon fuel together with air as the oxidizer. If this effort is successful, the future of fuel cell systems could be quite promising.

1.5 Relevancy of Other Engineering Sciences

Engineering thermodynamics is useful, but it has its limitations. Let us illustrate with a few simple examples. We can, making use of thermodynamics, predict the maximum output of a turbine for each pound of fluid flowing through it. But the actual turbine output depends to a great extent on the mechanism of momentum transfer, which is studied in the discipline of fluid mechanics. We can, making use of thermodynamics, predict the maximum outlet temperature of air when it is being heated in a heat exchanger with a given heat source. But the actual outlet temperature depends on the mechanism of heat transfer, which is studied in the discipline of heat transfer. We can, making use of thermodynamics, predict the maximum temperature of the products of combustion in the burning of fuel with air. But the actual temperature depends largely on the mechanism of chemical reaction, which is studied in the discipline of chemical kinetics. The actual temperature also depends on how well the fuel is mixed with the air, which is studied in the subject of mass transfer. The study of heat transfer, mass transfer, momentum transfer, and chemical kinetics, known as *rate processes*, are most important in the determination of the size of the hardware involved. The economic feasibility of any system thus depends much on these rate processes. Consequently, the complete design of a system to perform a given task requires knowledge of engineering thermodynamics as well as rate processes, as indicated in Fig. 1.1.

1.6 Macroscopic and Microscopic Views of Matter

We can study the behavior of matter from either a macroscopic or a microscopic point of view. Advantages and disadvantages exist in each case. *Classical thermodynamics*, which is presented in this book, is based on the macroscopic point of view. It is one of several important macroscopic engineering sciences (the others being mechanics of solids, mechanics of fluids, heat and mass transfer, and electromagnetics) which provide us with the fundamental knowledge for the analysis and synthesis of a wide variety of engineering problems.

The structure of classical thermodynamics is relatively simple. Making use of only a few basic concepts, we are able to deduce the laws of thermodynamics, which, to paraphrase the words of Bridgman,[9] are more palpably verbal and show more signs of their human origin than other laws of physics. Because the basic ideas used in macroscopic thermodynamics are more intuitively acceptable, they are fairly easy to grasp. The variables involved being few in number, the mathematics required for this branch of science is rather simple.

The macroscopic approach requires no hypotheses on the detailed structure of matter. Consequently, the laws of thermodynamics are not subject to change as new knowledge on the nature of matter is discovered. We must pay a price for generality, however—which is, restriction in scope. Even though classical thermodynamics can predict many relationships between properties of matter, it cannot explain why such relationships have their particular form. For example, we shall see that through the use of thermodynamic arguments, the specific heat at constant pressure of a pure substance (such as nitrogen, water, ammonia, or carbon dioxide) is always greater than the specific heat at constant volume of the same pure substance at the same state condition. But the explanation of such behavior cannot be obtained from classical thermodynamics. This disadvantage, fortunately, is not a serious one. In many engineering applications, it is more important to know the relations between the properties of matter than to have a clear understanding of the behavior of matter in terms of its atomic structure.

To study matter from the microscopic point of view, we must accept the atomic model of matter and know the physics (quantum mechanics) of the atoms, molecules, protons, electrons, and so on, that make up its structure. In the discipline of *statistical thermodynamics*, the microscopic point of view is used. It must begin, therefore, with a higher level of abstraction than in classical thermodynamics. In addition, statistics and probability theory must be used so that information on a very large number of particles (of the order of 10^{20} cm^{-3}) may be "averaged" out into a few quantities of importance in engineering applications. The mathematics required for statistical thermodynamics is quite a bit more difficult than for classical thermodynamics. But the reward for such sophistication is great: with statistical thermodynamics we can predict and interpret the macroscopic behavior of matter. Statistical thermodynamics has become

---

[9] P. W. Bridgman, *The Nature of Thermodynamics*, Harper & Row, Publishers, Inc., New York, 1961, p. 3.

more important in recent years. In certain areas of applications, such as in the study of high-temperature plasmas in connection with magnetohydrodynamic power generation, we must resort to the use of statistical thermodynamics, because macroscopic methods are no longer adequate or even feasible. Statistical thermodynamics is indeed useful, although it has its limitations. At the present state of development, statistical thermodynamics is not as general and fruitful in its applicability as is classical thermodynamics.

1.7 Dimensions and Units

Dimensions are names given to physical quantities. Some familiar examples of dimensions are length, time, mass, force, volume, and velocity. In engineering analysis, it is most important to check the dimensional homogeneity of any equation relating physical quantities. This means that the dimensions of terms on one side of the equation must equal those on the other side.

To reduce the number of dimensions, certain physical quantities can be expressed in terms of others. For example, if length and time are chosen as primary quantities, velocity would have the dimensions of length divided by time. Velocity is then a derived dimension. Any consistent set of dimensions may be chosen as the primary quantities. We shall examine four-dimensional systems commonly in use.

Whereas a dimension is a name, a *unit* is a definite standard or measure of a dimension. For example, foot, meters, and angstroms are all different units with the common dimension of length. Each of the primary scales of measure is based on a carefully chosen standard. The definition and the unit for each of the primary quantities of measurement are established through international agreement. The international standard of length used to be the distance between two marks on a platinum–iridium bar, but in 1960 an atomic standard based on the wavelength of the orange–red line in the spectrum of krypton 86 was adopted by international agreement. The present international standard of length is the meter defined as 1,650,763.73 times this wavelength. The international standard of time is the second. Until 1960 it was defined as 1/86,400 of a mean solar day. Since October 1967, the second has been given an atomic definition. It is now defined as the duration of 9,192,631,770 periods of the radiation corresponding to the transition between the two hyperfine levels of the ground state of the cesium 133 atom. The international standard of mass is the kilogram. It is equal to the mass of a particular cylinder of platinum–iridium alloy kept at the International Bureau of

Table 1.2 Primary Dimensions and Units in the SI System of Units

Physical Quantity	Basic Unit	Symbol
Length	meter	m
Mass	kilogram	kg
Time	second	s
Temperature	kelvin	K
Electric current	ampere	A
Luminous intensity	candela	cd

Weights and Measures in France. The definitions of other physical quantities may be found in the report by Mechtly.[10]

A very important system of units is the SI (Système International) or International System of Units. The primary or basic dimensions and their corresponding units in this system are given in Table 1.2. The units of all other physical quantities are then given in terms of these basic units. The derived units of important thermodynamic quantities given in SI units are given in Table 1.3.

In any dimensional system, the units of length, time, mass, and force are related through Newton's second law of motion, which states that the total force acting on a body is proportional to the product of the mass and the acceleration in the direction of the force. Thus

$$F \propto ma. \tag{1.1}$$

Introducing a constant of proportionality of $1/g_c$, Eq. 1.1 becomes

$$F = \frac{1}{g_c} ma. \tag{1.2}$$

In the SI system of units, force is a secondary quantity and is given in newtons (N). One newton is defined as the force necessary to accelerate 1 kilogram mass at a rate of 1 meter per second per second. Thus

$$1 \text{ N} = 1 \text{ kg} \cdot \text{m/s}^2$$

and

$$g_c = 1 \frac{(\text{kg})(\text{m})}{(\text{N})(\text{s}^2)}.$$

[10] E. A. Mechtly, *The International System of Units*, National Aeronautics and Space Administration Report SP-7012, 1969.

Table 1.3 Derived Units of Important Thermodynamic Quantities in the SI System of Units

Physical Quantity	Derived Unit	Symbol
Force	newton (kg-m/s^2)	N
Area	square meter	m^2
Volume	cubic meter	m^3
Pressure	newton per square meter	N/m^2
Density	kilogram per cubic meter	kg/m^3
Energy, work, quantity of heat	joule (newton-meter)	J
Power	watt (J/s)	W
Entropy	joule per kelvin	J/K

It is important to note that as a result of the definition of force, the value of the constant g_c is unity in the SI system of units.

Another important system of units is the English engineering system of units. In this dimensional system, both force and mass plus length and time are chosen as the primary quantities. Force and mass are assigned the units of pound-force (lbf) and pound-mass (lbm), respectively. The standard gravitational force of the earth is defined as such that a force of 1 lbf will accelerate a mass of 1 lbm at a rate of 32.174 ft/s^2. Thus

$$1 \text{ lbf} = \frac{(1 \text{ lbm})(32.174 \text{ ft/s}^2)}{g_c}$$

and

$$g_c = 32.174 \frac{(\text{ft})(\text{lbm})}{(\text{lbf})(\text{s}^2)}.$$

This means that a body having a weight of 1 lbf on the surface of the earth will have a mass of approximately 1 lbm. (For many engineering calculations, it is accurate enough to let $g_c = 32.2$ ft-lbm/lbf · s^2.)

It is important to note that in the English engineering system of units the constant g_c not only has a numerical value of 32.174 but also has units. Since both the SI system of units as well as the English engineering system of units will be used in this book, it is important to carry the constant g_c along in all equations in which the conversion between mass and force is involved.

A third system of units in use at the present time is the absolute engineering system of units. In this dimensional system, length, time, and force are chosen as the primary quantities; mass is a secondary quantity. The unit of force is the pound-force (lbf) and the unit of mass is the slug. One slug is defined as the mass that is accelerated at a rate of 1 ft/s² when acted upon by a force of 1 lbf. Thus

$$1 \text{ slug} = 1 \text{ lbf} \cdot \text{s}^2/\text{ft}$$

and

$$g_c = 1 \, \frac{(\text{slug})(\text{ft})}{(\text{lbf})(\text{s}^2)}.$$

The constant g_c has the value unity as a result of the definition of mass in this system of unit.

A fourth system of units still in use is the absolute metric (cgs) system. In this system, as in the SI case, mass, time, and length are chosen as primary quantities. The gram is the unit of mass, the second is the unit of time, and the centimeter is the unit of length. The unit of force, the dyne, is defined as that force required to accelerate a mass of 1 gram at a rate of 1 centimeter per second per second. Thus

$$1 \text{ dyne} = 1 \text{ g} \cdot \text{cm}/\text{s}^2$$

and

$$g_c = 1 \, \frac{(\text{g})(\text{cm})}{(\text{dyne})(\text{s}^2)}.$$

Table 1.4 Units of Different Dimensional Systems

Name of System	Unit of Mass	Unit of Length	Unit of Time	Unit of Force	g_c in $F = \dfrac{1}{g_c} ma$	Definition of Force
SI	kg	m	s	N	$1.0 \, \dfrac{\text{kg} \cdot \text{m}}{\text{N} \cdot \text{s}^2}$	1.0 N is the force needed to accelerate a mass of 1.0 kg at 1.0 m/s²
English engineering	lbm	ft	s	lbf	$32.174 \, \dfrac{\text{lbm} \cdot \text{ft}}{\text{lbf} \cdot \text{s}^2}$	1.0 lbf is the force needed to accelerate a mass of 1.0 lbm at 32.174 ft/s²
Absolute engineering	slug	ft	s	lbf	$1.0 \, \dfrac{\text{slug} \cdot \text{ft}}{\text{lbf} \cdot \text{s}^2}$	1.0 lbf is the force needed to accelerate a mass of 1.0 slug at 1.0 ft/s²
Absolute metric (cgs)	g	cm	s	dyne	$1.0 \, \dfrac{\text{g} \cdot \text{cm}}{\text{dyn} \cdot \text{s}^2}$	1.0 dyne is the force needed to accelerate a mass of 1.0 g at 1.0 cm/s²

Table 1.5 Unit Prefixes

Multiplier	Prefix	Symbol	Multiplier	Prefix	Symbol
10^9	giga	G	10^{-3}	milli	m
10^6	mega	M	10^{-6}	micro	μ
10^3	kilo	k	10^{-9}	nano	n

The results of our discussion are summarized in Table 1.4. We must keep in mind that the constant g_c has the numerical value of unity in the SI, cgs, and absolute engineering systems. But in the English engineering system, g_c has the value of 32.174 ft · lbm/lbf · s². We should also not confuse the term g_c with the local acceleration of gravity g. While g_c and g are numerically equal in many engineering calculations involving the use of the English engineering system of units, their dimensions are different.

The basic units in the SI system at times are either too large or too small to be convenient. In these cases we may use the prefixes given in Table 1.5. It should be pointed out that these prefixes may also be used with units in other systems.

Example 1.1

Convert 1.0 cal/g · K to

(a) SI units.
(b) English engineering units.

Solution

Making use of the various conversion factors given in the front of the book, we have

(a) 1.0 cal/g · K = (1.0 cal/g · K) × (4.1868 J/cal) × (1000 g/kg)

$$= 4186.8 \text{ J/kg} \cdot \text{K}.$$

(b) 1.0 cal/g · K = (1.0 cal/g · K) × (4.1868 J/cal) · $\left(\dfrac{1}{1.055 \times 10^3} \text{ Btu/J} \right)$

$$\times (453.59 \text{ g/lbm}) \times \left(\frac{1}{1.8} \text{ K/°R} \right)$$

$$= 1.0 \text{ Btu/lbm} \cdot \text{°R}.$$

Comments

The units of certain thermodynamic properties, such as specific entropy and specific heat, may be given in cal/g · K, J/kg · K, or Btu/lbm · °R. It is important to note that dimensions are an essential part of the answer to any engineering problems. Students are encouraged to acquire the good habit of always writing down units in a numerical problem.

1.8 Thermodynamic Systems

In mechanics, if we want to study the motion of a body, we must draw the "free body" and identify all the forces exerted on it by other bodies before we proceed to apply the governing equations of motion. In other words, we must first understand how other bodies interact with the body that we want to study.

We have an analogous situation in thermodynamics. If we want to study the behavior of a particular thermodynamic system, we must be able to identify the other thermodynamic systems that interact with the system in question before we apply the governing equations in thermodynamics.

In thermodynamics, a *system* is defined as any collection of matter or any region in space bounded by a closed surface or wall. The wall may be a real one, like that of a tank enclosing a certain amount of fluid. The wall may also be imaginary, like the boundary of a certain amount of fluid flowing along a pipe. All other systems outside the wall that interact with the system in question are known as the *surroundings*.

Depending on the nature of the wall involved, we can classify a thermodynamic system as a closed system, an open system, or an isolated system. In a *closed system*, the wall involved is impermeable to matter. That is, a closed system can have no material exchange with its surroundings, and consequently its mass must remain constant. On the other hand, a closed system can exchange energy with its surroundings in terms of heat and work. In an *open system*, there will be material flow across the boundary. In addition, there could also be heat flow and work flow across the boundary. In an *isolated system*, there can be absolutely no interaction with its surroundings. The wall involved is not only impermeable to matter; it is also impermeable to any form of energy. An isolated system may therefore be defined as an assembly of subsystems with any possible interaction between matter and energy restricted to the subsystems within the assembly. Any system plus its surroundings taken together would constitute an isolated system. Examples of a closed system, an open system, and an isolated system are shown in Fig. 1.11. We thus see

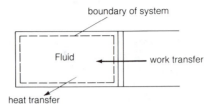

(a) Example of a closed system—fluid inside
the cylinder of a piston-cylinder apparatus

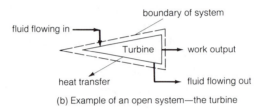

(b) Example of an open system—the turbine

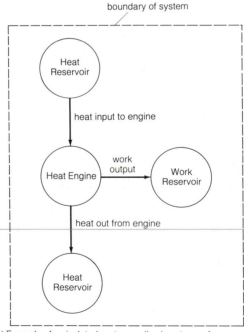

(c) Example of an isolated system—all subsystems of a power-
producing system

Figure 1.11 Closed, Open, and Isolated Systems

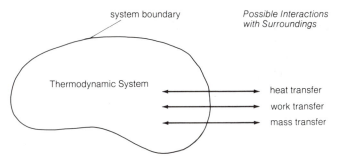

Figure 1.12 Schematic Representation of Interactions Between System and Surroundings

that a thermodynamic system may have in general only three types of interactions with its surroundings: heat interaction, work interaction, and mass interaction. This important idea is represented schematically in Fig. 1.12.

Other types of walls that we could use to separate the system from its surroundings are rigid walls, diathermal walls, and adiabatic walls. A *rigid wall*, as the name implies, is one that will not permit the volume of the system to change. A *diathermal wall* is one that will make it possible for the system to communicate thermally with its surroundings. Two systems separated by a diathermal wall are said to be in *thermal contact*. The walls that we encounter in the real world are generally diathermal walls. An *adiabatic wall* is one that is impermeable to thermal energy. Such a wall will cut off thermal interaction between a system and its surroundings. Although this is an idealized situation, a system may be assumed to be separated by an adiabatic wall if the amount of heat transfer involved is very small compared to the other kinds of energy exchange. In recent years, multilayer insulations[11] (called *superinsulations*) developed for use in the cryogenic industry may for all practical purposes be employed as adiabatic walls.

1.9 Thermodynamic Properties

The distinguishing characteristics of a system are called the *properties* of the system. They are quantities that we must specify to give a macroscopic description of a system. Many such quantities are

[11] Randall Barron, *Cryogenic Systems*, McGraw-Hill Book Company, New York, 1966, pp. 493–497.

familiar to us in other branches of science, such as mass, energy, pressure, volume, density, electric field, magnetic field, and magnetization of matter. Two other properties—temperature and entropy—are unique to thermodynamics. Together with energy, they play a most important role in the structure of thermodynamics. These three important properties will be considered in detail shortly.

A *property* is either a directly observable or an indirectly observable characteristic of a system. Any combination of such characteristics, such as, for example, the product of pressure and volume, is also a property. That is, we can obtain new properties by defining them in terms of other properties. We shall find that among the many possible derived properties, three of them—enthalpy, Gibbs function, and Helmholtz function—are particularly useful.

The definition of a property in thermodynamics has a unique meaning. Let us illustrate this by considering the property *pressure*. When a system has a pressure of p_1 at one instant and a pressure of p_2 at another instant, the change in pressure is simply given by $p_2 - p_1$, regardless of how the change is accomplished. This means that we must have

$$\text{pressure change} = \int_{p_1}^{p_2} dp = p_2 - p_1, \qquad (1.3)$$

where dp represents a differential change in pressure. Mathematically speaking, Eq. 1.3 indicates that dp is an exact differential, and the integral is absolutely independent of the particular path in which the pressure was changed. This is why thermodynamic properties are called *point functions* or *state functions*. A quantity whose value depends on the particular path followed in going from one state to another is called a *path function*. The differential of such a quantity is not exact.

All properties of a system may be divided into two types: intensive and extensive. Those properties which are independent of the amount of material in the system are called *intensive properties*. They are not additive. Pressure, temperature, and density are examples of intensive properties. Those properties which are proportional to the mass of a system are called *extensive properties*. They are additive. Volume, energy, and entropy are examples of extensive properties. Frequently, we find it convenient to obtain certain intensive properties from their corresponding extensive properties. For example, specific volume (volume per unit mass) is obtained by dividing the volume of the system by its mass. Similarly, specific entropy (units of entropy per unit mass) is obtained by dividing the entropy content of the system by its mass. If we use a capital letter for an extensive property and the same lowercase letter for the corresponding specific value, specific volume and specific entropy are given as

$$v = \frac{V}{m} \qquad s = \frac{S}{m},$$

where v is specific volume, V is volume, s is specific entropy, S is entropy, and m is mass of a system.

Density (ρ) is defined as the mass of a substance divided by its volume, or the mass per unit volume. Thus $\rho \equiv 1/v$, and it is an intensive property.

1.10 Thermodynamic Equilibrium and Equilibrium States

It is our observation that everything in our physical world is continuously changing. It is also our observation that under certain conditions a collection of matter can experience changes that are negligibly small. For example, the temperature and pressure of a tank of gas exposed to the atmospheric temperature long enough will be observed to be essentially constant as the instruments used to measure these properties are insensitive to fluctuations on the microscopic level.[12] If the gas is the product of combustion, chemical analysis will show that the relative amount of each species will also be constant when the temperature and pressure are constant. When a collection of matter experiences no more changes in all its properties, we say that it is at a state of *thermodynamic equilibrium*. Equilibrium states of a given system are then characterized by the unique values of all its properties.

The concept of equilibrium in classical thermodynamics is an important and primitive one. It is really an abstraction, as real systems are never strictly in equilibrium. However, we postulate that any thermodynamic system can be in equilibrium and that any isolated system will reach a state of thermodynamic equilibrium after it has been left to itself long enough. An important implication of this postulate is that eventually there can be no more tendencies for any macroscopic change in an isolated system. We shall see that this idea will lead to some far-reaching conclusions.

---

[12] The study of fluctuations belongs in statistical thermodynamics. See Joseph Kestin and J. R. Dorfman, *A Course in Statistical Thermodynamics*, Academic Press, Inc., New York, 1971, Chap. 14.

When a system has no unbalanced force within it and when the force it exerts on its boundary is balanced by external force, the system is said to be in *mechanical equilibrium*. When the temperature of a system is uniform throughout and is equal to the temperature of the surroundings, the system is said to be in *thermal equilibrium*. When the chemical composition of a system will remain unchanged, the system is said to be in *chemical equilibrium*. In order to have thermodynamic equilibrium, we must satisfy the conditions of mechanical equilibrium, thermal equilibrium, and chemical equilibrium.

1.11 Thermodynamic Processes

When a collection of matter experiences a change from one equilibrium state to another equilibrium state, it is said to have undergone a *process*. Since an engineering device or system in thermodynamics is simply a creation of man making use of various kinds of processes for the purpose of carrying out controlled interaction among energy and matter, it is essential that we develop a logical structure and methodology for the evaluation of changes in the equilibrium states of matter. A necessary part of this methodology is that we must be able to recognize what process actually takes place.

We shall encounter many processes in our course of study. The special features of certain processes may be recognized from the names given to them. For examples, an *isothermal process* is a constant-temperature process, an *isobaric process* is a constant-pressure process, and an *isometric process* is a constant-volume process. On the other hand, the significance of some processes may be recognized only if we fully understand the definitions involved. Examples of this type are adiabatic process, cyclic process, quasi-static process, and reversible process.

Adiabatic Process

A process in which no heat crosses the system boundary in either direction is called an *adiabatic process*.

Cyclic Process or Cycle

A *cycle* is simply a sequence of processes that a system undergoes in such a manner that its initial state and its final state are identical. In other words, the net change in any property of the system is zero for a cycle. Mathematically, this is

$$\oint dX = 0, \tag{1.4}$$

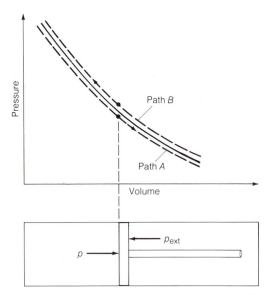

Figure 1.13 Quasi-static Expansion and Compression of a Gas

where X is any property and the symbol $\oint$ indicates integration around a cycle.

Quasi-static Process

If a process is carried out in such a manner that at every instant the system departs only infinitesimally from an equilibrium state, the process is called *quasi-static* (sometimes called *quasi-equilibrium*). For such a process, the path followed by the system may be represented by a succession of equilibrium states. If there are finite departures from equilibrium, the process is non-quasi-static.

Let us consider a gas in a cylinder provided with a movable piston as shown in Fig. 1.13. If the external pressure p_{ext} is maintained infinitesimally less than the gas pressure p, the gas will expand quasi-statically following path A. If the external pressure p_{ext} is maintained infinitesimally greater than the gas pressure p, the gas will be compressed quasi-statically following path B. In the limit these two processes follow the same path in opposite directions. Thus a *quasi-static process* is reversible, or more correctly, *internally reversible*.

A quasi-static process is an ideal process. It is approximately realized by making the change very slowly. All real processes are not quasi-static because they take place with finite differences of pressure, temperature, and so on, between system and surroundings.

Reversible Process

A process is *reversible* if, after it has been carried out, it is possible by any means whatsoever to restore the system and the surroundings involved in the interaction to exactly the same states they were in before the process. This implies that, if a process is reversible, it is possible to undo it in such a manner that there will be no trace anywhere of the fact that the process occurred. Thus a *reversible process* must be *internally reversible* as well as *externally reversible*. It is perfection from the thermodynamic point of view. More elaborate discussion of this ideal process will be given in connection with our study of the property entropy. Real processes are all *irreversible* processes, but some are less irreversible than others. An important part of our study of engineering thermodynamics is to recognize the factors that contribute to irreversibility so that we may select or create the best possible processes for a given problem.

Irreversibility is encountered when there is a lack of equilibrium during the process. For example, if heat is added to a system through a finite temperature difference, the system will undergo a non-quasistatic process. Thus the process is at least internally irreversible.

Irreversibility is also encountered when there is friction of any kind, be it mechanical friction, fluid friction, or electrical resistance. Frictional effects are known as dissipative effects in which the work-producing ability of the system and surroundings involved has been decreased due to the irreversible process. We shall see that the effect of irreversibility may be quantified through the use of the property entropy.

1.12 Temperature, the Zeroth Law of Thermodynamics, and Thermal Equilibrium

The concept of temperature, like that of force, originated in man's perceptions of hot and cold. It is a primitive concept resulting from the accepted scientific fact known as the *zeroth law of thermodynamics*, which states: When two systems are each in thermal equilibrium with a third system, they are also in thermal equilibrium with each other.

We say the temperature of system A is the same as the temperature of system B if system A and system B are in thermal equilibrium with each other. The concept of temperature will be shown to have additional physical significance after we have formulated the first and second laws of thermodynamics.

Through the use of the zeroth law, the temperature of a system may be determined by bringing it into thermal equilibrium with a thermometer, which is nothing but a system with numerical values assigned to a temperature-dependent property of the system. In general, the temperature of a system given by one kind of thermometer (for example, a mercury thermometer) is not exactly the same as that given by another kind of thermometer (for example, an electrical-resistance thermometer) except at their common fixed points. Empirical temperature scales are dependent on the nature of the thermometric substance used. We shall see that as a consequence of the second law of thermodynamics we can establish a thermodynamic temperature scale that is independent of the nature of any substance.

The concept of temperature will be explored in more detail later. We conclude our discussion of it in this section by giving the units of temperature and their conversion factors that we use in this book:

$$°R = °F + 459.67$$

$$K = °C + 273.15$$

$$1.0\ K = 1.8°R,$$

where °R is degree Rankine (absolute degree Fahrenheit); °F is degree Fahrenheit; K is degree Kelvin (absolute degree Celsius); °C is degree Celsius (formerly known as degree centigrade). In many engineering calculations, it is accurate enough to let

$$°R = °F + 460$$

$$K = °C + 273.$$

Since K differs from °C only by a constant, a 1.0-degree change in K is identical with a 1.0-degree change in °C. By the same token, a 1.0-degree change in °R is identical with a 1.0-degree change in °F.

It will be shown that the thermodynamic temperature scale corresponds to the absolute temperature scales. Therefore, we must use absolute temperature in our thermodynamic calculations.

If a system is separated from its surrounding by a diathermal wall, there will be heat interaction if there is a temperature gradient across the wall. The property temperature may thus be looked upon as the driving force for heat transfer. This consequence of the zeroth law of thermodynamics is illustrated in Fig. 1.14.

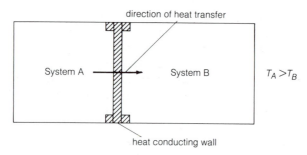

Figure 1.14 Temperature as a Driving Force for Heat Transfer

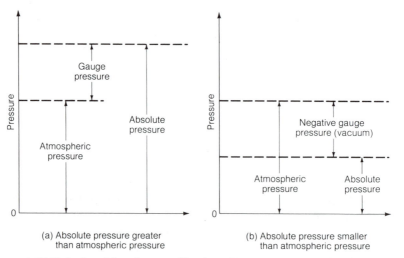

Figure 1.15 Relationships Among Absolute Pressure, Atmospheric Pressure, Gauge Pressure, and Vacuum

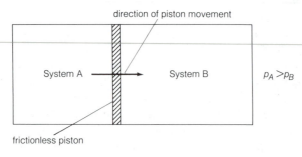

Figure 1.16 Pressure as a Driving Force for Volume Change

1.13 Pressure and Mechanical Equilibrium

The *thermodynamic pressure p* is defined as the total normal force per unit area exerted by the system on its boundary. It is also called the *absolute pressure*. In engineering calculations, it is the absolute pressure that we must use.

The pressure of a system is quite often measured with a gauge by using the atmospheric pressure as the reference points. If this is the case, the absolute pressure is related to the gauge pressure in the following manner:

absolute pressure = atmospheric pressure + gauge pressure.   (1.5)

For pressure below atmospheric, the gauge pressure would be negative. It is common practice to apply the term "vacuum" to the magnitude of the gauge pressure in this case. When the absolute pressure is zero, we say that we have a perfect vacuum. The relationship among absolute pressure, gauge pressure, atmospheric pressure, and vacuum is shown graphically in Fig. 1.15.

In engineering units, pressure is often given in lbf/in^2 (psi). To distinguish absolute pressure from gauge pressure, we shall use the abbreviations of psia for the former, and psig for the latter. In SI units, all pressures are absolute unless otherwise noted.

When two systems are separated by a frictionless piston in a cylinder, the piston will move in the direction of lower pressure due to the unbalanced force acting on the piston. The property pressure may thus be looked upon as the driving force for volume change. This concept is illustrated in Fig. 1.16. The piston will stop moving when the pressure in both systems are equal. Just as uniform temperature in a system is necessary to have thermal equilibrium, uniform pressure in a system is necessary to have mechanical equilibrium.

Example 1.2

A simple manometer is shown in Fig. 1.17 when the right arm is exposed to the atmosphere, p is the pressure of the system being measured, and the distance Z gives an indication of the pressure difference between p and the atmosphere. Determine the absolute pressure of the system if the manometer liquid is mercury and if the distance Z is 250 mm. The atmospheric pressure is 101 kPa. The density of mercury may be taken as $13.6 \times 10^3 \ kg/m^3$.

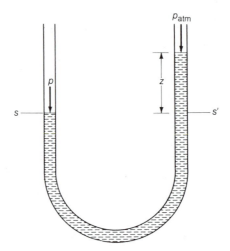

Figure 1.17 Simple Manometer

Solution

We first observe that the total force acting down at S is the same as the total force acting down at S' since we have mechanical equilibrium. From force balance we have

$$pA = p_{atm} A + \text{weight of column of liquid with distance } Z,$$

where A is the cross-sectional area of the manometer tube, and

$$\text{weight of column of liquid} = \frac{(\text{mass of column of liquid}) \, g}{g_c}$$

$$= \frac{\rho A Z g}{g_c}.$$

Solving, we have

$$p - p_{atm} = \frac{\rho g Z}{g_c},$$

where ρ is the density of the manometer liquid. Assuming that g has the value of 9.81 m/s², we have

$$p - p_{atm} = \frac{(13.6 \times 10^3 \text{ kg/m}^3)(9.81 \text{ m/s}^2)(0.250 \text{ m})}{1.0 \text{ kg} \cdot \text{m/s}^2 \cdot \text{N}}$$

$$= 3.34 \times 10^4 \text{ N/m}^2 = 33.4 \text{ kPa}$$

$$p = 101 + 33.4 = 134.4 \text{ kPa}.$$

Comments

We observe that the manometer measures the gauge pressure in terms of the distance Z and the density ρ of the manometer liquid. If the gauge pressure to be measured is very small, we should use a light liquid instead of mercury to give a greater distance Z so that the reading can be obtained more accurately.

1.14 Chemical Potential and Chemical Equilibrium

Suppose that we have system A and system B both initially containing only fresh water. If they are separated by a semipermeable membrane that will permit only pure water to flow through, the liquid level would be the same in both systems, as shown in Fig. 1.18a. If we now dissolve some salt in system B, we will observe that some fresh water will flow from system A to system B as shown in Fig. 1.18b. This phenomenon is known as osmosis. It occurs because a driving force exists. This driving force is related to a property called chemical potential (usually designated by the symbol μ). As long as the chemical potential of water in system A is greater than that of water in the mixture, flow will continue. This phenomenon of mass transfer will stop only when the chemical potential of water is the same in both systems (that is, when system B has reached chemical equilibrium). The pressure difference between system B and system A at equilibrium is known as the osmotic pressure of the salt–water mixture. The osmotic pressure will depend on the salt concentration. For seawater at 25°C, the osmotic pressure is about 25 atm.

We thus see that the property chemical potential is analogous to temperature and pressure. Each is a form of driving force. While tem-

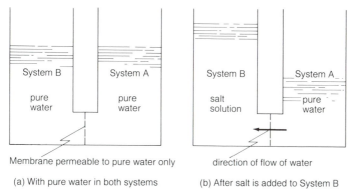

Membrane permeable to pure water only   direction of flow of water

(a) With pure water in both systems   (b) After salt is added to System B

Figure 1.18 Chemical Potential as a Driving Force for the Flow of Matter

perature difference would cause heat transfer and pressure difference would cause change in volume, difference in chemical potential would result in the flow of matter. The concept of chemical potential plays a very important role in the study of chemical thermodynamics.

1.15 Simple Structure of Engineering Thermodynamics

In order to design and evaluate the performance of a complex thermodynamic system or the components of the system such as those we have described earlier in this chapter, we must develop a logical structure so that we may apply it systematically. This logical structure of engineering thermodynamics turns out to be relatively simple. Specifically, we must in general satisfy only four equations in any design. These equations are:

1. Energy conservation equation.
2. Mass conservation equation.
3. Entropy accounting equation.
4. Property equation.

We shall provide a brief presentation of each of these four equations in this section. These equations are then developed in detail in the chapters to follow. The simple structure of engineering thermodynamics is then repeatedly employed in the analysis and synthesis of many practical and useful engineering applications.

The Energy Conservation Equation

Formulation of this equation is based on the observation of nature that energy is a commodity that can be neither destroyed nor created. Expressed in words, we have, for any thermodynamic system,

$$\text{energy entering} - \text{energy leaving} = \text{change of energy within system} \tag{1.6}$$

or

change of energy within system

$$+ \text{ (energy leaving} - \text{energy entering)} = 0 \tag{1.6a}$$

The left-hand side of Eq. 1.6a may be interpreted as the energy creation in the universe due to a given process. The first term gives us the energy change in the system selected for study, and the terms within the parentheses represent the energy change in the surroundings that interact with the system in question. For an isolated system, there can be no interactions with any surrounding. Thus Eq. 1.6a is reduced to the following statement:

The energy content of an isolated system is a constant,

which is the general statement of the first law of thermodynamics.

We thus see that the application of the energy conservation equation is just a matter of energy bookkeeping. We must account for the different forms of energy that enter and leave at the boundary of the system. We must also account for the different forms of energy that a system may possess.

Mass Conservation Equation

Although the energy E of a system is related to its mass m according to Einstein's famous equation $E = mc^2$, in this book we consider mass (as with energy) to be a commodity that can be neither destroyed nor created. This is reasonable because for all reactions involving energy, with the exception of nuclear reactions, the amount of mass converted to energy is extremely small. The mass conservation equation may thus be formulated in exactly the same manner as the energy conservation equation. That is, we have, for any thermodynamic system,

$$\text{mass entering} - \text{mass leaving} = \text{change of mass within system}$$

$$(1.7)$$

or

$$\text{change of mass within system} + (\text{mass leaving} - \text{mass entering}) = 0$$

$$(1.7a)$$

The principle of mass conservation, parallel to the principle of energy conservation, may thus be given by the following statement:

The mass content of an isolated system is a constant.

We thus see that the application of the mass conservation equation is just a matter of mass bookkeeping.

Entropy Accounting Equation

Formulation of this equation is based on the observation of nature that entropy is not a conservative property. In fact, entropy is being created, produced, or generated whenever a real process is carried out. Expressed in words in the same format as the conservation of mass and energy equations, we have, for any thermodynamic system,

change of entropy within system

$$+ \text{ (entropy leaving } - \text{ entropy entering)}$$

$$= \text{ entropy creation} > 0, \quad (1.8)$$

where entropy creation (also known as entropy production or entropy generation) is a positive, nonzero quantity for any real process. The magnitude of entropy creation will depend on the process involved. The more irreversible is the process, the greater will be the magnitude of entropy creation. Entropy change of the surroundings that interact with the system in question is given by the terms within the parentheses on the left-hand side of Eq. 1.8. For an isolated system this would be identically zero and Eq. 1.8 is reduced to the following statement:

The entropy content of an isolated system can never decrease,

which is the most general formulation of the second law of thermodynamics.

We thus see that the application of the second law of thermodynamics is just a matter of entropy accounting. But we must always keep in mind that entropy is not a conservative property.

Property Equation

Thermodynamics is an empirical science. Much of the information we need for engineering design depends on experimental data. But some properties, such as energy and entropy, are not directly measurable. We must develop expressions to relate them to other hopefully measurable properties. Any equation that relates one property to other properties may be called a property equation. Since such an equation gives the conditions of all equilibrium states, a property equation is also known as an equation of state.

The number of properties that can be given arbitrary values in any equation of state will vary from one substance to another. It has been found experimentally, however, that this number is relatively small and depends on the number of ways that we can independently change the energy of matter. This observation of nature has been formulated in a general rule on the behavior of matter known as the state postulate:

> The equilibrium states of a given system are completely specified by specifying $n + 1$ independent thermodynamic properties, where n is the number of relevant quasi-static work modes for the system.

The reference to "relevant quasi-static work modes" means that we count only the important work modes for the system involved. For example, mechanical work of expansion or compression is an important work mode for an ordinary gas such as air. On the other hand, mechanical work and magnetic work are both important for an ionized gas (plasma). For most substances used in engineering design, the number of relevant work modes turns out to be just one or two.

To apply the principles of engineering thermodynamics successfully to the design of processes, devices, and systems, we need a great amount of information pertaining to the properties of matter. Consequently, the more knowledge we have about matter, the more versatile we would be as designers. This is why much of the material in this book is devoted to the study of the behavior of different substances.

1.16 General Methodology for Problem Solving in Engineering Thermodynamics

The fundamentals of engineering thermodynamics are applicable to a wide variety of problems. In order that we may maximize the results and rewards of our effort, it is appropriate in applying these fundamentals that we develop a routine which we can follow systematically. Such a general procedure for problem solving is given below:

1. Read the problem carefully. Understand the task that is to be accomplished.
2. Since a sketch almost always aids in visualization, draw a simple diagram of all the components of the system involved. This could be a pump, a heat exchanger, gas inside a tank, or an entire power plant.
3. Select the system whose behavior we want to study by properly and clearly locating the boundary of the system. Do we have an isolated system, a closed system, or an open system?

4. Make use of the appropriate thermodynamic diagrams to locate the state points, and possibly the path of the process. These diagrams are extremely helpful as visual aids in our analysis.
5. Show all interactions (work, heat, and mass) across the boundary of the selected system.
6. Extract from the statement of the problem the unique features of the process and list them. Is the process isothermal, constant pressure, constant volume, adiabatic, isentropic, or constant enthalpy?
7. List all the assumptions that one might need to solve the problem. Are we neglecting a change of kinetic energy and change of potential energy?
8. Apply the first-law equation appropriate to the system that we have selected.
9. Apply the principle of mass conservation appropriate to the system that we have selected.
10. Apply the second-law equation appropriate to the system we have selected.
11. Apply the appropriate property relations. That is, bring in data from tables, charts, or appropriate property equations.
12. Try to work with general equations as long as possible before substituting in numbers.
13. Watch out for units. For example, when we use $h = u + pv$, h, u, and pv must all have the same units.
14. Make sure that the absolute temperature, in degrees Rankine or kelvin, is used in calculations.
15. Develop the habit of asking this question: Is the answer to our problem reasonable in terms of both magnitude and units? For example, we will see later that something is wrong if the thermal efficiency of a practical power plant is found to be 80%. Something is also wrong if the units of our answer are that of entropy, yet the physical quantity involved is energy.

Problems

1.1 The acceleration of a body falling freely in a vacuum is 9.8066 m/s^2. Express this quantity in ft/s^2.

1.2 An astronaut weighs in at 850 N on the surface of the earth, where g is 9.806 m/s^2.
    (a) What is the mass of the astronaut?
    (b) What is the weight of the astronaut on the surface of the moon, where g is one-sixth of the earth's gravity?

1.3 How much would a 190-lbm person weigh inside a space station in which an artificial gravity of 5.0 ft/s^2 is induced by rotation?

1.4 The weight of an average apple is about 1.0 N in California, where g is 9.807 m/s^2. What is the mass of this apple after it is shipped to New York?

1.5 If you are allowed to carry only 20 kg of luggage on board an airplane, what is the weight limit on your luggage in newtons when you check in at the airport, where g is 9.81 m/s^2? What is the weight limit in lbf?

1.6 If the speed limit on a highway is 100 km/h, what is it in miles per hour?

1.7 What force in newtons will accelerate a mass of 250 kg at the rate of 15 m/s^2?

1.8 What is the weight in newtons of a 2.0-kg mass at a location where g is 9.81 m/s^2?

1.9 What is the weight in dynes of a 20-g mass at a location where g is 981 cm/s^2?

1.10 What is the weight in lbf of a 2.0-slug mass at a location where g is 32.2 ft/s^2?

1.11 The weight of a body is 125 N in a location where g is 9.81 m/s^2. What is the mass of the body in kilograms? What is it in slugs?

1.12 The value of g at the equator and sea level is 32.088 ft/s^2, and this value decreases about 0.001 ft/s^2 for each 1000 ft of ascent. What is the weight of a person of 200 lbm at 5000 ft above sea level?

1.13 A person having a mass of 80 kg is being subjected to a deceleration (such as in an automobile accident) of 20 g's where 1 g is 9.807 m/s^2. What is the force in newtons that acts on the person? What is this force in lbf?

1.14 The universal gas constant $\bar{R}$ in SI units is 8314.3 J/kg mol · K.
(a) Convert it to Btu/lb mol · °R.
(b) Convert it to cal/g mol · K.

1.15 The density of water at room temperature and atmospheric pressure is quite often taken as 1.0 g/cc.
(a) Convert it to SI units.
(b) Convert it to lbm/ft^3.

1.16 The heat of vaporization of water at atmospheric pressure is about 2300 kJ/kg. Express this quantity in Btu/lbm.

1.17 The weight of 1 liter of a particular gasoline is found to be 7.0 N at a location where g is 9.81 m/s². Determine the density of this gasoline in kg/m³.

1.18 The absolute entropy of oxygen gas at 298 K and 1 atm is given in the literature as 49.004 cal/g mol · K.
(a) Express this quantity in SI units.
(b) Express this quantity in Btu/lbm mol · °R.

1.19 A cylindrical tank 30 m in length and 1.5 m in diameter contain 25 kg of air. Determine the specific volume and density of the air in the tank.

1.20 The energy consumption in the world is about 65 gJ per person per year. For a world population of 5.0 billion, what is the total annual energy consumption in barrels of crude oil equivalent? The energy content of one barrel of crude oil may be taken as 5.8×10^6 Btu.

1.21 The consumption of energy in the United States is about 75×10^{15} Btu per year. What is this consumption in gigajoules? What is it in barrels of crude oil equivalent? (For the energy content of crude oil, see Problem 1.20.)

1.22 A rigid vessel having a volume of 1.5 m³ initially holds 5.0 kg of air in a vacuum. Due to leakage, the mass of air inside the vessel is increased by 10%. What are the final density and specific volume of the air in the vessel?

1.23 One kilogram of a liquid having a density of 1200 kg/m³ is mixed with 2 kg of another liquid having a density of 2000 kg/m³. If the volume of the mixture is the sum of the initial volumes, what is the density of the mixture?

1.24 Two liters of water at 25°C is mixed with an unknown liquid to form a liquid mixture. The volume and mass of the liquid mixture are found to be 4000 cc and 4 kg, respectively. If the mixture volume is the sum of the initial volumes, determine the density of the unknown liquid.

1.25 Air in a storage tank is cooled from an initial temperature of 250°C to a final temperature of 25°C by exchanging heat with the atmosphere, which is at a temperature of 25°C. Does the air in the tank undergo a quasi-static process?

1.26 Heat, coming from a 100°C source, is added very slowly to an equilibrium mixture of ice and water at 0°C.
(a) Does the mixture undergo a quasi-static process?
(b) Is the heat-addition process reversible?

1.27 The normal human body temperature is 98.6°F. What is this temperature in °C, K, and °R?

1.28 Two thermometers, one reading in °C and the other in kelvin, are in thermal equilibrium with the same system. What is the temperature of the system if both thermometers have the same numerical value?

1.29 The temperature of a fluid is measured with a Celsius thermometer as well as a Fahrenheit thermometer. If the numerical reading is the same for both thermometers, what is the temperature of the fluid in K and °R?

1.30 The same as Problem 1.29, except that the Fahrenheit reading is numerically twice that of the Celsius reading.

1.31 In a constant-volume gas thermometer, it is found experimentally that regardless of the gas used, the following result is valid:

$$\frac{T_s}{T_i} = 1.3661$$

where T_s is the steam point (the equilibrium temperature of pure liquid water in contact with its vapor at atmospheric pressure) and T_i is the ice point (the equilibrium temperature of ice and air-saturated water at atmospheric pressure). Show that if we let $T_s - T_i = 100$, T_s and T_i are exactly the boiling point and freezing point for water on the Kelvin temperature scale.

1.32 Show that a pressure of 101.325 kPa is the same as 14.696 psi.

1.33 The gauge pressure of a fluid measured with a manometer is equivalent to a 25-cm column of mercury. What is the absolute pressure in pascal if the atmospheric pressure is 101.3 kPa? The density of mercury is 13,600 kg/m³.

1.34 A column of fluid in an open container is 1.5 m high. If the density of the fluid is 1500 kg/m³, what is the absolute pressure of the fluid at the bottom of the container? The atmospheric pressure is 101.325 kPa.

1.35 A pressure gauge reads 50 lbf/in². The atmospheric pressure is 14.7 lbf/in². What is the absolute pressure in the tank to which the pressure gauge is connected? Express the results in psia and SI units.

1.36 A vacuum gauge reads 20 mm Hg when the atmospheric pressure is 760 mm Hg. Determine the absolute pressure in pascal. The density of mercury is 13,600 kg/m³.

1.37 A skin diver descends to a depth of 25 m in the ocean. If the density of the seawater is 1020 kg/m³, what is the pressure exerted on the diver's body?

1.38 The pressure of high-vacuum systems is quite often given in torr. (One torr is defined as 1.0 mm Hg.) If the pressure of such a system is 1.0×10^{-8} torr, what is the pressure in atmospheres, psi, and SI units?

1.39 The height of a mercury column in a manometer used to measure a vacuum is 500 mm. The atmospheric pressure is 101.325 kPa. Calculate the absolute pressure of the vacuum system. The density of mercury is 13,600 kg/m³.

1.40 A barometer reads 101.3 kPa at the base of a mountain. If a barometer reads 80 kPa at the top of the mountain, determine the height of the mountain if we assume the average air density to be 1.21 kg/m³.

1.41 A mercury manometer attached to a tank registers a pressure of 2 in. Hg. If a manometer filled with oil having a density of 50 lbm/ft³ were substituted for the mercury manometer, what would it read in inches of oil?

1.42 The gauge pressure of the gas shown in Fig. P1.42 is 30 psig. The atmospheric pressure is 14.7 lbf/in². The piston has an area of 1 ft². What is the mass of the piston at a location where $g = 32.174$ ft/s²?

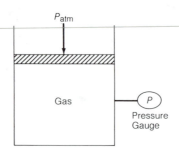

Figure P1.42

1.43 The pressure of air in a tank is measured with a U-tube manometer as shown in Fig. P1.43. If the manometer liquid is mercury, calculate the absolute pressure of the air for $Z = 760$ mm. The atmospheric pressure is 101.325 kPa. The density of mercury is 13,600 kg/m^3. Express your result in kPa and atm.

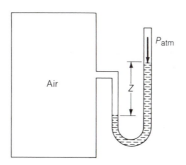

Figure P1.43

1.44 Same as Problem 1.43 except that $Z = 30$ in. What is the absolute pressure of the air in lbf/in^2?

1.45 A venturi tube, shown in Fig. P1.45, measures the pressure difference on the up- and downstream sides of a contraction through which a fluid is flowing. The pressure difference is often measured by means of a U-tube with mercury in the bottom of the tube and the fluid flowing on top of each column of mercury. If $Z = 125$ mm Hg and the fluid flowing is low-pressure air at 25°C, what is the pressure difference in pascal?

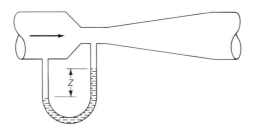

Figure P1.45

1.46 Same as Problem 1.45 except that $Z = 5$ in. Hg and fluid flowing is water at 70°F. What is the pressure difference in lbf/in^2?

1.47 A fluid inside a cylinder fitted with a frictionless piston, as shown in Fig. P1.47, is in mechanical equilibrium. The atmospheric pressure is 101.325 kPa and the piston area is 2500 mm². The linear spring has a constant of 5000 N/m. What is the absolute pressure of the fluid in pascal if the spring is compressed 150 mm beyond its free length?

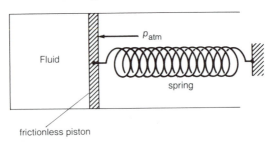

Figure P1.47

1.48 Same as Problem 1.47 except that the spring constant is 50 lbf/in. and the spring is compressed 6.0 in. beyond its free length. What is the gauge pressure of the fluid in lbf/in²?

1.49 A fluid inside a vertical cylinder fitted with a frictionless piston, as shown in Fig. P1.49, is in mechanical equilibrium. The atmospheric pressure is 101.325 kPa. The piston area is 3000 mm². The linear spring, having a spring constant of 1500 N/m, is compressed 250 mm beyond its free length. What is the mass of the piston if the absolute pressure of the fluid is 400 kPa?

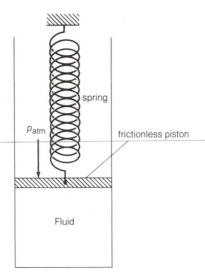

Figure P1.49

1.50 Same as Problem 1.49 except that the gauge pressure of the fluid is 20 psig. What is the mass of the piston in lbm?

1.51 The annual energy consumption of the world from 1950 through 1979 are given below:

Year	Annual Energy Consumption (10^6 terajoules)
1950	75
1955	95
1960	125
1965	160
1970	210
1975	243
1979	281

If we plot these data on a semilogarithmic graph (with the year along the horizontal linear axis), a straight line may be fitted. That is, the annual energy consumption of the world may be represented by the following equation:

$$\ln E = m\tau + B$$

where E is the annual energy consumption of the world
τ is the time (i.e., the year)
m is the slope of the straight line
B is the vertical axis intercept.

(a) Determine the approximate values of m and B from a semilog plot of the given data.
(b) Make an estimate of the annual energy consumption of the world in the year 2000. What is your answer in barrels of crude oil? The energy content of a barrel of crude oil may be taken as 6×10^6 kJ.

1.52 Write a computer program to solve Problem 1.51 using the least squares method for data that may be fitted by a straight line. Note: For data that may be fitted by a straight line, the equation of condition is

$$y = a_0 + a_1 x$$

and the normal equations are

$$\sum y_i = a_0 N + a_1 \sum x_i$$

$$\sum x_i y_i = a_0 \sum x_i + a_1 \sum x_i^2$$

N is the number of experimental points.

CHAPTER 2

Energy and the First Law of Thermodynamics

Through the careful study of a great amount of accumulated experimental evidence, scientists and engineers have been able to describe and predict the behavior of our physical world by using a few basic and primitive concepts and a few postulates or laws that we accept as scientific facts. One such basic concept is energy, and we accept its existence as a matter of fact. The scientific fact that has direct bearing on the concept of energy is that energy in our universe is something that cannot be destroyed or created. This principle of conservation of energy, the *first law of thermodynamics*, for any system may simply be stated as

$$\text{energy input} - \text{energy output} = \text{change in stored energy.} \quad (2.1)$$

Our job now is to identify the different forms of energy that are stored in matter as well as the different forms of energy that may be transferred into or out of a particular system.

2.1 Energy Content

Energy is a primitive property. We postulate that it is something that all matter has. We further define it as capacity to do work.[1] For convenience, we shall divide the total energy content, that is, the stored energy, of a system into two groups: energies that are expressed in terms of parameters measured with respect to a reference frame outside the system, and energies that are functions of the molecular configuration of matter and microscopic modes of motion.

---

[1] The word "energy" is formed by combining two words in Greek, meaning "capacity" and "work." See Ryogo Kubo, *Thermodynamics*, North-Holland Publishing Company, Amsterdam, 1968, p. 147.

Potential Energy

The concept of potential energy is familiar to us from our previous study in physics. It will be recalled that any two masses, regardless of size, exert an attraction on one another. If this force of attraction is multiplied by the distance of separation, the resultant energy is known as *potential energy*. That is, potential energy is the kind of energy that a body has because of its position in a potential field. By this definition, there is potential energy in the sun–earth system or in a system of molecules. In thermodynamics, the potential energies within a system of molecules are functions of the molecular configuration of matter and will be accounted for in a different manner. The only kind of potential energy that is relevant to our development is that associated with a mass above an arbitrary datum plane when the force of attraction is that due to the earth's gravitational field. That is, when we speak of potential energy in thermodynamics, we mean gravitational potential energy.

In raising a body of mass m to some elevation Z above the datum plane, the earth's gravitational field will exert on the body a force equal to $(1/g_c)mg$ according to Newton's second law of motion. Thus

$$dE_{pot} = F \, dZ = \frac{1}{g_c} mg \, dZ.$$

Integrating yields

$$\int_{(E_{pot})_1}^{(E_{pot})_2} dE_{pot} = \frac{m}{g_c} \int_{Z_1}^{Z_2} g \, dZ.$$

Assuming that the local acceleration of gravity g does not vary with Z, we obtain

$$(E_{pot})_2 - (E_{pot})_1 = \frac{mg}{g_c} (Z_2 - Z_1).$$

This means that the potential energy of a system having a mass m and an elevation Z above the datum plane in a gravitational field with constant g is given by

$$E_{pot} = \frac{mgZ}{g_c}. \tag{2.2}$$

Example 2.1

How much does the gravitational potential energy in a 200 kg package change when it is carried from the ground floor of a building to another floor that is 1000 m above the ground floor?

Solution

Let us select the ground floor as the datum plane. Using Eq. 2.2 and taking g to be 9.81 m/s^2, we have

$$\Delta E_{\text{pot}} = \frac{(200 \text{ kg})(9.81 \text{ m/s}^2)(1000 \text{ m})}{1.0 \dfrac{\text{kg} \cdot \text{m}}{\text{N} \cdot \text{s}^2}}$$

$$= 1{,}962{,}000 \text{ N} \cdot \text{m} = 1{,}962{,}000 \text{ J.} \qquad \blacksquare$$

Comments

1. Positive change in potential energy simply means that the potential energy of the package has been increased since it has been raised from one level to a higher level.
2. Neglecting friction, the potential energy increase in the package represents the work that must be done by the carrier against the gravitational force.

Kinetic Energy

The concept of *kinetic energy* is familiar to us from our study of mechanics. It is the kind of energy that a body has because of its bulk motion and is defined in terms of the relative motion of two bodies. Just as the arbitrary datum plane was assigned a height of zero when we were calculating potential energy, we normally assume that one of the two bodies is at rest. As a matter of convenience, we assume the earth to have zero velocity and measure velocities of bodies relative to the earth.

Consider a body that is initially at rest relative to the earth. Let this body be acted upon by an external force F and let dL be the differential distance through which the force F acts in the direction of motion. Then

$$F \, dL = \frac{1}{g_c} ma \, dL.$$

But

$$a = \frac{d\bar{V}}{d\tau}$$

and

$$dL = \bar{V} \, d\tau.$$

Thus

$$F \, dL = \frac{m}{g_c} \, \bar{V} \, d\bar{V}$$

and

$$\int_{L_1}^{L_2} F \, dL = \frac{m}{g_c} \int_{\bar{V}_1}^{\bar{V}_2} \bar{V} \, d\bar{V}.$$

The kinetic energy of a system having a mass m with a velocity $\bar{V}$ is found by integrating the above equation from a velocity of zero to $\bar{V}$,

$$E_{\text{kin}} = \frac{m \bar{V}^2}{2g_c}. \tag{2.3}$$

Example 2.2

Calculate the kinetic energy of a 4000-lbm car traveling at 5.0 miles per hour.

Solution

$$5.0 \text{ miles/h} = \frac{(5.0 \text{ miles/h})(5280 \text{ ft/mile})}{3600 \text{ s/h}} = 7.333 \text{ ft/s}$$

$$E_{\text{kin}} = \frac{(4000 \text{ lbm})(7.333 \text{ ft/s})^2}{(2)(32.174 \text{ ft} \cdot \text{lbm/lbf} \cdot \text{s}^2)}$$

$$= 3343 \text{ ft} \cdot \text{lbf.} \qquad \blacksquare$$

Comments

1. An energy-absorbing system designed for stopping a car having a mass of 4000 lbm and traveling at 5.0 miles/h must be capable of absorbing 3343 ft · lbf of energy.
2. Since kinetic energy is proportional to velocity squared, an energy-absorbing system designed for stopping a high-speed car would be bulky and costly.

Kinetic energy and potential energy are two forms of energy that are familiar to us from our study of mechanics. They are also known as *mechanical energies*. It will be recalled that in the absence of friction these mechanical energies are completely interchangeable; that is, one unit of potential energy can be ideally converted into one unit of kinetic energy, and vice versa.

Internal Energy

Internal energy, to which we give the symbol U, includes all forms of energy in a system other than kinetic energy and potential energy. It represents energy modes on the microscopic level, such as energy associated with nuclear spin, molecular binding, magnetic-dipole moment, molecular translation, molecular rotation, molecular vibration, and so on.[2] We do not know how to determine the absolute values of U. We can, however, determine changes in U and, fortunately, this is all we need in the solution of practical problems.

The total energy of a system, to which we give the symbol E, may now be expressed as

$$E = U + E_{kin} + E_{pot}. \tag{2.4}$$

Thus a change in the stored energy of a system is given as

$$\Delta E = \Delta U + \Delta E_{kin} + \Delta E_{pot}. \tag{2.5}$$

2.2 Conservation of Mass

Einstein in 1905 postulated and it was later proved that the energy, E, of a system is related to its mass, m, by the well-known equation

$$E = mc^2, \tag{2.6}$$

where

$$c = \text{velocity of light}$$

$$= 2.998 \times 10^8 \text{ m/s}$$

$$= 9.83 \times 10^8 \text{ ft/s}.$$

Thus the mass of a system is a measure of its energy content, and a change in mass will accompany a change in energy from any cause whatsoever.

Using Eq. 2.6, we may calculate a change in m due to a change in E as

$$\Delta m = \frac{\Delta E}{c^2} g_c.$$

[2] W. C. Reynolds and H. C. Perkins, *Engineering Thermodynamics*, McGraw-Hill Book Company, New York, 1970, p. 25.

For a ΔE of 2250 kJ,

$$\Delta m = \frac{2250 \times 1000 \text{ J} \times 1.0 \dfrac{\text{kg} \cdot \text{m}}{\text{N} \cdot \text{s}^2}}{(2.998 \times 10^8)^2 \dfrac{\text{m}^2}{\text{s}^2}}$$

$$= 2.50 \times 10^{-11} \text{ kg.}$$

A change in mass of this magnitude cannot be detected by the finest balance. Besides, this very small change in mass is insignificant compared to the mass of the systems that we deal with in engineering calculations. For example, it was found from experiment that 2250 kJ is approximately the amount of energy input required to vaporize 1.0. kg of water at ordinary conditions, say 101.3 kPa and 100°C. Therefore, we shall in this book consider the law of conservation of mass to be independent of the law of conservation of energy. Thus, for any system, we have, analogous to Eq. 2.1,

mass added − mass removed = change in mass storage.   (2.7)

2.3 First Law of Thermodynamics for a Closed System

The energy content of a closed (constant-mass) system may be changed, without involving a transfer of mass, by thermal conduction, radiation, mechanical compression or expansion, electromagnetic fields, gravitational fields, and similar means. We shall divide all such possible energy transfer into two categories: work and heat. This means that a change of state for a closed system may be brought about by work done to it or by it, and by heat addition to it or rejection by it. We shall use the symbol W to designate work and the symbol Q to designate heat. The symbols W_{12} and Q_{12} would represent the amount of work and heat involved for the process, respectively, when the system undergoes a change of state from state 1 with E_1 to state 2 with E_2. Now, heat added to a system is energy input and work done by a system is energy output. Thus, according to Eq. 2.1 and adopting the convention that heat addition is positive and work done by a system is positive, the first law of thermodynamics for a closed system may now be expressed mathematically as

$$Q_{12} - W_{12} = E_2 - E_1$$

or

$$Q_{12} = E_2 - E_1 + W_{12}. \tag{2.8}$$

In using Eq. 2.8, we must always remember the important sign convention that we have adopted on heat and work. In this equation, heat entering a system has the same positive sign as the work leaving a system. It follows that heat removed from a system represents negative heat input while work done on a system represents negative work output.

The validity of Eq. 2.8 is not affected by the scale of operations. It applies equally well to a finite transfer of energy or to a differentially small one. For the latter case, the first law of thermodynamics for a closed system is given as

$$\bar{d}Q = dE + \bar{d}W. \tag{2.8a}$$

The bar over d is used to indicate that $\bar{d}Q$ and $\bar{d}W$ are not exact differentials.

In problems in which changes in kinetic energy as well as changes in potential energy are zero, we have

$$Q_{12} = U_2 - U_1 + W_{12} \tag{2.9}$$

and

$$\bar{d}Q = dU + \bar{d}W. \tag{2.9a}$$

Equation 2.9 is sometimes known as the *energy equation for nonflow processes*.

Example 2.3

A certain amount of air is compressed inside a cylinder. The internal energy change of the air is $+16$ kJ; work required for the compression process is 300 kJ. What is the amount of heat transfer involved?

Solution

System selected for study: air inside cylinder
Assumptions: process is nonflow

From the first law for given process,

$$Q_{12} = +16 - 300 = -284 \text{ kJ.} \qquad \blacksquare$$

Comments

1. The minus sign for Q_{12} indicates that heat must be removed from the air.
2. Work is done on the air, hence W_{12} is negative.

Example 2.4

A certain amount of steam undergoes a change of state inside a cylinder-piston apparatus. Its internal energy increases by 900 Btu. If heat addition for the process is 950 Btu, what is the amount of work involved expressed in ft · lbf?

Solution

System selected for study: steam inside cylinder-piston apparatus
Assumption: process is nonflow

From the first law for given process,

$$950 = 900 + W_{12}$$

or

$$W_{12} = +50 \text{ Btu}$$

$$= +(50 \text{ Btu})(778.16 \text{ ft} \cdot \text{lbf/Btu}) = +38,900 \text{ ft} \cdot \text{lbf}. \qquad \blacksquare$$

Comments

1. The positive sign for W_{12} indicates that work is done by the steam.
2. We have made use of the conversion factor of

$$1 \text{ Btu(British thermal unit)} = 778.16 \text{ ft} \cdot \text{lbf}.$$

(For engineering calculations, it is adequate to use 1 Btu = 778 ft · lbf.)
3. We wish to point out that the historical development of the first law of thermodynamics is all wrapped up in this conversion factor, also known as the *mechanical equivalent of heat*.[3]

2.4 First Law of Thermodynamics for Cyclic Processes

Let a system operate in a cycle consisting of processes $1 \to 2$, $2 \to 3$, $3 \to 4$ and $4 \to 1$, as shown in Fig. 2.1. Applying Eq. 2.8 to each of the

[3] See E. Mendoza, "A Sketch for a History of Early Thermodynamics," *Physics Today*, Vol. 14, No. 2 (1961), p. 32. See also M. Zemansky and H. C. Van Ness, *Basic Engineering Thermodynamics*, McGraw-Hill Book Company, New York, 1966, pp. 73–78; P. S. Epstein, *Textbook of Thermodynamics*, John Wiley & Sons, Inc., New York, 1937, pp. 27–34.

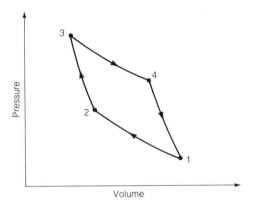

Figure 2.1 Cyclic Process

four processes involved, we have

$$Q_{12} = E_2 - E_1 + W_{12}$$

$$Q_{23} = E_3 - E_2 + W_{23}$$

$$Q_{34} = E_4 - E_3 + W_{34}$$

$$Q_{41} = E_1 - E_4 + W_{41}.$$

For the cycle,

$$Q_{12} + Q_{23} + Q_{34} + Q_{41} = W_{12} + W_{23} + W_{34} + W_{41}$$

or

$$\oint \bar{d}Q = \oint \bar{d}W, \tag{2.10}$$

where $\oint$ means summation around the cycle. In words, Eq. 2.10 says that in a work-producing cycle the net heat addition is equal to the net work output, and in a work-absorbing cycle the net heat rejection is equal to the net work input. This equation has been used as the first law of thermodynamics by some authors.

Example 2.5

An inventor claims to have developed a work-producing cycle which receives 1000 kJ of heat from a heat source, rejects 300 kJ of heat to a heat sink, and produces a net work of 700 kJ. How do we evaluate this claim?

Solution:

We have a cyclic process. Let us see if the first law of thermodynamics is satisfied. Using given data, we have

$$\text{net heat addition for cycle} = \oint \bar{d}Q = 1000 - 300 = 700 \text{ kJ}$$

$$\text{net work output for cycle} = \oint \bar{d}W = 700 \text{ kJ}$$

Since $\oint \bar{d}Q = \oint \bar{d}W$, the claim is valid as far as the first law is concerned.   ∎

Comments

In our study of work-producing cycles, we wish to compare the net work output (what you get) with the total amount of heat addition (what you pay). This ratio will be given the name "*thermal efficiency* of the cycle." For the given cycle, the thermal efficiency is 700/1000, or 70%. To evaluate whether or not this efficiency is possible, we must have more information and make use of another physical law of nature, the second law of thermodynamics. According to our present-day technology, a thermal efficiency of 70% is too good to be true. (Our most modern steam power plant operates at a thermal efficiency of only about 40%.)

Example 2.6

Consider the design of a work-absorbing cycle. Let the heat received from a low-temperature heat source be 100 Btu and the net work absorbed be 10 Btu. What is the amount of heat rejection of the cycle?

Solution

We have a cyclic process. Using the first law of thermodynamics for a cycle, we have

$$\oint \bar{d}Q = 100 \text{ Btu} + Q_{\text{out}} = -10 \text{ Btu}$$

and

$$Q_{\text{out}} = -110 \text{ Btu.}$$

The minus sign indicates that the cycle must reject 110 Btu of heat.   ∎

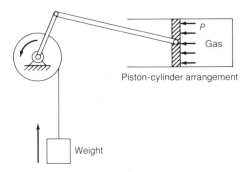

Piston-cylinder arrangement

Weight

Figure 2.2 Lifting of a Weight Due to Work Done by a Gas in an Expansion Process

Comments

In the study of work-absorbing cycles, we wish to compare the amount of heat received from the low-temperature heat source (what you get) with the net work required (what you pay). This ratio shall be given the name of *coefficient of performance*, defined as a positive number. For the given cycle, the coefficient of performance is 100/10, or 10. As in the case of Example 2.5, we must have more information before we can determine whether or not a coefficient of performance of 10 is possible. We note at this point that the coefficient of performance of a good industrial refrigeration plant is only about 4.

2.5 Thermodynamic Definition of Work

We have already noted that work and heat are the only two mechanisms by which the energy of a closed system can be changed. It is therefore essential for us to have a clear understanding of what is work interaction and what is heat interaction. We first take up the idea of work.

The concept of work is a primitive one. It has its origin in the study of mechanics, where it is defined as the product of a force and the distance through which this force acts. In thermodynamics, the force and the distance involved are sometimes not so easily recognized. To give a broader interpretation, we shall define work as follows: *work* is energy transferred, without transfer of mass, across the boundary of a system, and work is said to be done by a system on its surroundings if the sole effect external to the system could be the raising of a weight.

To illustrate, let us consider the expansion of a gas inside a cylinder. As we can see from Fig. 2.2, we can, through the proper selection of linkages, make use of the expansion process to lift a weight. Therefore, work will be done by the gas.

It should be emphasized that work, as defined, is energy in transit. Once this form of energy crosses the boundary of a system, it "disappears" and becomes part of the energy content of the system or the surroundings, as the case may be. Consequently, we cannot speak of a system having a certain amount of work at a given state; that is, the quantity of work is not a thermodynamic property. We can have work interaction only when a change of state has occurred.

2.6 Sign Convention for Work

In applying the principle of conservation of energy, we must set up a proper bookkeeping procedure for energy. It is most important that we have a common understanding as to what is meant by positive and negative work. Since energy leaves a system when work is done by it, it would seem natural to regard work done by a system as negative work. But the common convention is to regard work done by a system as positive work. We shall adopt this common convention. (It will be recalled that we used this convention in expressing the first law of thermodynamics mathematically.) A positive value (say $+50$ kJ) for W_{12} would mean that the system has performed 50 kJ of work on its surroundings in undergoing a change from state 1 to state 2. A negative value (say -25 kJ) for W_{12} would mean that in undergoing a change from state 1 to state 2 the system has absorbed 25 kJ of work from its surroundings.

2.7 Quasi-static Work of Expansion

Let us consider the expansion of a gas inside a cylinder from state 1 to state 2, as shown in the upper part of Fig. 2.3. If the expansion process is carried out quasi-statically, we may then plot the path on a pressure–volume diagram, as shown in the lower part of Fig. 2.3.

At any instant during the expansion process, the force exerted on the gas by the piston is only infinitesimally smaller than the force exerted on the piston by the gas, which is simply the product of the pressure of the gas and the area of the piston. Since this force acts in the direction of motion of the piston, the work done by the gas on the piston while the piston moves a distance dx is

$$\bar{d}W = pA \; dx,$$

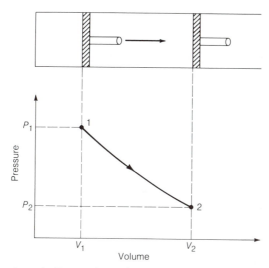

Figure 2.3 Quasi-static Expansion of a Gas Inside a Cylinder

where p is the pressure of the gas and A is the area of the piston. But $A\,dx = dV$, the change in volume of the gas as the piston travels the distance dx. Therefore,

$$\bar{d}W = p\,dV. \tag{2.11}$$

From Eq. 2.11, we see that $\bar{d}W$ is positive when dV is positive, while $\bar{d}W$ is negative when dV is negative. Thus work is done by the gas as it expands, and work is done to the gas if it is being compressed.

By integrating Eq. 2.11, we find that the total amount of work done by the gas as it expands quasi-statically is

$$W_{12} = \int_1^2 p\,dV. \tag{2.12}$$

Now, the integral $\int_1^2 p\,dV$ is simply the area under the path on the pressure–volume diagram. Since we can go from state 1 to state 2 along many different quasi-static paths, this integral will have a different value for each path. That is, work is a path function. This is consistent with what has been pointed out previously: work is not a thermodynamic property. In mathematical language, $\bar{d}W$ is an inexact differential. The bar over d is the notation we shall use in this book to indicate an inexact differential. The notation d will be used with point functions or exact differentials.

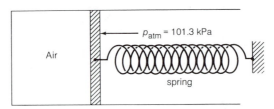

Figure 2.4 Expansion of Air Against Atmosphere and Spring

Equation 2.11 has been obtained based on the expansion of a gas inside a cylinder. It should be pointed out that this expression applies also to any substance undergoing a change of state quasi-statically, in which volume change is relevant.

Example 2.7

Consider the expansion of air inside a cylinder as shown in Fig. 2.4. Let the initial volume be 0.025 m^3 and the initial pressure be 10 MPa. Let the expansion process be quasi-static and let the path be given by $pV^{1.4}$ = constant. If the final volume of the gas is 0.20 m^3, determine

(a) the total amount of work done by the gas.
(b) the amount of work done by the gas against the spring.

Solution

System selected for study: air inside cylinder
Unique feature of process: quasi-static

Since the process is quasi-static,

$$W_{12} = \int_1^2 p \, dV. \tag{1}$$

Since $pV^{1.4}$ = constant,

$$p = \frac{\text{constant}}{V^{1.4}}$$

$$= \frac{p_1 V_1^{1.4}}{V^{1.4}}$$

$$= \frac{p_2 V_2^{1.4}}{V^{1.4}}.$$

Substituting into Eq. 1, we have

$$W_{12} = \text{constant} \int_1^2 \frac{dV}{V^{1.4}}$$

$$= \frac{\text{constant}}{1 - 1.4} (V_2^{1-1.4} - V_1^{1-1.4})$$

$$= \frac{p_1 V_1^{1.4}}{1 - 1.4} (V_2^{1-1.4} - V_1^{1-1.4})$$

$$= \frac{10 \times 10^6 (0.025)^{1.4}}{1 - 1.4} (0.20^{1-1.4} - 0.025^{1-1.4}) \text{ N} \cdot \text{m}$$

(a) $= 3.526 \times 10^5 \text{ J} = 352.6 \text{ kJ}.$

(b) We observe that the work done against the spring, $(W_{12})_s$, is simply the total work done by the gas, W_{12}, less the work done against the atmosphere, $(W_{12})_{atm}$. Since the pressure of the atmosphere is constant, the work done against the atmosphere is given by

$$(W_{12})_{atm} = p_{atm}(V_2 - V_1)$$

$$= 101.3 \times 10^3 (0.2 - 0.025) \text{ N} \cdot \text{m}$$

$$= 17.73 \text{ kJ}.$$

Thus the work done against the spring is

$$(W_{12})_s = 352.6 - 17.73 = 334.9 \text{ kJ}.$$ ∎

Comments

1. The process considered in this example is an important ideal process that we shall study in some detail in connection with some important cycles.
2. A path on the pressure–volume diagram for gas and vapor may in general be given by $pV^n = \text{constant}$, where n will have a given value for a given process. For example, we shall see that the value for n is 1.0 in the case of an ideal gas undergoing a change of state quasi-statically and isothermally.
3. The work done against the atmosphere is not useful. Thus the useful work done by the gas in this problem is simply the work done against the spring. If the spring is perfectly elastic, all the work done against it may be recovered as energy to do useful work. A perfect spring is an ideal energy storage device.

Example 2.8

Consider the compression of a liquid inside a cylinder (such as a hydraulic press). Let the compression process be quasi-static as well as isothermal. The path of such a process may be given as

$$\ln \frac{V}{V_0} = -A(p - p_0),$$

where A, V_0, and p_0 are positive constants. Derive an expression for the total amount of work required.

Solution

System selected for study: liquid inside cylinder
Unique features of process: quasi-static; isothermal

Since the process is quasi-static,

$$W_{12} = \int_1^2 p \, dV.$$

Since $\ln (V/V_0) = -A(p - p_0)$,

$$dV = -AV \, dp.$$

Therefore,

$$W_{12} = -A \int_1^2 Vp \, dp.$$

To carry out the integration, we should have the expression for V as a function of p. Now, in general the volume of a liquid is not sensitive to a change in pressure. Hence, assuming constant V in the integration, we have

$$W_{12} = -AV \int_1^2 p \, dp$$

$$= -\frac{AV}{2} (p_2^2 - p_1^2). \tag{2.13}$$

Since A and V are positive quantities, W_{12} will be negative if p_2 is greater than p_1. This is consistent with our convention that work is done to a system in a compression process. The values for A and V will be different for different substances. ∎

Comments

1. It may be shown later, by using thermodynamic arguments and reasonable assumptions, that the actual compression process for a liquid as well as for a solid is essentially isothermal.
2. Equation 2.13 is also valid for the quasi-static isothermal compression of a solid, since the path used for a liquid may also be used for a solid.
3. It is to be noted that the process is not constant volume.

2.8 Other Quasi-static Work Modes

In the preceding section, we considered one possible mode of work interaction in which the thermodynamic parameters involved are pressure and volume. These are what we call *mechanical parameters*, and the corresponding work is known as *mechanical work of expansion*. For a system in general, there could be other modes of work interaction in which the parameters involved are also mechanical in nature or where they are not mechanical in nature. Detailed discussions of many other modes of work are given in some excellent books.[4] We shall briefly mention a few of these modes below.

Mechanical Work of Stretching

Let us consider a wire as our system (Fig. 2.5). Two relevant parameters we may use to describe such a system are tension (τ) and length (L). To stretch a wire, work must be done to it. If the stretching process is being carried out quasi-statically, the work involved is

$$\bar{d}W = -\tau \, dL. \tag{2.14}$$

The tension is taken as positive. Since dL will be positive in stretching, dW due to stretching will be negative. The minus sign is used to be consistent with the convention for work that we have adopted in this book.

Electrical Work

Two relevant parameters we may use to describe an electrolytic cell (such as the battery in our automobile) are the *electromotive force* ($\mathscr{E}$) and the *quantity of electric charge* (Z) of such a system. Work is done by the cell when it is discharging, and work is done to it when it

[4] M. Zemansky, *Heat and Thermodynamics*, McGraw-Hill Book Company, New York, 1957, Chap. 3; M. Tribus, *Thermostatics and Thermodynamics*, Van Nostrand Reinhold Company, Inc., New York, 1961, pp. 9–21; T. A. Bruzustowski, *Introduction to the Principles of Engineering Thermodynamics*, Addison-Wesley Publishing Co., Inc., Reading, Mass., 1969, pp. 27–39.

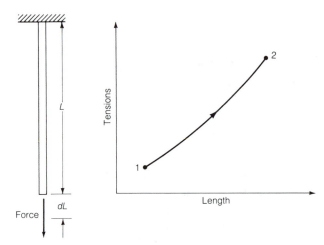

Figure 2.5 Quasi-static Stretching of a Wire

is being charged. If the charging or discharging process is being carried out quasi-statically, the work involved is

$$\bar{d}W = -\mathscr{E}\ dZ. \tag{2.15}$$

The electromotive force is positive by definition. Since dZ will be negative in a discharging process, $\bar{d}W$ due to discharging will be positive. The minus sign again is used because of the convention for work.

Magnetic Work

For a magnetic substance, two relevant parameters that we may use to describe it are the *external field* $\mathscr{H}$ and its *magnetization M*. If the magnetization or demagnetization process is being carried out quasi-statically, the magnetic work involved is

$$\bar{d}W = -\mathscr{H}\ dM. \tag{2.16}$$

The external magnetic field is taken as positive. Work must be done to the substance in a magnetization process—hence the minus sign following our adopted convention on work.

A special kind of magnetic substance is the paramagnetic material. A most fascinating application of a paramagnetic substance is the use of it in the design of a magnetic refrigerator that can produce temperatures below 1.0 K.[5]

[5] Randall Barron, *Cryogenic Systems*, McGraw-Hill Book Company, New York, 1966, pp. 347–359.

2.9 Summary of Work Interactions

We have seen that there are many kinds of quasi-static work modes. If all work modes are possible, then the quasi-static work in general is given as

$$\bar{d}W = p\ dV - \tau\ dL - \mathscr{E}\ dZ - \mathscr{H}\ dM + \cdots . \qquad (2.17)$$

We observe that each kind of work mode is given as the product of an intensive thermodynamic property and the differential of an extensive thermodynamic property. Analogous to the definition of work in mechanics, in which work done is the result of a displacement dX effected by a force $\mathscr{F}$, we may call each of the intensive properties a *generalized force* and its corresponding extensive property a *generalized displacement*. Thus a difference in pressure provides a generalized force for a volume change, and a difference in external magnetic field provides a generalized force for a change in magnetization.[6]

We shall see that identification of the work modes involved is a very important part in the development of the structure of thermodynamics. However, it is strictly a matter of experience to determine how many work modes should be taken into account. This means that the more we understand the behavior of various thermodynamic substances, the better we shall be equipped to solve practical problems.

To conclude our discussion on work, it should be noted that many types of work are also associated with non-quasi-static processes. In the study of lubrication, shear work due to viscous friction is involved, causing the lubricating oil to become hot. By turning the paddle wheel inside a vessel containing water, we observe that the temperature of the water will rise due to the transfer of shaft work. The rapid burning of gasoline inside the cylinder of our car results in work of expansion, which in turn becomes shaft work. An important point to remember is that, regardless of whether we are dealing with quasi-static processes or not, work transfer is involved when there will be energy crossing the boundary of the system without involving any mass transfer and when the sole effect external to the system could be the raising or lowering of a weight. Work is done by a system on its surroundings if the sole effect external to the system could be the raising of a weight. Work is done on a system by its surroundings if the sole effect external to the system could be the lowering of a weight.

---

[6] H. B. Callen, *Thermodynamics*, John Wiley & Sons, Inc., New York, 1960, pp. 43–46.

2.10 Thermodynamic Definition of Heat

Heat is an important concept. It is unique to thermodynamics. Although the true nature of work was understood rather early, heat was a concept subject to much misunderstanding. Actually, it was not until the beginning of the twentieth century that a clear, unequivocal definition of heat was given. *Heat* is simply energy transferred, without transfer of mass, across the boundary of a system when such energy is not accountable for as work. We have already introduced it into the first law of thermodynamics for a closed system. It will be recalled that

$$Q_{12} = E_2 - E_1 + W_{12} \qquad (2.8)$$

and

$$\bar{d}Q = dE + \bar{d}W. \qquad (2.8a)$$

One may wish to use this equation as the defining equation of heat.

By accepting this definition of heat, we tacitly accept heat, as in the case of work, as energy in transit. Once this form of energy crosses the boundary of a system, as in work, it "disappears" and becomes part of the energy content of the system or the surroundings. Positive heat transfer is heat addition to a system, and negative heat transfer is heat rejection by a system.

Since dE is an exact differential and $\bar{d}W$ is an inexact differential, $\bar{d}Q$ must be an inexact differential. Similar to work, heat is also a path function. That is, heat is not a thermodynamic property.

We have mentioned previously that when a system is enclosed by an adiabatic wall, there will be no heat transfer crossing its boundary. Applying the first law to a closed system undergoing a change of state adiabatically, we have

$$dE = -\bar{d}W_{\text{ad}}, \qquad (2.18)$$

where $\bar{d}W_{\text{ad}}$ means adiabatic work.

In words, Eq. 2.18 says that the work is the same for all possible paths when a closed system undergoes an adiabatic process between two given equilibrium states. This equation is used by some authors as the starting point in establishing the first law of thermodynamics.

Equation 2.8a, together with Eq. 2.18, indicates that from an operational point of view, heat is measured in a process by measuring the work W_{12} performed during the process under consideration and by measuring the adiabatic work $W_{\text{ad}} = -(E_2 - E_1)$ during an adiabatic process between the same states.

We have noted that pressure is a "driving force" associated with pressure–volume work. We shall see that the property "temperature" is the "driving force" associated with heat transfer. When two bodies at different temperatures are brought into thermal contact, heat interaction will occur.

2.11 Enthalpy—A Thermodynamic Property

Let us consider a differential change of state for a stationary closed system. The first law in general is given by Eq. 2.9a as

$$\bar{d}Q = dU + \bar{d}W.$$

If the process is carried out quasi-statically, $\bar{d}W$ is given by Eq. 2.17 in general as

$$\bar{d}W = p\,dV - \tau\,dL - \mathscr{E}\,dZ - \mathscr{H}\,dM + \cdots.$$

Therefore, $\bar{d}Q$ for a stationary closed system undergoing a change of state quasi-statically in general is given by

$$\bar{d}Q = dU + p\,dV - \tau\,dL - \mathscr{E}\,dZ - \mathscr{H}\,dM + \cdots. \qquad (2.19)$$

Let us limit our discussion to a substance for which only $p\,dV$ work is relevant. Then Eq. 2.19 becomes

$$\bar{d}Q = dU + p\,dV. \qquad (2.20)$$

If the process is constant pressure as well as quasi-static, Eq. 2.20 may be written as

$$\bar{d}Q = d(U + pV). \qquad (2.21)$$

Let us define a new function, called the *enthalpy*, by

$$H \equiv U + pV. \qquad (2.22)$$

Then Eq. 2.21 becomes

$$\bar{d}Q = dH. \qquad (2.23)$$

From Eq. 2.23, we see that ΔH is a measure of the heat effect in a quasi-static constant-pressure process for a substance for which only $p\,dV$ work is relevant. We wish to point out that the significance and use of enthalpy is not limited to this special process. For another process, enthalpy change would represent work transfer. Under still different constraints, zero enthalpy change is the criterion for thermodynamic equilibrium. Basically, the significance of the quantity enthalpy is purely that of mathematical convenience. It was devised simply as a convenient way to combine thermodynamic properties that occur together repeatedly in thermodynamic equations.

2.12 First Law of Thermodynamics for an Open System

We wish now to extend the basic statement of the first law (Eq. 2.1) to open systems across the boundary of which there will be mass flow in addition to the possibility of heat and work transfer. To study such systems, we shall introduce the concept of a *control volume*, which is simply a region in space upon which we are focusing our attention. The boundary of a control volume is known as the *control surface*. A control volume, shown in Fig. 2.6, will be analyzed as an open system.

With only one incoming stream and one outgoing stream, we have, according to Eq. 2.1,

$$E_{in} + Q_{cv} = (E_2 - E_1)_{cv} + W_{cv} + E_{out}, \qquad (2.24)$$

where

$$
\begin{aligned}
E_{in} &= \text{energy added to the control volume due to the} \\
&\quad \text{incoming stream at the inlet conditions} \\
E_{out} &= \text{energy removed from the control volume due to the} \\
&\quad \text{outgoing stream at the outlet conditions} \\
Q_{cv} &= \text{heat transfer across the control surface} \\
W_{cv} &= \text{all work transfer across the control surface} \\
(E_2 - E_1)_{cv} &= \text{change in stored energy within the control volume.}
\end{aligned}
$$

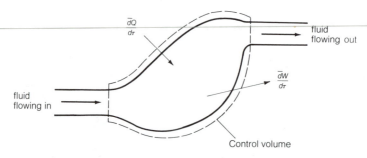

Figure 2.6 Open System

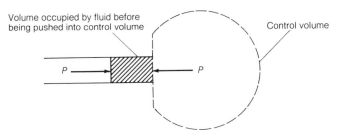

Figure 2.7 Flow Work

To make room in the control volume for the incoming fluid, a hole must be created. This is accomplished by the performance of *flow work*. To evaluate the amount of flow work involved, let us consider 1.0 kg of fluid about to enter the control volume, as shown in Fig. 2.7. Let p be the pressure at the point of entrance to the system. For the fluid to get into the control volume, work must be done on it in the amount sufficient to move it against the resistance offered by the control volume. This amount of work is equal to pAL, where A is the area on the control surface across which the fluid enters and L is the distance through which the force pA must act. But the product AL equals the specific volume, v, of the fluid at the entrance. Hence the flow work required to push 1.0 kg of fluid into the control volume is given by pv, evaluated at inlet conditions. In a similar manner, the flow work required to push 1.0 kg of fluid out of the control volume is also given by pv, evaluated at the outlet conditions. Separating the flow work from the work-transfer term in Eq. 2.24, we have

$$E_{in} + Q_{cv} = (E_2 - E_1)_{cv} + W_s + m_{out}(pv)_{out} - m_{in}(pv)_{in} + E_{out}, \quad (2.25)$$

where

W_s = all work transfer across the control surface except flow work
m_{out} = mass of outgoing stream
m_{in} = mass of incoming stream.

Note that we have introduced the negative sign for the flow work due to the incoming stream according to the sign convention on work that we use in this book.

Now E_{in} is simply given as

$$E_{in} = m_{in}\left(u + \frac{\bar{V}^2}{2g_c} + \frac{g}{g_c}Z\right)_{in} \quad (2.26)$$

and E_{out} is given as

$$E_{out} = m_{out}\left(u + \frac{\bar{V}^2}{2g_c} + \frac{g}{g_c}Z\right)_{out}.$$ (2.27)

Substituting into Eq. 2.25, the first law for an open system with one incoming stream and one outgoing stream becomes

$$Q_{cv} = (E_2 - E_1)_{cv} + W_s + m_{out}\left(u + pv + \frac{\bar{V}^2}{2g_c} + \frac{g}{g_c}Z\right)_{out}$$

$$- m_{in}\left(u + pv + \frac{\bar{V}^2}{2g_c} + \frac{g}{g_c}Z\right)_{in}.$$ (2.28)

Making use of the definition of enthalpy, $h = u + pv$, Eq. 2.28 may be written as

$$Q_{cv} = (E_2 - E_1)_{cv} + W_s + m_{out}\left(h + \frac{\bar{V}^2}{2g_c} + \frac{g}{g_c}Z\right)_{out}$$

$$- m_{in}\left(h + \frac{\bar{V}^2}{2g_c} + \frac{g}{g_c}Z\right)_{in}.$$ (2.29)

Equation 2.29 is the first-law equation for an open system in general with one stream in and one stream out. If more than one stream is involved, a summation is performed over all streams supplying energy to or removing energy from the control volume. In differential form, we have

$$dQ_{cv} = dE_{cv} + dW_s + dm_{out}\left(h + \frac{\bar{V}^2}{2g_c} + \frac{g}{g_c}Z\right)_{out}$$

$$- dm_{in}\left(h + \frac{\bar{V}^2}{2g_c} + \frac{g}{g_c}Z\right)_{in}.$$ (2.29a)

Equation 2.29a can be transformed readily into a rate equation. The rate at which the changes occur per unit time is obtained by dividing by the time differential $d\tau$. Thus

$$\frac{\bar{d}Q_{cv}}{d\tau} = \frac{dE_{cv}}{d\tau} + \frac{\bar{d}W_s}{d\tau} + \frac{dm_{out}}{d\tau}\left(h + \frac{\bar{V}^2}{2g_c} + \frac{g}{g_c}Z\right)_{out}$$

$$- \frac{dm_{in}}{d\tau}\left(h + \frac{\bar{V}^2}{2g_c} + \frac{g}{g_c}Z\right)_{in} \quad (2.30)$$

or, in equivalent notation,

$$\dot{Q}_{cv} = \dot{E}_{cv} + \dot{W}_s + \dot{m}_{out}\left(h + \frac{\bar{V}^2}{2g_c} + \frac{g}{g_c}Z\right)_{out}$$

$$- \dot{m}_{in}\left(h + \frac{\bar{V}^2}{2g_c} + \frac{g}{g_c}Z\right)_{in}. \quad (2.31)$$

In engineering application, many open systems may be modeled as operating under what is known as steady-state steady-flow conditions. These conditions are as follows:

1. The rate of heat transfer across the control surface is constant.
2. The rate of work transfer across the control surface is constant.
3. The state and velocity of each incoming stream are constant.
4. The state and velocity of each outgoing stream are constant.
5. The mass-flow rate of each incoming stream is constant.
6. The mass-flow rate of each outgoing stream is constant.
7. The total mass-flow rate of incoming streams is equal to the total mass-flow rate of outgoing streams.

With steady-state steady flow, $\dot{E}_{cv}$ will be identically equal to zero. Consequently, the energy equation for an open system operating under steady-state steady-flow conditions with one stream in and one stream out is given as

$$\dot{Q}_{cv} = \dot{W}_s + \dot{m}\left(h + \frac{\bar{V}^2}{2g_c} + \frac{g}{g_c}Z\right)_{out} - \dot{m}\left(h + \frac{\bar{V}^2}{2g_c} + \frac{g}{g_c}Z\right)_{in}. \quad (2.32)$$

Note that $\dot{m} = \dot{m}_{out} = \dot{m}_{in}$. Equation 2.32 is very important. Beginning in Chapter 9, we shall make use of it repeatedly in the study of many practical engineering processes, devices, and systems.

We wish to point out that the steady-state steady-flow equation is referred to by some authors simply as the steady-flow equation. We should realize that we could have steady-flow by just satisfying conditions 5, 6, and 7. But in order to have steady-state steady-flow, we must satisfy all seven conditions.

Example 2.9

Steam at a pressure of 101.3 kPa enters a condenser with a velocity of 50 m/s and leaves with a velocity of 15 m/s. The fluid leaves the heat exchanger at a point 2.0 m below the inlet point. Determine, for each kilogram of fluid flow under stead-state steady-flow conditions,

(a) the change of kinetic energy.
(b) the change of potential energy.

Solution

(a) $\Delta KE = \dfrac{\bar{V}_{out}^2 - \bar{V}_{in}^2}{2g_c}$

$$= \frac{[(15)^2 - (50)^2]\ \text{m}^2/\text{s}^2}{2 \times 1.0\ (\text{kg} \cdot \text{m})/(\text{N} \cdot \text{s}^2)} = -1138\ \text{J/kg} = -1.138\ \text{kJ/kg}.$$

(b) $\Delta PE = \dfrac{g(Z_{out} - Z_{in})}{g_c}$

$$= \frac{9.81\ \text{m/s}^2(-2.0)\ \text{m}}{1.0\ (\text{kg} \cdot \text{m})/(\text{N} \cdot \text{s}^2)} = -19.6\ \text{J/kg} = -0.0196\ \text{kJ/kg}. \qquad ▮$$

Comments

In the condensation of steam at 101.3 kPa, the change in enthalpy is about -2250 kJ, which is much greater than the changes in kinetic energy and potential energy. This would mean that when we applied the energy equation (Eq. 2.32) to this problem, we could have neglected the kinetic energy and potential energy terms without introducing too much of an error. This is true in general for other heat exchangers as well as devices such as pumps, compressors, and turbines.

2.13 Conservation of Mass for an Open System

With only one incoming stream and one outgoing stream, we have for an open system, according to the principle of conservation of mass

(Eq. 2.7),

$$\dot{m}_{\text{in}} - \dot{m}_{\text{out}} = \frac{dm}{d\tau}, \tag{2.33}$$

where $dm/d\tau$ is the rate of change in the mass content within the control volume. If more than one stream is involved, a summation is performed over all incoming and outgoing streams.

For an open system modeled as operating under steady-state steady-flow conditions,

$$\dot{m}_{\text{in}} = \dot{m}_{\text{out}} \tag{2.34}$$

and $dm/d\tau$ is identically equal to zero.

With one stream in and one stream out, Eq. 2.34 may be written as

$$(\rho \bar{V} A)_{\text{in}} = (\rho \bar{V} A)_{\text{out}}$$

or

$$\rho \bar{V} A = \text{constant}, \tag{2.35}$$

where ρ is the density and $\bar{V}$ is the average velocity of fluid at the section where the flowing area is A. Equation 2.35 is commonly known as the continuity equation in fluid mechanics.

2.14 Thermodynamic Reservoirs

In the development of thermodynamics, we shall find it convenient to introduce the idea of a *thermodynamic reservoir*. Since a system can have in general three types of interactions (work interaction, heat interaction, and mass interaction) with its surroundings, we shall assume that there will exist three kinds of thermodynamic reservoirs: work reservoir, heat reservoir, and matter reservoir.[7]

Work Reservoir

A *work reservoir* is a device that we may employ to keep track of the amount of work done by or done to a given thermodynamic system. It is a body in which every unit of energy crossing its boundary is work energy. A work reservoir might be visualized as a perfectly elastic spring that is compressed by the work done on it by a

[7] See Ryogo Kubo, *Thermodynamics*, North-Holland Publishing Co., Inc., Amsterdam, 1968, pp. 2–3. See also H. C. Weber and H. P. Meissner, *Thermodynamics for Chemical Engineers*, John Wiley & Sons, Inc., New York, 1957, p. 9.

system, or as a weight that is raised as the system does work upon the reservoir and lowered as the reservoir does work on the system. By definition, a work reservoir is a closed system with no heat interaction. It is an ideal system in which every unit of work transferred into it may be recovered completely to do work for us.

Heat Reservoir

A *heat reservoir* is a thermodynamic system that serves as a heat source or sink in the analysis of thermodynamic problems. We define it as a body with a very large energy capacity so that its temperature remains constant when heat flows into or out of it. The atmosphere around the earth and the ocean may be considered as heat reservoirs in many engineering applications. By definition, a heat reservoir is a closed system with no work interaction. It is an ideal system in which every unit of heat that enters the reservoir at the temperature of the reservoir may be recovered completely at the same temperature of the reservoir.

Matter Reservoir

Since matter as well as heat and work can cross the boundary of an open system, the surroundings of an open system may be imagined to contain not only heat and work reservoirs but also one or more *matter reservoirs* to supply and receive matter. A matter reservoir is considered to be sufficiently larger than the system so that the reservoir itself remains in a given equilibrium state. The atmosphere around the earth may be considered as a matter reservoir supplying air to the engines of our automobiles and to air-separation plants.

2.15 Fundamental Nature of Energy

The nature of energy may be summarized as follows:

1. Energy is a primitive concept.
2. Every system has energy.
3. Energy is an extensive property. That is, it is additive.
4. The quality of energy in a system is measured by its potential capacity to do work.
5. Energy is something that can be neither destroyed nor created. That is, the energy content of a system is conserved if we isolate the system from its surroundings.

Expressed in words, we have, for any system,

$$\text{energy input} - \text{energy output} = \text{change in stored energy} \quad (2.36)$$

or

$$\text{(change in stored energy)} + \text{(energy output} - \text{energy input)} = 0.$$

$$(2.36a)$$

The left-hand side of Eq. 2.36a may be interpreted as the energy creation in the universe due to a given process. The first term gives us the energy change in the system selected for study, and the terms within the second set of parentheses represents the energy change in the surroundings that interact with the system in question.

Equation 2.36 is the general statement of the first law of thermodynamics.

Problems

2.1 A pile driver, having a mass of 500 kg, is released from rest. It falls for a distance of 25 m when it strikes the piling. What is the ideal velocity of the pile driver just before impact? Let $g = 9.81$ m/s^2.

2.2 An elevator, having a mass of 10,000 kg, is to be raised a distance of 150 m at a location where the acceleration of gravity is 9.81 m/s^2. What is the minimum work required?

2.3 A jet of water is being delivered in a vertical direction from a nozzle with a velocity of 95 ft/s at a location where $g = 32.174$ ft/s^2. What is the ideal velocity of the jet at a point that is 100 ft directly above the nozzle?

2.4 A hydraulic turbine receives water from a reservoir at an elevation of 100 m above it. What is the minimum water flow, in kg/s, to produce a steady turbine output of 50,000 kW?

2.5 Water flows over a waterfall at the rate of 5.0×10^6 gal/h and drops 200 ft. Determine the ideal amount of power (in kilowatts) that can be generated by a power plant located at the bottom of the fall.

2.6 The mass of an automobile and its occupants is 2000 kg. What is the minimum energy needed to bring the automobile from rest to a speed of 88 km/h?

2.7 A linear spring is initially at its free length. It is then compressed 25 cm. If the average applied force is measured to be 125 N, what is the spring constant in N/m?

2.8 A linear spring has a spring constant of 100 N/m. How much work is needed to stretch it from a free length of 1.5 m to a final length of 2.0 m?

2.9 A 25-kg mass falls from rest for a distance of 5.0 m until it hits a linear spring that is initially at its equilibrium position. If all the energy of the falling body goes into compressing the spring, how far will the spring be deflected? The spring constant is 500 N/m.

2.10 A 12-V storage battery delivers current at the rate of 40 A. In a 30-min period, the heat lost from the battery is 90 kJ. What is the change in internal energy of the battery for the discharging period?

2.11 In charging a battery, it is supplied with 45 A at 12 V for a period of 30 min. During this time, the heat loss from the battery is 100 Btu. Determine the change in internal energy of the battery in this period.

2.12 A closed system undergoes a process during which 150 kJ of heat is added to it. The system is then restored to its initial state. If heat transfer and work transfer for the second process are -50 kJ and $+75$ kJ, respectively, what is the work transfer for the first process?

2.13 The stored energy of a closed system decreases by 5 kJ while 50 kJ of work is done to it. Determine the amount of heat transfer to or from the system.

2.14 A closed system undergoes a cycle made up of four processes. Fill in the missing data in the table.

Process	Q (kJ)	W (kJ)	ΔE (kJ)
$1 \rightarrow 2$	1040	0	
$2 \rightarrow 3$	0	142	
$3 \rightarrow 4$	-900	0	
$4 \rightarrow 1$	0		

2.15 The pressure, specific volume, and internal energy of a fluid at a given state are 100 kPa, 1.8 m³/kg, and 2500 kJ/kg, respectively. Calculate the enthalpy (in kJ/kg) of the fluid at this state.

2.16 The pressure, specific volume, and enthalpy of a fluid at a certain state are 50 kPa, 3.0 m³/kg, and 1560 kJ/kg, respectively. Calculate the internal energy (in kJ/kg) of the fluid at this state.

2.17 Calculate the enthalpy of 1 lbm of fluid which occupies a volume of 10 ft³ if the internal energy is 450 Btu/lbm and the pressure is 35 psia.

2.18 One kilogram of fluid undergoes a quasi-static process inside a cylinder at a constant pressure of 250 kPa. If 75 kJ of heat is removed while its volume changes from 0.5 m³ to 0.2 m³, determine
(a) the change of internal energy of the fluid in kilojoules.
(b) the change of enthalpy of the fluid in kilojoules.

2.19 Fifty pounds of air are compressed quasi-statically in a cylinder at a constant pressure of 25 psia from an initial volume of 415 ft³ to a final volume of 400 ft³. It is found that the amount of heat removed from the air is 243 Btu. Determine
(a) the change of internal energy of the air.
(b) the change of enthalpy of the air.

2.20 A piston, having a diameter of 25 cm, moves a distance of 1.0 m when acted on by a constant gas pressure of 500 kPa. Neglecting friction, determine the work done by the gas.

2.21 A gas expands quasi-statically inside a cylinder from 5000 kPa to 500 kPa according to the equation $pv = 100$, where p is the pressure in kPa and v is the specific volume in m³/kg. Determine the work done by the gas in kJ/kg.

2.22 A gas is compressed quasi-statically in a cylinder according to the equation $pv = 30{,}000$, where p is the pressure in lbf/ft² absolute and v is the specific volume in ft³/lbm. Determine the work done to the gas (in Btu/lbm) if the initial volume is five times the final volume.

2.23 A gas is compressed quasi-statically in a piston–cylinder arrangement. Show that the work done to the gas is given by

$$W_{12} = \frac{p_2 V_2 - p_1 V_1}{1 - n}$$

if the relationship between p and V for the process is

$$pV^n = \text{constant},$$

where n is any constant but is not equal to unity.

2.24 A gas is compressed quasi-statically in a piston–cylinder arrangement. We may make use of the following two paths to reach the final pressure starting with the same initial state:
(a) $pV = $ constant.
(b) $pV^{1.4} = $ constant.
Making use of a p–V plot, explain why one process will require less work than the other.

2.25 A gas expands quasi-statically in a piston–cylinder arrangement against the atmosphere and a spring. Initial pressure and volume are 400 kPa and 0.2 m³, respectively. The final volume of the gas is 0.6 m³. Determine the work done by the gas if the spring force is proportional to the volume of the gas. The atmospheric pressure is 101 kPa.

2.26 A gas in a rigid vessel is initially at 100 kPa and 25°C. Heat is then added to it in the amount of 30 kJ. If the increase in enthalpy of the gas is 40 kJ, what is the final pressure of the gas in kPa? The volume of the vessel is 0.50 m³.

2.27 Show in Fig. P2.27 is a cycle consisting of three processes executed by a closed system. Process $1 \rightarrow 2$ is quasi-static at constant volume, process $2 \rightarrow 3$ is quasi-static at constant pressure,

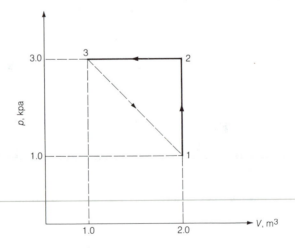

Figure P2.27

and process $3 \rightarrow 1$ is not quasi-static. Determine the net heat transfer for the cycle if work done by the system along the non-quasi-static process is 200 kJ.

2.28 The following expression is an equation of state for an elastic substance such as a strip of ideal rubber:

$$\tau = CT\left[\frac{L}{L_0} - \left(\frac{L_0}{L}\right)^2\right],$$

where τ is tension in the elastic substance, C is a constant, T is temperature in K, and L_0 is the length L at zero tension. Determine the mechanical work required to stretch such a substance from $L_1 = L_0$ to $L_2 = 3L_0$ quasi-statically and isothermally if $L_0 = 1.0$ m and $C = 1.5 \times 10^{-2}$ N/K. Temperature is 300 K.

2.29 The following expression is an equation of state for a paramagnetic substance:

$$M = \frac{C\mathcal{H}}{T},$$

where C is a constant (known as the Curie constant), M is magnetization, $\mathcal{H}$ is the external field, and T is temperature in kelvin. If we want to increase the magnetization of such a substance from M_1 to M_2 quasi-statically and isothermally, show that the work transfer is given as

$$W_{12} = -\frac{T}{2C}[(M_2)^2 - (M_1)^2].$$

2.30 If a quasi-static process may be represented by the expression $pV^n = $ constant, where n is a constant, show that the exponent n may be obtained as the slope of a straight line in a log-log plot of the p–V data.

2.31 Water at 25°C flows through a pipe with an inside diameter of 5.0 cm. Calculate the mass-flow rate of water (in kg/s) through the pipe if the average velocity of water is 3.0 m/s. The density of water is 1000 kg/m³.

2.32 Water at 25°C is being drained through a circular hole at the bottom of a tank with an average velocity of 15 m/s. The diameter of the hole is 10 cm. What is the rate of change of water in the tank?

2.33 Air flows through a compressor under steady-state steady-flow conditions. At the compressor suction, the specific volume and velocity of air are 12 ft³/lbm and 100 ft/s, respectively. At the compressor discharge, the specific volume and velocity of air are 4 ft³/lbm and 150 ft/s, respectively. Determine the ratio of the diameter of the discharge pipe to that of the suction pipe.

2.34 Crude oil flows into a storage tank at the steady rate of 5 m³/min. At the same time, crude oil is pumped out of the same tank at the steady rate of 250,000 kg/h. After 2 h of operation, what is the change of crude oil (in kilograms) in the storage tank? The density of the crude oil is 700 kg/m³.

2.35 A water pump takes suction from a cylindrical tank having a diameter of 1.5 m. If the mass-flow rate of water through the pump is 500 kg/s, what is the rate of change of the liquid level in the tank?

2.36 Water flowing in a stream contains 1000 ppm salt. The salt concentration is found to be 1100 ppm after 10 lbm/h of salt has been dumped into the stream by industrial plants. What is the flow rate of water in the stream? Give your answer in lbm/h and gal/min. (*Note:* ppm = parts of salt per million parts of water, by weight.)

2.37 A vapor is compressed in a water-jacketed compressor under steady-state steady-flow conditions. The enthalpy of the vapor at the inlet and at the outlet is 185 kJ/kg and 210 kJ/kg, respectively. If heat removed from the vapor is 10% of the work input, what is the power input in kilowatts for a flow rate of 75 kg/min? Changes in kinetic energy and potential energy are negligible.

2.38 A fluid enters a steady-state steady-flow work-producing device at a rate of 1.0 lbm/s. The enthalpy and velocity of the fluid at the entrance are 100 Btu/lbm and 300 ft/s, respectively. At the exit, the enthalpy and velocity are 75 Btu/lbm and 50 ft/s, respectively. If the process is adiabatic, and if the change in potential energy is negligible, what is the horsepower output of the device?

2.39 A fluid expands in a turbine under steady-state steady-flow conditions. It enters the turbine with a velocity of 3.0 m/s through a 2.5-cm-diameter pipe. At the entrance, the enthalpy of the fluid is 2400 kJ/kg and the specific volume is 0.34 m³/kg. At the exit, the enthalpy is 2100 kJ/kg. Neglecting heat transfer and changes in kinetic and potential energies, what is the power output of the turbine in kilowatts?

2.40 A nozzle is to be designed to produce an outlet velocity of 300 m/s under steady-state steady-flow conditions. Neglecting inlet velocity and heat transfer, what must be the drop in enthalpy (in kJ/kg) of the working fluid?

2.41 Steam enters a nozzle steadily with a velocity of 100 ft/s and an enthalpy of 1280 Btu/lbm and leaves with an enthalpy of 1150 Btu/lbm. Neglecting heat transfer, what is the velocity of steam at the nozzle outlet?

2.42 If we want to design a diffuser so that the enthalpy of the working fluid will be increased by 40 kJ/kg, what must be the velocity of the fluid at the inlet of the diffuser? Heat transfer and velocity at the outlet of the diffuser are both negligible.

2.43 Air enters a diffuser with a velocity of 1000 ft/s and temperature of 60°F and leaves with a velocity of 100 ft/s. What is the temperature of air at the diffuser outlet? Neglect heat transfer and assume that the relationship between enthalpy and temperature for air is

$$h = 0.24T + h_0 \qquad \text{Btu/lbm,}$$

where T is in °R and h_0 is a constant.

2.44 Air enters a diffuser with a velocity of 300 m/s and leaves with a velocity of 15 m/s and a temperature of 100°C. What is the temperature of air at the diffuser inlet? Neglect heat transfer and assume that the relationship between enthalpy and temperature for air is

$$h = 1.00T + h_0 \qquad \text{kJ/kg,}$$

where T is in kelvin and h_0 is a constant.

2.45 A steady flow of steam enters a condenser with an enthalpy of 2330 kJ/kg and a velocity of 350 m/s. The condensate leaves the condenser with an enthalpy of 140 kJ/kg and a velocity of 6 m/s. What is the amount of heat given to the cooling medium by each kilogram of steam condensed?

2.46 Steam enters a radiator steadily at 15 psia. At the entrance, the specific volume and enthalpy of the steam are 26.29 ft³/lbm and 1150.9 Btu/lbm, respectively. After condensing in the radiator, the condensate leaves at a pressure of 15 psia. At the exit, the specific volume and enthalpy are 0.01673 ft^3/lbm and 181.2 Btu/lbm, respectively. If changes in kinetic and potential energies are negligible, what is the heat transfer from the radiator per lbm of steam?

2.47 A gas expands quasi-statically inside a cylinder from an initial pressure of 1000 kPa and an initial volume of 5000 m^3 to a final pressure of 215 kPa and a final volume of 15,000 m^3. The variation of pressure with volume along the quasi-static path is as follows:

p, kPa	1000	775	625	520	440	380
V, m^3	5000	6000	7000	8000	9000	10000

p, kPa	330	295	260	240	215
V, m^3	11000	12000	13000	14000	15000

Write a computer program to obtain the quasi-static work of expansion by numerical integration using Simpson's Rule.

2.48 A vapor is compressed quasi-statically inside a cylinder from an initial pressure of 1000 kPa and an initial volume of 6000 m^3 to a final pressure of 6000 kPa and a final volume of 1000 m^3. The variation of pressure with volume along the quasi-static path is as follows:

p, kPa	1000	1090	1200	1330	1500	1710
V, m^3	6000	5500	5000	4500	4000	3500

p, kPa	2000	2400	3000	4000	6000
V, m^3	3000	2500	2000	1500	1000

Write a computer program to obtain the quasi-static work of compression by numerical integration using Simpson's Rule.

2.49 A vapor expands quasi-statically inside a cylinder from an initial pressure of 500 kPa and an initial volume of 50 m^3 to a final pressure of 100 kPa. The quasi-static process may be represented by the following expression:

$$pV^n = \text{constant}$$

where n is any constant.

(a) Write a computer program to generate the quasi-static paths in the p–V diagram for $n = 1.1, 1.2, 1.3, 1.4, 1.667$.

(b) In order to obtain more work output, should n be large or small?

2.50 A vapor is compressed quasi-statically inside a cylinder from an initial pressure of 100 kPa and an initial volume of 150 m³ to a final pressure of 1000 kPa. The quasi-static process may be represented by the following expression:

$$pV^n = \text{constant}$$

where n is any constant.

(a) Write a computer program to generate the quasi-static paths in the p–V diagram for $n = 1.1, 1.2, 1.3, 1.4, 1.667$.

(b) Should n be large or small if we wish to require less work input?

2.51 The variation of pressure p with volume V for a gas or vapor undergoing a change of state quasi-statically inside a cylinder may be represented by the following expression:

$$pV^n = C = \text{constant} \tag{1}$$

where n is any constant. This expression may be linearized as

$$y = a_0 + a_1 x \tag{2}$$

where

$$y = \ln p$$

$$x = \ln V$$

$$a_0 = \ln C$$

$$a_1 = -n.$$

(a) Assuming linearization is valid, write a computer program using the least square method to determine the exponent n in Eq. (1) for data given in Problem 2.47.

(b) Making use of the value of n determined in part (a), calculate the quasi-static work of the process.

2.52 Same as Problem 2.51 except for data given in Problem 2.48.

CHAPTER 3

Entropy and the Second Law of Thermodynamics

From a large accumulation of experimental evidence, scientists and engineers have deduced the scientific fact that energy in an isolated system must be conserved. But the first law of thermodynamics cannot explain physical phenomena such as the following: the conversion of heat into work cannot be carried out on a continuous basis with a conversion efficiency of 100%; heat cannot flow spontaneously from a region of low temperature to one of high temperature; air will rush into a vacuum chamber spontaneously; water and salt will mix spontaneously to form a solution, but the separation of such a solution cannot be made without some external means; a ball dropped on the floor will ultimately stop bouncing; and a vibrating spring will eventually come to rest all by itself. As the result of scientific detective work of the highest order in the study of such diverse processes, we now have accepted another important scientific fact: In an isolated system, the quality of energy, defined in terms of ability to do work, cannot be conserved. This is sometimes known as the *principle of degradation of energy*. We shall formulate the second law of thermodynamics on the basis of this primitive concept.

3.1 Entropy and Quality of Energy

Since energy has been defined as the capacity to do work, it would be reasonable for us to express the quality of energy in terms of its potential ability to do work. We may thus say that the quality of each unit of energy in system A is higher than the quality of each unit of energy in system B if each unit of energy in system A can potentially do more work than each unit of energy in system B. This means that energy in a work reservoir would have the highest quality. The portion of the energy content in a given system that can potentially do work for us is known as the available work or available energy of the system. The available work of a system may be determined in principle by having it undergo a change of state quasi-statically and interact with a standard reservoir. If we designate E_{AV} as the *available work of a system* having a total energy content of E, we may write

$$E = E_{AV} + E_{UA} ,$$

where E_{UA} may be called the *unavailable energy of the system.*

When a closed system undergoes a change of state in general, there will be a change in its total energy content as well as in its available work. Following Hatsopoulos and Keenan,[1] we shall define a primitive property, entropy S, so that the change in entropy dS in a change of state is related to the change in energy dE and the change in available work dE_{AV} evaluated with respect to one standard reservoir by the relation

$$dS = C(dE - dE_{AV}) , \tag{3.1}$$

where C is an arbitrary positive constant whose value depends on the standard reservoir. This definition implies that entropy is proportional to the unavailable energy of a system and that it is an extensive property.

For an isolated system, $dE = 0$ according to the first law of thermodynamics, and dE_{AV} must be negative according to the principle of energy degradation. Therefore, from Eq. 3.1, we must have, for an isolated system,

$$(dS)_{isol} \geq 0. \tag{3.2}$$

Equation 3.2 is the mathematical statement of the second law of thermodynamics. It states that the entropy of an isolated system can never decrease. Although the second law has been stated in many forms,[2] all statements of it can be shown to be equivalent. Moreover, all statements lead to the important result embodied in Eq. 3.2.

---

[1] G. N. Hatsopoulos and J. H. Keenan, *Principles of General Thermodynamics*, John Wiley & Sons, Inc., New York, 1965, p. 382. See also "Basic Thermodynamic Considerations for a Relativistic System" by J. H. Keenan and G. N. Hatsopoulos, contained in *A Critical Review of Thermodynamics*, edited by E. B. Stuart, B. Gal-Or, and A. J. Brainard, Mono Book Corp., Baltimore, 1970, p. 418; Huang, F. F., Clothier, R. F., "Let Us De-Mystify the Concept of Entropy: Formulation of the Second Law of Thermodynamics Based on Defining Entropy as a Measure of Unavailable Energy According to Hatsopoulos and Keenan", 1979 ASEE Annual Conference Proceedings, pp. 339–344.

[2] Nine statements of the second law of thermodynamics are given in S. R. Montgomery, *Second Law of Thermodynamics*, Pergamon Press Ltd., London, 1966.

Since a system plus its surroundings in any given interaction constitutes an isolated system, we may also express the second law as

$$(dS)_{\text{system}} + (dS)_{\text{surroundings}} \geq 0. \tag{3.3}$$

It will be recalled that the difference between a reversible process and an irreversible process is that in the former, there will be no dissipative effects whatsoever, whereas in the latter, there will always be dissipative effects. Consequently, the inequality sign in Eqs. 3.2 and 3.3 is for the case when any one of the processes carried out in an isolated system is an irreversible process. The equality sign in Eqs. 3.2 and 3.3 is thus for the case when all processes carried out in an isolated system are reversible processes.

In general, the right-hand side of Eqs. 3.2 and 3.3 must be a positive nonzero quantity. It must have the units of entropy. We may thus introduce the concept of entropy generation, production, or creation and express the second law of thermodynamics as

$$(dS)_{\text{isol}} = (dS)_{\text{system}} + (dS)_{\text{surroundings}} = \bar{d}S_{\text{irr}}, \tag{3.4}$$

where $\bar{d}S_{\text{irr}}$ is known as entropy generation due to irreversibility. It is always a positive, nonzero quantity for any real process. Its magnitude will depend on the process involved. That is, it is path dependent. It is zero only when the process is reversible.

Since entropy is a measure of unavailable energy, we would create unavailable energy (that is, destroy available energy) whenever we generate, produce, or create entropy. To conserve our natural resources, we must minimize entropy production in all our designs. How to develop processes and systems that will minimize entropy production and yet be consistent with economic, environmental, and other constraints is indeed a challenge to us all.

3.2 Entropy Change of a Work Reservoir

It will be recalled that a work reservoir is defined as a body in which every unit of energy crossing its boundary is work energy. From this definition, we have, according to the first law of thermodynamics,

$$[dE = -\bar{d}W]_{\text{work reservoir}}. \tag{3.5}$$

From the definition of available work, we have

$$[\bar{d}W = -dE_{\text{AV}}]_{\text{work reservoir}}. \tag{3.6}$$

Combining Eqs. 3.5 and 3.6, we get

$$[dE = dE_{AV}]_{\text{work reservoir}} \cdot \tag{3.7}$$

Substituting this expression into the definition of entropy, we have

$$[dS = 0]_{\text{work reservoir}} \cdot \tag{3.8}$$

This idea—that there can never be any entropy change in a work reservoir—is very useful in our study of engineering thermodynamics.

3.3 Entropy Change of a Heat Reservoir

A *heat reservoir* has been defined as a body with a very large energy capacity so that its temperature remains constant when heat, the only form of energy that can cross its boundary, flows into or out of it. From this definition, we have

$$[\bar{d}Q = dE]_{\text{heat reservoir}}, \tag{3.9}$$

which, in combination with the definition of entropy, gives us

$$[dS = C(\bar{d}Q - dE_{AV})]_{\text{heat reservoir}} \cdot$$

Since dE_{AV} is only a fraction of $\bar{d}Q$, we have

$$[dS = C_1\, \bar{d}Q]_{\text{heat reservoir}}, \tag{3.10}$$

where C_1 is a positive constant depending on the state of the heat reservoir.

It will be shown that C_1 is the reciprocal of the temperature of the heat reservoir. Therefore, we get

$$\left[dS = \frac{\bar{d}Q}{T} \right]_{\text{heat reservoir}} \cdot \tag{3.11}$$

Since the temperature of the heat reservoir remains constant,

$$\left[S_2 - S_1 = \frac{Q_{12}}{T} \right]_{\text{heat reservoir}} \cdot \tag{3.11a}$$

The Q_{12} will be positive when heat is added to the heat reservoir; the Q_{12} will be negative when heat is removed from it. Therefore, the entropy content of a heat reservoir will be increased when heat is added to it, and its entropy content will be decreased if heat is removed from it. Equation 3.11 is also very useful in our study of engineering thermodynamics.

Equation 3.11 would also imply that as temperature of the heat reservoir increases, so will the available energy content of the reservoir. This would mean that a heat reservoir becomes a work reservoir when its temperature is very, very large. That is, the temperature of a heat reservoir is an index of the quality of its energy.

3.4 Heat Transfer Between Two Heat Reservoirs

Let us investigate the heat-transfer phenomenon between two heat reservoirs. To transfer Q units of heat continuously from one heat reservoir to another heat reservoir, we may operate a device (such as a piece of metal or a heat pipe) in a cycle. During part of the cycle it will absorb heat from one heat reservoir, and during another part of the cycle it will reject the amount of heat it absorbed into another heat reservoir, as shown in Fig. 3.1.

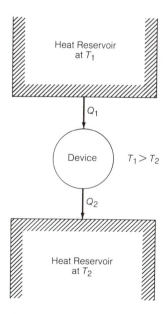

Figure 3.1 Heat Transfer Between Two Heat Reservoirs

Taking the device as our system, we have, according to the first law,

$$Q_1 - Q_2 = 0,$$

where Q_1 and Q_2 are absolute values. Note that $\Delta E = 0$ for a cyclic process.

From the second law,

$$(\Delta S)_{\text{systems}} + (\Delta S)_{\text{surroundings}} \geq 0.$$

Since the device operates in a cycle,

$$(\Delta S)_{\text{system}} = 0.$$

The only surroundings are the two reservoirs. Using Eq. 3.11a, we find

$$(\Delta S)_{\text{surroundings}} = (\Delta S)_{\text{heat reservoir at } T_1} + (\Delta S)_{\text{heat reservoir at } T_2}$$

$$= -\frac{Q_1}{T_1} + \frac{Q_2}{T_2}.$$

According to the first law, however, $Q_1 = Q_2$. Therefore, we get

$$(\Delta S)_{\text{system}} + (\Delta S)_{\text{surroundings}} = \frac{Q_1}{T_2} - \frac{Q_1}{T_1} = Q_1\left(\frac{1}{T_2} - \frac{1}{T_1}\right) \geq 0. \quad (3.12)$$

Equation 3.12 must be satisfied if any heat-transfer process is possible. It can be satisfied only if $T_1 > T_2$. This result is in agreement with experimental evidence that heat will flow by itself from a hot body to a cold body. It is also in agreement with the following classical Clausius statement of the second law, which is used by some authors as the starting point in deducing the property entropy:

It is impossible to operate a cyclic device in such a manner that the sole effect is the transfer of heat from one heat reservoir to another at a higher temperature.

3.5 Efficiencies of Heat Engines

A *heat engine* is a device that, operating in a cycle, produces work and exchanges heat with heat sources and sinks. The study of heat engines is of historical importance in the annals of science. In our present energy crisis, the study of heat engines is becoming more important than ever. The word "crisis" in Chinese is represented by the characters for two words, meaning danger (危) and opportunity (機). The danger of our energy crisis is well known. But there will also be greater opportunity and challenge for the scientists and engineers to develop new and improved heat engines so that we may utilize our energy resources more effectively.

To measure the quality of performance of a heat engine, we shall make use of the concept of *thermal efficiency* of a heat engine. It is defined as the ratio of net work produced by an engine (what you get) to the amount of heat added to the engine (what you pay). If we let

Q_H = amount of heat added to an engine

Q_L = amount of heat rejected by an engine

W = net amount of work produced by an engine,

the thermal efficiency η_{th} of a heat engine is then given by

$$\eta_{th} = \frac{W}{Q_H}.$$ (3.13)

From the first law, $Q_H - Q_L = W$, where Q_H, Q_L, and W are absolute values. Therefore,

$$\eta_{th} = 1 - \frac{Q_L}{Q_H}.$$ (3.13a)

Equation 3.13a is valid for any heat engine, reversible or irreversible.

Example 3.1

Let us consider a heat engine operating between a high-temperature reservoir at T_H and a low-temperature heat reservoir at T_L (Fig. 3.2). Derive an expression for the thermal efficiency of the heat engine if it is reversible.

Solution

We shall consider the heat engine as our system. From the first law of thermodynamics, we have

$$Q_H - Q_L = W,$$ (3.14)

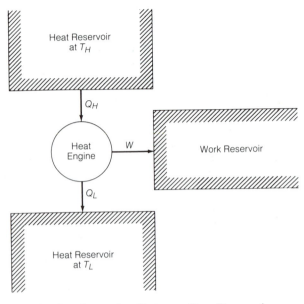

Figure 3.2 Heat Engine Operating Between Two Reservoirs

where Q_H, Q_L, and W are absolute values. From the second law of thermodynamics, we have, for a reversible system,

$$(\Delta S)_{\text{system}} + (\Delta S)_{\text{surroundings}} = 0.$$

But

$$(\Delta S)_{\text{system}} = 0 \qquad \text{since the process is cyclic}$$

and

$$(\Delta S)_{\text{surroundings}} = (\Delta S)_{\text{high-temp. heat reservoir}} + (\Delta S)_{\text{low-temp. heat reservoir}}$$

$$+ (\Delta S)_{\text{work reservoir}}.$$

Using Eq. 3.11a for the heat reservoirs and Eq. 3.8 for the work reservoir, we have

$$(\Delta S)_{\text{surroundings}} = -\frac{Q_H}{T_H} + \frac{Q_L}{T_L} + 0.$$

Therefore,

$$-\frac{Q_H}{T_H} + \frac{Q_L}{T_L} = 0$$

or

$$\frac{Q_L}{Q_H} = \frac{T_L}{T_H}.$$ (3.15)

Combining Eqs. 3.14 and 3.15 gives

$$(\eta_{th})_{ideal} = 1 - \frac{T_L}{T_H}.$$ ▮ (3.16)

We thus see that even the ideal heat engine could not have a thermal efficiency of 100% unless $T_L = 0$. This is in agreement with experimental evidence. It is also in agreement with the following classical Kelvin–Planck statement of the second law, which is used by some authors as the starting point in deducing the property entropy:

It is impossible to operate a cyclic device in such a manner that the sole effect is the exchange of heat with a single heat reservoir and the performance of an equivalent amount of work.

Comments

1. Students who have had some exposure to thermodynamics in physics should be able to recognize that Eq. 3.16 is the thermal efficiency of a Carnot cycle (a reversible cycle) operating between the temperature limits of T_H and T_L.
2. We have deduced Eq. 3.16 without using a Carnot cycle. We should not therefore jump to the conclusion that Eq. 3.16 is only valid for a Carnot cycle. We shall see that an ideal Stirling cycle and an ideal Ericsson cycle could also achieve the same kind of thermal efficiency as that given by Eq. 3.16.

Example 3.2

Consider the design of a power plant operating between a high-temperature heat reservoir at 3000°F and a low-temperature heat reservoir at 60°F.

(a) What is the ideal thermal efficiency of such a power plant?

(b) For a production of 1,000,000 kW of power, what is the rate of heat addition (energy cost) if the plant is ideal? What is the rate of heat rejection (*thermal discharge*) of such a plant? Express the results in Btu/h.

(c) For a production of 1,000,000 kW of power, what is the rate of heat addition if the actual thermal efficiency is only 40%? What is the rate of heat rejection of such a plant? Express the results in Btu/h.

Solution

(a) The ideal thermal efficiency is given by Eq. 3.16. Therefore,

$$\eta_{th} = 1 - \frac{60 + 460}{3000 + 460} = 1 - \frac{520}{3460} = 85\%.$$

(b) By definition,

$$\eta_{th} = \frac{W}{Q_{add}}.$$

For

$$\eta_{th} = 0.85$$

and

$$W = 1{,}000{,}000 \text{ kW} = 3.413 \times 10^9 \text{ Btu/h}$$

$$Q_{add} = \frac{3.413 \times 10^9}{0.85} = 4.015 \times 10^9 \text{ Btu/h}.$$

From the first law for a cycle,

$$\oint \bar{d}Q = \oint \bar{d}W.$$

Therefore,

$$|Q_{rej}| = 4.015 \times 10^9 - 3.413 \times 10^9$$

$$= 0.602 \times 10^9 \text{ Btu/h}.$$

(c) For $\eta_{th} = 0.40$,

$$Q_{add} = \frac{3.413 \times 10^9}{0.40} = 8.533 \times 10^9 \text{ Btu/h}$$

$$|Q_{rej}| = 8.533 \times 10^9 - 3.413 \times 10^9$$

$$= 5.120 \times 10^9 \text{ Btu/h}.$$

Comments

1. The maximum thermal efficiency of a practical modern power plant operating between given heat reservoirs is on the order of 40%. This means that there is considerable room for improvement, at least theoretically, since the ideal thermal efficiency is 85%.
2. Thermal discharge from power plants is one of the causes of our thermal pollution problems. More efficient power plants will help in reducing the seriousness of our environmental problems and our energy crisis.

3.6 Efficiencies of Refrigerators and Heat Pumps

Refrigerators and heat pumps are simply heat engines operating in reverse. To measure the quality of performance of these devices, we use the concept of *coefficient of performance*. It is defined as the ratio of "what you get" to "what you pay". A device may be called a refrigerator or a heat pump, depending on "what you get" from it. If the objective is to remove heat from a refrigerated space (a cold body), the device is known as a *refrigerator*. If the objective is to supply heat to another system (such as heating a house), the device is known as a *heat pump* (see Fig. 3.3).

If we let

Q_L = amount of heat transferred from a low-temperature body
Q_H = amount of heat transferred to a warmer body
W = net amount of work required,

the coefficient of performance of a refrigerator is given by

$$\beta_R = \frac{Q_L}{W} = \frac{Q_L}{Q_H - Q_L} = \frac{1}{Q_H/Q_L - 1} \qquad (3.17)$$

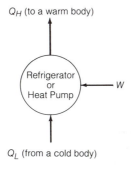

Figure 3.3 Refrigerator or Heat Pump

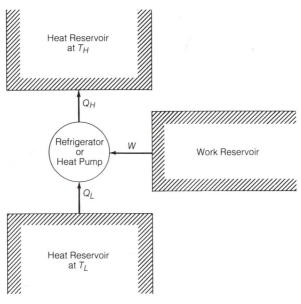

Figure 3.4 Refrigerator or Heat Pump Operating Between Two Heat Reservoirs

and the coefficient of performance of a heat pump is given by

$$\beta_{\text{HP}} = \frac{Q_H}{W} = \frac{Q_H}{Q_H - Q_L} = \frac{1}{1 - Q_L/Q_H}, \tag{3.18}$$

where Q_L, Q_H, and W are all absolute quantities. The coefficient of performance is defined as a positive quantity.

For a reversible refrigerator or heat pump operating between a low-temperature heat reservoir at T_L and a high-temperature heat reservoir at T_H, it may be shown by the method used in Example 3.1 that

$$(\beta_R)_{\text{ideal}} = \frac{T_L}{T_H - T_L} = \frac{1}{T_H/T_L - 1} \tag{3.19}$$

$$(\beta_{\text{HP}})_{\text{ideal}} = \frac{T_H}{T_H - T_L} = \frac{1}{1 - T_L/T_H}. \tag{3.20}$$

The derivations of Eqs. 3.19 and 3.20 are left as exercises (see Fig. 3.4).

Example 3.3

Determine the ideal coefficient of performance of a refrigerator operating between a low-temperature heat reservoir at 0°C and a high-temperature heat reservoir at 21°C.

Solution

The ideal coefficient of performance of a refrigerator is given by Eq. 3.19. Therefore,

$$(\beta_R)_{\text{ideal}} = \frac{1}{\dfrac{21 + 273}{0 + 273} - 1} = \frac{273}{294 - 273} = 13.0. \qquad \blacksquare$$

Comments

The two temperatures used in this example are approximately those used in the design of a home refrigerator. However, a good home refrigerator operates with a coefficient of performance of only about 4. Here is another case where there is considerable room for improvement, at least theoretically.

3.7 Examples of Irreversible Processes

It will be recalled that we have previously defined a reversible process as one that is internally reversible (quasi-static) as well as externally reversible. This means that, if a process is reversible, it is possible to undo it in such a manner that there will be no trace anywhere of the fact that the process occurred. Owing to our inability to eliminate all dissipative effects, all real processes are irreversible processes. Since dissipative effects imply that there will be a loss of opportunity to do work, there will be entropy creation in the universe whenever an irreversible process is carried out. In this section we shall make use of a few examples to demonstrate the one-to-one correspondence between irreversibility and entropy creation in the universe.

Heat Transfer Between Two Heat Reservoirs

Let Q units of energy flow from a heat reservoir at T_H to a heat reservoir at T_L, which is smaller than T_H by a finite amount. This is a natural phenomenon that we have all observed. As T_H is greater than T_L, each unit of energy located in the high-temperature reservoir will have a greater potential to do work than each unit of energy located in the low-temperature reservoir. Therefore, when Q units of energy has been relocated from the heat reservoir at T_H to the heat reservoir at T_L, the available work of this Q units of energy will have been reduced. According to our definition of entropy, the heat-transfer process will have resulted in an entropy increase in our universe. Since heat flow from a lower- to a higher-temperature heat reservoir does not occur, the given heat-transfer process is an irreversible one. The extent of irreversibility is quantified in terms of entropy creation in the universe.

Example 3.4

Determine the rate of entropy creation in the universe when the rate of heat transfer between a heat reservoir at 3000 K and another reservoir at 1000 K is 8.50×10^9 kJ/h.

Solution

System selected for study: as shown in Fig. 3.1.

Entropy creation in the universe is given by Eq. 3.12 for this problem.

$$(\Delta \dot{S})_{\text{creation}} = \dot{Q}\left(\frac{1}{T_L} - \frac{1}{T_H}\right)$$

$$= 8.50 \times 10^9 \left(\frac{1}{1000 + 273} - \frac{1}{3000 + 273}\right)\frac{\text{kJ}}{\text{K} \cdot \text{h}}$$

$$= 8.50 \times 10^9 \left(\frac{1}{1273} - \frac{1}{3273}\right)\frac{\text{kJ}}{\text{K} \cdot \text{h}}$$

$$= 4.08 \times 10^6 \frac{\text{kJ}}{\text{K} \cdot \text{h}}. \qquad \blacksquare$$

Comments

1. The positive sign for entropy creation indicates that the heat-transfer process is an irreversible one.
2. We have used absolute temperatures in our calculations, as we must always do.
3. For a given heat-source temperature, irreversibility in heat transfer can be reduced by reducing the difference between the temperature of the heat source and the temperature of the heat sink.
4. Reversible heat transfer can only be approached by making the temperature of the heat sink differ from the temperature of the heat source by an infinitesimal amount.

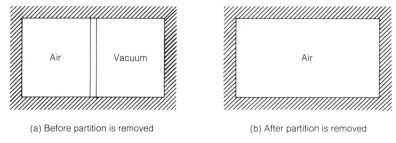

(a) Before partition is removed   (b) After partition is removed

Figure 3.5 Free Expansion—An Irreversible Process

Free-Expansion Process

Let us consider a rigid adiabatic vessel separated into two components by a partition. Let us fill one compartment with air and have the other compartment evacuated, as shown in Fig. 3.5a. Now let us carry out a free-expansion process by removing the partition. It is an experimental fact that almost instantaneously the air will fill the entire volume of the vessel, as shown in Fig. 3.5b. The process is definitely not quasi-static. Since the air is not interacting with any surroundings, the free expansion must be irreversible. A rigorous proof to show that this process is irreversible has been given by Hatsopoulos and Keenan.[3]

Let us now examine the quality of the energy content of the air. For a free expansion, the air constitutes an isolated system. Consequently, the energy content of the air will remain constant according to the first law. But since the final volume occupied by the air will be greater than its initial volume, the final pressure of the air will be found to be lower than its initial pressure. The available work of the air can be determined in principle by placing the air in a frictionless piston–cylinder arrangement and letting it expand quasi-statically against a standard reservoir. It is clear that the air will have a greater potential to do work at its initial state than at its final state. From our definition of entropy, then, there will be an entropy increase in our universe as a result of the free-expansion process. Again, entropy is created because of an irreversible process.

According to the second law for an irreversible process, the entropy creation in the universe is given by

$$(\Delta S)_{\text{creation}} = (\Delta S)_{\text{system}} + (\Delta S)_{\text{surroundings}} > 0. \tag{3.3}$$

[3] G. N. Hatsopoulos and J. H. Keenan, *Principles of General Thermodynamics*, John Wiley & Sons, Inc., New York, 1965, pp. 133–134.

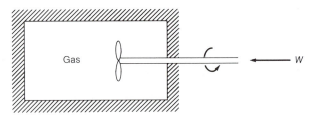

Figure 3.6 Stirring—An Irreversible Process

But $(\Delta S)_{\text{surroundings}} = 0$ for a free-expansion process. Thus we have, for our free-expansion process,

$$(\Delta S)_{\text{creation}} = (\Delta S)_{\text{air}} > 0.$$

We thus see that the entropy content of the air at the final state will be greater than its entropy content at the initial state.

Stirring Process

Let us consider a rigid adiabatic vessel containing a given amount of gas. Let us change the state of the gas by stirring it with a paddle wheel (Fig. 3.6). Since work is supplied to turn the paddle wheel, the temperature of the gas will increase. If this process were reversible, we would first be able to remove from the gas a quantity of heat exactly equal to the amount of work supplied, and then use this quantity of heat as the heat addition to a heat-engine cycle. Since it is not possible to have a 100% conversion efficiency from heat to work in a cyclic process, the work energy supplied to the gas cannot be completely recovered as work. The stirring process must be irreversible. There will be a decrease in available energy as a result of the stirring process. That is, an increase of entropy in the universe will occur because of the irreversible process of stirring.

Example 3.5

Let 100 kg of a given fluid undergo a change of state inside a rigid adiabatic vessel through the use of a stirring process. Determine the entropy increase in the universe caused by the stirring process.

Solution

The entropy increase in the universe is given by Eq. 3.3.

$$(\Delta S)_{\text{isol}} = (\Delta S)_{\text{system}} + (\Delta S)_{\text{surroundings}}.$$

Taking the fluid as our system, we have

$$(\Delta S)_{isol} = (\Delta S)_{fluid} + (\Delta S)_{work\ reservoir}.$$

But $(\Delta S)_{work\ reservoir} = 0$, according to Eq. 3.8. Therefore,

$$(\Delta S)_{isol} = (\Delta S)_{fluid} = m(s_2 - s_1) = 100(s_2 - s_1)\ kJ/K$$

where s_1 is the specific entropy of the fluid at state 1 in kJ/kg·K and s_2 is the specific entropy of the fluid at state 2 in kJ/kg·K.   ∎

Comments

1. To go further with this problem, we must know what kind of fluid is involved.
2. In addition, we must know how the states of this fluid are determined. That is, we must know how the entropy change is functionally related to changes in other thermodynamic properties.
3. We shall begin our study of the relations among various thermodynamic properties in Chapter 4.

3.8 Second Law of Thermodynamics for a Closed System

For a closed system, only heat and work can cross its boundary. As a consequence, the second law for such a system may be expressed as

$$\Delta S_{sys} + \Delta S^{HR} + \Delta S^{WR} \geq 0, \tag{3.21}$$

where

ΔS^{HR} = entropy change of the heat reservoir involved in the interaction

ΔS^{WR} = entropy change of the work reservoir involved in the interaction.

Now according to Eq. 3.11a, we have

$$\Delta S^{HR} = \frac{Q^{HR}}{T^{HR}}$$

and, according to Eq. 3.8, we have

$$\Delta S^{WR} = 0.$$

Therefore, for a closed system, we have

$$\Delta S_{\text{sys}} + \frac{Q^{\text{HR}}}{T^{\text{HR}}} \geq 0. \tag{3.22}$$

Since $Q^{\text{HR}} = -Q_{\text{sys}}$ according to the first law, Eq. 3.22 may be given as

$$\Delta S_{\text{sys}} - \frac{Q_{\text{sys}}}{T^{\text{HR}}} \geq 0. \tag{3.22a}$$

For a differential change of state, $T^{\text{HR}} = T + dT$, and T^{HR} may be replaced by T of the system as the second-order differential $(dS)(dT)$ may be neglected. Thus for a closed system, we have

$$dS \geq \frac{\bar{d}Q}{T}. \tag{3.23}$$

Equation 3.23 is the mathematical expression of the second law for a closed system. Note that the entropy change of a closed system can be positive, negative, or zero. However, if the process is adiabatic, it is always positive, and, in the limit of a reversible process, is equal to zero. Consequently, a reversible adiabatic process is the same as an isentropic (constant-entropy) process.

Equation 3.23 may be written as

$$dS = \frac{\bar{d}Q}{T} + \bar{d}S_{\text{irr}}, \tag{3.23a}$$

where $\bar{d}S_{\text{irr}}$ is entropy generation due to internal irreversibility. This equation shows that the change in entropy of a closed system is the result of two effects: heat transfer and internal irreversibility. Note that there is no entropy change associated with the work transfer effect since there is no entropy associated with work transfer.

Let us now make use of Eq. 3.23a to investigate the work obtainable from a closed system by following a reversible path and compare the work obtainable from the system by following an irreversible path for identical change of state. From the first law, we have

$$\bar{d}Q_{\text{rev}} = dE + \bar{d}W_{\text{rev}} \tag{1}$$

and

$$\bar{d}Q_{\text{irr}} = dE + \bar{d}W_{\text{irr}}. \tag{2}$$

From the second law, we have

$$dS = \frac{\bar{d}Q_{rev}}{T} \tag{3}$$

and

$$dS = \frac{\bar{d}Q_{irr}}{T} + \bar{d}S_{irr}. \tag{4}$$

From Eqs. 1, 2, 3, and 4 we have, for an identical change of state,

$$\bar{d}W_{rev} - \bar{d}W_{irr} = T\, \bar{d}S_{irr} > 0. \tag{3.24}$$

From Eq. 3.23a we observe that in a work-producing device (such as an engine) the reversible work is always larger than the actual work, while in a work-absorbing device (such as a compressor) the magnitude of the actual work required will be greater than that of the reversible work.

Defining a quantity known as irreversibility by

$$\text{irreversibility} = I = W_{rev} - W_{irr} \tag{3.25}$$

the second law for a closed system may also be given as

$$dS = \frac{\bar{d}Q}{T} + \frac{dI}{T}. \tag{3.26}$$

The term "irreversibility" is also known as "lost work". It is, just like entropy generation, a measure of the extent of irreversibility within the system. It is a positive quantity for an irreversible process. It is zero for a reversible process. The concept of lost work will be explored in more detail in Section 3.10.

3.9 Second Law of Thermodynamics for an Open System

For an open system, there will be mass transfer across its boundary. In our entropy accounting for such a system, we must take into consideration the entropy content of the incoming and outgoing streams. Thus the entropy change of an open system with one stream in and one stream out may be given as

$$dS = \frac{\bar{d}Q}{T} + \bar{d}S_{irr} + (s\, dm)_{in} - (s\, dm)_{out} \tag{3.27}$$

or

$$\bar{d}S_{\text{irr}} = dS + (s\ dm)_{\text{out}} - (s\ dm)_{\text{in}} - \frac{\bar{d}Q}{T} > 0, \qquad (3.28)$$

where

$$(s\ dm)_{\text{out}} = \text{entropy content of the outgoing stream}$$
$$(s\ dm)_{\text{in}} = \text{entropy content of the incoming stream.}$$

In the case of more than one stream in and more than one stream out, we have

$$\Delta S_{\text{irr}} = \Delta S + \sum_{\text{out}} ms - \sum_{\text{in}} ms - \sum \frac{Q_k}{T_k} > 0, \qquad (3.29)$$

where the heat-transfer quantities, Q_k, are being transferred at temperature T_k. Equation 3.29 is the mathematical expression of the second law for an open system. The left-hand side of this equation represents the entropy creation in the universe. As a rate equation, we have

$$\left(\frac{\bar{d}S}{d\tau}\right)_{\text{irr}} = \frac{dS}{d\tau} + \sum_{\text{out}} \dot{m}s - \sum_{\text{in}} \dot{m}s - \sum \frac{\dot{Q}_k}{T_k} > 0, \qquad (3.29a)$$

where $dS/d\tau$ is the rate of entropy change within the open system or control volume.

In the case of steady-state steady flow, $dS/d\tau$ will be identically equal to zero. Equation 3.29a then becomes

$$\dot{S}_{\text{gen}} = \sum_{\text{out}} \dot{m}s - \sum_{\text{in}} \dot{m}s - \sum \frac{\dot{Q}_k}{T_k} > 0, \qquad (3.30)$$

where $\dot{S}_{\text{gen}}$ is entropy generation due to internal irreversibility. Entropy generation is also known as entropy production or entropy creation. For steady-state steady flow involving a reversible adiabatic process, we have, according to Eq. 3.30,

$$\sum_{\text{out}} \dot{m}s = \sum_{\text{in}} \dot{m}s,$$

which means that the total entropy content of the outgoing streams must be equal to the total entropy content of the incoming streams. For this special and ideal case involving only one stream of fluid, the specific entropy of the fluid at the outlet state is identical to the specific entropy of the fluid at the inlet state, in which case one might say that the fluid has undergone an isentropic process.

For steady-state steady flow involving an irreversible adiabatic process, we have, according to Eq. 3.30,

$$\sum_{\text{out}} \dot{m}s > \sum_{\text{in}} \dot{m}s,$$

which means that the total entropy content of the outgoing streams must be greater than the total entropy content of the incoming streams. If there is only one stream for this special irreversible case, the specific entropy of the fluid at the outlet state will be greater than the specific entropy of the fluid at the inlet state.

We should realize that the entropy content of the outgoing streams could be greater or smaller than, or equal to, the entropy content of the incoming streams if there is any heat transfer involved in a steady-state steady-flow process.

3.10 The Concept of Lost Work

We have introduced the concept of entropy generation to quantify the amount of thermodynamic irreversibility. We could also make use of a concept known as lost work[4] to do the same thing.

By definition, we have

$$\bar{d}S_{\text{irr}} = \frac{\bar{d}\text{LW}}{T_R} \tag{3.31a}$$

or

$$\text{LW} = T_R S_{\text{gen}}, \tag{3.31b}$$

where LW is known as lost work, with T_R being a reference temperature. Since S_{gen}, entropy generation, is always positive, lost work is always positive for any real process. This is just another way of stating that entropy generation is a measure of lost opportunity to do work.

We should note that entropy generation S_{gen} depends solely on the degree of thermodynamic irreversibility of the process involved. On the other hand, the amount of lost work is a relative quantity that

---

[4] N. De Nevers and J. D. Seader, "Lost Work: A Measure of Thermodynamic Efficiency," *Energy*, Vol. 5 (1980), pp. 757–769.

depends on the choice of the reference temperature T_R. A designer may at times find the use of a particular reference temperature more relevant to the system being studied.

If we let the reference temperature be the absolute temperature of the environment T_0, we call the lost work the thermodynamic lost work to distinguish it from lost work using a different reference temperature. In this text, whenever we use the term "lost work," thermodynamic lost work is implied unless otherwise indicated. That is, thermodynamic lost work LW_0, by definition, is given by

$$LW_0 = T_0 S_{gen} \tag{3.32}$$

Equation 3.32 is also known as the Gouy–Stodola theorem, the result named after the first two thermodynamicists who recognized its importance.[5] It will be shown that the thermodynamic lost work is identical to the concept of lost availability or exergy destruction introduced in Chapter 11.

Example 3.6

If 1000 kJ of energy is transferred from a work reservoir to a heat reservoir at 373 K, determine:

(a) the amount of entropy generation.
(b) the amount of lost work with the environment at 300 K.

Solutions

Taking the work reservoir and the heat reservoir as our system, we have

(a)
$$S_{gen} = \Delta S^{WR} + \Delta S^{HR}$$

$$= 0 \quad + \frac{1000 \text{ kJ}}{373 \text{ K}} = 2.681 \text{ kJ.}$$

(b) Making use of Eq. 3.32, we have

$$LW_0 = T_0 S_{gen} = 300 \times 2.681 \text{ kJ} = 804.3 \text{ kJ.}$$

Comments

With the environment at 300 K, the available energy in 1000 kJ in the heat reservoir at 373 K is $(1 - 300/373) \times 1000$ kJ or 195.7 kJ, which is the differ-

---

[5] Jan Szargut, "International Progress in Second Law Analysis," *Energy*, Vol. 5 (1980), pp. 709–718; Adrian Bejan, *Entropy Generation Through Heat and Fluid Flow*, John Wiley & Sons, Inc., New York, 1982, Chap. 2.

ence between 1000 kJ and 804.3 kJ. We thus see that the amount of lost work is simply the amount of lost available energy since every unit of energy in the work reservoir is available energy. This problem is the same as boiling water at 373 K using electrical energy. Even though energy is conserved, the process is very inefficient or ineffective on the basis of available energy accounting. We may say that the efficiency of our process is 100% according to the first law of thermodynamics, but it is only 195.7/1000 or 19.57% according to the second law of thermodynamics.

Example 3.7

If we transfer 1000 kJ of energy from a heat reservoir at 1000 K to a heat reservoir at 500 K, determine:

(a) the amount of entropy generation.
(b) the amount of lost work with the environment at 300 K.

Solution

(a) Taking the two heat reservoirs as our system, we have

$$S_{gen} = \Delta S_{1000}^{HR} + \Delta S_{500}^{HR}$$

$$= -\frac{1000 \text{ kJ}}{1000 \text{ K}} + \frac{1000 \text{ kJ}}{500 \text{ K}} = +1.0 \text{ kJ/K}.$$

(b) The amount of lost work is given by Eq. 3.32 as

$$LW_0 = T_0 S_{gen} = 300 \times 1.0 \text{ kJ} = 300 \text{ kJ}. \qquad \blacksquare$$

Comment

With the environment at 300 K, the amount of available energy in 1000 kJ in the 1000-K heat reservoir is $(1 - 300/1000) \times 1000$ kJ or 700 kJ, and the amount of available energy in 1000 kJ in the 500-K heat reservoir is $(1 - 300/500) \times 1000$ kJ or 400 kJ. We thus see that we have lost $(700 - 400)$ or 300 kJ of available energy due to the irreversible heat-transfer process.

Example 3.8

A heat engine is designed to operate between a high-temperature heat reservoir at 2000 K and a low-temperature heat reservoir at 300 K. If the actual power output of the engine is 1.0×10^6 kW when heat addition is at the rate of 2.5×10^6 kJ/s, determine the rate of lost work production (lost power) due to irreversibility.

Solution

From Eq. 3.32 we have

$$\dot{LW}_0 = T_0 \dot{S}_{gen}.$$

From the second law for a heat engine, we have

$$\dot{S}_{gen} = -\frac{|\dot{Q}_H|}{T_H} + \frac{|\dot{Q}_L|}{T_L}.$$

From the first law for a heat engine, we have

$$|\dot{Q}_H| - |\dot{Q}_L| = \dot{W}_{net}.$$

Thus

$$|\dot{Q}_L| = 2.5 \times 10^6 - 1.0 \times 10^6 \text{ kJ/s} = 1.5 \times 10^6 \text{ kJ/s}$$

$$\dot{S}_{gen} = -\frac{2.5 \times 10^6}{2000} + \frac{1.5 \times 10^6}{300}$$

$$= 3750 \text{ kJ/K} \cdot \text{s}$$

$$\dot{LW}_0 = 300 \times 3750 \text{ kJ/s} = 1.125 \times 10^6 \text{ kW}. \qquad \blacksquare$$

Comments

1. The ideal power output is $(1 - 300/2000) \times 2.5 \times 10^6$ or 2.125×10^6 kW, which is the sum of the actual power output and the lost power. That is, $\dot{W}_{ideal} = \dot{W}_{act} + \dot{LW}_0 = 1.0 \times 10^6 + 1.125 \times 10^6 = 2.125 \times 10^6$ kW.
2. The actual thermal efficiency of the engine is 40%. On the other hand, the ratio of actual power output to ideal power output is $(1.0 \times 10^6)/(2.125 \times 10^6)$ or 47.06%. That is, our design is 47.06% perfect from the thermodynamics point of view, a more meaningful measure than the actual thermal efficiency.

Example 3.9

A heat pump is designed to deliver energy to a heat reservoir at 450 K at the rate of 1500 kJ/s. If power consumption of the heat pump is 1000 kW when it receives energy from the environment at 300 K, what is the rate of lost work production due to irreversibility?

Solution

From Eq. 3.32, we have

$$L\dot{W}_0 = T_0 \dot{S}_{gen}.$$

From the second law for a heat pump, we have

$$\dot{S}_{gen} = +\frac{|\dot{Q}_H|}{T_H} - \frac{|\dot{Q}_L|}{T_L}.$$

From the first law for a heat pump, we have

$$|\dot{Q}_H| = |\dot{W}_{net}| + |\dot{Q}_L|.$$

Thus

$$|\dot{Q}_L| = 1500 - 1000 = 500 \text{ kJ/s}$$

$$\dot{S}_{gen} = \frac{1500}{450} - \frac{500}{300} = 1.666 \text{ kJ/s} \cdot \text{K}$$

$$L\dot{W}_0 = 300 \times 1.666 \text{ kW} = 500 \text{ kW}. \qquad \blacksquare$$

Comments

1. The ideal power input is $[(450 - 300) \times 1500]/450$ or 500 kW, which is the algebraic sum of the actual power input and the lost power generated. That is, $\dot{W}_{ideal} = \dot{W}_{act} + L\dot{W}_0 = -1000 + 500 = -500$ kW.
2. The actual coefficient of performance is 1500/1000 or 1.5. The ratio of ideal power needed to actual power consumed is 500/1000 or 50%. That is, our design is 50% perfect from the thermodynamics point of view, a more meaningful measure than the actual coefficient of performance.

Example 3.10

A refrigerator is designed to remove energy from a heat reservoir at 480°R at the rate of 12,000 Btu/h. If power needed to run the refrigerator is 4000 Btu/h when heat is rejected to the environment at 520°R, what is the rate of lost work production due to thermodynamic imperfection?

Solution

From Eq. 3.32, we have

$$L\dot{W}_0 = T_0 \dot{S}_{gen}.$$

From the second law for a refrigerator, we have

$$\dot{S}_{gen} = +\frac{|\dot{Q}_H|}{T_H} - \frac{|\dot{Q}_L|}{T_L}.$$

From the first law for a refrigerator, we have

$$|\dot{Q}_H| = |\dot{W}_{net}| + |\dot{Q}_L|.$$

Thus

$$|\dot{Q}_H| = 12{,}000 + 4000 = 16{,}000 \text{ Btu/h}$$

$$\dot{S}_{gen} = \frac{16{,}000}{520} - \frac{12{,}000}{480} = 5.769 \text{ Btu/h} \cdot \text{R}$$

$$L\dot{W}_0 = 520 \times 5.769 \text{ Btu/h} = 3000 \text{ Btu/h}. \qquad \blacksquare$$

Comments

1. The ideal power input is $[(520 - 480) \times 12{,}000]/480$ or 1000 Btu/h, which is the algebraic sum of the actual power input and the lost power generated. That is, $\dot{W}_{ideal} = \dot{W}_{act} + L\dot{W}_0 = -4000 + 3000 = -1000$ Btu/h.
2. The actual coefficient of performance is 12,000/4000 or 3.0. The ratio of ideal power needed to actual power consumed is 1000/4000 or 25%. That is, our design is 25% perfect from the thermodynamics point of view, a more meaningful measure than the actual coefficient of performance.

3.11 Entropy, Equilibrium, and Direction of Change

The mathematical statement of the second law of thermodynamics for an isolated system is given by Eq. 3.2,

$$dS_{isol} \geq 0.$$

We thus see that the entropy of an isolated system tends to increase. This means that changes of state in an isolated system can occur only in the direction of increasing entropy. This also implies that an isolated system will have attained equilibrium only when its entropy has attained the maximum value. Thus the general equilibrium condition of an isolated system is given by

$$dS_{isol} = 0. \tag{3.33}$$

Now, an isolated system is simply a system under the constraint of no interaction with any surroundings. Let us now consider a system under different kinds of constraints. Let us consider a closed system constrained to constant temperature and constant pressure. We may consider such a system to be in contact with a heat reservoir at a temperature equal to that of the system and also in contact with a pressure reservoir[6] at a pressure equal to that of the system. For a change of state for this system, we have, from the first law,

$$\bar{d}Q = dU + p\,dV,$$

and from the second law, we have

$$dS \geq \frac{\bar{d}Q}{T}.$$

Combining, we have

$$dU + p\,dV - T\,dS \leq 0. \tag{3.34}$$

Since T and p are constant, Eq. 3.34 may be written as

$$d(U + pV - TS)_{T,\,p} \leq 0. \tag{3.35}$$

Let us define a new function, called the *Gibbs function*, by

$$G \equiv U + pV - TS \equiv H - TS. \tag{3.36}$$

Then Eq. 3.35 becomes

$$dG_{T,\,p} \leq 0. \tag{3.37}$$

Equation 3.37 gives the direction of change for a closed system constrained to constant temperature and pressure. It implies that a

---

[6] A pressure reservoir, such as the atmosphere, may be considered as a system of very large volume so that the only kind of energy that crosses its boundary is $p\,dV$ work. Thus a pressure reservoir is a form of work reservoir. See H. B. Callen, *Thermodynamics*, John Wiley & Sons, Inc., New York, 1960, p. 66.

change of state of such a system is only possible if its Gibbs function decreases. Thus, at equilibrium, the Gibbs function of such a system must have attained the minimum value. Important applications of this result are found in the study of vapor pressure, phase equilibrium, osmotic pressure, and chemical reactions. We wish to point out that the significance and use of the Gibbs function is not limited to the special case just studied. As in the case of the enthalpy function, the definition of the Gibbs function is purely a matter of mathematical convenience.

For a chemical reaction carried out at constant temperature and pressure, we have, according to Eq. 3.37,

$$\Delta G = \Delta H - T\Delta S \leq 0.$$

This means that such a chemical reaction is possible only if the Gibbs function for the products is less than the Gibbs function for the reactants.

In a chemical reaction, ΔH is known as the *heat of reaction at constant pressure*. It is negative for an exothermic reaction, and positive for an endothermic reaction. For a given reaction, the tendency to occur depends on the sign as well as the magnitude of ΔH and ΔS. As an example, let us consider the oxidation of solid carbon according to the following two chemical equations:

(A) $$C + O_2 \rightarrow CO_2$$

(B) $$2C + O_2 \rightarrow 2CO.$$

It is an experimental fact that much more heat is released in the first reaction than in the second. Therefore, ΔH is more negative in the first. However, ΔS may be shown to be much greater in the second reaction than in the first. We thus see that there are two opposing tendencies at work. Since the ΔS term is multiplied by the temperature T, the entropy effect is expected to be less at low temperature. We must conclude that to minimize the production of carbon monoxide, we must burn carbon at low temperature.

Let us consider another problem that is of great importance to our quality of life. It may be shown that both of the following reactions are possible:

(C) $$2NO + O_2 \rightarrow 2NO_2$$

(D) $$2NO_2 \rightarrow 2O_2 + N_2.$$

The first reaction shows that nitric oxide can react with oxygen to form nitrogen dioxide, which is harmful to our health. Since tons of nitrogen oxides are pouring into the atmosphere from our automobiles, this reaction is of major concern to us all. Looking at the second reaction, we see that nitrogen dioxide can decompose into nitrogen and oxygen, both of which are acceptable constituents of the atmosphere. Thus a way to solve our air pollution problem is to find the conditions that would accelerate the second reaction. This will require a good understanding of thermodynamics, chemistry, and chemical kinetics.

Let us now consider a closed system constrained to constant temperature and volume. For a change of state for this system, we have, from the first law,

$$\bar{d}Q = dU$$

and from the second law,

$$dS \geq \frac{\bar{d}Q}{T}.$$

Combining, we obtain

$$dU - T\,dS \leq 0. \tag{3.38}$$

But T is constant. Therefore, Eq. 3.38 may be written as

$$d(U - TS)_{T,V} \leq 0. \tag{3.39}$$

Let us define a new function, called the *Helmholtz function*, by

$$A \equiv U - TS. \tag{3.40}$$

Then Eq. 3.40 becomes

$$dA_{T,V} \leq 0. \tag{3.41}$$

Equation 3.41 gives the direction of change for a closed system constrained to constant temperature and volume. It implies that a change of state of such a system is only possible if its Helmholtz function decreases. Thus, at equilibrium, the Helmholtz function of such a system must have attained the minimum value. As in the case of H and G, the Helmholtz function was devised as a mathematical convenience. This function is of primary usefulness in problems involving chemical and electrochemical processes which operate essentially at constant temperature and volume. It is also of general use in the study of various thermodynamic properties and their numerical formulation.

3.12 Fundamental Nature of Entropy

The nature of entropy may be summarized as follows:

1. Entropy is a primitive concept.
2. Every system has entropy. That is, with the exception of a work reservoir, entropy change is always possible for any system.
3. Entropy is an extensive property.
4. One interpretation of entropy is that it is an index of that portion of the energy content in a system that is not available to do work.
5. Entropy content of an isolated system is not conserved. In fact, the entropy of an isolated system is a monotonically increasing function of time. In this respect, entropy differs fundamentally from energy: energy is a conservative property, entropy is not.

Expressed in words, we have for any system,

entropy creation in universe

= entropy change in system

+ entropy change in surroundings

$$\geq 0. \qquad (3.42)$$

Equation 3.42 is the general statement of the second law of thermodynamics. The equality sign is for the case when the process is carried out reversibly.

Analogous to Eq. 2.36a, the second law of thermodynamics for any system may also be given as entropy creation in universe

= (entropy change in system)

+ (entropy outflow − entropy inflow)

$\geq 0.$ (3.42a)

Problems

3.1 Heat is added to a heat engine at the rate of 4.5×10^7 kJ/h. If the power output of the engine is 5000 kW, what is the thermal efficiency of the engine? What is the rate of heat rejection by the engine?

3.2 The power output of a heat engine is 12,500 kW. If the thermal efficiency of the engine is 40%, what is the rate of heat addition to the engine? What is the rate of heat rejection by the engine?

3.3 A refrigerator removes heat from a cold body at the rate of 7.6×10^5 kJ/h. If the coefficient of performance of the refrigerator is 4.0, what is the power input to the refrigerator in kilowatts? What is the amount of heat rejected by the refrigerator?

3.4 Heat is added to the refrigerator at the rate of 12,000 kJ/min. If the power input to the refrigerator is 60 kW, what is the coefficient of performance of the refrigerator?

3.5 A refrigerator, having a coefficient of performance of 4, removes heat from a cold body at the rate of 12,000 Btu/min. What is the horsepower input to the refrigerator?

3.6 A heat pump delivers heat to a room at the rate of 2.0×10^5 kJ/h. What is the power, in kilowatts, required to run the heat pump if the coefficient of performance of the heat pump is 4.0?

3.7 If a heat engine of 30% thermal efficiency is used to drive a refrigerator having a coefficient of performance of 4, what is the heat input into the engine for each kilojoule removed from the cold body by the refrigerator?

3.8 A reversible heat engine, operating in a cycle, receives energy from a high-temperature reservoir at 1800 K and rejects heat to a low-temperature reservoir at 320 K. Determine the entropy change of the two heat reservoirs when 5000 kJ is added to the heat engine in each cycle. What is the entropy change of the universe?

3.9 A reversible heat engine, operating in a cycle, receives energy from a high-temperature reservoir at 3000°R. If the thermal efficiency of the heat engine is 40%, what is the temperature of the low-temperature reservoir that will accept the heat rejected by the engine?

3.10 A heat engine, operating in a cycle, receives heat from a high-temperature reservoir at T_H and rejects heat to a low-temperature reservoir at T_L, as shown in Fig. P3.12. Determine whether this machine is reversible, irreversible, or impossible for the following cases:

(a) $Q_H = 1000$ J   (b) $Q_H = 2000$ J   (c) $W_{net} = 1500$ J
$W_{net} = 900$ J. $Q_L = 300$ J. $Q_L = 500$ J.

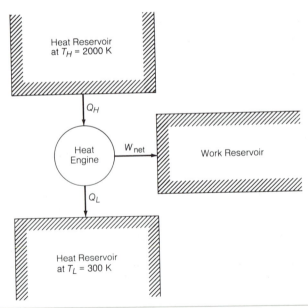

Figure P3.10

3.11 A refrigerator, operating in a cycle, removes heat from a low-temperature reservoir at T_L and rejects heat to a high-temperature reservoir at T_H, as shown in Fig. P3.11. Determine whether this machine is reversible, irreversible, or impossible for the following cases:

(a) $Q_L = 1000$ J   (b) $Q_L = 2000$ J   (c) $Q_H = 3000$ J
$W_{net} = 250$ J. $Q_H = 2400$ J. $W_{net} = 500$ J.

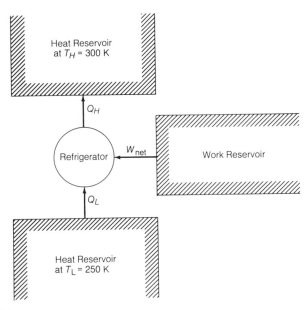

Figure P3.11

3.12 It is proposed to produce power by taking advantage of the thermal gradients that are present in certain parts of the ocean. If maximum ocean temperature available near the surface is 30°C and a minimum temperature near the ocean bottom is 5°C, what is the maximum possible thermal efficiency of such a heat engine?

3.13 A person claims to have developed a heat engine that, operating in a cycle between 30 and 5°C, will produce 100 kJ of work at the expense of 1000 kJ of heat addition for each cycle. Verify this claim by
(a) comparing the actual thermal efficiency with the maximum possible thermal efficiency.
(b) calculating the amount of entropy generation for each cycle.

3.14 A heat pump delivers 10 kJ/s of heat to a room maintained at 25°C, and receives heat from a reservoir at −10°C. If the actual coefficient of performance is 50% of that of an ideal heat pump operating between the same temperature limits, what is the actual power, in kilowatts, required to run the heat pump?

3.15 A refrigerator operates between 100 and 20°F. It requires twice as much work for a given amount of refrigeration as does an ideal refrigerator operating between the same temperature limits. What is the actual horsepower required to produce refrigeration of 24,000 Btu/min?

3.16 Derive Eq. 3.19.

3.17 Derive Eq. 3.20.

3.18 Show that if it is possible to violate the Kelvin–Planck statement of the second law of thermodynamics, it is possible to construct a device that operates in a cycle and produces no effect other than the transfer of heat from a cooler body to a hotter body.

3.19 Show that if we violate the Clausius statement of the second law, we would also violate the Kelvin–Planck statement of the second law.

3.20 Consider the design of a heat engine for use in outer space. The working substance of the system can reject heat to space only by radiation. Assume that the rate at which heat can be radiated to outer space is given by $\dot{Q} = \sigma A T_L^4$, where σ is the Stefan–Boltzmann constant, A is the area of the radiator, and T_L is the radiator temperature. Because the radiator is expensive to launch, such a heat engine is usually designed for minimum radiator area. Show that for a given power output $\dot{W}$ and a given T_H,
(a) the maximum thermal efficiency of such a heat engine is 25%.
(b) the radiator area is given by

$$A = \frac{256\,\dot{W}}{27\sigma T_H^4}.$$

3.21 Show that $(\beta_{HP})_{ideal} = 1 + (\beta_R)_{ideal}$, where $(\beta_{HP})_{ideal}$ is the coefficient of performance of an ideal heat pump operating between a high-temperature reservoir at T_H and a low-temperature reservoir at T_L, while $(\beta_R)_{ideal}$ is the coefficient of performance of a reversible refrigerator operating between the same temperature limits.

3.22 A reversible heat engine receives heat from a high-temperature reservoir at T_H and rejects heat at 1000 K. A second reversible heat engine receives the heat rejected by the first heat engine at 1000 K and rejects heat to a low-temperature reservoir at 300 K. If we want the same thermal efficiency for both engines, what is T_H?

3.23 Same as Problem 3.22 except that we want equal net work for both engines.

3.24 Two reversible refrigerators are connected in series. The first one removes heat from a cold reservoir at T_L and discharges heat at 10°C. The second refrigerator absorbs the heat discharged by the first at 10°C, and it in turn discharges heat to the environment at 25°C. If we want the same coefficient of performance for both refrigerators, what is T_L?

3.25 Same as Problem 3.24 except that we want T_L for same net work input for both refrigerators.

3.26 A reversible heat engine is used to drive a reversible refrigerator. The heat engine takes energy from a high-temperature reservoir at T_H and rejects heat to the surroundings at T_0. The refrigerator removes heat from a cold body at a temperature of T_L and discharges heat to the same surroundings at T_0. Show that the ratio of Q_L (heat removed from the cold body) to Q_H (heat supplied to the heat engine) approaches $T_L/(T_0 - T_L)$ if T_H is very much greater than T_0.

3.27 The scheme shown in Fig. P3.27 may be used to produce high-temperature process heat by making use of a low-temperature heat source. If the thermal efficiency of the real engine is only equal to 60% of that of the reversible engine operating between the same temperature limits, and the coefficient of performance of the real heat pump is only equal to 60% of that of the perfect

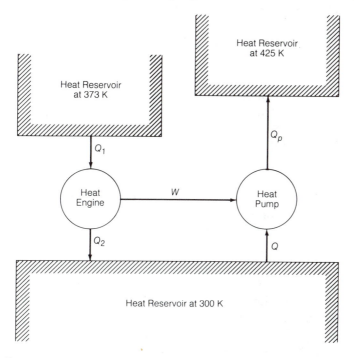

Figure P3.27

heat pump for given conditions, determine for the production of each unit of heat at 425 K the amount of heat we need from the heat source at 373 K.

3.28 A heat pump operates between a low-temperature reservoir at 40°F and a high-temperature reservoir at 70°F. If the heat pump has a coefficient of performance of 2.5 and delivers heat at the rate of 10,000 But/h, calculate the rate of lost work generated. What is the ideal power needed to deliver the same amount of heat? Let $T_0 = 500°R$

3.29 A heat engine receives heat reversibly in the amount of 200 kJ from a heat reservoir at 600 K. A reversible adiabatic expansion process next reduces the temperature of the system to 300 K. Heat in the amount of 150 kJ is then reversibly transferred to a heat sink at 300 K. The cycle is closed by an adiabatic compression process.
(a) Show that the adiabatic compression process is irreversible.
(b) What is the lost work for the cycle if the environmental temperature is 300 K?

3.30 A refrigerator removes heat from a cold body at T_L and discharges heat to the environment at T_0. Show that the lost work of the actual refrigeration cycle may be given by

$$LW_0 = Q_L\left(\frac{1}{\beta_{act}} - \frac{1}{\beta_{ideal}}\right),$$

where Q_L is the amount of heat removed from the low-temperature body, β_{act} is the coefficient of performance of the actual refrigerator, and β_{ideal} is the coefficient of performance of the ideal refrigerator operating between the same temperature limits.

3.31 Show that two constant-entropy lines cannot intersect. (*Hint:* If constant-entropy lines do intersect, we can create a cycle consisting of two reversible adiabatic processes and a reversible isothermal process that will violate the second law of thermodynamics.)

3.32 A closed system undergoes a reversible process. Work done by the system is 15 kJ. Heat added to the system is 5 kJ. The entropy change of the system
(a) is positive.
(b) is negative.
(c) can be either positive or negative.

3.33 A closed system undergoes an adiabatic process. Work done to the system is 15 kJ/kg. The entropy change of the system
(a) is positive.
(b) is negative.
(c) can be either positive or negative.

3.34 A closed system undergoes an irreversible process. Work done by the system is 15 Btu. Heat rejected by the system is 5 Btu. The entropy change of the system
(a) is positive.
(b) is negative.
(c) can be either positive or negative.

3.35 A closed system undergoes a process in which the entropy change of the system is $+25$ J/k. During the process, the system receives 6000 J from a heat reservoir at 300 K. Is the process reversible, irreversible, or impossible?

3.36 A closed system receives 10 kJ of heat from a heat reservoir and produces 20 kJ of work in changing from state 1 to state 2. Can we return the system to its initial state by an adiabatic process? Justify your answer.

3.37 A closed system undergoes a process in which heat is removed from it. Can we return the system to its initial state by an adiabatic process? Justify your answer.

3.38 A gas is compressed in a piston–cylinder assembly. The internal energy and entropy changes of the gas are 24 Btu/lbm and -0.07 Btu/lbm·°R, respectively. The work input to the gas is 80 Btu/lbm. Heat transfer to or from the gas is with the surroundings at a temperature of 70°F. Determine the amount of entropy generation for each pound of gas compressed.

3.39 A 50-Ω resistor carrying a constant direct current of 20 A is kept at a constant temperature of 100°C. The energy dissipated by the resistor is received by the air of the surroundings, which remains at the constant temperature of 25°C. In a time interval of 2 h, what is the amount of entropy created in the universe, in J/K?

3.40 The volume of a closed system in a cylinder is doubled in a reversible isothermal process at 25°C with no change in its internal energy. Determine the work done by the system in kJ/kg if the relationship between entropy and volume for the closed system may be given by

$$s = s_0 + 0.29 \ln (v) \qquad \text{kJ/kg} \cdot \text{K},$$

where v is specific volume in m^3/kg, and s_0 is a constant.

3.41 A stream of fluid undergoes an irreversible process adiabatically in a steady-state steady-flow device. The work produced by the device is 24 kJ. The entropy change of the fluid
(a) is positive.
(b) is negative.
(c) can be either positive or negative.

3.42 A stream of fluid undergoes a reversible process in a steady-state steady-flow device. The work produced by the device is 50 Btu. The heat removed from the device is 6 Btu. The entropy change of the fluid
(a) is positive.
(b) is negative.
(c) can be either positive or negative.

3.43 A stream of fluid undergoes an adiabatic process in a steady-state steady-flow work-producing device. The entropy change of the fluid
(a) is positive.
(b) is negative.
(c) can be either positive or negative.

3.44 The pressure of a gas expands reversibly and isothermally at 25°C in a steady-state steady-flow device with no change in its enthalpy. The pressure of the gas at the outlet is half that at the inlet. Changes in potential and kinetic energies are negligible. For a mass-flow rate of 50 kg/s, what is the power output in kilowatts of the device? The entropy–pressure relationship of the gas may be given as

$$s = s_0 - 0.29 \ln (p) \qquad \text{kJ/kg} \cdot \text{K}$$

where p is in kPa and s_0 is a constant.

3.45 A fluid is being cooled under steady-state steady-flow conditions in a heat exchanger by rejecting heat to the surrounding air, which is at 25°C. Fluid enters with an enthalpy value of 2326.1 kJ/kg and an entropy value of 7.508 kJ/kg · K. Fluid leaves with an enthalpy value of 162.5 kJ/kg and an entropy value of 0.555 kJ/kg · K. Changes in potential and kinetic energies are negligible. Show that this heat-transfer process is irreversible.

CHAPTER 4

Equations of State of a Pure Substance in Graphical and Tabular Forms

In the engineering design of thermodynamic processes, devices, and systems, we encounter many different types of thermodynamic substances. Since system design parameters (such as pressure and temperature) and system performance characteristics depend on the properties of the working fluid used, it is essential that we have a good understanding of the thermodynamic behavior of matter. In this chapter we first establish a general rule on such behavior. This general rule, known as the state postulate, will be applied to specific substances in the rest of the book. It provides us with the general framework to study the relationships among thermodynamic properties of a particular substance through the use of various equations of state. Unfortunately, these equations of state in general are rather complicated and cumbersome to handle, particularly when more than one relevant reversible work mode is involved. However, for a very special but extremely important class of substances, known as *pure substances*, there is only one relevant quasi-static work mode, with the consequence that the equilibrium states of such simple substances are completely specified by specifying two independent intensive properties. This important feature of a pure substance makes it possible for us to represent the equilibrium state of such a substance as a point on a diagram and to represent quasi-static processes as lines on a diagram. By the same token, we can in principle tabulate all thermodynamic properties of a pure substance on a sheet of paper, since paper is two-dimensional. It is our primary objective in this chapter to exploit the usefulness of such thermodynamic diagrams and tables so that we may acquire a qualitative feeling for the relationships among some of the important thermodynamic properties of pure substances.

4.1 The State Postulate

Although the number of properties that we can use to describe a system is many, it has been found by experiment that for a majority of the systems of interest to engineers and scientists, only a few of the properties can be given arbitrary values. That is, the number of properties that may be independently varied is quite small.

Since thermodynamics deal with energy and energy transformations, all thermodynamic properties are directly or indirectly related to energy. It would be reasonable to expect the number of independent properties for a given system to be governed by the number of ways we can independently change the energy of the system. It will be recalled that for a closed system the energy content can be changed by various kinds of quasi-static work modes and heat transfer. Each kind of quasi-static work mode provides us with an independent way to change the energy. Thus, if n is the number of relevant quasi-static work modes, we should have $n + 1$ independent variables for energy. This idea has been formalized and is given as the *state postulate*[1]:

> The equilibrium states of a given system are completely specified by specifying $n + 1$ independent thermodynamic properties, where n is the number of relevant quasi-static work modes for the system.

It is important to note that this rule does not say which $n + 1$ properties will constitute an independent set. Internal energy and the extensive variables corresponding to the various quasi-static work modes would constitute an independent set, however.

It will be recalled that there are many kinds of quasi-static work modes. Four kinds—mechanical work of expansion or compression, mechanical work of stretching, electrical work, and magnetic work— were given in Chapter 2. A common feature for all work modes is that each has a generalized force (intensive property) associated with a generalized displacement (extensive property). For mechanical work of expansion or compression, the generalized force and displacement are the pressure (p) and volume (V), respectively. For mechanical work of stretching, the generalized force and displacement are the tension (τ) and length (L). For electrical work, the generalized force and displacement are the electromotive force ($\mathscr{E}$) and quantity of electric charge (Z). For magnetic work, the generalized force and displacement are the external field ($\mathscr{H}$) and magnetization (M). Detailed

---

[1] S. J. Kline and F. O. Koenig, *Journal of Applied Mechanics*, Vol. 24 (1957), p. 29. See also H. B. Callen, *Thermodynamics*, John Wiley & Sons, Inc., New York, 1960, pp. 12, 192.

discussions of many quasi-static work modes are given in some excellent books.[2]

4.2 Definition of a Pure Substance

A pure substance is one that possesses the following two characteristics:

1. It is a simple substance having only one relevant quasi-static work mode. A pure substance is known as a *simple compressible substance* when the only relevant quasi-static work mode is the mechanical work of expansion or compression ($p\,dV$ work). A pure substance is known as a *simple magnetic substance* when the only relevant quasi-static work mode is the magnetic work. A pure substance is known as a *simple dielectric substance* when the only relevant quasi-static work mode is that due to electric polarization. We shall concentrate on the simple compressible substance, which is often termed merely a *simple substance*. It is to be noted that no real substance is really that simple. The use of the simple compressible substance model implies that we are neglecting surface effects, magnetic effects, and electrical effects.
2. It is a substance that has a homogeneous and invariable chemical composition. This means that if we examine a sample of a system of a pure substance at any time, we should find that the relative amount of each chemical species in the system will remain the same. Obviously, a system undergoing any chemical reaction is not a pure substance.[3]

According to our definition, water is a pure substance even if it exists as a mixture of liquid and vapor, or as a mixture of liquid, vapor, and solid, since its chemical composition is the same in all phases.[4] Atmospheric air, which is essentially a mixture of nitrogen and oxygen, may be treated as a pure substance as long as it remains in the gaseous state. On the other hand, a mixture of gaseous air in equilibrium with liquid air is not a pure substance since the chemical

[2] M. Zemansky, *Heat and Thermodynamics*, McGraw-Hill Book Company, New York, 1957, Chap. 3; M. Tribus, *Thermostatics and Thermodynamics*, Van Nostrand Reinhold Company, New York, 1961, pp. 9–21; T. A. Bruzustowski, *Introduction to the Principles of Engineering Thermodynamics*, Addison-Wesley Publishing Co., Inc., Reading, Mass., 1969, pp. 27–39.

[3] W. C. Reynolds and H. C. Perkins, *Engineering Thermodynamics*, McGraw-Hill Book Company, New York, 1970, p. 72.

[4] It will be recalled that a *phase* is defined as a homogeneous system having uniform intensive properties throughout.

composition in the gaseous state is not the same as that in the liquid phase. A mixture of oil and water is also not a pure substance since the chemical composition is not homogeneous throughout the system. However, a solution of ammonia and water may be treated as a pure substance if the solution strength stays the same.

We shall concentrate mostly on one-component simple compressible substances. We should, however, keep in mind that multi-component systems may be treated as pure substances if they satisfy the definition given above.

4.3 State Postulate of a Simple Compressible Substance

According to the state postulate, the following explicit functions in the case of a simple compressible substance of N moles will exist:

$$S = S(U,\ V,\ N) \tag{4.1}$$

$$U = U(S,\ V,\ N). \tag{4.2}$$

Each of these two equations is an equation of equilibrium states. Each, then, is known as an equation of state. We shall encounter many more equations of state, but these two are among the most important. In fact, they are often called the fundamental equations of state, because they contain all the thermodynamic information of a given substance. It will be shown that the following explicit functions will also exist for a simple compressible substance:

$$H = H(S,\ p,\ N) \tag{4.3}$$

$$A = A(T,\ V,\ N) \tag{4.4}$$

$$G = G(T,\ p,\ N). \tag{4.5}$$

These five equations deal with any amount of a simple compressible substance. It is often convenient to work with a unit mass of the pure substance. On this basis, we then have

$$s = s(u,\ v) \tag{4.1a}$$

$$u = u(s,\ v) \tag{4.2a}$$

$$h = h(s,\ p) \tag{4.3a}$$

$$a = a(T,\ v) \tag{4.4a}$$

$$g = g(T,\ p), \tag{4.5a}$$

where $s = S/N$, $u = U/N$, $h = H/N$, $a = A/N$, and $g = G/N$ are the specific values of entropy, internal energy, enthalpy, Helmholtz function, and Gibbs function, respectively. Since these are all intensive properties, the state postulate of a simple compressible substance may now be expressed as follows:

The equilibrium states of a simple compressible substance are completely specified by specifying two independent intensive properties. This is also known as the *two-property rule*.[5]

The key word in this rule is the word "independent," but the rule does not tell us *which* two properties may be taken as independent. We must determine this by experimental observation. Let us illustrate this point with the p–v–T data of a single-component simple compressible substance.

When a simple compressible substance exists in the liquid phase, we have the following experimental facts:

T is specified if p and v are specified.
p is specified if T and v are specified.
v is specified if T and p are specified.

We have the same experimental facts if the simple compressible substance exists in the vapor phase. This is also true when the pure substance exists in the solid phase.[6] On the other hand, if we have a mixture of liquid and vapor in equilibrium, the experimental facts are somewhat different:

T is still specified if p and v are specofoed;
p is still specified if T and v are specified; but
v is not specified if only T and p are given.

We thus see that p and v, T and v, or T and p may be taken as independent in general, while T and p may not be taken as independent in the case of a liquid–vapor mixture in equilibrium (see Fig. 4.1).

---

[5] D. B. Spalding and E. H. Cole, *Engineering Thermodynamics*, McGraw-Hill Book Company, New York, 1959, pp. 95–96.

[6] This implies that the solid is isotropic. Some solids can have pressures that differ according to the direction of measurement, but these kinds of solids would not satisfy our definition of a simple compressible substance, as more than two intensive properties must be specified in order to specify their states.

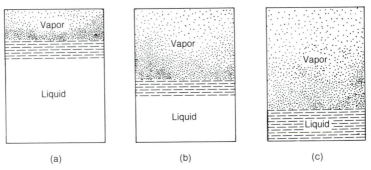

(a)   (b)   (c)

Figure 4.1 Various Combinations of Liquid and Vapor in Equilibrium at the Same T and p

4.4 Some Important General Thermodynamic Relations for a Simple Compressible Substance

The important contribution of classical thermodynamics is that by using a few basic laws and some relatively simple mathematical tools, thermodynamics provides us with the means to deduce many general thermodynamic relations among the many thermodynamic properties. These general thermodynamic relations, in turn, make it possible for us to identify the experimental data we need in order for us to correlate, tabulate, and calculate the properties that we use for engineering design. Several such general thermodynamic relations for a simple compressible substance may now be deduced.

Since a thermodynamic property is a state function, the change of any thermodynamic property between two equilibrium states is independent of the path connecting the same pair of equilibrium states. Suppose that we want to evaluate the internal energy change between a pair of equilibrium states. From the first law of thermodynamics for a closed system, we have, on the basis of unit mass,

$$\bar{d}q = du + \bar{d}w. \tag{1}$$

Let us make use of a reversible path to determine du. Then

$$\bar{d}w = p\, dv. \tag{2}$$

From the second law of thermodynamics,

$$\bar{d}q = T\, ds. \tag{3}$$

Combining Eqs. 1, 2, and 3, we have

$$T\, ds = du + p\, dv. \tag{4.6}$$

Equation 4.6 relates ds, du, and dv between a pair of equilibrium states for any process regardless of whether the process is reversible or irreversible.

On the basis of unit mass, enthalpy is defined as

$$h \equiv u + pv. \tag{4.7}$$

In differential form, we have

$$dh = du + p\,dv + v\,dp, \tag{4.8}$$

which, in combination with Eq. 4.6, gives

$$T\,ds = dh - v\,dp. \tag{4.9}$$

Equation 4.9 relates ds, dh, and dp between a pair of equilibrium states for any process regardless of whether the process is reversible or irreversible.

On the basis of unit mass, the Helmholtz function is defined as

$$a \equiv u - Ts. \tag{4.10}$$

Using the same procedure, we get

$$da = du - T\,ds - s\,dT, \tag{4.11}$$

which, in combination with Eq. 4.6, gives

$$da = -s\,dT - p\,dv. \tag{4.12}$$

Equation 4.12 relates da, dT, and dv between a pair of equilibrium states for any process, regardless of whether the process is reversible or irreversible.

On the basis of unit mass, the Gibbs function is defined as

$$g \equiv h - Ts. \tag{4.13}$$

In a similar manner, we get

$$dg = dh - T\,ds - s\,dT, \tag{4.14}$$

which, in combination with Eq. 4.9, gives

$$dg = -s\,dT + v\,dp. \tag{4.15}$$

Equation 4.15 relates dg, dT, and dp between a pair of equilibrium states for any process, regardless of whether the process is reversible or irreversible.

Equations 4.6 and 4.9, also known as $T\,ds$ equations, are two of the most important property equations in thermodynamics. They provide the means for the calculations of entropy changes of a simple compressible substance from measurable quantities. We shall have the opportunity to apply them in Chapter 5 in our study of ideal gases, which constitute a very special kind of simple compressible substance; but detailed study of Eqs. 4.6 and 4.9, as well as Eqs. 4.12 and 4.15, and their usefulness is deferred until Chapter 14.

4.5 The p–v–T Surface of a Simple Compressible Substance

We are now acquainted with eight thermodynamic properties: p, v, T, u, h, s, a, and g. The last two, a and g, turn out to be of only minor importance in the study of simple compressible substances. We shall concentrate our discussion on the first six properties.

Of the properties p, v, T, u, h, and s, the first three are directly measurable. It is not surprising that experimental p–v–T data are among the most plentiful. Figures 4.2 and 4.3 represent two possible experiments that one can perform to obtain p, v, and T data.

In the experiment shown in Fig. 4.2, any process taking place will be at constant pressure if the weight on the piston remains constant. Addition of heat to the substance will produce an increase in the temperature and a change in the volume. When this heat-addition process is repeated for various different loads on the piston, we would have obtained the necessary data for the construction of a p–v–T surface.

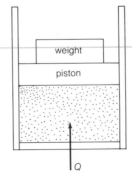

Figure 4.2 Determination of the Variation of T with v at Constant p

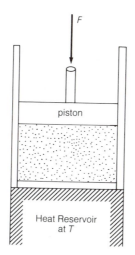

Figure 4.3 Determination of the Variation of p with v at Constant T

In the experiment in Fig. 4.3, any process taking place will be at constant temperature since the system is in contact with a heat reservoir. By varying the force applied to the piston, we would obtain a change in the volume corresponding to a change in pressure for each constant-temperature operation. When this process is repeated using heat reservoirs at different temperatures, we would also have obtained the necessary data for the construction of a p–v–T surface.

Experimental p–v–T data may be plotted on a rectangular coordinate system. The equilibrium states would then be represented in a three-dimensional space as a surface, known as the p–v–T surface. This surface is simply a graphical representation of the p–v–T equation of state, also known as the *thermal equation of state*, since p–v–T data are obtainable from a heat-transfer experiment such as the one shown in Fig. 4.2.

Figures 4.4 and 4.5 are schematic diagrams of the p–v–T surface for a simple compressible substance. Figure 4.4 is for a substance, such as carbon dioxide, that contracts on freezing. Figure 4.5 is for a substance, such as water, that expands on freezing. Most simple compressible substances contract on freezing.

The p–v–T surface, although not very useful for engineering analysis and design, does have the advantage of pictorially exhibiting some of the basic structure of a simple compressible substance. It clearly shows that a simple compressible substance can exist only in the vapor or gas phase,[7] liquid phase, or solid phase for certain

---

[7] We shall use the terms "vapor" and "gas" interchangeably.

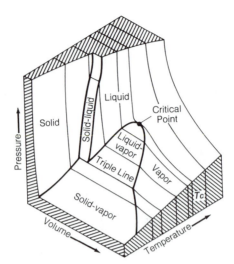

Figure 4.4 p–v–T Surface for a Substance That Contracts on Freezing

ranges of the variables. It also gives the regions in which both phases can exist simultaneously, and along a line called the *triple line* all three phases can coexist.[8] In addition, this surface contains the *critical point*, or more correctly, the *critical state*. The pressure and tem-

---

[8] All simple compressible substances possess a triple point with the exception of helium, which remains as a liquid even at the lowest attainable temperature.

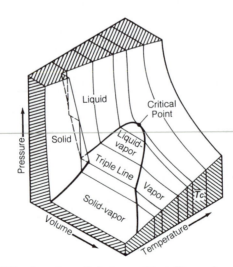

Figure 4.5 p–v–T Surface for a Substance That Expands on Freezing

Table 4.1 Critical-Point Data for Some Substances

Substance		p_c (MPa)	T_c (K)	v_c (m³/kmol)
Helium	He	0.23	5.2	0.0578
Hydrogen	H_2	1.30	33.3	0.0649
Nitrogen	N_2	3.40	126.2	0.0895
Oxygen	O_2	5.04	154.6	0.0734
Air	—	3.77	132.4	0.0905
Carbon monoxide	CO	3.50	133	0.0930
Carbon dioxide	CO_2	7.39	304.2	0.0943
Ammonia	NH_3	11.63	406.8	0.0717
Methane	CH_4	4.64	191.1	0.0993
Water	H_2O	22.12	647.3	0.0571
Propane	C_3H_8	4.26	370.0	0.2000
Ethylene	C_2H_4	5.12	282.4	0.1242

perature at the critical state are the highest pressure and highest temperature beyond which a liquid vapor transformation is not possible. The pressure, temperature, and specific volume at the critical point are known as the *critical pressure, p_c*, the *critical temperature, T_c*, and the *critical specific volume, v_c*, respectively. The critical constants for several substances are given in Table 4.1. To liquefy a particular gas, we must have hardware operating at or below its critical temperature. Study of Table 4.1 will show why the creation of a practical liquefaction process for helium gas ($T_c = 5.2$ K) is such a great challenge to engineers and scientists.[9]

In the liquid–vapor region, the vapor in an equilibrium mixture is called a *saturated vapor* and the liquid a *saturated liquid*. The line separating the liquid region from the liquid–vapor region is called the *saturated-liquid line*. The line separating the vapor region from the liquid–vapor region is called the *saturated-vapor line*. The two lines meet at the critical point. Every point on the saturated liquid line is a saturated-liquid state, and every point on the saturated-vapor line is a saturated-vapor state. The pressure at which a liquid vaporizes (or a vapor condenses) is called the *saturation pressure* corresponding to a given temperature. It is also known as the vapor pressure, as it is the pressure exerted by a saturated vapor or liquid on its boundary. The temperature at which these phenomena occur is called the *saturation*

[9] Randall Barron, *Cryogenic Engineering*, McGraw-Hill Book Company, New York, 1966, pp. 4, 120–128.

temperature corresponding to a given pressure. Consequently, there is one and only one value of saturation temperature corresponding to a given pressure. Similarly, there is one and only one value of saturation pressure corresponding to a given temperature.

4.6 Thermodynamic Diagrams for a Simple Compressible Substance

Thermodynamic surfaces, of which the p–v–T surface is the most familiar, are not convenient for engineering analysis and design. However, the projections of certain surfaces on certain planes are quite useful. These projections, known as *thermodynamic diagrams*, are useful in two respects. First, they offer a compact method of presenting thermodynamic data. Second, they may be used as "visual aids" for showing quasi-static processes as lines on them. Although there are many possible thermodynamic diagrams, only six of them are of interest to us: p–T, p–v, T–v, T–s, p–h, and h–s. In general, a thermodynamic diagram could include data for all regions of a thermodynamic surface. However, with the exception of our discussion of the p–T diagram, we shall have no need in this book to refer to behavior in the solid phase. We shall therefore only study the vapor region, the liquid region, and the liquid–vapor two-phase region.

Pressure–Temperature Diagram

The p–T diagram represents the projection of the p–v–T surface on the p–T plane. Figures 4.6 and 4.7 are two such diagrams for a simple compressible substance: Fig. 4.6 for a substance, such as carbon dioxide, that contracts on freezing (see Fig. 4.4), and Fig. 4.7 for a

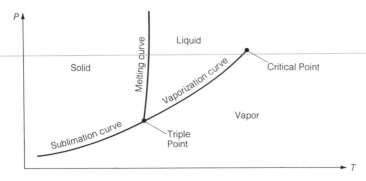

Figure 4.6 p–T Diagram for a Substance That Contracts on Freezing

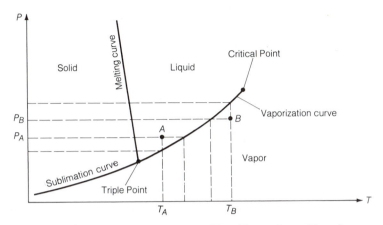

Figure 4.7 p–T Diagram for a Substance That Expands on Freezing

substance, such as water, that expands on freezing (see Fig. 4.5). In these diagrams we have three curves separating the solid, liquid, and vapor phases. The p–T diagram is therefore also known as a *phase diagram*. These three curves meet at the *triple point*, which is simply the projection of the triple line in the p–v–T surface on the p–T plane. It is important to note that the triple point is not a state point. Triple points for a few substances are given in Table 4.2.

The curve separating the liquid phase from the vapor phase is called the *vaporization curve*. This curve terminates at the upper end at the critical point. It is also known as the *boiling-point curve*, and represents the projection of the liquid–vapor region in the p–v–T surface on the p–T plane. Consequently, any point on this curve, with the exception of the critical point, is not a defined state point. The slope of the curve is positive for all simple compressible substances. This means that the temperature at which a liquid will boil increases

Table 4.2 Triple-Point Data for Some Substances

Substance		Triple-Point Pressure (kPa)	Triple-Point Temperature (K)
Water	H_2O	0.6112	273.16
Nitrogen	N_2	12.52	63.15
Oxygen	O_2	0.15	54.36
Carbon dioxide	CO_2	517.77	216.7
Methane	CH_4	10.031	88.7
Silver	Ag	0.0101	1233
Copper	Cu	0.000079	1356

with an increase in pressure, a feature that makes the pressure cooker so appealing to the busy homemaker. This phenomenon is also responsible for the increase in operating pressure of modern steam power plants,[10] as more efficient utilization of our energy resources requires that we operate such plants at high temperature.

The curve separating the solid phase from the liquid phase is called the *melting curve*. It is also known as the *fusion curve* or the *freezing-point curve*, and is the projection of the solid–liquid region in the $p–v–T$ surface on the $p–T$ plane. The slope of this curve for a substance that contracts on freezing is positive. For a substance that expands on freezing, such as water, the slope of the curve is negative, which means that the melting point of ice (solid water) decreases as pressure increases. Ice-skating enthusiasts can be thankful for this behavior of water since the melting ice beneath the blades of the skates (resulting from high pressure as a consequence of a large weight applied to a very small area) provides the lubrication that makes skating possible. Persons who admire the great beauty of the Yosemite Valley in California can also be thankful for this behavior, because melting ice also provided the lubrication for the motion of glaciers that resulted in the creation of one of the great wonders of nature.

The curve separating the solid phase from the vapor phase is called the *sublimation curve*. It is the projection of the solid–vapor region in the $p–v–T$ surface on the $p–T$ plane. The slope of this curve is positive for all simple compressible substances. We note that a solid can sublime and change to the vapor phase directly only at a pressure and temperature below that of its triple-point pressure and temperature. The sublimation process is responsible for the formation of frost (the reverse of sublimation) on our lawns on very cold days. One practical application of the sublimation process is in the deposit of very thin films of metals on other surfaces for the production of integrated circuits in the electronics industry. Another practical application of this process is in connection with the freeze-drying technique, used to produce dehydrated food. The use of solid carbon dioxide (called *dry ice*) to keep food and drinks refrigerated on a summer outing is of course familiar. In this application, the dry ice is, in essence, simply a disposable refrigerator.

We may also use the $p–T$ diagram advantageously to introduce the compressed liquid state and the superheated vapor state. State A, shown in Fig. 4.7 and defined by T_A and p_A in the liquid region, is

---

[10] The peak pressure for such plants using fossil fuel as the energy source is about 34 MPa.

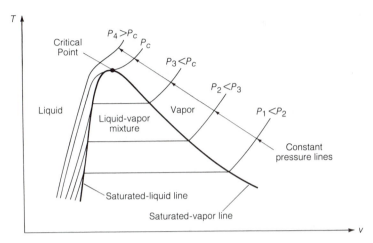

Figure 4.8 Vapor Dome on a T–v Diagram

called a compressed liquid state. Its actual pressure, p_A, is greater than the saturation pressure corresponding to its actual temperature, T_A. A compressed liquid is also known as a *subcooled liquid*, because its actual temperature, T_A, is lower than the saturation temperature corresponding to its actual pressure, p_A. State B, also shown in Fig. 4.7 and defined by T_B and p_B in the vapor region, is known as a *superheated vapor*, because its actual temperature, T_B, is higher than the saturation temperature corresponding to its actual pressure of p_B. For a superheated vapor, we also observe that its actual pressure is lower than the saturation pressure corresponding to its actual temperature. Note that the term "superheated" does not necessarily imply a high temperature. For example, oxygen gas at 1 atm and 100 K is a superheated vapor because the saturation temperature at 1 atm is only 90.19 K.

Temperature–Specific Volume Diagram

Figure 4.8 is a T–v diagram showing the saturation curve (the saturated liquid line plus the saturated vapor line) and constant-pressure lines. Such a diagram may be obtained by projecting the p–v–T surface on the T–v plane. The data used for the construction of the diagram may be obtained by performing the constant-pressure heat-addition experiment shown in Fig. 4.2. (It should be pointed out that the spacings of the constant-pressure lines in the liquid region have been greatly exaggerated; these spacings should be much smaller.) This is why a compressed liquid state may be approximated as the

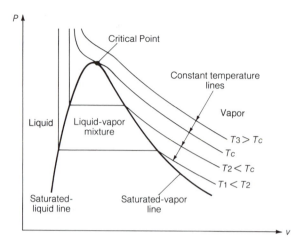

Figure 4.9 Vapor Dome on a p–v Diagram

saturated liquid state at its temperature.[11] Because vaporization at constant pressure also occurs at constant temperature, a constant-pressure line inside the liquid–vapor region is also a constant-temperature line. The volume change due to a change of phase alone is represented by the length of the horizontal line inside the liquid–vapor region. The length of the horizontal line decreases with increasing pressure, and it vanishes at the critical state. This means that the specific volume of saturated liquid at a given temperature or pressure approaches the specific volume of saturated vapor at the same temperature or pressure as we approach the critical point. At the critical state, specific volume of saturated liquid has the same value as the specific volume of saturated vapor. From the T–v diagram, we also observe that at constant temperature the volume change due to change in pressure is much greater for vapor than for liquid. This is because vapor is much more compressible than liquid.

Pressure–Specific Volume Diagram

Figure 4.9 is a p–v diagram showing the saturation curve and several constant-temperature lines. It may be obtained by projecting the p–v–T surface on the p–v plane. The data used for the construction of this diagram may be obtained by performing the constant-temperature compression experiment shown in Fig. 4.3. Alternatively, we can make use of data in the T–v diagram and obtain the p–v diagram through cross plotting. The behavior of the substance inside

---

[11] Approximation usually implies that we can tolerate the error involved. Under certain conditions we may not want to make this approximation.

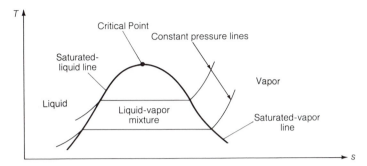

Figure 4.10 Vapor Dome on a T–s Diagram

the liquid–vapor region is similar to that for the T–v diagram. Because liquid is much less compressible than vapor, the slope of the constant-temperature lines is much steeper in the liquid region than in the vapor region. For a closed system involving a simple compressible substance, the quasi-static work is given by (Eq. 2.11)

$$\bar{d}W = p\ dV.$$

Hence the area under the quasi-static process path in the p–v diagram is simply the quasi-static work. For engineering analysis and design, the p–v diagram is very useful. Since we shall make use of this diagram quite often in the rest of the book, one should have a good understanding of the general trends of various curves in it.

Temperature–Entropy Diagram

Figure 4.10 is a T–s diagram showing the saturation curve and several constant-pressure lines. Lines of constant enthalpy are usually also plotted on such a T–s diagram designed for use in engineering analysis. To see how entropy and enthalpy are related to p–v–T data, let us apply the first law and the second law of thermodynamics to the constant-pressure heat-addition experiment shown in Fig. 4.2.

From the first law for a closed system (Eq. 2.9a) written for unit mass, we have

$$\bar{d}q = du + \bar{d}w.$$

If the system is a simple compressible substance, and if the process is quasi-static, the work-transfer term is given by Eq. 2.11:

$$\bar{d}w = p\ dv.$$

Therefore,

$$\bar{d}q = du + p\,dv. \tag{4.16}$$

If the process is also constant pressure, Eq. 4.16 may be written as

$$\bar{d}q = d(u + pv). \tag{4.17}$$

Making use of the definition of enthalpy ($h = u + pv$), we have

$$\bar{d}q = dh. \tag{4.18}$$

Equation 4.18 shows that enthalpy change for each constant-pressure operation may be determined by simply making measurement of the heat transfer involved. From the second law for a closed system undergoing a change of state quasi-statically, we have, according to Eq. 3.23,

$$\bar{d}q = T\,ds.$$

We thus see that entropy change for each constant-pressure operation is related to the heat transfer involved. Consequently, the constant-pressure heat-addition experiment in principle will yield the data for the construction of a T–s diagram having constant-pressure lines as well as constant-enthalpy lines.

Since $\bar{d}q = T\,ds$, the area under the quasi-static path in the T–s diagram is simply the heat-transfer quantity. For a reversible cycle made up of several reversible processes, the net heat transfer for the cycle is then given by the area enclosed by the processes involved. According to the first law (Eq. 2.10) for a cycle, this enclosed area is also the net work for the cycle. We thus see that the T–s diagram is very useful for the study of cycles.

The *latent heat of vaporization at constant pressure* is given by the area under the constant-pressure line, which is also a constant-temperature line, in the liquid–vapor region. The length of the horizontal lines inside the two-phase region becomes smaller and smaller as we approach the critical state. This is because the latent heat of vaporization becomes smaller and smaller as the pressure is increased. From Eq. 4.18, we also see that the latent heat of vaporization at constant pressure is given by the difference between the enthalpy of saturated vapor and the enthalpy of saturated liquid at the same pressure.

Enthalpy–Entropy Diagram
Figure 4.11 is an h–s diagram showing the saturation curve, constant-pressure lines, and constant-temperature lines. It is known

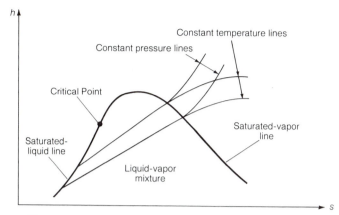

Figure 4.11 Vapor Dome on an *h–s* (Mollier) Diagram

as a *Mollier diagram*, in honor of its originator. The data used for the construction of this diagram are those which we used for the construction of the *T–s* diagram. Since this diagram contains information that we encounter in most engineering analysis, it has been used extensively. The diagram makes it possible for us to obtain directly the heat and work effects of some important processes. It will become quite clear that the availability of such diagrams will facilitate our study of such devices as nozzles, diffusers, compressors, turbines, and heat exchangers.

It will be recalled that a general thermodynamic relation for a simple compressible substance is given by Eq. 4.9:

$$T \, ds = dh - v \, dp.$$

This equation is valid for any process. If the process is constant pressure, it is reduced to

$$T \, ds_p = dh_p, \tag{4.19}$$

in which we have used the subscript p to denote that the entropy change and the enthalpy change are for a constant-pressure process. But $dh_p/ds_p = (\partial h/ds)_p$. Therefore, we have

$$\left(\frac{\partial h}{\partial s}\right)_p = T. \tag{4.20}$$

The quantity $(\partial h/\partial s)_p$ is simply the slope at a given state point on a constant-pressure line. We thus observe that constant-pressure lines

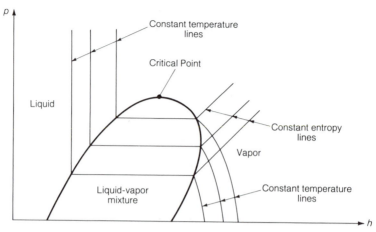

Figure 4.12 Vapor Dome on a *p–h* (Mollier) Diagram

will have a slope, at any point, equal to the temperature at that point. Inside the liquid–vapor region, constant-pressure lines are straight lines, since they are also constant-temperature lines in this region. These lines slope upward more steeply as the pressure is increased, because the boiling point increases with increase in pressure. In the vapor region, the constant-pressure lines start with a slope determined by the saturation temperature and then curve upward as the temperature increases.

Constant-temperature lines in the vapor region tend toward the horizontal. This is due to the fact that the enthalpy becomes more and more dependent on temperature alone as the vapor becomes more superheated. We shall come back to this point when we study the behavior of an ideal gas.

Pressure–Enthalpy Diagram

The skeleton of a typical *p–h* diagram is shown on Fig. 4.12. This diagram is useful in the study of refrigeration cycles, as several ideal processes of such cycles are represented by lines in it.

The lengths of the horizontal lines inside the two-phase region gives us the latent heat of vaporization according to Eq. 4.18. These lines become shorter and shorter as we approach the critical state, because the latent heat of vaporization becomes smaller as the pressure is increased.

The constant-temperature lines are practically vertical in the liquid region because the effect of pressure on the enthalpy is negligible. In the vapor region, these lines fall steeply and approach the vertical in the low-pressure region. This is due to the fact that the effect of pressure on the enthalpy of dilute vapor is also negligible.

4.7 Constant-Volume Specific Heat and Constant-Pressure Specific Heat

We have mentioned that Eqs. 4.6 and 4.9 are two of the most important equations in thermodynamics. We wish to make use of them now to arrive at two very important thermodynamic properties: constant-volume specific heat and constant-pressure specific heat.

From Eq. 4.6, we have

$$T \, ds = du + p \, dv.$$

For a constant-volume process, this equation is reduced to

$$T \, ds_v = du_v, \tag{4.21}$$

in which we have used the subscript v to denote that the entropy change and the internal energy change are for a constant-volume process.

Dividing Eq. 4.21 across by dT_v, the temperature change for the same constant-volume process, we have

$$\frac{T \, ds_v}{dT_v} = \frac{du_v}{dT_v}.$$

But

$$\frac{ds_v}{dT_v} = \left(\frac{\partial s}{\partial T}\right)_v \quad \text{and} \quad \frac{du_v}{dT_v} = \left(\frac{\partial u}{\partial T}\right)_v.$$

Therefore, we have

$$T\left(\frac{\partial s}{\partial T}\right)_v = \left(\frac{\partial u}{\partial T}\right)_v. \tag{4.22}$$

The quantity $(\partial u/\partial T)_v$, by definition, is known as the constant-volume specific heat. It will be given the symbol c_v. It is a positive quantity. Since it is a partial derivative involving properties, it is itself a thermodynamic property. Consequently, it is a function of two independent intensive properties. Substituting c_v for $(\partial u/\partial T)_v$ in Eq.

4.22, we have

$$c_v = T\left(\frac{\partial s}{\partial T}\right)_v \qquad (4.23)$$

and

$$\left(\frac{\partial T}{\partial s}\right)_v = \frac{T}{c_v}. \qquad (4.23a)$$

Equation 4.23 is a very useful and important general thermodynamic relation for a simple compressible substance. From this equation, we see that c_v must have the same dimensions as specific entropy, such as Btu/lbm · °R or J/kg · K.

Making use of Eq. 4.9, we could in a similar manner derive the following:

$$T\left(\frac{\partial s}{\partial T}\right)_p = \left(\frac{\partial h}{\partial T}\right)_p. \qquad (4.24)$$

The quantity $(\partial h/\partial T)_p$, by definition, is known as the *constant-pressure specific heat*. We shall give it the symbol c_p. It, too, is a positive quantity. It is also a thermodynamic property, since it is a derivative involving other thermodynamic properties. It is a function of two independent intensive properties. Substituting c_p for $(\partial h/\partial T)_p$ in Eq. 4.24, we have

$$c_p = T\left(\frac{\partial s}{\partial T}\right)_p \qquad (4.25)$$

and

$$\left(\frac{\partial T}{\partial s}\right)_p = \frac{T}{c_p}. \qquad (4.25a)$$

Equation 4.25 is another very useful and important general thermodynamic relation for a pure substance. The dimensions of c_p must also be that of specific entropy.

The quantity $(\partial T/\partial s)_v$ represents the slope of a constant-volume line on the T–s diagram. The quantity $(\partial T/\partial s)_p$ represents the slope of a constant-pressure line on the T–s diagram. Since it is an experimental fact that at a given state point, c_p is always greater than c_v, comparison between Eq. 4.23a and Eq. 4.25a will show that the slope of constant-volume lines must be steeper than that of constant-pressure lines. This fact will be useful to us in our study of energy-conversion cycles using a gaseous substance as the working fluid.

4.8 Tables of Thermodynamic Properties for a Simple Compressible Substance

We have discussed how thermodynamic properties may be presented in the form of diagrams. For engineering calculations, however, a more useful way of presenting such data is in the form of tables. Typically, these tables give a list of values for p, T, v, u, h, and s. From our study of thermodynamic diagrams, the data we need may be given in the following tables:

1. *Table for saturated liquid.* Since a saturated-liquid state is fixed when the value of one intensive property is given, we may for saturated liquid tabulate specific volume (v_f), specific internal energy (u_f), specific enthalpy (h_f), and specific entropy (s_f) for each value of pressure or temperature. Sometimes either u_f or h_f is not tabulated because each of these is related by the definition of enthalpy in the following manner:

$$h_f = u_f + pv_f.$$

2. *Table for saturated vapor.* Since a saturated-vapor state is fixed when the value of one intensive property is given, we may for saturated vapor tabulate specific volume (v_g), specific internal energy (u_g), specific enthalpy (h_g), and specific entropy (s_g) for each value of pressure or temperature. Although sometimes either u_g or h_g is also not tabulated, each is related by the following equation:

$$h_g = u_g + pv_g.$$

For convenience, saturated-liquid data and saturated-vapor data are always combined in one table, known as the *saturation table*. In this table, the differences of $(v_g - v_f)$, $(u_g - u_f)$, $(h_g - h_f)$, and $(s_g - s_f)$ are also given sometimes, and are represented by v_{fg}, u_{fg}, h_{fg}, and s_{fg}, respectively.

With data in the saturation table we can evaluate the properties of a liquid–vapor mixture at equilibrium. Let the volume occupied by the saturated vapor in the mixture be V_g and let the volume occupied by the saturated liquid in the mixture be V_f. Since volume is an extensive property, the volume of the mixture V is then given by

$$V = V_f + V_g.\tag{4.26}$$

From the definition of specific volume, the volumes V_f and V_g are then given by

$$V_f = m_f v_f$$

$$V_g = m_g v_g,$$

where m_f is the mass of the saturated liquid and m_g is the mass of the saturated vapor. Therefore,

$$V = m_f v_f + m_g v_g.\tag{4.27}$$

By the same token, the relations for the other extensive properties are given by

$$U = m_f u_f + m_g u_g\tag{4.28}$$

$$H = m_f h_f + m_g h_g\tag{4.29}$$

$$S = m_f s_f + m_g s_g.\tag{4.30}$$

The ratio of the mass of vapor to the total mass of liquid plus vapor is known as the *quality* of the mixture. It will be given the symbol x. Thus

$$x = \frac{m_g}{m_f + m_g} = \frac{m_g}{m}.\tag{4.31}$$

The specific volume of the mixture is obtained by dividing the volume of the mixture by the mass of the mixture; that is,

$$v = \frac{V}{m}.\tag{4.32}$$

Combining Eqs. 4.27, 4.31, and 4.32, we have

$$v = v_f + x(v_g - v_f) = v_f + xv_{fg}\tag{4.33}$$

or

$$v = v_g - (1 - x)(v_g - v_f) = v_g - (1 - x)v_{fg}.\tag{4.33a}$$

In a similar manner, other specific values of the mixture are given by

$$u = u_f + x(u_g - u_f) = u_f + xu_{fg} \tag{4.34}$$

or

$$u = u_g - (1 - x)(u_g - u_f) = u_g - (1 - x)u_{fg} \tag{4.34a}$$

$$h = h_f + x(h_g - h_f) = h_f + xh_{fg} \tag{4.35}$$

or

$$h = h_g - (1 - x)(h_g - h_f) = h_g - (1 - x)h_{fg} \tag{4.35a}$$

$$s = s_f + x(s_g - s_f) = s_f + xs_{fg} \tag{4.36}$$

or

$$s = s_g - (1 - x)(s_g - s_f) = s_g - (1 - x)s_{fg}. \tag{4.36a}$$

If the state of a liquid–vapor mixture is close to the saturated vapor line (that is, if its quality is high), we should use Eqs. 4.33a, 4.34a, 4.35a, and 4.36a to calculate the specific values of the mixture. If the quality of the mixture is small, we should use Eqs. 4.33, 4.34, 4.35, and 4.36.

The ratio of the mass of liquid to the total mass of liquid plus vapor is known as the *moisture content* of the mixture. It will be given the symbol y. Thus

$$y = \frac{m_f}{m_f + m_g} = \frac{m_f}{m} \tag{4.37}$$

$$y = 1 - x. \tag{4.38}$$

In the liquid–vapor region, we may use either quality or moisture content as an independent intensive property. A state in the liquid–vapor region may therefore be defined by temperature or pressure and quality.

3. *Table for superheated vapor.* In the vapor region, a given value of temperature and a given value of pressure will fix a state. Therefore, both temperature and pressure are usually chosen as the independent variables for the tabulation of v, u, h, and s. However, the internal energy u is often not given, but it can be found from $(h - pv)$.

4. *Table for compressed liquid.* In the liquid region, a given value of temperature and a given value of pressure will also fix a state. Consequently, compressed liquid data are usually tabulated in the same way as for superheated vapor. However, while properties of superheated vapor are quite sensitive to pressure in general, properties of compressed liquid change very little with pressure. As a result of this experimental fact, the properties v, u, h, and s may be approximated as the saturation values of v_f, u_f, h_f, and s_f at the temperature of the compressed liquid.

It should be pointed out that the tabulated values of u, h, and s in all the tables are not absolute values. Each is the difference between the value at any state and the value of the respective property at a reference state. But it makes no difference as far as engineering calculations are concerned, since we are only interested in changes of u, h, and s.

We should also appreciate the fact that tables of thermodynamic properties represent the hard work of many devoted engineers and scientists. The available tables save engineers time and energy to spend on more creative activities. It is therefore most important for engineers to become thoroughly familiar with the use of these tables.

4.9 Interpolation of Tabulated Data

The tables of thermodynamic properties provided in this book are of necessity in abbreviated form. It will be necessary to make an interpolation to obtain numerical values of certain properties in the solution of problems. If the increments of the independent variables (usually, temperature and pressure) are small, linear interpolation in general can be used to give fairly accurate values. That is, if X is the property we want to determine, we may make use of the following expression:

$$\frac{X - X_1}{X_2 - X_1} = \frac{T - T_1}{T_2 - T_1}, \tag{4.39}$$

where the subscripts 1 and 2 are the endpoints of the interval for which properties are given in the table, and T is the corresponding temperature.

If the increments of the independent variables are not small (such as in the superheated vapor tables), certain properties (such as specific volume v and entropy s) obtained by linear interpolation may not be as accurate as we like.

At a given value of pressure, v is proportional to temperature. Consequently, the use of linear interpolation for the determination of v would be appropriate. But entropy is proportional to the logarithm of the temperature ratio at a given value of pressure. Thus interpolation with the following expression would yield a more accurate value for entropy:

$$\frac{s - s_1}{s_2 - s_1} = \frac{\ln (T/T_1)}{\ln (T_2/T_1)}. \tag{4.40}$$

At a given value of temperature, v is proportional to the reciprocal of pressure, and s is proportional to the logarithm of the pressure ratio. Thus for interpolation at a given value of temperature, the following expressions would yield a more accurate value for v and s than would using linear interpolation:

$$\frac{v - v_1}{v_2 - v_1} = \frac{(1/p) - (1/p_1)}{(1/p_2) - (1/p_1)} \tag{4.41}$$

$$\frac{s - s_1}{s_2 - s_1} = \frac{\ln (p/p_1)}{\ln (p_2/p_1)}. \tag{4.42}$$

Example 4.1

Determine the specific volume v and entropy s for superheated steam at 600°C and 15 kPa.

Solution

From tabulated data for superheated steam, we have, at 600°C, the following data:

p (kPa)	v (m³/kg)	h (kJ/kg)	s (kJ/kg · K)
10	40.295	3705.5	10.1616
20	20.146	3705.4	9.8416

If we use linear interpolation,

$$v = v_1 + (v_2 - v_1)\left(\frac{p - p_1}{p_2 - p_1}\right)$$

$$= 40.295 + (20.146 - 40.295)\left(\frac{15 - 10}{20 - 10}\right) \text{ m}^3/\text{kg}$$

$$= 40.295 - 20.149 \times 0.5 = 30.221 \text{ m}^3/\text{kg}.$$

Interpolating with Eq. 4.41, we have

$$v = v_1 + (v_2 - v_1)\left(\frac{\frac{1}{15} - \frac{1}{10}}{\frac{1}{20} - \frac{1}{10}}\right)$$

$$= 40.295 - 20.149 \times 0.666 = 26.866 \text{ m}^3/\text{kg}.$$

The correct value is 26.864 m³/kg.

If we use linear interpolation for entropy, we have

$$s = s_1 + (s_2 - s_1)\left(\frac{p - p_1}{p_2 - p_1}\right)$$

$$= 10.1616 + (9.8416 - 10.1616)\left(\frac{15 - 10}{20 - 10}\right) \text{ kJ/kg} \cdot \text{K}$$

$$= 10.1616 - 0.3200 \times 0.5 = 10.0016 \text{ kJ/kg} \cdot \text{K}.$$

Interpolating with Eq. 4.42, we have

$$s = s_1 + (s_2 - s_1)\left[\frac{\ln (p/p_1)}{\ln (p_2/p_1)}\right]$$

$$= 10.1616 - 0.3200\left[\frac{\ln (15/10)}{\ln (20/10)}\right] \text{ kJ/kg} \cdot \text{K}$$

$$= 10.1616 - 0.3200 \times 0.5850 = 9.9744 \text{ kJ/kg} \cdot \text{K}.$$

The correct value is 9.9744 kJ/kg · K.   ∎

Comments

This example shows that specific volume is proportional to the reciprocal of pressure, and entropy is proportional to the logarithm of the pressure ratio for a given value of temperature. The data show that enthalpy is not sensitive to pressure. Consequently, it would be appropriate to use linear interpolation for the interpolation of enthalpy.

Example 4.2

Determine specific volume v, enthalpy h, and entropy s for superheated steam at 1000 kPa and 520°C.

Solution

From tabulated data for superheated steam, we have, at 1000 kPa:

t (C)	v (m³/kg)	h (kJ/kg)	s (kJ/kg · K)
500	0.3540	3478.3	7.7627
550	0.3775	3587.1	7.8991

Using linear interpolation for specific volume, we have

$$v = v_1 + (v_2 - v_1)\left(\frac{T - T_1}{T_2 - T_1}\right)$$

$$= 0.3540 + (0.3775 - 0.3540)\left(\frac{520 - 500}{550 - 500}\right) \text{ m}^3/\text{kg}$$

$$= 0.3540 + 0.0235 \times 0.4 \text{ m}^3/\text{kg} = 0.3634 \text{ m}^3/\text{kg},$$

which is the correct value.
Using linear interpolation for enthalpy, we have

$$h = h_1 + (h_2 - h_1)\left(\frac{T - T_1}{T_2 - T_1}\right)$$

$$= 3478.3 + (3587.1 - 3478.3)\left(\frac{520 - 500}{550 - 500}\right) \text{ kJ/kg}$$

$$= 3478.3 + 108.8 \times 0.4 \text{ kJ/kg} = 3521.8 \text{ kJ/kg},$$

compared to the correct value of 3521.6 kJ/kg.
Using linear entropy for entropy, we have

$$s = s_1 + (s_2 - s_1)\left(\frac{T - T_1}{T_2 - T_1}\right)$$

$$= 7.7627 + (7.8991 - 7.7627)\left(\frac{520 - 500}{550 - 500}\right) \text{ kJ/kg} \cdot \text{K}$$

$$= 7.7627 + 0.1364 \times 0.4 \text{ kJ/kg} \cdot \text{K} = 7.8173 \text{ kJ/kg} \cdot \text{K}.$$

Interpolating with Eq. 4.40, we have

$$s = s_1 + (s_2 - s_1)\left[\frac{\ln (T/T_1)}{\ln (T_2/T_1)}\right]$$

$$= 7.7627 + 0.1364\left[\frac{\ln (793.15/773.15)}{\ln (823.15/773.15)}\right] \text{kJ/kg} \cdot \text{K}$$

$$= 7.7627 + 0.1364 \times 0.4075 \text{ kJ/kg} \cdot \text{K} = 7.8183 \text{ kJ/kg} \cdot \text{K}.$$

The correct value is 7.8181 kJ/kg · K.   ∎

Comments

This example shows that, at a given value of pressure, interpolating specific volume and enthalpy linearly would be appropriate. Even in the case of entropy, linear interpolation gives a very good result. This is because the temperature ratios involved are not much greater than unity. However, Eq. 4.40 gives essentially the correct value.

4.10 Examples of Property Evaluation and of Property Changes Due to a Change of State

In this section we devote our discussion to the use of thermodynamic diagrams, property relations, and tabulated data. We shall evaluate property changes due to a change of state. In Chapters 8 and 9 we study some of the changes of state that we study in this section.

Example 4.3

Determine the specific volume of water when temperature and pressure are

(a) 40°C and 1.0 MPa, respectively.
(b) 40°C and 10 MPa, respectively.
(c) 320°C and 15.0 MPa, respectively.
(d) 320°C and 50.0 MPa, respectively.

Solution

All the given state points are those of a compressed liquid, as the saturated pressure is only 7.375 kPa at 40°C and only 11.289 MPa at 320°C. From the compressed water table, we have

(a) $v = 0.0010074$ m^2/kg.
(b) $v = 0.0010034$ m^3/kg.
(c) $v = 0.0014736$ m^3/kg.
(d) $v = 0.0013406$ m^3/kg.

If we had approximated the specific volume as that of a saturated liquid at the actual temperature, we would have introduced the following error:

$$\frac{v - v_f}{v} = \frac{0.0010074 - 0.0010078}{0.0010074} = 0.04\% \text{ for part (a)}$$

$$\frac{v - v_f}{v} = \frac{0.0010034 - 0.0010078}{0.0010034} = 0.44\% \text{ for part (b)}$$

$$\frac{v - v_f}{v} = \frac{0.0014736 - 0.0014995}{0.0014736} = 1.76\% \text{ for part (c)}$$

$$\frac{v - v_f}{v} = \frac{0.0013406 - 0.0014995}{0.0013406} = 11.85\% \text{ for part (d)}.$$

Comment

This example shows that water is not exactly incompressible, particularly when the actual pressure is much greater than the saturated pressure at the actual temperature, and when the actual temperature is high relative to the critical temperature of the substance.

Example 4.4

A storage tank for oxygen has a volume of 100 m³. Determine the amount of oxygen in the tank

(a) when the temperature and pressure of the fluid are 300 K and 101.325 kPa,

(b) when the fluid is a saturated liquid at 101.325 kPa.

Solution

Since $m = V/v$, we must first find the specific volume for each case.

(a) The given state is a superheated vapor, as 300 K is greater than 90.19 K, the saturated temperature corresponding to the actual pressure of 101.325 kPa. From the superheated vapor table for oxygen, we have

$$v = 0.768758 \text{ m}^3/\text{kg}.$$

Therefore,

$$m = \frac{100}{0.768758} \text{ kg} = 130 \text{ kg}.$$

(b) From the saturated oxygen table, v_f at 101.325 kPa is 0.000876 m³/kg. Therefore,

$$m = \frac{100}{0.000876} \text{ kg} = 114{,}155 \text{ kg.}$$ ∎

Comment

This example shows that we can store much more liquid oxygen than gaseous oxygen in a given volume. But because of the very low temperature, a liquid oxygen storage tank would require a very complicated insulation scheme to prevent heat leakage.

Example 4.5

Making use of enthalpy–entropy data, determine approximately the slope of a constant-pressure line in the h–s diagram for superheated steam at 1.0 MPa and 500°C.

Solution

The slope we want is simply the partial derivative $(\partial h/\partial s)_p$. For an approximation, $(\partial h/\partial s)_p = (\Delta h/\Delta s)$ for a constant-pressure line. From the superheated steam table at 1.0 MPa, we have:

T (°C)	h (kJ/kg)	s (kJ/kg · K)
450	3370.8	7.6190
500	3478.3	7.7627
550	3587.1	7.8991

Thus

$$\left(\frac{\partial h}{\partial s}\right)_p \approx \frac{3587.1 - 3370.8}{7.8991 - 7.6190} \text{ K} = 772.2 \text{ K.}$$ ∎

Comment

Since $(\partial h/\partial s) = T$ according to Eq. 4.20, a property relation, the slope we what is really the thermodynamic temperature at the given state. For this problem, the correct answer is $(500 + 273.15)$ K or 773.15 K.

Example 4.6

Nitrogen changes from an initial state at 200 K and 2.0 MPa to a final state defined by 300 K and a final pressure equal to the initial pressure. Determine the change in the property pv due to the change of state.

Solution

We want $p_2 v_2 - p_1 v_1$. But $p_2 = p_1$. Thus

$$p_2 v_2 - p_1 v_1 = p(v_2 - v_1).$$

Since the saturated temperature at 2.0 MPa is only 115.58 K, the initial state as well as the final state is a superheated vapor. From the superheated nitrogen table, we have

$$v_1 = 0.028438 \text{ m}^3/\text{kg}$$

$$v_2 = 0.044396 \text{ m}^3/\text{kg}.$$

Therefore,

$$p_2 v_2 - p_1 p_1 = 2 \times 1000(0.044396 - 0.028438) \text{ kJ/kg}$$

$$= 31.916 \text{ kJ/kg}. \qquad ▮$$

Comment

An example of the change of state just discussed could be that of the work done by the fluid in a cylinder due to a quasi-static constant-pressure process.

Example 4.7

Water changes from a saturated liquid at 40°C to a final state defined by 5000 kPa and a final value of entropy equal to the value of entropy at the initial state. Assuming that water is incompressible, determine the change of enthalpy due to the change of state.

Solution

Equation 4.9 relates ds, dh, and dp:

$$T \, ds = dh - v \, dp.$$

If $ds = 0$, $dh = v \, dp$. If fluid is incompressible, v is a constant. Then we have

$$h_2 - h_1 = v(p_2 - p_1).$$

With

$$v = v_f \text{ at } 40°C = 0.0010078 \text{ m}^3/\text{kg},$$

$$p_1 = \text{saturated pressure at } 40°C = 7.375 \text{ kPa},$$

$$h_2 - h_1 = 0.0010078(5000 - 7.375) \text{ kJ/kg} = 5.03 \text{ kJ/kg}. \qquad ▮$$

Comment

An example of the change of state just discussed is in the study of pumping liquid reversibly under steady-state steady-flow conditions.

Example 4.8

Nitrogen changes from 20 MPa and 160 K to a final pressure of 101.325 kPa and a final value of enthalpy equal to that of the enthalpy value at the initial state. Determine the mass fraction that is saturated liquid at the final state.

Solution

The initial state is in the gaseous region, as 160 K is greater than the critical temperature of 154.58 K for nitrogen. From the superheated nitrogen table, we have

$$h_1 = 53.625 \text{ kJ/kg.}$$

The final state is defined by

$$p_2 = 101.325 \text{ kPa}$$

and

$$h_2 = h_1 = 53.625 \text{ kJ/kg.}$$

At 101.325 kPa, we have

$$h_g = 76.689 \text{ kJ/kg} > 53.625 \text{ kJ/kg}$$

$$h_f = -122.150 \text{ kJ/kg} < 53.625 \text{ kJ/kg.}$$

Thus state 2 is a liquid–vapor mixture. Making use of Eq. 4.35 and data from the saturated nitrogen table, we have

$$h_2 = (h_f + xh_{fg}) \text{ at } 101.325 \text{ kPa and } x$$

$$53.625 = -122.150 + x(76.689 + 122.150)$$

$$x_2 = \frac{53.625 + 122.150}{76.689 + 122.150} = 0.884.$$

Thus the mass fraction that is saturated liquid is given by

$$y_2 = 1 - x_2 = 1 - 0.884 = 0.116. \qquad \blacksquare$$

Comment

An example of the change of state just discussed is in the study of a gas-liquefaction system.

Example 4.9

Freon-12 changes from a saturated liquid at 80°F to a final state defined by 10°F and a final value of h equal to the value of h at the initial state.

(a) Locate the initial and final state points on a T–s diagram.
(b) Determine the entropy change due to the change of state.

Solution

(a) The initial state is on the saturated-liquid line per the given information. To locate the final state point, we must determine s_2 and compare it with s_1. From the saturation table for Freon-12, we have the following data:

$$h_1 = h_f \text{ at } 80°\text{F} = 26.365 \text{ Btu/lbm}$$

$$s_1 = s_f \text{ at } 80°\text{F} = 0.05475 \text{ Btu/lbm} \cdot °\text{R}.$$

At 10°F,

$$h_f = 10.684 \text{ Btu/lbm} < 26.365$$

$$h_g = 78.335 \text{ Btu/lbm} > 26.365.$$

It can be seen that the final state is in the liquid–vapor region. Using Eq. 4.35, Eq. 4.36, and data from the saturation table for Freon-12, we have

$$h_2 = [h_f + x(h_g - h_f)] \text{ at } 10°\text{F and } x$$

$$26.365 = 10.684 + x(78.335 - 10.684)$$

and

$$x = \frac{26.365 - 10.684}{78.335 - 10.684} = 0.2318$$

and

$$s_2 = [s_f + x(s_g - s_f)] \text{ at } 10°\text{F and } x = 0.2318$$

$$= 0.02395 + 0.2318(0.16798 - 0.02395) = 0.05734 \frac{\text{Btu}}{\text{lbm} \cdot °\text{R}}.$$

Having found the value for s_2, we now can locate the two state points as shown:

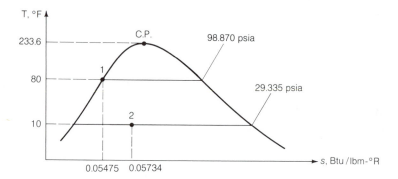

(b) $$s_2 - s_1 = 0.05734 - 0.05475 \text{ Btu/lbm} \cdot {}^{\circ}\text{R}$$

$$= 0.00259 \text{ Btu/lbm} \cdot {}^{\circ}\text{R}.$$

Comment

An example of the change of state just discussed is in the study of the expansion valve used in a mechanical refrigeration system.

Example 4.10

Oxygen changes from 500 kPa and 300 K to a final pressure of 101.325 kPa and a final value of entropy equal to the value of entropy at the initial state. Determine the final temperature of the fluid.

Solution

The initial state is in the gaseous region, as 300 K is greater than the critical temperature of 126.19 K for oxygen. From the superheated oxygen table, we have

$$s_1 = 5.9950 \text{ kJ/kg} \cdot \text{K}.$$

The final state is defined by

$$p_2 = 101.325 \text{ kPa}$$

$$s_2 = s_1 = 5.9950 \text{ kJ/kg} \cdot \text{K},$$

which is in the superheated region. From the superheated oxygen table, we have the following data:

	T (K)	s (kJ/kg · K)
A	180	5.9447
2	?	5.9950
B	200	6.0413

Thus the final temperature is between 180 and 200 K. Using linear interpolation, we have

$$T_2 = T_A + (T_B - T_A)\left(\frac{s_2 - s_A}{s_B - s_A}\right)$$

$$= 180 + (200 - 180)\left(\frac{5.9950 - 5.9447}{6.0413 - 5.9447}\right) K$$

$$= 180 + (20)\left(\frac{0.0503}{0.0966}\right) K = 190.4 \text{ K.}$$

∎

Comment

An example of the change of state just discussed is in the study of the expansion of oxygen gas reversibly and adiabatically in a nozzle under steady-state steady-flow conditions.

Example 4.11

Water changes state from 25°C and 100 kPa to 50°C 100 kPa. Determine the change of enthalpy due to change of state

(a) making use of tabulated enthalpy data.
(b) assuming constant c_p of 4.187 kJ/kg · K.

Solution

(a) For given states, water is only slightly compressed. We will assume the actual states as saturated liquid at actual temperature. That is,

$$h_2 - h_1 \approx h_f \quad \text{at } 50°C \qquad - h_f \quad \text{at } 25°C.$$

Making use of data from the saturated water table, we have

$$h_2 - h_1 \approx 209.26 - 104.77 \text{ kJ/kg} = 104.49 \text{ kJ/kg}.$$

(b) By definition, $(\partial h/\partial T)_p = c_p$. Hence $dh_p = c_p \, dT$. Assuming constant c_p, we have

$$h_2 - h_1 = c_p(T_2 - T_1)$$

$$= 4.187(50 - 25) \text{ kJ/kg}$$

$$= 104.68 \text{ kJ/kg.} \qquad \blacksquare$$

Comment

This example of change of state could be that of heating water in a steady-state steady-flow heat exchanger. Since no work is involved, $q = h_2 - h_1$ according to the first law of thermodynamics if we neglect changes in kinetic and potential energies.

Example 4.12

Nitrogen gas changes from an initial state defined by 14.7 psia and 540°R to a final state defined by 3000 psia and 540°R. Determine

(a) the enthalpy change due to change of state.
(b) the entropy change due to change of state.

Solution

(a) From the superheated vapor data for nitrogen, we have

$$h_1 = h \text{ at } 14.7 \text{ psia and } 540°R$$

$$= 133.862 \text{ Btu/lbm}$$

$$h_2 = h \text{ at } 3000 \text{ psia and } 540°R$$

$$= 119.735 \text{ Btu/lbm.}$$

Therefore

$$h_2 - h_1 = 119.735 - 133.862 \text{ Btu/lbm}$$

$$= -14.127 \text{ Btu/lbm.}$$

(b) From the superheated vapor data for nitrogen, we have

$$s_1 = s \text{ at } 14.7 \text{ psia and } 540°R$$

$$= 1.63523 \text{ Btu/lbm} \cdot °R$$

$$s_2 = s \text{ at } 3000 \text{ psia and } 540°R$$

$$= 1.23089 \text{ Btu/lbm} \cdot °R.$$

Therefore,

$$s_2 - s_1 = 1.23089 - 1.63523 \text{ Btu/lbm} \cdot °R$$

$$= -0.40434 \text{ Btu/lbm} \cdot °R. \qquad \blacksquare$$

Comment

An example of the change of state just discussed could be that of carrying out an ideal isothermal compression process such as that used in an ideal system for the liquefaction of nitrogen.

Example 4.13

Superheated steam changes from an initial state defined by $p_1 = 5.0$ psia and $T_1 = 700°F$ to a final state defined by $v_2 = 5v_1$ and $u_2 = u_1$. Determine

(a) the temperature of steam at the final state.
(b) the entropy change due to change of state.

Solution

(a) From superheated steam tables, we have

$$v_1 = v \text{ at } 5.0 \text{ psia and } 700°F$$

$$= 138.08 \text{ ft}^3/\text{lbm}$$

$$h_1 = 1384.3 \text{ Btu/lbm}.$$

Since u for superheated steam is not tabulated, we must make use of the definition of $h = u + pv$. Thus

$u_1 = (h - pv)$ at 5.0 psia and 700°F

$$= 1384.3 \text{ Btu/lbm} - \frac{(5.0 \text{ lbf/in}^2)(144 \text{ in}^2/\text{ft}^2)(138.08 \text{ ft}^3/\text{lbm})}{778 \text{ ft} \cdot \text{lbf/Btu}}$$

$$= 1384.3 \text{ Btu/lbm} - 127.79 \text{ Btu/lbm}$$

$$= 1256.5 \text{ Btu/lbm.}$$

Therefore,

$$v_2 = 5 \times 138.08 \text{ ft}^3/\text{lbm} = 690.4 \text{ ft}^3/\text{lbm}$$

$$u_2 = u_1 = 1256.5 \text{ But/lbm}$$

and T_2 is defined by

$$v_2 = 690.4 \text{ ft}^3/\text{lbm}$$

$$u_2 = 1256.5 \text{ Btu/lbm.}$$

To find T_2 from tables, we must resort to interpolation. But we note that for steam at 1.0 psia and 700°F, we have, from tables,

$$v = 690.7 \text{ ft}^3/\text{lbm}$$

$$h = 1384.5 \text{ Btu/lbm.}$$

Thus at 1.0 psia and 700°F, we have

$u = (h - pv)$ at 1.0 psia and 700°F

$$= 1384.5 \text{ Btu/lbm} - \frac{(1.0 \text{ lbf/in}^2)(144 \text{ in}^2/\text{ft}^2)(690.7 \text{ ft}^3/\text{lbm})}{778 \text{ ft} \cdot \text{lbf/Btu}}$$

$$= 1384.5 \text{ Btu/lbm} - 127.84 \text{ Btu/lbm}$$

$$= 1256.7 \text{ Btu/lbm.}$$

This means that we have $T = 700°F$ at

$$v = 690.7 \text{ ft}^3/\text{lbm}$$

$$u = 1256.7 \text{ Btu/lbm.}$$

We may thus conclude that the temperature corresponding to the final state of our problem is very close to 700°F. That is,

$$T_2 \approx 700°F.$$

(b) From the superheated steam tables, we have

$$s_1 = s \text{ at } 5.0 \text{ psia and } 700°F = 2.1369 \text{ Btu/lbm} \cdot °R$$

$$s_2 = s \text{ at } v = 690.4 \text{ ft}^3/\text{lbm and } u_2 = 1256.5 \text{ Btu/lbm}$$

$$\approx s \text{ at } 1.0 \text{ psia and } 700°F$$

$$= 2.3144 \text{ Btu/lbm} \cdot °R.$$

Therefore,

$$s_2 - s_1 = 2.3144 - 2.1369 \text{ Btu/lbm} \cdot °R = 0.1775 \text{ Btu/lbm} \cdot °R. \quad \blacksquare$$

Comments

An example of the change of state just discussed could be that of carrying out a free-expansion process (see Fig. 3.5) in which the fluid involved would constitute an isolated system. For a free-expansion process, the energy content of the fluid must remain constant according to the first law, and the entropy content must increase according to the second law. Our results are in agreement with what the two laws predict.

The reason the temperature remains constant in this particular case is that steam at low pressure and high temperature behaves very much like an ideal gas (which we shall soon study in detail).

Example 4.14

Ammonia changes state from saturated liquid to saturated vapor at 70°F. Calculate the change in the specific Gibbs function.

Solution

By definition, $g = h - Ts$. Thus

$$g_f = h_f - Ts_f$$

$$g_g = h_g - Ts_g.$$

Using data from saturation tables for ammonia, we have, at 70°F,

$$g_f = 120.5 - (459.67 + 70)(0.2537) \text{ Btu/lbm}$$

$$= 120.5 - 134.4 = -13.9 \text{ Btu/lbm}$$

$$g_g = 629.1 - (459.67 + 70)(1.214) \text{ Btu/lbm}$$

$$= 629.1 - 643.0 = -13.9 \text{ Btu/lbm.}$$

Therefore,

$$g_g - g_f = (-13.9) - (-13.9) = 0. \qquad \blacksquare$$

Comments

We see that the change in the specific Gibbs function for ammonia due to a phase change at 70°F is identically equal to zero. Actually this is true for any simple compressible substance having a change of state at constant temperature and pressure. According to Eq. 4.15, a general thermodynamic relation for a simple compressible substance, $dg = -s\,dT + v\,dp$. Since dT and dp are both zero for a phase change, dg must also be zero. In fact, this is one way to check the internal consistency of the tabulated data.

Example 4.15

A fluid changes state from 50°C to 100°C at constant volume. Assuming constant c_v of 0.716 kJ/kg · K, determine the change of entropy due to change of state.

Solution

From Eq. 4.23, we have

$$c_v = T\left(\frac{\partial s}{\partial T}\right)_v$$

$$ds_v = \frac{c_v\,dT_v}{T}.$$

Assuming constant c_v, we have

$$s_2 - s_1 = c_v \int_{T_1}^{T_2} \frac{dT}{T}$$

$$= c_v \ln \frac{T_2}{T_1}$$

$$= 0.716 \ln \frac{373.15}{323.15} \text{ kJ/kg} \cdot \text{K}$$

$$= 0.103 \text{ kJ/kg} \cdot \text{K}. \qquad \blacksquare$$

Comments

The result shows that the entropy at the final state is greater than the entropy at the initial state. Interpreting entropy as a measure of disorder, this would mean that the fluid is more disorders at 100°C than at 50°C occupying the same volume. This is reasonable, as the fluid at 100°C is more energetic than at 50°C.

Example 4.16

Calculate the property pv/T for oxygen gas at 320 K and 101.325 kPa.

Solution

Since p, v, and T are thermodynamic properties, the quantity pv/T is also a thermodynamic property. The given state is a superheated vapor. From the superheated oxygen table, we have, at 320 K and 101.325 kPa,

$$v = 0.820210 \text{ m}^3/\text{kg}.$$

Thus

$$\frac{pv}{T} = \frac{101.325 \times 0.820210}{320} \text{ kJ/kg} \cdot \text{K}$$

$$= 0.2597 \text{ kJ/kg} \cdot \text{K}. \qquad \blacksquare$$

Comments

We recall from previous courses in physics and chemistry that the gas constant R of a given gas is defined as the universal gas constant $\bar{R}$ divided by the molecular weight M of the gas. For oxygen, the gas constant R is given by

$$R = \frac{\bar{R}}{M} = \frac{8.3143 \text{ kJ/kg} \cdot \text{K}}{31.999 \text{ kg/kmol}}$$

$$= 0.2598 \text{ kJ/kg} \cdot \text{K}.$$

which is practically the same as the property pv/T in this example. This is because the state of oxygen gas in this example is in the region where oxygen may be modeled as an ideal gas, a concept that we study in detail in Chapter 5.

Problems

4.1 Determine whether H_2O at each of the following states is a subcooled liquid, a superheated vapor, or an equilibrium mixture of liquid and vapor.
(a) $T = 330$ K; $p = 50$ kPa.
(b) $p = 1.0$ MPa; $v = 1.5$ m^3/kg.

4.2 Determine whether Freon-12 at each of the following states is a subcooled liquid, a superheated vapor, or an equilibrium mixture of liquid and vapor.
(a) $T = -5°C$; $p = 20.0$ kPa.
(b) $T = 25°C$; $s = 0.50$ kJ/kg $\cdot$ K.
(c) $p = 1.0$ MPa; $h = 220$ kJ/kg.

4.3 Determine whether nitrogen at each of the following states is a subcooled liquid, a superheated vapor, or an equilibrium mixture of liquid and vapor.
(a) $T = 80$ K; $s = 2.5$ kJ/kg $\cdot$ K.
(b) $T = 120$ K; $h = -50$ kJ/kg.
(c) $p = 0.5$ MPa; $T = 100$ K.

4.4 Determine whether oxygen at each of the following states is a subcooled liquid, a superheated vapor, or an equilibrium mixture of liquid and vapor.
(a) $T = 80$ K; $s = 2.5$ kJ/kg $\cdot$ K.
(b) $T = 120$ K; $h = -50.0$ kJ/kg.
(c) $T = 100$ K; $p = 0.5$ MPa.

4.5 One kilogram of oxygen occupies a volume of 0.02 m^3 at a temperature of 100 K. What is the pressure of the fluid in kPa?

4.6 Ten kilograms of nitrogen occupies a volume of 0.05 m^3 at a pressure of 10.0 MPa. What is the temperature of the fluid?

4.7 A rigid tank having a volume of 10 m^3 is filled with nitrogen at a pressure and temperature of 3.0 MPa and 300 K. Determine the mass of nitrogen in the tank.

4.8 Ten kilograms of water is contained in a steam drum at a pressure of 1.0 MPa and a quality of 50%. What is the volume of the drum?

4.9 A rigid vessel having a volume of 5 m^3 is filled with oxygen at a pressure and temperature of 10.0 MPa and 250 K. Determine the mass of oxygen in the vessel.

4.10 A rigid tank having a volume of 10 ft^3 is filled with H$_2$O at a pressure and temperature of 100 psia and 400°F. Determine the mass of steam in the tank.

4.11 A rigid tank having a volume of 1.0 m^3 contains 60% saturated liquid water by volume and 40% saturated vapor at a temperature of 75°C. Determine
(a) the quality of the mixture.
(b) the total mass of fluid in the tank.

4.12 A rigid tank contains 1 kg of liquid ammonia and 0.5 kg of ammonia vapor at 20°C. Determine
(a) the pressure of the ammonia in kPa.
(b) the volume occupied by the saturated liquid.
(c) the total volume of the tank.

4.13 A closed vessel is filled with water at a pressure and temperature of 14.7 psia and 70°F. If exposure to the sun caused the water temperature to rise to 150°F, what will the water pressure be
(a) if the vessel is truly rigid?
(b) if the volume of the vessel increases by 1.0%?

4.14 Complete the following table for Freon-12.

State	(a)	(b)	(c)
p (kPa)		400	
T (°C)	20	90	−10
x (%)			
v (m³/kg)			0.2500
h (kJ/kg)	150		
u (kJ/kg)			

4.15 A rigid tank having a volume of 0.2 m³ contains equal volumes of liquid and vapor nitrogen at 120 K. Twenty-five kilograms of saturated liquid are then withdrawn from the tank. If the temperature in the tank remains at 120 K, what is the final volume of the saturated liquid?

4.16 A rigid tank having a volume of 0.25 m³ contains equal masses of saturated liquid and saturated vapor of oxygen at 130 K. Ten kilograms of saturated vapor are then withdrawn from the tank. If the temperature in the tank remains at 130 K, what is the final mass of the saturated vapor?

4.17 A rigid vessel having a volume of 0.5 m³ contains 80% by volume of saturated liquid water and 20% by volume of saturated steam at 200°C. Water in the amount of 140 kg is then pumped into the vessel. If the final temperature of the fluid in the vessel is 80°C, what is the final pressure in kPa?

4.18 A rigid vessel having a volume of 2.0 m³ contains dry saturated steam at 1.5 MPa initially. After heat loss to the surrounding, the pressure of the fluid drops to 500 kPa. What is the total change in the internal energy of the fluid due to the change of state?

4.19 One pound of nitrogen undergoes a change of state isothermally at 80°F from 1 atm to 200 atm. Determine
(a) the change in enthalpy in Btu/lbm.
(b) the change in entropy in Btu/lbm · °R.

4.20 One pound of oxygen undergoes a change of state isothermally at 80°F from 1 atm to 200 atm. Determine
(a) the change in enthalpy in Btu/lbm.
(b) the change in entropy in Btu/lbm · °R.

4.21 One kilogram of nitrogen undergoes a change of state isothermally at 300 K from 50.0 MPa to 101.325 kPa. Determine

(a) the change in enthalpy in kJ/kg.

(b) the change in entropy in kJ/kg · K.

4.22 One kilogram of oxygen undergoes a change of state isothermally at 300 K from 50.0 MPa to 101.325 kPa. Determine

(a) the change in enthalpy in kJ/kg.

(b) the change in entropy in kJ/kg · K.

4.23 What must be the proportions, by mass, of saturated liquid and vapor at 50 psia if H_2O in a rigid vessel is to pass through the critical state when it is heated?

4.24 The temperature of a liquid–vapor mixture of oxygen is increased from 120 K to 150 K in a rigid vessel. If we want 100% saturated liquid at the final temperature, what must be the initial liquid fraction?

4.25 The temperature of a liquid–vapor mixture of oxygen is increased from 120 K to 150 K in a rigid vessel. If we want 100% saturated vapor at the final temperature, what must be the initial vapor fraction?

4.26 One kilogram of water at 120°C and 200 kPa is converted into steam at 300°C under constant pressure. Determine the change in enthalpy of the fluid.

4.27 Freon-12 enters a device as a saturated liquid at a temperature of 100°F and leaves at a temperature of 20°F. If the enthalpy of the fluid at the outlet of the device is the same as that at the inlet, what is the quality of Freon-12 at the outlet?

4.28 A rigid vessel having a volume of 0.25 m^3 contains 50 kg of nitrogen at 80 K. Due to heat leakage, the temperature of the fluid reaches 300 K. Determine

(a) the final pressure of the fluid.

(b) the total change in internal energy due to the change of state.

4.29 Superheated steam at a pressure of 3000 psia and a temperature of 1000°F undergoes a constant-enthalpy process in which the pressure drops to 2000 psia. Find

(a) the final temperature.

(b) the change in entropy in Btu/lbm · °R.

4.30 Freon-12 is condensed from an initial state of 100 psia and 160°F to a final state of saturated liquid at the same pressure. Determine the change in enthalpy in Btu/lbm.

4.31 Water changes state isentropically from a saturated liquid at 40°C to a final state with a final pressure greater than the initial pressure. Determine the final temperature when the final pressure is
(a) 1.0 MPa. (b) 10.0 MPa. (c) 50.0 MPa.

4.32 Same as Problem 4.31 except that the initial state is saturated liquid at 320°C.

4.33 Water is compressed isentropically from an initial state of saturated liquid at 100°F to a final pressure of 3000 psia.
(a) What is the final temperature of the water?
(b) What is the change in enthalpy in Btu/lbm?

4.34 Freon-12 is compressed isentropically from an initial state of saturated vapor at −25°C to a final pressure of 800 kPa.
(a) What is the final temperature?
(b) What is the change in enthalpy?

4.35 Ammonia is compressed isentropically from an initial state of saturated vapor at −10°C to a final pressure of 1.0 MPa.
(a) What is the final temperature?
(b) What is the change in internal energy in Btu/lbm?

4.36 Ammonia is compressed isentropically from an initial state of saturated vapor at 20°F to a final pressure of 100 psia. Determine the change in enthalpy in Btu/lbm.

4.37 Steam expands from an initial state of 20.0 MPa and 550°C to a final pressure of 5.0 kPa. If the quality of steam at the final pressure is 90%, determine
(a) the change in enthalpy.
(b) the change in entropy.

4.38 Steam expands isentropically from an initial state of 2500 psia and 1000°F to a final pressure of 1.0 psia. Determine the change in enthalpy in Btu/lbm.

4.39 Determine the slope of the constant-temperature line in the enthalpy–entropy diagram in the liquid–vapor mixture region for the following fluids:
(a) nitrogen at 101.325 kPa.
(b) oxygen at 101.325 kPa.
(c) water at 100 kPa.

4.40 Nitrogen changes state from 120 K with a quality of 10% to a saturated vapor at 120 K.
(a) Locate the initial and final state points on a T–s diagram.

(b) What is the change in the specific Gibbs function due to the change of state?

4.41 Oxygen changes state from a dry saturated vapor state at 120 K to a final temperature of 120 K and a quality of 10%.
(a) Locate the initial and final state points on a T–s diagram.
(b) What is the change in the specific Gibbs function due to the change of state?

4.42 Making use of the tabulated data, complete the following table for saturated oxygen at the given temperature.

T (K)	$h_g - h_f$ (kJ/kg)	$T(s_g - s_f)$ (kJ/kg)
75		
100		
125		

4.43 To check the consistency of tabulated data for saturated vapor and saturated liquid, calculate $T(s_g - s_f)$ and compare with $(h_g - h_f)$ for the following cases:
(a) H_2O at 100 psia.
(b) Freon-12 at 80°F.
(c) nitrogen at 200°R.

4.44 Making use of h–T–p data from the nitrogen tables and the definition of $c_p = (\partial h/\partial T)_p$, determine approximately the constant-pressure specific heat for nitrogen at the following state points:
(a) 101.325 kPa and 180 K.
(b) 101.325 kPa and 300 K.
(c) 10 MPa and 180 K.
(d) 10 MPa and 300 K.

4.45 Making use of h–T–p data from the oxygen tables and the definition of $c_p = (\partial h/\partial T)_p$, determine approximately the constant-pressure specific heat for oxygen at the following state points:
(a) 101.325 kPa and 180 K.
(b) 101.325 kPa and 300 K.
(c) 10.0 MPa and 180 K.
(d) 10.0 MPa and 300 K.

4.46 Making use of h–T–p data from the steam tables and the definition of $c_p = (\partial h/\partial T)_p$, determine approximately the constant-pressure specific heat for steam at the following state points:
(a) 40 kPa and 250°C.

(b) 40 kPa and 500°C.
(c) 1.5 MPa and 250°C.
(d) 1.5 MPa and 500°C.

4.47 Making use of s–T–p data from the nitrogen tables and the property relation of $c_p = T(\partial s/\partial T)_p$, determine approximately the constant-pressure specific heat for nitrogen at the following state points:
(a) 101.325 kPa and 180 K.
(b) 101.325 kPa and 300 K.
(c) 10 MPa and 180 K.
(d) 10 MPa and 300 K.

4.48 Making use of s–T–p data from the oxygen tables and the property relation of $c_p = T(\partial s/\partial T)_p$, determine approximately the constant-pressure specific heat for oxygen at the following state points:
(a) 101.325 kPa and 180 K.
(b) 101.325 kPa and 300 K.
(c) 10 MPa and 180 K.
(d) 10 MPa and 300 K.

4.49 Making use of s–T–p data from the steam tables and the property relation of $c_p = T(\partial s/\partial T)_p$, determine approximately the constant-pressure specific heat for steam at the following state points:
(a) 40 kPa and 250°C.
(b) 40 kPa and 500°C.
(c) 1.5 MPa and 250°C.
(d) 1.5 MPa and 500°C.

4.50 The vapor pressure of water at a given temperature may be determined quite accurately over a fairly wide temperature range (0 to 200°C) by the following expression:

$$\ln p_g = 70.4346943 - \frac{7362.6981}{T} + 0.006952085T - 9.00000 \ln T$$

where T is in degrees Kelvin and p_g is vapor pressure in atm.
(a) Write a computer program for this expression and calculate the vapor pressure of water to three decimal places in units of kPa from 0 to 100°C in 10°C steps.
(b) Compare the calculated vapor pressures with those tabulated in Table A.1.1 (SI) and calculate the percentage variation in each case.

4.51 The following simple expression has been obtained by P. E. Liley for the determination of vapor pressure of water at a given temperature:

$$\ln p_g = 14.43509 - \frac{5333.3}{T}$$

where T is in degrees Kelvin and p_g is vapor pressure in bar. This expression, which is not as accurate as that given in Problem 4.50 is quite useful in certain applications, such as air conditioning.
(a) Write a computer program for this expression and calculate the vapor pressure of water to three decimal places in units of kPa from 0 to 100°C in 10°C steps.
(b) Compare the calculated vapor pressures with those tabulated in Table A.1.1 (SI) and calculate the percentage variation in each case.

4.52 Freon-12 undergoes a change of state isentropically from a saturated vapor to a higher pressure.
(a) Write a computer program involving the use of Eq. 4.40 for interpolation of entropy to determine the final temperature.
(b) If the initial state is saturated vapor at $-25°C$, calculate the final temperature corresponding to a final pressure of 0.80, 0.90, 1.00, 1.20, and 1.40 MPa.

4.53 Steam undergoes a change of state isentropically from a superheated vapor to a liquid–vapor mixture.
(a) Write a computer program to determine the quality of steam corresponding to a final pressure.
(b) If the initial state of steam is defined by 15.0 MPa and 500°C, calculate the quality of steam corresponding to a final pressure of 100, 200, 300, 400, and 500 kPa.

4.54 Freon-12 undergoes a change of state isentropically from a saturated vapor to a higher temperature.
(a) Write a computer program involving the use of Eq. 4.42 for interpolation of entropy to determine the final pressure.
(b) If the initial state is saturated vapor at $-25°C$, calculate final pressure corresponding to a final temperature of 20, 30, 40, 50, and 60°C.

4.55 Same as Problem 4.52 except that the initial state is saturated vapor at $-35°C$.

4.56 Same as Problem 4.53 except that the initial temperature is 450°C.

4.57 Same as Problem 4.54 except that the initial state is saturated vapor at −35°C.

CHAPTER 5

Behavior and Thermodynamic Properties of Ideal Gases

In our study of simple compressible substances in Chapter 4, it was mentioned that at very high temperatures or at very low pressures the enthalpy of vapor is essentially independent of pressure. This is simply one of the consequences of the unique behavior of the vapor under certain conditions. If we examine the $p-v-T$ relation for vapor in the regions of high temperature and low pressure, we would find that the quantity pv/T approaches a constant. When the vapor of a simple compressible substance behaves in this fashion, we say that it approaches the behavior of a hypothetical simple compressible substance called the *ideal gas*. As we shall see very shortly, this simplification reduces considerably the effort required for the determination of changes in various thermodynamic properties in engineering analysis and design.

No vapor or real gas behaves exactly like an ideal gas, of course. When we can use the ideal-gas model depends on how much deviation from real-gas behavior we are willing to accept. It is generally recognized that in the region where the temperature is in excess of twice the critical temperature, or in the region where the pressure is 1.0 atm or less, the ideal-gas assumption is quite valid. Since many engineering processes are concerned with vapor and real gases operating in regions of high temperature and low pressure, the ideal-gas concept is a very useful one in thermodynamics. In this chapter we explore the behavior of an ideal gas in fairly great detail. In fact, an essentially complete picture of the ideal-gas model is contained in this chapter, because of the manner in which we are structuring thermodynamic fundamentals in this book.

5.1 Definition of an Ideal Gas

An *ideal gas* is a simple compressible substance, a very unique kind of pure substance. It is defined as one whose thermal equation of

Table 5.1 Values of the Universal Gas Constant in Various Units

Value of $\bar{R}$	Units
1.986	Btu/lb mol $\cdot$ °R
1.986	cal/g mol $\cdot$ K
8314.3	J/kg mol $\cdot$ K
1545.3	ft $\cdot$ lbf/lb mol $\cdot$ °R
82.057	cm$^3 \cdot$ atm/g mol $\cdot$ K

state (that is, the p–v–T relation) is given by

$$p\bar{v} = \bar{R}T, \tag{5.1}$$

where $\bar{v}$ is the molal specific volume (volume per mole) and $\bar{R}$ is the universal gas constant. The value for $\bar{R}$ depends on the units chosen for p, $\bar{v}$, and T. Several values of $\bar{R}$ are given in Table 5.1.

Equation 5.1 may be written in the following alternative forms:

$$pV = n\bar{R}T \tag{5.1a}$$

$$pV = mRT \tag{5.1b}$$

$$pv = RT, \tag{5.1c}$$

where n is the amount of gas in moles, m is the mass, and R is the specific gas constant or simply the gas constant of a particular gas. The R is related to the universal gas constant $\bar{R}$ through its molecular weight M (that is, its molal mass) in the following manner:

$$R = \frac{\bar{R}}{M}. \tag{5.2}$$

We wish to point out that the thermal equation of state of an ideal gas may be deduced from kinetic theory of gases, a microscopic approach, on the basis of some simplifying assumptions. Two of the most important microscopic assumptions are that the gas molecules are point masses and that there are no intermolecular forces. The implication of these assumptions is that for an ideal gas the molecules would be far apart and the density of the gas would be low. Low density would correspond to the conditions of high temperatures or low pressures, of course.

For real gases, there will be intermolecular forces. It would be reasonable to expect that any deviation from ideal-gas behavior on the macroscopic level would represent the effect of intermolecular forces on the microscopic level and possibly could yield some information on the nature of the intermolecular forces involved. This observation indeed has been and is still being exploited. Many outstanding engineers and scientists, in fact, have made this kind of study a lifetime endeavor.[1]

Example 5.1

Determine the molal specific volume in SI units for an ideal gas at 1.0 atm and 0°C.

Solution

From the definition of an ideal gas,

$$\bar{v} = \frac{\bar{R}T}{p}.$$

In SI units,

$$\bar{R} = 8314.3 \text{ J/kg mol} \cdot \text{K} \quad (8314.3 \text{ N} \cdot \text{m/kmol} \cdot \text{K})$$

$$T = (0 + 273.15) \text{ K} = 273.15 \text{ K}$$

$$p = 1 \text{ atm}$$

$$= 1.01325 \times 10^5 \text{ N/m}^2.$$

Therefore,

$$\bar{v} = \frac{(8314.3 \text{ N} \cdot \text{m/kmol} \cdot \text{K})(273.15 \text{ K})}{1.01325 \times 10^5 \text{ N/m}^2}$$

$$= 22.41 \text{ m}^3/\text{kmol}. \qquad\blacksquare$$

Comment

The combination of 1.0 atm and 0°C is known as the *standard condition* (STP) for gas measurements. It is commonly stated in chemistry books that 1.0 g mol of ideal gas at STP would occupy 22.41 liters. Students should show that 22.41 m^3/kmol is indeed the same as 22.41 liters/g mol.

---

[1] See J. O. Hirschfelder, C. F. Curtiss, and R. B. Bird, *Molecular Theory of Gases and Liquids*, John Wiley & Sons, Inc., New York, 1964.

Example 5.2

A cylinder with a capacity of 2.0 m³ contained oxygen gas at a pressure of 500 kPa and 25°C initially. Then a leak developed and was not discovered until the pressure dropped to 300 kPa while the temperature stayed the same. Assuming ideal-gas behavior, determine how much oxygen had leaked out of the cylinder by the time the leak was discovered.

Solution

From the definition of an ideal gas,

$$m_1 = \frac{p_1 V_1}{R T_1}$$

$$m_2 = \frac{p_2 V_2}{R T_2}.$$

For oxygen,

$$R = \frac{8.3143 \text{ kJ/kmol} \cdot \text{K}}{31.999 \text{ kg/kmol}} = 0.2598 \text{ kJ/kg} \cdot \text{K}.$$

At $p_1 = 500$ kPa, $T_1 = (25 + 273.15)$ K $= 298.15$ K, and $V_1 = 2.0 \text{ m}^3$,

$$m_1 = \frac{500 \text{ kPa} \times 2.0 \text{ m}^3}{0.2598 \text{ kJ/kg} \cdot \text{K} \times 298.15 \text{ K}}$$

$$= 12.910 \text{ kg}.$$

At $p_2 = 300$ kPa, $T_2 = T_1 = 298.15$ K, and $V_2 = V_1 = 2.0 \text{ m}^3$,

$$m_2 = \frac{300 \text{ kPa} \times 2.0 \text{ m}^3}{0.2598 \text{ kJ/kg} \cdot \text{K} \times 298.15 \text{ K}}$$

$$= 7.746 \text{ kg}.$$

With the same V and T, m_2 may also be obtained from

$$m_2 = \left(\frac{p_2}{p_1}\right) m_1 = \frac{300}{500} \times 12.910 \text{ kg} = 7.746 \text{ kg}.$$

Therefore,

$$m_2 - m_1 = (7.746 - 12.910) \text{ kg}$$

$$= -5.164 \text{ kg}.$$

Comment

The negative sign in our answer simply means that the final amount of oxygen in the cylinder is less than the initial amount in the cylinder. That is, the amount of oxygen that leaked out was 5.164 kg.

Example 5.3

The temperature and pressure of an unknown ideal gas contained in a tube of 100-cm^3 capacity are found to be 25°C and 1.0 atm, respectively. The mass is found to be 0.1227 g. What is the molecular weight of the unknown gas?

Solution

From the definition of an ideal gas,

$$pV = mRT.$$

But

$$M = \frac{\bar{R}}{R}.$$

Therefore,

$$M = \frac{m\bar{R}T}{pV}.$$

Using

$$m = 0.1227 \text{ g} = 0.1227 \times 10^{-3} \text{ kg}$$

$$\bar{R} = 8314.3 \text{ J/kmol} \cdot \text{K}$$

$$p = 1 \text{ atm} = 101.325 \text{ kPa} = 101.325 \times 10^3 \text{ Pa}$$

$$V = 100 \text{ cm}^3 = 1 \times 10^{-4} \text{ m}^3$$

$$T = (25 + 273.15) \text{ K} = 298.15 \text{ K}$$

$$M = \frac{0.1227 \times 10^{-3} \times 8314.3 \times 298.15}{101.325 \times 10^3 \times 1 \times 10^{-4}}$$

$$= 30 \text{ kg/kmol.} \qquad \blacksquare$$

Comment

The molecular weight of a substance is simply the molal mass of the substance. Thus 30 kg/kmol is the same as 30 lbm/lb mol or 30 g/g mol.

Example 5.4

Air at 101 kPa and 25°C is supplied to the furnace of a steam power plant which burns coal at the rate of 240,000 kg/h. If we need 13.5 kg of air for each kilogram of coal burned, what must be the volume rate of airflow in m^3/h?

Solution

From the definition of specific volume, the volume rate of airflow is given by

$$\dot{V}_{air} = \dot{m}_{air} \times v_{air}.$$

With

$$\dot{m}_{air}/\dot{m}_{coal} = 13.5 \text{ kg/kg},$$

$$\dot{V}_{air} = 13.5 \, \dot{m}_{coal} \times v_{air}.$$

It is reasonable to assume ideal-gas behavior for air at given conditions. Then the specific volume of air at furnace inlet is given by

$$v_{air} = \frac{RT}{p}.$$

Using 28.97 for the molecular weight of air, we have

$$v_{air} = \frac{8.3143 \text{ kJ/kmol} \cdot \text{K} \times (25 + 273.15) \text{ K}}{28.97 \text{ kg/kmol} \times 101 \text{ kPa}}$$

$$= 0.8472 \text{ m}^3/\text{kg}.$$

Thus

$$\dot{V}_{air} = 13.5 \times 240,000 \times 0.8472 \text{ m}^3/\text{h}$$

$$= 2.745 \times 10^6 \text{ m}^3/\text{h}. \qquad\blacksquare$$

Comment

The amount of air needed in this example is quite typical for the design of large steam power plants.

Example 5.5

Find the specific volume of steam at 1.0 psia and 120°F by using

(a) the ideal-gas equation of state.

(b) data from superheated steam tables (that is, real-gas data).

Solution

(a) $v = \dfrac{RT}{p} = \dfrac{1545 \times 580}{18 \times 1.0 \times 144}$ ft^3/lbm $= 345.7$ ft^3/lbm.

(b) From the steam tables, $v = 344.6$ ft^3/lbm.   ∎

Comments

The saturation temperature for steam at 1.0 psia is 101.7°F. This means that the given state in this example is only slightly superheated. Generally speaking, vapor does not obey the ideal-gas equation of state too well if the state point is near the saturation curve. However, if the pressure is low enough, such as in this example, the ideal-gas relation is still quite valid.

5.2 Important Consequence of the Ideal-Gas Model

A very important consequence of the ideal-gas model is that the internal energy of an ideal gas is a function of temperature only. This may be deduced by starting with Eq. 4.6, a most important property relation for a simple compressible substance:

$$ds = \frac{1}{T}\, du + \frac{p}{T}\, dv.$$

Since ds is an exact differential, we have, according to the condition for exactness (see Eqs. 1.10, 1.10a, and 1.13),

$$\left[\frac{\partial(1/T)}{\partial v}\right]_u = \left[\frac{\partial(p/T)}{\partial u}\right]_v$$

or

$$\left(\frac{\partial T}{\partial v}\right)_u = -T^2\left[\frac{\partial(p/T)}{\partial u}\right]_v. \tag{5.3}$$

Equation 5.3 is valid for any simple compressible substance. In fact, it is one of many similar general thermodynamic relations for a simple compressible substance that we can derive.

The quantity on the left-hand side of Eq. 5.3 is a partial derivative (thermodynamic property) involving the properties T, v, and u. From the calculus of several variables, this thermodynamic property may also be given, according to Eq. 1.18, in terms of other partial derivatives involving the same three variables in the following manner:

$$\left(\frac{\partial T}{\partial v}\right)_u \left(\frac{\partial v}{\partial u}\right)_T \left(\frac{\partial u}{\partial T}\right)_v = -1$$

or

$$\left(\frac{\partial u}{\partial v}\right)_T = -\left(\frac{\partial T}{\partial v}\right)_u \left(\frac{\partial u}{\partial T}\right)_v . \tag{5.4}$$

Equations 5.3 and 5.4 may be used to study the behavior of any simple compressible substance. Let us apply them now to an ideal gas. Applying the equation of state of an ideal gas to the right-hand side of Eq. 5.3, we find that

$$\left[\frac{\partial(p/T)}{\partial u}\right]_v = \left[\frac{\partial(R/v)}{\partial u}\right]_v = 0.$$

Therefore, for an ideal gas, we have

$$\left(\frac{\partial T}{\partial v}\right)_u = 0. \tag{5.5}$$

Making use of this result, it follows from Eq. 5.4 that for an ideal gas, we must have

$$\left(\frac{\partial u}{\partial v}\right)_T = 0 \tag{5.6}$$

since $(\partial u/\partial T)_v$ (constant-volume specific heat) is always positive.

This means that for an ideal gas, u is independent of v along any line of constant T. It is left for the reader to show that for an ideal gas, u is independent of any thermodynamic property along any line of constant T. Consequently, u is a function only of temperature. That is, $u = u(T)$ for an ideal gas.

5.3 Internal Energy Change of an Ideal Gas

From the definition of constant-volume specific heat, the internal energy change of a real gas may be calculated from

$$du_v = c_v \, dT_v . \tag{5.7}$$

To integrate Eq. 5.7, we must have available the function relating c_v and T along the particular line of constant v. For an ideal gas, Eq. 5.7 becomes

$$du = c_v \, dT \tag{5.8}$$

and

$$u_2 - u_1 = \int_{T_1}^{T_2} c_v \, dT \tag{5.9}$$

since u depends only on T. This means that c_v must be a function only of temperature. That is, $c_v = c_v(T)$ for an ideal gas.

5.4 Enthalpy Change of an Ideal Gas

Since the enthalpy is defined as $(u + pv)$, and $pv = RT$ for an ideal gas, it follows that for an ideal gas

$$h = u + RT. \tag{5.10}$$

Since u is a function only of temperature, the enthalpy h for an ideal gas must also be a function only of temperature. That is, $h = h(T)$. Consequently, we must also have, from the definition of constant-pressure specific heat, for an ideal gas

$$dh = c_p \, dT \tag{5.11}$$

$$h_2 - h_1 = \int_{T_1}^{T_2} c_p \, dT \tag{5.12}$$

and

$$c_p = c_p(T).$$

5.5 Specific Heats of an Ideal Gas

Since $h = u + RT$ for an ideal gas, we have for a differential change of state,

$$dh = du + R \, dT \tag{5.13}$$

Combining Eqs. 5.8, 5.11, and 5.13, we have

$$c_p = c_v + R. \tag{5.14}$$

This shows that although c_p and c_v are functions of temperature, the difference of $(c_p - c_v)$ is a constant at a given state. This also shows that it is necessary to determine c_p or c_v, but not both, for any ideal gas. Since c_p can be measured more easily, most effort is generally devoted to the determination of c_p.

Defining $k = c_p/c_v$, known as the *specific-heat ratio*, we would also have, for an ideal gas,

$$c_p = \frac{Rk}{k-1} \tag{5.15}$$

$$c_v = \frac{R}{k-1}. \tag{5.16}$$

The specific-heat ratio for an ideal gas naturally is a function only of temperature. Equations 5.15 and 5.16 show that if we can determine k as a function of temperature, we would have obtained c_p and c_v as a function of temperature. It may be shown that one way to obtain k as a function of temperature is to measure the velocity of sound in low-density gas.

The molal specific heats of ideal gases at a given temperature are found, from experiment as well as from the kinetic theory of gases, to be dependent upon the complexity of the molecular structure of the gases—i.e., the more complex the molecules, the larger the molal specific heats. This means that at a given temperature, diatomic gases (such as hydrogen, nitrogen, and oxygen) would have larger $\bar{c}_p$ than monatomic gases (such as helium, argon, and mercury vapor), and polyatomic gases (such as water vapor, carbon dioxide, and propane) would have larger $\bar{c}_p$ than diatomic gases.

For monatomic gases at any temperature, molal specific heats are essentially constant, and $\bar{c}_p$ may be taken as $\frac{5}{2}\bar{R}$, and $\bar{c}_v$ may be taken as $\frac{3}{2}\bar{R}$. Consequently, the specific-heat ratio for a monatomic gas is a constant of $\frac{5}{3}$, or 1.667.

For diatomic gases at room temperature, $\bar{c}_p$ may be approximated as $\frac{7}{2}\bar{R}$ and $\bar{c}_v$ may be approximated as $\frac{5}{2}\bar{R}$, with $k = 1.40$.

Specific heats for several gases at room temperature are given in Table 5.2. Expressions giving $\bar{c}_p$ as a function of temperature for various gases at low pressure may be found in standard handbooks.

A study of Table 5.2 will show that the specific-heat ratio, k, decreases with an increase in the complexity of the molecular struc-

Table 5.2 Specific Heats for Gases at Low Pressure and 300 K

Gas		Molecular Weight	$\bar{c}_p$ (kJ/kmol·K)	c_p (kJ/kg·K)	c_v (kJ/kg·K)	$k = c_p/c_v$
Helium	He	4.003	20.786	5.1926	3.1156	1.667
Argon	Ar	39.948	20.786	0.5203	0.3122	1.667
Mercury	Hg	200.6	20.786	0.1037	0.0622	1.667
Air		28.97	29.0792	1.0038	0.7168	1.400
Nitrogen	N_2	28.013	29.0784	1.0380	0.7412	1.400
Oxygen	O_2	31.999	29.4315	0.9198	0.6599	1.394
Hydrogen	H_2	2.016	28.8720	14.3214	10.1972	1.404
Carbon monoxide	CO	28.011	29.0860	1.0384	0.7416	1.400
Water vapor	H_2O	18.015	33.6693	1.8690	1.4074	1.328
Methane	CH_4	16.043	35.8068	2.2319	1.7137	1.302
Carbon dioxide	CO_2	44.01	37.2539	0.8465	0.6576	1.287
Propane	C_3H_8	44.097	74.1090	1.6806	1.4920	1.126

ture of the gases. This trend is predicted by the kinetic theory of gases, with k varying from a maximum value of 1.667 for monatomic gases to a value approaching 1.0 for very complex molecules.

5.6 Entropy Change of an Ideal Gas

We have pointed out that the two $T\,ds$ equations (Eqs. 4.6 and 4.9) provide the means for the calculation of entropy change of a simple compressible substance from measurable quantities. In the case of an ideal gas, these two important equations take the form of

$$ds = \frac{du}{T} + \frac{p}{T}\,dv$$

$$= \frac{c_v\,dT}{T} + R\,\frac{dv}{v} \tag{5.17}$$

$$ds = \frac{dh}{T} - \frac{v}{T}\,dp$$

$$= \frac{c_p\,dT}{T} - R\,\frac{dp}{p}. \tag{5.18}$$

Integrating Eqs. 5.17 and 5.18, we get

$$s_2 - s_1 = \int_{T_1}^{T_2} \frac{c_v \, dT}{T} + R \ln \frac{v_2}{v_1} \qquad (5.19)$$

$$s_2 - s_1 = \int_{T_1}^{T_2} \frac{c_p \, dT}{T} - R \ln \frac{p_2}{p_1}. \qquad (5.20)$$

We thus see that the determination of entropy change for an ideal gas is a relatively simple matter if we have specific-heat data expressed as a function of temperature.

Example 5.6

An ideal gas has a molecular weight of 20. If its specific heat ratio at a given temperature is observed to be 1.25, determine c_p and c_v for the gas at the same temperature.

Solution

The gas constant R of the given gas is given by

$$R = \frac{\bar{R}}{M}$$

$$= \frac{8.3143 \ \text{kJ/kmol} \cdot \text{K}}{20 \ \text{kg/kmol}}$$

$$= 0.4157 \ \text{kJ/kg} \cdot \text{K}.$$

Using Eq. 5.15, we have

$$c_p = \frac{Rk}{k-1}$$

$$= \frac{0.4157 \times 1.25 \ \text{kJ/kg} \cdot \text{K}}{1.25 - 1}$$

$$= 2.0785 \ \text{kJ/kg} \cdot \text{K}.$$

Using Eq. 5.16, we have

$$c_v = \frac{R}{k-1}$$

$$= \frac{0.4157 \ \text{kJ/kg} \cdot \text{K}}{1.25 - 1}$$

$$= 1.6628 \ \text{kJ/kg} \cdot \text{K}. \qquad \blacksquare$$

Comment

It may be shown that the specific heat ratio k of an ideal gas is related to its velocity of sound. Thus one way to determine c_p and c_v of an ideal gas is to design an experiment to measure the velocity of sound within the gas.

Example 5.7

Five grams of argon gas undergoes a change of state at constant internal energy. Initial pressure and temperature are 6.0 atm and 300 K, respectively. The final volume occupied by the gas is three times that occupied initially. Assuming ideal-gas behavior, determine

(a) the final temperature of the gas.
(b) the final pressure of the gas.
(c) the entropy change of the gas due to the change of state.

Express the answers in SI units.

Solution

(a) For an ideal gas, internal energy is a function only of temperature. If internal energy is constant, temperature is constant. Therefore,

$$T_2 = T_1 = 300 \text{ K.}$$

(b) For an ideal gas,

$$p_1 V_1 = mRT_1$$

$$p_2 V_2 = mRT_2 .$$

If m is fixed, $T_2 = T_1$, and $V_2 = 3V_1$,

$$p_2 = \frac{P_1}{3} = \frac{6.0}{3} \text{ atm} = 2.0 \text{ atm}$$

$$= 2.0 \text{ atm} \times 1.0133 \times 10^5 \frac{\text{N/m}^2}{\text{atm}}$$

$$= 2.0266 \times 10^5 \text{ N/m}^2.$$

(c) Since argon is a monatomic gas, we may calculate entropy change using Eq. 5.19 or 5.20 with constant specific heats. Using Eq. 5.20, we have

$$s_2 - s_1 = c_p \ln \frac{T_2}{T_1} - R \ln \frac{p_2}{p_1}.$$

With $T_2 = T_1$, we have

$$s_2 - s_1 = -R \ln \frac{p_2}{p_1}$$

$$= -\frac{8314.3}{39.95} \ln \frac{2.0}{6.0} \text{ J/kg} \cdot \text{K}$$

$$= +\frac{8314.3}{39.95} \ln \frac{6.0}{2.0} \text{ J/kg} \cdot \text{K}$$

$$= +228.64 \text{ J/kg} \cdot \text{K}.$$

For 5 g of gas, the total entropy change is then

$$S_2 - S_1 = m(s_2 - s_1) = \frac{5 \times 228.64}{1000} \frac{\text{J}}{\text{K}} = 1.1432 \text{ J/K}. \qquad \blacksquare$$

Comment

The reader should show that the same entropy change may be obtained by using Eq. 5.19.

5.7 Thermodynamic Diagrams for an Ideal Gas

We have already seen how the ideal-gas model makes it possible to calculate property changes in a fairly straightforward manner. This simple model also makes it possible for us to construct thermodynamic diagrams for an ideal gas with relative ease. For one thing, an ideal gas, by definition, can exist only in the gaseous state. There is no saturation curve to worry about, because there is no phase change to consider. Although there are many possible thermodynamic diagrams that we can construct, we shall only discuss the p–v and the T–s diagrams, because these are the most useful ones as far as this book is concerned.

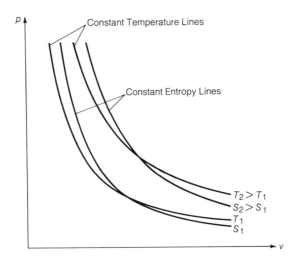

Figure 5.1 p–v Diagram for an Ideal Gas

Pressure–Specific Volume Diagram

Figure 5.1 is a p–v diagram for an ideal gas showing constant-temperature and constant-entropy lines. The constant-temperature lines are rectangular hyperbolas, since $pv = RT$. The slope of a constant-temperature line may be obtained in the following manner. From $pv = RT$, we have

$$p\,dv + v\,dp = R\,dT. \tag{5.21}$$

Equation 5.21 is simply the thermal equation of state for an ideal gas in differential form. Using this equation, we can form

$$\left(\frac{\partial p}{\partial v}\right)_T = -\frac{p}{v}. \tag{5.22}$$

Equation 5.22 is recognized as the slope of a constant-temperature line on the p–v diagram. Since both p and v are positive quantities, these slopes are always negative, as shown.

For completeness, we can also form the following partial derivatives for an ideal gas:

$$\left(\frac{\partial v}{\partial T}\right)_p = \frac{R}{p} \tag{5.23}$$

$$\left(\frac{\partial T}{\partial p}\right)_v = \frac{v}{R}. \tag{5.24}$$

The slope of the constant-entropy line, that is, $(\partial p/\partial v)_s$, may be shown to be equal to $-kp/v$. This means that these slopes are also always negative and are steeper than the slopes of the constant-temperature lines on the same p–v diagram.

The p–v diagram will be quite useful to us in our study of cycles using an ideal gas as the working fluid.

Temperature–Entropy Diagram

Figure 5.2 is a T–s diagram for an ideal gas showing constant-pressure lines and constant-volume lines. The slopes of these lines are given by Eqs. 5.17 and 5.18, respectively, as

$$\left(\frac{\partial T}{\partial s}\right)_v = \frac{T}{c_v}$$

$$\left(\frac{\partial T}{\partial s}\right)_p = \frac{T}{c_p}.$$

Since $c_p > c_v$, the constant-volume lines must have steeper slopes, as shown. These slopes are always positive, since T, c_v, and c_p are positive quantities.

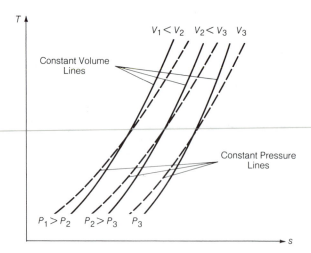

Figure 5.2 T–s Diagram for an Ideal Gas

The shifting of the constant-pressure lines to the right-hand side as the pressure decreases may be seen from an examination of Eq. 5.18,

$$ds = c_p \frac{dT}{T} - R \frac{dp}{p}.$$

From this equation, we observe that for a constant-temperature process,

$$ds = -R \frac{dp}{p}$$

and

$$s_2 - s_1 = -R \ln \frac{p_2}{p_1}.$$

Thus constant-pressure lines must shift to the right as pressure decreases, because entropy increases in that direction. Using this kind of reasoning, the reader should be able to show why the shifting of the constant-volume lines is to the left as volume decreases.

This diagram will also be useful to us in our study of cycles using an ideal gas as the working fluid.

5.8 Tabulation of Thermodynamic Properties for an Ideal Gas

p–v–T Data
For obvious reasons, there is no need to tabulate this kind of information.

Internal Energy Data
Since $u = u(T)$ and $du = c_v \, dT$, the tabulation of internal energy using the temperature as the argument can be made if we have the function of c_v as a function of temperature. This has been done for a number of common gases. A widely used source of such tabulated data is Keenan and Kaye's *Gas Tables*.[2]

Enthalpy Data
Since $h = h(T)$ and $dh = c_p \, dT$, the tabulation of enthalpy using the temperature as the argument has also been made and is given by Keenan and Kaye.

[2] J. H. Keenan and J. Kaye, *Gas Tables*, John Wiley & Sons, Inc., New York, 1948.

Entropy Data

For an ideal gas, the entropy change between two state points may be obtained from Eq. 5.20:

$$s_2 - s_1 = \int_{T_1}^{T_2} \frac{c_p}{T} \, dT - R \ln \frac{p_2}{p_1},$$

which may be rewritten as

$$s_2 - s_1 = \int_{T_0}^{T_2} \frac{c_p}{T} \, dT - \int_{T_0}^{T_1} \frac{c_p}{T} \, dT - R \ln \frac{p_2}{p_1}. \tag{5.25}$$

Defining a property ϕ by

$$\phi(T) = \int_{T_0}^{T} \frac{c_p}{T} \, dT, \tag{5.26}$$

where T_0 is some reference temperature, Eq. 5.25 may be expressed as

$$s_2 - s_1 = \phi_2 - \phi_1 - R \ln \frac{p_2}{p_1}, \tag{5.25a}$$

where ϕ_2 is a function of T_2 and ϕ_1 is a function of T_1.

The tabulation of ϕ using temperature as the argument is also given in the gas tables. With this kind of tabulated data, the evaluation of entropy change can be made quite easily by using Eq. 5.25a. Note that s and ϕ must have the same units.

Relative-Pressure Data

For an ideal gas undergoing a change state isentropically, we have, from Eq. 5.25a,

$$0 = \phi_2 - \phi_1 - R \ln \frac{p_2}{p_1}$$

or

$$\ln \frac{p_2}{p_1} = \frac{\phi_2 - \phi_1}{R}. \tag{5.27}$$

We may define relative pressure, P_r, by

$$\ln P_r = \frac{\phi}{R}. \tag{5.28}$$

Since ϕ is a function of temperature only, P_r is also a function of temperature only. This quantity is also given in the gas tables. Combining Eqs. 5.27 and 5.28, we have, for an isentropic process,

$$\ln \frac{p_2}{p_1} = \ln \frac{P_{r2}}{P_{r1}}$$

or

$$\frac{p_2}{p_1} = \frac{P_{r2}}{P_{r1}}, \tag{5.29}$$

where P_{r2} is the tabulated value at T_2 and P_{r1} is the value at T_1.

Relative-Volume Data

If we define a relative volume, v_r, by

$$\ln v_r = -\frac{1}{R} \int_{T_0}^{T} \frac{c_v \, dT}{T}, \tag{5.30}$$

we can show that for an isentropic process we would have

$$\frac{v_2}{v_1} = \frac{v_{r2}}{v_{r1}}, \tag{5.31}$$

where v_{r2} is the value at T_2 and v_{r1} is the value at T_1. The quantity v_r is also given in the gas tables.

Summary of Tabulated Data

The quantities u, h, ϕ, P_r, and v_r are tabulated in the gas tables with T as the argument. Table A.6 in the Appendix gives these data for air on the basis of ideal-gas behavior. Table 5.3 is a skeleton of

Table 5.3 Some Properties of Air at Low Pressures

T (°R)	h (Btu/lbm)	u (Btu/lbm)	ϕ (Btu/lbm · °R)	P_r	v_r
500	119.48	85.20	0.58233	1.0590	174.90
1000	240.98	172.43	0.75042	12.298	30.12
1500	369.17	266.34	0.85416	55.86	9.948

Table A.6. It is presented to remind the reader that the tabulated values of u, h, and ϕ may be used irrespective of the processes, while the tabulated values of P_r and v_r may be used only for isentropic processes.

Example 5.8

Air undergoes a change of state isentropically. The initial temperature and pressure are 1500°R and 75 psia, respectively. The final pressure is 16.5 psia.

(a) Sketch the process on a T–s diagram.
(b) Determine the temperature of the air at the final state.
(c) Determine the enthalpy change due to the change of state.
(d) Determine the entropy change due to the change of state.
(e) Determine the specific volume of air at the final state.

Solution

(a)

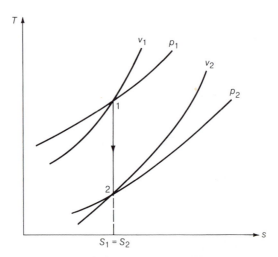

(b) For an isentropic process, we have, from Eq. 5.29,

$$\frac{p_2}{p_1} = \frac{Pr_2}{Pr_1}.$$

From Table 5.3, $P_{r1} = 55.86$ at 1500°R. Therefore,

$$P_{r2} = \frac{16.5}{75} \times 55.86 = 12.29.$$

From Table 5.3 the final temperature is about 1000°R.

(c) From Table 5.3,

$$h_1 = h \text{ at } 1500°R = 369.16 \text{ Btu/lbm}$$

$$h_2 = h \text{ at } 1000°R = 240.98 \text{ Btu/lbm.}$$

Therefore,

$$h_2 - h_1 = 240.98 - 369.17 \text{ Btu/lbm}$$

$$= -128.19 \text{ Btu/lbm.}$$

(d) From Table 5.3,

$$\phi_1 = \phi \text{ at } 1500°R = 0.85416 \text{ Btu/lbm} \cdot °R$$

$$\phi_2 = \phi \text{ at } 1000°R = 0.75042 \text{ Btu/lbm} \cdot °R.$$

Therefore,

$$s_2 - s_1 = \phi_2 - \phi_1 - R \ln \frac{p_2}{p_1}$$

$$= 0.75042 - 0.85416 - \frac{1.986}{28.9} \ln \frac{16.5}{75}$$

$$= -0.104 + 0.104$$

$$= 0,$$

which is as expected since the process is isentropic. Nevertheless, these calculations do show the consistency of the tabulated data.
(e) Since the process is isentropic, we have, from Eq. 5.31,

$$\frac{v_2}{v_1} = \frac{v_{r2}}{v_{r1}}.$$

From Table 5.3,

$$v_{r1} = v_r \text{ at } 1500°R = 9.948$$

$$v_{r2} = v_r \text{ at } 1000°R = 30.12,$$

Therefore,

$$v_2 = \frac{30.12}{9.948} v_1 = 3.028v_1,$$

which means that v_2 is greater than v_1. This is in agreement with our discussion of the T–s diagram for ideal gas. From $pv = RT$,

$$v_1 = \frac{RT_1}{p_1} = \frac{1545 \times 1500}{28.9 \times 75 \times 144} \frac{\text{Btu}}{\text{lbm}} = 7.425 \frac{\text{ft}^3}{\text{lbm}}.$$

Thus

$$v_2 = 3.028 \times 7.425 \frac{\text{ft}^3}{\text{lbm}} = 22.48 \frac{\text{ft}^3}{\text{lbm}}. \qquad \blacksquare$$

5.9 Isentropic Process for an Ideal Gas

In Example 5.8, involving an isentropic change of state, we have seen how the gas tables may be used advantageously. However, we do not have gas tables for all gases. Fortunately, p–v–T relations for an isentropic process of an ideal gas may without too much effort be deduced from basic equations.

Starting with Eq. 5.18, we have for an isentropic change of state,

$$\frac{c_p}{R} \frac{dT}{T} - \frac{dp}{p} = 0. \tag{5.32}$$

Making use of Eq. 5.15, Eq. 5.32 may be rewritten as

$$\frac{k}{k-1} \frac{dT}{T} - \frac{dp}{p} = 0. \tag{5.32a}$$

Using Eq. 5.17, we have, for an isentropic change of state,

$$\frac{c_v}{R} \frac{dT}{T} + \frac{dv}{v} = 0. \tag{5.33}$$

Making use of Eq. 5.16, Eq. 5.33 may be rewritten as

$$\frac{1}{k-1} \frac{dT}{T} + \frac{dv}{v} = 0. \tag{5.33a}$$

Combining Eqs. 5.32a and 5.33a, we would obtain the isentropic path on the p–v diagram as

$$k \frac{dv}{v} + \frac{dp}{p} = 0. \tag{5.34}$$

Equations 5.32a, 5.33a, and 5.34 may now be used to deduce the p–v–T relations for an isentropic change of state involving an ideal gas when functions of specific heats as functions of temperature are known. In the case of constant specific heats, these equations become, respectively,

$$\frac{T_2}{T_1} = \left(\frac{p_2}{p_1}\right)^{(k-1)/k} \tag{5.35}$$

$$\frac{T_2}{T_1} = \left(\frac{v_1}{v_2}\right)^{k-1} \tag{5.36}$$

$$p_1 v_1^k = p_2 v_2^k. \tag{5.37}$$

Example 5.9

Air undergoes a change of state isentropically. The initial temperature and pressure are 300 K and 100 kPa, respectively. The final pressure is 500 kPa. Determine the final temperature assuming ideal-gas behavior and constant specific heats, with $k = 1.40$.

Solution

Using Eq. 5.35 gives us

$$T_2 = T_1 \left(\frac{p_2}{p_1}\right)^{(k-1)/k}$$

$$= 300 \left(\frac{5}{1}\right)^{(1.4-1)/1.4} \quad \text{K}$$

$$= 475 \text{ K.} \qquad \blacksquare$$

Comment

Using data from the gas tables (that is, we still assume ideal-gas behavior but no longer assume constant specific heats), T_2 is 474 K. We obtain good agreement because k for air at low pressures does not vary too much from 1.4 in the temperature range of our problem, with $k = 1.400$ at 300 K and $k = 1.390$ at 475 K.

5.10 Reversible Polytropic Process for an Ideal Gas

Quasi-static, that is internally reversible, processes on the p–v diagram may be represented in general by the expression

$$pv^n = \text{constant.} \tag{5.38}$$

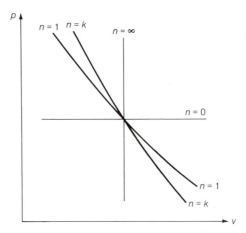

Figure 5.3 Polytropic Processes on a p–v Diagram

Equation 5.38 is the path on the p–v diagram of a *polytropic process* in which n, the polytropic exponent, has a particular value for a particular type of process. Thus for

$n = 0$, we have a constant-pressure process

$n = \infty$, we have a constant-volume process

$n = 1$, we have an isothermal process for an ideal gas

$n = k$, we have an isentropic process for an ideal gas with constant specific heats.

These four common processes are shown in Fig. 5.3, the p–v diagram, and in Fig. 5.4, the T–s diagram.

Combining Eq. 5.38 with the definition of an ideal gas, we have the following relations for a polytropic process for an ideal gas:

$$\frac{T_2}{T_1} = \left(\frac{p_2}{p_1}\right)^{(n-1)/n} \tag{5.39}$$

$$\frac{T_2}{T_1} = \left(\frac{v_1}{v_2}\right)^{n-1}. \tag{5.40}$$

Equations 5.38, 5.39, and 5.40 may be used to relate the initial and final states of a polytropic process. This means that they are useful to us in the determination of property changes for such a process.

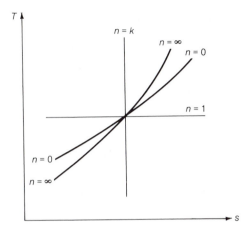

Figure 5.4 Polytropic Processes on a T–s Diagram

Example 5.10

Derive an expression giving the entropy change for a polytropic process of an ideal gas with constant specific heats.

Solution

Since we are dealing with an ideal gas with constant specific heats, we have from Eq. 5.20,

$$s_2 - s_1 = c_p \ln \frac{T_2}{T_1} - R \ln \frac{p_2}{p_1}.$$

For a polytropic process, we have, from Eq. 5.39,

$$\frac{T_2}{T_1} = \left(\frac{p_2}{p_1}\right)^{(n-1)/n}.$$

Combining Eqs. 5.20 and 5.39, we would have

$$s_2 - s_1 = \left[c_p \left(\frac{n-1}{n}\right) - R\right] \ln \frac{p_2}{p_1}. \tag{5.41}$$

Since $c_p/R = k/k - 1$ for an ideal gas (Eq. 5.15), Eq. 5.41 may be rewritten as

$$s_2 - s_1 = \frac{(n-k)R}{n(k-1)} \ln \frac{p_2}{p_1}. \tag{5.41a}$$

If n is not equal to 1, the entropy change may be shown to be given by the following expression in terms of temperatures:

$$s_2 - s_1 = \frac{(n-k)R}{(n-1)(k-1)} \ln \frac{T_2}{T_1} .$$ ▮ (5.42)

Comment

Let us make use of Eq. 5.41a to analyze the compression process of an ideal gas with constant specific heats from an initial state to a final pressure by three paths, path I, path II, and path III, as shown on Fig. 5.5.

For path I, s_2 is greater than s_1 and p_2 is greater than p_1. This means that the polytropic exponent, n, must be greater than k. Since the area under a line on the $T-s$ diagram represents heat transfer, the compression process for path I would involve heat addition to the gas.

For path II, s_2 equals s_1. This is possible only when n equals k, which is in agreement with the fact that this process is reversible and adiabatic for an ideal gas with constant specific heats.

For path III, s_2 is less than s_1 but p_2 is greater than p_1. This is possible only if n is less than k. For this process, heat must be removed from the gas, which is why the final temperature in this case must be smaller than in the other two cases.

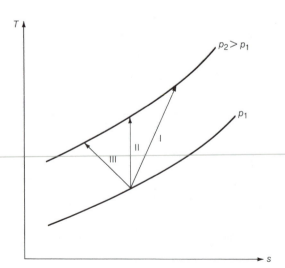

Figure 5.5 Three Compression Processes on a $T-s$ Diagram for an Ideal Gas with Constant Specific Heats

Example 5.11

Air changes state polytropically inside a cylinder from 100 kPa and 300 K to a final pressure of 1000 kPa according to the path given by

$$pv^{1.35} = \text{constant.}$$

Determine the change of entropy in kJ/kg·K. Assume ideal-gas behavior with constant specific heats of $c_p = 1.0038$ kJ/kg·K and $c_v = 0.7168$ kJ/kg·K.

Solution

Since we have an ideal gas undergoing a change of state polytropically with constant specific heats, the change of entropy may be evaluated with Eq. 5.41a:

$$s_2 - s_1 = \frac{(n - k)R}{n(k - 1)} \ln \frac{p_2}{p_1}$$

With given data,

$$k = \frac{c_p}{c_v} = \frac{1.0038}{0.7168} = 1.4$$

$$R = c_p - c_v = 1.0038 - 0.7168 \text{ kJ/kg·K}$$

$$= 0.2870 \text{ kJ/kg·K}$$

$$n = 1.35$$

Substituting into Eq. 5.41a, we have

$$s_2 - s_1 = \frac{(1.35 - 1.4) \times 0.287}{1.35(1.4 - 1)} \ln \frac{1000}{100} \text{ kJ/kg·K}$$

$$= -0.06119 \text{ kJ/kg·K.} \qquad \blacksquare$$

Comment

Since the entropy of the gas decreases for this internally reversible process, heat must be removed from the gas. This is why the cylinder used for such compression process is usually water jacketed.

Example 5.12

Air changes state isentropically. Its final volume is $\frac{1}{15}$ of its initial volume. Its initial temperature is 80°F. Assuming ideal-gas behavior and constant specific heats with $k = 1.4$, determine the final temperature.

Solution

Using Eq. 5.36,

$$T_2 = T_1\left(\frac{v_1}{v_2}\right)^{k-1}$$

$$= (80 + 460)(15)^{(1.4-1)\circ}\text{R}$$

$$= 540(15)^{0.4\circ}\text{R} = 1600^\circ\text{R}. \qquad \blacksquare$$

Comment

This is one of four processes making up the ideal Diesel cycle, which we may use to simulate the design and operation of a Diesel engine. The high-temperature air makes it possible to operate such an engine without spark plugs, since the injected fuel will burn when it meets the hot air. This is why a Diesel engine is known as a compression ignition engine.

Problems

5.1   A balloon is to be charged with a gas until the volume is 50 m³, the pressure is 200 kPa, and the temperature is 20°C.
   (a) What is the mass of gas contained in the balloon if helium is used?
   (b) What is the mass of gas contained in the balloon if hydrogen is used?
   (c) How many moles of each gas will the balloon contain?

5.2   Complete the following table for ideal gases.

Ideal Gas	Molecular Weight	p (kPa)	v (m³/kg)	R (kJ/kg·K)	T (K)
A	29	120			300
B			0.550	0.2200	500
C		300	0.5940	0.2376	

5.3 Complete the following table for ideal gases.

Ideal Gas	Molecular Weight	R (kJ/kg·K)	c_p (kJ/kg·K)	c_v (kJ/kg·K)	k
A	30				1.30
B		0.1386	0.4850		
C	20			1.0393	

5.4 A rigid tank having a volume of 1.5 m³ contains 2.0 kg of air initially at 120 kPa. Heat is added to the air until its final temperature exceeds its initial temperature by 50°C. Determine the final pressure.

5.5 A 2-m³ rigid tank contains oxygen gas at 50 kPa and 50°C. Another rigid tank of the same volume contains oxygen gas at 30 kPa and 25°C. The two tanks are then connected. If the final equilibrium temperature of the gas in both tanks is 25°C, what is the final pressure in both tanks?

5.6 A rigid tank contains a certain amount of nitrogen gas at 120 kPa and 20°C initially. Three kilograms of nitrogen gas is then added to the tank. If the final pressure and final temperature of the gas in the tank are found to be 240 kPa and 20°C, what is the volume of the tank?

5.7 A rigid tank contains oxygen gas at 500 kPa and 25°C initially. After 2.0 kg of the gas have been withdrawn for use in an experiment, the pressure and temperature of the gas in the tank are found to be 250 kPa and 25°C. What is the mass of oxygen in the tank initially?

5.8 Oxygen gas is heated from 25°C to 125°C. What is the increase in its internal energy in kJ/kg? What is its increase in enthalpy? Assume constant specific heats with $k = 1.4$.

5.9 A monatomic ideal gas with a molecular weight of 30 is cooled from 200°C to 100°C inside a rigid tank. What is the change in its internal energy in kJ/kg? What is the change in its enthalpy in kJ/kg?

5.10 Argon gas undergoes a change of state from 1 atm and 80°F to 5 atm and 200°F. Assuming ideal-gas behavior, determine the changes in its internal energy and enthalpy, in Btu/lbm.

5.11 The pressure of nitrogen gas in a rigid vessel is increased from 200 kPa to 400 kPa. If its initial temperature is 25°C, determine the changes in internal energy and enthalpy of the gas in kJ/kg. Assume ideal-gas behavior and constant specific heats with $k = 1.4$.

5.12 Air at low pressure is heated inside a rigid vessel from 50°C to 125°C. What is the change in its entropy in kJ/kg·K? Assume constant specific heats with $k = 1.4$.

5.13 An unknown ideal gas is cooled inside a rigid vessel from 200°C to 120°C. Assuming constant specific heats with $c_v = 0.5500$ kJ/kg·K and $k = 1.4$, determine
(a) the molecular weight of the gas.
(b) the change in entropy of the gas in kJ/kg·K.

5.14 An unknown ideal gas is cooled under constant pressure from 200°C to 50°C. Assuming constant specific heats with $c_p = 1.000$ kJ/kg·K, determine
(a) the change in entropy of the gas in kJ/kg·K.
(b) the ratio of the final volume to the initial volume of the gas.

5.15 Argon gas expands isentropically from 75 psia and 140°F to a pressure of 15 psia. Assuming ideal-gas behavior, determine the changes in its internal energy and enthalpy in Btu/lbm.

5.16 It may be shown from thermodynamics or from fluid mechanics that the velocity of sound, c, in an ideal gas may be given by the following expression:

$$c = \sqrt{kg_c\, RT}$$

If the velocity of sound in an ideal gas with a molecular weight of 29 is measured to be 400 m/s at 100°C, determine c_p and c_v in kJ/kg·K of the gas at 100°C.

5.17 A rigid tank of 20 ft³ capacity contains air at 100 psia and 70°F. The tank is equipped with a relief valve that opens at a pressure of 125 psia and remains open until the pressure drops to 120 psia. If a fire causes the valve to operate as described,
(a) what is the air temperature when the valve opens?
(b) what is the quantity of air lost due to the fire?
Assume that the temperature of the air in the tank remains constant during the discharge.

5.18 An ideal gas with a molecular weight of 40 is heated at constant pressure from 300°C to 600°C. If the change in its enthalpy is found to be 300 kJ/kg, determine the change in its internal energy. Assume constant specific heats.

5.19 An ideal gas having a molecular weight of 30 undergoes a change of state at constant volume from 540°F to 1540°F. If the change in its internal energy is 260 Btu/lbm, what is the change in its enthalpy, in Btu/lbm? Assume constant specific heats.

5.20 The pressure and density of air entering a compressor are 100 kPa and 1175 kg/m³, respectively. The pressure and density at the compressor outlet are 500 kPa and 5875 kg/m³, respectively. Determine the enthalpy difference between the state of air at the compressor outlet and that at the inlet. Assume constant specific heats with $c_p = 1.0038$ kJ/kg·K.

5.21 Methane gas enters a pipeline compressor at 250 psia and 80°F at the rate of 10,000 ft³/min. Assuming ideal-gas behavior, determine the mass rate of flow in lbm/s.

5.22 An ideal gas is compressed isentropically from state 1 to state 2. It is then cooled reversibly and at constant pressure from state 2 to state 3 with $T_3 = T_1$. Sketch the two processes
   (a) on a pressure–volume diagram showing the appropriate constant-temperature and constant-entropy lines.
   (b) on a temperature–entropy diagram showing the appropriate constant-pressure lines.

5.23 An ideal gas is compressed isentropically from state 1 to state 2. It is then expanded reversibly and isothermally with $V_3 = V_1$. Sketch the two processes
   (a) on a pressure–volume diagram showing the appropriate constant-temperature and constant-entropy lines.
   (b) on a temperature–entropy diagram showing the appropriate constant-pressure and constant-volume lines.

5.24 An ideal gas is first heated from state 1 to state 2 at constant volume and is then further heated from state 2 to state 3 at constant pressure. If the changes of states are carried out reversibly, sketch the processes
   (a) on a pressure-volume diagram showing the appropriate constant-temperature lines.
   (b) on a temperature–entropy diagram showing the appropriate constant-pressure and constant-volume lines.

5.25 Nitrogen gas is compressed reversibly and isothermally from 100 kPa and 25°C to a final pressure of 300 kPa.
   (a) Locate the state points on a T–s diagram showing the appropriate constant-pressure lines.

(b) Calculate the change in entropy, in kJ/kg·K. Assume ideal-gas behavior and constant specific heats with $k = 1.4$.

5.26 Nitrogen gas is compressed from 300 K and 100 kPa to a final pressure of 500 kPa. Assuming ideal-gas behavior and constant specific heats with $k = 1.4$, determine the change in enthalpy, in kJ/kg,
(a) if the process is reversible isothermal.
(b) if the process is isentropic.
(c) Sketch both processes on a pressure–volume diagram showing the appropriate constant-temperature and constant-entropy lines.

5.27 Oxygen gas expands from 300 K and 500 kPa to a final pressure of 100 kPa. Assuming ideal-gas behavior and constant specific heats with $k = 1.4$, determine the change in internal energy, in kJ/kg
(a) if the process is reversible isothermal.
(b) if the process is isentropic.
(c) Sketch both processes on a pressure–volume diagram showing the appropriate constant-temperature and constant-entropy lines.

5.28 An ideal gas having a molecular weight of 29 is compressed isentropically from a pressure and temperature of 15 psia and 540°R to a pressure p. The process is repeated with isothermal compression. The entropy difference between the two terminal states at pressure p is 0.14 Btu/lbm·°R. Determine the pressure p.

5.29 Air undergoes a change of state isentropically from 300 K and 110 kPa to a final pressure of 550 kPa. Assuming ideal-gas behavior, determine the change in enthalpy, in kJ/kg,
(a) assuming constant specific heats with $k = 1.4$.
(b) using data from Keenan and Kaye's *Gas Tables*.

5.30 Air undergoes a change of state isentropically from a pressure and temperature of 15 psia and 80°F to a pressure of 75 psia. Assuming ideal-gas behavior, determine the final temperature
(a) assuming constant specific heats of $c_p = 0.240$ Btu/lbm·°R and $c_v = 0.171$ Btu/lbm·°R.
(b) using data from Keenan and Kaye's *Gas Tables*.

5.31 Air expands isentropically from a pressure and temperature of 500 kPa and 800°C to a final pressure of 100 kPa. Assuming ideal-gas behavior, determine the change in enthalpy, in kJ/kg,
(a) assuming constant specific heats with $k = 1.4$.
(b) using data from Keenan and Kaye's *Gas Tables*.

5.32 Nitrogen gas is compressed polytropically from 100 kPa and 300 K to a final pressure of 600 kPa according to the path $pv^{1.25}$ = constant. Assuming ideal-gas behavior and constant specific heats with $k = 1.4$, determine the change in entropy, in kJ/kg·K, using
(a) Eq. 5.42.
(b) Eq. 5.20.

5.33 Show that the p–v–T relation of an ideal gas may be expressed in differential form as

$$\frac{dp}{p} + \frac{dv}{v} = \frac{dT}{T}$$

and

$$\frac{dp}{p} = \frac{d\rho}{\rho} + \frac{dT}{T}.$$

5.34 Prove the following relation for an ideal gas:

$$ds = \frac{c_p \, dv}{v} + \frac{c_v \, dp}{p},$$

from which we get $(\partial p/\partial v)_s = -kp/v$, which is the slope of a constant-entropy line in the p–v diagram.

5.35 Show that for an ideal gas,

$$\left(\frac{\partial u}{\partial p}\right)_T = 0.$$

5.36 Making use of the result in Problem 5.34, show that for an ideal gas undergoing a change of state isentropically with constant specific heats, we have

$$pv^k = \text{constant}.$$

5.37 The molal heat capacity of a gas at zero pressure may be expressed as a function of temperature as

$$\bar{c}_p^* = a + bT + cT^2 + dT^3$$

where
$\bar{c}_p^*$ is in kJ/kgmol·K
T is in degrees Kelvin

a, b, c, and *d* are constants which are different for different gases.

(a) Write a computer program for the calculation of $\bar{c}_p^*$ of a given gas at a given temperature.

(b) Calculate constant pressure specific heat c_p^*, constant volume specific heat c_v^*, and specific heat ratio k for air for the temperature range of 300 to 1000 K, in 100 K steps. Express c_p^* and c_v^* in units of kJ/kg · K.

The values of the constant for air in the given equation are:

$a = 28.11$
$b = 0.1967 \times 10^{-2}$
$c = 0.4802 \times 10^{-5}$
$d = -1.966 \times 10^{-9}$

5.38 Same as Problem 5.37 except that the gas is nitrogen. The values of the constants for nitrogen in the given equation are:

$a = 28.90$
$b = -0.1571 \times 10^{-2}$
$c = 0.8081 \times 10^{-5}$
$d = -2.873 \times 10^{-9}$

5.39 Same as Problem 5.37 except that the gas is carbon dioxide. The values of the constants for carbon dioxide in the given equation are:

$a = 22.26$
$b = 5.981 \times 10^{-2}$
$c = -3.501 \times 10^{-5}$
$d = 7.469 \times 10^{-9}$

5.40 (a) Write a computer program to calculate the molal enthalpy of an ideal gas at a given temperature, using 298 K as the datum, and using the following expression for its molal heat capacity as a function of temperature:

$$\bar{c}_p^* = a + bT + cT^2 + dT^3$$

which is the same equation given in Problem 5.37.

(b) Making use of this computer program, calculate the molal enthalpy of nitrogen gas in the temperature range of 300 to 1000 K, in 100 K steps.

See Problem 5.38 for the values of the constants *a, b, c,* and *d* for nitrogen.

5.41 Same as Problem 5.40 except that the gas is carbon dioxide. See Problem 5.39 for the values of the constants *a, b, c,* and *d* for carbon dioxide.

5.42 (a) Write a computer program to calculate the molal entropy of an ideal gas at 1 atm and a temperature T using its entropy at 298 K and 1 atm as the datum, and using the following expression for its molal heat capacity at a given temperature:

$$\bar{c}_p^* = a + bT + cT^2 + dT^3$$

which is the same equation given in Problem 5.37.

(b) Making use of this computer program, calculate the molal entropy for nitrogen gas in the temperature range of 300 to 1000 K, in 100 K steps. See Problem 5.38 for the values of the constants a, b, c, and d for nitrogen.

5.43 Same as Problem 5.42 except that the gas is carbon dioxide. See Problem 5.39 for the values of the constants a, b, c, and d for carbon dioxide.

5.44 (a) Write a computer program to calculate enthalpy change for an ideal gas between T_1 and T_2, for various values of T_2
   (1) taking into consideration the variation of specific heat with temperature;
   (2) assuming constant specific heat having a value at T_1;
   (3) assuming average specific heat between T_1 and T_2.

(b) Making use of this program, calculate the enthalpy change for nitrogen gas, in kJ/kg, with

$$T_1 = 300 \text{ K}$$

$$T_2 = 300 \text{ K to } 1000 \text{ K, in } 100 \text{ K steps.}$$

The molal heat capacity as a function of temperature for nitrogen is given by the equation given in Problem 5.37 with the appropriate constants given in Problem 5.38.

5.45 Same as Problem 5.44 except that the gas is carbon dioxide. The molal heat capacity as a function of temperature for carbon dioxide is given by the equation given in Problem 5.37 with the appropriate constants given in Problem 5.39.

5.46 (a) Write a computer program to calculate the final temperature of an ideal gas when it undergoes a change of state isentropically from an initial pressure p_1 and an initial temperature T_1 to a final pressure p_2

      (1) assuming constant specific heat having a value at T_1;

      (2) taking into consideration the variation of specific heat with temperature.

(b) Making use of this program, calculate the final temperature T_2 when nitrogen gas undergoes a change of state isentropically from 1 atm and 300 K to a final pressure of 10 atm. The molal heat capacity as a function of temperature for nitrogen is given by the equation given in Problem 5.37 with the appropriate constants given in Problem 5.38.

5.47 Same as Problem 5.46 except that the gas is carbon dioxide. The molal heat capacity as a function of temperature for carbon dioxide is given by the equation given in Problem 5.37 with the appropriate constants given in Problem 5.39.

CHAPTER 6

Some Consequences of the Second Law of Thermodynamics with an Introduction to Multicomponent Systems

This chapter is a continuation of the chapter on entropy and the second law of thermodynamics, in which the main objective was the introduction of the concept of entropy as a measure of unavailable energy and the development of the working equations of the second law that we need for the solution of engineering problems. In this chapter we devote our effort to the discussion of related ideas, such as absolute entropy, absolute zero and the third law of thermodynamics, and the concept of availability or exergy. We also give a brief presentation of the thermodynamic behavior of multicomponent systems, with particular reference to the concept of chemical potential and its physical significance in connection with chemical equilibrium. With this chapter we complete our discussion of all the fundamentals necessary for the engineering design of processes, devices, and systems involving the effective utilization of energy and matter.

6.1 Carnot Cycle and Thermodynamic Temperature Scale

The Carnot cycle is of historical importance. It was first introduced by Sadi Carnot in 1824 and led to the development of the second law of thermodynamics.[1] It is a reversible cycle, consisting of

[1] S. Carnot, *Reflections on the Motive Power of Fire* (E. Mendoza, ed.), Dover Publications, Inc., New York, 1960.

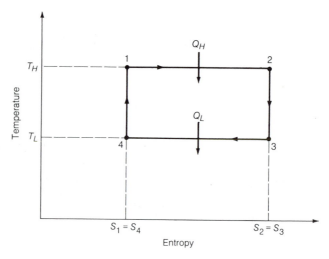

Figure 6.1 Carnot Cycle on a Temperature–Entropy Diagram

two reversible isothermal processes connected by two reversible adiabatic processes.

Let us carry out the Carnot cycle with a piston-cylinder arrangement (that is, using a closed-system approach). Making use of Eq. 3.23, we show the Carnot cycle on the temperature–entropy diagram of Fig. 6.1: process $1 \to 2$ is reversible isothermal; process $2 \to 3$ is reversible adiabatic (isentropic); process $3 \to 4$ is reversible isothermal; process $4 \to 1$ is reversible adiabatic (isentropic). For this reversible cycle,

$$\oint \bar{d}Q = Q_{12} + Q_{34}.$$

Making use of Eq. 3.23, we have, for reversible processes,

$$Q_{12} = \int_1^2 T \, dS = T_1(S_2 - S_1)$$

$$Q_{34} = \int_3^4 T \, dS = T_4(S_4 - S_3)$$

$$= -T_4(S_2 - S_1)$$

since $S_4 = S_1$ and $S_3 = S_2$. From the first law of thermodynamics for a cycle,

$$\oint \bar{d}Q = \oint \bar{d}W.$$

Therefore, the net work for the cycle is

$$W_{net} = \oint \bar{d}W = \oint \bar{d}Q = T_1(S_2 - S_1) - T_4(S_2 - S_1).$$

Making use of the definition of thermal efficiency, we have

$$(\eta_{th})_{Carnot} = \frac{T_1(S_2 - S_1) - T_4(S_2 - S_1)}{T_1(S_2 - S_1)}$$

$$= 1 - \frac{T_4}{T_1} = 1 - \frac{T_L}{T_H}. \qquad (6.1)$$

The definition of the thermal efficiency of a cycle is

$$\eta_{th} = 1 - \frac{|Q_L|}{|Q_H|}.$$

Combining this definition with Eq. 6.1, we have for a Carnot engine,

$$\frac{|Q_H|}{|Q_L|} = \frac{T_H}{T_L}. \qquad (6.2)$$

Equation 6.2 implies that the amount of work produced by a Carnot engine is dependent only on the temperature difference between the heat source and the heat sink. As proposed by Lord Kelvin, we may conceptually make use of Eq. 6.2 to define a thermodynamic temperature scale that is independent of the properties of any real substance.

To visualize Kelvin's concept, let us operate n Carnot engines in series as shown on Fig. 6.2. Further, let the work output be the same for all the reversible engines. Referring to Fig. 6.2, the work output of the first engine is

$$W = |Q_1| - |Q_2| = |Q_1|\left(1 - \frac{|Q_2|}{|Q_1|}\right) = \frac{|Q_1|}{T_1}(T_1 - T_2).$$

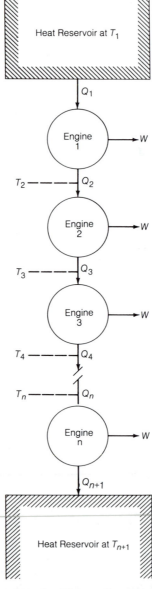

Figure 6.2 Kelvin's Concept of a Thermodynamic Temperature Scale with Carnot Engines Operating in Series

Similarly, the work output of the second engine is

$$W = |Q_2| - |Q_3| = \frac{|Q_2|}{T_2}(T_2 - T_3).$$

The output of the nth engine is

$$W = |Q_n| - |Q_{n+1}| = \frac{|Q_n|}{T_n}(T_n - T_{n+1}).$$

But according to Eq. 6.2, we have $|Q_1|/T_1 = |Q_2|/T_2 = |Q_n|/T_n$. Thus

$$T_1 - T_2 = T_2 - T_3 = T_n - T_{n+1}.$$

This equation states that the reversible engines in the series will deliver equal work output when the temperature differences across them are equal. If n is very large, the heat rejection of the last engine would approach zero, implying that the lowest temperature on the Kelvin scale is zero according to Eq. 6.2. This is why the zero temperature on the Kelvin scale is also known as the absolute zero temperature.

Although Kelvin's idea is important in giving us the concept of thermodynamic temperature and absolute zero temperature, it provides no practical way of measuring temperature. We shall see that if we use an ideal gas as the working fluid for a Carnot cycle operating between the temperature limits of T_H and T_L, we would reproduce Eq. 6.1. This means that the temperature used in the ideal-gas equation of state is consistent with the thermodynamic temperature defined in this section. This consistency is based on the comparison of the ratio of two temperatures only. Consequently, the thermodynamic temperature scale still could differ from the empirical temperature scale by an arbitrary multiplicative constant. However, by proper selection of the constant in the defining equation for the entropy, the two scales can be made identical. From now on we assume that the thermodynamic temperature is the same as the empirical absolute temperature.

6.2 Carnot Cycle Using an Ideal Gas as the Working Fluid

We have shown that the thermal efficiency of a Carnot cycle operating between the temperature limits of T_H and T_L is independent of the kind of substance used as the working fluid. Let us now carry out the Carnot cycle using an ideal gas as the working fluid.

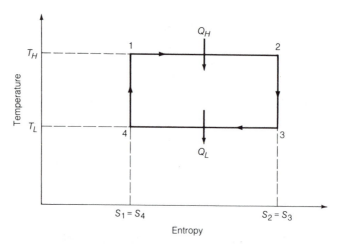

Figure 6.3 Carnot Cycle on a Temperature–Entropy Diagram

With reference to Fig. 6.3, a Carnot cycle shown on a T–s diagram, we have

$$Q_{12} = T_1(s_2 - s_1)$$

$$Q_{34} = T_3(s_4 - s_3).$$

For convenience, let us assume constant specific heats. Then, since processes $2 \to 3$ and $4 \to 1$ are both isentropic, we must also have

$$\frac{T_2}{T_3} = \left(\frac{p_2}{p_3}\right)^{(k-1)/k}$$

$$\frac{T_1}{T_4} = \left(\frac{p_1}{p_4}\right)^{(k-1)/k}.$$

Now, $T_1 = T_2$ and $T_3 = T_4$. Therefore,

$$\frac{p_2}{p_3} = \frac{p_1}{p_4}$$

and

$$\frac{p_1}{p_2} = \frac{p_4}{p_3}.$$

Thus the heat-transfer quantities Q_{34} and Q_{12} may now be given as

$$Q_{34} = -T_3 R \ln \frac{p_1}{p_2}$$

$$Q_{12} = T_1 R \ln \frac{p_1}{p_2}.$$

Applying the first law of thermodynamics to the cycle, we have

$$W_{net} = T_1 R \ln \frac{p_1}{p_2} - T_3 R \ln \frac{p_1}{p_2}.$$

The only heat addition to the cycle is Q_{12}. Making use of the definition of thermal efficiency of a cycle, we have

$$(\eta_{th})_{Carnot} = \frac{T_1 R \ln (p_1/p_2) - T_3 R \ln (p_1/p_2)}{T_1 R \ln (p_1/p_2)}$$

$$= 1 - \frac{T_3}{T_1} = 1 - \frac{T_L}{T_H},$$

which is as expected.

By defining an ideal gas by $pv = RT$, we have tacitly assumed that the empirical ideal-gas temperature scale, θ, is a linear function of the thermodynamic temperature scale. From our result just obtained, this means that we have shown that

$$\frac{\theta_1}{\theta_3} = \frac{T_1}{T_3}.$$

We have thus demonstrated that the empirical temperature scale could differ from the thermodynamic temperature scale by a multiplicated constant. In the next section, we shall show how this constant may be made equal to unity, thereby making the empirical temperature scale identical to the thermodynamic temperature scale.

6.3 Equivalence of the Thermodynamic and Ideal-Gas Empirical Temperature

On the basis of the experiments of Boyle, we may express the p–v–T relation of an ideal gas in terms of the empirical temperature scale, θ, as follows:

$$pv = A\theta, \tag{6.3}$$

where A is a constant that we can choose arbitrarily. Then the internal energy of the ideal gas is given by

$$u = f(\theta), \tag{6.4}$$

where $f(\theta)$ is a function of temperature.

For a simple compressible substance in general, we have, according to Eq. 4.6,

$$ds = \frac{du}{T} + \frac{p}{T} \, dv.$$

For an ideal gas, this may be given as

$$ds = \frac{df(\theta)}{T \, dT} \, dT + \frac{p}{T} \, dv.$$

Since ds is an exact differential, we have, according to the condition for exactness,

$$\frac{\partial}{\partial v} \left[\frac{df(\theta)}{T \, dT} \right]_T = \frac{\partial}{\partial T} \left(\frac{p}{T} \right)_v. \tag{6.5}$$

The left-hand side of Eq. 6.5 must be equal to zero. Consequently,

$$\frac{p}{T} = f(v), \tag{6.6}$$

where $f(v)$ is a function of volume alone. Combining Eqs. 6.3 and 6.6, we have

$$vf(v) = A \frac{\theta}{T}. \tag{6.7}$$

Since the left-hand side in Eq. 6.7 is only a function of v while the right-hand side is only a function of temperature, both sides must be equal to the same constant K. Hence

$$T = \frac{A}{K} \theta. \tag{6.8}$$

Since A can be chosen arbitrarily, A/K can be made equal to unity. This is indeed what is done, thereby making the thermodynamic temperature scale identical to the empirical temperature scale.

6.4 Absolute Zero, Absolute Entropy, and the Third Law of Thermodynamics

Let us investigate the temperature of a system as heat is being removed from it according to the scheme shown in Fig. 6.4. Let T be the temperature of the system at any instant. Let $\bar{d}Q_L$ be the amount of heat removed from the system at any instant. Let $\bar{d}W$ be the amount of work supplied corresponding to $\bar{d}Q_L$.

From the definition of the coefficient of performance of a refrigerator,

$$\beta_R = \frac{|\bar{d}Q_L|}{|\bar{d}W|}. \tag{1}$$

If the refrigerator is perfect, we have, according to Eq. 3.19,

$$(\beta_R)_{\text{ideal}} = \frac{T}{T_0 - T}. \tag{2}$$

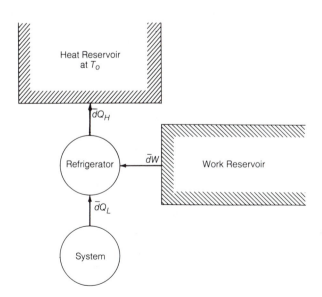

Figure 6.4 Scheme to Reduce the Temperature of a System

Combining Eqs. 1 and 2, we have

$$|\bar{d}W|_{\min} = |\bar{d}Q_L| \frac{T_0 - T}{T}. \tag{3}$$

We see that as T becomes smaller and smaller, $\bar{d}W_{\min}$ becomes larger and larger. It is true that $\bar{d}Q_L$ will also become smaller and smaller. But for $\bar{d}Q_L$ to become zero is quite unlikely, as this would require a perfect thermal insulator around the system. One may conclude that there will always be a finite value for $\bar{d}Q_L$. If this is the case, then $\bar{d}W_{\min}$ would approach infinity as T approaches zero. Thus the lowest attainable temperature for the system is greater than zero. In other words, a thermodynamic temperature of zero or less is unattainable. A temperature of zero on the thermodynamic scale is properly called *absolute zero*.

Generalizing from experience, we may now accept as true the following statement:

> It is impossible by any procedure, no matter how idealized, to reduce any system to the absolute zero of temperature in a finite number of operations.

This is known as the unattainability statement of the *third law of thermodynamics*, which may be given in the following form:

> The entropy of any substance approaches a constant value as its temperature approaches absolute zero. For a perfect crystalline substance, its entropy is zero at the absolute zero of temperature.[2]

From statistical thermodynamics, the entropy of any substance is given as

$$S = k \ln \Omega$$

where k is Boltzmann's constant and Ω is the thermodynamic probability or the number of microscopic arrangements. For a perfect crystal at absolute zero, it has only one microscopic configuration. That is, it is in the most ordered state when it is at absolute zero. Thus the third law is consistent with our previous comment that entropy is a measure of disorder in the microscopic level.

[2] C. J. Adkins, *Equilibrium Thermodynamics*, McGraw-Hill Book Company, New York, 1968, Chap. 12; M. W. Zemansky, *Heat and Thermodynamics*, McGraw-Hill Book Company, New York, 1957, pp. 396–401; J. D. Fast, *Entropy*, McGraw-Hill Book Company, New York, 1962, pp. 83–87.

With the third law, we can determine absolute entropies, which are necessary information required for the study of problems involving chemical reactions. However, further discussion of the third law and absolute entropy is beyond the scope of this book.

Example 6.1

Determine the absolute entropy of oxygen gas at 101.325 kPa and 100°C if absolute entropy of the gas at 101.325 kPa and 25°C is 6.4118 kJ/kg · K.

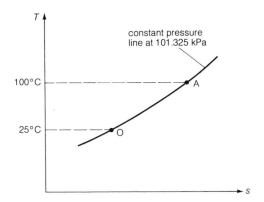

Solution

We want to find the absolute entropy at point A. Now

$$s_A = s_0 + (s_A - s_0).$$

With s_0 given, we need to evaluate the change of entropy between point 0 and point A. Since entropy is a point function, we may calculate the entropy change between two state points by selecting any convenient path connecting the state points involved.

Let us select a reversible path. Then from Eq. 3.23 for a reversible path, we have

$$ds = \frac{\bar{d}q}{T}. \tag{1}$$

We must now substitute for $\bar{d}q$ of the reversible path in terms of convenient parameters.

For given state points, a convenient path is a reversible constant-pressure process. Thus, we have, from the first and second laws,

$$\bar{d}q = du + p \, dv = d(u + pv) = dh.$$

Substituting into (1), we have

$$ds = \frac{dh}{T}$$

or

$$s_A - s_0 = \int_{101.325 \text{ kPa, } 25°C}^{101.325 \text{ kPa, } 100°C} \frac{dh}{T}. \tag{2}$$

Assuming an ideal gas with constant heats and $c_p = 0.9198$ kJ/kg $\cdot$ K, Eq. (2) becomes

$$s_A - s_0 = \int_{T_0}^{T_A} \frac{c_p \, dT}{T} \qquad \text{as } dh = c_p \, dT \text{ for ideal gas}$$

$$= 0.9198 \ln \frac{100 + 273.15}{25 + 273.15} \text{ kJ/kg} \cdot \text{K}$$

$$= 0.2064 \text{ kJ/kg} \cdot \text{K}.$$

Thus

$$s_A = 6.4118 + 0.2064 = 6.6182 \text{ kJ/kg} \cdot \text{K}. \qquad \blacksquare$$

Note that we must always use absolute temperatures in our calculations.

Comment

This problem shows that at a given pressure, the entropy of oxygen gas at a higher temperature is greater than its entropy at a lower temperature. From the microscopic point of view, this would mean that hot gas is more disorderly than the cool gas at the same pressure. The fact that solids have lower entropies than liquids, and liquids have lower entropies than gases, simply reflects the relative disorder of molecular motion in each of these forms of matter.

6.5 The Inequality of Clausius

The inequality of Clausius, the classical theorem through which the concept of entropy is deduced historically, may be obtained in different ways.[3] We shall take advantage of Eq. 3.23 (the second law of thermodynamics for a closed system that we have already established), which is given as

$$dS \geq \frac{\bar{d}Q}{T} \qquad (3.23)$$

where $\bar{d}Q$ is the differential amount of heat flowing into the system at temperature T.

Making use of Eq. 3.23, we have the following expression when a closed system undergoes a cyclic process:

$$\oint dS \geq \oint \frac{\bar{d}Q}{T},$$

where $\oint$ means summation around the cycle.

But $\oint dS$ must be equal to zero since entropy is a thermodynamic property. Consequently, we have

$$\oint \frac{\bar{d}Q}{T} \leq 0, \qquad (6.9)$$

which is the inequality of Clausius. The unequal sign applies when the cycle is irreversible. The equal sign applies when the cycle is reversible.

Example 6.2

Suppose that we have a cycle consisting of the following processes:

Process $1 \rightarrow 2$: adiabatic.
Process $2 \rightarrow 3$: heat addition at constant temperature.
Process $3 \rightarrow 4$: adiabatic.
Process $4 \rightarrow 1$: heat rejection at constant temperature.

Determine whether the inequality of Clausius is satisfied for the following cases:

(a) 5000 kJ are added at 1000 K; 3000 kJ are rejected at 300 K.
(b) 5000 kJ are added at 1000 K; 1500 kJ are rejected at 300 K.

[3] Gordon J. Van Wylen and Richard E. Sonntag, *Fundamentals of Classical Thermodynamics*, 3rd ed., SI version, John Wiley & Sons, Inc., New York, 1985, pp. 184–188. See also Franzo H. Crawford and William D. Van Vorst, *Thermodynamics for Engineers*, Harcourt Brace & World, Inc., New York, 1968, pp. 216–219.

Solution

(a) With the given information,

$$\oint \frac{\bar{d}Q}{T} = \int_2^3 \frac{\bar{d}Q}{T} + \int_4^1 \frac{\bar{d}Q}{T}$$

$$= + \frac{5000}{1000} - \frac{3000}{300} \text{ kJ/K}$$

$$= 5.0 - 10.0 = -5.0 \text{ kJ/K} < 0.$$

The inequality of Clausius is satisfied.

(b) With the given information,

$$\oint \frac{\bar{d}Q}{T} = \int \frac{\bar{d}Q}{T} + \int \frac{\bar{d}Q}{T}$$

$$= + \frac{5000}{1000} - \frac{1500}{300} \text{ kJ/K}$$

$$= 5.0 - 5.0 = 0.$$

The inequality of Clausius is satisfied, but the cycle must be reversible. ∎

Comment

We should realize that, when we make use of the inequality of Clausius to study cycles, we are making use of the second law of thermodynamics.

6.6 Quasi-static Heat Transfer for a Closed System

For a closed system, the quasi-static work is given by Eq. 2.17:

$$\bar{d}W = p\, dV - \tau\, dL - \mathscr{E}\, dZ - \mathscr{H}\, d\mu + \cdots$$

or

$$\bar{d}W = p\, dV + \sum_i \mathscr{F}_i\, dX_i, \tag{6.10}$$

where p and the $\mathscr{F}$'s are the generalized forces of the relevant quasi-static work modes, and V and the X's are the generalized displacements of the relevant quasi-static work modes.

Combining Eq. 6.10 with the first law for a closed system given by Eq. 2.9a, we find that the *quasi-static* (also known as *reversible*) *heat transfer* for a closed system is given by

$$\bar{d}Q_{rev} = dU + p \, dV + \sum_i \mathscr{F}_i \, dX_i \tag{6.11}$$

or

$$dU = \bar{d}Q_{rev} - p \, dV - \sum_i \mathscr{F}_i \, dX_i. \tag{6.11a}$$

According to the state postulate, the following explicit functions will exist:

$$S = S(U, V, X_1, \ldots, X_r) \tag{6.12}$$

$$U = U(S, V, X_1, \ldots, X_r), \tag{6.13}$$

where the V and the X's are the extensive parameters corresponding to the relevant quasi-static work modes. Each of these two equations is an equation of equilibrium states. Each, then, is known as an *equation of state*. We shall encounter many more equations of state, but these two are among the most important. In fact, they are often called the *fundamental* equations of state, because they contain all the thermodynamic information of a given substance.

Although a system could have many kinds of quasi-static work modes, we shall see in the rest of this book that the number of relevant quasi-static work modes is very small. This means that the number of X's in Eq. 6.12 or 6.13 is very small, usually one or two. If we limit our discussion to two X's only, the system we shall study will have only three kinds of relevant quasi-static work modes, one of which is the mechanical work of expansion or compression.

With two X's only, the independent variables in Eq. 6.12 are U, V, X_1, and X_2, and the independent variables in Eq. 6.13 are S, V, X_1, and X_2. From Eq. 6.13, we have

$$dU = \left(\frac{\partial U}{\partial S}\right)_{V, X_1, X_2} dS + \left(\frac{\partial U}{\partial V}\right)_{S, X_1, X_2} dV$$

$$+ \left(\frac{\partial U}{\partial X_1}\right)_{S, V, X_2} dX_1 + \left(\frac{\partial U}{\partial X_2}\right)_{S, V, X_1} dX_2. \tag{6.14}$$

We shall define the thermodynamic temperature T of a given substance by

$$T \equiv \left(\frac{\partial U}{\partial S}\right)_{V,X_1,X_2} \tag{6.15}$$

and the thermodynamic pressure p of a given substance by

$$p \equiv -\left(\frac{\partial U}{\partial V}\right)_{S,X_1,X_2} \tag{6.16}$$

The definitions for T and p will be shown to be consistent with our concepts of thermal equilibrium and mechanical equilibrium. Combining these definitions with Eq. 6.14 we have

$$dU = T\,dS - p\,dV + \left(\frac{\partial U}{\partial X_1}\right)_{S,V,X_2} dX_1 + \left(\frac{\partial U}{\partial X_2}\right)_{S,V,X_1} dX_2. \tag{6.17}$$

In general we would have

$$dU = T\,dS - p\,dV + \sum_i \left(\frac{\partial U}{\partial X_i}\right)_{S,V,X_{j\pm i}} dX_i. \tag{6.18}$$

For a simple compressible substance, we have only $p\,dV$ work. Thus Eq. 6.18 would become

$$dU = T\,dS - p\,dV,$$

which is Eq. 4.6 for unit mass.

Equation 6.18 must be compatible with Eq. 6.11a. Therefore, we get the following expression, which is valid for any closed system undergoing a change of state quasi-statically:

$$\bar{d}Q_{\text{rev}} = T\,dS$$

or

$$dS = \frac{\bar{d}Q_{\text{rev}}}{T}. \tag{6.19}$$

Equation 6.19 is the same as Eq. 3.23 for a reversible process. It is used by some authors as part of the definition of entropy.[4]

---

[4] M. M. Abbott and H. C. Van Ness, *Thermodynamics*, McGraw-Hill Book Company, New York, 1972, p. 36.

By integrating Eq. 6.19, we find that the total amount of heat transfer for a closed system undergoing a change of state quasi-statically is

$$(Q_{12})_{rev} = \int_1^2 T \, dS. \tag{6.20}$$

Now the integral $\int_1^2 T \, dS$ is simply the area under the path on the temperature–entropy diagram. Since we can go from state 1 to state 2 along many different quasi-static paths, this integral will have a different value for each path. That is, heat is a path function, as we pointed out previously.

Since entropy is a point function, Eq. 6.19 makes it possible for us to calculate the entropy change between two state points by selecting any convenient reversible path between the two state points and substituting for $\bar{d}Q_{rev}$ of this reversible path in terms of convenient parameters into the expression

$$S_2 - S_1 = \int_1^2 \frac{\bar{d}Q_{rev}}{T}. \tag{6.21}$$

In fact, we have made use of this idea in Example 6.1.

6.7 Another Look at the Entropy Change of a Heat Reservoir

From the definition of a heat reservoir, since no work transfer is allowed, we must have

$dU = T \, dS$ according to the state postulate

$\bar{d}Q = dU$ according to the first law of thermodynamics.

Therefore,

$$\left[dS = \frac{\bar{d}Q}{T} \right]_{\text{heat reservoir}}$$

(which is Eq. 3.11 and is the same as Eq. 6.19). This implies that any interaction within a heat reservoir must be quasi-static, and is consistent with our definition that a heat reservoir is a body of very large heat capacity.[5]

We have thus demonstrated that for a logical and consistent structure of principles, the constant C_1 in Eq. 3.10 is indeed the reciprocal of the temperature of the heat reservoir.

6.8 The Concept of Availability

We have stated in Chapter 3 that the portion of energy content in a given system that can potentially do work for us is known as the available work or available energy of the system. We have also stated that the available energy of a given system may be determined in principle by having it undergo a change of state ideally and interact with a standard reservoir. If we are interested in the maximum work that a system at a given state can produce, we would want to change its state reversibly until it is in equilibrium with the environment at p_0 and T_0. It is the common practice to select the environment as the standard reservoir for such determination.

Let us now investigate the available energy of a system at a state having the values of p, T, u, v, and s, using the environment as the reference. From the first law of thermodynamics for a closed system, we have

$$\bar{d}q = du + \bar{d}w. \tag{1}$$

From the second law of thermodynamics for a closed system undergoing a change of state reversibly, we have

$$ds + ds^{\text{HR}} = 0. \tag{2}$$

But

$$ds^{\text{HR}} = \frac{\bar{d}q^{\text{HR}}}{T^{\text{HR}}} \tag{3}$$

and

$$\bar{d}q^{\text{HR}} = -\bar{d}q. \tag{4}$$

---

[5] H. B. Callen, *Thermodynamics*, John Wiley & Sons, Inc., New York, 1960, p. 65. See also Ryogo Kubo, *Thermodynamics*, North-Holland Publishing Company, Amsterdam, 1968, p. 3.

Combining Eqs. 1, 2, 3, and 4, we have

$$T^{\text{HR}}\, ds = du + \bar{d}w_{\text{max}}. \tag{5}$$

If the heat reservoir is the environment at T_0, then

$$\bar{d}w_{\text{max}} = T_0\, ds - du. \tag{6}$$

From Eq. 6, the maximum work is given by

$$w_{\text{max}} = (u - T_0 s) - (u_0 - T_0 s_0), \tag{6.22}$$

where T_0, s_0 and u_0 are properties of the system when it is in equilibrium with the environment.

But part of the maximum work is done against the environment, which is not useful. Since the work done against the environment is $p_0(v_0 - v)$, the maximum useful work is given by

$$w_{\text{max useful}} = w_{\text{max}} - p_0(v_0 - v)$$

$$= (u - T_0 s) - (u_0 - T_0 s_0) - p_0(v_0 - v)$$

$$= (u + p_0 v - T_0 s) - (u_0 + p_0 v_0 - T_0 s_0). \tag{6.23}$$

The maximum useful work given by Eq. 6.23 is known as the availability or nonflow exergy. We use these two terms interchangeably in this book.

Introducing the symbol ϕ for availability or nonflow exergy per unit mass, we have

$$\phi = (u + p_0 v - T_0 s) - (u_0 + p_0 v_0 - T_0 s_0). \tag{6.24}$$

We observe that the availability or nonflow exergy is a function of the state properties of the system as well as a function of the properties of the environment. However, once the conditions of the environment have been specified, the availability is just like any other thermodynamic property. It can then be used as either a dependent or an independent variable in an equation of state. We will find the concept of availability very useful, particularly in the study of complex thermodynamic systems.

For a closed system undergoing a change from state 1 to state 2, we would have

$$w_{\text{max useful}} = \phi_1 - \phi_2. \tag{6.25}$$

It is important to note that there are no restrictions in Eq. 6.25.

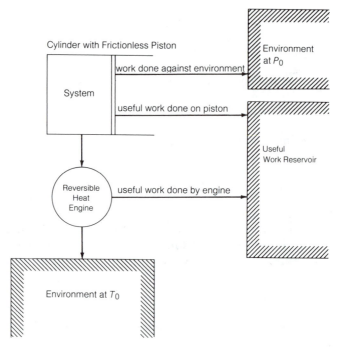

Figure 6.5 Conceptual Arrangement for the Determination of Availability

Let us now visualize, with the help of Fig. 6.5, how to determine the availability of a given system. Let the system be placed inside a cylinder. Since the processes involved must be reversible, the piston must be frictionless. Since the heat-transfer process must be reversible, we must employ a reversible heat engine to operate between the system and the environment at T_0. Part of the work done by the system on the piston is done against the environment at p_0. We thus see that the availability, as defined, is simply the maximum useful work done against the piston plus the work produced by the reversible heat engine. We should realize that if the temperature of the system is lower than the temperature of the environment, heat will be delivered to the system by the engine. If the pressure of the system is lower than the pressure of the environment, work will be done to the system by the environment.

Example 6.3

Calculate the availability (nonflow exergy) for nitrogen at the following state points:

(a) 10 MPa and 320 K.

(b) 0. 020 MPa and 300 K.
(c) 0.080 MPa and 200 K.

The pressure and temperature of the environment are 101.325 kPa and 25°C, respectively.

Solution

The nitrogen data we need are given in the following table:

State Point					$u = (h - pv)$
p (MPa)	T (K)	h (kJ/kg)	v (m³/kg)	s (kJ/kg · K)	(kJ/kg)
0.101325	298.15	309.237	0.873153	6.8353	220.765
10	320	315.553	0.009650	5.4932	219.053
0.020	300	311.341	4.451086	7.3241	222.319
0.080	200	207.068	0.740770	6.4898	147.806

By definition, availability or nonflow exergy is given by Eq. 6.24:

$$\phi = (u + p_0 v - T_0 s) - (u_0 + p_0 v_0 - T_0 s_0)$$

$$= (u - u_0) + p_0(v - v_0) - T_0(s - s_0).$$

For case (a), we have

$$\phi = (219.053 - 220.765) + 101.325(0.009650 - 0.873153)$$

$$- 298.15(5.4932 - 6.8353) \text{ kJ/kg}$$

$$= -1.712 - 87.494 + 400.147 \text{ kJ/kg}$$

$$= +310.941 \text{ kJ/kg}.$$

For case (b), we have

$$\phi = (222.319 - 220.765) + 101.325(4.451086 - 0.871353)$$

$$- 298.15(7.3241 - 6.8353)$$

$$= 1.554 + 362.534 - 145.736 \text{ kJ/kg}$$

$$= +218.352 \text{ kJ/kg}.$$

For case (c), we have

$$\phi = (147.806 - 220.765) + 101.325(0.740770 - 0.873153)$$

$$- 298.15(6.4898 - 6.8353) \text{ kJ/kg}$$

$$= -72.959 - 13.414 + 103.011 \text{ kJ/kg}$$

$$= +16.638 \text{ kJ/kg}. \qquad \blacksquare$$

Comment

This problem shows that nonflow exergy is positive even if both pressure and temperature of the fluid are lower than the pressure and temperature of the environment. This is true for all substances. Nonflow exergy or availability is always a positive quantity.

Example 6.4

A piece of metal with constant specific heats of 0.400 kJ/kg · K is being cooled from 200°C to 100°C. Heat removed from the metal is fed to a heat engine operating with the environment as the heat sink. Pressure and temperature of the environment are 101.325 kPa and 300 K, respectively. Calculate

(a) the heat removed from the metal, in kJ/kg.
(b) the change in the availability (nonflow exergy) of the metal, in kJ/kg.
(c) the useful work output of the engine if it is reversible.
(d) the thermodynamic lost work if the engine output is 10 kJ/kg.

Solution

(a) Applying the first law of thermodynamics to the metal, we have

$$\bar{d}q = du$$

Assuming that metal is incompressible, it would be reasonable to make use of the following property relation:

$$du = c \, dT$$

where c is the specific heat of the metal. Thus

$$\bar{d}q = c \, dT$$

and

$$q_{12} = c(T_2 - T_1) \text{ with constant specific heat}$$

$$= 0.400(100 - 200) \text{ kJ/kg}$$

$$= -40 \text{ kJ/kg,}$$

where the minus sign simply means that heat is removed from the metal, and that the heat added to the engine is 40 kJ/kg.

(b) From the definition of availability (Eq. 6.24), the change in the availability of the metal is given by

$$\phi_2 - \phi_1 = (u_2 - u_1) + p_0(v_2 - v_1) - T_0(s_2 - s_1). \tag{1}$$

Since the metal is assumed to be incompressible, $v_2 = v_1$. Using $du = c\,dT$ with constant specific heat, we obtain

$$u_2 - u_1 = c(T_2 - T_1). \tag{2}$$

According to Eq. 4.6,

$$T\,ds = du + p\,dv.$$

But dv is zero for our case. Thus $T\,ds = du$. With $du = c\,dT$,

$$ds = \frac{c\,dT}{T}$$

$$s_2 - s_1 = c \ln \frac{T_2}{T_1} \tag{3}$$

with constant specific heat. Substituting into (1), we have

$$\phi_2 - \phi_1 = c(T_2 - T_1) - (T_0 c) \ln \frac{T_2}{T_1}$$

$$= 0.400(100 - 200) - 300 \times 0.400 \ln \frac{373.15}{473.15} \text{ kJ/kg}$$

$$= -40 + 28.49 \text{ kJ/kg}$$

$$= -11.51 \text{ kJ/kg.}$$

(c) According to the second law of thermodynamics, entropy generation must be zero if the engine is reversible. Thus we have

$$s_{\text{gen}} = (\Delta s)_{\text{metal}} + (\Delta s)^{\text{HR}} = 0. \tag{4}$$

For the metal,

$$(\Delta s)_{metal} = c \ln \frac{T_2}{T_1}. \tag{5}$$

For a heat reservoir, we have

$$(\Delta s)^{HR} = \frac{q^{HR}}{T^{HR}}$$

$$= \frac{q^{HR}}{T_0}. \tag{6}$$

Combining Eqs. 4, 5, and 6, we have

$$q^{HR} = (T_0 c) \ln \frac{T_1}{T_2}$$

$$= 300 \times 0.400 \ln \frac{473.15}{373.15} \text{ kJ/kg}$$

$$= 28.49 \text{ kJ/kg},$$

which is the heat rejected by the engine. Thus the maximum work produced by the engine is

$$w_{max\ useful} = 40 - 28.49 \text{ kJ/kg}$$

$$= 11.51 \text{ kJ/kg},$$

which is exactly the nonflow exergy given up by the metal.
(d) By definition, the thermodynamic lost work is given by Eq. 3.32:

$$LW_0 = T_0 S_{gen}. \tag{7}$$

According to the second law, entropy generation is given by

$$S_{gen} = (\Delta S)_{metal} + (\Delta S)^{HR}$$

$$= (\Delta S)_{metal} + \frac{Q^{HR}}{T^{HR}}$$

$$= (\Delta s)_{metal} + \frac{q^{HR}}{T^{HR}} \text{ per unit mass of metal.} \tag{8}$$

Applying the first law to the heat engine, we have

$$|q_{out}| = |q_{in}| - |w|$$

$$= 40 - 10 \text{ kJ/kg}$$

$$= 30 \text{ kJ/kg}.$$

Thus

$$q^{HR} = 30 \text{ kJ/kg}. \tag{9}$$

Now

$$(\Delta s)_{metal} = c \ln \frac{T_2}{T_1}. \tag{10}$$

Combining Eqs. 7, 8, 9, and 10, we have

$$LW_0 = 300\left(0.400 \ln \frac{373.15}{473.15} + \frac{30}{300}\right) \text{ kJ/kg}$$

$$= 1.51 \text{ kJ/kg},$$

which indicates that the second law is satisfied. We note that the thermodynamic lost work is exactly the difference between the nonflow exergy given up by the metal and the actual work output of the engine. ∎

Comments

We have shown in this problem that the thermodynamic lost work is identical to the amount of exergy destroyed. This is always the case. We explore this relationship in more detail in Chapter 11. With heat input of 40 kJ/kg and work output of 10 kJ/kg, the thermal efficiency of the heat engine is 25%, which seems low. On the other hand, if we compare the engine output of 10 kJ/kg to the exergy input of 11.51 kJ/kg, the ratio is 10/11.51 or 86.9%, which is quite good. In fact, it may be too good to be true. If we compare the output in exergy to input in exergy, the ratio is usually called the second-law efficiency, an idea that we will also study in Chapter 11. We will see that the second-law efficiency is a much better measure of thermodynamic performance.

6.9 Availability for Steady-State Steady Flow

For a stream of fluid entering a steady-state steady-flow device at a given state, what is the maximum useful work that we can extract from it? Obviously, it depends on the state of the fluid leaving the device and on the heat interaction. Suppose that we want the outlet state to be in equilibrium with the environment and that only heat interaction with the environment is allowed. For maximum useful work, we must use a reversible process, of course.

From the first law for steady-state steady flow with no changes in kinetic and potential energies, we have

$$\bar{d}q = dh + \bar{d}w_s, \tag{1}$$

where $\bar{d}w_s$ is shaft work, which is all useful work.

From the second law for a steady-state steady-flow reversible process, we have

$$ds + ds^{HR} = 0. \tag{2}$$

But

$$ds^{HR} = \frac{\bar{d}q^{HR}}{T^{HR}} \tag{3}$$

and

$$\bar{d}q^{HR} = -\bar{d}q. \tag{4}$$

Combining Eqs. 1, 2, 3, and 4, we have

$$T^{HR}\, ds = dh + \bar{d}w_{\text{max useful}}. \tag{5}$$

If the heat reservoir is the environment at T_0, then

$$\bar{d}w_{\text{max useful}} = T_0\, ds - dh. \tag{6}$$

From Eq. 6, the maximum useful work is given by

$$w_{\text{max useful}} = T_0(s_0 - s) - (h_0 - h)$$

$$= (h - T_0 s) - (h_0 - T_0 s_0), \tag{6.26}$$

where T_0, s_0, and h_0 are properties of the flowing fluid at the outlet of the device and are in equilibrium with the environment.

The maximum useful work given by Eq. 6.26 is known as the flow availability or flow exergy of the flowing fluid. We will use the two terms interchangeably in this book.

Introducing the symbol ψ for the flow availability or flow exergy per unit mass, we have

$$\psi = (h - T_0 s) - (h_0 - T_0 s_0). \qquad (6.27)$$

We observe that this flow availability, as in the case of availability or nonflow exergy, is also a composite quantity. It is a function of the properties of the flowing fluid as well as a function of the properties of the environment. However, once the conditions of the environment have been specified, the flow availability or flow exergy is just like any other thermodynamic property. We will find the concept of flow availability extremely useful in engineering design.

With the definitions of ϕ, ψ, and h,

$$\phi = (u + p_0 v - T_0 s) - (u_0 + p_0 v_0 - T_0 s_0)$$

$$\psi = (h - T_0 s) - (h_0 - T_0 s_0)$$

$$h = u + pv,$$

we may obtain the following expression:

$$\phi = \psi - (p - p_0)v, \qquad (6.28)$$

which shows that the nonflow exergy ϕ and the flow exergy ψ are simply related. If we can determine one, we can determine the other using this expression. Although we have mentioned that the nonflow exergy is always a positive quantity, the flow exergy could be negative according to Eq. 6.28.

For a stream of fluid entering a steady-state steady-flow device at state 1 and leaving at state 2, the maximum useful work we can extract from the fluid is given by

$$w_{\text{max useful}} = \psi_1 - \psi_2 = (h_1 - T_1 s_1) - (h_2 - T_2 s_2). \qquad (6.29)$$

The quantity $(h - T_0 s)$ is sometimes referred to as the availability function for steady flow. Introducing b for $(h - T_0 s)$, per unit mass, Eq. 6.29 may be written as

$$w_{\text{max useful}} = b_1 - b_2. \qquad (6.29a)$$

It is important to note that Eq. 6.29 is valid as long as the process is reversible and that heat transfer is limited to that between the system and a heat reservoir at T_0. Therefore, it follows that for a

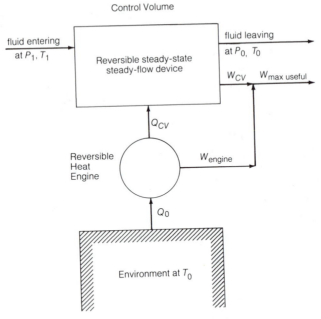

Figure 6.6 Conceptual Arrangement for the Determination of Flow Availability

reversible isothermal process, we would have

$$w_{\text{max useful}} = (h_1 - T_1 s_1) - (h_2 - T_2 s_2). \qquad (6.30)$$

Since the Gibbs function is defined as

$$g = h - Ts,$$

Eq. 6.30 may also be written as

$$w_{\text{max useful}} = (g_1 - g_2)_T$$

$$= -(g_2 - g_1)_T. \qquad (6.31)$$

Equation 6.31 states that the maximum work obtainable from a steady-state steady-flow device is given by the decrease in the Gibbs function if the process is reversible and isothermal. An application of this equation is in the study of fuel cells.

The flow availability of a flowing stream may be determined conceptually using the scheme shown in Fig. 6.6. To provide reversible heat interaction with the environment, we must operate a reversible heat engine between the environment and the system or control volume. The work output of the control volume is shaft work, which is all useful work. Consequently, the maximum useful work, that is, the flow availability or flow exergy, of the flowing stream is simply the sum of the reversible work output of the control volume and the work output of the reversible heat engine.

Example 6.5

Determine the flow availability for nitrogen gas at the following state points $(p_0 = 101.325 \text{ kPa}; \ T_0 = 298.15 \text{ K})$:

(a) 10 MPa and 320 K.
(b) 20 kPa and 300 K.

Solution

(a) From Example 6.3 we have the following data:

At 10 MPa and 320 K:	At 101.325 kPa and 298.15 K:
$h = 315.553 \text{ kJ/kg}$	$h_0 = 309.237 \text{ kJ/kg}$
$v = 0.009650 \text{ m}^3/\text{kg}$	$v_0 = 0.873153 \text{ m}^3/\text{kg}$
$s = 5.4932 \text{ kJ/kg} \cdot \text{K}$	$s_0 = 6.8353 \text{ kJ/kg} \cdot \text{K}$

Using Eq. 6.27, we have

$$\psi = (h - T_0 s) - (h_0 - T_0 s_0)$$

$$= (h - h_0) - T_0(s - s_0)$$

$$= (315.553 - 309.237) - 298.15(5.4932 - 6.8353) \text{ kJ/kg}$$

$$= 406.463 \text{ kJ/kg}.$$

Using Eq. 6.30 and the value of ϕ from Example 6.3 yields

$$\psi = \phi + (p - p_0)v$$

$$= 310.941 + (10 \times 1000 - 101.325) \times 0.009650 \text{ kJ/kg}$$

$$= 406.463 \text{ kJ/kg}.$$

(b) From Example 6.3, we have the following data at 20 kPa and 300 K:

$$h = 311.341 \text{ kJ/kg}$$

$$v = 4.451086 \text{ m}^3/\text{kg}$$

$$s = 7.3241 \text{ kJ/kg} \cdot \text{K.}$$

Using Eq. 6.27 gives us

$$\psi = (h - h_0) - T_0(s - s_0)$$

$$= (311.341 - 309.237) - 298.15(7.3241 - 6.8353) \text{ kJ/kg}$$

$$= 2.104 - 145.736 = -143.632 \text{ kJ/kg.} \qquad \blacksquare$$

Comment

As we have pointed out, the flow availability could indeed be a negative quantity.

Example 6.6

Dry saturated steam at 100°C is condensed in a heat exchanger to saturated liquid at 100°C. Determine the maximum useful work obtainable from this operation in kJ/kg. The pressure and temperature of the environment are 101.325 kPa and 25°C, respectively.

Solution

The maximum useful work is given by Eq. 6.29:

$$w_{\text{max useful}} = (h_1 - T_0 s_1) - (h_2 - T_0 s_2)$$

$$= (h_1 - h_2) - T_0(s_1 - s_2). \qquad (1)$$

From steam tables,

$$h_1 = h_g \text{ at } 100°C = 2676.0 \text{ kJ/kg}$$

$$s_1 = s_g \text{ at } 100°C = 7.3554 \text{ kJ/kg} \cdot \text{K}$$

$$h_2 = h_f \text{ at } 100°C = 419.06 \text{ kJ/kg}$$

$$s_2 = s_f \text{ at } 100°C = 1.3069 \text{ kJ/kg} \cdot \text{K.}$$

Substituting data into Eq. (1), we have

$$w_{\text{max useful}} = (2676.0 - 419.06) - 298.15(7.3554 - 1.3069) \text{ kJ/kg}$$

$$= 453.5 \text{ kJ/kg.} \qquad \blacksquare$$

Comment

The heat removed from the steam is $(2676.0 - 419.06)$ kJ/kg or 2256.9 kJ/kg according to the first law of thermodynamics. Since the steam is condensed at a constant temperature of 100°C, the steam may be visualized as a heat reservoir at 100°C. The maximum useful work for this problem is simply the work output of a Carnot engine operating between $T_H = 373.15$ K and $T_L = T_0 = 298.15$ K with a heat input of 2256.9 kJ/kg. That is,

$$w_{\text{max useful}} = \left(1 - \frac{T_0}{T_H}\right) Q_{\text{in}}$$

$$= \left(1 - \frac{298.15}{373.15}\right) \times 2256.9 \text{ kJ/kg}$$

$$= 453.5 \text{ kJ/kg.}$$

6.10 Chemical Potential and the Gibbs Function

For a system consisting of more than one component, the mole numbers of all the constituents might change in general. The relevant quasi-static work mode associated with the changes in composition of a multicomponent system is known as the *chemical work*.[6] This means that some of the X's in Eqs. 6.12 and 6.13 could be the mole numbers of the different components of such a system.

Let us now be specific about the X's in Eqs. 6.12 and 6.13. Let the X's be the mole numbers of a multicomponent system. Then

$$S = S(U, V, N_1, \ldots, N_r) \tag{6.32}$$

$$U = U(S, V, N_1, \ldots, N_r). \tag{6.33}$$

From Eq. 6.33 we have

$$dU = \left(\frac{\partial U}{\partial S}\right)_{V,N_i} dS + \left(\frac{\partial U}{\partial V}\right)_{S,N_i} dV + \sum \left(\frac{\partial U}{\partial N_i}\right)_{S,V,N_{j\pm i}} dN_i, \tag{6.34}$$

where $(\partial U/\partial S)_{V,N_i}$ is defined as the thermodynamic temperature T and $-(\partial U/\partial V)_{S,N_i}$ is defined as the thermodynamic pressure p.

---

[6] A simple example on chemical work is that performed by the common automobile storage battery. When the battery is doing work (discharging), lead, lead oxide, and sulfuric acid are consumed to form lead sulfate, thereby changing the mole numbers of lead, lead oxide, sulfuric acid, and lead sulfate in the system.

Let us now introduce the definition of the chemical potential of the ith species, μ_i, as

$$\mu_i \equiv \left(\frac{\partial U}{\partial N_i}\right)_{S,V,N_{j\pm i}}. \tag{6.35}$$

Then Eq. 6.34 becomes

$$dU = T\,dS - p\,dV + \sum \mu_i\,dN_i. \tag{6.36}$$

It is apparent from the form of Eq. 6.36 that the chemical potentials, like T and p, are intensive properties. The concepts of temperature and pressure as potentials or driving force for the transport of energy and volume change, respectively, are relatively familiar to us. The chemical potential as the driving force for chemical changes is not so easy to visualize. More than anyone else, Josiah Willard Gibbs, one of the greatest intellects America ever produced, is responsible for our present understanding of this important quantity. The symbol μ_i was first used by Gibbs and called by him the *chemical potential*. The term $\sum \mu_i\,dN_i$ in Eq. 6.36 represents the quasi-static chemical work of the system.

Enthalpy is defined as

$$H \equiv U + pV.$$

In differential form, we have

$$dH = dU + p\,dV + V\,dp,$$

which, in combination with Eq. 6.36 gives

$$dH = T\,dS + V\,dp + \sum \mu_i\,dN_i. \tag{6.37}$$

The Helmholtz function is defined as

$$A \equiv U - TS.$$

Using the same procedure, we get

$$dA = dU - T\,dS - S\,dT,$$

which, in combination with Eq. 6.36 gives

$$dA = -S\,dT - p\,dV + \sum \mu_i\,dN_i. \tag{6.38}$$

The Gibbs function is defined as

$$G \equiv H - TS.$$

In a similar fashion, we get

$$dG = dH - T \, dS - S \, dT,$$

which, in combination with Eq. 6.37, gives

$$dG = -S \, dT + V \, dp + \sum \mu_i \, dN_i. \tag{6.39}$$

Equation 6.37 implies that we have the explicit function of

$$H = H(S, p, N_1, \ldots, N_r). \tag{6.40}$$

Equation 6.38 implies that we have the explicit function of

$$A = A(T, V, N_1, \ldots, N_r). \tag{6.41}$$

Equation 6.39 implies that we have the explicit function of

$$G = G(T, p, N_1, \ldots, N_r). \tag{6.42}$$

From Eqs. 6.37, 6.38, and 6.39, it follows that μ_i can be related to H, A, and G by the equations

$$\mu_i = \left(\frac{\partial H}{\partial N_i} \right)_{S,p,N_{j \pm i}} \tag{6.43}$$

$$= \left(\frac{\partial A}{\partial N_i} \right)_{T,V,N_{j \pm i}} \tag{6.44}$$

$$= \left(\frac{\partial G}{\partial N_i} \right)_{T,p,N_{j \pm i}}. \tag{6.45}$$

Equations 6.36, 6.37, 6.38, and 6.39 are four of the most important equations in thermodynamics. They are extremely useful in the study of chemical thermodynamics.

Let us increase every extensive property of a multicomponent system in the same proportion. Then, for a differential increase, we have

$$dU = U \, d\varepsilon$$

$$dS = S \, d\varepsilon$$

$$dV = V \, d\varepsilon$$

$$dN_i = N_i \, d\varepsilon,$$

where $d\varepsilon$ is the proportionality factor.

Substituting into Eq. 6.36, we have

$$U \, d\varepsilon = TS \, d\varepsilon - pV \, d\varepsilon + \sum \mu_i N_i \, d\varepsilon$$

or

$$U = TS - pV + \sum \mu_i N_i.$$

But $U + pV - TS = G$, by definition. Therefore, we have the following expression relating the Gibbs function to the chemical potentials:

$$G = \sum \mu_i N_i \tag{6.46}$$

Equation 6.46 shows that the chemical potentials are indeed intensive properties, as we observed previously. In the case of a single-component system, Eq. 6.46 becomes

$$G = \mu N$$

or

$$\mu = \frac{G}{N} = g. \tag{6.47}$$

In this simple but not trivial case, the chemical potential is identical to the specific Gibbs function. We must keep in mind that this is not true in the case of a multicomponent system.

We have mentioned that Eqs. 6.12 and 6.13 (same as Eqs. 6.32 and 6.33) are fundamental equations of state. We wish to point that Eqs. 6.40, 6.41, and 6.42 are also known as fundamental equations of state, because they also contain all the thermodynamic information of a given substance.

Let us use the Helmholtz function as an example. From Eq. 6.41, we have

$$A = A(T, V, N_1, \ldots, N_r)$$

$$dA = \left(\frac{\partial A}{\partial T}\right)_{V,N_j} dT + \left(\frac{\partial A}{\partial V}\right)_{T,N_j} dV + \sum \left(\frac{\partial A}{\partial N_i}\right)_{T,V,N_{j \neq i}} dN_i. \quad (6.48)$$

Comparing Eq. 6.48 with Eq. 6.38, we must have

$$S = -\left(\frac{\partial A}{\partial T}\right)_{V,N_j}. \quad (6.49)$$

Equation 6.49 implies that if we have the function of A expressed as a function of $T, V, N_1, \ldots, N_r$, we should be able to use it to calculate entropy simply by differentiation. Unfortunately, this is not possible in classical thermodynamics. However, it is possible in statistical thermodynamics, at least for some simple substances. Similar comments can be made with respect to the functions $U, S, H,$ and G. A study of these functions from the microscopic point of view will give us a better insight into the behavior of matter. Any serious student of thermodynamics should follow a course on classical thermodynamics with a course on statistical thermodynamics.

6.11 Temperature as the Driving Force for Heat Transfer

Let system A be separated from system B in a rigid and thermally insulated chamber by a rigid, nonpermeable, but heat-conducting (diathermal) wall of negligible mass, as shown in Fig. 6.7. According

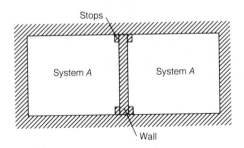

Figure 6.7 Two Systems Separated by a Heat-Conducting Wall

to Eq. 6.36, the entropy change of system A is given by

$$dS^A = \frac{dU^A}{T^A} \tag{1}$$

since its volume and mole number are both constant. In a similar manner, the entropy change of system B is obtained as

$$dS^B = \frac{dU^B}{T^B}. \tag{2}$$

Taking system A and system B together, we have

$$dS^A + dS^B = \frac{dU^A}{T^A} + \frac{dU^B}{T^B}. \tag{3}$$

When equilibrium is established, we have, according to the second law for an isolated system,

$$dS_{\text{isol}} = dS^A + dS^B = \frac{dU^A}{T^A} + \frac{dU^B}{T^B} = 0. \tag{4}$$

From the first law for an isolated system, we must have

$$dU^A = -dU^B. \tag{5}$$

Combining Eqs. 4 and 5, we have

$$\left(\frac{1}{T^A} - \frac{1}{T^B}\right) dU^A = 0, \tag{6}$$

which must be valid for all variations in dU^A. Clearly, we must have

$$T^A = T^B, \tag{7}$$

which is in agreement with our concept of thermal equilibrium. Our definition of temperature as $(\partial U/\partial S)_{V,N_j}$ is thus justified.

When equilibrium has not been reached, either dU^A or dU^B will be negative. Let dU^B be negative. Then dU^A will be positive. If equilibrium has not been reached, the second law for an isolated system says that we must have

$$dS_{\text{isol}} = dS^A + dS^B = \left(\frac{1}{T^A} - \frac{1}{T^B}\right) dU^A > 0. \tag{8}$$

If dU^A is positive, then T^A must be less than T^B. This means that as long as T^B is greater than T^A, system A will be gaining energy while system B will be losing. The property temperature may thus be looked upon as the driving force for heat transfer.

6.12 Pressure as the Driving Force for Volume Change

Let system A be separated from system B in a rigid and thermally insulated chamber by a diathermal, nonpermeable, but movable frictionless piston as shown in Fig. 6.8. According to Eq. 6.36, we have

$$dS^A = \frac{dU^A}{T^A} + \frac{p^A \, dV^A}{T^A} \tag{1}$$

$$dS^B = \frac{dU^B}{T^B} + \frac{p^B \, dV^B}{T^B}. \tag{2}$$

But we have the following constraints for our problem:

$$dU^A = -dU^B \tag{3}$$

and

$$dV^A = -dV^B. \tag{4}$$

Combining Eqs. 1, 2, 3, and 4, we have

$$dS^A + dS^B = \left(\frac{1}{T^A} - \frac{1}{T^B}\right) dU^A + \left(\frac{p^A}{T^A} - \frac{p^B}{T^B}\right) dV^A. \tag{5}$$

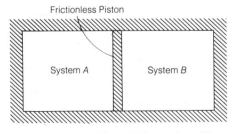

Figure 6.8 Two Systems Separated by a Frictionless Piston

When equilibrium is established, we have, according to the second law,

$$dS_{isol} = dS^A + dS^B = \left(\frac{1}{T^A} - \frac{1}{T^B}\right) dU^A + \left(\frac{p^A}{T^A} - \frac{p^B}{T^B}\right) dV^A = 0. \quad (6)$$

Equation 6 must be valid for arbitrary and independent values of dU^A and dV^A. Clearly, we must have

$$\frac{1}{T^A} - \frac{1}{T^B} = 0 \qquad (7)$$

$$\frac{p^A}{T^A} - \frac{p^B}{T^B} = 0. \qquad (8)$$

Consequently, we have at equilibrium,

$$T^A = T^B \qquad (9)$$

and

$$P^A = P^B. \qquad (10)$$

The equality of the temperatures is just our previous result for equilibrium with a diathermal wall. The equality of the pressures is because the wall is movable without friction, and this result is one we would expect on the basis of mechanics. Pressure may thus be looked upon as the driving force for volume change.

6.13 Chemical Potential as the Driving Force for Matter flow

Let system A be separated from system B in a rigid and thermally insulated chamber by a diathermal, rigid, but semipermeable membrane as shown in Fig. 6.9. Let system A contain only species 1. Let system B contain a mixture of species 1 and 2. Let the membrane be permeable to species 1 only. Then, for our problems, we have the fol-

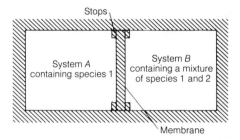

Figure 6.9 Two Systems Separated by a Semipermeable Membrane

lowing constraints:

$$U^A + U^B = \text{constant}$$

or

$$dU^A = -dU^B \tag{1}$$

and

$$N_1^A + N_1^B = \text{constant}$$

or

$$dN_1^A = -dN_1^B. \tag{2}$$

From Eq. 6.36, we have

$$dS^A = \frac{dU^A}{T^A} - \frac{\mu_1^A \, dN_1^A}{T^A} \tag{3}$$

$$dS^B = \frac{dU^B}{T^B} - \frac{\mu_1^B \, dN_1^B}{T^B}. \tag{4}$$

Combining Eqs. 1, 2, 3, and 4, we have

$$dS^A + dS^B = \left(\frac{1}{T^A} - \frac{1}{T^B} \right) dU^A - \left(\frac{\mu_1^A}{T^A} - \frac{\mu_1^B}{T^B} \right) dN_1^A. \tag{5}$$

When equilibrium is established, we have, from the second law,

$$dS_{\text{isol}} = dS^A + dS^B = \left(\frac{1}{T^A} - \frac{1}{T^B} \right) dU^A - \left(\frac{\mu_1^A}{T^A} - \frac{\mu_1^B}{T^B} \right) dN_1^A = 0, \tag{6}$$

which must be valid for arbitrary and independent values of dU^A and dN_1^A. Therefore, the necessary conditions for equilibrium are

$$T^A = T^B \tag{7}$$

and

$$\mu_1^A = \mu_1^B. \tag{8}$$

The equality of the chemical potential of species 1 is because the membrane is permeable to this particular species. If the membrane is also permeable to species 2, we must also have the equilibrium condition of

$$\mu_2^A = \mu_2^B. \tag{9}$$

Thus the chemical potential may be looked upon as the driving force for matter flow.

6.14 Gibbs Phase Rule for a Nonreacting System

It is an experimental fact that matter may exist in the gaseous, liquid, or solid forms or phases. A phase is defined as a *homogeneous system* having uniform intensive properties throughout. A multiphase system is known as a *heterogeneous system*. At the phase boundaries of a multiphase system, there will be sudden or abrupt changes in properties. For example, the density of liquid is quite different from the density of ice in a two-phase mixture of liquid water and ice. Certain important relationships for a multiphase system of several components or species are given in the celebrated phase rule derived by J. Willard Gibbs in 1875.[7] This *phase rule* says that for a multi-component nonreacting system, we have

$$F = C - P + 2, \tag{6.50}$$

where

F = number of intensive properties that may be varied arbitrarily (also called the degrees of freedom)

C = number of components

P = number of phases.

[7] *Collected Works of Josiah Willard Gibbs*, Yale University Press, New Haven, Conn., 1948.

The phase rule was derived by Gibbs from general thermodynamic considerations. We wish to show that it is consistent with the structure of thermodynamics presented in this text.

In a heterogeneous system, each phase may be imagined as separated from another phase by a diathermal, deformable, and permeable membrane. From our study of equilibrium between systems in the last three sections, we observe that for a heterogeneous system at equilibrium, its temperature, pressure, and chemical potentials of each species must be the same in all phases. Now, from the fundamental equation of state in the Gibbs function representation we have

$$G = G(T, p, N_1, \ldots, N_r). \tag{6.42}$$

Denoting the mole fraction by x, Eq. 6.42 may be written as

$$G = G(T, p, x_1 \ldots, x_{r-1}), \tag{6.51}$$

keeping in mind that the mole fractions must add up to unity.

Equation 6.51 says that in a homogeneous system, the number of intensive properties required to fix the state is qual to $(C-1) + 2$, where C is the number of components in the system. Consequently, the number of intensive properties that must be fixed in order to define a state for a heterogeneous system is equal to $P(C-1) + 2$, where P is the number of phases. But these are not all independent. At equilibrium, we must have, for a system of C components and P phases,

$$\mu_1^1 = \mu_1^2 = \cdots = \mu_1^P$$
$$\mu_2^1 = \mu_2^2 = \cdots = \mu_2^P$$
$$\mu_C^1 = \mu_C^2 = \cdots = \mu_C^P,$$

from which we get $C(P-1)$ equilibrium conditions. Thus the number of intensive properties that may be varied arbitrarily is given by

$$F = P(C-1) + 2 - C(P-1)$$

or

$$F = C - P + 2,$$

which is the Gibbs phase rule.

The phase rule summarizes important limitations on the various types of systems that can be encountered. Let us illustrate with the following examples:

1. Consider a one-component system such as water, carbon dioxide, or oxygen; the phase rule says that no more than three phases are permitted.
2. Consider a one-component one-phase system, such as liquid water, solid carbon dioxide, or gaseous oxygen; the phase rule says that $F = 2$. This means that the state of such a system is fixed by specifying any two intensive properties, such as temperature and pressure.
3. Consider a one-component two-phase system, such as an equilibrium mixture of liquid water and steam; the phase rule says that $F = 1$. This means that one one intensive property may be varied arbitrarily.
4. Consider a two-component one-phase system, such as a water–ammonia solution; the phase rule says that $F = 3$. Thus the state of such a system is fixed by specifying any three intensive properties, such as temperature, pressure, and mole fraction of ammonia in the solution.

Problems

6.1 A simple compressible substance inside a cylinder undergoes a change of state quasi-statically, isothermally, and at constant pressure. If the enthalpy change of the system is 2257 kJ at 100°C, what is the corresponding change in its entropy?

6.2 A gas undergoes a change of state reversibly and isothermally in a piston-cylinder assembly. Work done by the gas is 50 J. Internal energy change of the gas is zero. What is the entropy change of the gas? Temperature of the gas is 100°C.

6.3 A closed system undergoes a change of state isothermally at 25°C. Work done to the system is 150 J. Internal energy change of the system is zero. If the process is reversible in addition to being isothermal, what is the change in the Helmholtz function of the system?

6.4 A closed system undergoes a change of state reversibly, isothermally, and at constant pressure. If the only relevant quasi-static work mode for the system is the mechanical work of expansion or compression, what is the change in the Gibbs function of the system?

6.5 The enthalpy change of water when it freezes at constant pressure at 32°F is about -143 Btu/lbm. How much does the entropy of each pound of water change when it turns into ice at 32°F?

6.6 The total quasi-static work of a closed system may be divided into its mechanical work of expansion or compression and a second term known as the *useful work*. That is, for a closed system, the quasi-static work may be given as

$$dW = p\,dV + d W_{useful}.$$

(a) Show that for a closed system, we have, in general,

$$d W_{useful} = T\,dS - dU - p\,dV.$$

(b) Show that for a closed system undergoing a change of state isentropically and at constant pressure, we have

$$d W_{useful} = -dH.$$

(c) Show that for a closed system undergoing a change of state reversibly and at constant temperature and constant volume, we have

$$d W_{useful} = -dA.$$

(d) Show that for a closed system undergoing a change of state reversibly and at constant temperature and constant pressure, we have

$$d W_{useful} = -dG.$$

6.7 For part (a) of Problem 6.6, show that if the closed system is a simple compressible substance, the useful work is identically zero.

6.8 For part (d) of Problem 6.6, show that if the closed system is a simple compressible substance, the useful work is identically zero.

6.9 A closed system undergoes a change of state reversibly, isothermally, inside a rigid vessel. If the only relevant quasi-static work mode of the system is the mechanical work of expansion or compression, what is the change in the Helmholtz function of the system?

6.10 The absolute entropy of methane at 101.325 kPa and 25°C is 186.271 kJ/kmol · K. Determine the absolute entropy of methane at 500 kPa and 25°C. Assume ideal-gas behavior. Is your result consistent with the microscopic interpretation of entropy as a measure of molecular disorder? Explain.

6.11 The absolute entropy of nitrogen is 6.8407 kJ/kg · K at 101.325 kPa and 25°C. Determine the absolute entropy of nitrogen at 100°C with a specific volume of 0.8733 m^3/kg. Is your result consistent with the microscopic interpretation of entropy as a measure of molecular disorder? Explain. Assume ideal-gas behavior and constant specific heats with $k = 1.4$.

6.12 The absolute entropy of oxygen at 101.325 kPa and 25°C is 6.4118 kJ/kg · K. Determine the absolute entropy of oxygen at 200 kPa and 250 K. Is your result consistent with the microscopic interpretation of entropy as a measure of molecular disorder? Explain. Assume ideal-gas behavior and constant specific heats with $k = 1.4$.

6.13 Determine the availability (nonflow exergy) for oxygen gas at the following states (use $p_0 = 101.325$ kPa, and $T_0 = 300$ K):
(a) 10 MPa and 320 K.
(b) 1.0 MPa and 120 K.

6.14 A tank with a volume of 2.0 m^3 contains air at 500 kPa and 100°C. What is the nonflow exergy of the air in the tank, in kilojoules? Assume ideal-gas behavior and constant specific heats with $k = 1.4$. The pressure and temperature of the environment are 101.325 kPa and 300 K, respectively.

6.15 A tank with a volume of 2.0 m^3 contains oxygen at 6.0 MPa and 320 K. What is the nonflow exergy of the oxygen in the tank, in kilojoules? The pressure and temperature of the environment are 101.325 kPa and 300 K, respectively.

6.16 A tank with a volume of 5.0 m^3 contains dry saturated steam at 100°C. What is the nonflow exergy of the steam in the tank, in kilojoules? The pressure and temperature of the environment are 101.325 kPa and 300 K, respectively.

6.17 Air is stored in a rigid tank at 200 kPa and 200°C. If we want its nonflow exergy content to be 5000 kJ, what is the size of the tank, in m^3? Pressure and temperature of the environment are 101.325 kPa and 300 K. Assume ideal-gas behavior and constant specific heats with $k = 1.4$.

6.18 Air in a rigid tank is cooled from 500 K to 300 K, which is the temperature of the environment. The initial pressure of the air is 1000 kPa. What is the maximum useful work obtainable from this cooling process, in kilojoules, if the volume of the tank is 5.0 m³? Assume ideal-gas behavior and constant specific heats with $k = 1.4$.

6.19 Carbon monoxide gas is heated in a rigid tank from 300 K to 500 K with the use of a reversible heat pump operating between the tank and the environment at 300 K. For each kilogram of gas, determine

(a) the change in nonflow exergy of the gas due to the heating process.

(b) the work needed to drive the heat pump.

Assume ideal-gas behavior and constant specific heats with $k = 1.4$.

6.20 Bodies A and B in Fig. P6.20 are two identical blocks of solid each having a mass of 100 kg. They are used as reservoirs for a heat engine. Body A is initially at a temperature of T_1 kelvin, which is greater than the initial temperature of T_2 kelvin of body B. Show that for maximum engine output, the final temperature T_f of the two bodies is given by

$$T_f = \sqrt{T_1 T_2} \quad \text{K.}$$

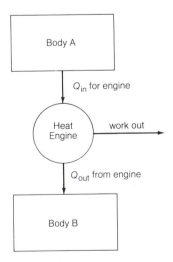

Figure P6.20

Assume that the solids are incompressible with constant specific heat of $c_v = 0.500$ kJ/kg · K.

6.21 For Problem 6.20 with the final temperature $T_f = \sqrt{T_1 T_2}$ kelvin,
(a) determine the amount of heat given up by body A.
(b) determine the amount of heat absorbed by body B.
(c) show that with the results from parts (a) and (b) the maximum engine output is given by

$$W_{\text{max useful}} = 50(T_1 + T_2 - 2\sqrt{T_1 T_2}) \quad \text{kJ.}$$

6.22 For Problem 6.20 with the final temperature $T_f = \sqrt{T_1 T_2}$ kelvin,
(a) calculate the amount of nonflow exergy given up by body A.
(b) calculate the amount of nonflow exergy gained by body B.
(c) show that with the results from parts (a) and (b) the maximum engine output is given by

$$W_{\text{max useful}} = 50(T_1 + T_2 - 2\sqrt{T_1 T_2}) \quad \text{kJ.}$$

(*Hint:* For a reversible process exergy is conserved.)

6.23 Bodies A and B in Fig. P6.23 are two identical blocks of solid each having a mass of 100 kg. Each block is also initially at the same temperature T_1 kelvin. A refrigerator operates between these two bodies until body A is cooled to a temperature of T_2 kelvin. Show that if the refrigerator is a reversible machine, the final temperature $(T_f)_B$ of body B is given by $(T_1)^2/T_2$ kelvin.

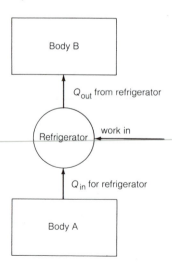

Figure P6.23

Assume that the solids are incompressible with constant specific heats.

6.24 For Problem 6.23 with $(T_f)_B = (T_1)^2/T_2$ kelvin and $c_v = 0.400$ kJ/ kg $\cdot$ K,

(a) calculate the amount of heat removed from body A.

(b) calculate the amount of heat added to body B.

(c) show that with the results of parts (a) and (b), the minimum work needed to operate the refrigerator is given by

$$| W_{min} | = 40\left[\frac{(T_1)^2}{T_2} + T_2 - 2T_1\right] \quad \text{kJ.}$$

6.25 For Problem 6.23 with $(T_f)_B = (T_1)^2/T_2$ kelvin and $c_v = 0.400$ kJ/ kg $\cdot$ K,

(a) calculate the nonflow exergy given up by body A.

(b) calculate the nonflow exergy gained by body B.

(c) show that with the results of parts (a) and (b), the minimum work needed to operate the refrigerator is given by

$$W_{min} = 40\left[\frac{(T_1)^2}{T_2} + T_2 - 2T_1\right] \quad \text{kJ.}$$

(*Hint:* For a reversible process, exergy is conserved.)

6.26 Determine the flow availability (flow exergy) in kJ/kg for H_2O at the following states:

(a) saturated liquid at 100°C.

(b) saturated vapor at 100°C.

The pressure and temperature of the environment are 101.325 kPa and 300 K, respectively.

6.27 Determine the flow exergy for nitrogen at the following states (let $p_0 = 101.325$ kPa, and $T_0 = 300$ K):

(a) 10 MPa and 320 K.

(b) 1.0 MPa and 120 K.

6.28 If geothermal steam is available at the wellhead at 100 psia and 400°F, what is the maximum useful power that we can obtain from such an energy source for a flow rate of 250,000 lbm/h? The pressure and temperature of the environment are 14.7 psia and 70°F, respectively.

6.29 Air is cooled in a heat exchanger from 100°C to 27°C. If the rate of airflow is 75 kg/min, determine the loss of flow exergy (in kilowatts) in the air due to the cooling process. The pressure and temperature of the environment are 101.325 kPa and 300 K, respectively.

6.30 In a steam condenser, the cooling water enters at 25°C and leaves at 38°C. What is the increase in the flow exergy of the water, in kJ/kg? Assume that water is incompressible with a constant specific heat of $c_p = 4.1868$ kJ/kg · K. The pressure and temperature of the environment are 101.325 kPa and 300 K, respectively.

6.31 Nitrogen gas enters a steady-state steady-flow compressor at 300 K and 101.325 kPa and leaves at 300 K and 10 MPa. The flow rate is 10,000 kg/h.
(a) What is the rate of flow exergy increase in the nitrogen due to the compression process?
(b) What is the minimum power needed to operate the compressor?
The pressure and temperature of the environment are 101.325 kPa and 300 K, respectively.

6.32 In the heat recovery steam generator of a cogeneration system, the hot gas enters steadily at 1000°F and leaves at 300°F. Water enters steadily as liquid at 100 psia and 80°F and leaves as dry saturated vapor at 100 psia. The hot gas may be assumed to be an ideal gas with a constant specific heat of $c_p = 0.25$ Btu/lbm · °R. For a flow rate of 50,000 lbm/h of hot gas,
(a) determine the decrease in the flow exergy content of the hot gas in Btu/h.
(b) determine the increase in the flow exergy content of the water in Btu/h.
(c) compare the result of part (a) with that of part (b). Are they equal? If not, why?
The pressure and temperature of the environment are 14.7 psia and 70°F, respectively.

6.33 Same as Problem 6.32 except that water enters as liquid at 200 psia and 80°F and leaves as steam at 200 psia and 700°F.

6.34 If the fundamental equation of state of a fluid in the enthalpy function representation is

$$H = H(S, p, N),$$

show that

$$T = \left(\frac{\partial H}{\partial S}\right)_{p,N}$$

$$V = \left(\frac{\partial H}{\partial p}\right)_{S,N}.$$

6.35 If the fundamental equation of state of a fluid in the internal energy representation is

$$U = U(S, V, N),$$

show that

$$\frac{1}{T} = \left(\frac{\partial S}{\partial U}\right)_{V,N}$$

$$\frac{p}{T} = \left(\frac{\partial S}{\partial V}\right)_{U,N}.$$

6.36 If the fundamental equation of state of a fluid in the Helmholtz function representation is

$$A = A(T, V, N),$$

show that

$$p = -\left(\frac{\partial A}{\partial V}\right)_{T,N}$$

$$S = -\left(\frac{\partial A}{\partial T}\right)_{V,N}.$$

6.37 If the Helmholtz function for the fluid in Problem 6.36 is given by the expression

$$A = -N\bar{R}T\left\{\ln\left[\frac{V}{Nh^3}(2\pi mkT)^{3/2}\right] + 1\right\},$$

where N is the number of moles, $\bar{R}$ is the universal gas constant, T is the thermodynamic temperature, V is the volume, and all the other quantities are constants, show that the p–V–T equation of state is given by

$$pV = N\bar{R}T,$$

which is that of an ideal gas. (*Note:* The fundamental equation of state in the Helmholtz function representation given in this problem is that of a monatomic ideal gas obtained from statistical thermodynamics. The quantity h is the Planck constant, the quantity k is the Boltzmann constant, and m is the mass of a particle of the fluid.)

6.38 If the fundamental equation of state of a fluid in the Gibbs function representation is

$$G = G(T, p, N),$$

show that

$$S = -\left(\frac{\partial G}{\partial T}\right)_{p,N}$$

$$V = \left(\frac{\partial G}{\partial p}\right)_{T,N}.$$

6.39 For a batch of blackbody radiation inside an enclosure, the fundamental equations of state may be given as

$$U = U(S, V)$$

and

$$dU = T\, dS - p\, dV,$$

which implies that the Gibbs function G of blackbody radiation is identically equal to zero. For blackbody radiation, we also have the familiar relation

$$p = \frac{1}{3}\frac{U}{V}.$$

Making use of this information, show that for blackbody radiation,

(a) $S\, dT - V\, dp = 0.$
(b) p is proportional to the fourth power of the absolute temperature, which is the *Stefan–Boltzmann relation*.

6.40 It may be shown from statistical thermodynamics that the fundamental equation of state in the entropy representation for a monatomic ideal gas is given by

$$S = S(U, V, N) = AN + N\bar{R} \ln \frac{BVU^{3/2}}{N}.$$

where A and B are constants and $\bar{R}$ is the universal gas constant. Show that
(a) $U = \frac{3}{2}N\bar{R}T$.
(b) $pV = N\bar{R}T$.
[*Hint:* Make use of $1/T = (\partial S/\partial U)_{V,N}$ and $p/T = (\partial S/\partial V)_{V,N}$.]

6.41 For N moles of a substance in which only $p\,dV$ work is relevant, show that we have

$$\left(\frac{\partial T}{\partial V}\right)_{S,N} = -\left(\frac{\partial p}{\partial S}\right)_{V,N}$$

$$\left(\frac{\partial T}{\partial p}\right)_{S,N} = \left(\frac{\partial V}{\partial S}\right)_{p,N}$$

$$\left(\frac{\partial S}{\partial V}\right)_{T,N} = \left(\frac{\partial p}{\partial T}\right)_{V,N}$$

$$\left(\frac{\partial S}{\partial p}\right)_{T,N} = -\left(\frac{\partial V}{\partial T}\right)_{p,N},$$

which are known as *Maxwell relations.* (*Hint:* Make use of the mathematical fact that $\partial^2 x/\partial y\,\partial z = \partial^2 x/\partial z\,\partial y$.)

6.42 Show that for a multicomponent system in which only mechanical work of expansion or compression and chemical work are relevant, we have

$$S\,dT - V\,dp + \sum N_i\,d\mu_i = 0.$$

(*Note:* This expression is known as the *Gibbs–Duhem equation,* relating the intensive properties of such a system. It is of considerable importance in the study of chemical thermodynamics.)

6.43 A mixture of nitrogen, oxygen, and argon can exist as a two-phase mixture. For such a mixture, how many intensive properties must be specified in order to specify a state?

6.44 If the solid phase, liquid phase, and the vapor phase of a simple compressible substance exist as an equilibrium mixture, what is the degree of freedom of such a mixture?

6.45 In heating, ventilating, and air-conditioning design, we usually consider the atmospheric air as a mixture of dry air and water vapor, a two-component single-phase system. For such a system, how many intensive properties must be specified in order to determine a state?

6.46 (a) Write a computer program to calculate the absolute molal entropy of an ideal gas at 1 atm and a given temperature, assuming the absolute molal entropy at 1 atm and 298 K is known, and that the molal heat capacity as a function of temperature is given by

$$\bar{c}_p^* = a + bT + cT^2 + dT^3$$

   where

$$\bar{c}_p^* \text{ is in kJ/Kgmol} \cdot \text{K}$$

$$T \text{ is in degrees Kelvin}$$

$$a, b, c, \text{ and } d \text{ are constants.}$$

(b) Making use of this computer program, calculate the absolute entropy for nitrogen at 1 atm for the temperature range of 300 to 1000 K, in 100 K steps. Express your results in kJ/kg · K. The values of the constants in the molal heat capacity equation for nitrogen are:

$$a = 28.90$$

$$b = -0.1571 \times 10^{-2}$$

$$c = 0.8081 \times 10^{-5}$$

$$d = -2.873 \times 10^{-9}$$

   The absolute entropy for nitrogen at 1 atm and 298 K is 6.8407 kJ/kg · K.

6.47 Same as Problem 6.46 except that the gas is oxygen. The values of the constants in the molal heat capacity equation for oxygen are:

$$a = 25.48$$

$$b = 1.520 \times 10^{-2}$$

$$c = -0.7155 \times 10^{-5}$$

$$d = 1.312 \times 10^{-9}$$

The absolute entropy for oxygen at 1 atm and 298 K is 6.4118 kJ/kg · K.

6.48 (a) Write a computer program to calculate the nonflow exergy of an ideal gas at a given pressure p and a given temperature T.

(b) Making use of this computer program, calculate the nonflow exergy for nitrogen gas, in kJ/kg, at 200 kPa and for the temperature range of 300 to 1000 K, in 100 K steps. The molal heat capacity equation for nitrogen and the values for the constants in this equation are given in Problem 6.46.
Pressure and temperature of the environment are 101.325 kPa and 300 K, respectively.

6.49 Same as Problem 6.48 except that the gas is oxygen. The molal heat capacity equation for oxygen is given in Problem 6.46 and the values of the constants in this equation are given in Problem 6.47.

CHAPTER 7

Engineering Analysis of Processes for Closed Systems

Processes encountered in engineering are of many varieties. Some are complex, some are relatively simple. In many cases, however, even a complex process may be reduced to a simple one or broken up into a set of simple processes through the use of fairly reasonable assumptions or idealizations. Consequently, it is necessary for one first to acquire the ability to analyze simple processes before he can acquire the ability to analyze complex processes. In this chapter, we shall apply all the fundamentals that we have discussed to the study of several common types of processes for closed systems. In Chapter 8, we shall devote our attention to processes for open systems. It will be seen that although the basic principles are identical for all cases, the details of the solutions will depend on the kind of working substance involved as well as on the form of the property data available.

7.1 Constant-Volume Process

For a closed system, the mass is constant. If the volume for such a system is also constant, the final specific volume of the system must be the same as its initial specific volume. This is a unique feature of this kind of process. If the process is quasi-static as well as constant volume, the specific volume of the system is the same at any point along the process path.

Example 7.1

A rigid tank of volume 1.0 m^3 contains dry saturated steam initially at 200°C. Due to heat transfer to the environment, the steam temperature drops to 100°C. The temperature of the environment is 300 K.

(a) Calculate the amount of heat transfer involved.
(b) Calculate the amount of thermodynamic lost work involved.

Solution

System selected for study: all the fluid inside rigid tank
Unique feature of process: $v_2 = v_1$
Idealization: H_2O is a simple compressible substance

(a) Since no work is involved, we have, from the first law,

$$Q_{12} = m(u_2 - u_1), \tag{1}$$

where $m = V/v$. We have dry saturated steam initially. Thus

$$u_1 = u_g \text{ at } 200°C$$

$$= (h_g - pv_g) \text{ at } 200°C$$

$$= (2790.9 - 1554.9 \times 0.12716) \text{ kJ/kg}$$

$$= 2593.2 \text{ kJ/kg}$$

$$v_1 = v_2 = v_g \text{ at } 200°C = 0.12716 \text{ m}^3/\text{kg}.$$

Then $m = 1.0/0.12716 \text{ kg} = 7.8641 \text{ kg}$.

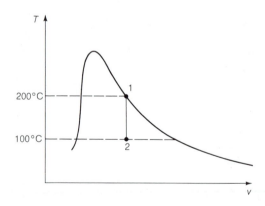

From the T–v diagram, it can be seen that state 2 is a liquid–vapor mixture. Thus

$$v_2 = (v_f + x_2 v_{fg}) \text{ at } 100°C$$

and

$$x_2 = \frac{0.12716 - 0.0010437}{1.6720} = 0.07543$$

$$u_2 = (u_f + x_2 u_{fg}) \text{ at } 100°C,$$

where

$$u_f = (h_f - pv_f) \text{ at } 100°C$$

$$= 419.06 - 101.33 \times 0.0010437 \text{ kJ/kg} = 418.95 \text{ kJ/kg}$$

$$u_g = (h_g - pv_g) \text{ at } 100°C$$

$$= 2676.0 - 101.33 \times 1.6730 \text{ kJ/kg} = 2506.5 \text{ kJ/kg}.$$

Then

$$u_2 = 418.95 + 0.07543 \times (2506.5 - 418.95) \text{ kJ/kg}$$

$$= 576.41 \text{ kJ/kg}.$$

Substituting data into Eq. 1, we have

$$Q_{12} = 7.8641(576.41 - 2593.2) \text{ kJ} = -15860.2 \text{ kJ}.$$

The minus sign of our result simply means that heat is removed from the steam.

(b) By definition, thermodynamic lost work is given by

$$\text{LW}_0 = T_0 S_{\text{gen}}. \tag{2}$$

From the second law, we have

$$S_{\text{gen}} = \Delta S + \Delta S^{\text{HR}} + \Delta S^{\text{WR}} = \Delta S + \Delta S^{\text{HR}} \qquad \text{since } \Delta S^{\text{WR}} \text{ is zero.}$$

Now

$$\Delta S = m(s_2 - s_1)$$

$$\Delta S^{\text{HR}} = \frac{Q^{\text{HR}}}{T^{\text{HR}}} = \frac{Q^{\text{HR}}}{T_0}$$

$$Q^{\text{HR}} = -Q_{12}.$$

$$Q_{12} = m(u_2 - u_1).$$

Substituting into Eq. 2, we have

$$\text{LW}_0 = m[(u_1 - u_2) - T_0(s_1 - s_2)]. \tag{3}$$

Now

$$s_1 = s_g \text{ at } 200°C$$

$$= 6.4278 \text{ kJ/kg} \cdot \text{K}$$

$$s_2 = (s_f + x_2 s_{fg}) \text{ at } 100°C$$

$$= 1.3069 + 0.07543 \times 6.0485 \text{ kJ/kg} \cdot \text{K}$$

$$= 1.7631 \text{ kJ/kg} \cdot \text{K}.$$

Thus

$$\text{LW}_0 = 7.8641[(2593.2 - 576.41) - 300(6.4278 - 1.7631)] \text{ kJ}$$

$$= +4855.1 \text{ kJ}. \qquad\qquad \blacksquare$$

Comments

Since LW_0 is positive, the heat-transfer process is an irreversible one. We also wish to point out that the right-hand side of Eq. 3 is simply the exergy given up by the steam. Thus thermodynamic lost work is identical to exergy destruction.

Example 7.2

A rigid container encloses 1500 lbm of air at 15 psia and 500°R. We wish to increase the temperature to 540°R. Assume ideal-gas behavior and $c_v = 0.171$ Btu/lbm · °R.

(a) Determine the requirement of energy input to the air for such a change of state.
(b) Determine the entropy creation in the universe if the change of state is accomplished by using energy from a heat reservoir at 300°F alone.
(c) Determine the entropy creation in the universe if the change of state is accomplished by using energy from a work reservoir alone.

Solution

System selected for study: all the air inside a rigid container
Unique feature of process: $v_2 = v_1$
Assumptions: ideal-gas behavior with constant specific heat

(a) From the first law, we have

$$\text{net energy input} = Q_{12} - W_{12} = U_2 - U_1 = m(u_2 - u_1). \quad (1)$$

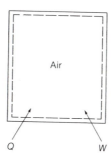

Air

Q   W

From the property relation for an ideal gas with constant specific heat,

$$u_2 - u_1 = c_v(T_2 - T_1). \quad (2)$$

Substituting Eq. 2 into Eq. 1, we have

$$\text{net energy input} = mc_v(T_2 - T_1)$$

$$= 1500 \times 0.171(540 - 500) \text{ Btu}$$

$$= 10{,}260 \text{ Btu.}$$

(b) If no work is used, net energy input $= Q_{12} = 10{,}260$ Btu. Therefore,

$$Q^{HR} = -10{,}260 \text{ Btu.}$$

From the second law,

$$S_{\text{gen}} = \Delta S_{\text{air}} + \Delta S^{HR}. \quad (3)$$

From the property relation for an ideal gas with constant specific heat,

$$s_2 - s_1 = c_v \ln \frac{T_2}{T_1} + R \ln \frac{v_2}{v_1},$$

which reduces to $s_2 - s_1 = c_v \ln (T_2/T_1)$ as $v_2 = v_1$. Then

$$\Delta S_{air} = m c_v \ln \frac{T_2}{T_1}$$

$$= 1500 \times 0.171 \times \ln \frac{540}{500} \text{ Btu/°R}$$

$$= 19.74 \text{ Btu/°R}.$$

For a heat reservoir,

$$\Delta S^{HR} = \frac{Q^{HR}}{T^{HR}}$$

$$\Delta S^{HR} = -\frac{10{,}260}{760} \text{ Btu/°R}$$

$$= -13.5 \text{ Btu/°R}.$$

Substituting into Eq. 3, we have

$$S_{gen} = 19.74 - 13.5 = 6.24 \text{ Btu/°R}.$$

(c) From second law, if no heat is used,

$$S_{gen} = \Delta S_{air} + \Delta S^{WR} = \Delta S_{air}$$

$$= 19.74 \text{ Btu/°R}$$

since $\Delta S^{WR} = 0$ according to Eq. 3.8. ∎

Comments

This simple problem brings out a most important lesson which one can get out of thermodynamics. To do a given job, the process will be less irreversible if we use heat instead of work, which is high-quality energy. One who prefers to heat his house in the winter with a fire rather than to use electric panels not only has a smaller heating bill to pay, but he can also claim that he is doing it according to sound thermodynamic principles.

Example 7.3

A rigid tank is completely filled with 50 lbm of water at 14.7 psia and 70°F. If the temperature of the bottled-up water might be caused to go to 100°F due to an accidental energy input (such as a flash fire),

(a) determine the amount of energy input that could have caused such a temperature rise.

(b) what pressure must the tank be good for in case of such an emergency?

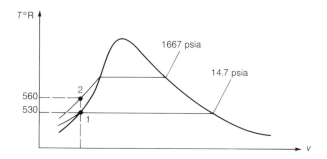

Solution

System selected for study: 50 lbm of water inside a rigid tank
Unique feature of process: $v_2 = v_1$
Idealization: H_2O is a simple compressible substance

From the first law, we have

$$\text{energy input } = U_2 - U_1 = m(u_2 - u_1). \tag{1}$$

Since water at 14.7 psia and 70°F is only slightly compressed, we may let

$$u_1 \approx u_f \text{ at } 70°F \approx h_f \text{ at } 70°F = 38.05 \text{ Btu/lbm}$$

$$v_1 \approx v_f \text{ at } 70°F = 0.01605 \text{ ft}^3/\text{lbm}.$$

State 2 is defined by $T_2 = 100°F$ and $v_2 = v_1 = 0.01605$ ft^3/lbm. Therefore, from compressed liquid tables for water, we have

$$p_2 = 1667 \text{ psia} \quad \text{and} \quad h_2 = 72.39 \text{ Btu/lbm}.$$

Thus

$$u_2 = h_2 - p_2 v_2$$

$$= 72.39 \text{ Btu/lbm} - \frac{(1667 \text{ lbf/in}^2)(144 \text{ in}^2/\text{ft}^2)(0.01605 \text{ ft}^3/\text{lbm})}{778 \text{ ft} \cdot \text{lbf/Btu}}$$

$$= 67.44 \text{ Btu/lbm}.$$

Substituting into Eq. 1, we find that

$$\text{energy input} = 50(67.44 - 38.05) \text{ Btu} = 1470 \text{ Btu.}$$

Since the water pressure could reach 1667 psia, the tank must be good for at least 1667 psia. ∎

Comments

We see from this problem that the pressure of bottled-up water increases rapidly due to a small increase in temperature. In fact, this is true for liquid in general. A good engineer would not design the tank to withstand 1667 psia. Instead, he probably would design it at a much lower pressure, say 200 psia, and provide the tank with a safety valve.

7.2 Constant-Pressure Process

A unique feature of a constant-pressure process is that the final pressure of a system is the same as its initial pressure. For a simple compressible substance undergoing a change of state quasi-statically and at constant pressure, the work-transfer quantity may be conveniently evaluated as

$$W_{12} = \int_1^2 p \, dV = p(V_2 - V_1) = mp(v_2 - v_1). \tag{7.1}$$

Example 7.4

Twenty-five kilograms of nitrogen gas, confined inside a cylinder equipped with a piston, are initially at 140 K and 101.325 kPa. If the nitrogen undergoes a change of state quasi-statically at constant pressure until its volume is doubled, determine

(a) the total amount of work transfer for the nitrogen.
(b) the final temperature of the gas.

Solution

System selected for study: 25 kg of nitrogen
Unique feature of process: $p_2 = p_1$; quasi-static
Idealization: nitrogen is a simple compressible substance

(a) Since we have a closed system undergoing a change of state at constant pressure quasi-statically, the work transfer is given by Eq. 7.1:

$$W_{12} = mp(v_2 - v_1). \tag{1}$$

For nitrogen at 140 K and 101.325 kPa, we have

$$v_1 = 0.40709 \text{ m}^3/\text{kg}.$$

Thus

$$v_2 = 2v_1$$

$$= 2 \times 0.40709 \text{ m}^3/\text{kg} = 0.81418 \text{ m}^3/\text{kg}$$

and

$$W_{12} = 25 \times 101.325(0.81418 - 0.40709) \text{ kJ}$$

$$= +1031.2 \text{ kJ}.$$

(b) State 2 is defined by $p_2 = p_1 = 101.325$ kPa and $v_2 = 0.81418 \text{ m}^3/\text{kg}$. From nitrogen tables, we have the following data at 101.325 kPa:

T (K)	260	280
v (m³/kg)	0.761145	0.819883

Using linear interpolation, we have

$$T_2 = 260 + (280 - 260)\left(\frac{0.81418 - 0.761145}{0.819883 - 0.761145}\right) \text{ K}$$

$$= 278 \text{ K.} \qquad\qquad ▮$$

Comments

We have positive work. This means that work is done by the gas. This is consistent with the fact that the gas expands. If we had assumed ideal gas behavior, the final temperature would be 280 K. Consequently, it would be quite reasonable to use the ideal-gas model for this problem.

Example 7.5

Twenty kilograms of liquid–vapor mixture of H_2O initially at 100°C and a quality of 60% are being cooled quasi-statically until they are 100% saturated liquid at 100°C. They can exchange heat with the environment at 25°C only.

(a) Determine the amount of heat transfer for the fluid.
(b) Show that the heat-transfer process is irreversible.

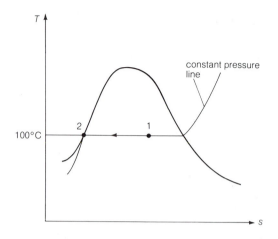

Solution

System selected for study: 20 kg of H_2O
Unique features of process: $T_2 = T_1$; $p_2 = p_1$; quasi-static
Idealization: H_2O is a simple compressible substance

(a) Since process is quasi-static and of constant pressure, the heat-transfer term may be evaluated as

$$Q_{12} = m(h_2 - h_1). \tag{1}$$

For H_2O at 100°C and $x = 0.60$, we have

$$h_1 = (h_f + x_1 h_{fg}) \text{ at } 100°C$$

$$= (419.06 + 0.60 \times 2256.9) \text{ kJ/kg} = 1773.2 \text{ kJ/kg}.$$

Now state 2 is saturated liquid at 100°C. Thus

$$h_2 = h_f \text{ at } 100°C$$

$$= 419.06 \text{ kJ/kg}.$$

Substituting data into Eq. 1, we have

$$Q_{12} = 20(419.06 - 1773.2) \text{ kJ}$$

$$= -27{,}083 \text{ kJ}.$$

Since the process is quasi-static and of constant temperature, we could also determine the heat-transfer quantity as

$$Q_{12} = mT(s_2 - s_1), \tag{2}$$

where

$$s_1 = (s_f + x_1 s_{fg}) \text{ at } 100°C$$

$$= (1.3069 + 0.60 \times 6.0485) \text{ kJ/kg} \cdot \text{K}$$

$$= 4.9360 \text{ kJ/kg} \cdot \text{K}$$

$$s_2 = s_f \text{ at } 100°C$$

$$= 1.3069 \text{ kJ/kg} \cdot \text{K}.$$

Substituting data into Eq. 2, we have

$$Q_{12} = 20 \times 373.15(1.3069 - 4.9360) \text{ kJ}$$

$$= -27,084 \text{ kJ}.$$

(b) From the second law, we have

$$S_{gen} = \Delta S + \Delta S^{HR}.$$

Now

$$\Delta S = m(s_2 - s_1)$$

$$\Delta S^{HR} = \frac{Q^{HR}}{T^{HR}}$$

$$= -\frac{Q_{12}}{T_0}.$$

Thus

$$S_{gen} = m(s_2 - s_1) - \frac{Q_{12}}{T_0} = m\left[(s_2 - s_1) - \frac{T(s_1 - s_2)}{T_0}\right] \qquad (3)$$

$$= 20\left[(1.3069 - 4.9360) - \frac{373.15(1.3069 - 4.936)}{298.15}\right] \text{ kJ/K}$$

$$= 20(-3.6291 + 4.5420) \text{ kJ/K}$$

$$= +18.258 \text{ kJ/K}.$$

Comments

Since entropy generation is positive, the heat-transfer process is irreversible. We observe from Eq. 3 that the only way we could have reversible heat transfer is to have $T = T_0$. Our process is not reversible because it is externally irreversible even though it is quasi-static or internally reversible.

7.3 Isothermal Process

A unique feature of an isothermal process is that the final temperature of a system is the same as its initial temperature. For any substance undergoing a change of state quasi-statically and isothermally, the heat-transfer quantity may be conveniently evaluated according to the second law of thermodynamics as

$$Q_{12} = \int_1^2 T \, dS = T(S_2 - S_1) = mT(s_2 - s_1). \tag{7.2}$$

Combining Eq. 7.2 with the first law, we have for a closed system undergoing a change of state quasi-statically and isothermally,

$$W_{ideal} = (U_1 - T_1 S_1) - (U_2 - T_2 S_2). \tag{7.3}$$

From the definition of the Helmholtz function,

$$A = U - TS.$$

Therefore, Eq. 7.3 may also be written as

$$W_{ideal} = (A_1 - A_2)_T = -(A_2 - A_1)_T. \tag{7.3a}$$

Thus the decrease in the Helmholtz function of a system represents the maximum work that can be delivered by the system in a work-producing device in an isothermal process. In a work-absorbing device, the decrease in the Helmholtz function would represent the minimum work required for an isothermal process. This is why the Helmholtz function is also known as a *potential function*.

Example 7.6

Two kilograms of nitrogen gas, confined inside a cylinder equipped with a piston, undergoes a change of state quasi-statically from 300 K and 101.325 kPa to a final state of 300 K and 20,000 kPa. Heat transfer can occur between the nitrogen gas and a heat reservoir at 300 K.

(a) Determine the total amount of work transfer for the nitrogen gas.
(b) Determine the total amount of heat transfer for the nitrogen gas.
(c) Show that the process is both internally reversible as well as externally reversible.

Solution

System selected for study: 2.0 kg of nitrogen gas
Unique features of process: quasi-static; $T_2 = T_1$
Idealization: nitrogen is a simple compressible substance

(a) Since process is quasi-static and isothermal, the amount of work transfer may be evaluated as

$$W_{12} = m[(u_1 - u_2) - T(s_1 - s_2)]. \tag{1}$$

Using data from nitrogen tables, we have

$$u_1 = h_1 - p_1 v_1$$

$$= (311.163 - 101.325 \times 0.878604) \text{ kJ/kg}$$

$$= 222.138 \text{ kJ/kg}$$

$$s_1 = 6.8418 \text{ kJ/kg} \cdot \text{K}$$

$$u_2 = h_2 - p_2 v_2$$

$$= (279.010 - 20{,}000 \times 0.004704) \text{ kJ/kg}$$

$$= 184.93 \text{ kJ/kg}$$

$$s_2 = 5.1630 \text{ kJ/kg} \cdot \text{K}.$$

Substituting data into Eq. 1, we have

$$W_{12} = 2[(222.138 - 184.93) - 300(6.8418 - 5.1630)] \text{ kJ}$$

$$= -932.864 \text{ kJ}.$$

The minus sign means that work is done to the gas.
(b) Since process is quasi-static and isothermal, we have, from the second law,

$$Q_{12} = mT(s_2 - s_1)$$

$$= 2 \times 300(5.1630 - 6.8418) \text{ kJ}$$

$$= -1007.28 \text{ kJ}.$$

The minus sign means that heat is removed from the gas.
(c) From the second law,

$$S_{gen} = \Delta S + \Delta S^{HR}, \qquad (2)$$

where

$$\Delta S = m(s_2 - s_1)$$

$$\Delta S^{HR} = \frac{Q^{HR}}{T^{HR}}$$

$$= -\frac{Q_{12}}{T^{HR}}.$$

Thus

$$S_{gen} = m(s_2 - s_1) - \frac{mT(s_2 - s_1)}{T^{HR}}. \qquad (3)$$

Since $T = T^{HR} = 300$ K, we have, from Eq. 3,

$$S_{gen} = m(s_2 - s_1) - m(s_2 - s_1)$$

$$= 0,$$

which means that the process is indeed internally reversible (quasi-static) as well as externally reversible. ▮

Comment

If we had assumed ideal gas behavior, we would have

$$Q_{12} = W_{12} = -mRT \ln \frac{p_2}{p_1} = -941.3 \text{ kJ},$$

which is, compared with the correct results, quite adequate for the purpose of preliminary design. We should realize that the ideal gas model in general is not valid for gas or vapor at high pressures.

Example 7.7

One kilogram of air is compressed isothermally and quasi-statically in a cylinder–piston apparatus from 101 kPa and 300 K to 505 kPa and 300 K. Assuming ideal-gas behavior, determine

(a) the amount of heat transfer for the air.
(b) the amount of work transfer for the air.

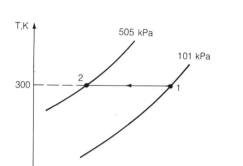

Solution

System selected for study: air
Unique features of process: quasi-static; $T_2 = T_1$
Idealization: air is an ideal gas

(a) From the second law,

$$q_{12} = T(s_2 - s_1). \tag{1}$$

From the property relation for an ideal gas,

$$s_2 - s_1 = \int_1^2 \frac{c_p \, dT}{T} - R \ln \frac{p_2}{p_1}.$$

For our problem, $T_2 = T_1$. Therefore,

$$s_2 - s_1 = -R \ln \frac{p_2}{p_1}.$$

Using 28.97 for the molecular weight of air, we have

$$s_2 - s_1 = -\frac{8.3143}{28.97} \ln \frac{505}{101} \ \text{kJ/kg} \cdot \text{K}$$

$$= -0.4619 \ \text{kJ/kg} \cdot \text{K}.$$

Substituting into Eq. 1, we find that

$$q_{12} = 300(-0.4619) \ \text{kJ/kg}$$

$$= -138.57 \ \text{kJ/kg}.$$

The negative sign means that heat is removed from the gas.

(b) From the first law,

$$q_{12} = u_2 - u_1 + w_{12}.$$

For an ideal gas, $u_2 = u_1$ if $T_2 = T_1$. Therefore,

$$w_{12} = q_{12} = -138.57 \text{ kJ/kg}.$$

The negative sign for work transfer means that work is done to the gas, which is as expected. ∎

7.4 Adiabatic Process

The unique feature of this process is that heat transfer is prevented from crossing the boundary of a system. If the process is quasi-static as well as adiabatic, it is isentropic, according to the second law for a closed system.

Example 7.8

One kilogram of air is compressed quasi-statically and adiabatically from 101 kPa and 300 K to 505 kPa. Assuming ideal-gas behavior with $c_v = 0.7168$ kJ/kg · K and $c_p = 1.0038$ kJ/kg · K, determine the amount of work transfer for the air.

Solution

System selected for study: air
Unique features of process: quasi-static; adiabatic
Idealization: air is an ideal gas with constant specific heats

Since there is no heat transfer, we have, from the first law,

$$w_{12} = -(u_2 - u_1). \tag{1}$$

From the property relation for an ideal gas with constant specific heats,

$$u_2 - u_1 = c_v(T_2 - T_1).$$

Therefore,

$$w_{12} = -c_v(T_2 - T_1). \tag{2}$$

Since we have an ideal gas undergoing a change of state isentropically with constant specific heats, T_2 may be obtained from Eq. 5.35:

$$\frac{T_2}{T_1} = \left(\frac{p_2}{p_1}\right)^{(k-1)/k}.$$

from which we get

$$T_2 = T_1 \left(\frac{p_2}{p_1}\right)^{(k-1)/k}$$

$$= 300 \left(\frac{505}{101}\right)^{(1.4-1)/1.4}$$

$$= 475 \text{ K.}$$

Substituting in Eq. 2, we have

$$w_{12} = -0.7168(475 - 300) \text{ kJ/kg}$$

$$= -125.4 \text{ kJ/kg.} \qquad\qquad \blacksquare$$

Comments

Comparing this with Example 7.7, we see that the work required is less in the case of isentropic compression than in the case of quasi-static isothermal compression. This is confirmed by an examination of the two paths shown on the p–v diagram, as the quasi-static work for a closed system is represented by the area under the process path. Path I is isothermal, and path II is isentropic. It will be recalled that the slope of isentropic lines on the p–v diagram is steeper than those of the isothermal lines.

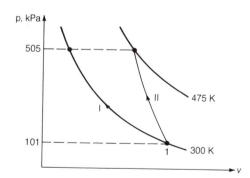

7.5 Constant-Internal-Energy Process

When a stationary closed system undergoes a change of state involving no heat transfer and no work transfer, its internal energy must remain constant.

Example 7.9

One kilogram mole of an ideal gas at 300 K and 1.0 atm pressure is expanded adiabatically into a vacuum (see Fig. 3.5). The final pressure of the gas is 0.1 atm. Show that this free-expansion process is irreversible.

Solution

System selected for study: 1 kg mol of ideal gas
Unique features of process: no heat transfer; no work transfer

From the first law for given process,

$$u_2 = u_1.$$

This means that $T_2 = T_1 = 300$ K, as the internal energy of an ideal gas is a function of temperature only.
From the second law for given process,

$$S_{gen} = \Delta S_{ideal\ gas}.$$

From the property relation for an ideal gas for given process,

$$\bar{s}_2 - \bar{s}_1 = -\bar{R} \ln \frac{p_2}{p_1}$$

$$= +8314.3 \ln 10 \text{ J/kg mol} \cdot \text{K}$$

$$= +19144.0 \text{ J/kg mol} \cdot \text{K}.$$

For 1 kg mol of an ideal gas,

$$S_{gen} = +19144.0 \text{ J/K}.$$

Since entropy generation or creation is positive, the process is irreversible.

Problems

7.1 A rigid vessel has a total volume of 0.5 m³. Initially, it contains 90% by volume of saturated liquid nitrogen and 10% of saturated vapor at 101.325 kPa. Due to heat leak into the vessel from the surroundings, the pressure and temperature of the fluid will increase. If the heat added to the fluid is a constant of 90 kJ/h, how long will it take for the fluid to reach a temperature of 100 K?

7.2 Oxygen stored in a rigid vessel is initially at 101.325 kPa and a quality of 20%. Due to heat leak into the vessel from the surroundings, its pressure will increase. Determine, when the final pressure is 200 kPa,
(a) the final temperature of the fluid.
(b) the amount of heat added to the fluid.

7.3 A rigid vessel, having a volume of 2.0 m³, contains helium gas at a pressure of 300 kPa. If we transfer heat in the amount of 500 kJ to the gas,
(a) show that the final pressure is independent of the initial temperature of the gas.
(b) calculate the final pressure. Assume ideal-gas behavior.

7.4 Fifty kilograms of radioactive water (saturated liquid at 200°C) is stored in a pressure vessel surrounded by a secondary containment structure. The space between the pressure vessel and the containment structure is normally a vacuum. The containing chamber must be large enough so that in case of pressure vessel failure, the pressure of the fluid would not exceed the design pressure of the containing structure. How large must the containing chamber be if the containing structure design pressure is 150 kPa?

7.5 A rigid vessel consists of compartment A separated from compartment B by a rigid and adiabatic partition. Compartment A, having a volume of 0.25 m³, is filled with air at 30°C and 100 kPa. Compartment B, having a volume of 0.50 m³, is filled with air at 60°C and 300 kPa. If the partition is removed,
(a) determine the final equilibrium pressure and temperature of the air.
(b) show that the mixing process is irreversible.
Assume that the mixing process is carried out adiabatically. Also consider the air to be an ideal gas with constant specific heats with $k = 1.4$.

7.6 Water in a rigid tank is initially a liquid–vapor mixture at 100°C and a quality of 50%. Heat is then transferred from a heat reservoir to the fluid until all the liquid is vaporized.
(a) Determine the heat added to the fluid, in kJ/kg.
(b) What is the minimum temperature of the heat reservoir?

7.7   A rigid vessel having a volume of 50 ft^3 is filled with ammonia at 100 psia and 250°F. Heat is transferred from the ammonia until it exists as saturated vapor. Determine the amount of heat transferred from the ammonia.

7.8   Steam in a rigid vessel exists as dry saturated steam at 100 psia initially. It loses heat to the surroundings until its pressure is 14.7 psia. What is the amount of heat given up by each pound of steam?

7.9   Dry saturated steam at a pressure of 14.7 psia is heated in a closed rigid tank to a pressure of 20 psia. For each pound of steam, calculate
(a) the amount of heat added to the steam.
(b) the entropy increase in the universe if the heat source is a heat reservoir at 1000°F.

7.10 An ideal gas with a molecular weight of 30 expands inside a cylinder quasi-statically and isothermally from an initial state defined by $p_1 = 500$ kPa and $T_1 = 20°C$ to a final pressure of $p_2 = 250$ kPa. Determine
(a) the work done by the gas, in kJ/kg.
(b) the work done against the spring (Fig. P7.10) alone.
The pressure of the environment is 101.325 kPa.

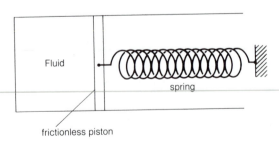

Figure P7.10

7.11 A cylinder, as in Fig. P7.10, contains 0.5 kg of dry saturated steam at 120°C initially. Heat is then added to it until its final

pressure is 300 kPa. The linear spring has a constant of 5000 N/m. The piston cross-sectional area is 0.05 m². Determine
(a) the final temperature of the steam.
(b) the amount of work done by the steam.
(c) the amount of heat transferred to the steam.

7.12 Argon, assumed to be an ideal gas, undergoes the following two processes inside a cylinder. The gas first changes state quasi-statically and adiabatically from 505 kPa and 300 K to a pressure of 101 kPa. It is then followed by a change of state quasi-statically at constant pressure until its final temperature is 100 K.
(a) Sketch the two processes on a T–S diagram showing the appropriate constant-pressure lines.
(b) Determine the amount of heat that must be removed from the gas, in kJ/kg.

7.13 Five kilograms of air undergo a quasi-static process in two steps: they are first expanded at constant pressure from 300 kPa and 50°C until their volume doubles; then they are heated at constant volume until its pressure doubles. Assuming air to be an ideal gas with constant specific heats and $k = 1.4$, determine
(a) the amount of work transfer for the entire process.
(b) the amount of heat transfer for the entire process.

7.14 In a closed system, steam at 100 psia and 500°F is cooled quasi-statically at constant pressure until it exists as saturated vapor. It is then cooled quasi-statically at constant volume to a pressure of 50 psia. For the overall process,
(a) what is the amount of work done to or done by each pound of steam?
(b) what is the amount of heat transferred from each pound of steam?

7.15 Two kilograms of nitrogen, initially at $T_1 = 100$ K and $v_1 = 0.015$ m³/kg, undergo a quasi-static process at constant pressure until they are a saturated liquid. Determine
(a) the amount of work transfer for the process.
(b) the amount of heat transfer for the process.

7.16 Steam expands quasi-statically and isothermally from 800 kPa and 200°C to a final pressure of 200 kPa. Determine the work done, in kJ/kg,
(a) on the basis of real vapor.
(b) on the basis of the ideal-gas model.

7.17 A rigid vessel that is thermally insulated has two compartments. Initially, each compartment has a volume of 1.0 m³, with one compartment containing helium and the other compartment containing argon gas. The compartments are separated by a frictionless piston which is also non-heat conducting. Both gases are initially at 150 kPa and 20°C. Work is transferred very, very slowly to the helium gas until the pressure of both gases is 300 kPa. On the basis of ideal-gas behavior for both gases, determine
(a) the final temperature of argon.
(b) the final temperature of helium.
(c) the amount of entropy generation for the process.

7.18 The same as Problem 7.17 except that heat input is used instead of work input to the helium gas. Assume that the heat reservoir supplying the heat is at the minimum temperature that is needed for the process to be possible.

7.19 One pound of air initially at 100°F expands quasi-statically inside a cylinder at a constant pressure of 75 psia until its volume is doubled. Assuming air to be an ideal gas with constant specific heats of $c_p = 0.240$ Btu/lbm · °R and $c_v = 0.171$ Btu/lbm · °R, find
(a) the final temperature.
(b) the work done by the air.
(c) the heat transferred to or from the air.

7.20 A certain amount of a monatomic ideal gas expands quasi-statically inside a cylinder at a constant pressure. If the work done by the gas is 5000 Btu, what is the amount of heat transferred to or from the gas?

7.21 Helium gas is compressed quasi-statically inside a cylinder from 110 kPa and 27°C to a pressure of 550 kPa. The compression process follows the path given by

$$pv^{1.20} = \text{constant.}$$

On the basis of ideal-gas behavior, determine
(a) the final temperature of the gas.
(b) the amount of heat transfer for the gas, in kJ/kg.

7.22 Argon gas is being compressed very, very slowly inside a cylinder. The process on the p–V diagram is represented by the expression

$$pv^{1.8} = \text{constant.}$$

The pressure and temperature of the gas at the beginning of the process are 200 kPa and 20°C, respectively. The final pressure is 800 kPa. On the basis of ideal-gas behavior,
(a) determine the final temperature of the gas.
(b) show the compression process on a T–S diagram with the appropriate constant-pressure lines.
(c) explain why there will be heat added to or removed from the gas. (No calculations are necessary.)

7.23 Helium gas is being compressed quasi-statically inside a cylinder. The process on the p–V diagram is represented by the expression

$$pV^{1.4} = \text{constant.}$$

The pressure and temperature of the gas at the beginning of the compression process are 300 kPa and 25°C, respectively. The final pressure is 900 kPa. On the basis of ideal-gas behavior,
(a) determine the final temperature of the gas.
(b) show the compression process on a T–S diagram with the appropriate constant-pressure lines.
(c) explain why there will be heat added to or removed from the gas. (No calculations are necessary.)

7.24 The frictionless piston in Fig. P7.24 is restrained by a linear spring that has a constant of 25 kN/m. The cylinder has a diameter of 0.3 m and contains nitrogen gas. At the position shown, the spring exerts no force on the piston and the nitrogen gas is saturated vapor at the atmospheric pressure of 101.325 kPa, occupying a volume of 1.4 m³. Heat is then added very slowly to the gas until the piston has moved 0.6 m to the right. Determine
(a) the final pressure of the gas.
(b) the final volume of the gas.
(c) the amount of work done by the gas.
(d) the amount of heat added to the gas.

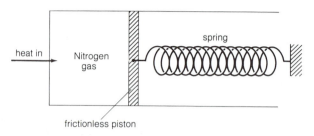

Figure P7.24

7.25 A frictionless piston divides a cylinder into two compartments as shown in Fig. P7.25. Compartment B is a vacuum. Compartment A contains 1.0 kg of H_2O. The piston may slide against a linear spring that has a constant of 8.5 kN/m. Before the stop is removed, the fluid in compartment A is saturated liquid at 3000 kPa. When the stop is removed and final equilibrium is observed, the pressure of the fluid in compartment A is found to be 300 kPa. The spring is initially at its free length. Assuming the process to be quasi-static and adiabatic, determine
(a) the work done by the fluid.
(b) the distance between the final position and the initial position of the piston.

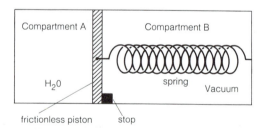

Figure P7.25

7.26 A rigid cylinder that is well insulated has a total volume of 0.2 m^3. It is fitted with a frictionless but heat-conducting piston which divides the cylinder into compartment A and compartment B. Initially, the piston is clamped in the center with 0.1 m^3 of argon gas at 300 K and 300 kPa in compartment A, and 0.1 m^3 of argon gas at 300 K and 100 kPa in compartment B. The clamp is then removed and the system reaches equilibrium with the piston at a new position. Assuming ideal-gas behavior and neglecting any heat transfer to or from the cylinder,

(a) show that the final temperature of the gas is 300 K.
(b) determine the final pressure of the gas in each compartment.
(c) determine the amount of entropy generation due to the process.

7.27 N kilomoles of an ideal gas are contained inside a cylinder at an initial pressure of p_1 and an initial volume of V_1. They are cooled quasi-statically at constant volume until their pressure is reduced to one-half of p_1. It is then compressed quasi-statically and isothermally until their final pressure is the same as their initial pressure of p_1.
   (a) Sketch the processes involved in the p–V diagram, showing the appropriate lines.
   (b) Calculate the work transfer involved, in kilojoules, if $p_1 = 300$ kPa and $V_1 = 0.2$ m³.

7.28 Five kilograms of air are cooled quasi-statically from 150°C to 25°C at a constant pressure of 500 kPa. They are then followed by a quasi-static and adiabatic expansion process until their final pressure is 100 kPa. Assuming air to be an ideal gas with constant specific heats and $k = 1.4$, determine, for the entire process,
   (a) the amount of heat transfer involved.
   (b) the amount of work transfer involved.

7.29 Steam is compressed adiabatically and quasi-statically inside a cylinder. At the beginning of the process, the temperature and pressure of the steam are 212°F and 14.7 psia. At the end of the process, the steam pressure is 60 psia. Determine the compression work required for each pound of steam.

7.30 Freon-12 is compressed inside a cylinder. At the beginning of the process, the fluid exists as a dry saturated vapor at -10°F. At the end of the process, the pressure and temperature of the fluid are 100 psia and 140°F, respectively. The work required for compression is 40 Btu/lbm.
   (a) Determine the amount of heat removed from each pound of Freon-12.
   (b) Show that the compression process is irreversible if the heat removed from the fluid during compression flows to a heat reservoir at 80°F.

7.31 Five kilograms of Freon-12 are compressed quasi-statically inside a cylinder from 500 kPa and 80°C to a final pressure of 1.0

MPa. The path of the process is given by

$$pV^{1.2} = \text{constant}.$$

Determine
(a) the amount of work transfer for the fluid.
(b) the amount of heat transfer for the fluid.

7.32 Five kilograms of Freon-12 expand quasi-statically inside a cylinder from 1.0 MPa and 100°C to a final pressure of 0.50 MPa. The path of the expansion process may be taken as

$$pV = \text{constant}.$$

Determine
(a) the amount of work transfer for the fluid.
(b) the amount of heat transfer for the fluid.

7.33 Air is compressed adiabatically inside a cylinder from a pressure and temperature of 15 psia and 80°F to a pressure and temperature of 75 psia and 500°F. Making use of data from Keenan and Kaye's *Gas Tables*,
(a) calculate the compression work required for each pound of air.
(b) show that the compression process is irreversible.

7.34 An ideal gas expands adiabatically and quasi-statically inside a cylinder. Its temperature falls from 100°F to −40°F while its volume is doubled. The gas does 25 Btu/lbm of work in the process. Assuming constant specific heats, find the values of the specific heats c_p and c_v.

7.35 Five kilograms of air are compressed polytropically inside a cylinder. The initial pressure and temperature of the gas are 150 kPa and 20°C. The final pressure and temperature of the gas are 750 kPa and 200°C. Assuming ideal-gas behavior with constant specific heats and $k = 1.4$, determine
(a) the amount of work transfer for the air.
(b) the amount of heat transfer for the air.

7.36 Nitrogen is compressed polytropically from 110 kPa and 25°C to 770 kPa inside a cylinder. The polytropic exponent for the process is 1.30. Assuming ideal-gas behavior with constant specific heats and $k = 1.4$, determine
(a) the amount of work transfer for the nitrogen, in kJ/kg.
(b) the amount of heat transfer for the nitrogen, in kJ/kg.

7.37 Five kilograms of nitrogen are heated in a rigid vessel from 100 K to 300 K. The initial pressure is 0.50 MPa. Determine
(a) the final pressure of the fluid.
(b) the amount of heat transfer for the fluid.

7.38 For Problem 7.37, if the heat source is the environment at $T_0 = 300$ K and $p_0 = 101.325$ kPa, determine for the process
(a) the amount of entropy generation.
(b) the amount of thermodynamic lost work.
(c) the change in the nonflow exergy of the fluid.
(d) Compare the result of part (c) with that of part (b).

7.39 One kilogram of oxygen expands inside a cylinder isothermally (but not quasi-statically) at 300 K from 10 MPa to 1.0 MPa. The actual work done by the gas is 100 kJ/kg. The pressure and temperature of the environment are 101.325 kPa and 300 K.
(a) Calculate the thermodynamic lost work for the process.
(b) Calculate the change in the nonflow exergy of the gas.
(c) Compare the result of part (b) with the sum of thermodynamic lost work and actual useful work done by the gas. Any comment?

7.40 Nitrogen is compressed isothermally (but not quasi-statically) at 300 K from 101.325 kPa to 20 MPa. The actual work done to the gas is 600 kJ/kg. The pressure and temperature of the environment are 101.325 kPa and 300 K.
(a) Calculate the thermodynamic lost work, in kJ/kg.
(b) Calculate the change in the nonflow exergy of the gas, in kJ/kg.
(c) Compare the result of part (b) with the sum of the thermodynamic lost work and the actual useful work involved. Any comment?

7.41 A rigid vessel, having a volume of 5.0 m³, contains a liquid–vapor mixture of water at 100°C and a quality of 10% initially. An electric heater inside the vessel is then turned on until the final temperature is 180°C. Assuming the process to be adiabatic,
(a) determine the amount of work transfer for the fluid.
(b) determine the amount of thermodynamic lost work for the process.
(c) calculate the change in the nonflow exergy of the fluid.
(d) Compare the result of part (c) with the sum of thermodynamic lost work and work input. Any comment?
Let $T_0 = 300$ K.

7.42 A rigid tank of 50 ft^3 capacity contains air at 30 psia and 500°F. If a heat-engine system may be developed to utilize the air in the tank as the heat source, what is the maximum useful work obtainable from the air when the temperature of the air drops to 80°F, which is also the temperature of the surroundings?

7.43 An ideal gas is to be heated inside a rigid vessel from an initial temperature T_1 to a final temperature T_2. The molal heat capacity as a function of temperature is given by

$$\bar{c}_p^* = a + bT + cT^2 + dT^3$$

where $\bar{c}_p^*$ is in kJ/kgmol · K
   T is in degrees Kelvin
   a, b, c, and d are constants.
Write a computer program to calculate the amount of heat input required in kJ/kg to heat nitrogen gas from 300 K to 1000 K
(a) assuming constant specific heat at T_1;
(b) using average specific heat between T_1 and T_2;
(c) taking into consideration of the variation of specific heat with temperature.
The values of the constants in the molal heat capacity equation for nitrogen are:

$$a = 28.90$$

$$b = -0.1571 \times 10^{-2}$$

$$c = 0.8081 \times 10^{-5}$$

$$d = -2.873 \times 10^{-9}$$

7.44 Same as Problem 7.43 except that the gas is carbon dioxide. The molal heat capacity as a function of temperature for carbon dioxide is given by the equation given in Problem 7.43. The values of the constants in the molal heat capacity equation for carbon dioxide are:

$$a = 22.26$$

$$b = 5.981 \times 10^{-2}$$

$$c = -3.501 \times 10^{-5}$$

$$d = 7.469 \times 10^{-9}$$

7.45 An ideal gas is to be heated inside a rigid vessel from an initial temperature T_1 to a final temperature T_2. The source of heat is a heat reservoir at T^{HR}. The molal heat capacity as a function of temperature is given in Problem 7.43.

Write a computer program to determine the amount of entropy generation due to irreversibility in kJ/kg when nitrogen gas is heated from 300 K to 1000 K with heat coming from a heat reservoir at 1000 K, 2000 K, and 3000 K

(a) assuming constant specific heat at T_1;

(b) using average specific heat between T_1 and T_2;

(c) taking into consideration of the variation of specific heat with temperature.

The values of the constants in the molal heat capacity equation for nitrogen are given in Problem 7.43.

7.46 Same as Problem 7.45 except that the gas is carbon dioxide. The values of the constants in the molal heat capacity equation for carbon dioxide are given in Problem 7.44.

7.47 Nitrogen gas is compressed quasi-statically and adiabatically inside a cylinder from an initial pressure of 1 atm and an initial temperature of 300 K to a final pressure of 10 atm. On the basis of ideal gas behavior, determine the quasi-static work involved, in kJ/kg,

(a) assuming constant specific heat at T_1;

(b) taking into consideration the variation of specific heat with temperature.

Write a computer program to solve this problem.

The molal heat capacity equation and the values of the constants in this equation for nitrogen are given in Problem 7.43.

7.48 Same as Problem 7.47 except that the gas is carbon dioxide. The molal heat capacity as a function of temperature is given in Problem 7.43. The values of the constants in this equation are given in Problem 7.44.

CHAPTER 8

Engineering Analysis of Processes for Open Systems

In Chapter 7 we studied several common engineering problems by focusing attention on a fixed quantity of matter. In principle, all problems may be analyzed in this fashion. In many engineering applications, however, it is much more convenient to draw the system boundary around a fixed region in space through which material may flow. In this chapter we study many engineering applications using the open-system approach. Since flow of material across a system boundary (control surface) is possible, it is necessary to make use of the principle of conservation of mass to keep track of the mass within the open system (control volume). Consequently, the tools available to us in the study of flow processes are four in number:

1. The first law of thermodynamics (the energy-balance equation).
2. The second law of thermodynamics (the entropy-creation equation).
3. The state postulate (property relations).
4. The principle of conservation of mass (the mass-balance equation).

Since the majority of engineering devices may be modeled as operating under steady-state steady-flow conditions, the major effort of this chapter is devoted to the study of hardware and systems with steady-state steady flow. A few examples are given of non-steady-state steady-flow processes to illustrate the usefulness of the general equations and to strengthen our understanding of basic principles.

8.1 Unique Characteristics of Open Systems with Steady-State Steady Flow

It will be recalled that steady-state steady flow involves the assumption that all properties at each point in the control volume are constant with respect to time, with the consequence that there can be no change in mass, no change in energy, and no change in entropy at any given point within the control volume. It will also be recalled that this assumption necessitates the following operating conditions:

1. The rate of heat transfer across the control surface is constant.
2. The rate of work transfer across the control surface is constant.
3. The state and velocity of each incoming stream are constant.
4. The state and velocity of each outgoing stream are constant.
5. The mass-flow rate of each incoming stream is constant.
6. The mass-flow rate of each outgoing stream is constant.
7. The total mass-flow rate of incoming streams is equal to the total mass-flow rate of outgoing streams.

In solving problems assuming open systems operating under steady-state steady-flow conditions, it is necessary for us to remember these unique characteristics of such systems.

8.2 Functional Classification of Steady-State Steady-Flow Devices

Many complex engineering systems designed for continuous operations, such as modern power plants, refrigeration plants, air-separation plants, gas-liquefaction plants, and water-desalination plants, are usually formed by joining together several well-known components through which material will flow. These well-known steady-state steady-flow devices may be conveniently grouped into the following six classes, according to their functions:

1. Work-absorbing devices, such as pumps and compressors.
2. Work-producing devices, such as turbines and expansion engines.
3. Nozzles and diffusers.
4. Heat exchangers, such as boilers and condensers.
5. Throttling devices, such as expansion valves and control valves.
6. Piping.

Since the performance of a complex system depends on the performance of its parts, it is essential for us to have a good understanding of the behavior of the components of the system. The capability we gain in analyzing steady-flow devices in this section will greatly enhance our ability to analyze and synthesize the behavior of complex systems.

8.3 Ideal Shaft Work for a Steady-State Steady-Flow Device

The work transmitted into or out of a steady-state steady-flow device through a rotating shaft is known as the shaft work of the device. It is generally of interest to the engineer to know what is the ideal shaft work of a steady-state steady-flow device. From a thermodynamics point of view, ideal implies that the process must be reversible. We shall investigate such a process involving one stream in and one stream out.

Neglecting change in kinetic energy and change in potential energy, we have from the first law for steady-state steady flow and working on the basis of unit mass-flow rate,

$$\bar{d}q = dh + \bar{d}w_{\text{shaft}}, \tag{8.1}$$

which is valid for irreversible processes as well as for reversible processes.

For a reversible process, we have from the second law for steady-state steady flow,

$$\bar{d}q = T\,ds. \tag{8.2}$$

Combining Eqs. 8.1 and 8.2, we have

$$T\,ds = dh + \bar{d}w_{\text{shaft}}. \tag{8.3}$$

If the fluid is a simple compressible substance, we have, from the property relation,

$$T\,ds = dh - v\,dp. \tag{8.4}$$

Combining Eqs. 8.3 and 8.4, we find

$$\bar{d}w_{\text{shaft}} = -v\,dp. \tag{8.5}$$

Equation 8.5 gives the ideal shaft work of a steady-state steady-flow device. For a work-absorbing device, this is the minimum work required. For a work-producing device, this is the maximum work that can be delivered. To integrate Eq. 8.5, we must know the function relating v to p for the particular reversible process used, which could be reversible isothermal, reversible adiabatic (isentropic), or reversible polytropic. This equation brings out once again the concept that work is a path function.

8.4 Pumps and Compressors

The purpose of a pump or a compressor is usually to bring a fluid from an initial state to a specified final pressure that is greater than the initial pressure. We call the device a pump when the flowing fluid is a liquid. We call the device a compressor when the flowing fluid is a vapor or a gas.

In an actual compressor, when no effort is made to cool the gas during compression, the compression process may be modeled as adiabatic. This is a reasonable assumption because the surface area of the machine available for heat transfer is relatively small, and the length of time required for the gas to pass through the machine is quite short. Therefore, the ideal compression process is a reversible adiabatic or isentropic one. Comparison between the actual and the ideal compressor performance is given by the adiabatic compressor efficiency, η_c, as

$$\eta_c \equiv \frac{w_{rev}}{w_{act}},\tag{8.6}$$

where

w_{rev} = work of isentropic compression from the initial state to final pressure

w_{act} = work of actual adiabatic compression from the initial state to the same final pressure.

In the case of a pump, its adiabatic pump efficiency, η_p, is defined in the same manner. That is,

$$\eta_p \equiv \frac{w_{rev}}{w_{act}}.\tag{8.7}$$

Example 8.1

Water enters a steady-state steady-flow pump as saturated liquid at 40°C. It leaves the pump at 20 MPa. For an adiabatic pump efficiency of 70%, determine the actual pump work. The pumping process may be assumed to be adiabatic.

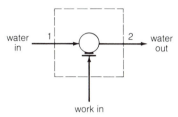

water in — 1 → ○ → 2 — water out

work in

Solution

System selected for study: pump
Unique features of process: steady-state steady flow; adiabatic
Idealization: water is incompressible; neglect ΔKE; neglect ΔPE

From the definition of pump efficiency, we have

$$w_{act} = \frac{w_{rev}}{\eta_p} \qquad (1)$$

where w_{rev} may be obtained from Eq. 8.5 as

$$w_{rev} = -\int_1^2 v \, dp. \qquad (2)$$

Assuming that water is incompressible,

$$w_{rev} \approx -v(p_2 - p_1). \qquad (3)$$

Using data for water from the tables, we have

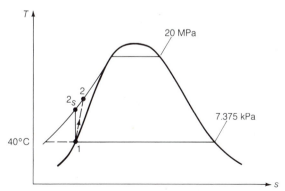

$p_1 =$ saturation pressure at $40°C = 7.375$ kPa

$v = v_f$ at $40°C = 0.0010078$ m^3/kg.

Substituting into Eq. 3, we find that

$$w_{rev} = -0.0010078(20 \times 1000 - 7.375) \text{ kJ/kg}$$

$$= -20.15 \text{ kJ/kg}.$$

Therefore,

$$w_{act} = -\frac{20.15}{0.70} = -28.79 \text{ kJ/kg}. \qquad ∎$$

Comments

Since $w_{rev} = -(h_{2s} - h_1)$ in which h_{2s} is defined by $p_{2s} = 20$ MPa and $s_{2s} = s_1 = s_f$ at $40°C = 0.5721$ kJ/kg · K, we could have obtained w_{rev} without neglecting the compressibility effect. To do it this way, however, we must have very accurate compressed water data. We wish to point out that the constant-pressure lines in the compressed-liquid region are very close to each other. The separation between the two constant-pressure lines in the T–S diagram has been exaggerated. The temperature of the water leaving the pump is only about $41°C$. Consequently, to measure the temperature difference between the outlet and the inlet would require a very accurate and sensitive instrument.

Example 8.2

Air enters a steady-state steady-flow fan at $25°C$ and 101.325 kPa with a volume rate of flow of 2.745×10^6 m³/h. The pressure rise of the air across the fan is 10 kPa. For a fan efficiency of 80%, determine the actual power needed to drive the fan. We may assume the process to be adiabatic.

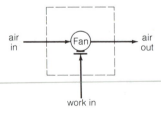

Solution

System selected for study: fan
Unique features of process: steady-state steady flow; adiabatic
Idealization: neglect ΔKE; neglect ΔPE; neglect the compressibility effect for air

A fan is simply a compressor in which the pressure rise of the fluid across the machine is very small. Thus we have

$$w_{\text{act}} = \frac{w_{\text{rev}}}{\eta_c}, \tag{1}$$

where w_{rev} may be obtained from Eq. 8.5 as

$$w_{\text{rev}} = -\int_1^2 v \, dp. \tag{2}$$

Because the pressure rise of the air across the fan is so small, Eq. 2 may be approximated as

$$w_{\text{rev}} \approx -v(p_2 - p_1) \tag{3}$$

$$\dot{W}_{\text{rev}} \approx -\dot{m}v(p_2 - p_1)$$

$$\approx -\dot{V}(p_2 - p_1).$$

Thus the actual power needed is given by

$$\dot{W}_{\text{act}} = -\frac{\dot{V}(p_2 - p_1)}{\eta_c}$$

$$= -\frac{2.745 \times 10^6 \times 10}{0.8} \text{ kJ/h}$$

$$= -3.431 \times 10^7 \text{ kJ/h} = -9530 \text{ kW.} \qquad \blacksquare$$

Comment

The fan power needed is quite typical for the design of large steam power plants.

8.5 Turbines

Turbines are work-producing devices operating at high rotating speeds. As a consequence, we can obtain large power from a relatively small turbine. Therefore, the unavoidable heat transfer between the turbine surface area and the surrounding is usually quite small compared to the work-transfer quantity. In general, we may model the expansion process in a turbine as adiabatic. The ideal expansion process used for comparison is then a reversible adiabatic or isentropic one. The adiabatic turbine efficiency, when comparing the actual performance of a turbine to its ideal performance, is defined as

$$\eta_T = \frac{w_{act}}{w_{rev}},$$
(8.8)

where

w_{act} = work of actual adiabatic expansion from the initial state to the final pressure

w_{rev} = work of isentropic expansion from the initial state to the same final pressure.

Example 8.3

A windmill or wind turbine is simply a device we employ to convert the kinetic energy of air into useful power. Suppose that the blade diameter D of a wind turbine is 30 m. What is the ideal power that we can generate when the wind is blowing steadily with a velocity of 50 km/h? Assume that the air is at 20°C and 101.325 kPa.

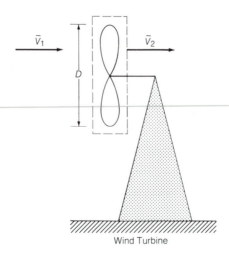

Wind Turbine

Solution

System selected for study: wind turbine
Unique feature of process: steady-state steady flow
Idealization: air is an ideal gas

From the first law we have

$$\bar{d}q = dh + d\left(\frac{\bar{V}^2}{2g_c}\right) + d\left(\frac{g}{g_c}\, Z\right) + \bar{d}w_s.$$

(1)

For ideal power output, the process must be reversible. Thus we have, from the second law,

$$\bar{d}q = T\, ds.$$

(2)

From the property relation we have

$$T\, ds = dh - v\, dp.$$

(3)

Combining Eqs. 1, 2, and 3, and neglecting change in potential energy, we have

$$\bar{d}w_{\text{rev}} = -v\, dp - d\left(\frac{\bar{V}^2}{2g_c}\right).$$

(4)

But $p_2 = p_1$. Thus

$$\bar{d}w_{\text{rev}} = -d\left(\frac{\bar{V}^2}{2g_c}\right)$$

(5)

$$w_{\text{rev}} = -\frac{\bar{V}_2^2 - \bar{V}_1^2}{2g_c}.$$

(6)

If we assume that $\bar{V}_2 = 0$, then the absolute maximum power we can generate is given by

$$\dot{W}_{\text{max}} = \frac{\dot{m}\bar{V}_1^2}{2g_c}.$$

(7)

From conservation of mass for steady flow,

$$\dot{m} = \rho \bar{V}A,$$

(8)

where

$$A = \frac{\pi D^2}{4}$$

(9)

and ρ is the density of air.

Combining Eqs. 7, 8, and 9, we have

$$\dot{W}^d_{max} = \frac{\pi D^2 \rho \bar{V}^3_1}{8 g_c}, \tag{10}$$

which shows that the maximum power output of a wind turbine is a function of the wind velocity cubed. For air as ideal gas at 20°C and 101.325 kPa,

$$\rho = \frac{p}{RT}$$

$$= \frac{101.325}{(8.3143/28.97)(293.15)} \, kg/m^3$$

$$= 1.204 \, kg/m^3.$$

With $D = 30$ m and $\bar{V}_1 = 50$ km/h = $(50 \times 1000)/3600$ m/s $= 13.89$ m/s,

$$\dot{W}_{max} = \frac{\pi \times (30)^2 (1.204)(13.89)^3}{8 \times 1} \, J/s$$

$$= 1,140,000 \, J/s$$

$$= 1140 \, kJ/s = 1140 \, kW.$$ ∎

Comments

We wish to point out that the maximum power output we obtain is the absolute maximum. It is deduced on the basis that the velocity of air leaving the wind turbine is zero. If we take the velocity of air leaving the turbine into consideration, it may be shown in fluid mechanics that the maximum power output is about 60% of what we have. Nonetheless, this is a source of pollution-free energy. It is also cost-free energy. However, the cost of installing such energy conversion devices is still very expensive.

Example 8.4

Steam enters a steady-state steady-flow turbine at 3000 psia and 1000°F. It leaves the turbine at 1.0 psia. For an adiabatic turbine efficiency of 85%, determine the actual turbine work. The expansion process may be assumed to be adiabatic.

Solution

System selected for study: turbine
Unique feature of process: steady-state steady flow
Assumptions: steam is a simple compressible substance; process is adiabatic; neglect ΔKE and ΔPE

From the definition of turbine efficiency, we have

$$w_{\text{act}} = \eta_T \, w_{\text{rev}}. \tag{1}$$

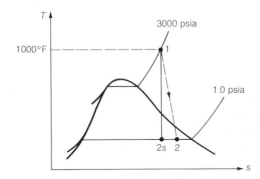

steam in

1

Turbine — work out

2

steam out

Since there is no heat transfer, we have, from the first law, neglecting change in kinetic energy and change in potential energy,

$$w_{\text{rev}} = h_1 - h_{2s}. \tag{2}$$

From the steam tables, we have

$$h_1 = 1440.2 \text{ Btu/lbm}$$

$$s_1 = 1.4976 \text{ Btu/lbm} \cdot {}^\circ\text{R}$$

$$s_g \text{ at } 1.0 \text{ psia} = 1.9781 \text{ Btu/lbm} \cdot {}^\circ\text{R} > 1.4976$$

$$s_f \text{ at } 1.0 \text{ psia} = 0.1326 \text{ Btu/lbm} \cdot {}^\circ\text{R} < 1.4976.$$

State 2s is a liquid–vapor mixture. Therefore,

$$s_{2s} = (s_f + x_{2s}s_{fg}) \text{ at } 1.0 \text{ psia}$$

and

$$1.4976 = 0.13266 + x_{2s}(1.9781 - 0.1326),$$

yielding $x_{2s} = 0.7396$. Then

$$h_{2s} = (h_f + x_{2s}h_{fg}) \text{ at } 1.0 \text{ psia}$$

and

$$h_{2s} = 69.73 + (0.7396)(1036.1) \text{ Btu/lbm}$$

$$= 836.0 \text{ Btu/lbm}.$$

Substituting into Eq. 2, we find that

$$w_{rev} = 1440.2 - 836.0 \text{ Btu/lbm} = 604.2 \text{ Btu/lbm}.$$

Substituting into Eq. 1, we find that

$$w_{act} = 0.85 \times 604.2 \text{ Btu/lbm} = 513.6 \text{ Btu/lbm}. \qquad \blacksquare$$

Comments

The turbine work obtained in this example and the pump work obtained in the previous example are typical values for a modern steam-power cycle. The algebraic sum of the turbine work and the pump work represents the net work of such a cycle. Since the turbine work is so much larger than the pump work, it is quite often accurate enough to consider the turbine work as the net work of the cycle.

8.6 Throttling Devices

A throttling process is said to have occurred when a flowing fluid suffers a loss of pressure in passing through some form of a restriction, such as a valve, an orifice, and similar devices. By the nature of this process, no shaft work is involved. In addition, the following assumptions are usually made in the study of throttling processes:

1. Neglect heat transfer.
2. Neglect change in kinetic energy.
3. Neglect change in potential energy.

With these assumptions, the energy equation for a throttling process becomes

$$H_{in} = H_{out}, \tag{8.9}$$

or, on the basis of unit mass-flow rate,

$$h_{in} = h_{out}.$$

We thus see that the enthalpy of the flowing fluid at the inlet of a throttling device and the enthalpy of the flowing fluid at the outlet are equal.

Example 8.5

We may use a throttling valve to liquefy nitrogen gas if the gas enters the valve at high pressure and low temperature. If we design the system so that the pressure of the fluid leaving the valve is 101.325 kPa, determine the fraction of the gas that is liquefied if the gas enters steadily

(a) at 160 K and 20 MPa.
(b) at 160 K and 30 MPa.
(c) at 170 K and 20 MPa.
(d) at 170 K and 30 MPa.

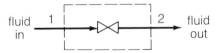

Solution

System selected for study: throttling valve
Unique features of process: steady-state steady flow; no work transfer
Assumptions: nitrogen is a simple compressible fluid; neglect heat transfer; neglect ΔKE and ΔPE

From the first law, we have

$$h_1 = h_2. \tag{1}$$

From tabulated data for nitrogen, we have

(a) $h_1 = h$ at 160 K and 20 MPa $= 53.625$ kJ/kg

h_f at 101.325 kPa $= -122.150$ kJ/kg < 53.625 kJ/kg

h_g at 101.325 kPa $= 76.689$ kJ/kg > 53.625 kJ/kg.

State 2 is a liquid–vapor mixture. Therefore,

$$h_2 = (h_f + x_2 h_{fg}) \text{ at } 101.325 \text{ kPa}$$

$$53.625 = -122.150 + x_2(76.689 + 122.150)$$

yielding

$$x_2 = 0.8840$$

and

$$y_2 = 1 - x_2 = 1 - 0.8840 = 0.1160.$$

That is, the fraction of the gas that is liquefied is 11.60%.

(b) From tabulated data for nitrogen, we have

$h_1 = h$ at 160 K and 30 MPa $= 52.522$ kJ/kg $> h_f$ at 101.325 kPa

$< h_g$ at 101.325 kPa.

Thus

$$h_2 = (h_f + x_2 h_{fg}) \text{ at } 101.325 \text{ kPa}$$

$$52.522 = -122.150 + x_2(76.689 + 122.150),$$

yielding

$$x_2 = 0.8785.$$

and

$$y_2 = 0.1215.$$

That is, the fraction of the gas that is liquefied is 12.15%.

(c) From tabulated data for nitrogen, we have

$$h_1 = h \text{ at 170 K and 20 MPa} = 73.323 \text{ kJ/kg} > h_f \text{ at 101.325 kPa}$$

$$< h_g \text{ at 101.325 kPa}.$$

Thus

$$h_2 = (h_f + x_2 \, h_{fg}) \text{ at 101.325 kPa}$$

$$73.323 = -122.150 + x_2(76.689 + 122.150),$$

yielding

$$x_2 = 0.9831$$

and

$$y_2 = 0.0169.$$

That is, the fraction of the gas that is liquefied is 1.69%.

(d) From tabulated data for nitrogen, we have

$$h_1 = h \text{ at 170 K and 30 MPa} = 70.26 \text{ kJ/kg} > h_f \text{ at 101.325 kPa}$$

$$< h_g \text{ at 101.325 kPa}.$$

Thus

$$h_2 = (h_f + x_2 \, h_{fg}) \text{ at 101.325 kPa}$$

$$70.26 = -122.150 + x_2(76.689 + 122.150),$$

yielding

$$x_2 = 0.9677$$

and

$$y_2 = 0.0323.$$

That is, the fraction of the gas that is liquefied is 3.23%. ▮

Comments

The results of this problem are summarized as follows:

State of Gas at Valve Inlet	160 K, 20 MPa	160 K, 30 MPa	170 K, 20 MPa	170 K, 30 MPa
Fraction of Gas Liquefied	0.1160	0.1215	0.0169	0.0323

From the results above, lowering the inlet temperature appears to be more effective than increasing the pressure at the inlet in the liquefaction of nitrogen gas. The decision maker will have to compare the cost of refrigeration with the cost of power needed for compression. It would be a matter of trade-offs, as in any real engineering design.

Example 8.6

A throttling steam calorimeter (see Fig. 8.1) is an instrument used for the determination of the quality of wet steam flowing in a steam main. It utilizes the fact that when wet steam is throttled sufficiently, superheated steam will form. If wet steam at 200 psia is throttled in a calorimeter to 15.0 psia and 300°F, determine the quality of the wet steam.

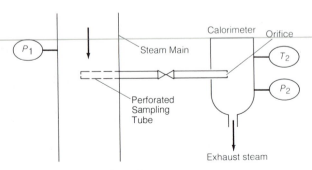

Figure 8.1 Throttling Steam Calorimeter

Solution

System selected for study: calorimeter
Unique features of process: steady-state steady flow; no work transfer
Assumptions: steam is a simple compressible substance; neglect heat transfer; neglect ΔKE and ΔPE

From the first law, we have

$$h_1 = h_2. \tag{1}$$

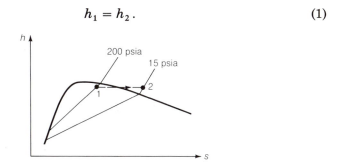

Using data from the steam tables, we have

$$h_2 = 1192.5 \text{ Btu/lbm}$$

$$h_f \text{ at 200 psia} = 355.5 \text{ Btu/lbm}$$

$$h_g \text{ at 200 psia} = 1198.3 \text{ Btu/lbm}.$$

Using $h_1 = h_f + x_1 h_{fg}$, we find that

$$1192.5 = 355.5 + x_1(1198.3 - 355.5),$$

yielding

$$x_1 = 0.993. \qquad \blacksquare$$

Comment

The reader should show that s_2 is greater than s_1, since the throttling process is an irreversible process, according to the second law.

8.7 Heat Exchangers

Any device the primary function of which is to make heat transfer possible between one fluid and another fluid, or between one fluid and a heat source or heat sink, or between one region and another region, is known as a *heat exchanger*. It is an essential component in many engineering systems. Its applications are extremely broad and wide, ranging from heat removal from a package of electronic gears to heat removal from a nuclear reactor core. Examples are indeed many, and they are often known by names that denote their applications. In a steam power plant, we have heat exchangers known as boilers, condensers, and feedwater heaters. In an internal combustion engine, we have water jackets and radiators. In a gas-liquefaction plant, we have evaporators, regenerators, and cold boxes.

The three physical mechanisms by which heat is transferred are conduction, convection, and radiation. The study of these physical mechanisms and related material belongs in the separate discipline of heat transfer.[1] We wish, however, to point out that a good understanding of the engineering science of heat transfer is essential to the proper design of a heat exchanger for a given application. We shall examine the behavior of heat exchangers primarily from the point of view of the first law and the second law, treating them as "black boxes."

By the nature of the process, no shaft work is involved in the study of heat exchangers. Since the change of kinetic energy and the change of potential energy are usually small compared to the heat-transfer quantity, the first law for heat exchangers becomes

$$\dot{Q} = \sum_{\text{out}} \dot{m}h - \sum_{\text{in}} \dot{m}h. \tag{8.10}$$

Example 8.7

Hot gas enters the heat recovery steam generator of a cogeneration system at 500°C and 101.325 kPa and leaves at 150°C. Water enters steadily at 100°C and 1.0 MPa and leaves as dry saturated steam at 1.0 MPa. The hot gas may be assumed to be an ideal gas with constant specific heat of $c_p = 1.05$ kJ/kg · K. For a flow rate of hot gas of 25,000 kg/h,

(a) determine the flow rate of water, in kg/h.
(b) determine the rate of generation of thermodynamic lost work.

[1] See Frank Kreith, *Principles of Heat Transfer*, International Textbook Company, Scranton, Pa., 1958; Max Jakob and G. A. Hawkins, *Elements of Heat Transfer*, 3rd ed., John Wiley & Sons, Inc., New York, 1957; W. H. McAdams, *Heat Transmission*, 3rd ed., McGraw-Hill Book Company, New York, 1954.

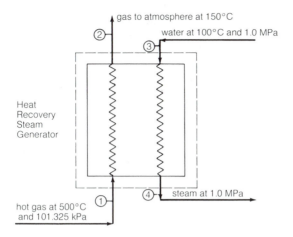

Solution

System selected for study: heat recovery steam generator
Unique features of process: steady-state steady flow; no work transfer
Assumptions: Both the gas and H_2O are simple compressible substances; neglect ΔKE; neglect ΔPE; neglect pressure drops; neglect heat loss to surroundings

(a) From the first law, we have

$$0 = \sum_{\text{out}} \dot{m}h - \sum_{\text{in}} \dot{m}h$$

or

$$\dot{m}_{\text{gas}}(h_1 - h_2) = \dot{m}_{H_2O}(h_4 - h_3)$$

and

$$\dot{m}_{H_2O} = \frac{\dot{m}_{\text{gas}}(h_1 - h_2)}{h_4 - h_3}. \tag{1}$$

On the basis of ideal gas with constant specific heats,

$$h_1 - h_2 = c_p(T_1 - T_2). \tag{2}$$

Then we have

$$\dot{m}_{H_2O} = \frac{\dot{m}_{\text{gas}}\, c_p(T_1 - T_2)}{h_4 - h_3}. \tag{3}$$

From steam tables, we have

$$h_3 \approx h_f \text{ at } 100°\text{C}$$

$$= 419.06 \text{ kJ/kg}$$

$$h_4 = h_g \text{ at } 1.0 \text{ MPa}$$

$$= 2776.2 \text{ kJ/kg}.$$

Substituting into Eq. 3, we have

$$\dot{m}_{\text{H}_2\text{O}} = \frac{25{,}000 \times 1.05(500 - 150)}{2776.2 - 419.06} \text{ kg/h}$$

$$= 3897.7 \text{ kg/h}.$$

(b) By definition, thermodynamic lost power is given by

$$L\dot{W}_0 = T_0 \dot{S}_{\text{gen}}. \tag{4}$$

From the second law, we have

$$\dot{S}_{\text{gen}} = \sum_{\text{out}} \dot{m}s - \sum_{\text{in}} \dot{m}s$$

$$= \dot{m}_{\text{gas}}(s_2 - s_1) + \dot{m}_{\text{H}_2\text{O}}(s_4 - s_3). \tag{5}$$

On the basis of ideal gas with constant specific heats,

$$s_2 - s_1 = c_p \ln \frac{T_2}{T_1} - R \ln \frac{p_2}{p_1}.$$

Neglecting pressure drops yields

$$s_2 - s_1 = c_p \ln \frac{T_2}{T_1}$$

$$= 1.05 \ln \frac{423.15}{773.15} \text{ kJ/kg} \cdot \text{K}$$

$$= -0.6329 \text{ kJ/kg} \cdot \text{K}.$$

From the steam tables, we have

$$s_3 \approx s_f \text{ at } 100°C$$

$$= 1.3069 \text{ kJ/ kg} \cdot \text{K}$$

$$s_4 = s_g \text{ at } 1.0 \text{ MPa}$$

$$= 6.5828 \text{ kJ/kg} \cdot \text{K}.$$

Substituting into Eq. 5, we have

$$\dot{S}_{gen} = 25{,}000(-0.6329) + 3897.7(6.5828 - 1.3069) \text{ kJ/K} \cdot \text{h}$$

$$= 4741.4 \text{ kJ/K} \cdot \text{h}.$$

Let temperature of the environment be 300 K. Then

$$L\dot{W}_0 = 300 \times 4741.4 \text{ kJ/h}$$

$$= 1.422 \times 10^6 \text{ kJ/h}.$$ ▮

Comment

The operating conditions of this heat recovery steam generator are quite typical of current applications. We could have reduced the amount of lost power generation by producing steam at a higher pressure. Why?

Example 8.8

Ten lbm/s of steam enters a steady-state steady-flow heat exchanger as saturated vapor at 300°F. Heat transfer is to the atmosphere at 70°F. Determine the entropy creation due to the heat-transfer process if the fluid leaves as saturated liquid at 300°F.

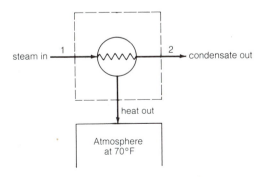

Solution

System selected for study: condenser
Unique features of process: steady-state steady flow; no work transfer
Assumptions: steam is a simple compressible substance; neglect ΔKE; neglect ΔPE

From the first law, we have

$$\dot{Q} = \dot{m}(h_2 - h_1), \tag{1}$$

and from the second law,

$$\dot{S}_{gen} = \dot{m}(s_2 - s_1) + \frac{\dot{Q}^{HR}}{T^{HR}}. \tag{2}$$

From the steam tables, we have

$$h_1 = h_g \text{ at } 300°F = 1179.7 \text{ Btu/lbm}$$

$$h_2 = h_f \text{ at } 300°F = 269.7 \text{ Btu/lbm}$$

$$s_1 = s_g \text{ at } 300°F = 1.6351 \text{ Btu/lbm} \cdot °R$$

$$s_2 = s_f \text{ at } 300°F = 0.4372 \text{ Btu/lbm} \cdot °R.$$

Substituting enthalpy data into Eq. 1, we have

$$\dot{Q} = 10(269.7 - 1179.7) \text{ Btu/s}$$

$$= -9100 \text{ Btu/s}.$$

Therefore,

$$\dot{Q}^{HR} = +9100 \text{ Btu/s}.$$

Substituting data into Eq. 1, we have

$$\dot{S}_{gen} = 10(0.4372 - 1.6351) + \frac{9100}{530} \text{ Btu/°R} \cdot s$$

$$= +5.191 \text{ Btu/°R} \cdot s.$$

Positive entropy generation means that the heat-transfer process is an irreversible one.

Comments

We have mentioned before that entropy generation is an index of measuring the loss of opportunity to do work, that is, of unavailable energy creation in the universe. Since steam is condensing at a constant temperature of 300°F, we may look upon it as a heat reservoir at 300°F. Then, if we operate a Carnot engine between this heat reservoir and the atmosphere at 70°F, the ideal work that we could have obtained from the condensation of the steam is simply given by

$$\dot{W} = \left(1 - \frac{530}{760}\right)(9100) \text{ Btu/s} = 2754 \text{ Btu/s}.$$

Since none of this amount of ideal work is extracted from the steam, it is all "lost." That is, this represents all the unavailable energy created by the condensation process. It may be shown that the unavailable energy is equal to the entropy generation of this process multiplied by the absolute temperature of the available sink at 530°R. In fact, unavailable energy creation is always equal to the product of entropy generation in the universe and the absolute temperature of the available sink.

Example 8.9

We want to cool 50 kg/s of air from 101 psia and 300 K to as low a temperature as possible in a steady-state steady-flow heat exchanger. If nitrogen gas at 101 kPa and 200 K is available, provide the heat-exchanger design engineer with the following information:

(a) the flow rate of the nitrogen gas.
(b) the temperature of the air at the outlet.

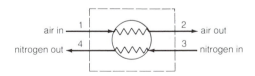

Solution

System selected for study: heat exchanger
Unique features of process: steady-state steady flow; no work transfer
Assumptions: nitrogen is an ideal gas with $c_p = 1.0380$ kJ/kg · K; air is an ideal gas with $c_p = 1.0038$ kJ/kg · K; neglect ΔKE and ΔPE; neglect heat transfer between the heat exchanger and the surroundings; neglect pressure drops

From the first law, we have

$$\sum_{\text{out}} \dot{m}h = \sum_{\text{in}} \dot{m}h$$

or

$$\dot{m}_{\text{air}}(h_1 - h_2) = \dot{m}_{\text{N}}(h_4 - h_3). \tag{1}$$

From the property relation for an ideal gas with constant specific heat,

$$\Delta h = c_p \Delta T.$$

Therefore, Eq. 1 becomes

$$\dot{m}_{\text{air}}(c_p)_{\text{air}}(T_1 - T_2) = \dot{m}_{\text{N}}(c_p)_{\text{N}}(T_4 - T_3). \tag{2}$$

From the second law, we have

$$\dot{S}_{\text{gen}} = \sum_{\text{out}} \dot{m}s - \sum_{\text{in}} \dot{m}s$$

$$= \dot{m}_{\text{air}}(s_2 - s_1) + \dot{m}_{\text{N}}(s_4 - s_3).$$

For the best result, we should have

$$\dot{S}_{\text{gen}} = 0.$$

Therefore, we want

$$\dot{m}_{\text{air}}(s_2 - s_1) + \dot{m}_{\text{N}}(s_4 - s_3) = 0. \tag{3}$$

For an ideal gas with constant specific heat and constant pressure, we have

$$s_2 - s_1 = (c_p)_{\text{air}} \ln \frac{T_2}{T_1}$$

and

$$s_4 - s_3 = (c_p)_{\text{N}} \ln \frac{T_4}{T_3}.$$

Then, Eq. 3 becomes

$$\dot{m}_{\text{air}}(c_p)_{\text{air}} \ln \frac{T_2}{T_1} + \dot{m}_{\text{N}}(c_p)_{\text{N}} \ln \frac{T_4}{T_3} = 0. \tag{4}$$

We now have established two equations (Eqs. 2 and 4) with three unknowns: $\dot{m}_N$, T_2, and T_4.

In order to get the lowest temperature of T_2, we should get the highest possible temperature for T_4, which can only be 300 K. Now that we have identified what T_4 should be, Eqs. 2 and 4 only involve two unknowns, $\dot{m}_N$ and T_2. Substituting data into Eqs. 2 and 4, we have

$$50 \times 1.0038(300 - T_2) = \dot{m}_N \times 1.0380(300 - 200) \tag{5}$$

$$50 \times 1.0038 \ln \frac{T_2}{300} + \dot{m}_N \times 1.0380 \ln \frac{300}{200} = 0 \tag{6}$$

Solving Eqs. 5 and 6 simultaneously, we find that

$$\dot{m}_N = 50 \times \frac{1.0038}{1.0380} \text{ kg/s} = 48.353 \text{ kg/s}$$

and

$$T_2 = 200 \text{ K.} \qquad\qquad \blacksquare$$

Comments

If $T_2 = 200$ K is specified for the design of this heat exchanger, the size of the heat exchange must be infinitely large, since the size of a heat exchanger is inversely proportional to the temperature difference between the cold stream and the hot stream. In fact, this temperature can only be obtained in a true counter flow heat exchanger. That is, the flowing streams must be parallel to each other but in opposite direction at all times.

Example 8.10

Steam enters a condenser at a pressure of 1.0 psia and 90 percent quality. It leaves the condenser as saturated liquid at 1.0 psia. Cooling water available at 60°F is used to remove heat from the steam. For a flow rate of 750,000 lb/h of steam, determine the flow rate of cooling water

(a) if the cooling water leaves the condenser at 75°F.
(b) if the cooling water leaves the condenser at 85°F.

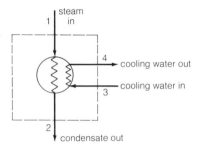

Solution

System selected for study: condenser
Unique features of process: steady-state steady flow; no work transfer
Assumptions: H_2O is a simple compressible substance; neglect ΔKE and ΔPE; neglect heat transfer between heat exchanger and surroundings

From the first law, we have

$$\sum_{\text{out}} \dot{m}h = \sum_{\text{in}} \dot{m}h$$

or

$$\dot{m}_s(h_1 - h_2) = \dot{m}_w(h_4 - h_3). \tag{1}$$

Using data from steam tables, we have

$$h_1 = (h_f + xh_{fg}) \text{ at } 1.0 \text{ psia and } x = 0.90$$

$$= 69.73 + 0.90 \times 1036.1 = 1002.2 \text{ Btu/lbm}$$

$$h_2 = h_f \text{ at } 1.0 \text{ psia} = 69.73 \text{ Btu/lbm}$$

$$h_3 \approx h_f \text{ at } 60°F = 28.06 \text{ Btu/lbm}$$

$$h_4 \approx h_f \text{ at } T_4 = 53.03 \text{ Btu/lbm if } T_4 \text{ is } 85°F$$

$$= 43.05 \text{ Btu/lbm if } T_4 \text{ is } 75°F.$$

Substituting into Eq. 1, we find that

(a) $$\dot{m}_w = \frac{750{,}000(1002.2 - 69.73)}{43.05 - 28.06} \text{ lbm/h} = 4.665 \times 10^7 \text{ lbm/h}$$

(b) $\dot{m}_w = \dfrac{750{,}000(1002.2 - 69.73)}{53.03 - 28.06}$ lbm/h $= 2.801 \times 10^7$ lbm/h.   ∎

Comments

The cooling water required for case (a) is about 1.667 times that for case (b). This would mean larger pipes, a larger water pump, and more pumping power for the cooling-water circuit for case (a). But one should not jump to the conclusion that we definitely should use 85°F as the outlet temperature of the cooling water because the size of the condenser in this case, from heat-transfer consideration, turns out to be about 1.45 that for case (a). The cost of the condenser, and other factors such as the space requirement and the quantity of available cooling water, must be taken into consideration in making the final decision. This is a typical engineering problem in which many trade-offs are made.

8.8 Steady-Flow Open Systems Consisting of More Than One Steady-State Steady-Flow Device

We have studied several problems involving only one steady-flow device in each case. In this section we study a few problems in which two or more devices are combined to perform an engineering task.

Example 8.11

We may make use of the following scheme to produce 350 kPa dry saturated steam and saturated water steadily.

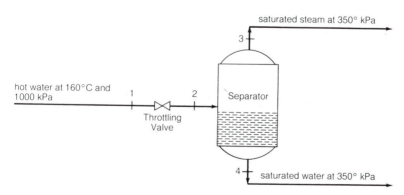

If we want a flow rate of 1000 kg/h of 350-kPa saturated steam, determine

(a) the flow rate of hot water.
(b) the rate of entropy generation for the overall process.

Solution

System selected for study: the entire system
Unique features of process: steady-state steady flow; no work transfer
Assumptions: H_2O is a simple compressible substance; neglect ΔKE and ΔPE; neglect heat transfer to or from system

(a) From the first law, we have

$$0 = \sum_{\text{out}} \dot{m}h - \sum_{\text{in}} \dot{m}h$$

or

$$0 = \dot{m}_3 h_3 + \dot{m}_4 h_4 - \dot{m}_1 h_1. \tag{1}$$

From conservation of mass, we have

$$\dot{m}_1 = \dot{m}_3 + \dot{m}_4. \tag{2}$$

Combining Eqs. 1 and 2 gives us

$$\dot{m}_1 = \frac{\dot{m}_3(h_3 - h_4)}{h_1 - h_4}. \tag{3}$$

From the steam tables, we have

$$h_1 = h \text{ at } 160°C \text{ and } 1000 \text{ kPa} = 675.7 \text{ kJ/kg}$$

$$h_3 = h_g \text{ at } 350 \text{ kPa} = 2731.6 \text{ kJ/kg}$$

$$h_4 = h_f \text{ at } 350 \text{ kPa} = 584.3 \text{ kJ/kg}.$$

Substituting into Eq. 3, we have

$$\dot{m}_1 = \frac{1000(2731.6 - 584.3)}{675.7 - 584.3} \text{ kg/h}$$

$$= 23{,}493.4 \text{ kg/h}$$

(b) From the second law, we have

$$\dot{S}_{\text{gen}} = \sum_{\text{out}} \dot{m}s - \sum_{\text{in}} \dot{m}s$$

$$= \dot{m}_3 s_3 + \dot{m}_4 s_4 - \dot{m}_1 s_1 = \dot{m}_3 s_3 + (\dot{m}_1 - \dot{m}_3)s_4 - \dot{m}_1 s_1. \tag{4}$$

From the steam tables, we have

$$s_1 = s \text{ at } 160°C \text{ and } 1000 \text{ kPa} = 1.9420 \text{ kJ/kg} \cdot K$$

$$s_3 = s_g \text{ at } 350 \text{ kPa} = 6.9392 \text{ kJ/kg} \cdot K$$

$$s_4 = s_f \text{ at } 350 \text{ kPa} = 1.7273 \text{ kJ/kg} \cdot K.$$

Substituting into Eq. 4, we have

$$\dot{S}_{gen} = 1000 \times 6.9392 + (23{,}493.4 - 1000) \times 1.7273$$

$$- 23{,}493.4 \times 1.9420 \text{ kJ/K} \cdot h$$

$$= 167.8 \text{ kJ/K} \cdot h.\qquad\blacksquare$$

Comments

The reader should check that the entire amount of entropy generation is the consequence of the throttling process.

Example 8.12

We may make use of the following scheme to produce high-velocity air. Air enters the compressor at 20°C and 100 kPa steadily and leaves with a pressure of 500 kPa. Air leaves the nozzle at 100 kPa. Assume that air is an ideal gas with constant specific heats of $c_p = 1.0038$ kJ/kg $\cdot$ K and $c_v = 0.7168$ kJ/kg $\cdot$ K. If we have a thermodynamically perfect nozzle but an imperfect compressor that has an adiabatic compressor efficiency of 85%, determine

(a) the temperature of air leaving the compressor.
(b) the velocity of air leaving the nozzle.

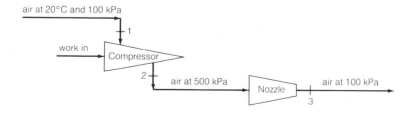

air at 20°C and 100 kPa

work in

Compressor

1

2

air at 500 kPa

Nozzle

air at 100 kPa

3

Solution

(a) *System selected for study:* compressor

Unique feature of process: steady-state steady flow
Assumptions: air is an ideal gas with constant specific heats; neglect ΔKE and ΔPE; neglect heat transfer to or from the compressor

From the definition of adiabatic compressor efficiency, we have

$$\eta_C = \frac{h_{2s} - h_1}{h_2 - h_1}. \tag{1}$$

For an ideal gas with constant specific heats, Eq. 1 becomes

$$\eta_C = \frac{T_{2s} - T_1}{T_2 - T_1}. \tag{2}$$

Thus

$$T_2 = T_1 + \frac{T_{2s} - T_1}{\eta_C}$$

$$= T_1 + \frac{T_1}{\eta_C}\left(\frac{T_{2s}}{T_1} - 1\right). \tag{3}$$

Making use of Eq. 5.35 for an ideal gas with constant specific heats, we obtain

$$\frac{T_{2s}}{T_1} = \left(\frac{p_2}{p_1}\right)^{(k-1)/k}.$$

Then Eq. 3 becomes

$$T_2 = T_1 + \frac{T_1}{\eta_C}\left[\left(\frac{p_2}{p_1}\right)^{(k-1)/k} - 1\right]$$

$$= 293.15 + \frac{293.15}{0.85}\left[\left(\frac{500}{100}\right)^{(1.4)-1/1.4} - 1\right] K$$

$$= 494.49 \text{ K.}$$

(b) *System selected for study:* nozzle

Unique features of process: steady-state steady flow; no work transfer

  Assumptions: air is an ideal gas with constant specific heats; neglect ΔPE; neglect KE at nozzle inlet; neglect heat transfer to or from the nozzle

From the first law, we have

$$0 = dh + d\left(\frac{\bar{V}^2}{2g_c}\right). \tag{1}$$

Since the nozzle is thermodynamically perfect, the process must be reversible and adiabatic or isentropic. That is,

$$ds = 0 \tag{2}$$

and dh in Eq. 1 must be evaluated along a constant-entropy line. Integrating Eq. 1 and neglecting velocity at the nozzle inlet, we have

$$\bar{V}_3 = \sqrt{2g_c(h_2 - h_{3s})}. \tag{3}$$

For an ideal gas with constant specific heats,

$$h_2 - h_{3s} = c_p(T_2 - T_{3s}) = c_p T_2\left(1 - \frac{T_{3s}}{T_2}\right). \tag{4}$$

Making use of Eq. 5.35 again gives us

$$\frac{T_{3s}}{T_2} = \left(\frac{p_3}{p_2}\right)^{(k-1)/k}$$

$$h_2 - h_{3s} = c_p T_2\left[1 - \left(\frac{p_3}{p_2}\right)^{(k-1)/k}\right]$$

$$= 1.0038 \times 494.49\left[1 - \left(\frac{100}{500}\right)^{(1.4-1)/1.4}\right] \text{kJ/kg}$$

$$= 182.96 \text{ kJ/kg} = 182{,}960 \text{ J/kg}$$

Substituting into Eq. 3, we have

$$V_3 = \sqrt{2 \times 1.0 \times 182{,}960} \text{ m/s}$$

$$= 605 \text{ m/s.} \qquad\blacksquare$$

Comments

It may be shown that with a velocity of 605 m/s at the nozzle outlet, we would have a supersonic nozzle. That is, 605 m/s is greater than the velocity of sound, corresponding to the temperature at nozzle outlet. Nozzles and diffusers are studied in Chapter 9.

Example 8.13

Air at 300 K and 100 kPa is to be compressed steadily to a final pressure of 20,000 kPa in a two-stage arrangement with intercooling as shown.

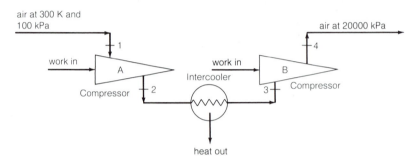

The design specifications include the following:

1. The temperature at the inlet of compressor B is also 300 K.
2. The pressure at the inlet of compressor B is the same as the pressure at the outlet of compressor A. That is, neglect pressure drops between the two compressors.

Assuming ideal gas behavior with constant specific heats of $c_p = 1.0038$ kJ/ kg · K and $c_v = 0.7168$ kJ/kg · K, and the same adiabatic compressor efficiency for both compressors, determine the optimum interstage pressure, p_2, for the minimum work needed for the entire system.

Solution

System selected for study: compressor A
Unique feature of process: steady-state steady flow
Assumptions: air is an ideal gas with constant specific heats; neglect ΔKE and ΔPE; neglect heat transfer to or from the compressor

From the first law, we have

$$w_A = h_1 - h_2. \tag{1}$$

For an ideal gas with constant specific heats,

$$h_1 - h_2 = c_p(T_1 - T_2), \tag{2}$$

and the adiabatic compressor efficiency is given as

$$\eta_C = \frac{T_{2s} - T_1}{T_2 - T_1}. \tag{3}$$

Combining Eqs. 1, 2, and 3, we have

$$w_A = \frac{c_p T_1}{\eta_C}\left(1 - \frac{T_{2s}}{T_1}\right). \tag{4}$$

Now

$$\frac{T_{2s}}{T_1} = \left(\frac{p_2}{p_1}\right)^{(k-1)/k}.$$

Thus

$$w_A = \frac{c_p T_1}{\eta_C}\left[1 - \left(\frac{p_2}{p_1}\right)^{(k-1)/k}\right]. \tag{5}$$

In a similar manner, we can obtain the work for compressor B as

$$w_B = \frac{c_p T_1}{\eta_C}\left[1 - \left(\frac{p_4}{p_2}\right)^{(k-1)/k}\right]. \tag{6}$$

In Eq. 6 we have applied the design specifications of $T_3 = T_1$ and $p_3 = p_2$. The total work needed for the entire system is then given by

$$w_{\text{total}} = w_A + w_B. \tag{7}$$

To get the optimum interstage pressure for the minimum work needed for the entire system, we must satisfy the following condition:

$$\frac{dw_{\text{total}}}{dp_2} = 0.$$

The solution of Eq. 7 would yield

$$(p_2)_{opt} = \sqrt{p_1 p_4} \tag{8}$$

$$= \sqrt{100 \times 20{,}000} \text{ kPa}$$

$$= 1414 \text{ kPa.} \qquad \blacksquare$$

Comment

The derivation of Eq. 8 may be found in books on fluid machinery.

Example 8.14

Water for irrigation is to be pumped from a lake to an elevation of 300 m as shown.

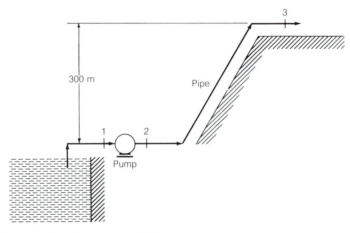

The design specifications are as follows:

1. The flow rate of water through the pump is 6.5 m^3/s.
2. The adiabatic pump efficiency is 75%.
3. Friction in the pipe may be neglected.

Determine

(a) the pressure of water at the pump discharge.
(b) the power, in kilowatts, needed to drive the pump.

Solution

(a) *System selected for study:* the entire section of pipe between pump discharge and outlet at the pipe

Unique features of process: steady-state steady flow; frictionless; no work transfer

Assumptions: water is incompressible; neglect ΔKE

From the first law, we have

$$\bar{d}q = dh + d\left(\frac{g}{g_c} Z\right). \tag{1}$$

From the second law for a reversible process,

$$\bar{d}q = T\ ds. \tag{2}$$

From the property relation,

$$T\ ds = dh - v\ dp. \tag{3}$$

Combining Eqs. 1, 2, and 3, we have

$$v\ dp = -d\left(\frac{g}{g_c} Z\right). \tag{4}$$

If water is incompressible, we can integrate Eq. 4 to give

$$v(p_3 - p_2) = -\left(\frac{g}{g_c}\right)(Z_3 - Z_2)$$

or

$$p_2 = p_3 + \frac{g\rho}{g_c}(Z_3 - Z_2), \tag{5}$$

where ρ is the density of water. Using $p_3 = 101.325$ kPa,

$$g = 9.81 \text{ m/s}^2,$$

and $\rho = 1000$ kg/m^3, we obtain

$$p_2 = 101{,}325 + \frac{9.81 \times 1000}{1.0}(300) \text{ Pa}$$

$$= 3{,}044{,}000 \text{ Pa} = 3044 \text{ kPa}.$$

(b) *System selected for study:* pump
   Unique feature of process: steady-state steady flow

Assumptions: water is incompressible; neglect ΔKE and ΔPE; neglect heat transfer to or from the pump

The pump work is given by

$$w_{act} = \frac{w_{rev}}{\eta_p}, \tag{6}$$

where

$$w_{rev} = -\int_1^2 v \, dp.$$

For an incompressible fluid,

$$w_{rev} \approx -v(p_2 - p_1).$$

Thus

$$w_{act} = -\frac{v(p_2 - p_1)}{\eta_p}$$

and

$$\dot{W}_{act} = -\frac{\dot{m}v(p_2 - p_1)}{\eta_p} = -\frac{\dot{V}(p_2 - p_1)}{\eta_p}. \tag{7}$$

Substituting data into Eq. 7, we have

$$\dot{W}_{act} = -\frac{6.5(3044 - 101.325)}{0.75} \text{ kJ/s}$$

$$= -25{,}500 \text{ kJ/s} = -25{,}500 \text{ kW.} \qquad \blacksquare$$

Comments

If we include the kinetic energy term in the first-law equation, then Eq. 4 may be written as

$$d\left(\frac{\bar{V}^2}{2g_c}\right) + d\left(\frac{g}{g_c} Z\right) + \frac{dp}{\rho} = 0, \tag{8}$$

which is useful for the study of flow of fluid in pipe with no friction involving either compressible or incompressible fluids. For the case of incompressible fluids, Eq. 8 becomes

$$\frac{V_2^2}{2g_c} + \frac{g}{g_c} Z_2 + \frac{p_2}{\rho} = \frac{\bar{V}_1^2}{2g_c} + \frac{g}{g_c} Z_1 + \frac{p_1}{\rho} = \text{constant}, \qquad (9)$$

which is the famous Bernoulli equation for the flow of incompressible fluid under steady-state steady-flow conditions with no friction. This equation is discussed further in Chapter 9.

Example 8.15

Figure 8.2 is a schematic diagram of a geothermal steam power plant. If such a power plant is to be designed according to the following specifications:

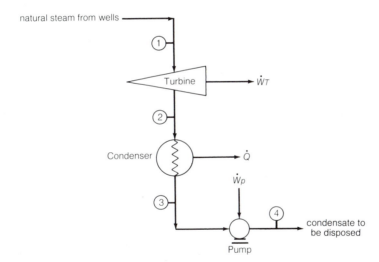

Figure 8.2 Schematic Diagram of a Geothermal Steam Power Plant

Natural steam is available (such as at the Geysers in northern California) at 100 psia and 400°F

Turbine output desired	= 15,000 kW
Steam pressure at turbine outlet =	2.0 psia
Adiabatic turbine efficiency =	70%
Discharge pressure at pump =	14.7 psia
Adiabatic pump efficiency =	60%

Determine the following design data for such a plant:

(a) steam-flow rate, in lbm/h.
(b) heat-transfer duty for condenser, in Btu/h.
(c) power required for pump, in kW.

Solution

(a) *System selected for study:* turbine
   Unique feature of process: steady-state steady flow
   Assumptions: natural steam is a simple compressible substance; neglect heat transfer; neglect ΔKE and ΔPE

From the first law, we have

$$\dot{m} = \frac{\dot{W}}{h_1 - h_2} \tag{1}$$

From the definition of adiabatic turbine efficiency, we have

$$\eta_T = \frac{h_1 - h_2}{h_1 - h_{2s}}. \tag{2}$$

Combining Eqs. 1 and 2, we find that

$$\dot{m} = \frac{\dot{W}}{\eta_T(h_1 - h_{2s})}. \tag{3}$$

From the property relations for steam, we have

$$h_1 = 1227.4 \text{ Btu/lbm}$$

$$s_1 = 1.6516 \text{ Btu/lbm} \cdot {}^\circ\text{R}$$

$$s_{2s} = s_1 = 1.6516 \text{ Btu/lbm} \cdot {}^\circ\text{R}.$$

At 2.0 psia,

$$s_f = 0.1750 \text{ Btu/lbm} \cdot {}^\circ\text{R} < 1.6516$$

$$s_g = 1.9200 \text{ Btu/lbm} \cdot {}^\circ\text{R} > 1.6516.$$

Therefore,

$$s_{2s} = (s_f + x_{2s} s_{fg}) \text{ at 2.0 psia}$$

and

$$1.6516 = 0.1750 + x_{2s}(1.9200 - 0.1750)$$

yielding

$$x_{2s} = 0.846$$

$$h_{2s} = (h_f + x_{2s} h_{fg}) \text{ at 2.0 psia}$$

$$= 94.03 + 0.846 \times 1022.1 \text{ Btu/lbm} = 958.7 \text{ Btu/lbm}.$$

Substituting into Eq. 3, we find that

$$\dot{m} = \frac{15,000 \times 3413}{0.7 \times (1227.4 - 958.7)} \text{ lbm/h}$$

$$= 272,200 \text{ lbm/h}.$$

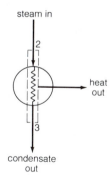

steam in

2

heat
out

3

condensate
out

(b) *System selected for study:* steam side of condenser
   Unique features of process: steady-state steady flow; no work transfer
   Assumptions: neglect ΔKE and ΔPE

From the first law, we have

$$\dot{Q} = \dot{m}(h_3 - h_2). \tag{4}$$

From Eq. 2, we get

$$h_2 = h_1 - \eta_T(h_1 - h_{2s})$$

$$= 1227.4 - 0.7(1227.4 - 958.7)$$

$$= 1039.3 \text{ Btu/lbm.}$$

Assuming that the condensate leaves as saturated liquid at 2.0 psia,

$$h_3 = h_f \text{ at 2.0 psia} = 94.03 \text{ Btu/lbm.}$$

Substituting into Eq. 4 we find that

$$\dot{Q} = 272,200(94.03 - 1039.3) \text{ Btu/h}$$

$$= -2.573 \times 10^8 \text{ Btu/h.}$$

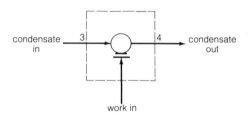

(c) *System selected for study:* pump
   Unique feature of process: steady-state steady flow
   Assumptions: neglect ΔKE and ΔPE; neglect heat transfer

From the definition of adiabatic pump efficiency,

$$w_{act} = \frac{w_{rev}}{\eta_p}. \tag{5}$$

where w_{rev} may be evaluated from

$$w_{rev} = -\int_3^4 v \, dp$$

$$\approx -v(p_4 - p_3)$$

neglecting the compressibility effect. From the property relations for water, we have

$$v = v_f \text{ at } 2.0 \text{ psia} = 0.01623 \text{ ft}^3/\text{lbm}.$$

Therefore,

$$w_{rev} = -\frac{0.01623(14.7 - 2.0)144}{778} \text{ Btu/lbm} = -0.038 \text{ Btu/lbm}$$

and

$$w_{act} = -\frac{0.038}{0.6} \text{ Btu/lbm} = -0.063 \text{ Btu/lbm}$$

$$\dot{W}_p = \dot{m} w_{act} = \frac{272,200(-0.063)}{3413} \text{ kW} = -5.02 \text{ kW}. \qquad \blacksquare$$

Comments

The condensate of natural steam usually contains some chemicals, such as boron and ammonia, which would pollute local streams if released into them. Consequently, a more acceptable method of disposing of them is to reinject them into the ground through deep wells.

Example 8.16

Suppose that a decision has been made to produce air at 20 psia and 200°R at the rate of 50 lbm/s by making use of (1) an air supply at 100 psia and 80°F, and (2) a heat sink at 139°R (such as a supply of liquid nitrogen). The following scheme has been proposed:

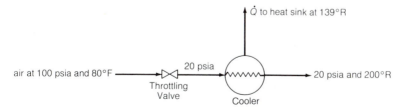

Determine for this scheme

(a) heat-transfer duty for the cooler.
(b) entropy generation for the entire process.

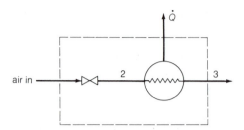

Solution

(a) *System selected for study:* valve and cooler as shown

Unique features of process: steady-state steady flow; no work transfer

Assumptions: air is an ideal gas; neglect ΔKE and ΔPE; neglect heat transfer from the valve and piping

From the first law, we have

$$\dot{Q} = \dot{m}(h_3 - h_1). \tag{1}$$

Using data from the gas tables, we have

$$h_1 = h \text{ at } 80°\text{F} = 129.06 \text{ Btu/lbm}$$

$$h_3 = h \text{ at } 200°\text{R} = 47.67 \text{ Btu/lbm}.$$

Substituting into Eq. 1, we find that

$$\dot{Q} = 50(47.67 - 129.06) \text{ Btu/s}$$

$$= -4069.5 \text{ Btu/s}.$$

(b) From the second law, we have

$$\dot{S}_{\text{gen}} = \dot{m}(s_3 - s_1) + \frac{\dot{Q}^{\text{HR}}}{T^{\text{HR}}}. \tag{2}$$

Using data from the gas tables, we have

$$s_3 = s_1 = \phi_3 - \phi_1 - R \ln \frac{p_3}{p_1}$$

$$= 0.36303 - 0.60078 - \frac{53.34}{778} \ln \frac{20}{100} \text{ Btu/lbm} \cdot °\text{R}$$

$$= -0.1274 \text{ Btu/lbm} \cdot °\text{R}.$$

Substituting into Eq. 2 we find that

$$\dot{S}_{gen} = 50(-0.1274) + \frac{4069.5}{139} \text{ Btu/}^{\circ}\text{R} \cdot \text{s}$$

$$= +22.91 \text{ Btu/}^{\circ}\text{R} \cdot \text{s}. \qquad \blacksquare$$

Example 8.17

The task to be performed is the same as in Example 8.16, but the following scheme is proposed instead:

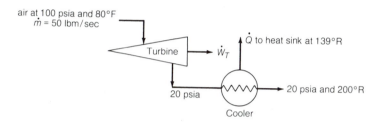

For an adiabatic turbine efficiency of 80%, determine for this scheme

(a) turbine-power output, in kW.
(b) heat-transfer duty for the cooler.
(c) entropy generation for the entire process.

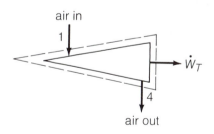

Solution

(a) *System selected for study:* turbine
  Unique feature of process: steady-state steady flow
  Assumptions: neglect ΔKE and ΔPE; neglect heat transfer; air is an ideal gas

From the first law, we have

$$\dot{W}_T = \dot{m}(h_1 - h_4), \qquad (1)$$

and from the definition of adiabatic turbine efficiency,

$$h_1 - h_4 = \eta_T(h_1 - h_{4s}). \tag{2}$$

Using data from the gas tables, we have

$$h_1 = 129.06 \text{ Btu/lbm}$$

$$P_{r4} = P_{r1}\left(\frac{p_4}{p_1}\right)$$

$$= 1.386\left(\frac{20}{100}\right)$$

$$= 0.2772,$$

yielding

$$h_{4s} = 81.33 \text{ Btu/lbm}$$

and

$$T_{4s} = 340.6°\text{R}.$$

Substituting into Eq. 2 we find that

$$h_1 - h_4 = 0.8(129.06 - 81.33) \text{ Btu/lbm}$$

$$= 38.18 \text{ Btu/lbm}.$$

Substituting into Eq. 1, we find that

$$\dot{W}_T = \frac{50 \times 38.18 \times 3600}{3413} \text{ kW} = 2014 \text{ kW}.$$

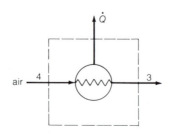

(b) *System selected for study:* cooler
   Unique features of process: steady-state steady flow; no work transfer
   Assumptions: neglect ΔKE and ΔPE; air is an ideal gas

From the first law, we have

$$\dot{Q} = \dot{m}(h_3 - h_4). \tag{3}$$

From Eq. 2, we have

$$h_4 = h_1 - \eta_T(h_1 - h_{4s})$$

$$= 129.06 - 0.8(129.06 - 81.33) \text{ Btu/lbm}$$

$$= 90.88 \text{ Btu/lbm}.$$

From the gas tables, $h_3 = 47.67$ Btu/lbm. Substituting into Eq. 3, we find that

$$\dot{Q} = 50(47.67 - 90.88) \text{ Btu/s}$$

$$= -2160.5 \text{ Btu/s}.$$

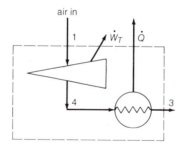

(c) *System selected for study:* turbine and cooler as shown
   Unique feature of process: steady-state steady flow
   Assumptions: neglect heat transfer from turbine and piping; air is an ideal gas

From the second law, we have

$$\dot{S}_{gen} = \Delta\dot{S} + \frac{\dot{Q}^{HR}}{T^{HR}}$$

$$= \dot{m}(s_3 - s_1) + \frac{\dot{Q}^{HR}}{T^{HR}}. \tag{4}$$

where $s_3 - s_1 = -0.1274$ Btu/lbm $\cdot$ °R, the same as in part (b) of Example 8.16. For this example,

$$\dot{Q}^{HR} + -\dot{Q} = 2160.5 \text{ Btu/s}.$$

Substituting into Eq. 4, we find that

$$\dot{S}_{gen} = 50(-0.1274) + \frac{2160.5}{139} \text{ Btu/}^\circ\text{R} \cdot \text{s}$$

$$= +9.17 \text{ Btu/}^\circ\text{R} \cdot \text{s.} \qquad \blacksquare$$

Comments

The results from Examples 8.16 and 8.17 are summarized as follows:

Scheme Used	Power Output (kW)	Heat-Transfer Duty for Cooler (Btu/s)	Entropy Generation (Btu/°R · s)
Example 8.16: pressure reduction done with a throttling valve	None	−4069.5	22.91
Example 8.17: pressure reduction done with a turbine	2014	−2160.5	9.17

We see that the scheme used in Example 8.17 is obviously superior to the scheme used in Example 8.16 from the point of view of thermodynamics and effective utilization of our energy resources. However, if simplicity is the criterion, the scheme used in Example 8.16 would be better.

8.9 Open Systems with Non-Steady-State Steady Flow

We have studied many processes for open systems modeled as steady-state steady flow. However, not all flow processes may be treated as such. Specifically, we must treat a process as non-steady-state steady flow when any one of the necessary conditions for a process to be steady-state steady flow is not satisfied. Some examples of non-steady-state steady flow are:

1. Discharging of air from a storage tank to operate pneumatic tools.
2. Filling of an oxygen gas bottle from a large reservoir of oxygen.
3. Leaking of air into an evacuated chamber.
4. Starting up of a steam turbine.

The first three examples are similar in nature; let us examine the behavior of just the first one. Taking the storage tank as our system, we have air leaving but no air coming in. Consequently, the content of mass, energy, and entropy in the tank will be changing with time. These are indeed the unique characteristics of non-steady-state steady-flow problems in general. For example 4, taking the turbine as our system, the content of mass, energy, and entropy within the control volume will also be changing with time in general. However, we could have the same amount of steam flowing in and out and the problem would still be nonsteady flow, since the state at any point of the turbine still could change with time.

Example 8.18

. A rigid vessel contains 100 kg of dry saturated steam at 300 kPa initially. Water at 25°C is then injected into the vessel until the fluid inside the vessel is saturated water at 40°C. Determine

(a) the amount of water injection needed.
(b) the amount of heat transfer involved.

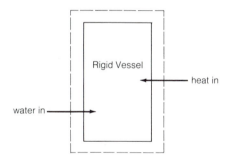

Solution

System selected for study: rigid vessel
Unique features of process: non-steady-state steady flow; no work transfer
Assumptions: H_2O is a simple compressible substance; neglect ΔKE and ΔPE

(a) Since the vessel is rigid,

$$V = m_1 v_1 = m_2 v_2$$

and

$$m_2 = \frac{m_1 v_1}{v_2}.$$

From the steam tables, we have

$$v_1 = v_g \text{ at 300 kPa}$$

$$= 0.6056 \text{ m}^3/\text{kg}$$

$$v_2 = v_f \text{ at } 40°C$$

$$= 0.0010078 \text{ m}^3/\text{kg}.$$

Thus

$$m_2 = \frac{100 \times 0.6056}{0.0010078} \text{ kg}$$

$$= 60,090 \text{ kg}.$$

From mass balance,

water injection needed $= m_2 - m_1 = 60,090 - 100 \text{ kg} = 59,990 \text{ kg}.$

(b) From the first law for non-steady-state steady flow, we have

$$Q_{cv} = m_2 u_2 - m_1 u_1 - \left[\int h \, dm\right]_{in}. \tag{1}$$

Since water is injected at a constant temperature of 25°C,

$$\left[\int h \, dm\right]_{in} = h_{in} m_{in}. \tag{2}$$

From mass balance,

$$m_{in} = m_2 - m_1. \tag{3}$$

Combining Eqs. 1, 2, and 3, we have

$$Q_{cv} = m_2 u_2 - m_1 u_1 - (m_2 - m_1)h_{in}. \tag{4}$$

From the steam tables,

$$u_1 = u_g \text{ at } 300 \text{ kPa}$$

$$= 2543.0 \text{ kJ/kg}$$

$$u_2 = u_f \text{ at } 40°C$$

$$= (h_f - pv_f) \text{ at } 40°C$$

$$= 167.45 - 7.375 \times 0.0010078 \text{ kJ/kg}$$

$$= 167.44 \text{ kJ/kg}$$

$$h_{in} = h_f \text{ at } 25°C$$

$$= 104.77 \text{ kJ/kg}.$$

Substituting into Eq. 4 yields

$$Q_{cv} = 60,090 \times 167.44 - 100 \times 2543.0 - 59,990 \times 104.77 \text{ kJ}$$

$$= 3.522 \times 10^6 \text{ kJ}.$$

The positive sign for Q_{cv} means that heat must be added. ∎

Comments

Let us make a quick check on the order of magnitude of heat addition. The 59,990 kg of water injected into the vessel changes the temperature from 25°C to 40°C. Using a c_p of 4.2 kJ/kg · K for water, this means that the energy we need to do this is $59,990 \times 4.2 \times (40 - 15)$ kJ or 3,780,000 kJ. Now each kilogram of steam inside the vessel gives up about 2200 kJ in becoming water. Thus heat to be added externally is about $(3,780,000 - 1000 \times 2200)$ kJ or 3.560×10^6 kJ, which is quite close to the correct answer.

Example 8.19

Suppose that the decision has been made to store liquid nitrogen at a pressure of 14.7 psia ($T = 139.224°R$). In order to maintain the liquid inside the storage vessel at 14.7 psia, we must provide a line to vent off the vapor formed due to heat leak. If the amount vented from a spherical vessel of 5.0-ft inside diameter is 600 lb in 100 h, determine the amount of heat leaked into the fluid in 100 h. Consider the vessel to be filled initially with saturated-liquid nitrogen.

Vent

Q

Solution

System selected for study: storage vessel
Unique features of process: non-steady-state steady flow; no work transfer
Assumptions: neglect ΔKE and ΔPE

From the first law for non-steady-state steady flow,

$$Q = \left[\int h\ dm\right]_{\text{out}} + U_2 - U_1$$

$$= \left[\int h\ dm\right]_{\text{out}} + m_2 u_2 + m_1 u_1, \tag{1}$$

where

m_1 = total mass of fluid inside the vessel initially

m_2 = total mass of fluid inside the vessel finally

u_1 = specific internal energy of all fluid inside the vessel initially

u_2 = specific internal energy of all fluid inside the vessel finally.

Assuming that the fluid leaving the vessel at any instant is saturated vapor at 14.7 psia, we have

$$\left[\int h\ dm\right]_{\text{out}} = m_{\text{out}}\, h_g. \tag{2}$$

Combining Eqs. 1 and 2, we get

$$Q = m_{\text{out}}\, h_g + m_2 u_2 - m_1 u_1, \tag{3}$$

where

$$m_{\text{out}} = 600 \text{ lbm, total mass vented}$$

$$h_g = \text{enthalpy of saturated vapor at 14.7 psia}$$

$$= 33.218 \text{ Btu/lbm.}$$

From mass balance, we have

$$m_{\text{out}} = m_1 - m_2, \tag{4}$$

and from property relations,

$$m_1 = \frac{V}{v_f} = \frac{\frac{1}{6}\pi(5)^3}{0.01981} \text{ lbm}$$

$$= 3303.9 \text{ lbm.}$$

Therefore,

$$m_2 = 3303.9 - 600 = 2703.9 \text{ lbm}$$

and

$$v_2 = \text{specific volume of all fluid inside the vessel finally}$$

$$= \frac{V}{m_2} = \frac{\frac{1}{6}\pi(5)^3}{2703.9} \text{ ft}^3/\text{lbm} = 0.02421 \text{ ft}^3/\text{lbm.}$$

Since $v_2 < v_g$ at 14.7 psia and $v_2 > v_f$ at 14.7 psia, we have a liquid–vapor mixture inside the vessel finally. For a liquid–vapor mixture, we have

$$v = v_f + xv_{fg}$$

$$u = u_f + xu_{fg}.$$

Thus

$$0.02421 = 0.01981 + x_2(3.4734 - 0.01981),$$

yielding

$$x_2 = 0.001274.$$

Using data from the nitrogen tables, we have, at 14.7 psia,

$$u_f = -52.297 \text{ Btu/lbm}$$

$$u_g = 23.767 \text{ Btu/lbm}.$$

Therefore,

$$u_2 = -52.297 + 0.001274(23.767 + 52.297)$$

$$= -52.200 \text{ Btu/lbm}$$

$$u_1 = u_f \text{ at 14.7 psia} = -52.297 \text{ Btu/lbm}.$$

Substituting into Eq. 3, we find that

$$Q = 600 \times 33.218 + 2703.9(-52.200) - 3303.9(-52.297)$$

$$= 51{,}570 \text{ Btu.} \qquad \blacksquare$$

Comments

The average rate of heat leak into the fluid for our problem is about 516 Btu/h. If the surrounding of the vessel is the atmosphere at 70°F, this heat-transfer rate is quite low, requiring a fairly sophisticated scheme of insulation. The study of insulation schemes for very low temperature vessels is an important topic in cryogenic engineering.[2]

Example 8.20

The scheme shown here is a power-producing system. Nitrogen gas inside the tank is at 5000 kPa and 20°C before the turbine is turned on. Pressure at turbine inlet is maintained at 505 kPa by a pressure regulator which operates as a throttling device. The nitrogen discharges from the turbine at 101 kPa.

[2] See Randall Barron, *Cryogenic Systems*, McGraw-Hill Book Company, New York, 1966, Chap. 7.

Determine the size of the tank for a turbine output of 36,000 kJ. Make calculations on the following basis:

1. The turbine operates adiabatically and has an adiabatic turbine efficiency of 100%.
2. Gas in the tank remains at a constant temperature of 20°C.
3. Power production ceases when the tank pressure reaches 505 kPa.
4. Assume ideal-gas behavior with constant specific heats of $c_p = 1.0380$ kJ/kg · K and $c_v = 0.7412$ kJ/kg · K.

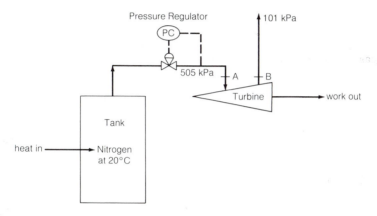

Solution

Let us first apply mass balance to the tank. The amount of gas left inside the tank during the operation is given by

$$m_e = m_1 - m_2. \tag{1}$$

Since the volume of the tank is a constant,

$$V = m_1 v_1 = m_2 v_2. \tag{2}$$

For ideal-gas behavior,

$$v_1 = \frac{RT_1}{p_1} \tag{3}$$

$$v_2 = \frac{RT_2}{p_2}. \tag{4}$$

Now

$$T_1 = T_2. \tag{5}$$

Combining Eqs. 1, 2, 3, 4, and 5, we have

$$V = \frac{m_e R T_1}{p_1 - p_2}, \tag{6}$$

in which the only unknown is m_e. Let us now select the turbine as our control volume, which may be considered as operating under steady-state steady-flow conditions. Neglecting changes in ΔKE and ΔPE, we have from the first law for the turbine,

$$W_T = m_e(h_A - h_B). \tag{7}$$

For ideal gas with constant specific heats,

$$h_A - h_B = c_p(T_A - T_B) \tag{8}$$

$$= c_p T_A \left(1 - \frac{T_B}{T_A}\right). \tag{9}$$

For a reversible adiabatic expansion,

$$\frac{T_B}{T_A} = \left(\frac{p_B}{p_A}\right)^{(k-1)/k}. \tag{10}$$

Combining Eqs. 7, 8, 9, and 10, we have

$$m_e = \frac{W_T}{c_p T_A[1 - (p_B/p_A)^{(k-1)/k}]}. \tag{11}$$

Substituting into Eq. 11, we have

$$m_e = \frac{36{,}000}{1.0380 \times 293.15[1 - (101/505)^{(1.4-1)/1.4}]} \text{ kg}$$

$$= 321 \text{ kg}.$$

Substituting into Eq. 6, we have

$$V = \frac{(321)(8.3143/28.013)(293.15)}{5000 - 505} \text{ m}^3$$

$$= 6.21 \text{ m}^3.$$

Comment

We could have solved this problem by applying the first law to the entire system and to the tank alone. Each equation would involve the same two unknowns: m_e and Q for the tank. Then solving these two equations simultaneously would yield m_e, which we need to obtain V.

Example 8.21

A rigid tank containing air at 14.7 psia and 70°F initially is connected to a supply main that furnishes air at the steady conditions of 100 psia and 70°F. The tank is initially isolated from the main by a closed valve. If a filling process is carried out until the pressure in the tank reaches 30 psia, determine, for a tank volume of 1.0 ft^3, the equilibrium temperature of the air inside the tank at the end of the filling process. Consider air to be an ideal gas with $c_v = 0.171$ Btu/lbm · °R and $c_p = 0.240$ Btu/lbm · °R. Assume that the filling is done adiabatically.

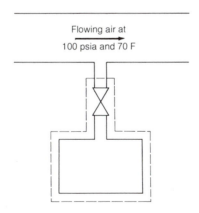

Flowing air at
100 psia and 70 F

Solution

System selected for study: tank and valve as shown
Unique feature of process: non-steady-state steady flow
Assumptions: adiabatic; neglect ΔKE and ΔPE; air is an ideal gas with constant specific heats; neglect the heat capacity of the tank and valve

From the first law (Eq. 2.29), we have

$$0 = -\left(\int h \, dm\right)_{in} + m_2 u_2 - m_1 u_1. \tag{1}$$

Since the air supply is at steady conditions, we have

$$\left(\int h \, dm \right)_{\text{in}} = m_{\text{in}} \, h_{\text{in}}. \tag{2}$$

Combining Eqs. 1 and 2, we find that

$$0 = -m_{\text{in}} \, h_{\text{in}} + m_2 \, u_2 - m_1 \, u_1, \tag{3}$$

where

m_{in} = total mass of air flow into tank from the main

m_1 = total mass of air inside tank initially

m_2 = total mass of air inside tank finally

h_{in} = specific enthalpy of air at the conditions of supply main

u_1 = specific internal energy of air inside tank initially

u_2 = specific internal energy of air inside tank finally.

From the mass balance, we have

$$m_{\text{in}} = m_2 - m_1. \tag{4}$$

Combining Eqs. 3 and 4, we find that

$$m_2(u_2 - h_{\text{in}}) - m_1(u_1 - h_{\text{in}}) = 0. \tag{5}$$

From the definition of enthalpy and the thermal equation of state of an ideal gas, we have

$$h = u + RT.$$

Therefore, Eq. 5 may be written as follows in the case of an ideal gas:

$$m_2(u_2 - u_{\text{in}} - RT_{\text{in}}) - m_1(u_1 - u_{\text{in}} - RT_{\text{in}}) = 0. \tag{6}$$

For an ideal gas with constant specific heats, we have

$$\Delta u = c_v \Delta T.$$

In addition, we have for an ideal gas in general,

$$c_p - c_v = R.$$

Consequently, Eq. 6 may be written as

$$m_2(c_v T_2 - c_p T_{in}) - m_1(c_v T_1 - c_p T_{in}) = 0. \tag{7}$$

From the thermal equation of state of an ideal gas, we have

$$m_2 = \frac{p_2 V_2}{RT_2}$$

$$m_1 = \frac{p_1 V_1}{RT_1}.$$

Substituting into Eq. 7 and simplifying, we have

$$T_2 = \frac{p_2 c_p T_{in}}{c_v(p_2 - p_1) + (p_1 c_p T_{in}/T_1)}, \tag{8}$$

yielding $T_2 = 621°R$ for our problem. ∎

Comments

Equation 8 shows that for an adiabatic filling process, the final temperature is independent of the volume of the tank. Equation 7 shows that for an adiabatic filling process with the tank being evacuated initially, the final temperature in the tank not only is independent of the size of the tank but is also independent of the amount of gas bled into the tank, having a value equal to $(c_p/c_v)T_{in}$. But we must keep in mind that our results are based on neglecting the heat capacity of the tank.

Problems

8.1 Water enters a boiler feed pump as saturated liquid at 220°C. The water ends up in a boiler that has an operating pressure of 20 MPa and is 30 m above the pump inlet. For a mass-flow rate of

2.5×10^6 kg/h and an adiabatic pump efficiency of 80%, determine the power needed, in kilowatts, to drive the pump
(a) assuming that water is incompressible.
(b) using compressed water data.
The process is steady-state steady flow. Heat transfer and changes in kinetic energy may be neglected.

8.2 A fan circulates air steadily in a ventilating system. The following operating data are collected:

Temperature of air at fan inlet $= 20°C$

Pressure of air at fan inlet $= 101.325$ kPa

Pressure difference between outlet and inlet air $= 0.75$ kPa

Volume flow rate $= 900$ m^3/h

Power input $= 0.25$ kW

What is the efficiency of the fan? Changes in kinetic and potential energies may be neglected.

8.3 Oxygen enters a pump as a saturated liquid at 14.7 psia and leaves at a pressure of 500 psia. The pumping process may be considered to be adiabatic. Neglecting change in kinetic and potential energies, determine the horsepower required to drive the pump for a flow rate of 100 lbm/s and an adiabatic pump efficiency of 70 percent.

8.4 For the steady-state steady-flow, reversible adiabatic process involving an ideal gas, show that if specific heats are constant and if changes in kinetic and potential energies are negligible, the work transfer per pound of gas is given by

$$w_{12} = \frac{kRT_1}{k-1}\left[1 - \left(\frac{p_2}{p_1}\right)^{(k-1)/k}\right].$$

8.5 Argon, assumed to be an ideal gas, is compressed adiabatically in a steady-state steady-flow compressor from 101 kPa and 25°C to 505 kPa. If the compressor work required is 175 kJ/kg, show that the compression process is irreversible because the actual work required is more than the reversible work given by the expression in Problem 8.4.

8.6 Air is to be compressed adiabatically in a steady-state steady-flow compressor from 101 kPa and 25°C to 707 kPa. Changes in kinetic and potential energies are negligible. If the actual compressor work required is 270 kJ/kg, show that the process is irreversible because entropy generation for the process is positive. Assume that air is an ideal gas with constant specific heats and $k = 1.4$.

8.7 Air is to be compressed reversibly and adiabatically in a steady-state steady-flow compressor from 15 psia and 80°F to 75 psia. Assuming ideal-gas behavior and neglecting change in kinetic and potential energies, calculate the work required in Btu/lbm and in SI units using
(a) data from Keenan and Kaye's *Gas Tables*.
(b) the expression in Problem 8.4 with $c_p = 0.240$ Btu/lbm · °R and $c_v = 0.171$ Btu/lbm · °R.

8.8 Helium, assumed to be an ideal gas, expands adiabatically in a steady-state steady-flow turbine from 25°C and 505 kPa to 101 kPa. Changes in kinetic and potential energies are negligible. If the acual turbine work is 600 kJ/kg, show that the process is irreversible because the actual turbine work is less than the reversible given by the expression in Problem 8.4.

8.9 A gas turbine operates with an inlet pressure and temperature of 75 psia and 1600°F and an outlet pressure of 15 psia. Assuming ideal-gas behavior and neglecting any change in kinetic and potential energies, calculate the isentropic turbine work in Btu/lbm and in SI units using
(a) data for air from Keenan and Kaye's *Gas Tables*.
(b) the expression in Problem 8.4 with $c_p = 0.275$ Btu/lbm · °R and $c_v = 0.206$ Btu/lbm · °R.

8.10 Oxygen gas is compressed reversibly and isothermally in a steady-state steady-flow compressor from 101 kPa at 25°C to 20 MPa. Changes in kinetic and potential energies are negligible. Determine the work required, in kJ/kg,
(a) using data from the oxygen tables.
(b) assuming ideal-gas behavior.

8.11 Steam is compressed in a steady-state steady-flow compressor from a saturated vapor at 50°C to 100 kPa and 250°C at a rate of 150 kg/s. Heat transferred from the vapor during the process is 2000 kJ/s. Neglecting changes in kinetic and potential energies, calculate the power input, in kilowatts, required to drive the compressor.

8.12 Water is drawn from a lake, pumped up to a city 500 ft above the water level in the lake, and forced through the nozzles of a fire-hose at 70 ft/s. What is the minimum horsepower needed to deliver 1000 lbm/s of water?

8.13 Figure P8.13 shows an integral part of a gas-liquefaction system. If we want 5% of each kilogram of nitrogen gas entering the throttling valve to be liquefied, what is the temperature of the nitrogen upstream of the valve, where the pressure is 20 MPa? Process is steady-state steady flow. Heat transfer and changes in kinetic and potential energies are negligible.

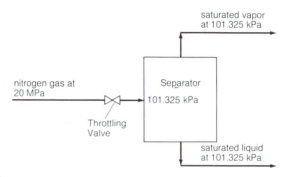

Figure P8.13

8.14 A throttling process is used to produce low-pressure steam steadily at 300 kPa from a high-pressure source. If the high-pressure steam enters the throttling valve at 200°C and a quality of 90%,
   (a) determine the temperature (if superheated) or the quality of steam (if a liquid–vapor mixture) leaving the valve.
   (b) calculate the thermodynamic lost work, in kJ/kg, for the process.
   The temperature of the environment is 300 K. Heat transfer and changes in kinetic and potential energies are negligible.

8.15 It has been proposed to produce very cold oxygen gas by reducing a high-pressure gas in a throttling device. If the desirable product must be at 101.325 kPa and 100 K, determine
   (a) the temperature of the gas at the valve inlet if its pressure is 20 MPa.
   (b) the pressure of the gas at the valve inlet if its temperature is 200 K.

8.16 Nitrogen at 3000 psia and 260°R is throttled to a pressure of 14.7 psia. What are its final temperature and quality?

8.17 What is the lowest quality that can be determined by the use of a throttling calorimeter (see Fig. 8.1) if the steam-line pressure is 300 psia, calorimeter pressure is 14.7 psia, and a minimum of 10°F of superheat is required in the calorimeter to establish the fact that the steam at that point is superheated?

8.18 In a space heater, steam enters the heating coils steadily at 101.325 kPa and a quality of 98% and the condensate leaves the coils at 70°C. The air that is to be heated flows over the heating coils. If the entering air is at 15°C and the leaving air is at 30°C, what must be the ratio of air mass flow rate to steam mass flow rate? The air may be treated as an ideal gas with constant specific heats and $k = 1.4$. Changes in kinetic and potential energies are negligible.

8.19 Atmospheric air is to be cooled in the evaporator of a refrigerator from 18°C to −5°C. Freon-12 is supplied to the evaporator as a liquid–vapor mixture at −10°C and a quality of 35%, and it leaves the evaporator as saturated vapor at −10°C. For a flow rate of 150 kg/h of air, determine
(a) the mass flow rate of Freon-12.
(b) the amount of entropy generation for the process.
Air may be treated as an ideal gas with constant specific heats and $k = 1.4$. Changes in kinetic and potential energies are negligible.

8.20 Steam in a modern power plant is condensed from $1\frac{1}{2}$ in. Hg absolute. We may assume that all the heat given up by the steam is completely transferred to the circulating cooling water. If we design the condenser on the basis that the temperature rise of the cooling water is 20°F, what is the cooling water requirement for each pound of steam entering the condenser with a quality of 90%?

8.21 Five pounds per minute of ammonia enters a water-jacketed compressor as saturated vapor at 5°F and leaves at 150°F and 180 psia. Power supplied to drive the compressor is 10 hp. If cooling water enters the water jacket of compressor at 70°F and leaves at 90°F, what is the flow rate of cooling water?

8.22 Figure P8.22 shows a scheme that we may use to produce high-velocity air. If heat added to each kilogram of air is 100 kJ, what is the maximum velocity that we can expect at the exit of the nozzle? Assume that air is an ideal gas with constant specific heats and $k = 1.4$.

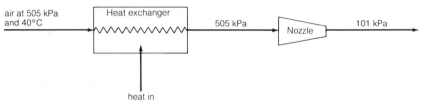

Figure P8.22

8.23 Superheated vapor of nitrogen may be produced by the scheme shown in Fig. P8.23. Assuming the pump to be perfect, determine for a mass flow rate of 100 kg/h of nitrogen
(a) the power, in kilowatts, required to operate the pump.
(b) the heat input to the fluid in the heat exchanger.
(c) the rate of entropy generation for the entire system. The temperature of the environment is 300 K.

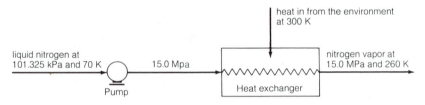

Figure P8.23

8.24 The combination of hardware shown in Fig. P8.24 is an integral part of a mechanical refrigeration system. Determine the amount of water that can be chilled by 1.0 lbm of Freon-12 in a steady-state steady-flow process.

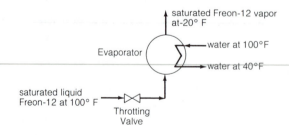

Figure P8.24

8.25 The combination of hardware shown in Fig. P8.25 is an arrange ment used in a steam power plant for preheating the feedwater.

The pump efficiency is 70%.

(a) Determine the extraction steam requirement for each pound of feedwater leaving the heater.

(b) Determine the entropy increase in the universe on the basis of 1 lb of feedwater leaving the heater.

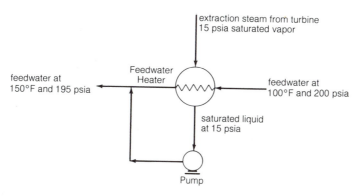

extraction steam from turbine
15 psia saturated vapor

Feedwater
Heater

feedwater at
150°F and 195 psia

feedwater at
100°F and 200 psia

saturated liquid
at 15 psia

Pump

Figure P8.25

8.26 The arrangement shown in Fig. P8.26 is a scheme that we may use to recover the kinetic energy from low-pressure high-velocity air for the production of useful work. If the turbine has an adiabatic efficiency of 90%, determine

(a) the temperature of air at the turbine inlet.

(b) the turbine output in kJ/kg.

Assume that air is an ideal gas with constant specific heats and $k = 1.4$.

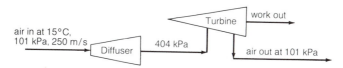

work out

Turbine

air in at 15°C,
101 kPa, 250 m/s

404 kPa

Diffuser

air out at 101 kPa

Figure P8.26

8.27 Air is to be expanded steadily in two stages with a reheater as shown in Fig. P8.27. The process is steady-state steady flow. The expansion process is reversible and adiabatic for both turbines. Design specifications include the following:

1. Neglect pressure drop in the reheater.

2. The temperature at the inlet of turbine B is the same as the temperature at the inlet of turbine A.

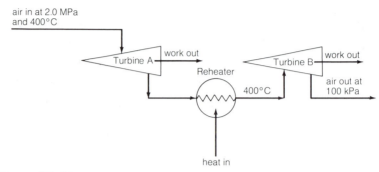

Figure P8.27

3. The work output of turbine A is the same as the work output of turbine B.
What must be the interstage pressure? Assume that air is an ideal gas with constant specific heats and $k = 1.4$.

8.28 An inventor claims that he has developed a steady-state steady-flow device to produce hot and cold air as shown in Fig. P8.28. No work or heat transfer are required. Could this device operate as shown?

Figure P8.28

8.29 A conceptual scheme that one could use to produce liquid oxygen continuously is shown in Fig. P8.29. What is the minimum amount of work input required to produce 1 kg of liquid oxygen?

8.30 A conceptual scheme that could be used to produce cold water continuously is shown in Fig. P8.30. What is the minimum amount of heat addition to the system required to produce 1 lb of cold water?

8.31 It has been proposed to produce saturated steam at 3.0 MPa steadily at the rate of 2000 kg/h by mixing superheated steam with water at 3.5 MPa and 40°C. If superheated steam is available at 3.50 MPa and 400°C,
(a) determine the flow rate of superheated steam required.
(b) determine the flow rate of water required.
(c) show that the mixing process is irreversible.

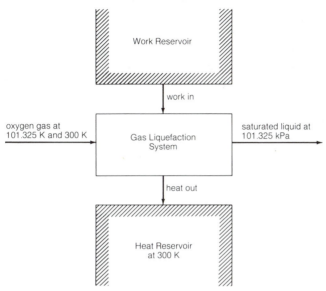

Figure P8.29

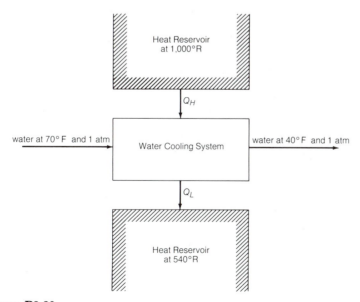

Figure P8.30

The mixing process may be considered to be carried out adiabatically. Changes in kinetic and potential energies are negligible.

8.32 The steam supplied to a turbine comes from two steam generators, A and B. Generator A produces steam at 1500 kPa and 400°C. Generator B produces steam at 1500 kPa and 500°C. The rate of steam coming from generator A is twice that coming from generator B. If the pressure of steam at turbine inlet is measured to be 1500 kPa, what is the temperature of steam entering the turbine? Assume that the two streams mix adiabatically. Changes in kinetic and potential energies are negligible.

8.33 Steam is condensed from a saturated-vapor state at 14.7 psia to a saturated-liquid state at 14.7 psia by giving up heat to the environment in a steady-state steady-flow process. The pressure and temperature of the environment are 14.7 psia and 537°R, respectively.
(a) Determine the thermodynamic lost work, in kJ/kg.
(b) Determine the change in the flow availability of the fluid, in kJ/kg.
(c) Compare the result of part (a) with that of part (b). Any comments?
Changes in kinetic and potential energies may be neglected.

8.34 Water is heated in a steady-state steady-flow heat exchanger from 25°C to 50°C with heat coming from a heat reservoir at 250°C. The mass flow rate of water is 0.5 kg/s. The temperature of the environment is 300 K. Calculate
(a) the amount of heat added to the water.
(b) the amount of exergy gained by the water, in kJ/s.
(c) the generation rate of thermodynamic lost work for the process, in kJ/s.
(d) Add the result of part (b) to that of part (c). Can you explain what this is?

8.35 The scheme shown in Fig. P8.35 is a cogeneration system in which we produce electricity and steam simultaneously using one source of energy. For a steam generation rate of 25,000 kg/h and a thermal efficiency of 30% for the heat engine, determine
(a) the amount of heat gained by the steam (this is sometimes called the process heat production of the system), in kJ/h.
(b) the amount of heat input to the heat engine, in kJ/h.
(c) the power output of the heat engine, in kJ/h.
(d) the generation rate of thermodynamic lost work.
(e) the gain in exergy by the steam, in kJ/h.
(f) Add the results of parts (c), (d), and (e). Can you explain what this is?

The pressure and temperature of the environment are 101.325 kPa and 300 K, respectively.

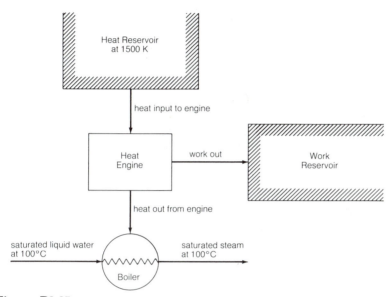

Figure P8.35

8.36 Nitrogen gas is compressed isothermally (but not reversibly) in a steady-state steady-flow compressor from 300 K and 101.325 kPa to 300 K and 20 MPa. Heat transfer is with the environment at 300 K. If the actual work required is 600 kJ/kg, calculate
(a) the thermodynamic lost work, in kJ/kg.
(b) the gain in exergy by the gas, in kJ/kg.
(c) Add the results of parts (a) and (b). Can you explain what this is?

8.37 One kilogram of fluid undergoes a change of state adiabatically from state 1 to state 2 in a steady-state steady-flow device in which changes in kinetic and potential energies are negligible. Show that the difference between the maximum useful work and the actual useful work for the same process is exactly equal to the thermodynamic lost work. The temperature of the environment is T_0.

8.38 Air is compressed from 15 psia and 80°F to 75 psia and 460°F. The actual work required is 92 Btu/lbm. Assuming ideal-gas behavior, show that this compression process is irreversible by showing that the actual work input is numerically greater than the minimum useful work required for the compression process. The pressure and temperature of the surroundings are 14.7 psia and 80°F.

8.39 A rigid tank of 0.5 m³ contains a mixture of liquid water and saturated vapor (Fig. P8.39). Initially, it is half-filled with saturated water and half-filled with saturated vapor at 3000 kPa. Heat is then added until one-half of the liquid is evaporated while a pressure controller lets saturated vapor escape at such a rate that the pressure in the tank remains at 3000 kPa. What is the amount of heat added to the tank?

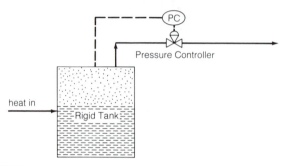

Figure P8.39

8.40 A rigid oxygen storage tank, having a total volume of 3.0 m³, initially contains 2.7 m³ of saturated liquid at 140 K and 0.3 m³ of saturated vapor at 140 K. Liquid is then withdrawn slowly from the bottom using a pressure controller to maintain constant pressure inside the tank (Fig. P8.40). Determine the amount of heat added to the tank when 1000 kg of oxygen have been withdrawn from the tank.

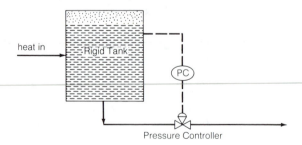

Figure P8.40

8.41 The arrangement shown in Fig. P8.41 is used to charge air to the rigid air storage tank, which has a volume of 3.0 m³. At the beginning of the charging process, air inside the tank is at 300 K

and 101 kPa. At the end of the charging process, air inside the tank is 300 K and 690 kPa. If work input to the adiabatic compressor is 8000 kJ for the entire charging process, determine
(a) the amount of heat exchange between the tank and the environment.
(b) the amount of thermodynamic lost work for the process.
Assume that air is an ideal gas with constant specific heats and $k = 1.4$. The temperature of the environment is 300 K.

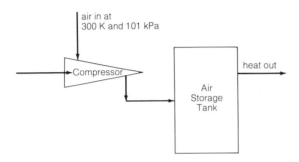

Figure P8.41

8.42 Two rigid tanks, A and B, are connected as shown in Fig. P8.42. Tank B, the connecting line, and the valve are all heavily insulated. Initially, tank A contains nitrogen gas at 1.0 MPa and 300 K, and tank B is evacuated. The valve is then opened slightly and then closed when the two tanks have come to the same pressure. During the process, the gas in tank A remains at a constant temperature of 300 K by exchanging heat with the environment, which is at a temperature of 300 K. The volume of each tank is 0.05 m^3.
(a) Calculate the final pressure and final temperature of the gas in tank B.

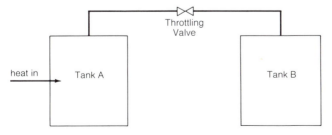

Figure P8.42

(b) Calculate the amount of heat exchange between the environment and tank A during the process.

(c) Show that the process is irreversible.

Assume ideal-gas behavior with constant specifics and $k = 1.40$.

8.43 A nitrogen cylinder of 30 ft^3 capacity is to be filled with a high-pressure nitrogen gas by connecting it to a line in which the pressure and temperature are maintained at 3000 psia and 80°F. At the beginning of the filling process, nitrogen inside the cylinder is at a pressure and temperature of 14.7 psia and 80°F. The filling process ends when the pressure of the gas inside the cylinder reaches 2000 psia. If the temperature of the gas inside the cylinder is to be kept at a constant 80°F, what is the amount of heat that must be removed from the cylinder for the filling process? Neglect the heat capacity of the tank.

8.44 A rigid tank of 3 ft^3 capacity contains air at 15 psia and 80°F. It is quickly filled adiabatically with air from a large reservoir at 100 psia and 80°F until the final pressure is 100 psia. Determine
(a) the final temperature of the air in the tank.
(b) the amount of air added to the tank.
Neglect the heat capacity of the tank and assume air to be an ideal gas with $c_p = 0.240$ Btu/lbm · °R and $c_v = 0.171$ Btu/lbm · °R.

8.45 Repeat Problem 8.44 for the case when the tank is initially evacuated.

8.46 A rigid tank of 5 ft^3 capacity is initially evacuated. It is to be filled adiabatically with steam from a large reservoir at 100 psia and 500°F until the final pressure is 100 psia.
(a) Determine the final temperature of the steam in the tank.
(b) Determine the amount of steam added to the tank.
(c) Show that the filling process is irreversible.

8.47 Nitrogen gas is compressed adiabatically from 300 K and 1 atm to a final pressure of 10 atm under steady-state steady-flow conditions. The adiabatic efficiency of the compressor is 85%. On the basis of ideal gas behavior, write a computer program to determine the compressor work, in kJ/kg,
(1) assuming constant specific heat at T_1;
(2) taking into consideration the variation of specific heat with temperature.
The molal heat capacity as a function of temperature for nitrogen is given by

$$\bar{c}_p^* = a + bT + cT^2 + dT^3$$

where $\bar{c}_p^*$ is in kJ/kg mol · K,

T is in degrees Kelvin,

a, b, c, and d are constants having the following values:

$a = 28.90$

$b = -0.1571 \times 10^{-2}$

$c = 0.8081 \times 10^{-5}$

$d = -2.873 \times 10^{-9}$

8.48 Air expands adiabatically from 1000 K and 10 atm to a final pressure of 1 atm under steady-state steady-flow conditions. The adiabatic turbine efficiency is 85%. On the basis of ideal gas behavior, write a computer program to determine the turbine work, in kJ/kg,
(1) assuming constant specific heat at T_1;
(2) taking into consideration the variation of specific heat with temperature.
 The molal heat capacity as a function of temperature for air is the same as that given in Problem 8.47, with the following values for the constants in this equation:

$a = 28.11$

$b = 0.19 \times 10^{-2}$

$c = 0.4802 \times 10^{-5}$

$d = -1.966 \times 10^{-9}$

8.49 For the steady-state steady-flow, reversible adiabatic process involving an ideal gas, the work transfer per unit mass of gas is given by the following expression (see Problem 8.4):

$$w_{12} = \frac{kRT_1}{k-1}\left[1 - \left(\frac{p_2}{p_1}\right)^{(k-1)/k}\right]$$

Making use of this equation, write a computer program to calculate the compressor work needed, in kJ/kgmol, for values of

p_2/p_1 ranging from 2 to 10 and values of $k = 1.2$, 1.3, 1.4, 1.667 with $T_1 = 300$ K. Graph your results and comment.

8.50 Making use of the expression in Problem 8.49, write a computer program to calculate the work output of a turbine, in kJ/kgmol, for values of p_1/p_2 ranging from 10 to 4 and values of $k = 1.2$, 1.3, 1.4, and 1.667 with $T_1 = 1000$ K.

8.51 Same as Problem 8.49 except that $T_1 = 200$ K.

8.52 Same as Problem 8.50 except that $T_1 = 1400$ K.

8.53 Write a computer program to determine the total compressor work, in kJ/kg, for Example 8.13 by varying the interstage pressure p_2 from 250 kPa to 2000 kPa. Graph your results. Make use of the plot to determine the optimum interstage pressure that will yield the minimum total compressor work needed. Compare your result with that obtained analytically in Example 8.13.

CHAPTER 9

Thermodynamics of One-Dimensional Steady Flow of Fluids

This chapter, also, is concerned with engineering analysis of processes for open systems; it is really a continuation of Chapter 8. There is, however, one important difference between the processes studied in Chapter 8 and the processes that we are going to study in this chapter. This is that the kinetic-energy terms were usually insignificant in the problems in Chapter 8, whereas the velocities of the flowing streams are quite often of primary importance in the processes we study now.

The flow of fluids is encountered in many engineering problems. A good understanding of its various aspects is essential to the design of pipes used for the transport of fluids; nozzles and flow passages in turbomachinery; jet engines for propulsion; the wings and fuselage of jet airliners; orifices for fluid metering; the capsules used to return astronauts from outer space; and many other interesting devices and systems.

Whenever we deal with any object in motion, Newton's second law of motion must be satisfied. Consequently, a complete analysis of fluid-flow problems involves not only the four basic tools we mentioned in the beginning of Chapter 8, but also the momentum principle. In addition, we must have a good understanding of such concepts as viscosity, turbulence, friction factor, and boundary layer. The study of fluid flow in general belongs in the discipline of fluid mechanics.[1]

[1] See R. M. Olson, *Essentials of Engineering Fluid Mechanics*, Intext Educational Publishers, New York, 1973; V. L. Streeter, *Fluid Mechanics*, McGraw-Hill Book Company, New York, 1958; A. H. Shapiro, *Compressible Fluid Flow*, Vols. I and II, Ronald Press, New York, 1953; Newman Hall, *Thermodynamics of Fluid Flow*, Prentice-Hall, Inc., Englewood Cliffs, N.J., 1951.

9.1 Energy Equation for One-Dimensional Steady Flow with No Work Transfer

The flow of fluids is quite complicated in its broadest aspect. For example, the fluid properties (pressure, temperature, velocity, and so on) could depend on three space coordinates, making the governing equations very complicated even in the case of simple fluids. However, for many practical engineering problems, the fluid properties may be assumed to be dependent on only one space coordinate. That is, a large class of fluid-flow problems may be approximated as one-dimensional in that all properties of the flowing fluid are uniform across each cross section normal to the direction of flow. In this chapter, we shall examine a few one-dimensional steady-flow problems based on thermodynamics principles alone, partly to demonstrate the usefulness of thermodynamics in fluid flow, and partly to give readers an introduction to the interesting and important subject of fluid mechanics.

By the nature of the process, no shaft work is involved in the study of fluid-flow problems. However, change of kinetic energy and change of potential energy may or may not be significant. The first law for the study of fluid-flow problems thus becomes

$$\dot{Q} = \sum_{\text{out}} \dot{m}h - \sum_{\text{in}} \dot{m}h + \sum_{\text{out}} \frac{\dot{m}\bar{V}^2}{2g_c} - \sum_{\text{in}} \frac{\dot{m}\bar{V}^2}{2g_c} + \sum_{\text{out}} \frac{\dot{m}gZ}{g_c} - \sum_{\text{in}} \frac{\dot{m}gZ}{g_c}. \qquad (9.1)$$

This equation, together with the second law, the continuity equation, and the property relations of fluids, are the tools we shall use in the study of one-dimensional steady-flow problems.

Example 9.1

Steam flows steadily through a horizontal pipe line 3.0 in in diameter. At a certain section in the pipe, the pressure and temperature of the steam are 20 psia and 400°F, respectively. At another section, the pressure and temperature are 15.0 psia and 390°F, respectively. The pipe is thoroughly insulated, so heat loss is negligible.

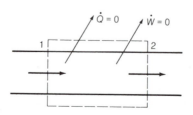

Determine

(a) velocity of steam at section 1.
(b) velocity of steam at section 2.
(c) mass-flow rate in lbm/s.
(d) rate of entropy creation in the universe.

Solution

System selected for study: pipe between section 1 and section 2
Unique features of process: steady-state steady flow; no work transfer; no change in PE
Assumption: one-dimensional flow

From the first law, on the basis of unit mass-flow rate,

$$h_1 + \frac{\bar{V}_1^2}{2g_c} = h_2 + \frac{\bar{V}_2^2}{2g_c}. \tag{1}$$

From the continuity equation (mass balance) for steady flow,

$$\dot{m}_1 = \dot{m}_2$$

$$\dot{m}_1 = \frac{\bar{V}_1 A_1}{v_1}$$

$$\dot{m}_2 = \frac{\bar{V}_2 A_2}{v_2}.$$

But $A_1 = A_2$ since the diameter of pipe is constant. Therefore,

$$\bar{V}_1 = \frac{v_1}{v_2} \bar{V}_2. \tag{2}$$

Combining Eqs. 1 and 2, we get

$$\bar{V}_1 = \sqrt{\frac{2g_c(h_1 - h_2)}{(v_2/v_1)^2 - 1}}. \tag{3}$$

From the property relations (steam tables), we have

$$h_1 = 1239.2 \text{ Btu/lbm}$$

$$v_1 = 25.428 \text{ ft}^3/\text{lbm}$$

$$s_1 = 1.8397 \text{ Btu/lbm} \cdot {}^\circ\text{R}$$

$$h_2 = 1235.2 \text{ Btu/lbm}$$

$$v_2 = 33.559 \text{ ft}^3/\text{lbm}$$

$$s_2 = 1.8665 \text{ Btu/lbm} \cdot {}^\circ\text{R}.$$

Substituting the appropriate data into Eq. 3, we find that

(a) $\bar{V}_1 = 519.6$ ft/s.

Using Eq. 2, we find that

(b) $\bar{V}_2 = 685.7$ ft/s.

Using the continuity equation, we find that

(c) $\dot{m} = 1.003$ lbm/s.

(d) From the second law for adiabatic steady flow, we have

$$\dot{S}_{gen} = \dot{m}(s_2 - s_1). \tag{4}$$

Substituting the appropriate data into Eq. 4, we determine that

$$\dot{S}_{gen} = 1.003(1.8665 - 1.8397) \text{ Btu/}^{\circ}\text{R} \cdot \text{s}$$

$$= +0.0269 \text{ Btu/}^{\circ}\text{R} \cdot \text{s.} \qquad\blacksquare$$

Comments

This example illustrates the simultaneous use of the energy equation and the continuity equation in solving problems in fluid flow. This is the basis of many flow-measuring devices.

The second law provides us the tool with which to check whether the flow conditions as specified are possible. The given flow process is irreversible, since we have positive entropy creation in the universe. The irreversibility involved is partly due to the shear stress in the fluid resulting from fluid viscosity, and partly to the relative velocity between the fluid and the pipe wall. In the case of a nonviscous fluid, the irreversibility due to shear stress will be zero.

Example 9.2

Water enters a nozzle steadily at 500 kPa and 20°C with a velocity of 2.0 m/s. It leaves the nozzle at 101 kPa with a velocity of 25 m/s. Show that the process is irreversible

(a) because the ideal velocity is greater than the actual velocity for the same inlet conditions and same outlet pressure.
(b) because entropy generation for the actual process is positive.

Solution

System selected for study: nozzle
Unique features of process: steady-state steady flow; no work transfer
Assumptions: one-dimensional flow; neglect heat transfer and ΔPE

(a) From the first law, we have

$$0 = dh + d\left(\frac{\bar{V}^2}{2g_c}\right). \tag{1}$$

For maximum velocity at the outlet, the process must be reversible and adiabatic. Thus the process is isentropic and

$$ds = 0. \tag{2}$$

From a property relation, we have

$$T\,ds = dh - v\,dp. \tag{3}$$

Combining Eqs. 1, 2, and 3, we have, for the reversible case,

$$0 = v\,dp + d\left(\frac{\bar{V}^2}{2g_c}\right). \tag{4}$$

Assuming water to be incompressible, we can integrate Eq. 4 to yield

$$v(p_2 - p_1) + \frac{\bar{V}_{2s}^2 - \bar{V}_1^2}{2g_c} = 0$$

and

$$\bar{V}_{2s} = \sqrt{2g_c\,v(p_1 - p_2) + \bar{V}_1^2}. \tag{5}$$

For water at 20°C and 500 kPa,

$$v \approx v_f \text{ at } 20°C$$

$$= 0.0010017 \text{ m}^3/\text{kg}.$$

Substituting into Eq. 5, we have

$$\bar{V}_{2s} = \sqrt{2 \times 0.0010017(500 - 101) \times 1000 + (2.0)^2} \text{ m/s}$$

$$= \sqrt{799.4 + 4} \text{ m/s}$$

$$= 28.3 \text{ m/s},$$

which is greater than the actual velocity of 25.0 m/s. Thus process is irreversible.

(b) Since the process is adiabatic, we have, from the second law,

$$S_{gen} = m(s_2 - s_1). \tag{6}$$

From the first law, we have

$$0 = dh + d\left(\frac{\bar{V}^2}{2g_c}\right).$$

Combining this with the property relation

$$T \, ds = dh - v \, dp,$$

we have

$$T \, ds = -v \, dp - d\left(\frac{\bar{V}^2}{2g_c}\right). \tag{7}$$

Since the temperature of the incompressible fluid is essentially a constant, we can integrate Eq. 7 to give

$$T(s_2 - s_1) \approx v(p_1 - p_2) - \frac{\bar{V}_2^2 - \bar{V}_1^2}{2g_c}. \tag{8}$$

But from part (a), we have

$$v(p_1 - p_2) = \frac{\bar{V}_{2s}^2 - \bar{V}_1^2}{2g_c}. \tag{9}$$

Thus

$$s_2 - s_1 = \frac{1}{2g_c \, T} [(\bar{V}_{2s}^2 - \bar{V}_1^2) - (\bar{V}_2^2 - \bar{V}_1^2)]. \tag{10}$$

Equation 10 shows that $(s_2 - s_1)$ must be positive since $\bar{V}_{2s}$, the ideal velocity found in part (a), is greater than the actual velocity of $\bar{V}_1$. Consequently, S_{gen} must be a positive quantity, according to Eq. 6.   ∎

Comments

We could have neglected the kinetic energy of the fluid at the nozzle inlet. This is indeed a reasonable assumption that we could make in the study of nozzles.

Example 9.3

Air enters a thermally insulated pipe at a temperature of 100°C and a velocity of 50 m/s. It leaves the pipe with a velocity of 20 m/s at an elevation of 100 m above the inlet. Determine the temperature of air at the outlet of the pipe. Assume that air is an ideal gas with constant specific heats of $c_p = 1.01$ kJ/kg · K. The process is steady-state steady flow.

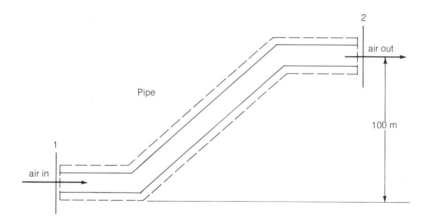

Solution

System selected for study: pipe
Unique features of process: steady-state steady flow; no work transfer
Assumption: air is an ideal gas with constant specific heats

From the first law we have, on the basis of unit mass,

$$0 = h_2 - h_1 + \frac{\bar{V}_2^2 - \bar{V}_1^2}{2g_c} + \frac{g}{g_c}(Z_2 - Z_1). \tag{1}$$

For an ideal gas with constant specific heats, we have

$$h_2 - h_1 = c_p(T_2 - T_1). \tag{2}$$

Combining Eqs. 1 and 2, we have

$$0 = c_p(T_2 - T_1) + \frac{\bar{V}_2^2 - \bar{V}_1^2}{2g_c} + \frac{g}{g_c}(Z_2 - Z_1). \tag{3}$$

Let $g = 9.81$ m/s^2. Then substituting into Eq. 3, we have

$$0 = 1.01(T_2 - 373.15) + \frac{(20)^2 - (50)^2}{2 \times 1} + \frac{9.81}{1}(100 - 0)$$

and

$$T_2 = 441.47 \text{ K} = 168.32°C. \qquad \blacksquare$$

Comment

The air heats up because part of the kinetic energy was converted into thermal energy.

9.2 Bernoulli Equation for Incompressible Fluid

Let us now consider one-dimensional flow involving one stream in and one stream out. Then Eq. 9.1 in differential form, and on the basis of unit mass-flow rate, becomes

$$\bar{d}q = dh + d\left(\frac{\bar{V}^2}{2g_c}\right) + d\left(\frac{g}{g_c} Z\right). \tag{1}$$

If the process is reversible, we have, from the second law,

$$\bar{d}q = T \, ds. \tag{2}$$

Combining Eqs. 1 and 2, we have for the one-dimensional reversible steady flow,

$$T \, ds = dh + d\left(\frac{\bar{V}^2}{2g_c}\right) + d\left(\frac{g}{g_c} Z\right). \tag{3}$$

If the flowing fluid is a simple compressible fluid, we have, from the property relation,

$$T \, ds = dh = v \, dp. \tag{4}$$

Combining Eqs. 3 and 4, we find that

$$v \, dp + d\left(\frac{\bar{V}^2}{2g_c}\right) + d\left(\frac{g}{g_c} Z\right) = 0. \tag{9.2}$$

If the fluid is incompressible, Eq. 9.2 may be integrated to give

$$v(p_2 - p_1) + \frac{\bar{V}_2^2 - \bar{V}_1^2}{2g_c} + \frac{g}{g_c}(Z_2 - Z_1) = 0. \tag{9.3}$$

Equation 9.3 is the famous Bernoulli equation for one-dimensional steady, reversible flow of an incompressible fluid. It may also be deduced from the momentum principle if flow is frictionless, as given in the next section. Note that this equation is valid regardless of whether there is heat transfer.

Equation 9.3 may also be written as

$$\frac{p_1}{\rho} + \frac{\bar{V}_1^2}{2g_c} + \frac{g}{g_c} Z_1 = \frac{p_2}{\rho} + \frac{\bar{V}_2^2}{2g_c} + \frac{g}{g_c} Z_2 = \text{constant}, \tag{9.3a}$$

where $\rho = 1/v$, density of the flowing fluid. Equation 9.3a simply says that all the mechanical energy of the fluid is conserved since there are no energy-dissipation mechanisms involved.

The Bernoulli equation for incompressible flow may also be written in the following two forms:

$$p_1 + \frac{\rho \bar{V}_1^2}{2g_c} + \frac{\rho g Z_1}{g_c} = p_2 + \frac{\rho \bar{V}_2^2}{2g_c} + \frac{\rho g Z_2}{gc} = \text{constant total pressure} \tag{9.3b}$$

and

$$\frac{g_c p_1}{g\rho} + \frac{\bar{V}_1^2}{2g} + Z_1 = \frac{g_c p_2}{g\rho} + \frac{\bar{V}_2^2}{2g} + Z_2 = \text{constant total head.} \tag{9.3c}$$

In Eq. 9.3b, the term $\rho \bar{V}^2/2g_c$ is known as the *dynamic pressure, p* is the *static pressure* (the pressure exerted on the wall parallel to the direction of flow), and $\rho g Z/g_c$ is the *potential pressure*. In Eq. 9.3c the term $\bar{V}^2/2g$ is known as the *velocity head*, $g_c p/g\rho$ is the *pressure head*, and Z is the *potential head*.

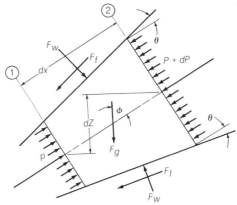

Figure 9.1 Forces Acting on an Element of a Fluid Flowing in a Duct

The Bernoulli equation states that in one-dimensional steady flow of an incompressible fluid without friction, the total pressure and the total head of the flowing fluid remain constant.

9.3 Derivation of the Bernoulli Equation from the Momentum Principle

Let us consider the one-dimensional flow of an element of a fluid in a duct from left to right, such as that shown in Fig. 9.1. The element of fluid is bounded by two planes separated by a distance dx. It is subject to the action of five external forces: the pressure force at plane 1, the pressure force at plane 2, the gravitational force, the pressure force exerted on the fluid by the sidewalls, and the sidewall frictional force.

The force due to the fluid pressure at plane 1 is

$$(F_p)_1 = pA. \tag{1}$$

The force due to the fluid pressure at plane 2 is

$$(F_p)_2 = -(p + dp)(A + dA) = -pA - p\,dA - A\,dp, \tag{2}$$

neglecting the second-order term $dp\,dA$.

If we assume that the mean density of the fluid equals $\rho + (d\rho/2)$, the gravitational force acting on the fluid in the direction of motion is

$$F_g = -\frac{g}{g_c}\left(\rho + \frac{d\rho}{2}\right)\left(A + \frac{dA}{2}\right)\cos\phi\ dx$$

$$= -\frac{g}{g_c}\rho A\ dZ, \tag{3}$$

neglecting second-order terms.

If we assume that the mean pressure of the fluid equals $p + (dp/2)$, the pressure force exerted by the sidewalls on the fluid in the direction of motion is

$$F_w = \left(p + \frac{dp}{2}\right)\sin\theta\ dA_w$$

$$= p\ dA, \tag{4}$$

neglecting second-order term.

The sidewall frictional force, in the direction of motion, is

$$F_f = A\ dp_f,$$

where the change in pressure dp_f because of friction is given in terms of the friction factor f as

$$dp_f = -\frac{f\rho \bar{V}^2\ dx}{2Dg_c},$$

in which D is the local diameter of the duct. Therefore, the sidewall frictional force, in the direction of motion, is

$$F_f = -\frac{fA\rho \bar{V}^2\ dx}{2Dg_c}. \tag{5}$$

According to Newton's second law of motion, summation of all the external forces acting on the fluid must be equal to the product of mass of the fluid and acceleration. Thus

$$pA - pA - p\,dA - A\,dp - \frac{g}{g_c}\rho A\,dZ + p\,dA - \frac{fA\rho\bar{V}^2\,dx}{2Dg_c}$$

$$= \frac{1}{g_c}\left(\rho + \frac{d\rho}{2}\right)\left(A + \frac{dA}{2}\right)(dx)\frac{d\bar{V}}{d\tau}$$

$$= \frac{1}{g_c}\rho A\bar{V}\,d\bar{V}, \tag{6}$$

neglecting second-order terms. We have also made use of the fact that

$$\frac{d\bar{V}}{d\tau} = \bar{V}\frac{d\bar{V}}{dx}.$$

Simplifying, the equation of motion becomes

$$dp + \frac{\rho\bar{V}\,d\bar{V}}{g_c} + \frac{g}{g_c}\rho\,dZ + \frac{f\rho\bar{V}^2\,dx}{2Dg_c} = 0 \tag{9.4}$$

or

$$\frac{dp}{\rho} + \frac{\bar{V}\,d\bar{V}}{g_c} + \frac{g}{g_c}\,dZ + \frac{f\bar{V}^2\,dx}{2Dg_c} = 0. \tag{9.4a}$$

If flow is frictionless, there will be no frictional force, and Eq. 9.4a is identical to Eq. 9.2, which is derived from thermodynamic considerations.

For the flow of an incompressible fluid with friction, we may integrate Eq. 9.4a to give

$$\frac{p_1}{\rho} + \frac{\bar{V}_1^2}{2g_c} + \frac{gZ_1}{g_c} = \frac{p_2}{\rho} + \frac{\bar{V}_2^2}{2g_c} + \frac{gZ_2}{g_c} + h_L \tag{9.5}$$

where h_L represents the portion of the mechanical energy that has been dissipated due to friction and is commonly known as "head" loss. For the flow of an incompressible fluid in a pipe, we have

$$h_L = f\frac{L}{D}\frac{\bar{V}^2}{2g_c}, \tag{9.6}$$

In Eq. 9.6, f is a dimensionless quantity known as the Fanning friction factor. The units for the other quantities in the SI and English Engineering systems are given in the following table:

Quantity in Eq. 9.6	Units in SI System	Units in English Engineering System
L = length of pipe	m	ft
D = pipe diameter	m	ft
$\bar{V}$ = average velocity	m/s	ft/s
g_c	1.0 kg·m/N·s²	32.174 lbm·ft/lbf·s²
h_L	N·m/kg	ft·lbf/lbm

The friction factor depends on the fluid properties, the pipe geometry, and the fluid velocity. It is an important topic in the study of fluid mechanics.

Example 9.4

In a chemical process plant, a large quantity of atmospheric air at 101 kPa and 25°C is to be delivered by a fan to a location that is 150 m above the outlet of the fan with a pressure of 101 kPa. The velocity in the pipe is to be kept low to minimize the power needed to drive the fan. Make an estimate of

(a) the discharge pressure for the design of the fan.
(b) the power needs for the fan for a volume flow rate of 5.0×10^4 m³/min if the fan efficiency is 75%.

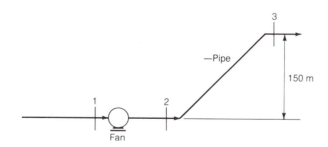

Solution

System selected for study: pipe
Unique features of process: steady-state steady flow; no work transfer
Assumptions: air is an ideal gas; air is incompressible; neglect friction; neglect ΔKE

(a) For steady flow of an incompressible fluid with no friction, Eq. 9.3 applies:

$$v(p_3 - p_2) + \frac{\bar{V}_3^2 - \bar{V}_2^2}{2g_c} + \frac{g}{g_c}(Z_3 - Z_2) = 0. \qquad (1)$$

Neglecting changes in kinetic energy, Eq. 1 becomes

$$p_2 = p_3 + \frac{\rho g}{g_c}(Z_3 - Z_2). \qquad (2)$$

Assuming air to be an ideal gas,

$$\rho = \frac{p}{RT}$$

$$= \frac{101 \times 1000}{(8314.3/28.97)(298.15)} \text{ kg/m}^3$$

$$= 1.1803 \text{ kg/m}^3.$$

Let $g = 9.81$ m/s^2. Substituting into Eq. 2, we have

$$p_2 = 101 + \frac{1.1803 \times 9.81(150 - 0)}{1.0 \times 1000} \text{ kPa}$$

$$= 101 + 1.7368 \text{ kPa} = 102.7368 \text{ kPa}.$$

(b) The actual power needed is simply the ideal power needed divided by the fan efficiency:

$$\dot{W}_{act} = \frac{\dot{W}_{rev}}{\eta_c}$$

$$= -\frac{\dot{m}v(p_2 - p_1)}{\eta_c}$$

$$= -\frac{\dot{V}(p_2 - p_1)}{\eta_c}$$

$$= -\frac{5 \times 10^4(102.7368 - 101)}{0.75} \text{ kJ/min}$$

$$= -1.93 \times 10^3 \text{ kJ/s}$$

$$= -1930 \text{ kW}. \qquad \blacksquare$$

Comment

Since we have neglected friction, our estimated power needed for the fan would be a little on the low side. However, our result should be adequate for the purpose of preliminary design in many instances.

Example 9.5

A pipe is to be selected to carry 20°C water at the rate of 10 kg/s. Using a Fanning friction factor of 0.02 and a friction pressure drop of 200 kPa for 100 m length, determine the diameter of the pipe.

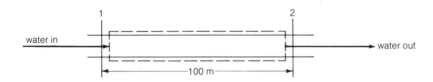

Solution

System selected for study: pipe
Unique features of process: steady-state steady flow; no work transfer
Assumptions: water is an incompressible fluid; neglect ΔKE and ΔPE

From the definition of the Fanning friction factor, we have

$$h_L = \frac{\Delta p}{\rho} = f\frac{L}{D}\frac{\bar{V}^2}{2g_c} \tag{1}$$

or

$$\frac{\Delta p}{L} = \frac{f\rho\bar{V}^2}{2Dg_c}. \tag{2}$$

From conservation of mass for steady flow,

$$\dot{m} = \rho\bar{V}A. \tag{3}$$

For a circular pipe,

$$A = \frac{\pi D^2}{4}. \tag{4}$$

Combining Eqs. 2, 3, and 4, we have

$$\frac{\Delta p}{L} = \frac{8f\dot{m}^2}{\pi^2 g_c \rho D^5}. \tag{5}$$

For water at 20°C,

$$v_f = 0.0010017 \text{ m}^3/\text{kg}$$

and

$$\rho = \frac{1}{0.0010017} \text{ kg/m}^3 = 998.3 \text{ kg/m}^3.$$

Substituting into Eq. 5, we have

$$\frac{200 \times 1000}{100} = \frac{8 \times 0.02 \times (10)^2}{\pi^2 \times 1.0 \times 998.3 \times D^5} \text{ Pa/m}.$$

Solving for the diameter, we have

$$D = 0.06 \text{ m}$$

$$= 6 \text{ cm}. \qquad \blacksquare$$

Comment

With the given flow rate and a 6-cm-diameter pipe, it may be shown in fluid mechanics that we would be operating in the turbulent flow region making the use of a Fanning friction factor of 0.02 a reasonable one.

Example 9.6

Water flows steadily with a velocity of 10 ft/s in a horizontal pipe having a diameter of 6.0 in. At one section of the pipe, the temperature and pressure of the water are 70°F and 100 psia, respectively. At a distance of 1000 ft downstream, the pressure is 75 psia. Determine the friction factor.

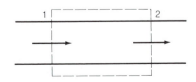

Solution

System selected for study: pipe between section 1 and section 2
Unique features of process: steady-state steady flow; no change in PE; no work transfer
Assumptions: one-dimensional flow; water is incompressible

Using Eq. 9.5, we have

$$h_L = \frac{p_1 - p_2}{\rho} + \frac{\bar{V}_1^2 - \bar{V}_2^2}{2g_c}. \tag{1}$$

From the continuity equation, we have

$$\rho_1 \bar{V}_1 A_1 = \rho_2 \bar{V}_2 A_2.$$

But $A_1 = A_2$ and $\rho_1 = \rho_2$; hence $\bar{V}_1 = \bar{V}_2$. Thus Eq. 1 becomes

$$h_L = \frac{p_1 - p_2}{\rho}. \tag{2}$$

By definition, "head" loss is given by Eq. 9.6,

$$h_L = f \frac{L}{D} \frac{\bar{V}^2}{2g_c}. \tag{3}$$

Combining Eqs. 2 and 3, we get

$$f = \frac{2Dg_c \, v(p_1 - p_2)}{L\bar{V}^2}. \tag{4}$$

For water at 70°F and 100 psia, we know that

$$v \approx v_f \text{ at } 70°F = 0.01605 \text{ ft}^3/\text{lbm}.$$

Substituting into Eq. 4, we find that

$$f = \frac{2 \times 6 \times 32.2 \times 0.01605 \times (100 - 75) \times 144}{12 \times 1000 \times (10)^2}$$

$$= 0.019.$$ ∎

Comment

It is shown in fluid mechanics that if we specify a certain value for the friction factor for a given situation, we are making the decision to use pipe of a certain roughness.

9.4 Sonic Velocity and Mach Number

A parameter of great importance in the study of fluid flow, especially in compressible flow, is the *sonic* or *acoustic velocity*, denoted by c. This is the velocity of propagation of an infinitely small disturbance relative to the fluid ahead of it.

Let us analyze the factors affecting the velocity of an acoustic wave by having an observer moving with the wave front (Fig. 9.2). Then, as far as the observer is concerned, a steady-state steady-flow situation exists. The fluid enters the control volume with sonic velocity and leaves with a reduced velocity of $c - d\bar{V}$. Because the acoustic wave is infinitesimal, the pressure change, the density change, and the enthalpy change across the wave front are given by dp, $d\rho$, and dh, respectively.

Since the process is adiabatic and no work transfer is involved, we have, from the first law for the control volume,

$$h + \frac{c^2}{2g_c} = (h + dh) + \frac{(c - d\bar{V})^2}{2g_c}. \tag{9.7}$$

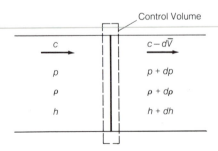

Figure 9.2 Observer Moving with Front of an Acoustic Wave

Neglecting higher-order differentials, Eq. 9.7 becomes

$$dh - \frac{c \, d\bar{V}}{g_c} = 0. \qquad (9.8)$$

From the continuity equation for steady flow, we have

$$\rho Ac = (\rho + d\rho)A(c - d\bar{V}). \qquad (9.9)$$

Neglecting higher-order differentials, Eq. 9.9 becomes

$$c \, d\rho - \rho \, d\bar{V} = 0. \qquad (9.10)$$

Combining Eqs. 9.8 and 9.10, we have

$$dh - \frac{c^2}{g_c} \frac{d\rho}{\rho} = 0. \qquad (9.11)$$

If the flowing fluid is a simple compressible substance, we have

$$T \, ds = dh - v \, dp.$$

Since the acoustic wave is infinitesimal, we can consider the process reversible. Since the process is also adiabatic, the process is isentropic. Thus $ds = 0$ and we have

$$dh = v \, dp = \frac{dp}{\rho}.$$

Substituting into Eq. 9.11, we get

$$\frac{dp}{\rho} - \frac{c^2}{g_c} \frac{d\rho}{\rho} = 0 \qquad (9.12)$$

or

$$c^2 = g_c \frac{dp}{d\rho} \qquad (9.13)$$

Since the process is isentropic, we have

$$\frac{dp}{d\rho} = \left(\frac{\partial p}{\partial \rho}\right)_s.$$

Thus

$$c^2 = g_c \left(\frac{\partial p}{\partial \rho} \right)_s$$

and

$$c = \sqrt{g_c \left(\frac{\partial p}{\partial \rho} \right)_s}. \tag{9.14}$$

Since $(\partial p / \partial \rho)_s$ is a thermodynamic property, c is also a thermodynamic property. This means that we have only one value of c at a particular state condition. Note that the expression for c can also be derived by applying the momentum and continuity equations to a weak pressure wave, as shown in textbooks on fluid mechanics. A sonic or acoustic wave is the same as a weak pressure wave.

For an ideal gas, it may be shown that by combining Eqs. 5.17 and 5.18, we obtain the following property relation:

$$ds = c_v \frac{dp}{p} + c_p \frac{dv}{v}, \tag{9.15}$$

from which we get

$$(\partial p / \partial \rho)_s = kRT. \tag{9.16}$$

Thus, for an ideal gas, the velocity of sound is given by

$$c = \sqrt{kg_c RT}. \tag{9.17}$$

We see that the velocity of sound in an ideal gas is only a function of temperature. Consequently, measuring of the velocity of sound provides an experimental method of determining the ratio of specific heat, k, of ideal gases.

In the study of compressible fluids, it is convenient to introduce a dimensionless parameter, called the *Mach number*, denoted by M. By definition,

$$M = \frac{\bar{V}}{c}, \tag{9.18}$$

where $\bar{V}$ is the actual velocity at a given location and c is the velocity of sound at the same location. Making use of the concept of the Mach number, we may classify flow regimes as follows:

When $M < 1$, flow is subsonic.

When $M = 1$, flow is sonic.

When $M > 1$, flow is supersonic.

Example 9.7

A jet plane travels at a speed of 1100 km/h. Determine the Mach number of flight

(a) for sea-level operation with an air temperature of 25°C.
(b) for high-altitude operation with an air temperature of $-55°C$.

Solution

Assuming air to be an ideal gas, the velocity of sound is then given by

$$c = \sqrt{kg_c\, RT}. \tag{1}$$

(a) Using $k = 1.4$ and a molecular weight of 28.97 for air, we have, at 25°C,

$$c = \sqrt{1.4 \times 1.0\ \frac{\text{kg} \cdot \text{m}}{\text{N} \cdot \text{s}^2} \times \frac{8314.3}{28.97}\ \frac{\text{J}}{\text{kg} \cdot \text{K}} \times 298.15\ \text{K}}$$

$$= 346.1\ \text{m/s}.$$

With $\bar{V} = 1100$ km/h $= (1100 \times 1000)/3600$ m/s $= 305.6$ m/s,

$$M = \frac{\bar{V}}{c} = \frac{305.6}{346.1} = 0.883,$$

which means that the flight is subsonic.
(b) At $-55°C$, we have

$$c = \sqrt{1.4 \times 1.0 \times (8314.3/28.97) \times 218.15}\ \text{m/s}$$

$$= 296.1\ \text{m/s}$$

With $\bar{V} = 305.6$ m/s,

$$M = \frac{\bar{V}}{c} = \frac{305.6}{296.1} = 1.03,$$

which means that the flight is supersonic.   ∎

Comment

For the same speed, the plane is flying subsonically at sea level but flying supersonically at high altitude. The plane must be carefully designed to meet these requirements.

Example 9.8

Determine the velocity of sound

(a) for air at 14.7 psia and 100°F.
(b) for water at 14.7 psia and 100°F.

Solution

(a) Assuming air to be an ideal gas, with $k = 1.4$ and a molecular weight of 28.9, we have

$$c = \sqrt{1.4 \times 32.2 \, \frac{\text{lbm} \cdot \text{ft}}{\text{lbf} \cdot \text{s}^2} \times \frac{1545}{28.9} \, \frac{\text{ft} \cdot \text{lbf}}{\text{lbm} \cdot {}^\circ\text{R}} \times 560^\circ\text{R}}$$

$$= 1162 \text{ ft/s}.$$

(b) Since $\rho = 1/v$, Eq. 9.14 may also be written as

$$c = \sqrt{-g_c \, v^2 \left(\frac{\partial p}{\partial v}\right)_s}. \tag{9.14a}$$

Water is only slightly compressible. Therefore,

$$\left(\frac{\partial p}{\partial v}\right)_s \approx \left(\frac{\Delta p}{\Delta v}\right)_s.$$

From property relations, we have, for saturated water at 100°F,

$$v_f = 0.01613 \text{ ft}^3/\text{lbm}$$

$$s_f = 0.1295 \text{ Btu/lbm} \cdot {}^\circ\text{R}$$

$$p = 0.9492 \text{ psia},$$

and at 500 psia and $s = 0.1295$ Btu/lbm $\cdot$ °R,

$$v = 0.01611 \text{ ft}^3/\text{lbm}.$$

Therefore,

$$\left(\frac{\partial p}{\partial v}\right)_s \approx \frac{500 - 0.9492}{0.01611 - 0.01613}$$

$$= -2.495 \times 10^7 \frac{\text{psi} \cdot \text{lbm}}{\text{ft}^3}.$$

Substituting into Eq. 9.14a, we find that

$$c \approx \sqrt{-32.2 \times (0.01613)^2 \times (-2.495 \times 10^7 \times 144)} \text{ ft/s}$$

$$= 5500 \text{ ft/s}.$$

Comments

A source of sound consists of a material substance in a state of vibration. From Eq. 9.14, it is seen that sound will propagate faster in a fluid that is less compressible. The fact that the velocity of sound in water at 14.7 psia and 100°F is much greater than the velocity of sound in air at 14.7 psia and 100°F is consistent with this observation. In fact, in a truly incompressible fluid, the velocity of sound would be infinite because there would be no adjustment in volume possible, and consequently any pressure change would be immediately transmitted and felt throughout the fluid.

9.5 Stagnation Properties

A concept that greatly simplifies the analysis of certain problems involving fluid flow is that of the isentropic stagnation state. Properties associated with this state are known as *stagnation properties*. The *isentropic stagnation state* is the state a flowing fluid would attain if it were brought to rest reversibly in a steady-state steady-flow device involving no heat transfer and no work transfer. We shall designate this state by the subscript zero.

From the first law of thermodynamics for steady-state steady flow, the stagnation enthalpy, h_0, is related to the enthalpy and velocity of the flowing fluid by

$$h_0 = h + \frac{\bar{V}^2}{2g_c}. \tag{9.19}$$

From the definition of an isentropic stagnation state, the stagnation entropy, s_0, is the same as the entropy of the flowing fluid. Thus all stagnation properties may be defined by h_0 and s_0. The isentropic stagnation process is shown in Fig. 9.3.

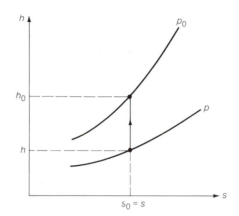

Figure 9.3 Isentropic Stagnation State on a T–s Diagram

Example 9.9

The airspeed of an aircraft is indicated as Mach 0.8 on board at atmospheric conditions of 44 kPa and $-15°C$. Determine the isentropic stagnation pressure and temperature recorded on the aircraft. Assume that air is an ideal gas with constant specific heats of $c_p = 1.0038$ kJ/kg $\cdot$ K and $c_v = 0.7168$ kJ/kg $\cdot$ K.

Solution

Using a molecular weight of 28.97 for air, we have, at $-15°C$,

$$c = \sqrt{1.4 \times 1.0(8314.3/28.97) \times 258.15} \text{ m/s}$$

$$= 322.1 \text{ m/s}$$

With $M = \bar{V}/c = 0.8$,

$$\bar{V} = 0.8 \times 322.1 \text{ m/s} = 257.7 \text{ m/s}.$$

Using Eq. 9.19, the stagnation enthalpy is given by

$$h_0 = h + \frac{\bar{V}^2}{2g_c}$$

and

$$h_0 - h = \frac{\bar{V}^2}{2g_c}. \tag{1}$$

For an ideal gas with constant specific heats,

$$h_0 - h = c_p(T_0 - T). \tag{2}$$

Combining Eqs. 1 and 2, we have

$$T_0 = T + \frac{\bar{V}^2}{2c_p g_c}. \tag{3}$$

Substituting into Eq. 3, the stagnation temperature is

$$T_0 = 258.15 + \frac{(257.7)^2}{2 \times 1.0038 \times 1000 \times 1.0} \, K = 291.2 \, K.$$

For an isentropic process involving an ideal gas with constant specific heats, the stagnation pressure may be obtained by using Eq. 5.35:

$$\frac{T_0}{T} = \left(\frac{p_0}{p}\right)^{(k-1)/k}$$

or

$$p_0 = p\left(\frac{T_0}{T}\right)^{k/(k-1)}. \tag{4}$$

Substituting into Eq. 4, we have

$$p_0 = 44\left(\frac{291.2}{258.15}\right)^{1.4/(1.4-1)} \, kPa$$

$$= 67.1 \, kPa. \qquad\blacksquare$$

9.6 Variations of Velocity and Pressure for Adiabatic Flow Through Passage with Varying Area

When a fluid flows through a passage of varying cross section, its velocity varies from point to point along the passage. If the velocity increases, the passage is called a *nozzle*. If the velocity decreases, the passage is called a *diffuser*. In this section, we shall examine the behavior of steady one-dimensional adiabatic flow of a compressible fluid through such devices.

From the continuity equation for steady flow, we have

$$\rho \bar{V} A = \text{constant.}$$

Taking the logarithmic differentiation, the continuity equation becomes

$$\frac{d\rho}{\rho} + \frac{d\bar{V}}{\bar{V}} + \frac{dA}{A} = 0. \tag{9.20}$$

Neglecting change in potential energy, we have, from the first law for steady-state steady flow with no heat transfer and no work transfer,

$$dh + \frac{\bar{V}\, d\bar{V}}{g_c} = 0. \tag{9.21}$$

Equations 9.20 and 9.21, together with second law of thermodynamics and property relations of the flowing fluid, are the basic tools that we shall use to study the behavior of flow through passages of varying cross section.

Let us now investigate the behavior of isentropic flow involving a simple compressible substance. For an isentropic process, we have, for a simple compressible substance,

$$dh = v\, dp.$$

Substituting into Eq. 9.21, we find that

$$\frac{dp}{\rho} + \frac{\bar{V}\, d\bar{V}}{g_c} = 0. \tag{9.22}$$

Combining Eqs. 9.20 and 9.22, we get

$$\frac{dA}{A} = g_c \frac{dp}{\rho}\left(\frac{1}{\bar{V}^2} - \frac{1}{g_c}\frac{d\rho}{dp}\right). \tag{9.23}$$

Since the process is isentropic, we have

$$\frac{dp}{d\rho} = \left(\frac{\partial p}{\partial \rho}\right)_s.$$

Thus, introducing the velocity of sound into Eq. 9.23, we have

$$\frac{dA}{A} = g_c \frac{dp}{\rho} \left(\frac{1}{\bar{V}^2} - \frac{1}{c^2} \right). \tag{9.24}$$

Introducing the concept of Mach number, Eq. 9.24 may be written as

$$\frac{dA}{A} = \frac{g_c}{\rho \bar{V}^2} (1 - M^2) \, dp. \tag{9.25}$$

Combining Eqs. 9.25 and 9.22 we find that

$$\frac{dA}{A} = -(1 - M^2) \frac{d\bar{V}}{\bar{V}}. \tag{9.26}$$

From Eqs. 9.25 and 9.26 we can make the following interesting observations:

1. When $M < 1$, that is, when the flow is subsonic,

$$\frac{dA}{dp} > 0 \quad \text{and} \quad \frac{dA}{d\bar{V}} < 0.$$

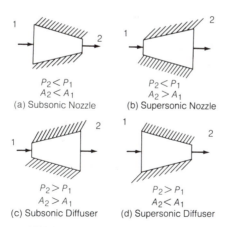

$P_2 < P_1$
$A_2 < A_1$
(a) Subsonic Nozzle

$P_2 < P_1$
$A_2 > A_1$
(b) Supersonic Nozzle

$P_2 > P_1$
$A_2 > A_1$
(c) Subsonic Diffuser

$P_2 > P_1$
$A_2 < A_1$
(d) Supersonic Diffuser

Figure 9.4 Nozzles and Diffusers

This means that a subsonic nozzle must have a converging section. That is, the cross-sectional area must decrease in the direction of increasing velocity. On the other hand, a subsonic diffuser must be a diverging section.

2. When $M > 1$, that is, when the flow is supersonic,

$$\frac{dA}{dp} < 0 \quad \text{and} \quad \frac{dA}{d\bar{V}} > 0.$$

This means that a supersonic nozzle must be a diverging section and a supersonic diffuser must be a converging section.

3. When $M = 1$, that is, when the flow is sonic,

$$\frac{dA}{dp} = 0 \quad \text{and} \quad \frac{dA}{d\bar{V}} = 0.$$

This means that the cross-sectional area must be either a maximum or a minimum. But a maximum is physically impossible. Therefore, the velocity of sound occurs only at the minimum cross section in a converging section.

These results are shown in Fig. 9.4.

9.7 Nozzle and Diffuser Efficiencies

The flow through actual nozzles and diffusers may be modeled as adiabatic, but it is not reversible. The effect of irreversibility, primarily due to friction of the wall, is to increase the entropy of the flowing fluid. In the case of nozzle flow, this effect is shown in Fig. 9.5. In the case of diffuser flow, it is shown in Fig. 9.6.

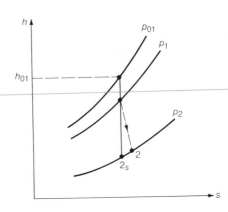

Figure 9.5 Ideal and Actual Process in Nozzle

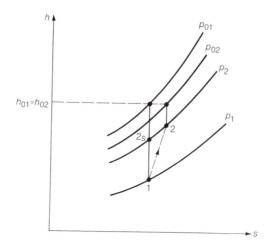

Figure 9.6 Ideal and Actual Process in Diffuser

Nozzle Efficiency

In a nozzle, the objective is to have the maximum kinetic energy leaving the nozzle for the given inlet conditions and exhaust pressure. Since the process may be considered adiabatic, the ideal process used for comparison is then a reversible adiabatic or isentropic one. The *nozzle efficiency* η_N is defined as the ratio of the actual kinetic energy at the nozzle outlet to the kinetic energy for an isentropic process between the same inlet conditions and exhaust pressure. That is,

$$\eta_N = \frac{(\bar{V}^2/2g_c)_{\text{act}}}{(\bar{V}^2/2g_c)_s}. \tag{9.27}$$

Applying the first law of thermodynamics to nozzle flow, we have

$$\left(\frac{\bar{V}^2}{2g_c}\right)_{\text{act}} = h_{01} - h_2$$

and

$$\left(\frac{\bar{V}^2}{2g_c}\right)_s = h_{01} - h_{2s},$$

where h_{01} is the stagnation enthalpy of the flowing fluid at the nozzle inlet. In terms of enthalpy changes, the nozzle efficiency may also be given as

$$\eta_N = \frac{h_{01} - h_2}{h_{01} - h_{2s}}. \tag{9.27a}$$

In nozzle flow the frictional effect is usually of minor importance. Consequently, nozzle efficiencies are generally quite high (on the order of 90 to 95%).

Diffuser Efficiency

In a diffuser the objective is to increase the fluid pressure with a corresponding decrease in velocity. A diffuser, then, is a nozzle in reverse. As in the case of nozzle flow, the ideal process for diffuser flow is isentropic. If the fluid is decelerated isentropically to zero velocity, it would attain a stagnation pressure of p_{01} and a stagnation enthalpy of h_{01}. In the actual flow, the fluid leaves the diffuser at state 2 with a corresponding stagnation pressure of p_{02}, which is seen (Fig. 9.6) to be lower than p_{01}. The stagnation enthalpy h_{02}, however, is equal to h_{01} as a consequence of the first law of thermodynamics. The *diffuser efficiency* is defined as the ratio of the outlet stagnation pressure to the inlet stagnation pressure. That is,

$$\eta_D = \frac{p_{02}}{p_{01}}. \tag{9.28}$$

Diffusers are inherently not as efficient as nozzles. This is because it is more "natural" for the fluid to expand (as in the case of a nozzle) than to be compressed (as in the case of a diffuser). The lower diffuser efficiencies are the result of boundary-layer growth and turbulence, which are two very important concepts in fluid mechanics.

Example 9.10

Air enters a diffuser with a velocity of 250 m/s at 120 kPa and 40°C. It leaves with a velocity of 90 m/s. If the process is isentropic, determine

(a) the Mach number at inlet.
(b) the temperature at exit.
(c) the pressure at exit.
(d) the ratio of the exit area to entrance area.

Assume that air is an ideal gas with constant specific heats with $c_p = 1.0038$ kJ/kg · K and $c_v = 0.7168$ kJ/kg · K.

Solution

(a) At 40°C, the velocity of sound is given by

$$c = \sqrt{1.4 \times 1.0 \times \frac{8314.3}{28.97} \times 313.15} \text{ m/s}$$

$$= 354.7 \text{ m/s}$$

Thus

$$M = \frac{\bar{V}}{c} = \frac{250}{354.7} = 0.705,$$

which means that we have subsonic flow at the entrance.
(b) From the first law, we have

$$0 = h_2 - h_1 + \frac{\bar{V}_2^2 - \bar{V}_1^2}{2g_c}. \tag{1}$$

For an ideal gas with constant specific heats,

$$h_2 - h_1 = c_p(T_2 - T_1). \tag{2}$$

Combining Eqs. 1 and 2 gives us

$$T_2 = T_1 + \frac{\bar{V}_1^2 - \bar{V}_2^2}{2c_p g_c}$$

$$= 313.15 + \frac{(250)^2 - (90)^2}{2 \times 1.0038 \times 1000 \times 1.0} \text{ K}$$

$$= 340.3 \text{ K} = 67.15°C.$$

(c) For an isentropic process involving an ideal gas with constant specific heats, we have

$$p_2 = p_1 \left(\frac{T_2}{T_1}\right)^{k/(k-1)}$$

$$= 120 \left(\frac{340.3}{313.15}\right)^{1.4/(1.4-1)} \text{ kPa}$$

$$= 160.5 \text{ kPa}.$$

(d) From conservation of mass for steady flow,

$$\dot{m} = \rho_1 \bar{V}_1 A_1 = \rho_2 \bar{V}_2 A_2$$

and

$$\frac{A_2}{A_1} = \frac{\rho_1 \bar{V}_1}{\rho_2 \bar{V}_2} \tag{3}$$

where

$$\rho_1 = \frac{p_1}{RT_1}$$

$$= \frac{120 \times 1000}{(8314.3/28.97) \times 313.15} \text{ kg/m}^3$$

$$= 1.335 \text{ kg/m}^3$$

$$\rho_2 = \frac{p_2}{RT_2}$$

$$= \frac{160.5 \times 1000}{(8314.3/28.97) \times 340.3} \text{ kg/m}^3$$

$$= 1.643 \text{ kg/m}^3.$$

Substituting into Eq. 3 yields

$$\frac{A_2}{A_1} = \frac{1.335 \times 250}{1.643 \times 90}$$

$$= 2.26. \qquad \blacksquare$$

Comment

Since flow is subsonic at entrance and $A_2/A_1 = 2.26$, we have a subsonic diffuser.

Example 9.11

Helium gas enters a nozzle from a reservoir at 100 psia and 80°F. If the nozzle efficiency is 95%, what is the velocity at the section of the nozzle where the static pressure is 20 psia? Assume ideal-gas behavior.

Solution

Refer to Fig. 9.5. From the definition of nozzle efficiency,

$$\eta_N = \frac{h_{01} - h_2}{h_{01} - h_{2s}}. \tag{1}$$

Helium is a monatomic gas and its specific heats are constant. Since $dh = c_p \, dT$ for an ideal gas, we have

$$h_{01} - h_2 = c_p(T_{01} - T_2)$$

and

$$h_{01} - h_{2s} = c_p(T_{01} - T_{2s}).$$

Substituting into Eq. 1, we have

$$T_2 = T_{01} - \eta_N(T_{01} - T_{2s}). \tag{2}$$

The temperature T_{2s} may be found from the familiar isentropic relation for an ideal gas with constant specific heats:

$$\frac{T_{01}}{T_{2s}} = \left(\frac{p_{01}}{p_{2s}}\right)^{(k-1)/k}$$

and

$$T_{2s} = \frac{T_{01}}{(p_{01}/p_{2s})^{(k-1)/k}}.$$

Now, for monatomic gas, $k = 1.667$. Thus

$$T_{2s} = \frac{540}{(\frac{100}{20})^{(1.667-1)/1.667}}$$

$$= 283.8°\text{R}.$$

Substituting into Eq. 2, the temperature T_2 is then given by

$$T_2 = 540 - 0.95(540 - 283.8)°\text{R}$$

$$= 296.6°\text{R}.$$

From the first law of thermodynamics for nozzle flow, we have

$$h_{01} = h_2 + \frac{\bar{V}_2^2}{2g_c}$$

and

$$\bar{V}_2 = \sqrt{2g_c(h_{01} - h_2)}.$$

Since $h_{01} - h_2 = c_p(T_{01} - T_2)$ for our problem, the velocity $\bar{V}_2$ may now be given as

$$\bar{V}_2 = \sqrt{2g_c \, c_p(T_{01} - T_2)}. \tag{3}$$

For monatomic ideal gas the constant-pressure specific heat c_p is related to the universal gas constant $\bar{R}$ and the molecular weight of the gas, MW, in the following manner:

$$c_p = \frac{\frac{5}{2}(\bar{R})}{\text{MW}},$$

which gives, for helium,

$$c_p = \frac{\frac{5}{2}(1545.3 \text{ ft} \cdot \text{lbf/lb mol} \cdot {}^\circ\text{R})}{(4.003 \text{ lbm/lb mol})(778 \text{ ft} \cdot \text{lbf/Btu})}$$

$$= 1.24 \text{ Btu/lbm} \cdot {}^\circ\text{R}.$$

Substituting into Eq. 3, we have

$$\bar{V}_2 = \sqrt{2 \times 32.174 \times 1.24 \times (540 - 296.6) \times 778} \text{ ft/s} = 3890 \text{ ft/s}. \quad \blacksquare$$

Comments

If the fluid involved in this problem is air instead of helium, then the specific-heat ratio k is about 1.4 and the constant-pressure specific heat c_p is about 0.240 Btu/lbm · °R, yielding a velocity of only 1510 ft/s instead of 3890 ft/s. It will be seen in the next section that for a fixed initial condition and a fixed outlet pressure, the ideal gas with the largest value of k and R will yield the highest velocity. This is why helium is often used as the working fluid in a high-velocity wind tunnel.

9.8 Mass Flow Through a Converging Nozzle

The equations developed in the previous section are valid for nozzles as well as for diffusers. They are also valid for any simple compressible substance. In this section, we wish to make use of some of the same basic equations to obtain an expression giving the mass-flow rate of an ideal gas flowing through a converging nozzle isentropically. Using Eq. 9.21, we have

$$h_1 + \frac{\bar{V}_1^2}{2g_c} = h_2 + \frac{\bar{V}_2^2}{2g_c} = h + \frac{\bar{V}^2}{2g_c}.$$ (9.29)

Introducing the concept of stagnation enthalpy into Eq. 9.29, we have

$$h_0 = h + \frac{\bar{V}^2}{2g_c}$$

and

$$\bar{V} = \sqrt{2g_c(h_0 - h)}.$$ (9.30)

In the case of ideal gas with constant specific heats, we have

$$h_0 - h = c_p(T_0 - T) = \frac{kR}{k-1}(T_0 - T).$$

Substituting into Eq. 9.30, we find that

$$\bar{V} = \sqrt{\frac{2kg_c\,RT_0}{k-1}\left(1 - \frac{T}{T_0}\right)}.$$ (9.31)

Combining Eq. 9.31 with the continuity equation, we get

$$\dot{m} = \rho A\sqrt{\frac{2kg_c\,RT_0}{k-1}\left(1 - \frac{T}{T_0}\right)}.$$ (9.32)

For an ideal gas undergoing a change of state isentropically, we have

$$\frac{T}{T_0} = \left(\frac{p}{p_0}\right)^{(k-1)/k}$$ (9.33)

and

$$\frac{v}{v_0} = \frac{\rho_0}{\rho} = \left(\frac{p_0}{p}\right)^{1/k}. \tag{9.34}$$

Substituting Eq. 9.33 into Eq. 9.31, we find that

$$\bar{V} = \sqrt{\frac{2kg_c\,RT_0}{k-1}\left[1 - \left(\frac{p}{p_0}\right)^{(k-1)/k}\right]}. \tag{9.35}$$

Equation 9.35 indicates that for a fixed initial condition and a fixed outlet pressure, the ideal gas with the largest value of k and R will yield the highest velocity. From the study of the properties of ideal gases, it is clear that a monatomic gas with low molecular weight provides the desirable combination. This is why helium is often used as the working fluid in a high-velocity wind tunnel.

Combining the continuity equation, Eq. 9.34, and Eq. 9.35, we get

$$\frac{\dot{m}}{A} = p_0\sqrt{\left(\frac{g_c}{RT_0}\right)\left(\frac{2k}{k-1}\right)\left[\left(\frac{p}{p_0}\right)^{2/k} - \left(\frac{p}{p_0}\right)^{(k+1)/k}\right]}. \tag{9.36}$$

Equation 9.36 indicates that for a given value of p_0, T_0, and k, the mass-flow rate per unit area varies as a function of the exit pressure. As the exit pressure decreases, the mass flow increases. When the exit pressure reaches a particular value p^*, there will be a maximum mass flow. This maximum value can be determined analytically by differentiating the left-hand side of Eq. 9.36 with respect to the pressure ratio and equating the result to zero. This pressure ratio corresponding to the maximum flow is called the *critical pressure ratio* and is given by

$$\frac{p^*}{p_0} = \left(\frac{2}{k+1}\right)^{k/(k-1)}. \tag{9.37}$$

The temperature ratio corresponding to the critical pressure ratio is

$$\frac{T^*}{T_0} = \frac{2}{k+1}. \tag{9.38}$$

Introducing the critical pressure ratio into Eq. 9.35, we have

$$\bar{V}^* = \sqrt{\frac{2g_c\,kRT_0}{k+1}}. \tag{9.39}$$

But

$$T^* = \frac{2}{k+1} T_0.$$

Thus

$$\bar{V}^* = \sqrt{kg_c RT^*}. \tag{9.40}$$

But this is just the sonic velocity c for the fluid at the critical temperature T^*. We thus have the interesting result that the maximum mass-flow rate per unit area in a converging nozzle is reached when the exit velocity equals the sonic velocity, or when the Mach number is equal to unity at the exit. Therefore, the pressure at the exit of a converging nozzle cannot fall below the critical value, no matter how low the back pressure (that is, the pressure in the space into which the nozzle is discharging).

The reason why the mass-flow rate cannot be increased above that corresponding to $M = 1$ at the nozzle exit is that when the fluid velocity at the exit is equal to the sonic velocity, pressure changes downstream from the exit cannot be transmitted upstream. The flow from the inlet to the critical section in a sense "does not know" what is happening downstream of the critical region. This phenomenon is called *choked flow*.

Example 9.12

Air enters a nozzle from a reservoir at 2200 kPa and 100°C. If the exit area is 3.25 cm², what is the maximum mass-flow rate that this nozzle can handle? The process is reversible and adiabatic. Assume that air is an ideal gas with constant specific heats with $c_p = 1.0038$ kJ/kg · K and $c_v = 0.7168$ kJ/kg · K.

Solution

To have maximum mass flow, we must have a critical condition at the exit. Now the critical pressure ratio is given by Eq. 9.37:

$$\frac{p^*}{p_0} = \left(\frac{2}{k+1}\right)^{k/(k-1)}$$

$$= \left(\frac{2}{1.4+1}\right)^{1.4/(1.4-1)}$$

$$= 0.528$$

and

$$p^* = 0.528 \times 2200 \text{ kPa}$$

$$= 1161.6 \text{ kPa}.$$

The critical temperature is given by Eq. 9.38:

$$T^* = T_0\left(\frac{2}{k+1}\right)$$

$$= 373.15\left(\frac{2}{1.4+1}\right) \text{K}$$

$$= 310.96 \text{ K}.$$

Then the critical density is given by

$$\rho^* = \frac{p^*}{RT^*}$$

$$= \frac{1161.6 \times 1000}{(8314.3/28.97) \times 310.96} \text{ kg/m}^3$$

$$= 13.0159 \text{ kg/m}^3.$$

The velocity at the exit is the sonic velocity, given as

$$\bar{V}^* = \sqrt{kg_c\, RT^*}$$

$$= \sqrt{1.4 \times 1.0(8314.3/28.97) \times 310.96} \text{ m/s}$$

$$= 353.47 \text{ m/s}$$

From continuity, we have

$$\dot{m} = \rho^* \bar{V}^* A^*$$

$$= 13.0159 \times 354.47 \times 3.25 \times 10^{-4} \text{ kg/s}$$

$$= 1.5 \text{ kg/s}$$

Comments

The reader should realize that a quicker way to obtain the answer of this problem is to make use of Eq. 9.36 once we recognize that the pressure ratio to be used is the critical pressure ratio. It may also be shown that the maximum mass-flow rate $\dot{m}_{max}$ of a nozzle is related to the stagnation pressure p_0, the stagnation temperature T_0, and the throat area A^* in the following manner:

$$\dot{m}_{max} = A^* \sqrt{\frac{g_c k}{R}} \frac{p_0}{\sqrt{T_0}} \left(\frac{2}{k+1}\right)^{(k+1)/2(k-1)}.$$

The reader should use this expression to check the consistency of the result obtained in this example.

Example 9.13

Determine the exit area of an ideal nozzle for a mass flow of air of 1.2 lbm/s flowing from a reservoir at 100 psia and 100°F and discharging into a reservoir maintained at a pressure of 50 psia. Let $k = 1.4$, and assume ideal-gas behavior.

Solution

$$\frac{p_2}{p_0} = \frac{50}{100} = 0.50$$

$$\frac{p^*}{p_0} = \left(\frac{2}{k+1}\right)^{k/(k-1)}$$

$$= \left(\frac{2}{1.4+1}\right)^{1.4/(1.4-1)} = 0.528.$$

Since $p^*/p_0 > p_2/p_0$, we have critical conditions at the exit. Thus

$$\dot{m} = \rho^* \bar{V}^* A^*.$$

To find ρ^* and $\bar{V}^*$, we need to determine T^*. Using Eq. 9.38, we find that

$$T^* = T_0\left(\frac{2}{k+1}\right) = 560\left(\frac{2}{1.4+1}\right) = 466.7°R.$$

Using Eq. 9.40, we find that

$$\bar{V}* = \sqrt{kg_c \, RT*}$$

$$= \sqrt{1.4 \times 32.2 \times \frac{1545}{28.9} \times 466.7} \text{ ft/s} = 1060 \text{ ft/s}.$$

For an ideal gas, we have

$$\rho* = \frac{p*}{RT*} = \frac{100 \times 0.528 \times 144}{(1545/28.9) \times 466.7} = 0.305 \text{ lbm/ft}^3.$$

Substituting the appropriate data into the continuity equation, we find that

$$A* = 0.534 \text{ in}^2. \qquad \blacksquare$$

9.9 Adiabatic Flow in Constant-Area Pipe

For adiabatic steady flow in a constant-area pipe, the energy equation, in terms of stagnation enthalpy, is given by

$$h_0 = h + \frac{\bar{V}^2}{2g_c} = \text{constant}.$$

From the continuity equation for flow in a constant-area pipe, we have

$$\frac{\dot{m}}{A} = \rho \bar{V} = \text{constant}.$$

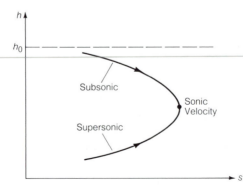

Figure 9.7 Fanno Line

Combining, we find that

$$h_0 = h + \frac{v^2}{2g_c}\left(\frac{\dot{m}}{A}\right)^2.$$ (9.41)

This equation shows that for a given initial condition, the relation between h and v is fixed. If we are dealing with a simple compressible substance, this means that the relation between any two properties of the flowing fluid is also fixed. Thus all the states that satisfy Eq. 9.41 can be plotted on an h–s diagram. The locus of these states on such a diagram is called the *Fanno line*. Figure 9.7 shows such a line for a certain value of h_0 and $\dot{m}/A$.

Since the process is adiabatic, the entropy of the fluid at any point along the pipe must be greater than that at the initial state. Therefore, processes represented by the Fanno line in Fig. 9.7 can proceed only in the directions shown by the arrows on the line.

To investigate the significance of the point at maximum entropy, let us write the energy equation in differential form as

$$dh + \frac{\bar{V}\,d\bar{V}}{g_c} = 0.$$ (9.42)

From the continuity equation, we have

$$\frac{d\rho}{\rho} + \frac{d\bar{V}}{\bar{V}} = 0.$$ (9.43)

Combining Eqs. 9.42 and 9.43, we get

$$dh - \frac{\bar{V}^2}{g_c}\frac{d\rho}{\rho} = 0.$$ (9.44)

For a simple compressible substance, we have

$$T\,ds = dh - v\,dp.$$

At maximum entropy, $ds = 0$ and $dh - v\,dp = 0$. Substituting this into Eq. 9.44, we find that

$$\bar{V}^2 = g_c\frac{dp}{d\rho}.$$ (9.45)

But

$$\frac{dp}{d\rho} = \left(\frac{\partial p}{\partial \rho}\right)_s$$

since $ds = 0$ at this point. Thus

$$\bar{V} = \sqrt{g_c\left(\frac{\partial p}{\partial \rho}\right)_s},$$

which is the velocity of sound.

Hence, at the point of maximum entropy, we have sonic velocity. For subsonic flow, the enthalpy decreases as the velocity increases in the direction of flow. For supersonic flow, the enthalpy increases as the velocity decreases in the direction of flow. Thus the upper part of the Fanno line represents the states of subsonic flow, while the lower part of the line represents the states of supersonic flow.

The physical significance of the point of maximum entropy may be illustrated by considering the flow in a pipe with friction. If in such a case the initial pressure and the discharge pressure are both maintained constant, there is a maximum pipe length that we can use for a given mass-flow rate. However, the variation of pressure, enthalpy, entropy, and velocity of the fluid as a function of pipe length can be predicted only if we know what friction factor to use. These kinds of problems are treated in books on fluid mechanics.

Example 9.14

Nitrogen flows steadily in a well-insulated constant-area pipe. At one section of the pipe, the pressure and temperature of nitrogen are found to be 10.0 MpA and 300 K, respectively. The velocity of nitrogen at this section is 100 m/s. At another section downstream, the pressure is 8 MPa. Determine the temperature and velocity at the second section.

Solution

System selected for study: pipe between section 1 and section 2
Unique features of process: steady-state steady flow; no work transfer
Assumptions: no heat transfer; no change in PE; one-dimensional flow

For this problem, Eq. 9.41 applies:

$$h_0 = h + \frac{v^2}{2g_c} \left(\frac{\dot{m}}{A}\right)^2 .$$

(1)

Now

$$h_0 = h_1 + \frac{\bar{V}_1^2}{2g_c}$$

$$\frac{\dot{m}}{A} = \rho_1 \bar{V}_1 .$$

From the nitrogen tables, we have

$$h_1 = 291.907 \text{ kJ/kg}$$

$$v_1 = 0.008952 \text{ m}^3/\text{kg}.$$

Thus

$$h_0 = 291.907 + \frac{(100)^2}{2 \times 1.0 \times 1000} \text{ kJ/kg} = 296.907 \text{ kJ/kg}$$

$$\frac{\dot{m}}{A} = \frac{1}{0.008952} \times 100 \text{ kg/s} \cdot \text{m}^2 = 11{,}170.7 \text{ kg/s} \cdot \text{m}^2.$$

Substituting into Eq. 1, we have

$$296.907 = h_2 + \frac{v_2^2}{2g_c} \frac{(11{,}170.7)^2}{1000} \text{ kJ/kg}.$$

(2)

where h_2 is in kJ/kg, v_2 is in m^3/kg, and $g_c = 1.0 \text{ kg} \cdot \text{m/N} \cdot \text{s}^2$.

Equation 2 shows that the relation between h_2 and v_2 is fixed. Since p_2 is known, there is only one combination of h_2 and v_2 at p_2 that will satisfy Eq. 2. That is, there is only value of T_2 at given p_2 that will make the right side of Eq. 2 equal to 296.907 kJ/kg. However, we must find T_2 by trial.

Before we begin our trial process, we must recognize whether T_2 is greater or smaller than T_1. Since it is reasonable to assume subsonic flow at section 1, T_2 should be smaller than T_1 according to the Fanno line shown in Fig. 9.7.

Assuming that $T_2 = 290$ K, then together with $p_2 = 8.0$ MPa, we have

$$h_2 = \frac{295.324 + 271.749}{2} = 283.537 \text{ kJ/kg}$$

$$v_2 = \frac{0.011136 + 0.010264}{2} = 0.0107 \text{ m}^3/\text{kg}.$$

With these values of h_2 and v_2, the right-hand side (RHS) of Eq. 2 is

$$\text{RHS of Eq. 2} = 283.537 + \frac{(0.0107)^2}{2} \times \frac{(11,170.7)^2}{1000} \text{ kJ/kg}$$

$$= 290.68 \text{ kJ/kg} < 296.907 \text{ kJ/kg}.$$

Assuming that $T_2 = 295$ K, then together with $p_2 = 8.0$ MPa, we have

$$h_2 = 271.749 + (295.324 - 271.749)\frac{295 - 280}{300 - 280} \text{ kJ/kg}$$

$$= 289.430 \text{ kJ/kg}$$

$$v_2 = 0.010264 + (0.011136 - 0.010264)\frac{295 - 280}{300 - 280} \text{ m}^3/\text{kg}$$

$$= 0.010918 \text{ m}^3/\text{kg}.$$

With these values of h_2 and v_2, the right side of Eq. 2 is

$$\text{RHS of Eq. 2} = 289.430 + \frac{(0.010918)^2}{2 \times 1.0} \times \frac{(11,170.7)^2}{1000} \text{ kJ/kg}$$

$$= 296.9 \text{ kJ/kg}$$

which is essentially what we have on the left-hand side of Eq. 2. Thus the temperature at section 2 is 295 K.

From continuity, we have, with constant area,

$$\bar{V}_2 = \frac{\rho_1 \bar{V}_1}{\rho_2} = \frac{v_2 \bar{V}_1}{v_1}$$

$$= \frac{0.010918 \times 100}{0.008952} \text{ m/s}$$

$$= 122 \text{ m/s.}$$

Comments

We may approximate the velocity of sound at section 2 as

$$c_2 \approx \sqrt{k g_c R T_2}$$

$$= \sqrt{1.4 \times 1.0 \times \frac{8314.3}{28.013} \times 295} \text{ m/s}$$

$$= 350 \text{ m/s.}$$

This means that we have not reached the maximum entropy point along the Fanno line. From tables, we have $s_1 = 5.4168$ kJ/kg · K and $s_2 = 5.4743$ kJ/kg · K. Since process is adiabatic, $(s_2 - s_1)$ is simply the entropy generation of the process. As this quantity is positive, the process is irreversible as expected.

Example 9.15

Air flows steadily in a constant-area pipe with a mass-flow rate of 40 lbm/s · ft². At one section of the pipe, the pressure and temperature of the air are found to be 50 psia and 80°F, respectively. At another section downstream, the pressure is 20 psia. Neglect heat transfer and assume that air is an ideal gas with constant specific heats of $c_p = 0.24$ Btu/lbm · °R and $c_v = 0.171$ Btu/lbm · °R. Determine the Mach number at the second section.

Solution

System selected for study: pipe between section 1 and section 2
Unique features of process: steady-state steady flow; no work transfer
Assumptions: no heat transfer; no change in PE; one-dimensional flow; ideal gas with constant specific heats

To find the Mach number at the second section, we must find the velocity and the temperature at the second section. From the first law and the continuity equation, we have, according to Eq. 9.41,

$$h_1 + \frac{v_1^2}{2g_c}\left(\frac{\dot{m}}{A}\right)^2 = h_2 + \frac{v_2^2}{2g_c}\left(\frac{\dot{m}}{A}\right)^2. \tag{1}$$

From property relations for an ideal gas, we have

$$h_1 - h_2 = c_p(T_1 - T_2) \tag{2}$$

$$v_1 = \frac{RT_1}{p_1} \tag{3}$$

$$v_2 = \frac{RT_2}{p_2}. \tag{4}$$

Combining Eqs. 1, 2, 3, and 4, we have

$$c_p(T_1 - T_2) = \left(\frac{\dot{m}}{A}\right)^2 \frac{R^2}{2g_c}\left[\left(\frac{T_2}{p_2}\right)^2 - \left(\frac{T_1}{p_1}\right)^2\right]. \tag{5}$$

Substituting the appropriate data into Eq. 5, we have

$$(T_2)^2 + 22{,}000\,T_2 - 11{,}910{,}000 = 0 \tag{6}$$

with T_2 in degrees Rankine. Solving Eq. 6, we get $T_2 = 529°R$. For an ideal gas, the velocity of sound is given by Eq. 9.17,

$$c = \sqrt{kg_c\,RT}.$$

At $T_2 = 529°R$, we find that

$$c_2 = \sqrt{\frac{1.4 \times 32.2 \times 1545 \times 529}{28.9}} \text{ ft/s}$$

$$= 1129 \text{ ft/s.}$$

From continuity, we have

$$\frac{\dot{m}}{A} = \rho_2 \bar{V}_2 = \frac{p_2}{RT_2} \bar{V}_2 ,$$

from which we get

$$\bar{V}_2 = \frac{40 \times 1545 \times 529}{28.9 \times 20 \times 144} \text{ ft/s}$$

$$= 393 \text{ ft/s.}$$

Thus the Mach number at section 2 is

$$M_2 = \frac{\bar{V}_2}{c_2} = \frac{393}{1129} = 0.348.$$

Comments

Since the velocity of the air has not reached sonic, the entropy of the air has not reached the maximum. It may be shown that for adiabatic flow in a constant-area pipe with a given mass-flow rate of ideal gas, the relation between M, p, and T at any point is given by

$$M = \frac{\dot{m}/A}{p} \sqrt{\frac{RT}{g_c k}}.$$

The reader should use this expression to check the consistency of the results obtained in this example.

Problems

9.1 Show that the stagnation pressure, p_0, for an incompressible fluid may be given by

$$p_0 = p + \frac{\rho \bar{V}^2}{2g_c}.$$

9.2 Determine the stagnation pressure for water at 100°C and 200 kPa with a velocity of 50 m/s. Assume that water is incompressible.

9.3 Determine the stagnation pressure and temperature on a jet plane traveling at 500 m/s through air at 12 kPa and −20°C. Assume ideal-gas behavior.

9.4 Steam at 200 psia has a stagnation enthalpy of 1270.8 Btu/lbm and a velocity of 300 ft/s. What is the static temperature and stagnation entropy?

9.5 The melting temperature of aluminum is about 660°C. An aluminum probe is inserted into a supersonic airstream. What is the minimum velocity at which the aluminum will melt if the static stream temperature is 30°C? Assume ideal-gas behavior with constant specific heats of $c_p = 1.004$ kJ/kg · K.

9.6 Air enters a constant area duct, 25 cm by 40 cm, at 110 kPa and 24°C with a velocity of 50 m/s. It leaves at 108 kPa and 240°C. The process is steady-state steady flow. Determine
  (a) the velocity of the air at the exit.
  (b) the mass-flow rate.
  Assume that air is an ideal gas with constant specific heats of $c_p = 1.004$ kJ/kg · K.

9.7 Water at 20°C flows steadily into a 4-cm-diameter pipe with a velocity of 5 m/s at 1000 kPa. The water leaves the pipe at 100°C. Neglecting friction, what is the static pressure at the exit?

9.8 The stagnation pressure and static pressure can be measured with an instrument known as the *pitot tube*. Information on these two types of pressure may then be used to determine the velocity of the fluid. Show that the velocity of an incompressible fluid may be expressed in terms of the stagnation pressure p_0 and the static pressure p as

$$\bar{V} = \sqrt{2g_c\, v(p_0 - p)},$$

where v is the specific volume.

9.9 Water flows in a horizontal duct that has a 6-in.-diameter section connected to an 8-in.-diameter section. If the stagnation pressure and static pressure in the 6-in. section are found to be

33 psia and 30 psia, respectively, what is the static pressure in the 8-in. section? Assume that the flow process is reversible.

9.10 Water at 70°F flows in a horizontal pipe. The friction factor is 0.008. The pressure drop due to friction is 1.0 psi for 100 ft of run. What is the mass-flow rate that a 6-in.-diameter pipe can deliver?

9.11 Steam leaves a nozzle with a velocity of 300 m/s, a stagnation pressure of 1000 kPa, and a stagnation temperature of 300°C. Determine
(a) the static pressure.
(b) the static temperature.

9.12 Helium is expanded reversibly and adiabatically in a horizontal duct from 750 kPa and 60°C to 150 kPa. The initial velocity is 60 m/s. Determine the stream velocity at the final state. Assume ideal-gas behavior.

9.13 Water enters a 4-cm-diameter pipe at 2000 kPa and 30°C, with a velocity of 2.0 m/s, and leaves at an elevation of 25 m above that at entrance. If the frictional pressure drop is 15 kPa, determine the pressure at exit.

9.14 Assuming that flow is reversible, what is the pressure in a 3-in.-diameter fire hose just upstream from a 1.0-in.-diameter nozzle for a discharge of 1.0 ft^3/s?

9.15 Assuming ideal-gas behavior, determine the velocity of sound for the following cases:
(a) air at 25°C.
(b) carbon monoxide at 25°C.
(c) argon at 25°C.

9.16 Assuming ideal-gas behavior, determine the velocity of sound for steam at 10 kPa and 200°C.

9.17 Air expands isentropically from 1500 kPa and 100°C to 500 kPa. Determine the ratio of the final to initial velocity of sound. Assume ideal-gas behavior with constant specific heats and $k = 1.4$.

9.18 Air flows into the test section of a wind tunnel at a Mach number of 1.5 and at a pressure and temperature of 14.7 psia and 70°F. What temperature rise would take place on a small object immersed in the flow at the test section?

9.19 For fluid flow involving an ideal gas with constant specific heats, show that the stagnation temperature T_0 is related to the static temperature T and the velocity $\bar{V}$ or the Mach number M in the following manner:

$$\frac{T_0}{T} = 1 + \frac{\bar{V}^2}{2g_c\,c_p\,T}$$

$$= 1 + \frac{k-1}{2}\,M^2.$$

9.20 For fluid flow involving an ideal gas with constant specific heats, show that the stagnation pressure p_0 is related to the static pressure p and the velocity $\bar{V}$ or the Mach number M in the following manner:

$$\frac{p_0}{p} = \left(1 + \frac{\bar{V}^2}{2g_c\,c_p\,T}\right)^{k/(k-1)}$$

$$= \left(1 + \frac{k-1}{2}\,M^2\right)^{k/(k-1)}.$$

9.21 Helium at 250 kPa and 100°C flows through a 40 cm by 50 cm rectangular duct with a velocity of 500 m/s. Determine the Mach number, stagnation temperature, stagnation pressure, and the mass-flow rate. Assume ideal-gas behavior.

9.22 Water enters a nozzle at 750 kPa and 25°C and leaves the nozzle at 150 kPa. For an inlet velocity of 3 m/s, what is the ideal velocity at the outlet?

9.23 Water enters a nozzle at 1000 kPa and 50°C and leaves at 150 kPa with a velocity of 35 m/s. Neglecting the inlet velocity, what is the temperature of water at the nozzle outlet?

9.24 An impact tube extending ahead of an airplane wing measures the stagnation pressure to be 19.5 psia when the undisturbed air has a pressure and temperature of 14.7 psia and 70°F. What is the speed of the airplane?

9.25 Oxygen flows steadily and isentropically through a passage. At section 1, where the cross-sectional area is 0.2 m², the oxygen is at 70 kPa and 5°C with a velocity of 600 m/s. At section 2 downstream, the gas is at a pressure of 340 kPa. Determine
(a) the cross-sectional area at section 2.
(b) the Mach number at section 2.
Assume ideal-gas behavior with constant specific heats and $k = 1.4$.

9.26 Nitrogen at 1500 kPa and 100°C expands adiabatically and reversibly in a nozzle. An exit velocity of 600 m/s is desired. Neglecting inlet velocity, determine
(a) the Mach number at exit.
(b) the pressure at exit.
Assume ideal-gas behavior with constant specific heats and $k = 1.4$.

9.27 Steam expands through a nozzle from stagnation conditions of 200 psia and 800°F to static conditions of 40 psia and 450°F. What is the nozzle efficiency?

9.28 Superheated steam flows from a boiler where the pressure is 1.0 MPa and the temperature is 500°C. Find the velocity in a nozzle at a point where the pressure is 100 kPa if process is reversible.

9.29 Steam at 6.0 MPa and 400°C expands to 4.0 MPa in a nozzle the efficiency of which is 95%. For a nozzle exit area of 6.5 cm², determine the mass-flow rate.

9.30 Water enters a nozzle steadily at 100°C and 2000 kPa with negligible velocity. It leaves at a pressure of 500 kPa. For a nozzle efficiency of 90%, determine
(a) the outlet velocity.
(b) the outlet temperature.

9.31 Show that for a nozzle involving the isentropic flow of an ideal gas with constant specific heats, the maximum mass-flow rate $\dot{m}_{max}$ is related to the stagnation pressure p_0, the stagnation temperature T_0, and the throat area A^* in the following manner:

$$\dot{m}_{max} = A^* \sqrt{\frac{g_c k}{R}} \frac{p_0}{\sqrt{T_0}} \left(\frac{2}{k+1}\right)^{(k+1)/2(k-1)}.$$

9.32 Helium expands reversibly and adiabatically in a nozzle from 2 MPa and 200°C to 0.5 MPa. Determine the maximum mass flow for a throat area of 0.02 m². Assume ideal-gas behavior.

9.33 Helium expands reversibly and adiabatically in a nozzle from 2 MPa and 200°C to 0.5 MPa. The mass-flow rate is 5 kg/s. Determine the throat and exit areas for the nozzle. Assume ideal-gas behavior.

9.34 Air enters a diffuser at a Mach number of 3.0 and at a pressure and temperature of 15 psia and 80°F. The flow rate is 5.0 lbm/s. Assuming isentropic flow, determine the throat area.

9.35 Helium, coming from a reservoir at 100 psia and 80°F, flows through a convergent nozzle with an outlet area of 1.50 in². Calculate the maximum discharge from the nozzle if flow is isentropic.

9.36 Air expands reversibly and adiabatically from 4 MPa and 300°C to 2 MPa. The mass-flow rate is 5 kg/s. Determine
(a) the exit area.
(b) the outlet velocity.
Assume ideal-gas behavior with constant specific heats and $k = 1.4$.

9.37 Helium expands isentropically in a convergent–divergent nozzle from stagnation conditions of 100 psia and 80°F to a pressure of 15 psia. For a flow rate of 1.0 lbm/s, determine
(a) the throat area.
(b) the outlet velocity.

9.38 Steam at a pressure of 1000 kPa and a temperature of 300°C expands to a pressure of 400 kPa in a nozzle having an efficiency of 90%. For a mass-flow rate of 5 kg/s, determine the nozzle exit area.

9.39 A convergent–divergent nozzle, having a throat area of 1.0 in², is required to discharge air at a pressure of 15 psia and a velocity of 1800 ft/s. The air supply is a reservoir at 100 psia and 80°F. Assuming isentropic flow, determine
(a) the mass-flow rate.
(b) the outlet area.
Consider air to be an ideal gas with constant specific heats of $c_p = 0.240$ Btu/lbm · °R and $c_v = 0.171$ Btu/lbm · °R.

9.40 A diffuser is attached to the outlet of the test section of a wind tunnel. At the diffuser inlet, the air is at 70 kPa and 5°C with a Mach number of 1.2. At the outlet of the diffuser, the air is 125 kPa and 75°C. Assuming that the process is adiabatic, determine
(a) the velocity of air leaving the diffuser.
(b) the entropy generation for the process, in kJ/kg · K.

9.41 Air flows in a well-insulated constant-area pipe. At section 1, its pressure and temperature are 500 kPa and 20°C, respectively. Its velocity is 500 m/s. At section 2 downstream, the pressure is 300 kPa. Will the temperature of the air at section 2 be greater or smaller than the temperature of the air at section 1? Explain.

9.42 Helium flows in a well-insulated constant-area duct. At section 1, its pressure and temperature are 500 kPa and 20°C. Its velocity is 800 m/s. At section 2 downstream, the pressure is 400 kPa. Will the temperature of helium at section 2 be greater or smaller than the temperature of helium at section 1? Explain.

9.43 Helium enters a well-insulated 6-cm-diameter pipe at the rate of 5 kg/s. If the temperature and pressure of the gas at the entrance are 100°C and 500 kPa, respectively, determine the minimum pressure and maximum velocity that can occur in the pipe.

9.44 Air enters a constant-area pipe with a velocity of 500 ft/s at 30 psia and 80°F and leaves with a velocity that is sonic. Assuming that flow is adiabatic, determine the temperature and pressure at the exit of the pipe. Assume ideal gas with constant specific heats of $c_p = 0.240 \, \text{Btu/lbm} \cdot {}^\circ\text{R}$ and $c_v = 0.171 \text{Btu/lbm} \cdot {}^\circ\text{R}$.

9.45 Steam at 200 psia and 600°F is expanded in a nozzle to 30 psia. The throat area is 1 in², and the inlet velocity is negligible. Assuming that the flow is isentropic and taking the critical-pressure ratio (the ratio of pressure at the throat to the stagnation pressure) to be 0.55, determine
   (a) the mass rate of flow, in lbm/s.
   (b) the area at the exit.

9.46 Steam expands isentropically through a nozzle from 200 psia and 600°F to 40 psia. The initial velocity is negligible. The mass-flow rate is 5 lbm/s. Determine, from a plot of nozzle area versus static pressure,
   (a) the throat area.
   (b) the critical-pressure ratio.
   Make calculations along the nozzle where the static pressure is 140, 130, 120, 110, 100, 90, and 80 psia.

9.47 An ideal gas with constant specific heat enters a nozzle at 500 kPa and 300 K. It leaves the nozzle at 100 kPa. The process is steady-state steady flow and isentropic. Write a computer

program to calculate the velocity, in m/s, at the nozzle outlet for
   (a) air
   (b) helium
   (c) argon
   (d) carbon dioxide.
   Graph your results and comment.
   Assume constant specific heat at 300 K. Neglect velocity at nozzle inlet.

9.48 Air enters a nozzle with negligible velocity. Temperature and pressure at nozzle inlet are 1000 K and 10 atm respectively. Pressure at nozzle outlet is 1 atm. Process is steady-state steady flow and isentropic. On the basis of ideal gas behavior, write a computer program to determine the velocity, in m/s,
   (a) assuming constant specific heat at 1000 K.
   (b) taking into consideration the variation of specific heat with temperature.
   The molal heat capacity as a function of temperature for air is given by

$$\bar{c}_p^* = a + bT + cT^2 + dT^3$$

where $\bar{c}_p^*$ is in kJ/kgmol $\cdot$ K,
   T is in degrees Kelvin,
   a, b, c, and d are constants having the following values:

$$a = 28.11$$

$$b = 0.196 \times 10^{-2}$$

$$c = 0.4802 \times 10^{-5}$$

$$d = -1.966 \times 10^{-9}.$$

9.49 An ideal gas enters a nozzle with specified pressure p_1 and temperature T_1. Process is steady-state steady flow and isentropic. Write a computer program to determine the cross-sectional area A, velocity $\bar{V}$, and temperature T as a function of the local pressure p, where p is any value less than the inlet pressure p_1.

Make calculations for the following design conditions:

ideal gas:	air
mass flowrate:	5.0 kg/s
pressure at nozzle inlet:	1000 kPa
temperature at nozzle inlet:	300 K
pressure at nozzle outlet:	100 kPa.

Graph your results. What is the pressure at the throat of the nozzle?
Assume constant specific heat at 300 K. Neglect velocity at nozzle inlet.

9.50 Same as Problem 9.49 except that the gas is helium.

9.51 An ideal gas enters a nozzle with negligible velocity and specified inlet pressure p_1 and inlet temperature T_1. Process is steady-state steady flow and isentropic. Assuming constant specific heat at T_1, write a computer program to calculate the mass flowrate per unit area, in kg/s · m², for different reservoir pressure.
Make calculations for the following design conditions:

ideal gas:	air
pressure at nozzle inlet:	1000 kPa
temperature at nozzle inlet:	300 K
pressure of reservoir:	800, 700, 600, 500, 400, 300 kPa.

Graph your results and comment.
Assume constant specific heat at 300 K.

9.52 Same as Problem 9.51 except that the gas is argon.

9.53 Write a computer program to calculate the pressure p, the density ρ, the temperature T, and entropy s as a function of the local Mach number M for the adiabatic flow of air along a constant area pipe under steady-state steady-flow conditions.
Let the initial conditions be as follows:

$$p_1 = 1.00 \text{ kPa}$$

$$T_1 = 100 \text{ K}$$

$$M_1 = 3.0$$

$$s_1 = 1.4143 \text{ kJ/kg} \cdot \text{K}.$$

Make calculations for M ranging from 2.5 to 0.3.
Assume air is an ideal gas with constant specific heats of

$$c_p = 1.0038 \text{ kJ/kg} \cdot \text{K}$$

and

$$c_v = 0.7168 \text{ kJ/kg} \cdot \text{K}.$$

Tabulate your results, plot the Fanno line in the T–s diagram, and comment.

CHAPTER 10

Availability, Irreversibility, and Availability Analysis of Engineering Processes

There are essentially two related approaches that we can use to quantify the irreversible nature of real processes and to perform analyses based on the combined first and second laws of thermodynamics commonly called second-law analysis. One approach makes use of the concept of entropy generation and lost work, which have been already introduced. The second approach makes use of the concept of availability (exergy) and irreversibility (loss of availability or exergy destruction). We concentrate on the second approach in this chapter.

To carry out second-law or availability analysis based on the second approach, we must develop the availability or exergy equation, which is simply the combined first and second laws of thermodynamics. We shall see that the thermodynamic lost work is indeed identical to the loss of availability (irreversibility) or exergy destruction, as we pointed out previously. By employing the concept of availability or exergy, we may introduce the concept of second-law or exergetic efficiency as a more accurate and more meaningful measure of thermodynamic performance.

The availability equation is similar to the energy equation except for one fundamental difference. While the energy equation is a statement of the law of conservation of energy, the availability or exergy equation may be looked upon as a statement of the principle of degradation of energy, which may be simply stated as

exergy in − exergy out

$$= \text{change in stored exergy} + \text{exergy destruction} \quad (10.1)$$

where exergy destruction is a positive nonzero quantity for any real process. Equation 10.1 is based on the observation of nature that useful energy (exergy) is not conserved except in a thermodynamically perfect process.

To make use of Eq. 10.1 effectively in engineering design, we must learn to identify and calculate the exergy content in different forms of energy. This chapter provides readers with some of the fundamental concepts involved. For detailed treatment of second-law analysis, however, other books on this subject should be consulted.[1]

10.1 Availability (Exergy) Equation for a Closed System

The first law of thermodynamics for a closed system with no changes in kinetic and potential energies is given by Eq. 2.9a as

$$dU = \bar{d}Q - \bar{d}W. \tag{2.9a}$$

The second law of thermodynamics for a closed system is given by Eq. 3.23a as

$$dS = \frac{\bar{d}Q}{T} + \bar{d}S_{irr}, \tag{3.32a}$$

where T is the temperature on the surface of the system where the heat transfer occurs, and $\bar{d}S_{irr}$ is entropy generation due to internal irreversibility.

Multiplying the entropy equation by T_0, the temperature of the environment, and combining it with the energy equation, we have

$$dU - T_0\,dS = \left(1 - \frac{T_0}{T}\right)\bar{d}Q - \bar{d}W - T_0\,\bar{d}S_{irr}. \tag{10.2}$$

[1] R. A. Gaggioli, ed., *Thermodynamics: Second Law Analysis*, ACS Symposium Series 122, American Chemical Society, Washington, D.C., 1980; R. A. Gaggioli, ed., *Efficiency and Costing: Second Law Analysis of Processes*, ACS Symposium Series 235, American Chemical Society, Washington, D.C., 1983; M. J. Moran, *Availability Analysis: A Guide to Efficient Energy Use*, Prentice-Hall, Inc., Englewood Cliffs, N.J., 1982; T. J. Kotas, *The Exergy Method of Thermal Plants Analysis*, Butterworth & Company (Publishers) Ltd., London, 1985; J. E. Ahern, *The Exergy Method of Energy Systems Analysis*, John Wiley & Sons, Inc., New York, 1980; A. Bejan, *Entropy Generation Through Heat and Fluid Flow*, John Wiley & Sons, Inc., New York, 1982; Ali B. Cambel, ed., "Second Law Analysis of Energy Devices and Processes," *ENERGY*, Vol. 5, No. 8–9 (1980).

Adding the term $p_0 \, dV$ to both sides of Eq. 10.2, we have

$$d(U + p_0 V - T_0 S) = \left(1 - \frac{T_0}{T}\right) dQ - (dW - p_0 \, dV) - T_0 \, dS_{irr},$$

(10.3)

where p_0 is the pressure of the environment.

Equation 10.3 may be used to study any process for a closed system in which changes in kinetic and potential energies are negligible.

Let us apply Eq. 10.3 to a unique reversible process in which heat interaction is limited to that with the environment and in which the final state of the system is in thermodynamic equilibrium with the environment. Then, from integrating Eq. 10.3 we would have the following expression for this unique reversible process:

$$W_{rev} - p_0(V_0 - V) = (U + p_0 V - T_0 S) - (U_0 + p_0 V_0 - T_0 S_0).$$

(10.4)

On the basis of unit mass, we have

$$w_{rev} - p_0(v_0 - v) = (u + p_0 v - T_0 s) - (u_0 + p_0 v_0 - T_0 s_0). \quad (10.4a)$$

Now Eq. 10.4a is simply the concept of availability or nonflow exergy, ϕ, obtained previously, that is,

$$d\phi = d(u + p_0 v - T_0 s).$$

Making use of the concept of nonflow exergy, Eq. 10.3 may now be written as

$$d\Phi = \left(1 - \frac{T_0}{T}\right) dQ - (dW - p_0 \, dV) - T_0 \, dS_{irr}, \quad (10.5)$$

where $\Phi = m\phi$.

Equation 10.5 is the availability or exergy equation for a closed system in differential form. It may be used to study any process in which changes in kinetic and potential energies are negligible. On the basis of unit mass, we have

$$d\phi = \left(1 - \frac{T_0}{T}\right) dq - (dw - p_0 \, dv) - T_0 \, ds_{irr}. \quad (10.5a)$$

Integrating Eq. 10.5, we have

$$\Phi_2 - \Phi_1 = \int_1^2 \left(1 - \frac{T_0}{T}\right) dQ - [W_{12} - p_0(V_2 - V_1)] - T_0 S_{gen}, \quad (10.6)$$

where

$$\Phi = m[(u + p_0 v - T_0 s) - (u_0 + p_0 v_0 - T_0 s_0)].$$

Equation 10.6 is the availability or exergy equation for a closed system in integral form. The quantity $T_0 S_{gen}$ is known as the irreversibility I, which we recognize as the thermodynamic lost work. That is,

$$T_0 S_{gen} = I, \text{ irreversibility}$$

$$= LW_0, \text{ thermodynamic lost work} \qquad (10.7)$$

From Eq. 10.6 we note that for a given change of state, the amount of exergy transfer associated with heat transfer is given by

$$\int_1^2 \left(1 - \frac{T_0}{T}\right) \bar{d}Q,$$

the amount of exergy transfer due to work transfer is given by

$$- [W_{12} - p_0(V_2 - V_1)],$$

and the quantity $T_0 S_{gen}$ is the amount of exergy destroyed due to thermodynamic imperfection.

If we had included the changes in kinetic and potential energies in the energy equation, the complete availability or exergy equation for a closed system would be given as

$$\Phi_2 - \Phi_1 = \int_1^2 \left(1 - \frac{T_0}{T}\right) \bar{d}Q - [W_{12} - p_0(V_2 - V_1)]$$

$$- m\left(\frac{\bar{V}_2^2 - \bar{V}_1^2}{2g_c}\right) - m\left[\frac{g}{g_c}(Z_2 - Z_1)\right] - T_0 S_{gen}. \qquad (10.8)$$

Equation 10.8 implies that kinetic and potential energies are pure exergies. This is in agreement with our study from mechanics that change in kinetic and potential energies can be completely converted into useful work in the absence of friction.

Now the first law for a closed system is

$$\bar{d}Q = dU + \bar{d}W. \qquad (2.9a)$$

Combining Eqs. 10.5 and 2.9a, we have

$$T_0 S_{gen} = -(\Phi_2 - \Phi_1) - T_0 \int_1^2 \frac{\bar{d}Q}{T} + p_0(V_2 - V_1) + (U_2 - U_1). \quad (10.9)$$

By definition,

$$\Phi_2 - \Phi_1 = (U_2 - U_1) + p_0(V_2 - V_1) - T_0(S_2 - S_1).$$

Substituting into Eq. 10.9, we have

$$T_0 S_{gen} = T_0(S_2 - S_1) - T_0 \int_1^2 \frac{\bar{d}Q}{T} \quad (10.10)$$

and

$$S_{gen} = (S_2 - S_1) - \int_1^2 \frac{\bar{d}Q}{T},$$

which is simply Eq. 3.23a, the second law for a closed system. This is consistent with our statement that the availability equation is really just the combined first and second laws of thermodynamics.

10.2 Second-Law Analysis of Closed Systems

We have now developed the necessary tools to make second-law analysis of closed systems. For convenience, these tools are summarized in this section.

1. From availability, the principle of degradation of energy, we have

$$I = T_0 S_{gen} = -(\Phi_2 - \Phi_1) - [W_{12} - p_0(V_2 - V_1)] + \sum \left(1 - \frac{T_0}{T_k}\right) Q_k \quad (10.11)$$

$$W_{rev,\,useful} = -(\Phi_2 - \Phi_1) + \sum \left(1 - \frac{T_0}{T_k}\right) Q_k \quad (10.12)$$

$$W_{12,\,useful} = W_{12} - p_0(V_2 - V_1) \quad (10.13)$$

$$I = T_0 S_{gen} = W_{rev,\,useful} - W_{12,\,useful}. \quad (10.14)$$

2. From the second law of thermodynamics, we have

$$S_{gen} = (S_2 - S_1) - \sum \frac{Q_k}{T_k}. \tag{10.15}$$

3. From the first law of thermodynamics, we have

$$W_{12} = Q_{12} - (U_2 - U_1), \tag{10.16}$$

where $Q_{12} = \sum Q_k$.

Example 10.1

Making use of the availability equation, determine the maximum thermal efficiency of a heat engine operating between a high-temperature heat reservoir at T_H and a low-temperature heat reservoir at T_L.

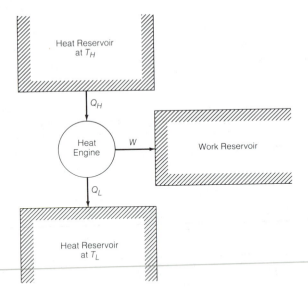

Solution

Applying Eq. 10.11 to a cyclic process, we have

$$\oint d\Phi = \oint \left(1 - \frac{T_0}{T}\right) \bar{d}Q - \oint \bar{d}W - P_0 \oint dV - T_0 S_{gen}. \tag{1}$$

Since Φ and V are properties of the system,

$$\oint d\Phi = 0$$

$$\oint dV = 0.$$

For maximum thermal efficiency, the process must be reversible and

$$T_0 S_{\text{gen}} = 0.$$

Substituting into Eq. 1, we have, for maximum thermal efficiency,

$$\oint \bar{d}W = \oint \left(1 - \frac{T_0}{T}\right) \bar{d}Q. \tag{2}$$

Now $\oint \bar{d}W = W$, net work for cycle, and

$$\oint \bar{d}Q = \left(1 - \frac{T_0}{T_H}\right) Q_H - \left(1 - \frac{T_0}{T_L}\right) Q_L, \tag{3}$$

where Q_H and Q_L in Eq. 3 are absolute values. From the first law for a cyclic process, we have

$$W = Q_H - Q_L. \tag{4}$$

Combining Eqs. 2, 3, and 4, we have

$$\frac{T_L}{T_H} = \frac{Q_L}{Q_H}. \tag{5}$$

Now the definition of thermal efficiency of a heat engine cycle is

$$\eta_{\text{th}} = 1 - \frac{|Q_{\text{out}}|}{|Q_{\text{in}}|}.$$

Thus the maximum thermal efficiency of our cycle is

$$(\eta_{\text{th}})_{\text{ideal}} = 1 - \frac{T_L}{T_H},$$

which is what we should expect from our previous study.

Comments

This problem shows that if the heat sink is at the temperature T_0 of the environment, then, for a given amount of Q_H, the maximum useful work obtainable from it is given by

$$W_{\text{max useful}} = \left(1 - \frac{T_0}{T_H}\right) Q_H .$$

The factor $(1 - T_0/T_H)$ is sometimes called the exergetic factor of the energy content in a heat reservoir at T_H.

Example 10.2

A perfectly elastic spring is compressed with an input of 5000 kJ of work. What is the change of the exergy content of the spring?

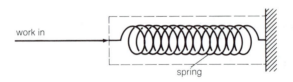

work in

spring

Solution

System selected for study: spring
Unique feature of process: work transfer
Assumptions: spring is perfectly elastic; neglect heat transfer; neglect volume change

Since the spring is perfectly elastic, there will be no entropy generation due to internal irreversibility. That is,

$$I = T_0 S_{\text{gen}} = 0.$$

Neglecting change of volume, we have

$$\Phi_2 - \Phi_1 = -(-5000) \text{ kJ}$$

$$= +5000 \text{ kJ}. \qquad \blacksquare$$

Comment

This problem shows that all exergy supplied to a perfectly elastic spring is conserved as exergy in the spring. This is why a perfectly elastic spring may be considered as a useful work reservoir in which all its energy content is 100% exergy.

Example 10.3

Oxygen is compressed inside a cylinder from 300 K and 1.0 MPa to 300 K and 10 MPa. The actual work required is 200 kJ/kg. Pressure and temperature of the environment are 101.325 kPa and 300 K, respectively. For the process, determine

(a) the actual useful work, in kJ/kg.
(b) the ideal useful work, in kJ/kg.
(c) the irreversibility per unit mass.

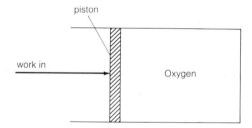

Solution

System selected for study: oxygen inside cylinder
Unique feature of process: isothermal
Assumption: oxygen is a simple compressible substance

(a)
$$w_{12,\,useful} = w_{12} - p_0(v_2 - v_1). \tag{1}$$

From oxygen tables,

$$v_2 = 0.007430 \text{ m}^3/\text{kg}$$

$$v_1 = 0.077457 \text{ m}^3/\text{kg}.$$

Substituting into Eq. 1, we have

$$w_{12,\,useful} = -200 - 101.325(0.007430 - 0.077457) \text{ kJ/kg}$$

$$= -200 + 7.095 \text{ kJ/kg}$$

$$= -192.905 \text{ kJ/kg}.$$

(b) Making use of Eq. 10.12, we have

$$w_{\text{rev, useful}} = -(\phi_2 - \phi_1) - \int_1^2 \left(1 - \frac{T_0}{T}\right) \bar{d}Q$$

$$= -(\phi_2 - \phi_1) \qquad \text{with } T = T_0$$

$$= -[(u_2 - u_1) + p_0(v_2 - v_1) - T_0(s_2 - s_1)]. \qquad (2)$$

From oxygen tables,

$$u_1 = h_1 - p_1 v_1$$

$$= 270.572 - 1000 \times 0.077457 \text{ kJ/kg}$$

$$= 193.115 \text{ kJ/kg}$$

$$u_2 = h_2 - p_2 v_2$$

$$= 249.609 - 10{,}000 \times 0.007430 \text{ kJ/kg}$$

$$= 175.309 \text{ kJ/kg}$$

$$s_1 = 5.8119 \text{ kJ/kg} \cdot \text{K}$$

$$s_2 = 5.1566 \text{ kJ/kg} \cdot \text{K}.$$

Substituting into Eq. 2, we have

$$w_{\text{rev, useful}} = -[(175.309 - 193.115) + 101.325(0.007430 - 0.077457)$$

$$- 300(5.1566 - 5.8119)] \text{ kJ/kg}$$

$$= -171.689 \text{ kJ/kg}.$$

(c) Making use of Eq. 10.14, we have

$$\frac{I}{m} = w_{\text{rev, useful}} - w_{12, \text{useful}}$$

$$= -171.689 - (-192.905) \text{ kJ/kg}$$

$$= +21.216 \text{ kJ/kg}.$$

This result could also be obtained using Eq. 10.11.   ∎

Comment

For the given process, we have, from the first law,

$$q_{12} = u_2 - u_1 + w_{12}$$

$$= 175.309 - 193.115 + (-200) \text{ kJ/kg}$$

$$= -217.806 \text{ kJ/kg},$$

and from the second law,

$$\frac{S_{gen}}{m} = s_2 - s_1 - \frac{q_{12}}{T}$$

$$= 5.1566 - 5.8119 - \frac{-217.806}{300} \text{ kJ/kg} \cdot \text{K}.$$

Then

$$\frac{I}{m} = 300(5.1566 - 5.8119 + \frac{217.806}{300} \text{ kJ/kg}$$

$$= +21.216 \text{ kJ/kg},$$

which is the same result as for part (c).

Example 10.4

Nitrogen in tank A is to be evacuated and charged to tank B by the compressor, as shown. Tank A has a volume of 5.0 m³ and initially contains nitrogen at 200 kPa and 300 K. Tank B is initially evacuated and its volume is such that it will hold all the gas originally in tank A at a pressure of 1000 kPa and 300 K. The temperature of the environment is 300 K. Determine the minimum work needed to drive the compressor.

Solution

System selected for study: tank A, tank B, and compressor
Unique features of process: reversible; isothermal
Assumption: nitrogen is a simple compressible substance

This problem may be interpreted as that of changing all the nitrogen from the state in tank A to a final state in tank B. Thus we may make use of Eq. 10.11:

$$I = -(\Phi_2 - \Phi_1) - [W_{12} - p_0(v_2 - v_1)] + \int_1^2 \left(1 - \frac{T_0}{T}\right) dQ. \qquad (1)$$

To get minimum work, $I = 0$. With $T = T_0$, Eq. 1 then becomes

$$W_{rev} = -(\Phi_2 - \Phi_1) + p_0(V_2 - V_1)$$

$$= -m[(u_2 - u_1) + p_0(v_2 - v_1) - T_0(s_2 - s_1)] + mp_0(v_2 - v_1)$$

$$= -m[(u_2 - u_1) - T_0(s_2 - s_1)]. \qquad (2)$$

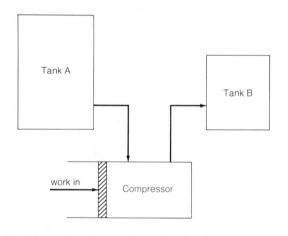

From the nitrogen tables,

$$v_1 = 0.445053 \text{ m}^3/\text{kg}$$

$$u_1 = h_1 - p_1 v_1$$

$$= 310.941 - 200 \times 0.445053 \text{ kJ/kg}$$

$$= 221.93 \text{ kJ/kg}$$

$$s_1 = 6.6394 \text{ kJ/kg} \cdot \text{K}$$

$$u_2 = h_2 - p_2 v_2$$

$$= 309.185 - 1000 \times 0.088893 \text{ kJ/kg}$$

$$= 220.292 \text{ kJ/kg}$$

$$s_2 = 6.1564 \text{ kJ/kg} \cdot \text{K}.$$

Thus

$$m = \frac{V_A}{v_1}$$

$$= \frac{5}{0.445053} \text{ kg}$$

$$= 11.2346 \text{ kg}$$

Substituting into Eq. 2, we have

$$W_{rev} = -11.2346[(220.292 - 221.930) - 300(6.1564 - 6.6394)] \text{ kJ}$$

$$= -1609.5 \text{ kJ}.$$

Comments

The reversible work is the total work input to the gas. The reversible useful work is given by

$$W_{rev, \, useful} = W_{rev} - p_0(V_2 - V_1)$$

$$= W_{rev} - mp_0(v_2 - v_1).$$

Using $p_0 = 101.325$ kPa gives us

$$W_{rev, \, useful} = -1609.5 - 11.2346 \times 101.325(0.088893 - 0.445053) \text{ kJ}$$

$$= -1609.5 + 405.4 \text{ kJ}$$

$$= -1204.1 \text{ kJ}.$$

This means that the minimum useful work needed is only 1204.1 kJ.

10.3 Availability (Exergy) Equation for an Open System

An open system will have mass interaction in addition to the possibility of heat and work interactions. We could thus deduce the availability equation for an open system simply by modifying the availability equation for a closed system to account for the exergy flow due to mass flow.

Making use of the concept of flow availability ψ deduced previously, the exergy flow per unit mass associated with mass flow is given by

$$\psi + \frac{\bar{V}^2}{2g_c} + \frac{g}{g_c} Z,$$

where

$$\psi = (h - T_0 s) - (h_0 - T_0 s_0).$$

We also make use of the fact that kinetic and potential energies are pure exergies. Together with Eq. 10.5, the availability equation for a closed system, the availability equation for an open system may then be given as

$$d\Phi_{cv} = \sum \left(1 - \frac{T_0}{T_k}\right) \bar{d}Q_k - (\bar{d}W_{useful} - p_0 \, dV_{cv})$$

$$+ \sum_{in} \left(\psi + \frac{\bar{V}^2}{2g_c} + \frac{g}{g_c} Z\right) dm$$

$$- \sum_{out} \left(\psi + \frac{\bar{V}^2}{2g_c} + \frac{g}{g_c} Z\right) dm - T_0 \, \bar{d}S_{irr}. \tag{10.17}$$

As a rate equation, we have

$$\dot{I} = T_0 \dot{S}_{gen} = -\frac{d\Phi_{cv}}{d\tau} + \sum \left(1 - \frac{T_0}{T_k}\right)\dot{Q}_k - W_{useful} + p_0 \frac{dV_{cv}}{d\tau}$$

$$+ \sum_{in} \dot{m}\left(\psi + \frac{\bar{V}^2}{2g_c} + \frac{g}{g_c} Z\right)$$

$$- \sum_{out} \dot{m}\left(\psi + \frac{\bar{V}^2}{2g_c} + \frac{g}{g_c} Z\right). \tag{10.17a}$$

Equation 10.17a may be used to study any process for an open system.

For an open system in general, the energy equation is given by Eq. 2.30:

$$\sum \dot{Q}_k = \frac{dU_{cv}}{d\tau} + \dot{W}_{useful} + \sum_{out} \dot{m}\left(h + \frac{\bar{V}^2}{2g_c} + \frac{g}{g_c} Z\right)$$

$$- \sum_{in} \dot{m}\left(h + \frac{\bar{V}^2}{2g_c} + \frac{g}{g_c} Z\right). \qquad (2.30)$$

Combining Eqs. 10.17a and 2.30, we have

$$\dot{I} = -\frac{d\Phi_{cv}}{d\tau} + \frac{dU_{cv}}{d\tau} + p_0 \frac{dV_{cv}}{d\tau} - \sum \frac{T_0}{T_k} \dot{Q}_k + \sum_{in} \dot{m}\psi - \sum_{out} \dot{m}\psi$$

$$+ \sum_{out} \dot{m}h - \sum_{in} \dot{m}h. \qquad (10.18)$$

By definition,

$$\Phi_{cv} = (U + p_0 V - T_0 S) - (U_0 + p_0 V_0 - T_0 S_0).$$

Thus

$$\frac{d\Phi_{cv}}{d\tau} = \frac{dU_{cv}}{d\tau} + p_0 \frac{dV_{cv}}{d\tau} - T_0 \frac{dS_{cv}}{d\tau} - (h_0 - T_0 s_0) \frac{dm_{cv}}{d\tau}. \qquad (10.19)$$

Combining Eqs. 10.18 and 10.19 and making use of the definition of $\psi = (h - T_0 s) - (h_0 - T_0 s_0)$, we have

$$\dot{I} = T_0 \frac{dS_{cv}}{d\tau} + T_0\left(\sum_{out} \dot{m}s - \sum_{in} \dot{m}s\right) - \sum \frac{T_0}{T_k} \dot{Q}_k$$

$$+ (h_0 - T_0 s_0)\left[\sum_{out} \dot{m} - \sum_{in} \dot{m} + \frac{dm_{cv}}{d\tau}\right]. \qquad (10.20)$$

Now from conservation of mass for an open system in general, we have

$$\frac{dm_{cv}}{d\tau} = \sum_{in} \dot{m} - \sum_{out} \dot{m}. \qquad (10.21)$$

Combining Eqs. 10.20 and 10.21, we have

$$\dot{I} = T_0 \frac{dS_{cv}}{d\tau} + T_0\left[\sum_{out} \dot{m}s - \sum_{in} \dot{m}s\right] - \sum \frac{T_0}{T_k} \dot{Q}_k. \qquad (10.22)$$

Since $\dot{I} = T_0 \dot{S}_{gen}$, we thus have

$$\dot{S}_{gen} = \frac{dS_{cv}}{d\tau} + \sum_{out} \dot{m}s - \sum_{in} \dot{m}s - \sum \frac{\dot{Q}_k}{T_k}, \qquad (10.23)$$

which is the familiar second-law equation for an open system in general. This is again in agreement with our statement that the availability equation is simply the combined first and second laws.

10.4 Second-Law Analysis of Open Systems

We have now at our disposal all the necessary tools we need to make second-law analysis of open systems. For convenience, these tools are summarized in this section.

1. From availability or the principle of degradation of energy, we have

$$\dot{I} = -\frac{d\Phi_{cv}}{d\tau} + \sum \left(1 - \frac{T_0}{T_k}\right)\dot{Q}_k - \left[\dot{W}_{useful} - p_0 \frac{dV_{cv}}{d\tau}\right]$$

$$+ \sum_{in} \dot{m}\left(\psi + \frac{\bar{V}^2}{2g_c} + \frac{g}{g_c}Z\right)$$

$$- \sum_{out} \dot{m}\left(\psi + \frac{\bar{V}^2}{2g_c} + \frac{g}{g_c}Z\right). \qquad (10.24)$$

For steady-state steady-flow processes, we have

$$\dot{I} = \sum \left(1 - \frac{T_0}{T_k}\right)\dot{Q}_k - \dot{W}_{useful} + \sum_{in} \dot{m}\left(\psi + \frac{\bar{V}^2}{2g_c} + \frac{g}{g_c}Z\right)$$

$$- \sum_{out} \dot{m}\left(\psi + \frac{\bar{V}^2}{2g_c} + \frac{g}{g_c}Z\right) \qquad (10.25)$$

$$\dot{W}_{rev} = \dot{W}_{rev, useful} = \sum \left(1 - \frac{T_0}{T_k}\right)\dot{Q}_k + \sum_{in} \dot{m}\left(\psi + \frac{\bar{V}^2}{2g_c} + \frac{g}{g_c}Z\right)$$

$$- \sum_{out} \dot{m}\left(\psi + \frac{\bar{V}^2}{2g_c} + \frac{g}{g_c}Z\right) \qquad (10.26)$$

$$\dot{I} = \dot{W}_{rev, useful} - \dot{W}_{useful}. \qquad (10.27)$$

2. From the second law of thermodynamics, we have

$$\dot{S}_{\text{gen}} = \frac{dS_{\text{cv}}}{d\tau} + \sum_{\text{out}} \dot{m}s - \sum_{\text{in}} \dot{m}s - \sum \frac{\dot{Q}_k}{T_k}, \qquad (10.28)$$

where $dS_{\text{cv}}/d\tau = 0$ if the process is steady-state steady flow.

3. From the first law of thermodynamics, we have

$$\dot{Q}_{\text{cv}} = \sum \dot{Q}_k = \frac{dE_{\text{cv}}}{d\tau} + \dot{W}_{\text{useful}} + \sum_{\text{out}} \dot{m}\left(h + \frac{\bar{V}^2}{2g_c} + \frac{g}{g_c} Z\right)$$

$$- \sum_{\text{in}} \dot{m}\left(h + \frac{\bar{V}^2}{2g_c} + \frac{g}{g_c} Z\right), \qquad (10.29)$$

where $dE_{\text{cv}}/d\tau = 0$ if the process is steady-state steady flow.

Example 10.5

Nitrogen is compressed reversibly and isothermally in a steady-state steady-flow compressor from 300 K and 101.325 kPa to 300 K and 1.0 MPa. Changes in kinetic and potential energies may be neglected. The pressure and temperature of the environment are 101.325 kPa and 300 K, respectively. For a mass-flow rate of 5.0 kg/s, calculate the minimum power needed for the process.

Solution

System selected for study: compressor
Unique features of process: steady-state steady flow; reversible; isothermal
Assumptions: nitrogen is a simple compressible substance; neglect ΔKE and ΔPE

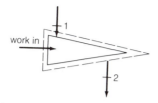

Making use of Eq. 10.26, we have

$$\dot{W}_{\text{rev, useful}} = \dot{m}(\psi_1 - \psi_2) + \int_1^2 \left(1 - \frac{T_0}{T}\right) d\dot{Q}$$

$$= \dot{m}(\psi_1 - \psi_2) \qquad \text{with } T = T_0$$

$$= -\dot{m}[(h_2 - h_1) - T_0(s_2 - s_1)]. \qquad (1)$$

From the nitrogen tables,

$$h_1 = 311.163 \text{ kJ/kg}$$

$$s_1 = 6.8418 \text{ kJ/kg} \cdot \text{K}$$

$$h_2 = 309.185 \text{ kJ/kg}$$

$$s_2 = 6.1564 \text{ kJ/kg} \cdot \text{K}.$$

Substituting into Eq. 1, we have

$$\dot{W}_{\text{rev, useful}} = -5[(309.185 - 311.163) - 300(6.1564 - 6.8418)] \text{ kJ/s}$$

$$= -1018.2 \text{ kJ/s}$$

$$= -1018.2 \text{ kW}. \qquad \blacksquare$$

Comments

For steady-state steady flow, $\dot{W}_{\text{rev}}$ is the same as $\dot{W}_{\text{rev, useful}}$ since there is no work done to or done by the environment. Since our process is isothermal at $T = T_0$, then $T_2 = T_1 = T_0$. Equation 1 may then be written as

$$\dot{W}_{\text{rev, useful}} = -\dot{m}[(h_2 - T_2 s_2) - (h_1 - T_1 s_1)]. \qquad (2)$$

Now $g = h - Ts$, the Gibbs function. Thus

$$\dot{W}_{\text{rev, useful}} = -\dot{m}(g_2 - g_1)$$

or

$$w_{\text{rev, useful}} = -(\dot{G}_2 - \dot{G}_1), \qquad (3)$$

which states that when a system undergoes a change of state reversibly and isothermally in a steady-state steady-flow device, the decrease in its Gibbs function is a measure of the maximum useful work that we can obtain from a work-producing device, or the minimum useful work we need for a work-absorbing device.

Example 10.6

Ammonia is compressed adiabatically in a steady-state steady-flow compressor from a saturated vapor at 10°F to 140 psia and 200°F. Neglect changes in kinetic and potential energies. Temperature of the environment is 77°F.

Determine

(a) the irreversibility per unit mass, in Btu/lbm.
(b) the minimum useful work needed for the given process.

Solution

System selected for study: compressor
Unique features of process: steady-state steady flow; adiabatic
Assumptions: ammonia is a simple compressible substance; neglect ΔKE and ΔPE

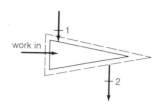

(a) Making use of Eq. 11.25,

$$\frac{I}{m} = -w_{\text{useful}} + (\psi_1 - \psi_2)$$

$$= -w_{\text{useful}} - [(h_2 - h_1) - T_0(s_2 - s_1)]. \tag{1}$$

From the first law,

$$w_{\text{useful}} = -(h_2 - h_1). \tag{2}$$

Combining Eqs. 1 and 2,

$$\frac{I}{m} = T_0(s_2 - s_1), \tag{3}$$

which is simply the second-law equation for the given process. From the ammonia tables,

$$s_1 = s_g \text{ at } 10°\text{F}$$

$$= 1.3157 \text{ Btu/lbm} \cdot °\text{R}$$

$$s_2 = 1.3418 \text{ Btu/lbm} \cdot °\text{R}.$$

Substituting into Eq. 3, we have

$$\frac{I}{m} = 537(1.3418 - 1.3157) \text{ Btu/lbm}$$

$$= 14.02 \text{ Btu/lbm},$$

which shows that the process is irreversible.

(b) Minimum work is given by Eq. 10.26:

$$w_{\text{rev, useful}} = \psi_1 - \psi_2$$

$$= -[(h_2 - h_1) - T_0(s_2 - s_1)]. \tag{4}$$

From the ammonia tables,

$$h_1 = h_g \text{ at } 10°\text{F}$$

$$= 614.9 \text{ Btu/lbm}$$

$$h_2 = 709.9 \text{ Btu/lbm}.$$

Substituting into Eq. 4 gives us

$$w_{\text{rev, useful}} = -[(709.9 - 614.9) - 537(1.3418 - 1.3157)] \text{ Btu/lbm}$$

$$= -(95 - 14.02) \text{ Btu/lbm}$$

$$= -80.98 \text{ Btu/lbm}, \qquad \blacksquare$$

which is the minimum work needed.

Comment

The actual useful work is given by

$$w_{\text{useful}} = -(h_2 - h_1)$$

$$= -(709.9 - 614.9) \text{ Btu/lbm}$$

$$= -95 \text{ Btu/lbm}.$$

Then

$$\frac{I}{m} = W_{\text{rev, useful}} - W_{\text{useful}}$$

$$= -80.98 - (-95) \text{ Btu/lbm} = +14.02 \text{ Btu/lbm},$$

as it should.

Example 10.7

Hot gas enters the heat recovery steam generator of a cogeneration system at 500°C and 101.325 kPa and leaves at 150°C. Water enters steadily at 100°C and 2.0 MPa and leaves as dry saturated steam at 2.0 MPa. The hot gas may be assumed to be an ideal gas with constant specific heats of $c_p = 1.05$ kJ/ kg · K. For a flow rate of hot gas of 25,000 kg/h, determine the irreversibility for the process, in kJ/h.

Solution

System selected for study: heat recovery steam generator
Unique features of process: steady-state steady flow; no work transfer
Assumptions: both hot gas and H_2O are simple compressible substances; neglect ΔKE and ΔPE; neglect pressure drops; neglect heat loss to the surroundings

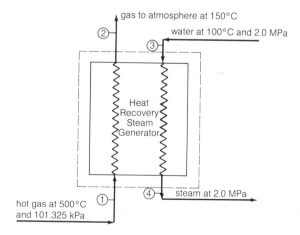

From the availability equation,

$$\dot{I} = \sum_{\text{in}} \dot{m}\psi - \sum_{\text{out}} \dot{m}\psi$$

$$= (\dot{m}_1 \psi_1 + \dot{m}_3 \psi_3) - (\dot{m}_2 \psi_2 + \dot{m}_4 \psi_4). \tag{1}$$

Now

$$\dot{m}_1 = \dot{m}_2 = \dot{m}_{gas}$$

$$\dot{m}_3 = \dot{m}_4 = \dot{m}_{H_2O}.$$

Thus Eq. 1 may be written as

$$\dot{I} = \dot{m}_{gas}[(h_1 - h_2) - T_0(s_1 - s_2)] + \dot{m}_{H_2O}[(h_3 - h_4) - T_0(s_3 - s_4)]. \quad (2)$$

From the first law,

$$0 = \sum_{out} \dot{m}h - \sum_{in} \dot{m}h$$

$$= (\dot{m}_2 h_2 + \dot{m}_4 h_4) - (\dot{m}_1 h_1 + \dot{m}_3 h_3)$$

$$= \dot{m}_{gas}(h_2 - h_1) + \dot{m}_{H_2O}(h_4 - h_3). \quad (3)$$

Combining Eqs. 2 and 3 gives us

$$\dot{I} = T_0[\dot{m}_{gas}(s_2 - s_1) + \dot{m}_{H_2O}(s_4 - s_3)]. \quad (4)$$

To find $\dot{m}_{H_2O}$ we make use of Eq. 3:

$$\dot{m}_{H_2O} = \dot{m}_{gas} \frac{h_1 - h_2}{h_4 - h_3}. \quad (5)$$

For an ideal gas with constant specific heats,

$$h_1 - h_2 = c_p(T_1 - T_2).$$

From steam tables,

$$h_3 \approx h_f \text{ at } 100°C$$

$$= 419.06 \text{ kJ/kg}$$

$$h_4 = h_g \text{ at } 2.0 \text{ MPa}$$

$$= 2797.2 \text{ kJ/kg}.$$

Substituting into Eq. 5 yields

$$\dot{m}_{H_2O} = 25{,}000 \times \frac{1.05(500 - 150)}{2797.2 - 419.06} \text{ kg/h}$$

$$= 3863.3 \text{ kg/h}$$

For an ideal gas with constant specific heats,

$$s_2 - s_1 = c_p \ln \frac{T_2}{T_1} - R \ln \frac{p_2}{p_1}.$$

Neglecting pressure drops, $p_2 = p_1$, and

$$s_2 - s_1 = c_p \ln \frac{T_2}{T_1}$$

$$= 1.05 \ln \frac{423.15}{773.15} \text{ kJ/kg} \cdot \text{K}$$

$$= -0.6329 \text{ kJ/kg} \cdot \text{K}.$$

From steam tables,

$$s_3 \approx s_f \text{ at } 100°\text{C} = 1.3069 \text{ kJ/kg} \cdot \text{K}$$

$$s_4 = s_g \text{ at } 2.0 \text{ MPa} = 6.3367 \text{ kJ/kg} \cdot \text{K}.$$

Substituting into Eq. 4, we have

$$I = 300[25{,}000(-0.6329) + 3863.3(6.3367 - 1.3069)] \text{ kJ/h}$$

$$= 300(-15{,}822.5 + 19{,}431.6) \text{ kJ/h}$$

$$= 1.083 \times 10^6 \text{ kJ/h}. \qquad \blacksquare$$

Comments

By neglecting heat loss to the surroundings, we are assuming that all the energy given up by the hot gas is absorbed by the water. That is, our first-law efficiency for this process is 100%, which is not very meaningful. Making use of the concept of exergy and exergy destruction, we will introduce the idea of a second-law efficiency, which is a more accurate and more meaningful measure of thermodynamic performance.

10.5 Second-Law Efficiency

To measure the performance of any process, device, or system, we make use of the concept of efficiency. The figure of merit widely used in thermodynamics is usually based on the concept of energy, in which no attempt is made to distinguish low-quality energy from high-quality energy. A simple example is the thermal efficiency of a heat engine, which is defined as the ratio of net work output to the amount of heat addition. In this definition, heat and work are given the same weight. Another example is the coefficient of performance of a refrigerator that compares the amount of refrigeration (heat quantity) to the amount of work input. Although these types of efficiencies do serve some useful purposes, they do not give an accurate measure of thermodynamic performance. Since it is exergy, not energy, that is consumed in causing changes, it would be more logical to have an efficiency based on the concept of exergy. In this section we introduce the concept of second-law efficiency, which is based on the concept of exergy and exergy destruction. It is also called exergetic efficiency.

In general, any kind of efficiency η may be expressed as

$$\eta = \frac{\text{output}}{\text{input}} \tag{1}$$

or

$$\eta = \frac{\text{what you get}}{\text{what you pay}}. \tag{2}$$

If all the quantities in Eq. 1 or 2 are expressed in units of exergy, we would have a second-law or exergetic efficiency. Because there could be different interpretations as to what constitutes output and what constitutes input, we could have different expressions of second-law efficiency for the same process. However, regardless of how one interprets output and input for a given process, it is important to note that the upper limit of any second-law efficiency is 100%, which corresponds to the ideal case with no exergy destruction. Expressions on second-law efficiency for some common devices and systems are presented in this section.

Heat Engine

For a heat engine (Fig. 10.1), "what you get" is work output W. "What you pay" is heat addition Q_H. Now work is pure exergy. The exergy content in Q_H is $(1 - T_0/T_H)Q_H$. Thus a second-law efficiency for a heat engine may be given as

$$\eta_{II} = \frac{W}{(1 - T_0/T_H)Q_H}.$$

Now the thermal efficiency of a heat engine is defined as

$$\eta_{th} = \frac{W}{Q_H}.$$

Thus we have

$$\eta_{II} = \frac{\eta_{th}}{1 - T_0/T_H}. \tag{10.30}$$

If $T_L = T_0$, then

$$1 - \frac{T_0}{T_H} = 1 - \frac{T_L}{T_H},$$

which is the Carnot cycle efficiency for a heat engine operating between the temperature limits of T_L and T_H.

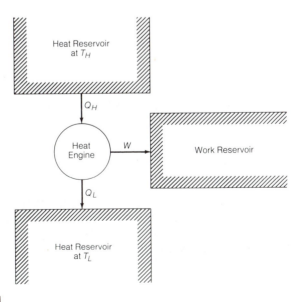

Figure 10.1

Since either $(1 - T_0/T_H)$ or $(1 - T_L/T_H)$ is less than unity, the second-law efficiency of a heat engine is always greater than the thermal efficiency, the first-law efficiency, of a heat engine.

Alternatively, we can say that the exergy input is given by

$$\left(1 - \frac{T_0}{T_H}\right)Q_H - \left(1 - \frac{T_0}{T_L}\right)Q_L,$$

where both Q_H and Q_L are absolute quantities.

Then the second-law efficiency of a heat engine may also be given as

$$\eta_{II} = \frac{W}{(1 - T_0/T_H)Q_H - (1 - T_0/T_L)Q_L}$$

$$= \frac{\eta_{th}}{(1 - T_0/T_H) - (1 - T_0/T_L)(Q_L/Q_H)}. \qquad (10.31)$$

Since $W = Q_H - Q_L$, Eq. 10.31 may also be given as

$$\eta_{II} = \frac{\eta_{th}}{(1 - T_0/T_H) - (1 - T_0/T_L)(1 - \eta_{th})}. \qquad (10.31a)$$

If $T_L = T_0$, Eqs. 10.30 and 10.31 would give the same result, as follows:

$$\eta_{II} = \frac{\eta_{th}}{1 - T_0/T_H}.$$

The difference between Eqs. 10.30 and 10.31 is simply that we are not charging the exergy content in Q_L as part of the exergy input in Eq. 10.31.

Refrigerator

The second-law efficiency for a refrigerator (Fig. 10.2) may be defined as the ratio of the minimum work needed to the actual work used for the production of the same amount of refrigeration Q_L under the same temperature limits. Then we have

$$\eta_{II} = \frac{W_{min}}{W_{act}}. \qquad (10.32)$$

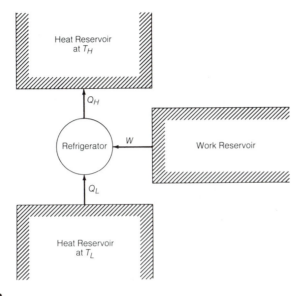

Figure 10.2

Now the coefficient of performance of a refrigerator is given as

$$\beta_R = \frac{Q_L}{W}.$$

Then Eq. 10.32 may be written as

$$\eta_{\text{II}} = \frac{(\beta_R)_{\text{act}}}{(\beta_R)_{\text{ideal}}}, \tag{10.32a}$$

where $(\beta_R)_{\text{ideal}} = T_L/(T_H - T_L)$, the coefficient of performance of a reversed Carnot cycle operating between the given temperature limits.

Now the availability equation for a refrigeration cycle may be written as

$$|W| = \left(1 - \frac{T_0}{T_H}\right)|Q_H| - \left(1 - \frac{T_0}{T_L}\right)|Q_L| + I. \tag{10.33}$$

We may interpret the input as W and the output as the first two terms on the right side of Eq. 10.33 since I is exergy destruction. Thus using the concept of output to input, the second-law efficiency of a refrigerator may also be given as

$$\eta_{\text{II}} = \frac{(1 - T_0/T_H)|Q_H| - (1 - T_0/T_L)|Q_L|}{|W|}. \tag{10.34}$$

Since $|Q_H| = |Q_L| + |W|$, Eq. 10.34 may be rearranged to give

$$\eta_{\text{II}} = T_0 \left(\frac{T_H - T_L}{T_L T_H} \right) (\beta_R)_{\text{act}} + \left(1 - \frac{T_0}{T_H} \right). \qquad (10.34\text{a})$$

If $T_0 = T_H$, Eq. 10.34a becomes

$$\eta_{\text{II}} = \frac{T_H - T_L}{T_L} (\beta_R)_{\text{act}} = \frac{(\beta_R)_{\text{act}}}{(\beta_R)_{\text{ideal}}},$$

which is the same as Eq. 10.32a. Since $(\beta_R)_{\text{ideal}}$ is always greater than $(\beta_R)_{\text{act}}$, the second-law efficiency is always less than unity, as it should be.

Heat Pump

As in the case of a refrigerator, the second-law efficiency for a heat pump (Fig. 10.3) may also be defined as the ratio of the minimum work needed to the actual work used for the production of the same amount of Q_H under the same temperature limits. Then we have

$$\eta_{\text{II}} = \frac{W_{\text{min}}}{W_{\text{act}}}. \qquad (10.35)$$

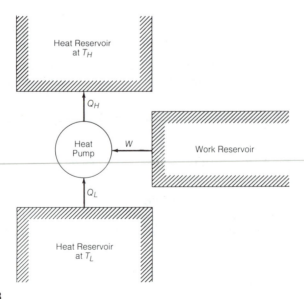

Figure 11.3

Now the coefficient of performance of a heat pump is given as

$$\beta_{HP} = \frac{Q_H}{W}.$$

Then Eq. 10.35 may be written as

$$\eta_{II} = \frac{(\beta_{HP})_{act}}{(\beta_{HP})_{ideal}}, \tag{10.35a}$$

where $(\beta_{HP})_{ideal} = T_H/(T_H - T_L)$, the coefficient of performance of a reversed Carnot cycle operating between the given temperature limits.

Now "what we get" from a heat pump is Q_H, the exergy content of which is $(1 - T_0/T_H)Q_H$. "What we pay" is W. Thus the second-law efficiency of a heat pump may also be given as

$$\eta_{II} = \frac{(1 - T_0/T_H)Q_H}{W}$$

$$= \frac{T_H - T_0}{T_H}(\beta_{HP})_{act}. \tag{10.36}$$

If $T_L = T_0$, Eq. 10.36 becomes

$$\eta_{II} = \frac{T_H - T_L}{T_H}(\beta_{HP})_{act} = \frac{(\beta_{HP})_{act}}{(\beta_{HP})_{ideal}},$$

which is the same as Eq. 10.35a. The second-law efficiency is less unity as $(\beta_{HP})_{ideal}$ is always greater than $(\beta_{HP})_{act}$.

Pumps and Compressors

Pumps and compressors are work-absorbing devices (see Fig. 10.4). Under steady-state steady-flow conditions, "what we pay" is the useful work supplied, "what we get" is increase in the exergy content of the flowing stream. Neglecting heat transfer and changes in kinetic and potential energies, the second-law efficiency of a pump or compressor may be expressed as

$$\eta_{II} = \frac{\text{"what you get"}}{\text{"what you pay"}}$$

$$= \frac{\psi_{out} - \psi_{in}}{|w_{in}|}. \tag{10.37}$$

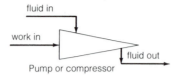

Figure 10.4

Turbines

Turbines or expanders are work-producing devices (see Fig. 10.5). Under steady-state steady-flow conditions, "what we get" is simply the turbine output, "what we pay" is the drop in the exergy content of the flowing stream. Neglecting heat transfer and changes in kinetic and potential energies, the second-law efficiency of a turbine may be expressed as

$$\eta_{II} = \frac{\text{``what we get''}}{\text{``what we pay''}}$$

$$= \frac{w_T}{\psi_{in} - \psi_{out}}. \tag{10.38}$$

Heat Exchangers

In a heat exchanger (Fig. 10.6), one flowing stream will give up exergy (this is "what we pay"), and another stream will gain exergy (this is "what we get"). If the heating stream gives up exergy and the heated stream gains exergy, and if changes in kinetic and potential energies are negligible, the second-law efficiency of a heat exchanger may be expressed as

$$\eta_{II} = \frac{\text{exergy gained by the heated stream}}{\text{exergy supplied by the heating stream}}$$

$$= \frac{\dot{m}_c(\psi_{out} - \psi_{in})_c}{\dot{m}_h(\psi_{in} - \psi_{out})_h}. \tag{10.39}$$

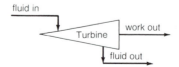

Figure 10.5

Figure 10.6

Throttling Valve

In the case of a throttling process operating under steady-state steady-flow conditions, the availability equation may be written as

$$\psi_{in} = \psi_{out} + \frac{I}{m}.$$

We may interpret the input as ψ_{in} of the incoming stream and the output as ψ_{out} of the outgoing stream. Thus the second-law efficiency of throttling process may be expressed as

$$\eta_{II} = \frac{\psi_{out}}{\psi_{in}}, \tag{10.40}$$

which may also be written as

$$\eta_{II} = 1 - \frac{I/m}{\psi_{in}}. \tag{10.40a}$$

Example 10.8

Determine the second-law efficiency of the compressor in Example 10.3.

Solution

This is a compression process carried out inside a cylinder. We may give the second-law efficiency as

$$\eta_{II} = \frac{\text{exergy gained by the fluid being compressed}}{\text{useful work input}}$$

$$= \frac{\phi_1 - \phi_2}{|w_{12,\,useful}|}.$$

From Example 10.3,

$$\phi_1 - \phi_2 = 171.689 \text{ kJ/kg}$$

$$|w_{12,\,useful}| = 192.905 \text{ kJ/kg}.$$

Thus

$$\eta_{II} = \frac{171.689}{192.905}$$

$$= 89.0\%.$$ ▮

Comment

The exergy gained by the fluid being compressed in the cylinder is also the reversible useful work (the minimum) needed. Thus our second-law efficiency in this case is simply the ratio of the reversible useful work needed to the actual useful work required.

Example 10.9

For the compressor in Example 10.6, determine

(a) the second-law efficiency.
(b) the adiabatic compressor efficiency.

Solution

    System selected for study: compressor
    Unique features of process: steady-state steady flow; adiabatic
    Assumptions: ammonia is a simple compressible substance; neglect ΔKE
    and ΔPE

(a) For a steady-state steady-flow compressor, the second-law efficiency may be given by Eq. 10.37:

$$\eta_{II} = \frac{\psi_2 - \psi_1}{|w_{useful}|}. \tag{1}$$

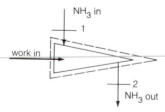

From Example 10.6,

$$\psi_2 - \psi_1 = 80.98 \text{ Btu/lbm}$$

$$|w_{useful}| = 95 \text{ Btu/lbm}.$$

Thus

$$\eta_{\mathrm{II}} = \frac{80.98}{95}$$

$$= 85.2\%.$$

(b) By definition, the adiabatic compressor efficiency η_c is given by

$$\eta_c = \frac{h_{2s} - h_1}{h_2 - h_1}. \tag{2}$$

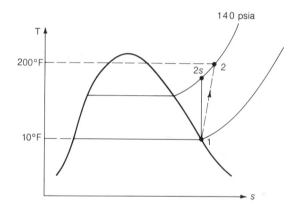

From the ammonia tables,

$$h_1 = h_g \text{ at } 10°F = 614.9 \text{ Btu/lbm}$$

$$h_2 = 709.9 \text{ Btu/lbm}$$

$$s_1 = s_g \text{ at } 10°F = 1.3157 \text{ Btu/lbm} \cdot °R$$

$$h_{2s} = h \text{ at } \begin{cases} s_{2s} = s_1 = 1.3157 \text{ Btu/lbm} \cdot °R \\ p_{2s} = p_2 = 140 \text{ psia}. \end{cases}$$

By interpolation,

$$h_{2s} = 679.9 + \frac{(709.9 - 679.9)(1.3157 - 1.2945)}{1.3418 - 1.2945}$$

$$= 693.3 \text{ Btu/lbm}$$

Substituting into Eq. 2,

$$\eta_c = \frac{693.3 - 614.9}{709.9 - 614.9}$$

$$= 82.5\%.$$ ∎

Comments

The second-law efficiency and the adiabatic compressor efficiency are different because they are defined differently. The adiabatic compressor efficiency may be looked upon as a quasi-second-law efficiency. In many cases, its value is not too different from that of the second-law efficiency.

Example 10.10

Determine the second-law efficiency of the heat recovery steam generator in Example 10.7.

Solution

The heat recovery steam generator is simply a heat exchanger. We may thus make use of Eq. 10.39 to determine its second-law efficiency:

$$\eta_{\text{II}} = \frac{\dot{m}_c(\psi_{\text{out}} - \psi_{\text{in}})_c}{\dot{m}_h(\psi_{\text{in}} - \psi_{\text{out}})_h}. \tag{1}$$

From Example 10.7,

$$\dot{m}_c = \dot{m}_{H_2O} = 3863.3 \text{ kg/h}$$

$$\dot{m}_h = \dot{m}_{\text{gas}} = 25{,}000 \text{ kg/h}$$

$$(\psi_{\text{out}} - \psi_{\text{in}})_c = \psi_4 - \psi_3$$

$$= (h_4 - h_3) - T_0(s_4 - s_3).$$

From the steam tables,

$$h_4 = h_g \text{ at } 2.0 \text{ MPa}$$

$$= 2797.2 \text{ kJ/kg}$$

$$h_3 = h_f \text{ at } 100°C$$

$$= 419.06 \text{ kJ/kg}$$

$$s_4 = s_g \text{ at 2.0 MPa}$$

$$= 6.3367 \text{ kJ/kg} \cdot \text{K}$$

$$s_3 \approx s_f \text{ at } 100°\text{C}$$

$$= 1.3069 \text{ kJ/kg} \cdot \text{K}.$$

Then

$$\psi_4 - \psi_3 = (2797.2 - 419.06) - 300(6.3367 - 1.3069) \text{ kJ/kg}$$

$$= 869.2 \text{ kJ/kg}.$$

Now

$$(\psi_{in} - \psi_{out})_h = \psi_1 - \psi_2$$

$$= (h_1 - h_2) - T_0(s_1 - s_2).$$

For an ideal gas with constant specific heats,

$$h_1 - h_2 = c_p(T_1 - T_2)$$

$$s_1 - s_2 = c_p \ln \frac{T_1}{T_2} - R \ln \frac{p_1}{p_2}$$

$$= c_p \ln \frac{T_1}{T_2} \qquad \text{neglecting pressure drop}$$

Thus

$$\psi_1 - \psi_2 = c_p(T_1 - T_2) - T_0 c_p \ln \frac{T_1}{T_2}$$

$$= 1.05(500 - 150) - 300 \times 1.05 \ln \frac{773.15}{423.15} \text{ kJ/kg}$$

$$= 177.63 \text{ kJ/kg}.$$

Substituting into Eq. 1 gives us

$$\eta_{II} = \frac{3863.3 \times 869.2}{25,000 \times 177.63} = 75.6\%.$$

∎

Comment

The cold stream did gain exergy as expected. Thus the hot stream was the exergy supplier. After all, the idea of the use of the heat recovery steam generator is to recover energy and exergy from the hot gas.

Example 10.11

For the simple steam power plant shown below,

(a) make a first-law analysis.
(b) make a second-law analysis.

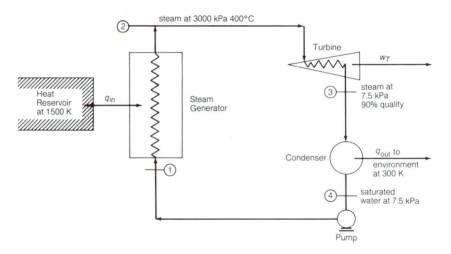

Solution

Assumptions: neglect pump work; neglect ΔKE and ΔPE; neglect heat transfer except in steam generator and condenser

The data we need are obtained from the steam tables:

$$h_1 \approx h_f \text{ at } 7.5 \text{ kPa}$$

$$= 168.77 \text{ kJ/kg}$$

$$h_2 = 3232.5 \text{ kJ/kg}$$

$$h_3 = (h_f + x_3 h_{fg}) \text{ at } 7.5 \text{ kPa and } 90\% \text{ quality}$$

$$= 168.77 + 0.90 \times 2406.2 \text{ kJ/kg}$$

$$= 2334.35 \text{ kJ/kg}$$

$$h_4 = h_f \text{ at } 7.5 \text{ kPa} = 168.77 \text{ kJ/kg}$$

$$s_1 \approx s_f \text{ at } 7.5 \text{ kPa} = 0.5763 \text{ kJ/kg} \cdot \text{K}$$

$$s_2 = 6.9246 \text{ kJ/kg} \cdot \text{K}$$

$$s_3 = (s_f + x_3 s_{fg}) \text{ at 7.5 kPa and 90\% quality}$$

$$= 0.5763 + 0.90 \times 7.6760 = 7.4847 \text{ kJ/kg} \cdot \text{K}$$

$$s_4 = s_f \text{ at 7.5 kPa} = 0.5763 \text{ kJ/kg} \cdot \text{K}.$$

(a) First-law analysis:

$$q_{in} = h_2 - h_1$$

$$= 3232.5 - 168.77 = 3063.73 \text{ kJ/kg}$$

$$w_T = h_2 - h_3$$

$$= 3232.5 - 2334.35 = 898.15 \text{ kJ/kg}$$

$$|q_{out}| = h_3 - h_4$$

$$= 2334.35 - 168.77 = 2165.58 \text{ kJ/kg}$$

$$w_{net} \approx w_T = 898.15 \text{ kJ/kg}$$

$$\eta_{th} = \frac{w_{net}}{q_{in}} \approx \frac{898.15}{3063.73} = 29.3\%.$$

(b) Second-law analysis:

$$\text{exergy input to plant} = \left(1 - \frac{T_0}{T^{HR}}\right) q_{in}$$

$$= \left(1 - \frac{300}{1500}\right) 3063.73 \text{ kJ/kg}$$

$$= 2450.98 \text{ kJ/kg}$$

exergy added to working fluid

$$= \psi_2 - \psi_1$$

$$= (h_2 - h_1) - T_0(s_2 - s_1)$$

$$= (3232.5 - 168.77) - 300(6.9246 - 0.5763) \text{ kJ/kg}$$

$$= 1159.24 \text{ kJ/kg}$$

exergy destroyed in heat addition process $= 2450.98 - 1159.24$ kJ/kg

$$= 1291.74 \text{ kJ/kg}$$

$$(\eta_{\text{II}})_{\text{cycle}} = \frac{898.15}{1159.24} = 77.5\%$$

$$(\eta_{\text{II}})_{\text{plant}} = \frac{898.15}{2450.98} = 36.6\%$$

exergy destroyed in turbine $= (\psi_2 - \psi_1) - w_T$

$$= (h_2 - h_3) - T_0(s_2 - s_3) - (h_2 - h_3)$$

$$= T_0(s_3 - s_2)$$

$$= 300(7.4847 - 6.9246) \text{ kJ/kg}$$

$$= 168.03 \text{ kJ/kg}$$

exergy destroyed in condenser

$$= \int \left(1 - \frac{T_0}{T}\right) dq + (\psi_3 - \psi_4)$$

$$= \psi_3 - \psi_4 \qquad \text{with } T = T_0$$

$$= (h_3 - h_4) - T_0(s_3 - s_4)$$

$$= (2334.35 - 168.77) - 300(7.4847 - 0.5763) \text{ kJ/kg}$$

$$= 93.06 \text{ kJ/kg.}$$

Exergy accounting for cycle:

$$\text{exergy in} - \text{exergy out} = \sum \text{exergy destruction}$$

$$1159.24 - 898.15 = 168.03 + 93.06$$

$$261.09 = 261.09 \text{ kJ/kg.}$$

We can also write the second-law efficiency for the cycle as

$$\eta_{II} = 1 - \frac{\sum \text{exergy destruction}}{\text{exergy input for cycle}}$$

$$= 1 - \frac{261.09}{1159.24}$$

$$= 77.5\%,$$

which is the same as before.

Exergy accounting for plant:

$$\text{exergy in} - \text{exergy out} = \sum \text{exergy destruction}$$

$$2450.98 - 898.15 = 1291.74 + 168.03 + 93.06$$

$$1552.83 = 1552.83 \text{ kJ/kg}.$$

We can also write the second-law efficiency for the plant as

$$\eta_{II} = 1 - \frac{\sum \text{exergy destruction}}{\text{exergy input to plant}}$$

$$= 1 - \frac{1552.83}{2450.98}$$

$$= 36.6\%,$$

which is the same as before.   ∎

Comments

The first-law analysis shows that 2165.58 kJ/kg, or 70.3% of total heat added to system, is rejected at the condenser. It gives the impression that the main cause of inefficiency lies in the condenser. On the other hand, the second-law analysis shows that the exergy destruction (that is, irreversibility) in the condenser is only 93.06 kJ/kg, or only 3.8% of the exergy input to the system. The second-law analysis pinpoints the main cause of thermodynamic imperfection: exergy destruction in the heat addition process is 1291.74 kJ/kg, which is 52.7% of the total exergy input to the system. The designer must find ways to reduce this amount. In the next example we show how this can be done.

Example 10.12

The steam power plant shown has one open feedwater heater (a heat exchanger). This makes it possible to increase the temperature of the condensate from the condenser before pumping into the steam generator. For this system

(a) make a first-law analysis.
(b) make a second-law analysis.

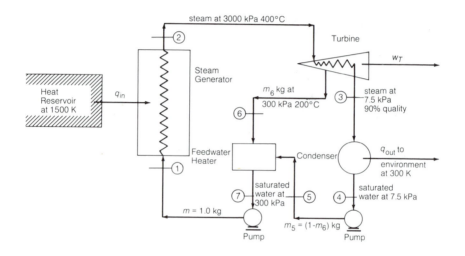

Solution

Assumptions: neglect pump work; neglect ΔKE and ΔPE; neglect heat transfer except in the steam generator, condenser, and feedwater heater

Calculations will be made on the basis of 1.0 kg of fluid entering the turbine. With m_6 kilograms extracted from turbine at point 6, we will have $(1.0 - m_6)$ kg of fluid entering the condenser. The data we need are obtained from the steam tables:

$$h_1 \approx h_f \text{ at } 300 \text{ kPa} = 561.4 \text{ kJ/kg}$$

$$h_2 = 3232.5 \text{ kJ/kg}$$

$$h_3 = (h_f + x_3 h_{fg}) \text{ at } 7.5 \text{ kPa and } 90\% \text{ quality}$$

$$= 168.77 + 0.90 \times 2406.2 = 2334.35 \text{ kJ/kg}$$

$$h_4 = h_f \text{ at } 7.5 \text{ kPa} = 168.77 \text{ kJ/kg}$$

$$h_5 \approx h_4 = 168.77 \text{ kJ/kg}$$

$$h_6 = 2865.5 \text{ kJ/kg}$$

$$h_7 = h_f \text{ at } 300 \text{ kPa} = 561.4 \text{ kJ/kg}$$

$$s_1 \approx s_f \text{ at } 300 \text{ kPa} = 1.6716 \text{ kJ/kg} \cdot \text{K}$$

$$s_2 = 6.9246 \text{ kJ/kg} \cdot \text{K}$$

$$s_3 = (s_f + x_3 s_{fg}) \text{ at } 7.5 \text{ kPa and } 90\% \text{ quality}$$

$$= 0.5763 + 0.90 \times 7.6760 = 7.4847 \text{ kJ/kg} \cdot \text{K}$$

$$s_4 = s_f \text{ at } 7.5 \text{ kPa} = 0.5763 \text{ kJ/kg} \cdot \text{K}$$

$$s_5 \approx s_4 = 0.5763 \text{ kJ/kg} \cdot \text{K}$$

$$s_6 = 7.3119 \text{ kJ/kg} \cdot \text{K}$$

$$s_7 = s_f \text{ at } 300 \text{ kPa} = 1.6716 \text{ kJ/kg} \cdot \text{K}.$$

We need to calculate m_6 before we can make any system analysis. Making an energy balance around the feedwater heater, we obtain

$$0 = \sum_{\text{out}} mh - \sum_{\text{in}} mh$$

$$= 1.0 \, h_7 - m_6 h_6 - (1 - m_6)h_5,$$

from which we get

$$m_6 = \frac{h_7 - h_5}{h_6 - h_5}$$

$$= \frac{561.4 - 168.77}{2865.5 - 168.77} \text{ kg}$$

$$= 0.1456 \text{ kg}.$$

(a) First-law analysis (on the basis of $m_1 = m_2 = 1.0$ kg):

$$q_{in} = h_2 - h_1$$

$$= 3232.5 - 561.4 \text{ kJ} = 2671.1 \text{ kJ}$$

$$w_T = m_2(h_2 - h_6) + (1.0 - m_6)(h_6 - h_3)$$

$$= 1.0(3232.5 - 2865.5) + (1.0 - 0.1456)(2865.5 - 2334.35) \text{ kJ}$$

$$= 820.8 \text{ kJ}$$

$$q_{out} = m_3(h_3 - h_4)$$

$$= (1.0 - 0.1456)(2334.35 - 168.77) \text{ kJ}$$

$$= 1850.3 \text{ kJ}$$

$$w_{net} \approx w_T = 820.8 \text{ kJ}$$

$$\eta_{th} = \frac{w_{net}}{q_{in}} \approx \frac{820.8}{2671.1} = 30.7\%.$$

(b) Second-law analysis (on the basis of $m_1 = m_2 = 1.0$ kg):

$$\text{exergy input to plant} = \left(1 - \frac{T_0}{T^{HR}}\right)q_{in}$$

$$= \left(1 - \frac{300}{1500}\right)2671.1 \text{ kJ}$$

$$= 2136.9 \text{ kJ}$$

exergy added to working fluid

$$= \psi_2 - \psi_1$$

$$= (h_2 - h_1) - T_0(s_2 - s_1)$$

$$= (3232.5 - 561.4) - 300(6.9246 - 1.6716) \text{ kJ}$$

$$= 1095.2 \text{ kJ}$$

exergy destroyed in heat addition process $= 2136.9 - 1095.2$ kJ

$$= 1041.7 \text{ kJ}$$

$$(\eta_{\text{II}})_{\text{cycle}} = \frac{820.8}{1095.2} = 74.9\%$$

$$(\eta_{\text{II}})_{\text{plant}} = \frac{820.8}{2136.9} = 38.4\%$$

exergy destroyed in turbine

$$= T_0 S_{\text{gen}}$$

$$= T_0[(1.0 - m_6)s_3 + m_6 s_6 - m_2 s_2]$$

$$= 300[(1.0 - 0.1456)7.4847 + 0.1456 \times 7.3119 - 1.0 \times 6.9246] \text{ kJ}$$

$$= 160.47 \text{ kJ}$$

exergy destroyed in condenser

$$= (1.0 - m_6)(\psi_3 - \psi_4)$$

$$= (1.0 - m_6)[(h_3 - h_4) - T_0(s_3 - s_4)]$$

$$= (1.0 - 0.1456)[(2334.35 - 168.77) - 300(7.4847 - 0.5763)] \text{ kJ}$$

$$= 79.51 \text{ kJ}$$

exergy destroyed in feedwater heater

$$= T_0 S_{\text{gen}}$$

$$= T_0[m_1 s_7 - (1.0 - m_6)s_5 - m_6 s_6]$$

$$= 300[1.0 \times 1.6716 - (1.0 - 0.1456) \times 0.5763 - 0.1456 \times 7.3119] \text{ kJ}$$

$$= 34.38 \text{ kJ.}$$

Exergy accounting for cycle:

$$\text{exergy in} - \text{exergy out} = \sum \text{exergy destruction}$$

$$1095.2 - 820.8 = 160.47 + 79.52 + 34.38$$

$$274.4 = 274.4 \text{ kJ.}$$

We can also write the second-law efficiency for the cycle as

$$\eta_{II} = 1 - \frac{\sum \text{exergy destruction}}{\text{exergy input for cycle}}$$

$$= 1 - \frac{274.4}{1095.2} = 74.9\%,$$

which is the same as before.

Exergy accounting for plant:

exergy in − exergy out = exergy destruction

$$2136.9 - 820.8 = 1041.7 + 160.47 + 79.51 + 34.38 \text{ kJ}$$

$$1316.1 = 1316.1 \text{ kJ}.$$

We express the second-law efficiency as

$$\eta_{II} = 1 - \frac{\sum \text{exergy destruction}}{\text{exergy input to plant}}$$

$$= 1 - \frac{1316.1}{2136.9} = 38.4\%,$$

the same as before. ∎

Comments

Some of the results for Examples 10.11 and 10.12 are shown below. (Calculations are made on the basis of 1.0 kg of fluid at the turbine inlet.)

	Heat Input to Plant (kJ)	Exergy Input to Plant (kJ)	Turbine Output (kJ)	$(\eta_{th})_{plant}$ (%)	$(\eta_{II})_{plant}$ (%)	$\sum$ Exergy Destructions for Plant (kJ)
Example 10.11	3063.73	2450.98	898.15	29.3	36.6	1552.83
Example 10.12	2671.1	2136.9	820.8	30.7	38.4	1316.1

From this table it can be seen that the scheme used in Example 10.12 is better thermodynamically since it has a higher thermal efficiency as well as a higher second-law efficiency. This improvement in thermodynamic performance is obtained at the expense of a more complicated arrangement, which would mean more capital cost. The designer must balance the savings in energy cost with the additional investment involved. Trade-offs again!

Problems

10.1 Argon gas is contained in a closed system at 5.0 MPa and 300°C. What is the maximum useful work that could be obtained from the gas? Assume that argon is an ideal gas. The pressure and temperature of the environment are 101.325 kPa and 300 K, respectively.

10.2 Five kilograms of steam is contained in a closed system at 5.0 MPa and 400°C. What is the maximum useful work that could be obtained from the steam, in kilojoules? The pressure and temperature of the environment are 101.325 kPa and 300 K, respectively.

10.3 A rigid tank having a volume of 2.0 m³ contains air at 1500 kPa and 100°C. Calculate the nonflow exergy of the air in the tank, in kilojoules. Assume that air is an ideal gas with constant specific heats of $c_p = 1.0038$ kJ/kg · K and $c_v = 0.7168$ kJ/kg · K. The pressure and temperature of the environment are 101.325 kPa and 300 K, respectively.

10.4 Five kilograms of air is heated quasi-statically at a constant pressure of 101.325 kPa from 25°C to 100°C. Determine the change in the availability of the air due to the change of state, in kilojoules. Assume that air is an ideal gas with constant specific heats of $c_p = 1.0038$ kJ/kg · K and $c_v = 0.7168$ kJ/kg · K. The pressure and temperature of the environment are 101.325 kPa and 300 K, respectively.

10.5 Two kilograms of air initially at 1000 kPa and 30°C undergo a free expansion to a final pressure of 100 kPa. Determine the irreversibility of the process. Assume that air is an ideal gas with constant specific heats of $c_p = 1.0038$ kJ/kg · K and $c_v = 0.7168$ kJ/kg · K. Pressure and temperature of the environment are 101.325 kPa and 300 K, respectively.

10.6 Two kilograms of metal having a temperature of 850°C is quenched by immersing it in a tank containing 100 kg of water with a temperature of 30°C. The specific heat of the metal is a constant of 0.465 kJ/kg · K. The specific heat of water may also be considered a constant of 4.187 kJ/kg · K. Determine the irreversibility of the process. Pressure and temperature of the environment are 101.325 kPa and 300 K, respectively.

10.7 A rigid tank contains steam at a pressure of 100 kPa and a temperature of 100°C. Heat is added to the steam until the pressure is doubled. Calculate
(a) the change in the availability of the steam.
(b) the irreversibility of the process if the heat source is a heat reservoir at 550°C.
The pressure and temperature of the environment are 101.325 kPa and 300 K, respectively.

10.8 Five kilograms of nitrogen is heated in a rigid vessel from 100 K to 300 K. The initial pressure is 0.50 MPa. The heat source is the environment at 300 K. Determine
(a) the change in the availability of the nitrogen, in kilojoules.
(b) the irreversibility of the process.
Pressure of the environment is 101.325 kPa.

10.9 Nitrogen is compressed polytropically from 100 kPa and 25°C to 700 kPa. The polytropic exponent for the process is 1.30. Heat transfer is with the environment at 25°C. Assume that nitrogen is an ideal gas with constant specific heats of $c_p = 1.0380$ kJ/kg · K and $c_v = 0.7412$ kJ/kg · K. Determine
(a) the actual useful work required, in kJ/kg.
(b) the reversible useful work, in kJ/kg.
(c) the thermodynamic lost work making use of results of parts (a) and (b).
The pressure and temperature of the environment are 101.325 kPa and 25°C, respectively.

10.10 Air is compressed adiabatically inside a cylinder from a pressure and temperature of 15 psia and 80°F to a pressure of 75 psia and a temperature of 500°F. Making use of data from Keenan and Kaye's *Gas Tables*, determine
(a) the actual useful work required, in Btu/lbm.
(b) the reversible useful work, in Btu/lbm.
(c) the irreversibility for the process making use of the results in parts (a) and (b).

10.11 One kilogram of oxygen expands inside a cylinder isothermally (but not quasi-statically) at 300 K from 10 MPa to 1.0 MPa. The actual work done by the gas is 100 kJ/kg. Determine
(a) the actual useful work performed, in kJ/kg.
(b) the reversible useful work, in kJ/kg.
(c) the irreversibility of the process making use of the results of parts (a) and (b).
The pressure and temperature of the environment are 101.325 kPa and 300 K, respectively.

10.12 Making use of the availability equation, deduce the coefficient of performance of an ideal refrigerator operating between a low-temperature heat reservoir at T_L and a high-temperature heat reservoir at T_H.

10.13 Making use of the availability equation, deduce the coefficient of performance of an ideal heat pump operating between a low-temperature heat reservoir at T_L and a high-temperature heat reservoir at T_H.

10.14 A heat engine operates between a high-temperature heat reservoir at 1000°C and a low-temperature heat reservoir at 200°C. If the thermal efficiency of the heat engine is 40%, what is its second-law efficiency? $T_0 = 300$ K.

10.15 A heat engine operates between a high-temperature heat reservoir at 2000°C and a low-temperature heat reservoir at 200°C. If the net work output of the engine is 2000 kJ when the heat input is 5000 kJ, what is the second-law efficiency of the engine? $T_0 = 300$ K.

10.16 A heat engine operates between a high-temperature heat reservoir at 1000°C and a low-temperature heat reservoir at 200°C. What is the thermal efficiency of the engine if its second-law efficiency is 100%? $T_0 = 300$ K.

10.17 A refrigerator operates between a low-temperature heat reservoir at 5°C and a high-temperature heat reservoir at 20°C. If the coefficient of performance of the refrigerator is 3.0, what is its second-law efficiency? $T_0 = 20$°C.

10.18 A refrigerator operates between a low-temperature heat reservoir at 5°C and a high-temperature heat reservoir at 20°C. If 1000 kJ of work input is required to produce 4000 kJ of refrigeration, what is the second-law efficiency of the refrigerator? $T_0 = 20$°C.

10.19 A heat pump operates between a low-temperature heat reservoir at 10°C and a high-temperature heat reservoir at 20°C. If the second-law efficiency is 100%, what is the coefficient of performance of the heat pump? $T_0 = 20$°C.

10.20 A heat pump operates between a low-temperature heat reservoir at 40°F and a high-temperature heat reservoir at 70°F. If the coefficient of performance is 3.5, what is the second-law efficiency of the heat pump? $T_0 = 70$°F.

10.21 Saturated liquid ammonia at 180 psia is throttled to 60 psia under steady-state steady-flow conditions. If heat transfer and changes in kinetic and potential energies are negligible, determine
(a) the reversible useful work for the process, in Btu/lbm.
(b) the irreversibility for the process, in Btu/lbm.
(c) What is the second-law efficiency of the process?
$p_0 = 14.7$ psia; $T_0 = 77$°F.

10.22 Saturated liquid Freon-12 at 2.0 MPa is throttled to 500 kPa under steady-state steady-flow conditions. If heat transfer and changes in kinetic and potential energies are negligible, determine
(a) the reversible useful work for the process, in kJ/kg.
(b) the thermodynamics lost work for the process, in kJ/kg.
(c) What is the second-law efficiency of the process?
$p_0 = 101.325$ kPa; $T_0 = 300$ K.

10.23 Water is heated in a steady-state steady-flow heat exchanger from 25°C to 50°C with heat coming from a heat reservoir at 250°C. If the mass flow rate of water is 0.5 kg/s, determine
(a) the reversible useful work of the process, in kJ/kg.
(b) the irreversibility of the process, in kJ/kg.
$p_0 = 101.325$ kPa; $T_0 = 300$ K.

10.24 Water enters a steam generator at 100°C and 2.2 MPa and leaves as steam at 2.0 MPa and 400°C. The heat source is a heat reservoir at 800°C. For a steady flow rate of 500 kg/h, determine
(a) the change in the availability of the flowing fluid, in kJ/h.
(b) the availability of the heat added to the steam, in kJ/h.
(c) the second-law efficiency of the process.
$p_0 = 101.325$ kPa; $T_0 = 300$ K.

10.25 Five kilograms of air is to be heated from 15°C and 100 kPa to 50°C in a steady-state steady-flow heat exchanger. If the heat source is steam flowing in at 200 kPa and 150°C and flowing out as water at 120°C, determine

(a) the change in the availability of the air, in kilojoules.

(b) the change in the availability of the steam, in kilojoules.

(c) the irreversibility of the process.

(d) the second-law efficiency of the process.

$p_0 = 101.325$ kPa; $T_0 = 300$ K.

10.26 A stream of steam enters an adiabatic mixing chamber at 1500 kPa and 300°C. Another stream of steam enters at 1500 kPa with a quality of 90%. The mass flow rates of the two streams are equal. The mixture leaves the chamber at 1500 kPa. Determine

(a) the temperature (or quality if it is a liquid–vapor mixture) of the mixture.

(b) the irreversibility of the process per unit mass of mixture.

$p_0 = 101.325$ kPa; $T_0 = 300$ K.

10.27 The air preheater is an important piece of hardware in a modern power plant in which air supply to the furnace is preheated by the products of combustion (flue gas). If the steady-state steady-flow operating conditions for such an air preheater are as shown in Fig. P10.27, determine

(a) the temperature of the air leaving the preheater.

(b) the change in the availability of the air, in Btu/h.

(c) the change in the availability of the flue gas, in Btu/h.

(d) the irreversibility of the process.

(e) the second-law efficiency of the process.

Make use of data from Keenan and Kaye's *Gas Tables* for the air as well as for the flue gas. $p_0 = 14.7$ psia; $T_0 = 77$°F.

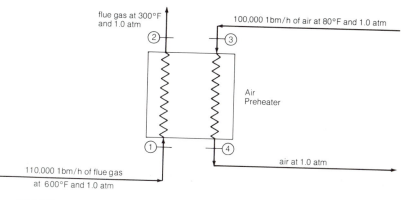

flue gas at 300°F and 1.0 atm

100,000 1bm/h of air at 80°F and 1.0 atm

② ③

Air Preheater

① ④

air at 1.0 atm

110,000 1bm/h of flue gas at 600°F and 1.0 atm

Figure P10.27

10.28 Air is compressed adiabatically in a steady-state steady-flow compressor from 100 kPa and 300 K to a final pressure of 500 kPa. What is the minimum useful power required if the mass flow rate is 2.0 kg/s? Assume that air is an ideal gas with constant specific heats of $c_p = 1.0038$ kJ/kg $\cdot$ K and $c_v = 0.7168$ kJ/kg $\cdot$ K. $p_0 = 101.325$ kPa; $T_0 = 300$ K.

10.29 Air is compressed isothermally in a steady-state steady-flow compressor from 100 kPa and 300 K to a final pressure of 500 kPa. What is the minimum power required for a mass flow rate of 2.0 kg/s? Assume that air is an ideal gas with constant specific heats with $c_p = 1.0038$ kJ/kg $\cdot$ K and $c_v = 0.7168$ kJ/kg $\cdot$ K. Heat transfer is with the environment at 300 K.

10.30 Argon gas expands adiabatically in a steady-state steady-flow turbine from 5000 kPa and 500°C to 500 kPa. If the turbine produces 10,000 kW of power when the mass flow rate is 50 kg/s, determine
(a) the adiabatic turbine efficiency.
(b) the second-law efficiency of the turbine.
Assume argon is an ideal gas. $p_0 = 101.325$ kPa; $T_0 = 300$ K.

10.31 Figure P10.31 may be used to produce high-velocity water operating under steady-state steady-flow conditions. The adiabatic pump efficiency is 70%. The nozzle may be assumed to be thermodynamically perfect. For a mass flow rate of 50 kg/s, determine
(a) the pressure of water at nozzle inlet, in kPa.
(b) the power required to drive the pump, in kilowatts.

(c) the irreversibility of the process.
$p_0 = 101.325$ kPa; $T_0 = 300$ K.

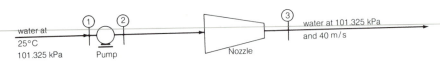

Figure P10.31

10.32 Figure P10.32 may be used to produce 1000-kPa steam if we have two sources of steam at 2500 kPa and 300 kPa. On the basis that both the turbine and the compressor will operate isentropically under steady-state steady-flow conditions, determine

(a) the ratio of the mass flow rate of 2500-kPa steam to the mass flow rate of 300-kPa steam.

(b) the temperature of the 1000-kPa steam formed from the discharge of the turbine and the discharge of the compressor.

(c) the irreversibility of the process per kg of 300 kPa steam.

Assume that the entire process is carried out adiabatically. $p_0 = 101.325$ kPa; $T_0 = 300$ K.

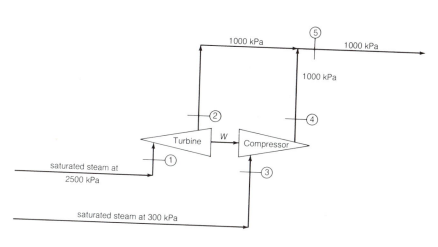

Figure P10.32

10.33 Ammonia is compressed in a steady-state steady-flow compressor which is driven by a steam turbine, as shown in Fig. P10.33. Assuming that the power output of the turbine matches the power needed for the compressor, determine, for a mass flow rate of 2500 lbm/hr of ammonia,

(a) the mass flow rate of steam, in lbm/hr.

(b) the adiabatic efficiency of the ammonia compressor.

(c) the irreversibility due to the compression process.

(d) the irreversibility due to the expansion process.

(e) the irreversibility for the entire process.

Assume that the entire process is carried out adiabatically. $p_0 = 101.325$ kPa; $T_0 = 300$ K.

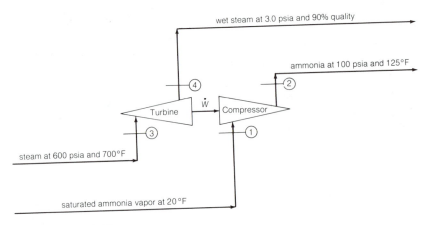

wet steam at 3.0 psia and 90% quality

ammonia at 100 psia and 125°F

Turbine $\dot{W}$ Compressor

steam at 600 psia and 700°F

saturated ammonia vapor at 20°F

Figure P10.33

10.34 For the indirect-fired (external combustion) air turbine power plant shown in Fig. P10.34,
(a) make a first-law analysis of the system.
(b) make a second-law analysis of the system.
Both compressor and turbine operate adiabatically. Neglect pressure drops in air heater. In addition to the specifications shown, make calculations on the basis of

$$\eta_C = \eta_T = 85\%$$

$$\frac{p_2}{p_1} = \frac{p_3}{p_4} = 7.$$

Assume that air is an ideal gas with constant specific heats of $c_p = 1.045$ kJ/kg $\cdot$ K and $k = 1.36$. $p_0 = 100$ kPa; $T_0 = 15°C$.

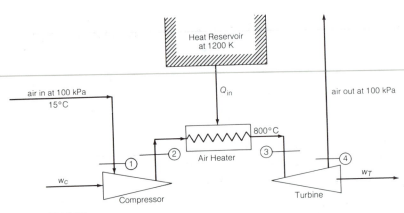

Heat Reservoir at 1200 K

air in at 100 kPa
15°C

Q_{in}

air out at 100 kPa

800°C

Air Heater

w_C

Compressor

w_T

Turbine

Figure P10.34

10.35 A rigid tank having a volume of 1.5 m³ initially contains a mixture of saturated vapor steam and saturated liquid water at 1.5 MPa. Of the total mass, 25% is vapor. Saturated liquid only is being withdrawn slowly through a valve from the bottom of the tank until the final mass in the tank is half of the initial total mass. During the process the temperature of the contents of the tank is kept constant through the use of an electrical heater immersed in the tank. Assuming that the process is carried out adiabatically, determine
(a) the energy input needed for the process.
(b) the irreversibility of the process.
$p_0 = 101.325$ kPa; $T_0 = 300$ K.

10.36 A rigid tank having a volume of 15 m³ contains air initially at 700 kPa and 40°C. When air is withdrawn from the tank, heat is added to the tank so that the outlet air and the tank air are always at the same temperature of 40°C. When the pressure of air in the tank has dropped to 175 kPa, determine
(a) the total amount of heat added to the contents in the tank.
(b) the irreversibility of the process.
Assume that air is an ideal gas with constant specific heats of $c_p = 1.0038$ kJ/kg · K and $c_v = 0.7168$ kJ/kg · K. $p_0 = 101.325$ kPa; $T_0 300$ K.

10.37 Ten kilograms of nitrogen gas will be heated in a rigid vessel from 300 K to 1000 K with energy coming from a heat reservoir at T^{HR}. Assuming ideal-gas behavior with variable specific heats, write a computer program to determine
(a) the exergy gained by the gas.
(b) the exergy supplied to the gas.
(c) the thermodynamic lost work for the process.
(d) the second-law efficiency for the process making use of results of parts (a) and (b).
(e) the second-law efficiency for the process making use of results of parts (b) and (c).
Make calculations for temperature of the heat reservoir at 3000 K, 2500 K, 2000 K, and 1000 K. Graph your results and comment. Let T_0 be 300 K. The molal heat capacity as a function of temperature for nitrogen is given by

$$\bar{c}_p^* = a + bT + cT^2 + dT^3$$

where $\bar{c}_p^*$ is in kJ/kgmol · K,
T is in degrees Kelvin,
a, b, c, and d are constants having the following values:

$a = 28.9$

$b = -0.1571 \times 10^{-2}$

$c = 0.8081 \times 10^{-5}$

$d = -2.873 \times 10^{-9}$.

10.38 Air is compressed adiabatically under steady-state steady-flow conditions from 100 kPa and 300 K to a final pressure of 1000 kPa. Write a computer program to calculate the second-law efficiency of the compressor corresponding to adiabatic compressor efficiency of 90%, 85%, 80%, and 75%. Assume air is an ideal gas with constant specific heats of

$$c_p = 1.0038 \text{ kJ/kg} \cdot \text{K}$$
$$c_v = 0.7168 \text{ kJ/kg} \cdot \text{K}.$$

Let T_0 be 300 K.

10.39 Argon expands adiabatically under steady-state steady-flow conditions from 300 K and 1000 kPa to a final pressure of 100 kPa. Write a computer program to calculate the second-law efficiency of the turbine corresponding to adiabatic turbine efficiency of 90%, 85%, 80%, and 75%. Assume ideal-gas behavior. Let T_0 be 300 K.

10.40 Air is being heated by a hot gas under steady-state steady-flow conditions in a counterflow heat exchanger (such as a regenerator in a regenerative gas turbine power plant). Let the inlet and outlet temperature of the air be T_1 and T_2, respectively. Let the inlet and outlet temperature of the hot gas be T_3 and T_4, respectively. Let the mass flowrate of the air be the same as that of the hot gas. Assume that specific heat c_p of the air is the same as that of the hot gas. Assuming ideal-gas behavior and neglecting pressure drops and changes in kinetic and potential energies, write a computer program to calculate the second-law efficiency of the heat transfer process for given values of T_1, T_3, and η_R, the heat exchanger effectiveness. Let T_0 be 300 K. Make calculations for

$$T_1 = 450 \text{ K}$$

and $\qquad$ $T_3 = 800 \text{ K}$

$$\eta_R = 0.50, 0.6, 0.7, 0.8, 0.9.$$

Plot η_{II} and η_R against T_2.

(*Note:* The heat exchanger effectiveness is a parameter commonly used to assess the performance of a heat exchanger. It is defined as the ratio of the actual amount of heat transfer to the maximum possible amount of heat transfer. For our problem, the heat exchanger effectiveness is given by

$$\eta_R = \frac{T_2 - T_1}{T_3 - T_1}.$$

11

Power-Producing, Combined-Cycle, and Cogeneration Systems

In the previous chapters, we studied and analyzed many devices that are important in engineering applications. We are now prepared to investigate the design of some important thermodynamic systems of which these devices are the major components. In this chapter we examine systems for the production of power. In Chapter 12 we examine systems for the production of refrigeration. In Chapter 16 we study systems with other objectives, such as gas-liquefaction systems and water-desalination systems.

11.1 Decision Making in Power-Plant Design

An industrial society consumes an enormous amount of energy. This is particularly true with advanced industrial societies. For example, the United States, with 6% of the world's population, consumes one-third of the world's energy.[1] Moreover, an advanced industrial society is further characterized by its increasing dependence on electricity, a trend that has direct effects on gross energy consumption and indirect effects on environmental quality. Consequently, the many decisions that we must make in the total system of energy-conversion-system design are intimately tied to the well-being of our economy and our quality of life.

Energy Source

Assuming the decision has been made that there is a need for a power plant of a certain capacity, we must next decide what the

---

[1] A. L. Hammond, W. D. Metz, and T. H. Maugh II, *Energy and the Future*, American Association for the Advancement of Science, Washington, D.C., 1973, p. v.

energy source should be. It could be the chemical energy stored in coal, oil, and natural gas. It could be the atomic energy stored in matter. It could be the solar radiation that we receive daily. The factors affecting this decision are many—some technical in nature, some economic, and some political. Discussion of all these factors is beyond the scope of this book. We shall limit our study to that of energy coming from a heat reservoir. This means that the energy input to our system is heat which may be released as the result of burning fossil fuels or a nuclear reaction. The heat could also be coming from the hot rocks deep underground.

Energy Sink

It is a general deduction from the second law of thermodynamics that we must have a heat sink in order to convert heat into useful work continuously. The second law also tells us that we should make use of the heat sink with the lowest temperature. Natural heat sinks are the rivers, the oceans, the lakes, and the atmosphere. With our present technology, we are rejecting about two-thirds of the heat input to the heat sink. This large quantity of waste heat could have considerable adverse effect on our environment. In the case of power plants burning fossil fuels, there is the additional problem of dumping into the atmosphere of a large amount of combustion products. It is clear that the problem associated with the siting of large power-producing plants deserves serious study.

Working Fluid

As shown in Fig. 11.1, the schematic representation of a power plant may be considered as consisting of four major subsystems: an energy source, an energy sink, a work reservoir, and an energy-conversion system. To produce power continuously, the energy-conversion system must operate in a cycle in such a fashion that during part of the cycle, it will absorb heat from the heat source; during another part of the cycle, it will reject heat to the heat sink; and during still another part of the cycle, it will produce net work output. To accomplish these effects, we employ a thermodynamic sub-stance commonly called the *working fluid* of the cycle. In modern steam power plants, the working fluid is water. In nuclear power plants proposed for long-range space missions, the working fluid is a liquid metal. In stationary gas-cooled nuclear power plants, a working fluid is helium. In a thermoelectric generator, the working "fluid" is the electron. Each working fluid has certain desirable features, but each also has undesirable features. One of the major challenges to engineers is the continual search for a better working fluid.

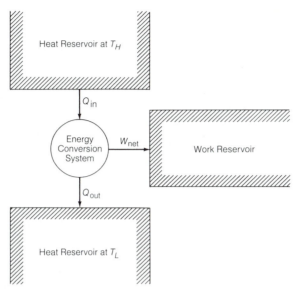

Figure 11.1 Four Major Subsystems of a Power Plant

Cycle Selection

Since the rate of producing work is called *power*, a cycle that provides for the continuous conversion of heat into work is called a *power cycle*. A power cycle in which the working fluid undergoes a change of phase, such as in a steam power plant, is known as a *vapor power cycle*. A power cycle in which the working fluid remains in the gaseous phase throughout is known as a *gas power cycle*. There are advantages and disadvantages in each kind of cycle using a particular working fluid. It is obvious that the decision regarding the cycle to be used depends on the decision regarding the working fluid to be used.

Components Selection

Once a cycle using a given working fluid has been decided upon, we must turn from conceptual design to hardware design. We must decide on what physical arrangement of components we should use to carry out the various processes of the cycle. It is at this stage of the design process that we shall arrive at the important parameters, such as temperature, pressure, and flow quantities, for the design of each piece of hardware. These design parameters will make it possible for the decision maker to determine the economic feasibility of such a system.

General Considerations

Engineers should always strive to increase the thermal efficiency of power-producing systems. From the thermodynamic point of view, this means that we must try to minimize irreversibilities in our systems. However, to do so usually requires more bulky and more expensive equipment, since reversibility can be achieved only at vanishingly small rates of energy transfer. Thus engineers must learn to compromise. A good power-producing system is one that has reasonably high thermal efficiency consistent with economic, environmental, and other constraints.

VAPOR POWER CYCLES

11.2 Carnot Vapor Cycle

Suppose that we have decided on the temperature range to be used for a power-cycle design. Suppose that we have also decided to use a simple compressible substance, such as water, as the working fluid. We must now select a cycle and then translate the cycle into hardware design.

Since the Carnot cycle has the maximum thermal efficiency for any given temperature range, it is natural that we should examine this cycle first. A Carnot cycle operating completely within the· liquid–vapor region is shown in Fig. 11.2a. The equipment arrangement that may be used to carry out the processes for this cycle is shown in Fig. 11.2b. The heat-addition process $1 \rightarrow 2$, a constant-pressure as well as a constant-temperature process, can be accomplished in the steam generator or boiler. The isentropic expansion process $2 \rightarrow 3$ can be simulated by a well-designed turbine. The heat-rejection process $3 \rightarrow 4$, also a constant-pressure as well as a constant-temperature process, can be carried out in a condenser, but it is difficult to control the final quality. The isentropic process $4 \rightarrow 1$ involves the compression of a liquid–vapor mixture. This is most difficult to simulate. Consequently, the actual compressor efficiency will be very low, thereby consuming a good portion of the turbine output.

In addition, we are not adding heat at the highest possible temperature, since this cycle operates at a top temperature below the critical temperature of the working fluid. For steam, the critical temperature is 374.15°C, which is well below the allowable operating temperature of about 566°C with available materials.

Figure 11.3a shows a version of the Carnot cycle that does not have the difficulties of the first one. Figure 11.3b shows equipment arrangement that may be used to carry out the processes of the cycle. In this version of the Carnot cycle, both the isentropic

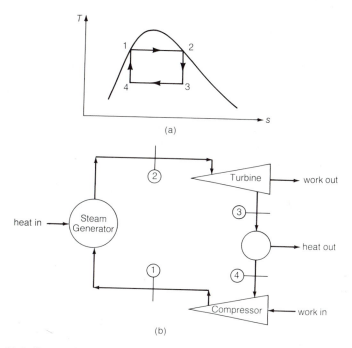

(a)

(b)

Figure 11.2 Carnot Vapor Cycle Operating Inside the Vapor Dome

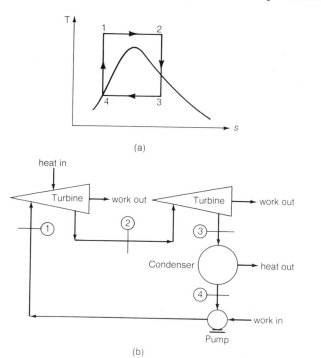

(a)

(b)

Figure 11.3 Carnot Cycle Operating Above the Critical Temperature

expansion process $2 \rightarrow 3$ and the heat-rejection process $3 \rightarrow 4$ can be simulated very closely in actual devices. But the heat-addition process $1 \rightarrow 2$ must be carried out in an isothermal turbine at high temperature. This is most difficult to accomplish. In addition, the final pressure at point 1 would be extremely high. For example, the discharge pressure of the pump must be about 90 MPa just to increase the water temperature from 38°C to 41°C isentropically. The pressure corresponding to a temperature above the critical temperature is utterly unreasonable. Thus there are difficulties with this cycle, also.

Because of the difficulties encountered in carrying out some of the processes in actual devices, the Carnot vapor cycle is simply not practical.

11.3 Simple Rankine Cycle

The difficulties with the Carnot vapor cycle may be eliminated by replacing the reversible isothermal heat-addition process with a reversible constant-pressure heat-addition process. We would then have a *Rankine cycle* consisting of the following processes:

1. Reversible constant-pressure heat addition.
2. Reversible adiabatic expansion.
3. Reversible constant-pressure heat rejection.
4. Reversible adiabatic compression.

The Rankine cycle is the base cycle for a vapor–liquid system such as the steam power plant. Such a cycle $(1 \rightarrow 2 \rightarrow 3 \rightarrow 4 \rightarrow 5 \rightarrow 1)$ is shown in the T–s diagram in Fig. 11.4a. The equipment arrangement that may be used to carry out the processes involved is shown in Fig. 11.4b. Note that the length of line 5—1 in the T–s diagram of Fig. 11.4a is greatly exaggerated since the temperature rise of liquid water resulting from compression is very small.

Since the heat-addition process is carried out reversibly, the area under the path $1 \rightarrow 2 \rightarrow 3$ in the T–s diagram of Fig. 11.4a represents the total amount of heat added to cycle. If we make this area equal to the area under the horizontal line 6—7, we would have

$$T_{m2}(s_7 - s_6) = Q_{in},$$

(11.1)

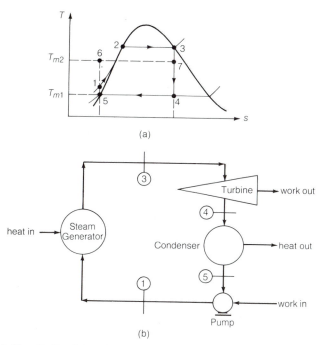

Figure 11.4 Simple Rankine Cycle

where T_{m2} is known as the *mean effective temperature of heat addition.*[2] By the same token, we can let

$$T_{m1}(s_4 - s_5) = |Q_{out}|, \tag{11.2}$$

where T_{m1} is known as the *mean effective temperature of heat rejection.*
   By definition, the thermal efficiency η_{th} of a cycle is given by

$$\eta_{th} = \frac{Q_{in} - |Q_{out}|}{Q_{in}}. \tag{11.3}$$

In terms of the mean effective temperature of heat addition and the mean effective temperature of heat rejection, the thermal efficiency of cycle $1 \rightarrow 2 \rightarrow 3 \rightarrow 4 \rightarrow 5 \rightarrow 1$ may now be written as

$$\eta_{th} = \frac{T_{m2}(s_7 - s_6) - T_{m1}(s_4 - s_5)}{T_{m2}(s_7 - s_6)}. \tag{11.4}$$

[2] J. F. Lee and F. W. Sears, *Thermodynamics,* Addison-Wesley Publishing Co., Inc., Reading, Mass., 1955, p. 350.

But $s_7 = s_4$ and $s_6 = s_5$. Thus Eq. 11.4 is reduced to

$$\eta_{th} = \frac{T_{m2} - T_{m1}}{T_{m2}}, \tag{11.5}$$

which is simply the thermal efficiency of a Carnot cycle operating between T_{m2} and T_{m1}. The thermal efficiency of a Carnot cycle operating between T_2 and T_1 is of course given by

$$\eta_{th} = \frac{T_2 - T_1}{T_2}.$$

Since T_1 is the same as T_{m1} while T_{m2} is less than T_2, the Rankine cycle is seen to have a lower thermal efficiency than that of a Carnot cycle operating between the same temperature limits.

Equation 11.5 says that for a fixed T_{m1} we must increase the mean effective temperature of heat addition in order to increase the thermal efficiency of a simple Rankine cycle. In the next several sections, we study the different ways of accomplishing this effect.

Performance calculations for a Rankine cycle are relatively simple. Assuming steady-state steady-flow processes throughout and neglecting changes in kinetic energy and potential energy across each piece of cycle component, we have, from the first law, the following results:

$$\text{Steam generator:} \quad Q_{in} = h_3 - h_1. \tag{11.6}$$

$$\text{Turbine:} \quad W_T = h_3 - h_4. \tag{11.7}$$

$$\text{Condenser:} \quad Q_{out} = h_5 - h_4. \tag{11.8}$$

$$\text{Pump:} \quad W_p = h_5 - h_1. \tag{11.9}$$

The cycle thermal efficiency may now be expressed as

$$\eta_{th} = \frac{W_{net}}{Q_{in}}$$

$$= \frac{(h_3 - h_4) + (h_5 - h_1)}{h_3 - h_1}$$

$$= \frac{(h_3 - h_4) - (h_1 - h_5)}{h_3 - h_1}. \tag{11.10}$$

Knowing the state condition at each point, we may determine the enthalpy terms by using available tabulated data. The pump work, however, may be obtained in another way. From the first law, we have

$$\bar{d}W_p = -dh.$$

But from the property relation, $T\ ds = dh - v\ dp$. Since the pumping process is isentropic $(ds = 0)$, the pump work may be given as

$$\bar{d}W_p = -v\ dp.$$

Assuming incompressible fluid, we have

$$W_p \approx -v_5(p_1 - p_5). \tag{11.11}$$

Equation 11.11 gives a fairly good approximation of the correct pump work. It will be seen that the pump work is quite small compared to the turbine work. Consequently, any error in the determination of pump work would introduce only a very small error in the cycle thermal efficiency.

It is important for us to know the cycle thermal efficiency, as it is a measure of the energy cost for power production. In practice, it is equally important for us to know the rate of fluid circulation in the cycle, as the construction cost of each piece of hardware depends on this quantity. It is common practice to express the fluid-flow rate on the basis of 1.0 kW of power output. In steam-power-plant design, this is known as the *steam rate* or specific steam consumption.

Since 1.0 kW = 3600 kJ/h, we have

$$\text{steam rate} = \frac{3600}{w_{\text{net}}}\ \text{kg/kWh}, \tag{11.12}$$

where w_{net} is in kJ/kg.

Since 1.0 kW = 3413 Btu/h, we also have

$$\text{steam rate} = \frac{3413}{w_{\text{net}}}\ \text{lbm/kWh}, \tag{11.12a}$$

where w_{net} is in Btu/lbm.

Example 11.1

A simple Rankine cycle using water as the working fluid operates between the pressure limits of 7.5 kPa and 1800 kPa. Determine

(a) the cycle thermal efficiency.
(b) the steam rate (specific steam consumption).

Solution (refer to Fig. 11.4)

For the turbine,

$$w_T = h_3 - h_4.$$

From the steam tables,

$$h_3 = h_g \text{ at } 1800 \text{ kPa} = 2794.8 \text{ kJ/kg}.$$

Since process $3 \to 4$ is isentropic, h_4 is defined by $p_4 = 7.5$ kPa and $s_4 = s_3 = s_g$ at 1800 kPa $= 6.3751$ kJ/kg $\cdot$ K. Since s_g at 7.5 kPa is 8.2523 kJ/kg $\cdot$ K and s_f at 7.5 kPa is 0.5763 kJ/kg $\cdot$ K, point 4 is inside the liquid–vapor region. Thus we may use the following expression to determine h_4:

$$h_4 = [h_g - (1 - x)_4 \, h_{fg}] \text{ at } 7.5 \text{ kPa}.$$

To find the quality at point 4, we use

$$s_3 = s_4 = [s_g - (1 - x)_4 \, s_{fg}] \text{ at } 7.5 \text{ kPa}$$

$$6.3751 = 8.2523 - (1 - x)_4 (7.6760),$$

from which we get $(1 - x)_4 = 0.2446$. Then

$$h_4 = 2574.0 - (0.2446)(2406.2) \text{ kJ/kg}$$

$$= 1986.3 \text{ kJ/kg}$$

and

$$w_T = 2794.8 - 1986.3 \text{ kJ/kg}$$

$$= 808.5 \text{ kJ/kg}.$$

For the condenser,

$$q_{out} = h_5 - h_4,$$

where $h_5 = h_f$ at 7.5 kPa = 168.77 kJ/kg. Thus

$$q_{out} = 168.77 - 1986.3 \text{ kJ/kg}$$

$$= -1817.5 \text{ kJ/kg}.$$

The negative sign simply means that heat is leaving the condenser according to our convention.

For the pump,

$$w_p = (h_5 - h_1) \approx -v(p_1 - p_5),$$

where v may be taken as v_f at 7.5 kPa, which is 0.0010079 m³/kg. Thus

$$w_p = -0.0010079(1800 - 7.5) \text{ kJ/kg}$$

$$= -1.81 \text{ kJ/kg},$$

where the negative sign means that work is supplied to the pump.

$$h_1 = h_5 - w_p$$

$$= 168.77 - (-1.81) \text{ kJ/kg} = 170.58 \text{ kJ/kg}.$$

For the steam generator,

$$q_{in} = h_3 - h_1$$

$$= 2794.8 - 170.58 \text{ kJ/kg} = 2624.22 \text{ kJ/kg}.$$

(a) The cycle thermal efficiency is

$$\eta_{th} = \frac{w_{net}}{q_{in}}$$

$$= (808.5 - 1.81)/2624.22 = 30.7\%.$$

It is instructive to determine the cycle thermal efficiency using Eq. 11.5. Now

$$T_{m2} = \frac{q_{in}}{s_3 - s_5} = \frac{2624.22}{6.3751 - 0.5763} \text{ K} = 452.55 \text{ K}$$

and

$$T_{m1} = T_1 = \text{saturation temperature at 7.5 kPa}$$

$$= (40.316 + 273.15) \text{ K} = 313.466 \text{ K}.$$

Therefore, we have

$$\eta_{th} = \frac{452.55 - 313.466}{452.55} = 30.7\%,$$

which is the same as before.

(b)
$$\text{steam rate} = \frac{3600}{w_{net}}$$

$$= \frac{3600}{808.5 - 1.81} \text{ kg/kWh}$$

$$= 4.46 \text{ kg/kWh.} \qquad \blacksquare$$

Comments

Note that the pump work is quite small compared to the turbine work. The ratio of net work to turbine work, known as the *net work ratio*, is practically unity for a Rankine cycle. This implies that the performance of the Rankine cycle is not sensitive to the efficiency of the pump. Also note that the maximum temperature of the cycle is 207.11°C (saturation temperature at 1800 kPa) and the minimum temperature of the cycle is 40.316°C (saturation temperature at 1.0 psia). The Carnot cycle efficiency between these temperature limits is 34.7 percent which is, as expected, greater than 30.7 percent.

Example 11.2

A simple Rankine cycle using water as the working fluid operates between the pressure limits of 7.5 kPa and 17.0 MPa. Determine

(a) the cycle thermal efficiency.
(b) the steam rate.

Solution (refer to Fig. 11.4)

For the turbine,

$$w_T = h_3 - h_4.$$

From the steam tables,

$$h_3 = 2551.6 \text{ kJ/kg}$$

$$s_3 = 5.1855 \text{ kJ/kg} \cdot \text{K.}$$

Thus

$$s_3 = s_4 = [s_g - (1 - x)_4 \, s_{fg}] \text{ at 7.5 kPa}$$

$$5.1855 = 8.2523 - (1 - x)_4(7.6760),$$

from which we get $(1 - x)_4 = 0.3995$. Then

$$h_4 = [h_g - (1 - x)_4 \, h_{fg}] \text{ at 7.5 kPa}$$

$$= 2574.9 - (0.3995)(2406.2) \text{ kJ/kg}$$

$$= 1613.6 \text{ kJ/kg}$$

and

$$w_T = (2551.6 - 1613.6) \text{ kJ/kg}$$

$$= 938.0 \text{ kJ/kg}.$$

For the condenser,

$$q_{\text{out}} = h_5 - h_4$$

$$= 168.77 - 1613.6 \text{ kJ/kg} = -1444.8 \text{ kJ/kg}.$$

For the pump,

$$w_p = (h_5 - h_1) \approx -v(p_1 - p_5)$$

$$= -0.0010079(17000 - 7.5) \text{ kJ/kg}$$

$$= -17.13 \text{ kJ/kg}$$

$$h_1 = h_5 - w_p$$

$$= 168.77 - (-17.13) \text{ kJ/kg} = 185.9 \text{ kJ/kg}.$$

For the steam generator,

$$q_{\text{in}} = h_3 - h_1$$

$$= 2551.6 - 185.9 \text{ kJ/kg} = 2365.7 \text{ kJ/kg}.$$

(a) The cycle thermal efficiency is

$$\eta_{th} = \frac{w_{net}}{q_{in}}$$

$$= \frac{938.0 - 17.13}{2365.7} = 38.9\%$$

or

$$\eta_{th} = \frac{q_{net}}{q_{in}}$$

$$= \frac{2365.7 - 1444.8}{2365.7} = 38.9\%$$

In terms of the mean temperature of heat addition and the mean temperature of heat rejection, we have

$$T_{m2} = \frac{q_{in}}{s_3 - s_5} = \frac{2365.7}{5.1855 - 0.5763} \text{ K} = 513.3 \text{ K}$$

$$T_{m1} = T_1 = (40.316 + 273.15) \text{ K} = 313.466 \text{ K}$$

$$\eta_{th} = \frac{513.3 - 313.466}{513.3} = 38.9\%.$$

(b)   steam rate $= \dfrac{3600}{w_{net}}$ kg/kWh $= \dfrac{3600}{938.0 - 17.13}$ kg/kWh

$$= 3.91 \text{ kg/kWh.} \qquad\blacksquare$$

Comments

The cycle in this example, compared to the cycle in Example 11.1, is better in thermal efficiency and in steam rate. But the quality of the steam leaving the turbine is only 60.05%, which is much too low. Since the presence of more than 10% of liquid content in the exhaust steam will cause serious erosion of turbine blades, this low quality of steam must be avoided to maintain long turbine life and high performance.

11.4 Rankine Cycle with Superheat

To prevent erosion of the low-pressure end of the turbine due to excessive moisture content at the end of the expansion process, we may operate a Rankine cycle with superheat, as shown in the *T–s*

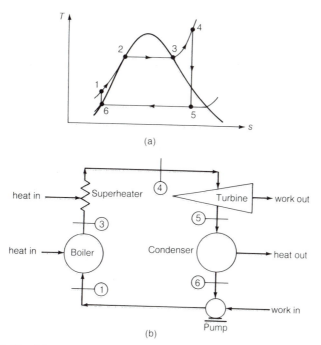

Figure 11.5 Rankine Cycle with Superheat

diagram of Fig. 11.5a. The equipment arrangement that may be used is shown in Fig. 11.5b.

By having the steam further heated in a superheater, we also achieve a higher mean temperature of heat addition without increasing the maximum pressure of the cycle. These benefits are gained, however, at a price: the addition of the superheater.

For this cycle, we have

$$Q_{in} = h_4 - h_1 \tag{11.13}$$

$$W_T = h_4 - h_5 \tag{11.14}$$

$$Q_{out} = h_6 - h_5 \tag{11.15}$$

$$W_p = h_6 - h_1 \approx -v_6(p_1 - p_6) \tag{11.16}$$

$$\eta_{th} = \frac{(h_4 - h_5) - (h_1 - h_6)}{h_4 - h_1} \tag{11.17}$$

$$\text{steam rate} = \frac{3600}{(h_4 - h_5) - (h_1 - h_6)} \text{ kg/kWh} \tag{11.18}$$

Example 11.3

A Rankine cycle using water as the working fluid operates between the pressure limits of 7.5 kPa and 17.0 MPa. The peak temperature of the cycle is 550°C. Determine

(a) the cycle thermal efficiency.
(b) the steam rate.

Solution (refer to Fig. 11.5)

For the turbine,

$$w_T = h_4 - h_5.$$

From the steam tables,

$$h_4 = 3428.0 \text{ kJ/kg}$$

$$s_4 = 6.4430 \text{ kJ/kg} \cdot \text{K}.$$

Thus

$$s_4 = s_5 = [s_g - (1 - x)_5 \, s_{fg}] \text{ at 7.5 kPa}$$

$$6.4430 = 8.2523 - (1 - x)_5(7.6760),$$

from which we get $(1 - x)_5 = 0.2357$. Then

$$h_5 = [h_g - (1 - x)_5 \, h_{fg}] \text{ at 7.5 kPa}$$

$$= 2574.9 - (0.2357)(2406.2) \text{ kJ/kg}$$

$$= 2007.8 \text{ kJ/kg}$$

and

$$w_T = 3428.0 - 2007.8 \text{ kJ/kg} = 1420.2 \text{ kJ/kg}.$$

For the condenser,

$$q_{out} = h_6 - h_5$$

$$= 168.77 - 2007.8 \text{ kJ/kg} = -1839.0 \text{ kJ/kg}.$$

For the pump,

$$w_p = h_6 - h_1 \approx -v(p_1 - p_6)$$

$$= -0.0010079(17{,}000 - 7.5) \text{ kJ/kg}$$

$$= -17.13 \text{ kJ/kg}$$

$$h_1 = h_6 - w_p$$

$$= 168.77 - (-17.13) \text{ kJ/kg} = 185.9 \text{ kJ/kg}.$$

For the boiler plus superheater,

$$q_{in} = h_4 - h_1$$

$$= 3428.0 - 185.9 \text{ kJ/kg} = 3242.1 \text{ kJ/kg}.$$

(a) The cycle thermal efficiency is

$$\eta_{th} = \frac{w_{net}}{q_{in}}$$

$$= \frac{1420.2 - 17.13}{3242.1} = 43.3\%.$$

In terms of the mean temperature of heat addition and the mean temperature of heat rejection, we have

$$T_{m2} = \frac{q_{in}}{s_4 - s_6} = \frac{3242.1}{6.4430 - 0.5763} \text{ K} = 552.6 \text{ K}$$

$$T_{m1} = T_1 = (40.316 + 273.15) \text{ K} = 313.466 \text{ K}$$

$$\eta_{th} = \frac{552.6 - 313.466}{552.6} = 43.3\%.$$

(b) $$\text{steam rate} = \frac{3600}{w_{net}} \text{ kg/kWh}$$

$$= \frac{3600}{1420.2 - 17.13} \text{ kg/kWh} = 2.57 \text{ kg/kWh}. \qquad \blacksquare$$

Comments

The moisture content of the exhaust steam leaving the turbine for this cycle is about 23.57%, which is still much larger than 10%. If we want to stay with the maximum cycle temperature of 550°C, we can shift to the drier region by operating at a lower maximum cycle pressure, in which case we would get a lower thermal efficiency and a larger steam rate. But we can retain the same maximum cycle temperature and maximum cycle pressure and still be able to shift into a drier region by the use of reheat, which is to be discussed in the next section.

Example 11.4

A steam power cycle operates between the same pressure limits of 7.5 kPa and 17.0 MPa and the same peak temperature of 550°C, as in Example 11.3. However, the adiabatic efficiency of the turbine is only 85%, and the adiabatic pump efficiency is only 80%. For this irreversible cycle determine

(a) the thermal efficiency.
(b) the steam rate.

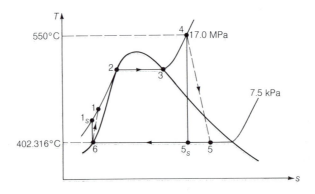

Solution (the cycle is as shown)

For the turbine,

$$(w_T)_{act} = \eta_T(h_4 - h_{5s}).$$

From Example 11.3, $(h_4 - h_{5s}) = 1420.2$ kJ/kg. Therefore,

$$(w_T)_{act} = 0.85 \times 1420.2 \text{ kJ/kg} = 1207.2 \text{ kJ/kg}.$$

Then

$$h_5 = h_4 - (w_T)_{act} = 3428.0 - 1207.2 \text{ kJ/kg} = 2220.8 \text{ kJ/kg}.$$

Since $h_5 = 2220.8 < h_g$ at 7.5 kPa, $h_5 > h_f$ at 7.5 kPa, state 5 is in the liquid–vapor region. Thus

$$h_5 = [h_g - (1 - x)_5 \, h_{fg}] \text{ at 7.5 kPa}$$

$$2220.8 = 2574.9 - (1 - x)_5(2406.2)$$

and

$$(1 - x)_5 = 0.1472.$$

For the condenser,

$$q_{out} = h_6 - h_5$$

$$= 168.77 - 2220.8 \text{ kJ/kg} = -2052.0 \text{ kJ/kg}.$$

For the pump,

$$(w_p)_{act} = h_6 - h_1 = \frac{h_6 - h_{1s}}{\eta_p}.$$

From Example 11.3, $(h_6 - h_{1s}) = -17.13$ kJ/kg. Therefore,

$$(w_p)_{act} = -\frac{17.13}{0.8} \text{ kJ/kg} = -21.41 \text{ kJ/kg}.$$

Then

$$h_1 = h_6 - (w_p)_{act} = 168.77 - (-21.41) \text{ kJ/kg} = 190.18 \text{ kJ/kg}.$$

For the boiler plus superheater,

$$q_{in} = h_4 - h_1 = 3428.0 - 190.18 = 3236.8 \text{ kJ/kg}.$$

(a) The cycle thermal efficiency is

$$\eta_{th} = \frac{w_{net}}{q_{in}}$$

$$= \frac{1207.2 - 21.41}{3236.8} = 36.6\%.$$

(b)   steam rate $= \dfrac{3600}{w_{net}}$ kg/kWh

$$= \frac{3600}{1207.2 - 21.41}\text{ kg/kWh} = 3.04\text{ kg/kWh.} \qquad \blacksquare$$

Comments

Comparing the results of this example with those of Example 11.3, we see that due to the irreversible effects in the turbine and pump, the cycle thermal efficiency is reduced from 43.3 to 36.6%, while the steam rate is increased from 2.57 kg/kWh to 3.04 kg/kWh. The reader should check that the reduction in cycle thermal efficiency and the increase in steam rate are primarily due to imperfections in the turbine. This is consistent, of course, with our previous comment that the performance of the Rankine cycle is not sensitive to the efficiency of the pump.

Example 11.5

The energy source for the cycle of Example 12.4 is a heat reservoir at 1400 K. For this irreversible cycle, determine

(a) the irreversibility per unit mass of working fluid for each of the processes involved.

(b) the second-law efficiency of the system. Let $T_0 = 300$ K.

Solutions

The irreversibility per unit mass for each process may be calculated by using

$$\frac{I}{m} = \frac{LW_0}{m} = \frac{T_0 S_{gen}}{m}$$

$$= T_0\left[(s_{out} - s_{in}) - \frac{q_k}{T_k}\right].$$

The data we need for analysis are as follows:

$$h_1 = 190.18\text{ kJ/kg}$$

$$h_4 = 3428.0\text{ kJ/kg}$$

$$h_5 = 2220.8\text{ kJ/kg}$$

$$h_6 = 168.77\text{ kJ/kg}$$

$$q_{in} = 3236.8\text{ kJ/kg}$$

$$q_{out} = -2052.0 \text{ kJ/kg}$$

$$s_1 \approx s_6 + \frac{h_1 - h_6}{T_6}$$

$$= 0.5763 + \frac{190.18 - 168.77}{313.466} = 0.6446 \text{ kJ/kg} \cdot \text{K}$$

$$s_4 = 6.4430 \text{ kJ/kg} \cdot \text{K}$$

$$s_5 = [s_g - (1 - x)_5 \, s_{fg}] \text{ at 7.5 kPa}$$

$$= 8.2523 - (0.1472)(7.6760) = 7.1224 \text{ kJ/kg} \cdot \text{K}$$

$$s_6 = 0.5763 \text{ kJ/kg} \cdot \text{K}.$$

(a) The irreversibility of the processes involved are as follows:

For process $1 \to 4$ (heat addition):

$$\frac{I}{m} = T_0\left(s_4 - s_1 - \frac{q_{in}}{T^{HR}} \right)$$

$$= 300\left(6.4430 - 0.6446 - \frac{3236.8}{1400} \right) \text{ kJ/kg} = 1045.92 \text{ kJ/kg}.$$

For process $4 \to 5$ (turbine):

$$\frac{I}{m} = T_0(s_5 - s_4)$$

$$= 300(7.1224 - 6.4430) \text{ kJ/kg} = 203.82 \text{ kJ/kg}.$$

For process $5 \to 6$ (condenser):

$$\frac{I}{m} = T_0\left(s_6 - s_5 - \frac{q_{out}}{T_0} \right)$$

$$= 300\left(0.5763 - 7.1224 - \frac{-2052.0}{300} \right) \text{ kJ/kg}$$

$$= 88.17 \text{ kJ/kg}.$$

For process $6 \rightarrow 1$ (pump):

$$\frac{I}{m} = T_0(s_1 - s_6)$$

$$= 300(0.6446 - 0.5763) \text{ kJ/kg} = 20.49 \text{ kJ/kg}.$$

Thus the total irreversibility for the cycle is

$$\sum \frac{I}{m} = 1045.92 + 203.82 + 88.17 + 20.49 \text{ kJ/kg} = 1358.4 \text{ kJ/kg}.$$

Now

$$\text{exergy input} = \left(1 - \frac{T_0}{T^{HR}}\right)q_{in}$$

$$= \left(1 - \frac{300}{1400}\right)(3236.8) \text{ kJ/kg} = 2543.2 \text{ kJ/kg}$$

$$\text{exergy output} = 1207.2 - 21.43 \text{ kJ/kg} = 1185.77 \text{ kJ/kg}.$$

Thus from exergy accounting, we have

$$\text{exergy in} - \text{exergy out} = \sum \frac{I}{m}$$

$$2543.2 - 1185.77 = 1358.4$$

$$1358.4 = 1358.4 \text{ kJ/kg}.$$

(b) The second-law efficiency of the system is given by

$$\eta_{II} = 1 - \frac{\sum(I/m)}{\text{exergy input}}$$

$$= 1 - \frac{1358.4}{2543.2} = 46.59\%. \qquad \blacksquare$$

Comments

First-law analysis shows that 2052.0/3236.8 or 63.34% of heat input is rejected in the condenser. But the second-law analysis shows that the exergy

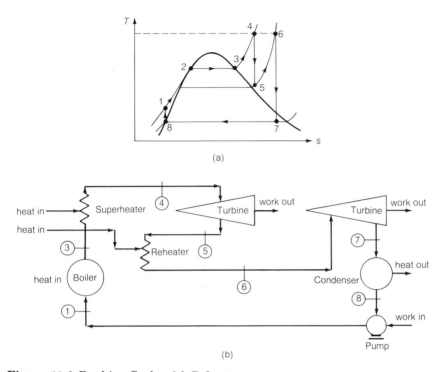

Figure 11.6 Rankine Cycle with Reheat

destroyed in the condenser is only 88.17/2543.2 or 3.47% of exergy input. The major irreversibility occurs in the steam generator, which is 1045.92/2543.2 or 41.11% of the exergy input. This is due to the large temperature difference between the heat source and the temperature of the working fluid in the steam generator. To improve the second-law efficiency, we must do a better job of matching the heat source with the temperature of the working fluid in the steam generator.

11.5 Rankine Cycle with Reheat

The reheat cycle is developed so that we can operate the Rankine cycle with high cycle pressure and temperature and still avoid excessive moisture in the low-pressure stages of the turbine. Such a cycle is shown in the T–s diagram of Fig. 11.6a. The equipment arrangement for this cycle is shown in Fig. 11.6b. In this cycle, the steam is first expanded to some intermediate pressure in the turbine and is then reheated at constant pressure in the reheater, after which it is expanded to the exhaust pressure. The turbine may be considered as having two stages: high pressure and low pressure. Heat is added at three places: in the boiler (process $1 \rightarrow 3$); in the superheater (process

$3 \rightarrow 4$); and in the reheater (process $5 \rightarrow 6$). Work is produced in two places: in the high-pressure turbine (process $4 \rightarrow 5$) and in the low-pressure turbine (process $6 \rightarrow 7$). Heat is rejected in the condenser, and work is supplied to the pump. Thus, for this cycle, we have

$$Q_{in} = (h_4 - h_1) + (h_6 - h_5) \tag{11.19}$$

$$W_T = (h_4 - h_5) + (h_6 - h_7) \tag{11.20}$$

$$Q_{out} = h_8 - h_7 \tag{11.21}$$

$$W_p = h_8 - h_1 \approx -v_8(p_1 - p_8) \tag{11.22}$$

$$\eta_{th} = \frac{(h_4 - h_5) + (h_6 - h_7) - (h_1 - h_8)}{(h_4 - h_1) + (h_6 - h_5)} \tag{11.23}$$

$$\text{steam rate} = \frac{3600}{(h_4 - h_5) + (h_6 - h_7) - (h_1 - h_8)} \text{ kg/kWh} \tag{11.24}$$

The thermal efficiency of a reheat cycle may or may not be higher than that of a cycle without reheat, depending on the pressure at which reheat is carried out. From Fig. 11.6a we observe that the average temperature of heat addition for the reheat process $5 \rightarrow 6$ is not much different than that of the heat-addition process $1 \rightarrow 2 \rightarrow 3 \rightarrow 4$. Consequently, the cycle thermal efficiency will not change very much due to reheat. The steam rate, however, will go down, since we are obtaining more work output from each pound of circulating fluid. The benefits of reheat are gained at the expense of the reheating equipment, a result that is economically justifiable in practice only in large power plants.

Example 11.6

A reheat Rankine cycle using water as the working fluid operates between the pressure limits of 7.5 kPa and 17.0 MPa. Steam is superheated to 550°C before it is expanded to the reheat pressure of 4.0 MPa. Steam is reheated to a final temperature of 550°C. Determine

(a) the cycle thermal efficiency.
(b) the steam rate.

Solution (refer to Fig. 11.6)

For the turbine,

$$w_T = (h_4 - h_5) + (h_6 - h_7).$$

From the steam tables, at 17.0 MPa and 550°C,

$$h_4 = 3428.0 \text{ kJ/kg}$$

$$s_4 = 6.4430 \text{ kJ/kg} \cdot \text{K}.$$

At $p_5 = 4.0$ MPa and $s_5 = s_4 = 6.4430$ kJ/kg · K,

$$h_5 = 3009.1 \text{ kJ/kg}.$$

At 4.0 MPa and 550°C,

$$h_6 = 3558.6 \text{ kJ/kg}$$

$$s_6 = 7.2333 \text{ kJ/kg} \cdot \text{K}$$

$$s_6 = s_7 = [s_g - (1 - x)_7 s_{fg}] \text{ at 7.5 kPa}$$

$$7.2333 = 8.2523 - (1 - x)_7(7.6760),$$

from which we get $(1 - x)_7 = 0.1328$ and

$$h_7 = [h_g - (1 - x)_7 h_{fg}] \text{ at 7.5 kPa}$$

$$= 2574.9 - (0.1328)(2406.2) \text{ kJ/kg} = 2255.4 \text{ kJ/kg}.$$

Then

$$w_T = (3428.0 - 3009.1) + (3558.6 - 2255.4) \text{ kJ/kg}$$

$$= 1722.1 \text{ kJ/kg}.$$

For the condenser,

$$q_{out} = h_8 - h_7$$

$$= 168.77 - 2255.4) = -2086.63 \text{ kJ/kg}.$$

For the pump,

$$w_p = h_8 - h_1 \approx -v_8(p_1 - p_8)$$

$$= -0.0010079(17{,}000 - 7.5) \text{ kJ/kg}$$

$$= -17.13 \text{ kJ/kg}$$

$$h_1 = h_8 - w_p = 168.77 - (-17.13) = 185.9 \text{ kJ/kg}.$$

For the boiler plus superheater plus reheater,

$$q_{in} = (h_4 - h_1) + (h_6 - h_5)$$

$$= (3428.0 - 185.9) + (3558.6 - 3009.1) \text{ kJ/kg}$$

$$= 3791.6 \text{ kJ/kg}.$$

(a) The cycle thermal efficiency is

$$\eta_{th} = \frac{w_{net}}{q_{in}} = \frac{1722.1 - 185.9}{3791.6} = 40.52\%.$$

(b) $\qquad$ Steam rate $= \dfrac{3600}{w_{net}} \text{ kg/kWh}$

$$= \frac{3600}{1722.1 - 185.9} = 2.34 \text{ kg/kWh.} \qquad \blacksquare$$

Comments

Our calculations show that the moisture content is 13.28 percent, which is still greater than 10%. However, an actual turbine with efficiency less than 100% will probably bring the moisture content down to the acceptable level. The reheat pressure of about 4.0 MPa is therefore quite suitable for the actual case.

11.6 Rankine Cycle with Regenerative Feedwater Heating

The use of superheat and the use of reheat are two methods by which we may increase the mean effective temperature of heat addition by increasing the amount of heat supplied at high temperatures.

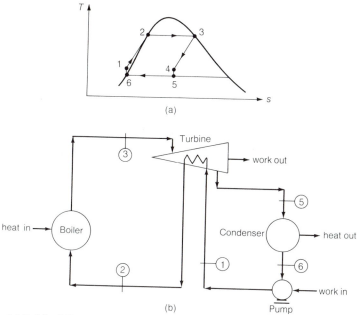

Figure 11.7 Ideal Regenerative Cycle

We may also increase the mean effective temperature of heat addition by decreasing the amount of heat supplied at low temperatures. Referring to Fig. 11.4, we observe that the average temperature of heat addition along path $1 \rightarrow 2$ is quite low. If the amount of heat required for this process is supplied internally instead of externally, the cycle thermal efficiency would approach that of a Carnot cycle. This could be done in a regenerative cycle in which the feedwater is preheated by the expanding steam.

An ideal regenerative cycle is shown in Fig. 11.7. If we could expand the steam in a reversible manner so that the area under path $3 \rightarrow 4$ would be exactly equal to the area under path $1 \rightarrow 2$, we would have all heat supplied externally at T_2 and all heat rejected at T_1. Consequently, the thermal efficiency of an ideal regenerative cycle is the same as that of a Carnot cycle operating between the same temperature limits.

To carry out the ideal regenerative process, we need in essence an infinite number of heat exchangers, known as *feedwater heaters*, to preheat the condensate with steam extracted from the steam turbine. This is physically not practical. In addition, thermodynamic analysis will show that the gain in thermal efficiency resulting from the addition of a heater drops as the number of heaters increases. Conse-

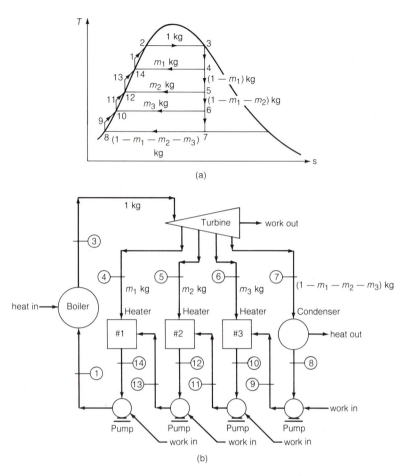

Figure 11.8 Regenerative Cycle with Three Feedwater Heaters

quently, a point is reached where any further increase in the number of heaters is no longer economically justified since the capital cost increases with the number of heaters used. In practice, six or seven heaters is the maximum number employed, and this quantity is used only in very large power plants. We illustrate by studying a regenerative cycle using three heaters, as shown in Fig. 11.8.

Referring to Fig. 11.8, we have, from the first law, on the basis of 1.0 kg of fluid at the turbine inlet,

$$Q_{in} = (1)(h_3 - h_1) \tag{11.25}$$

$$W_T = (1)(h_3 - h_4) + (1 - m_1)(h_4 - h_5)$$
$$+ (1 - m_1 - m_2)(h_5 - h_6)$$
$$+ (1 - m_1 - m_2 - m_3)(h_6 - h_7) \tag{11.26}$$

$$Q_{out} = (1 - m_1 - m_2 - m_3)(h_8 - h_7) \tag{11.27}$$

$$W_p = (1)(h_{14} - h_1) + (1 - m_1)(h_{12} - h_{13})$$
$$+ (1 - m_1 - m_2)(h_{10} - h_{11})$$
$$+ (1 - m_1 - m_2 - m_3)(h_8 - h_9) \tag{11.28}$$
$$\approx -v_{14}(p_1 - p_{14}) - (1 - m_1)v_{12}(p_{13} - p_{12})$$
$$- (1 - m_1 - m_2)v_{10}(p_{11} - p_{10})$$
$$- (1 - m_1 - m_2 - m_3)v_8(p_9 - p_8)$$

$$\eta_{th} = \frac{Q_{net}}{Q_{in}} = \frac{(h_3 - h_1) - (1 - m_1 - m_2 - m_3)(h_7 - h_8)}{h_3 - h_1}. \tag{11.29}$$

In Eq. 11.29, we may approximate h_1 with h_{14} since the temperature rise of liquid due to compression is very small. Thus η_{th} may be given as

$$\eta_{th} = \frac{(h_3 - h_{14}) - (1 - m_1 - m_2 - m_3)(h_7 - h_8)}{h_3 - h_{14}}. \tag{11.29a}$$

The rate of fluid flow at turbine inlet for 1.0 kW of net power output is given by

$$\text{steam rate} = \frac{3600}{W_{net}} \text{ kg/kWh} \tag{11.30}$$

where W_{net} is the algebraic sum of Eqs. 11.26 and 11.28 in kJ/kg.

To determine the various quantities of extraction, we make an energy balance around each feedwater heater. We shall assume that the energy given up by the extraction steam is completely absorbed by the feedwater. Thus, we have for the number 1 heater,

$$m_1 = \frac{h_{14} - h_{13}}{h_4 - h_{13}}. \tag{11.31}$$

Since $h_{13} \approx h_{12}$, we have

$$m_1 \approx \frac{h_{14} - h_{12}}{h_4 - h_{12}}. \tag{11.31a}$$

In a similar manner, we have for the number 2 heater,

$$m_2 = \frac{(1 - m_1)(h_{12} - h_{11})}{h_5 - h_{11}} \tag{11.32}$$

$$\approx \frac{(1 - m_1)(h_{12} - h_{10})}{h_5 - h_{10}} \tag{11.32a}$$

and for the number 3 heater,

$$m_3 = \frac{(1 - m_1 - m_2)(h_{10} - h_9)}{h_6 - h_9} \tag{11.33}$$

$$\approx \frac{(1 - m_1 - m_2)(h_{10} - h_8)}{h_6 - h_8}. \tag{11.33a}$$

In the case of an ideal regenerative cycle, the best result is obtained by heating the feedwater to a temperature equal to the saturation temperature corresponding to the boiler pressure. In the case of a finite number of heaters, the question naturally arises as to what should be the temperature of the feedwater entering the boiler in order to get optimum results. A detailed analysis of this problem is given by Salisbury.[3] We shall present an approximate method that yields results that are quite accurate. Let

T_B = saturation temperature corresponding to boiler pressure

T_C = saturation temperature corresponding to condenser pressure

T_E = temperature of feedwater entering boiler

$(\Delta T)_{\text{possible}} = T_B - T_C$ is the maximum temperature rise for feedwater in an ideal regenerative cycle

$(\Delta T)_{\text{act}} = T_E - T_C$ is the actual temperature rise of feedwater when a finite number of heaters is used.

[3] J. K. Salisbury, *Steam Turbines and Their Cycles*, John Wiley & Sons, Inc., New York, 1950; also J. K. Salisbury, "The Steam Turbine Regenerative Cycle—An Analytical Approach," *Transactions of the ASME*, April 1942.

Table 11.1 Optimum Feedwater Temperature Rise

Number of Feedwater Heaters	1	2	3	4	5	6	8	10	∞
Optimum $\dfrac{(\Delta T)_{act}}{(\Delta T)_{possible}}$	0.47	0.63	0.71	0.77	0.81	0.84	0.88	0.90	1.00

It may be shown that for a given number of feedwater heaters, we have a different optimum value for $(\Delta T)_{act}/(\Delta T)_{possible}$. Some of these optimum values are given in Table 11.1. Once $(\Delta T)_{act}$ is obtained, we may distribute it equally among the heaters. This information will make it possible for us to arrive at the extraction pressure at various points. Let us illustrate with the case of three heaters (Fig. 11.8).

$$(\Delta T)_{possible} = T_2 - T_8$$

$$(\Delta T)_{act} = T_{14} - T_8$$

$$\frac{(\Delta T)_{act}}{(\Delta T)_{possible}} = \frac{T_{14} - T_8}{T_2 - T_8}.$$

For the optimum result, we have, from Table 11.1,

$$\frac{(\Delta T)_{act}}{(\Delta T)_{possible}} = \frac{T_{14} - T_8}{T_2 - T_8} = 0.71.$$

Distributing $(\Delta T)_{act}$ equally, we have

$$T_{10} = T_8 + \tfrac{1}{3}(0.71)(T_2 - T_8)$$

$$T_{12} = T_{10} + \tfrac{1}{3}(0.71)(T_2 - T_8)$$

$$T_{14} = T_{12} + \tfrac{1}{3}(0.71)(T_2 - T_8).$$

Referring to Fig. 11.8 again, we see that the extraction pressure at point 4 is simply the saturation pressure at T_{14}, the extraction pressure at point 5 is the saturation pressure at T_{12}, and the extraction pressure at point 6 is the saturation pressure at T_{10}.

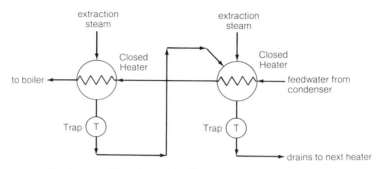

Figure 11.9 Feedwater Heating with Closed Heaters

We wish to conclude our study of the regeneration cycle by noting that feedwater heaters are of two types: open heaters (shown in Fig. 11.8) and closed heaters. Open heaters are also known as *direct-contact heaters*, because the extraction steam and the feedwater mix intimately at the same pressure, so that heat transfer is quite effective. Open heaters are simple to construct and consequently are relatively inexpensive, but the disadvantage is that we need to provide a pump for each open heater.

In closed heaters, the extraction steam is separated by tube walls from the high-pressure feedwater. Heat transfer would not be as effective as in the case of open heaters, since the process must take place through the walls of tubes. A cascaded system of closed heaters is shown in Fig. 11.9. In this system, the condensed extraction steam flows from one heater to another through a trap. In most feedwater heating systems, closed heaters are favored, but at least one open heater is used. The open heater used in the chain of closed heaters is known as a *deaerator*, because one function of it is to vent off dissolved and entrained gases such as oxygen and carbon dioxide which would cause a corrosion problem if allowed to enter the boiler.

Other arrangements of heaters and methods for disposing of the condensed extraction steam (also known as *drains* or *drips*) are used in practice. For details of these arrangements, books on the subject of steam-power-plant design should be consulted.[4]

[4] J. K. Salisbury, *Steam Turbines and Their Cycles*, John Wiley & Sons, Inc., New York, 1950, Chap. 10; B. G. A. Skrotzki and W. A. Vopat, *Power Station Engineering and Economy*, McGraw-Hill Book Company, New York, 1960, Chap. 14.

11.7 Alternative Working Fluids for Vapor Power Cycles

Although we have studied vapor power cycles using water as the working fluid in our examples, we do not wish to leave the impression that we should always use water. Depending on the energy source and other factors and constraints, we may find other working fluids more suitable. Some of the desirable characteristics of a good working fluid are given below:

1. High critical temperature—this would make it possible to vaporize the fluid at the maximum temperature of the cycle.
2. Large enthalpy of vaporization at maximum temperature of the cycle—this would make it possible to obtain a high mean temperature of heat addition and to minimize the mass-flow rate required for a given power output.
3. Positive gauge pressure at the condensing temperature—this would mean that we shall have no need of the auxiliary equipment to maintain a high vacuum in the condenser, such as in the case of steam power plants.
4. Steep slope for the saturated-vapor curve in the T–s diagram—this would make superheat and reheat unnecessary.
5. High density at operating temperature and pressure—this would minimize the size of equipment.
6. Nontoxic and noncorrosive.
7. Chemically stable.
8. Low cost and readily available in large quantities.

Of course, no working fluid could satisfy all the above requirements. The decision to use a particular working fluid is always based on a compromise. For example, water is rather good only in regard to items 2, 6, 7, and 8, but is still the most widely used working fluid because of its superiority overall.

GAS POWER CYCLES

11.8 Air-Standard Cycles

Heat engines that use gases as the working fluid may be classified as internal-combustion engines and external-combustion engines. In an external-combustion engine, heat is supplied from an external heat source and heat is rejected to an external heat sink. In addition, the working fluid is returned periodically to its initial state point. Thus an external combustion engine, such as a closed-cycle gas turbine power plant that utilizes the heat released in a nuclear reaction, operates in a thermodynamic cycle. External-combustion engines are currently receiving considerable attention in an effort to reduce the seriousness of the air pollution problem.

In an internal-combustion engine, as the name implies, the heat addition is accomplished through a chemical reaction between a fuel and air that takes place in the cylinder or combustion chamber itself. In this type of engine, the working fluid suffers a permanent chemical change and cannot be returned to its initial state. Although they operate in a mechanical cycle, internal-combustion engines do not operate in a thermodynamic cycle. In addition to the chemical changes of the working fluid, there are many other factors, such as friction, acceleration, and loss of heat to the cooling medium, that make an accurate analysis of an internal-combustion engine practically impossible. For a detailed treatment of internal-combustion engines, books of this subject should be consulted.[5]

In our study, we shall make use of a thermodynamic cycle to approximate the actual operation of an internal-combustion engine. Results obtained from this simple model are far from correct for most actual combustion engines. However, from the study of gas cycles we may gain considerable insight into the characteristics of internal-combustion engines. Specifically, this method of analysis will indicate the relative effect of the principal variables, such as compression ratio, cycle thermal efficiency, and the relative size of the apparatus.

To simplify our analysis, we shall make the following assumptions in our study of gas cycles:

1. The working fluid is an ideal gas.
2. The specific heats are constant.
3. Heat is added to the cycle from an external source.
4. Heat is rejected by the cycle to an external sink.
5. The amount of working fluid circulating in the cycle is constant.
6. The processes making up the cycle are all internally reversible.

With the above assumptions, the ideal-gas cycles are known as *air-standard cycles* if the working fluid is air.

[5] L. C. Lichty, *Combustion Engine Processes*, McGraw-Hill Book Company, New York, 1967; C. F. Taylor, *The Internal-Combustion Engine in Theory and Practice*, Vol. I, The MIT Press and John Wiley & Sons, Inc., New York, 1960.

11.9 Carnot Gas Cycle

A *Carnot gas cycle* operating in a given temperature range is shown in the T–s diagram in Fig. 11.10a. One way to carry out the processes of this cycle is through the use of steady-state steady-flow devices as shown in Fig. 11.10b. The isentropic expansion process $2 \rightarrow 3$ and the isentropic compression process $4 \rightarrow 1$ can be simulated quite well by a well-designed turbine and compressor, respectively, but the isothermal expansion process $1 \rightarrow 2$ and the isothermal compression process $3 \rightarrow 4$ are most difficult to approach. Because of these difficulties, a steady-flow Carnot gas cycle is not practical.

The Carnot gas cycle could also be achieved in a cylinder-piston apparatus (a reciprocating engine) as shown in Fig. 11.11b. The Carnot cycle in the p–v diagram is shown in Fig. 11.11a, in which processes $1 \rightarrow 2$ and $3 \rightarrow 4$ are isothermal while processes $2 \rightarrow 3$ and $4 \rightarrow 1$ are isentropic.

From our previous study, we know that the Carnot cycle efficiency is given by the expression

$$\eta_{th} = 1 - \frac{T_L}{T_H} = 1 - \frac{T_4}{T_1} = 1 - \frac{T_3}{T_2}.$$

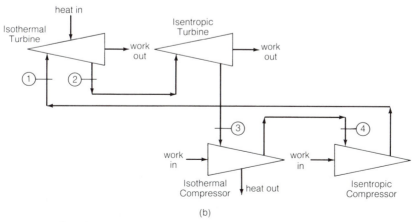

(a)

(b)

Figure 11.10 Steady-Flow Carnot Gas Cycle

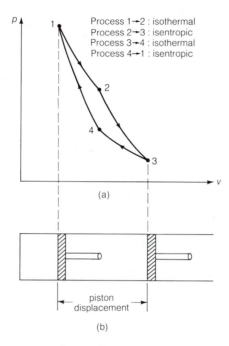

Figure 11.11 Reciprocating Carnot Engine

Since the working fluid is an ideal gas with constant specific heats, we have, for the isentropic processes,

$$\frac{T_1}{T_4} = \left(\frac{v_4}{v_1}\right)^{k-1}$$

$$\frac{T_2}{T_3} = \left(\frac{v_3}{v_2}\right)^{k-1}.$$

Now $T_1 = T_2$ and $T_4 = T_3$. Therefore,

$$\frac{v_4}{v_1} = \frac{v_3}{v_2}.$$

Denoting r_v, the isentropic compression or expansion ratio, as

$$r_v = \frac{v_4}{v_1} = \frac{v_3}{v_2},$$

the Carnot cycle efficiency may now be given as

$$\eta_{th} = 1 - \frac{1}{r_v^{k-1}}. \tag{11.34}$$

Equation 11.34 shows that the Carnot cycle thermal efficiency increases as r_v increases. This implies that the high thermal efficiency of a Carnot cycle is obtained at the expense of large piston displacement, that is, large cylinder size. For the isentropic processes, we also have

$$\frac{T_1}{T_4} = \left(\frac{p_1}{p_4}\right)^{(k-1)/k}$$

$$\frac{T_2}{T_3} = \left(\frac{p_2}{p_3}\right)^{(k-1)/k}.$$

Since $T_1 = T_2$ and $T_4 = T_3$, we have

$$\frac{p_1}{p_4} = \frac{p_2}{p_3}.$$

Denoting r_p, the pressure ratio, as

$$r_p = \frac{p_1}{p_4} = \frac{p_2}{p_3},$$

the Carnot cycle efficiency may also be given as

$$\eta_{th} = 1 - \frac{1}{r_p^{(k-1)/k}}. \tag{11.35}$$

Equation 11.35 indicates that the high Carnot cycle efficiency is obtained at the expense of a large r_p. This would mean that a Carnot cycle must operate with high peak cycle pressure to obtain high cycle efficiency.

A quantity of special interest in connection with reciprocating engines is the *mean effective pressure* (mep), which is defined as the constant pressure acting on the engine piston for its work stroke that would result in the net work of the cycle. It is a measure of the effectiveness of utilizing the cylinder volume. Thus we have

$$\text{mep} = \frac{\text{cycle net work}}{\text{piston displacement}}. \tag{11.36}$$

Referring to Fig. 11.11, we have for a Carnot cycle,

$$\text{mep} = \frac{W_{net}}{V_3 - V_1}. \tag{11.37}$$

Example 11.7

Assume that we are to design an air-standard Carnot cycle according to the following specifications:

$$\text{Maximum cycle temperature} = 1700 \text{ K}$$

$$\text{Minimum cycle temperature} = 340 \text{ K}$$

$$\text{Minimum cycle pressure} = 101 \text{ kPa}$$

$$\text{Heat added to cycle} = 250 \text{ kJ/kg}$$

Determine

(a) the cycle thermal efficiency.
(b) the isentropic compression ratio.
(c) the pressure ratio.
(d) the maximum cycle pressure.
(e) the mep.

Solution

(a)
$$\eta_{th} = 1 - \frac{T_L}{T_H} = 1 - \frac{340}{1700} = 0.80.$$

(b)
$$\eta_{th} = 1 - \frac{1}{r_v^{k-1}}.$$

Therefore,

$$0.80 = 1 - \frac{1}{r_v^{k-1}},$$

from which we get

$$r_v = (5)^{2.5} \qquad \text{with } k = 1.4$$

$$= 55.9.$$

(c)
$$\eta_{th} = \frac{1}{r_p^{(k-1)/k}}.$$

Therefore

$$0.80 = 1 - \frac{1}{r_p^{(k-1)/k}},$$

from which we get

$$r_p = 279.5 \qquad \text{with } k = 1.4.$$

(d) $\qquad r_p = \dfrac{p_2}{p_3} = 279.5$

$$p_2 = 279.5 p_3 = 279.5 \times 101 \text{ kPa} = 28.23 \text{ MPa}.$$

Now

$$q_{12} = T_H(s_2 - s_1).$$

From property relations for ideal gas with constant specific heats, we have

$$s_2 - s_1 = c_p \ln \frac{T_2}{T_1} - R \ln \frac{p_2}{p_1}$$

But $T_2 = T_1$. Thus

$$q_{12} = -T_H R \ln \frac{p_2}{p_1}.$$

For $q_{12} = 250$ kJ/kg, we have

$$250 = \frac{1700 \times 8.3143}{28.97} \ln \frac{p_1}{p_2},$$

from which we get $p_1/p_2 = 1.669$, and

$$p_1 = 1.669 \times 28.23 \text{ MPa} = 47.12 \text{ MPa}.$$

(e) On the basis of 1.0 kg of air,

$$w_{net} = \eta_{th} q_{12} = 0.80 \times 250 \text{ kJ/kg} = 200 \text{ kJ/kg}$$

$$v_3 = \frac{RT_3}{p_3} = \frac{8314.3 \times 340}{28.97 \times 101 \times 1000} \text{ m}^3/\text{kg} = 0.96613 \text{ m}^3/\text{kg}$$

$$v_1 = \frac{RT_1}{p_1} = \frac{8314.3 \times 1700}{28.97 \times 47120 \times 1000} \text{ m}^3/\text{kg} = 0.01035 \text{ m}^3/\text{kg}.$$

Therefore,

$$\text{mep} = \frac{w_{net}}{v_3 - v_1} = \frac{200}{0.96613 - 0.01035} \text{ kPa} = 209.3 \text{ kPa.} \qquad \blacksquare$$

Comments

This example shows that the Carnot cycle must operate with very high peak pressure as well as very large piston displacement, even with a heat addition of only 250 kJ/kg. The high Carnot cycle efficiency is indeed obtained at the expense of heavy and bulky machinery. In addition, the mep is much too small to be practical in reciprocating engines.

11.10 Otto Cycle

The cycle that may be used to approximate the operation of a spark-ignition internal-combustion engine is the *Otto cycle*. Such a cycle, shown in the T–s and p–v diagrams of Fig. 12.12, is made up of the following reversible processes:

1. Constant-volume heat addition (process $1 \to 2$).
2. Isentropic expansion (process $2 \to 3$).
3. Constant-volume heat rejection (process $3 \to 4$).
4. Isentropic compression (process $4 \to 1$).

On the basis of unit mass of gas, we have for the Otto cycle,

$$Q_{in} = Q_{12} = c_v(T_2 - T_1) \qquad (11.38)$$

$$Q_{out} = Q_{34} = c_v(T_4 - T_3) \qquad (11.39)$$

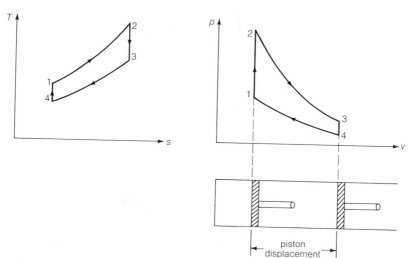

Figure 11.12 Otto Cycle

$$\eta_{\text{th}} = \frac{W_{\text{net}}}{Q_{\text{in}}}$$

$$= \frac{c_v(T_2 - T_1) - c_v(T_3 - T_4)}{c_v(T_2 - T_1)}$$

$$= 1 - \frac{T_3 - T_4}{T_2 - T_1} \qquad (11.40)$$

$$\text{mep} = \frac{c_v(T_2 - T_1) - c_v(T_3 - T_4)}{v_4 - v_1}. \qquad (11.41)$$

Since processes $2 \to 3$ and $4 \to 1$ are isentropic, we have

$$\frac{T_3}{T_2} = \left(\frac{v_2}{v_3}\right)^{k-1}$$

$$\frac{T_4}{T_1} = \left(\frac{v_1}{v_4}\right)^{k-1}.$$

But $v_3 = v_4$ and $v_2 = v_1$. Thus $T_3/T_2 = T_4/T_1$. Making use of this result, we may reduce the Otto cycle efficiency given in Eq. 11.40 to

$$\eta_{th} = 1 - \frac{T_3}{T_2} = 1 - \frac{T_4}{T_1}$$

$$= 1 - \frac{1}{r_v^{k-1}}. \tag{11.42}$$

where $r_v = v_3/v_2 = v_4/v_1$ is known as the *compression ratio* of the cycle.

This equation indicates that the Otto cycle efficiency increases with increased compression ratio. In actual internal-combustion engines, the compression ratio may not be increased, however, beyond a certain limit without encountering undesirable combustion phenomenon known as "knocking." Furthermore, high compression ratio has also been responsible for some air pollution problems.

Example 11.8

The pressure and temperature at the start of compression in an air-standard Otto cycle are 101 kPa and 300 K. The compression ratio is 8. The amount of heat addition is 2000 kJ per kilogram of air. Determine

(a) the cycle thermal efficiency.
(b) the maximum cycle temperature.
(c) the maximum cycle pressure.
(d) the mep.

Solution (refer to Fig. 11.12)

(a)
$$\eta_{th} = 1 - \frac{1}{r_v^{k-1}}.$$

Using $k = 1.4$, we have

$$\eta_{th} = 1 - \frac{1}{8^{1.4-1}} = 0.565.$$

(b)

$$\eta_{th} = 1 - \frac{T_4}{T_1}$$

$$0.565 = 1 - \frac{T_4}{T_1}$$

$$T_1 = \frac{300}{0.435} \text{ K} = 690 \text{ K}$$

$$q_{12} = c_v(T_2 - T_1)$$

$$2000 = 0.7168(T_2 - T_1)$$

and

$$T_2 - T_1 = \frac{2000}{0.7168} \text{ K} = 2790 \text{ K}$$

Thus

$$T_2 = 2790 + 690 \text{ K} = 3480 \text{ K} = T_{max}.$$

(c) Since process $4 \rightarrow 1$ is isentropic, we have

$$\frac{p_1}{p_4} = \left(\frac{v_4}{v_1}\right)^k$$

$$= 8^{1.4} = 18.4$$

and

$$p_1 = 18.4p_4 = 18.4 \times 101 \text{ kPa} = 1858.4 \text{ kPa}.$$

Since process $1 \rightarrow 2$ is constant volume, we have

$$\frac{p_2}{p_1} = \frac{T_2}{T_1} = \frac{3480}{690} = 5.043.$$

Thus

$$p_2 = 5.043p_1 = 5.043 \times 1858.4 \text{ kPa} = 9371.9 \text{ kPa} = p_{max}.$$

(d) $w_{net} = \eta_{th} q_{in} = 0.565 \times 2000 \text{ kJ/kg} = 1130 \text{ kJ/kg}$

$$v_4 = \frac{RT_4}{p_4} = \frac{8314.3 \times 300}{28.97 \times 101 \times 1000} \text{ m}^3/\text{kg} = 0.85247 \text{ m}^3/\text{kg}$$

$$v_1 = \frac{RT_1}{p_1} = \frac{8314.3 \times 690}{28.97 \times 1858.4 \times 1000} \text{ m}^3/\text{kg} = 0.10656 \text{ m}^3/\text{kg}.$$

Thus

$$\text{mep} = \frac{w_{net}}{v_4 - v_1}$$

$$= \frac{1130}{0.85247 - 0.10656} \text{ kPa} = 1515 \text{ kPa}. \qquad \blacksquare$$

Comments

The results of this example show that the Otto cycle has fairly high thermal efficiency and mean effective pressure. But these desirable features of the Otto cycle are accompanied by high maximum cycle temperature and pressure. In practical spark-ignition engines, the maximum temperature and pressure encountered, however, are much lower than those obtained in this example because the combustion process differs considerably from the ideal constant-volume process of heat addition.

11.11 Diesel Cycle

The cycle that may be used to approximate the operation of a compression-ignition internal-combustion engine is the *Diesel cycle*. Such a cycle, shown in the *T–s* and *p–v* diagrams of Fig. 11.13, is made up of the following reversible processes:

1. Constant-pressure heat addition (process $1 \rightarrow 2$).
2. Isentropic expansion (process $2 \rightarrow 3$).
3. Constant-volume heat rejection (process $3 \rightarrow 4$).
4. Isentropic compression (process $4 \rightarrow 1$).

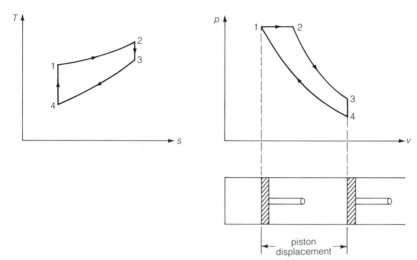

Figure 11.13 Diesel Cycle

On the basis of unit mass of gas, we have, for the Diesel cycle,

$$Q_{in} = Q_{12} = c_p(T_2 - T_1) \tag{11.43}$$

$$Q_{out} = Q_{34} = c_v(T_4 - T_3) \tag{11.44}$$

$$\eta_{th} = \frac{c_p(T_2 - T_1) - c_v(T_3 - T_4)}{c_p(T_2 - T_1)}$$

$$= 1 - \frac{T_3 - T_4}{k(T_2 - T_1)} \tag{11.45}$$

$$\text{mep} = \frac{c_p(T_2 - T_1) - c_v(T_3 - T_4)}{v_4 - v_1}. \tag{11.46}$$

Since the compression and expansion processes are isentropic, we have

$$\frac{T_4}{T_1} = \left(\frac{v_1}{v_4}\right)^{k-1}$$

and

$$\frac{T_3}{T_2} = \left(\frac{v_2}{v_3}\right)^{k-1}.$$

from which we get

$$\frac{T_3}{T_4} = \frac{T_2}{T_1}\left(\frac{v_2/v_3}{v_1/v_4}\right)^{k-1}.$$

Since the heat-addition process is constant pressure, we have

$$\frac{T_2}{T_1} = \frac{v_2}{v_1}.$$

We also have $v_3 = v_4$, since process $3 \rightarrow 4$ is a constant-volume one. Making use of these results, we may transform Eq. 11.45 into

$$\eta_{th} = 1 - \frac{1}{r_v^{k-1}} \frac{r_c^k - 1}{k(r_c - 1)}, \qquad (11.47)$$

where

$$r_v = \frac{v_4}{v_1} \text{ is known as the compression ratio}$$

$$r_c = \frac{v_2}{v_1} \text{ is known as the cutoff ratio.}$$

Equation 11.47 shows that the thermal efficiency of a Diesel cycle is increased by increasing the compression ratio r_v, by decreasing the cutoff ratio r_c, or by using a gas with a large value of k. Since the quantity $(r_c^k - 1)/k(r_c - 1)$ in Eq. 11.47 is always greater than unity, the efficiency of a Diesel cycle is always lower than that of an Otto cycle having the same compression ratio. However, engines using the Diesel cycle usually operate at higher compression ratios than engines using the Otto cycle. This is why the efficiencies of actual compression-ignition engines are not too different from those of actual spark-ignition engines.

Example 11.9

The pressure and temperature at the start of compression in an air-standard Diesel cycle are 101 kPa and 300 K. The compression ratio is 15. The amount of heat addition is 2000 kJ per kilogram of air. Determine

(a) the maximum cycle pressure.
(b) the maximum cycle temperature.

(c) the cycle thermal efficiency.
(d) the mep.

Solution (refer to Fig. 11.13)

(a) $\qquad v_4 = \dfrac{RT_4}{p_4} = \dfrac{8314.3 \times 300}{28.97 \times 101 \times 1000} \text{ m}^3/\text{kg} = 0.85247 \text{ m}^3/\text{kg} = v_3$

$\qquad v_1 = \dfrac{v_4}{r_v} = \dfrac{0.85247}{15} \text{ m}^3/\text{kg} = 0.05683 \text{ m}^3/\text{kg}$

$\qquad \dfrac{T_1}{T_4} = \left(\dfrac{v_4}{v_1}\right)^{k-1} = (15)^{1.4-1} = 2.954$

$\qquad T_1 = 2.954 \times 300 \text{ K} = 886 \text{ K}$

$\qquad \dfrac{p_1}{p_4} = \left(\dfrac{v_4}{v_1}\right)^{k} = (15)^{1.4} = 44.3$

$\qquad p_1 = 44.3 \times p_4 = 44.3 \times 101 \text{ kPa} = 4474.3 \text{ kPa} = p_{max}.$

(b) $\qquad q_{in} = q_{12} = c_p(T_2 - T_1)$

$\qquad T_2 - T_1 = \dfrac{2000}{1.0038} \text{ K} = 1992 \text{ K}$

$\qquad T_2 = 886 + 1992 \text{ K} = 2878 \text{ K} = T_{max}.$

$\qquad v_2 = \dfrac{RT_2}{p_2} = \dfrac{8314.3 \times 2878}{28.97 \times 4474.3 \times 1000} \text{ m}^3/\text{kg} = 0.18460 \text{ m}^3/\text{kg}$

$\qquad \dfrac{T_2}{T_3} = \left(\dfrac{v_3}{v_2}\right)^{k-1} = \left(\dfrac{0.85247}{0.18460}\right)^{1.4-1} = 1.844$

$\qquad T_3 = \dfrac{2878}{1.844} \text{ K} = 1561 \text{ K}.$

(c) $\qquad \eta_{th} = 1 - \dfrac{T_3 - T_4}{k(T_2 - T_1)}$

$\qquad\qquad = 1 - \dfrac{1561 - 300}{1.4(2878 - 886)} = 0.548.$

(d) $$w_{net} = \eta_{th}\, q_{in} = 0.548 \times 2000 \text{ kJ/kg} = 1096 \text{ kJ/kg}$$

$$\text{mep} = \frac{w_{net}}{v_4 - v_1} = \frac{1096}{0.85247 - 0.05683} \text{ kPa} = 1378 \text{ kPa.} \qquad \blacksquare$$

Comments

This example shows that the cycle efficiency and mep of the Diesel cycle are comparable to those of the Otto cycle if the Diesel cycle operates at a higher compression ratio. This is why actual Diesel engines having efficiencies comparable to those of the spark-ignition engines are bulkier.

For this example, the cutoff ratio r_c is 0.18460/0.05683, or 3.248. The reader should check that with $r_c = 3.248$, $r_v = 15$, and $k = 1.4$, the cycle thermal efficiency of 0.548 may also be obtained by using Eq. 11.47.

We wish to point out that combustion in an actual engine is never wholly at constant volume nor wholly at constant pressure. For a detailed study of internal-combustion engines, works on this subject should be consulted.

11.12 Wankel Rotary Combustion Engine

The spark-ignition internal-combustion engine operates according to the Otto cycle, which has four main events or phases: induction phase, compression phase, expansion or working phase, and exhaust phase. The engine can be designed so that different combinations of components may be used. In the familiar conventional four-stroke reciprocating piston engine, the four main events are carried out in a cylinder (see Fig. 11.14). During the intake stroke of such an engine, a fuel–air mixture is drawn into the cylinder. This is followed by the compression stroke, with the volume of the mixture reduced to about 10 percent of its original volume. The mixture is then ignited by the firing of a spark plug. The expanding gas does work on the piston, resulting in the power stroke. The cycle is then completed with the exhaust stroke, during which the products of combustion are pushed out from the cylinder.

The four-stroke reciprocating piston engine has enjoyed much commercial success for a long time, particularly as the power plant for our automobiles. However, it is now facing a big challenge from the *Wankel rotary combustion engine*, which is so designed that the four main events of the Otto cycle are carried out in a radically different fashion. First, the Wankel engine does not have the piston-connecting rod assemblies. Instead, it has an equilateral triangular

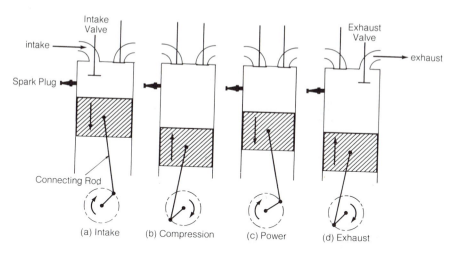

Figure 11.14 Sequence of Events in a Four-Stroke Reciprocating Piston Engine

rotor or rotary piston (see Fig. 11.15). The locus of every apex of the rotating triangular rotor is oval, with a slight necking in on the minor axis and is known as an *epitrochoid*.[6] Felix Wankel found in 1954 that three variable-volume chambers could be formed between the stationary epitrochoid-shaped housing and the rotating equilateral triangular rotor. Together with proper arrangement of intake, exhaust, and ignition mechanisms, this variation of volume in the three chambers makes it possible to carry out the four main events of the Otto cycle in each of the three chambers. A diagrammatic sketch of the Wankel engine is shown in Fig. 11.15. The three chambers are labeled *A, B,* and *C.* In the intake stroke for each chamber, a fresh charge of fuel–air mixture enters the chamber through the intake port, which is always open. The charge is then isolated and compressed as the volume in the chamber steadily decreases with rotation of the rotor. The mixture is then ignited by the firing of a spark plug when the volume in the chamber reaches a minimum. The expanding gases drive against the rotor, resulting in the power stroke. The products of combustion are then pushed out from the chamber when the exhaust port is uncovered.

---

[6] The term "epitrochoid" is given to the path described by a point within a circle rolling without slipping around another circle.

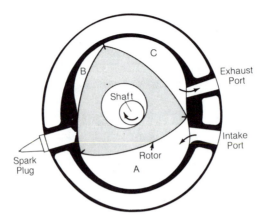

Figure 11.15 Wankel Engine

The rotor of the Wankel engine has an internal gear that is meshed with a stationary gear. The stationary gear, mounted concentrically with the main shaft, is rigidly attached to the housing and is used strictly to maintain the rotor in proper orientation with the housing. The crankshaft, from which all the power output is derived, carries the rotor on an eccentric, which is an integral part of the main shaft. It is the eccentric that turns the main shaft. When the rotor advances 30 degrees, the eccentric advances 90 degrees. Thus the crankshaft makes three revolutions for each revolution of the rotor. Since we have three variable-volume chambers to work with, the Wankel engine delivers one power stroke for each full rotation of the main shaft. This means that the power-stroke frequency for the Wankel engine is exactly twice that of a conventional four-stroke recipro-cating engine. In other words, the Wankel engine uses its piston-displacement volume twice as often as the four-stroke reciprocating engine does.

As we mentioned before, the Wankel engine does not have the piston-connecting rod assemblies. In addition, the Wankel does not need a camshaft, valve lifters, and associated mechanisms, because it has no valves. Consequently, the Wankel engine is a much simpler engine than the conventional four-stroke engine. According to Cole,[7] a typical V-8 engine of 195 hp has 388 moving parts, but one American-built two-rotor Wankel of 185 hp has only 154 moving parts. It is reasonable to expect that when Wankel engines are put into mass production, they should cost considerably less than the conventional engines. To what extent the Wankel engine will dis-place the reciprocating engine from the numerous areas of applica-

[7] D. E. Cole, "The Wankel Engine," *Scientific American*, Vol. 227, No. 2 (1972).

tions in which the latter is well established will depend upon how well it can measure up to such other factors as fuel economy, exhaust emissions, reliability, and serviceability. For a detailed treatment of the Wankel engine, books on this subject should be consulted.[8]

11.13 Stirling Cycle

A possible alternative to the internal-combustion engine is an external-combustion engine known as the *Stirling engine*. The basic cycle of the Stirling engine is the *Stirling cycle*, shown in the *T–s* and *p–v* diagrams of Fig. 11.16. From these diagrams, we observe that the Stirling cycle is made up of two isothermal processes and two constant-volume processes. Through the use of a reversible regenerator (an energy-storage device), it is theoretically possible to recover all the heat given up by the working fluid in the constant-volume cooling process (process $2 \rightarrow 3$) for the constant-volume heating process (process $4 \rightarrow 1$). Then all the heat received by the cycle from the external heat source is at T_H and all the heat rejected by the cycle to the external sink is at T_L. This means that the thermal efficiency of the Stirling cycle with perfect regeneration would be equal to that of the Carnot cycle for the same temperature range.

For a Stirling cycle with perfect regeneration, we have, on the basis of unit mass,

$$Q_{in} = Q_{12} = T_H(s_2 - s_1).$$

[8] R. F. Ansdale, *The Wankel RC Engine*, The Butterworth Group, London, 1968; see also J. P. Norbye, *The Wankel Engine*, Chilton Book Company, Philadelphia, 1971.

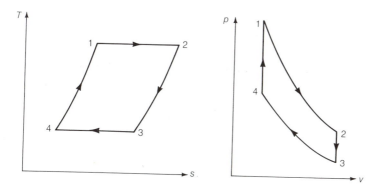

Figure 11.16 Stirling Cycle

If the working fluid is an ideal gas,

$$s_2 - s_1 = \int_1^2 c_p \frac{dT}{T} - R \ln \frac{p_2}{p_1}$$

$$= \int_1^2 c_v \frac{dT}{T} + R \ln \frac{v_2}{v_1}.$$

But $T_2 = T_1$. Thus

$$Q_{12} = RT_H \ln \frac{v_2}{v_1} = -RT_H \ln \frac{p_2}{p_1}. \tag{11.48}$$

Note that we have no need to assume constant specific heat here. In a similar manner, we have

$$Q_{out} = Q_{34} = RT_L \ln \frac{v_4}{v_3} = -RT_L \ln \frac{p_4}{p_3}. \tag{11.49}$$

The cycle thermal efficiency is given by

$$\eta_{th} = \frac{W_{net}}{Q_{in}}$$

$$= 1 - \frac{T_L \ln (v_3/v_4)}{T_H \ln (v_2/v_1)}.$$

But $v_2 = v_3$ and $v_4 = v_1$. Consequently, we have

$$\eta_{th} = 1 - \frac{T_L}{T_H}. \tag{11.50}$$

which is the same as that of the Carnot cycle operating between T_L and T_H.

While no practical machine for the Carnot cycle has ever been devised, the Stirling cycle has been approximated in a compact mechanical system operating over a reasonable pressure ratio. One promising practical version of the Stirling engine has been developed in recent years by the Philips Research Laboratories, The Netherlands.[9] A survey of current developments on the Stirling engine is

[9] B. D. Wood, *Applications of Thermodynamics*, Addison-Wesley Publishing Co., Inc., Reading, Mass., 1969, pp. 151–155.

given by Walker.[10] Anyone who has interest in this topic should also consult the *Proceedings* of the annual Intersociety Energy Conversion and Engineering Conference.

11.14 Brayton Cycle

The *Brayton cycle* is the basic cycle for the simple gas turbine power plant. It is shown in the p–v and T–s diagrams of Fig. 11.17. From these diagrams, we observe that the Brayton cycle consists of two reversible constant-pressure processes and two isentropic processes. For an air-standard Brayton cycle, we have, on the basis of unit mass,

$$Q_{in} = Q_{12} = c_p(T_2 - T_1) \tag{11.51}$$

$$Q_{out} = Q_{34} = c_p(T_4 - T_3) \tag{11.52}$$

$$\eta_{th} = 1 - \frac{T_3 - T_4}{T_2 - T_1}, \tag{11.53}$$

For the isentropic processes, we have

$$\frac{T_2}{T_3} = \left(\frac{p_2}{p_3}\right)^{(k-1)/k}$$

---

[10] Graham Walker, "The Stirling Engine," *Scientific American*, August 1973, and Graham Walker, *Stirling Engine*, Oxford University Press, Oxford, 1980.

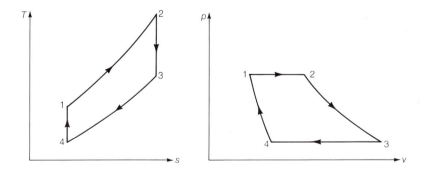

Figure 11.17 Brayton Cycle

and

$$\frac{T_1}{T_4} = \left(\frac{p_1}{p_4}\right)^{(k-1)/k}.$$

But $p_2 = p_1$ and $p_3 = p_4$. Thus $T_2/T_3 = T_1/T_4$. Making use of these results, we may transform Eq. 11.53 into

$$\eta_{\text{th}} = 1 - \frac{T_4}{T_1} = 1 - \frac{T_3}{T_2}$$

$$= 1 - \frac{1}{r_p^{(k-1)/k}}, \tag{11.54}$$

where $r_p = p_1/p_4 = p_2/p_3$ is known as the *pressure ratio*. It is seen that the efficiency of the air-standard Brayton cycle is only a function of the pressure ratio.

The net work of the cycle is given by

$$W_{\text{net}} = c_p(T_2 - T_1) - c_p(T_3 - T_4)$$

$$= c_p(T_2 - T_3) - c_p(T_1 - T_4). \tag{11.55}$$

In terms of the pressure ratio, Eq. 11.55 may be transformed into

$$W_{\text{net}} = c_p T_2 \left[1 - \frac{1}{r_p^{(k-1)/k}}\right] - c_p T_4 [r_p^{(k-1)/k} - 1]. \tag{11.56}$$

Equation 11.56 indicates that for an ideal Brayton cycle operating between two given temperature limits, the cycle net work is only a function of the pressure ratio. To find the maximum work per unit mass of working fluid, we must operate at the optimum pressure ratio, which is obtained by differentiating Eq. 11.56 with respect to r_p, equating to zero, and solving for r_p. The result is

$$(r_p)_{\text{opt}} = \left(\frac{T_2}{T_4}\right)^{k/2(k-1)}, \tag{11.57}$$

where T_2 is the maximum temperature of the cycle and T_4 is the minimum temperature of the cycle. In practice, T_2 is limited to about 1000°C at the present time, owing to metallurgical considerations, while T_4 is limited to the atmospheric temperature.

Two general types of gas-turbine power plants have been developed based on the Brayton cycle. They are the closed-cycle (Fig. 11.18) and

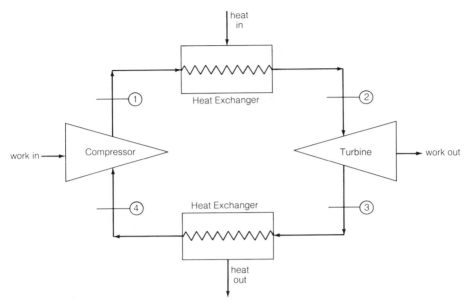

Figure 11.18 Closed Gas-Turbine Power Plant

open-cycle (Fig. 11.19) plants. In the closed-cycle plant, heat is supplied from an external heat source and is rejected to an external sink. This type is currently being developed to operate with energy input coming from a nuclear reactor. With increased interest and advancement in fluidized-bed combustion technology which would make it possible to burn coal and other low-cost fuels externally, the closed-cycle system has received a great deal of attention in recent years. In the open-cycle plant, energy input comes from the fuel that is injected into the combustion chamber.

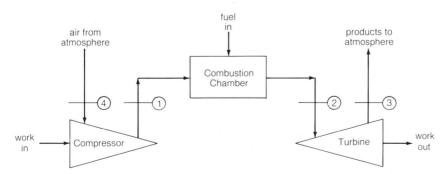

Figure 11.19 Open Gas-Turbine Power Plant

Example 11.10

A simple gas-turbine power plant operating on an air-standard Brayton cycle is to be designed according to the following specifications:

$$\text{Maximum cycle temperature} = 1150 \text{ K}$$

$$\text{Minimum cycle temperature} = 300 \text{ K}$$

$$\text{Maximum cycle pressure} = 505 \text{ kPa}$$

$$\text{Minimum cycle pressure} = 101 \text{ kPa}$$

Determine

(a) the cycle thermal efficiency.
(b) the compressor work.
(c) the turbine work.
(d) the air-flow rate for 1.0 kW of net power output.

Solution (refer to Fig. 11.17)

(a)
$$\eta_{th} = 1 - \frac{1}{r_p^{(k-1)/k}}$$

$$= 1 - \frac{1}{5^{(1.4-1)/1.4}} = 1 - \frac{1}{1.584} = 0.369.$$

(b) The compressor work is given by

$$W_c = c_p(T_4 - T_1).$$

With $r_p = 5$,

$$T_1 = T_4 \left(\frac{p_1}{p_4}\right)^{(k-1)/k}$$

$$= 300 \times 1.584 = 475 \text{ K}.$$

Thus

$$w_c = c_p(T_4 - T_1)$$

$$= 1.0038(300 - 475) \text{ kJ/kg} = -175.7 \text{ kJ/kg}.$$

(c) The turbine work is given by

$$w_T = c_p(T_2 - T_3)$$

where

$$T_3 = \frac{T_2}{(r_p)^{(k-1)/k}} = \frac{1150}{1.584} \text{ K} = 726 \text{ K}.$$

Thus

$$w_T = 1.0038(1150 - 726) \text{ kJ/kg} = 425.6 \text{ kJ/kg}.$$

(d) $$w_{net} = w_T + w_c = 425.6 - 175.7 = 249.9 \text{ kJ/kg}.$$

For 1.0 kW of net power output, we have

$$\dot{m} = \frac{3600}{w_{net}} \text{ kg/h with } w_{net} \text{ in kJ/kg}$$

$$= \frac{3600}{249.9} \text{ kg/h} = 14.4 \text{ kg/h}. \qquad \blacksquare$$

Comment

This example shows that even with a reversible compressor and turbine, a large portion of the turbine work is consumed by the compressor. This implies that in practice the gas-turbine power plant is very sensitive to turbine and compressor inefficiencies.

11.15 Simple Gas-Turbine Power Plant with Real Compressor and Turbine

Because the gas-turbine power cycle is very sensitive to inefficiencies in the compressor and the turbine, it is of interest to examine the effect on cycle performance of the irreversible compression and expansion experienced in the actual power plant. With reference to

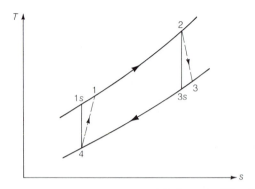

Figure 11.20 Simple Gas-Turbine Cycle with Irreversible Compression and Expansion

Fig. 11.20, we have, assuming ideal gas with constant specific heat,

$$w_T = h_2 - h_3 = c_p(T_2 - T_3) \qquad (11.58)$$

$$w_c = h_4 - h_1 = c_p(T_4 - T_1). \qquad (11.59)$$

From the definition of turbine efficiency, we have

$$\eta_T = \frac{h_2 - h_3}{h_2 - h_{3s}} = \frac{T_2 - T_3}{T_2 - T_{3s}}. \qquad (11.60)$$

From the definition of compressor efficiency, we have

$$\eta_c = \frac{h_{1s} - h_4}{h_1 - h_4} = \frac{T_{1s} - T_4}{T_1 - T_4}. \qquad (11.61)$$

Making use of Eq. 11.60, we may also give the turbine work as

$$w_T = \eta_T c_p T_2 \left[1 - \frac{1}{r_p^{(k-1)/k}} \right], \qquad (11.62)$$

where $r_p = p_1/p_4$ is the pressure ratio. Making use of Eq. 11.61, we may also give the compressor work as

$$w_c = - \frac{c_p T_4}{\eta_c} [r_p^{(k-1)/k} - 1]. \qquad (11.63)$$

The cycle net work is then given by

$$w_{net} = c_p(T_2 - T_3) - c_p(T_1 - T_4)$$

$$= \eta_T c_p T_2 \left[1 - \frac{1}{r_p^{(k-1)/k}} \right] - \frac{c_p T_4}{\eta_c} [r_p^{(k-1)/k} - 1]. \qquad (11.64)$$

For a given set of values for η_T, η_c, T_2, T_4, c_p, and k, there will be an optimum pressure ratio to give the maximum net cycle work. This optimum pressure ratio is obtained by differentiating Eq. 11.64 with respect to r_p, equating to zero, and solving for r_p. The result is

$$(r_p)_{opt} = \left(\frac{T_2}{T_4} \eta_T \eta_c \right)^{k/2(k-1)}, \qquad (11.65)$$

which is reduced to Eq. 11.57 when $\eta_T = 1$ and $\eta_c = 1$.
    Heat addition to the cycle is given by

$$q_{in} = q_{12} = c_p(T_2 - T_1), \qquad (11.66)$$

which may be transformed into

$$q_{in} = c_p \left\{ T_2 - T_4 \left[1 + \frac{r_p^{(k-1)/k} - 1}{\eta_c} \right] \right\}. \qquad (11.67)$$

The cycle thermal efficiency is given by

$$\eta_{th} = \frac{w_{net}}{q_{in}}$$

$$= \frac{(T_2 - T_3) - (T_1 - T_4)}{T_2 - T_1}$$

$$= \frac{\eta_T \dfrac{T_2}{T_4} \left[1 - \dfrac{1}{r_p^{(k-1)/k}} \right] - \dfrac{1}{\eta_c} [r_p^{(k-1)/k} - 1]}{\dfrac{T_2}{T_4} - \left[1 + \dfrac{r_p^{(k-1)/k} - 1}{\eta_c} \right]}. \qquad (11.68)$$

Example 11.11

A simple gas turbine power plant is to be designed according to the following specifications:

Maximum cycle temperature $= 1150$ K

Minimum cycle temperature $= 300$ K

Maximum cycle pressure $= 505$ kPa

Minimum cycle pressure $= 101$ kPa

Adiabatic turbine efficiency $= 85\%$

Adiabatic compressor efficiency $= 80\%$

Assuming an ideal gas with constant specific heat, determine

(a) the cycle thermal efficiency.
(b) the compressor work.
(c) the turbine work.
(d) the air-flow rate for 1.0 kW of net power output.
(e) the temperature of the working fluid at the turbine exhaust.

Solution (refer to Fig. 11.20)

(a) With $r_p = 5$, $\eta_T = 0.85$, $\eta_C = 0.80$, $T_2 = 1150$ K, and $T_4 = 300$ K, we have from Eq. 11.68,

$$\eta_{th} = 0.224.$$

(b)
$$w_C = -\frac{c_p T_4}{\eta_c} [r_p^{(k-1)/k} - 1]$$

$$= -\frac{1.0038 \times 300}{0.80} [5^{(1.4-1)/1.4} - 1]$$

$$= -219.8 \text{ kJ/kg}.$$

(c)
$$w_T = \eta_T c_p T_2 \left[1 - \frac{1}{r_p^{(k-1)/k}} \right]$$

$$= 0.85 \times 1.0038 \times 1150 \left[1 - \frac{1}{5^{(1.4-1)/1.4}} \right]$$

$$= 361.7 \text{ kJ/kg}.$$

(d)
$$w_{net} = w_T + w_C = 361.7 - 219.8 = 141.9 \text{ kJ/kg}.$$

For 1.0 kW of net power output, we have

$$\dot{m} = \frac{3600}{141.9} \text{ kg/h} = 25.37 \text{ kg/h}$$

$$w_T = c_p(T_2 - T_3)$$

$$T_2 - T_3 = \frac{361.7}{1.0038} \text{ K} = 360.3 \text{ K}.$$

Thus

$$T_3 = 1150 - 360.3 = 789.7 \text{ K}. \qquad \blacksquare$$

Comments

Comparing the above with Example 11.10, we note that the cycle efficiency is considerably reduced because of irreversible expansion and compression. The development of highly efficient compressors and turbines is therefore an important aspect of the development of efficient gas-turbine power plants. Since the temperature of turbine exhaust is quite high, we could also improve the cycle efficiency of making use of this hot gas for internal heating, thereby reducing the mean temperature of heat rejection. This method of improving cycle efficiency will be discussed in the section on the Brayton cycle with regenerative heating.

11.16 Closed-Cycle Gas-Turbine Power Plant with Real Compressor and Turbine and Pressure Drops in Heat Exchangers

The gas-turbine power cycle is very sensitive to inefficiencies in the compressor and the turbine. It is also sensitive to pressure drops in heat exchangers. The effect on cycle performance due to irreversibilities in the compressor and turbine was studied in the preceding section. We wish to extend the study to include the effect of pressure drops in heat exchanger.

With reference to Fig. 11.21, and assuming ideal gas with constant specific heats, the compression work is given by Eq. 11.63:

$$w_c = -\frac{c_p T_4}{\eta_C}[r_p^{(k-1)/k} - 1]. \qquad (11.63)$$

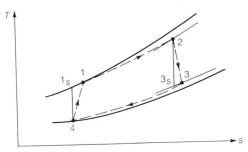

Figure 11.21 Closed-Cycle Gas-Turbine Cycle with Irreversible Compressor and Turbine and Pressure Drops in Heat Exchangers

The turbine expansion ratio, p_2/p_3, may be expressed in terms of the compressor compression ratio r_p and pressure drops to be used in each of the two heat exchangers. If p_{in} and p_{out} are the inlet pressure and outlet pressure of air for each heat exchanger, then

$$p_{out} = \beta p_{in} \tag{11.69}$$

and

$$\beta = 1 - \frac{p_{in} - p_{out}}{p_{in}}$$

$$= 1 - \frac{\Delta p}{p}. \tag{11.70}$$

The quantity $(\Delta p)/p$ is known as the relative pressure drop. β may be called the pressure drop factor.

From Fig. 11.18 and Fig. 11.21, we thus have

$$p_2 = \beta_{12} p_1 \tag{11.71}$$

$$p_4 = \beta_{34} p_3, \tag{11.72}$$

where β_{12} is the pressure drop factor for the heat exchanger with heat addition, and β_{34} is the pressure drop factor for the heat exchanger with heat rejection.

Combining Equations 11.71 and 11.72, we have

$$\frac{p_2}{p_1} \times \frac{p_4}{p_3} = \beta_{12} \beta_{34}$$

and

$$\frac{p_2}{p_3} = \beta_{12} \beta_{34} \left(\frac{p_1}{p_4}\right)$$

$$= \beta_{12} \beta_{34} r_p, \tag{11.73}$$

where p_2/p_3 is the turbine expansion ratio and r_p is the compressor compression ratio.

Making use of Eq.11.62, the turbine work may now be given as

$$w_T = \eta_T c_p T_2 \left[1 - \frac{1}{(\beta_{12} \beta_{34} r_p)^{(k-1)/k}}\right]. \tag{11.74}$$

The cycle net work is then given by

$$w_{net} = \eta_T c_p T_2 \left[1 - \frac{1}{(\beta_{12} \beta_{34} r_p)^{(k-1)/k}}\right] - \frac{c_p T_4}{\eta_C} [r_p^{(k-1)/k} - 1]. \tag{11.75}$$

Dividing Eq. 11.75 across by $c_p T_4$, the specific power output of the cycle is given by

$$\frac{w_{net}}{c_p T_4} = \eta_T \left(\frac{T_2}{T_4}\right) \left[1 - \frac{1}{(\beta_{12} \beta_{34} r_p)^{(k-1)/k}}\right] - \frac{1}{\eta_C} [r_p^{(k-1)/k} - 1]. \tag{11.76}$$

The quantity $w_{net}/(c_p T_4)$ is known as the specific power output of the cycle, which is a dimensionless quantity. It is an indicator of how compact the system is.

For a given set of values of η_T, η_C, T_2/T_4, c_p, and k, there will be an optimum pressure ratio r_p to give the maximum cycle specific power. This optimum pressure ratio is obtained by differentiating Eq. 11.76 with respect to r_p, equating to zero, and solving for r_p. The result is

$$(r_p)_{opt} = \left[\left(\frac{T_2}{T_4}\right) \frac{\eta_T \eta_C}{\phi}\right]^{k/2(k-1)}, \tag{11.77}$$

where

$$\phi = (\beta_{12} \beta_{34})^{(k-1)/k}.$$

Equation 11.77 is reduced to Eq. 11.65 when $\beta_{12} = \beta_{34} = 1$ (that is, when pressure drops are neglected).

Heat addition to the cycle is given by

$$q_{in} = q_{12} = c_p(T_2 - T_1),$$

which may be transformed into

$$q_{in} = c_p \left\{ T_2 - T_4 \left[1 + \frac{r_p^{(k-1)/k} - 1}{\eta_C} \right] \right\}, \tag{11.78}$$

which is also Eq. 11.67.

The cycle thermal efficiency is obtained by making use of Equations 11.75 and 11.78:

$$\eta_{th} = \frac{w_{net}}{q_{in}}$$

$$= \frac{\eta_T \left(\dfrac{T_2}{T_4} \right) \left[1 - \dfrac{1}{(\beta_{12}\beta_{34}r_p)^{(k-1)/k}} \right] - \dfrac{1}{\eta_C} [r_p^{(k-1)/k} - 1]}{\dfrac{T_2}{T_4} - \left[1 + \dfrac{r_p^{(k-1)/k} - 1}{\eta_C} \right]}. \tag{11.79}$$

For a given set of values of η_T, η_C, T_2/T_4, c_p, and k, there will be an optimum pressure ratio r_p to give the maximum cycle thermal efficiency. This optimum pressure ratio is obtained by differentiating Eq. 11.79 with respect to r_p, equating to zero, and solving for r_p. The result is

$$(r_p)_{opt} = \left[\frac{-b - \sqrt{b^2 - 4ac}}{2a} \right]^{k/(k-1)}, \tag{11.80}$$

where

$$a = \left(\frac{T_2}{T_4} \right) \frac{\eta_T}{\eta_C} - \frac{(T_2/T_4) - 1}{\eta_C} \tag{11.81}$$

$$b = - \frac{2\eta_T(T_2/T_4)}{\eta_C \phi} \tag{11.82}$$

$$c = \left(\frac{T_2}{T_4} - 1 + \frac{1}{\eta_C} \right) \frac{\eta_T}{\phi} \left(\frac{T_2}{T_4} \right) \tag{11.83}$$

$$\phi = (\beta_{12}\beta_{34})^{(k-1)/k}. \tag{11.84}$$

Example 11.12

A closed-cycle gas-turbine power plant is to be designed according to the following specifications:

Maximum cycle temperature	= 1150 K
Minimum cycle temperature	= 300 K
Adiabatic turbine efficiency	= 85%
Adiabatic compressor efficiency	= 80%
Relative pressure drop for each heat exchanger =	3%

Select pressure ratio for maximum power output. Assuming ideal gas with constant specific heats, determine

(a) the cycle thermal efficiency.
(b) the cycle net work.
(c) the airflow rate for 1.0 kW of net power output.
(d) the temperature of the working fluid at turbine exhaust.

Solution (refer to Fig. 11.21)

For maximum power output, the optimum pressure ratio is given by Eq. 11.77:

$$(r_p)_{opt} = \left[\left(\frac{T_2}{T_4} \right) \frac{\eta_T \eta_C}{\phi} \right]^{k/2(k-1)} \qquad \text{where } \phi = (\beta_{12}\beta_{34})^{(k-1)/k}.$$

With

$$\beta_{12} = 1 - 0.03 = 0.97$$

$$\beta_{34} = 1 - 0.03 = 0.97$$

$$\phi = (0.97 \times 0.97)^{(1.4-1)/1.4} = 0.983,$$

then

$$(r_p)_{\text{opt}} = \left(\frac{1150}{300} \times \frac{0.85 \times 0.80}{0.983} \right)^{1.4/2(1.4-1)}$$

$$= 5.5.$$

(a) The cycle thermal efficiency is given by Eq. 11.79:

$$\eta_{\text{th}} = \frac{\eta_T \left(\dfrac{T_2}{T_4} \right) \left[1 - \dfrac{1}{(\beta_{12}\, \beta_{34}\, r_p)^{(k-1)/k}} \right] - \dfrac{1}{\eta_C} [r_p^{(k-1)/k} - 1]}{\dfrac{T_2}{T_4} - \left[1 + \dfrac{r_p^{(k-1)/k} - 1}{C} \right]}$$

$$= \frac{0.85 \times \dfrac{1150}{300} \left[1 - \dfrac{1}{(0.97 \times 0.97 \times 5.5)^{(1.4-1)/1.4}} \right] - \dfrac{1}{0.8} [5.5^{(1.4-1)/1.4} - 1]}{\dfrac{1150}{300} - \left[1 + \dfrac{5.5^{(1.4-1)/1.4} - 1}{0.8} \right]}$$

$$= 21.32\%.$$

(b) The cycle net work is given by Eq. 11.75:

$$w_{\text{net}} = \eta_T\, c_p\, T_2 \left[1 - \frac{1}{(\beta_{12}\, \beta_{34}\, r_p)^{(k-1)/k}} \right] - \frac{c_p\, T_4}{\eta_C} [r_p^{(k-1)/k} - 1]$$

$$= 0.85 \times 1.0038 \times 1150 \left[1 - \frac{1}{(0.97 \times 0.97 \times 5.5)^{(1.4-1)/1.4}} \right]$$

$$- \frac{1.0038 \times 300}{0.80} [5.5^{(1.4-1)/1.4} - 1]$$

$$= 131.55 \text{ kJ/kg.}$$

(c) For 1.0 kW of net power output, we have

$$\dot{m} = \frac{3600}{131.55} \text{ kg/h} = 27.37 \text{ kg/h.}$$

(d) The turbine work is given by Eq. 11.74:

$$w_T = \eta_T c_p T_2 \left| 1 - \frac{1}{(\beta_{12} \beta_{34} r_p)^{(k-1)/k}} \right|$$

$$= 0.85 \times 1.0038 \times 1150 \left[1 - \frac{1}{(0.97 \times 0.97 \times 5.5)^{(1.4-1)/1.4}} \right]$$

$$= 367.76 \text{ kJ/kg}$$

$$= c_p(T_2 - T_3).$$

Therefore,

$$T_2 - T_3 = \frac{367.76}{1.0038} \text{ K} = 366.4 \text{ K}$$

$$T_3 = 1150 - 366.4 = 783.6 \text{ K.} \qquad \blacksquare$$

Comments

Because the pressure drop factor is very close to unity, the optimum pressure ratio is not very sensitive to these factors. But pressure drop will reduce the turbine output and consequently the cycle net work output. The high-temperature air leaving the turbine is a good heat source for additional power generation or process heat production. This is why the gas-turbine power plant has been used widely as the topping cycle for combined-cycle power generation or cogeneration applications.

Example 11.13

Repeat Example 12.14 except that the design will use the optimum pressure ratio for maximum thermal efficiency.

Solution (refer to Fig. 11.21)

For maximum cycle thermal efficiency, the optimum pressure ratio is given by Eq. 11.80:

$$(r_p)_{\text{opt}} = \left(\frac{-b - \sqrt{b^2 - 4ac}}{2a} \right)^{k/(k-1)},$$

where

$$a = \left(\frac{T_2}{T_4}\right)\frac{\eta_T}{\eta_C} - \frac{(T_2/T_4) - 1}{\eta_C}$$

$$= \frac{1150}{300} \times \frac{0.85}{0.80} - \frac{1150/300 - 1}{0.80} = 0.5313$$

$$b = -\frac{2\eta_T(T_2/T_4)}{\eta_C \phi}$$

$$= -\frac{2 \times 0.85 \times (1150/300)}{0.80 \times 0.983} \qquad (\phi = 0.983, \text{ from Example 11.12})$$

$$= -8.2866$$

$$c = \left(\frac{T_2}{T_4} - 1 + \frac{1}{\eta_C}\right)\frac{\eta_T}{\phi}\left(\frac{T_2}{T_4}\right)$$

$$= \left(\frac{1150}{300} - 1 + \frac{1}{0.80}\right)\frac{0.85}{0.983} \times \frac{1150}{300} = 13.5348.$$

Thus

$$(r_p)_{\text{opt}} = \left[\frac{-(-8.2866) - \sqrt{(-8.2866)^2 - 4 \times 0.5313 \times 13.5348}}{2 \times 0.5313}\right]^{1.4/(1.4-1)}$$

$$= 8.7.$$

(a) The cycle thermal efficiency is given by Eq. 11.79:

$$\eta_{\text{th}} = \frac{\eta_T\left(\frac{T_2}{T_4}\right)\left[1 - \frac{1}{(\beta_{12}\beta_{34}r_p)^{(k-1)/k}}\right] - \frac{1}{\eta_C}[r_p^{(k-1)/k} - 1]}{\frac{T_2}{T_4} - \left[1 + \frac{r_p^{(k-1)/k} - 1}{\eta_C}\right]}$$

$$= \frac{0.85 \times \frac{1150}{300}\left[1 - \frac{1}{(0.97 \times 0.97 \times 8.7)^{(1.4-1)/1.4}}\right] - \frac{1}{\eta_C}[8.7^{(1.4-1)/1.4} - 1]}{\frac{1150}{300} - \left[1 + \frac{8.7^{(1.4-1)/1.4} - 1}{0.80}\right]}$$

$$= 22.83\%.$$

(b) The cycle net work is given by Eq. 11.75:

$$w_{net} = \eta_T c_p T_2 \left[1 - \frac{1}{(\beta_{12} \beta_{34} r_p)^{(k-1)/k}} \right] - \frac{c_p T_4}{\eta_C} [r_p^{(k-1)/k} - 1]$$

$$= 0.85 \times 1.0038 \times 1150 \left[1 - \frac{1}{(0.97 \times 0.97 \times 8.7)^{(1.4-1)/1.4}} \right]$$

$$- \frac{1.0038 \times 300}{0.80} [8.7^{(1.4-1)/1.4} - 1]$$

$$= 443.02 - 321.96 = 121.06 \text{ kJ/kg}$$

(c) For 1.0 kW of net power output, we have

$$m = \frac{3600}{121.06} \text{ kg/h} = 29.74 \text{ kg/h}.$$

(d)
$$w_T = c_p(T_2 - T_3)$$

$$= 443.02 \text{ kJ/kg} \qquad \text{[from part (b)]}$$

$$T_2 - T_3 = \frac{443.02}{1.0038} = 441.3 \text{ K}$$

$$T_3 = 1150 - 441.3 = 708.7 \text{ K}.$$ ▮

Comments

The results from Examples 11.12 and 11.13 are summarized as follows:

Scheme Used	r_p	η_{th}	w_{net} (kJ/kg)	$\dot{m}$ (kg/kWh)	Temperature of Air Leaving Turbine (K)
Example 11.12 (design for maximum power output)	5.5	21.32%	131.55	27.35	783.6
Example 11.13 (design for maximum cycle thermal efficiency)	8.7	22.83%	121.06	29.74	708.7

It can be seen that the design for maximum power output will be slightly less efficient but more compact. The design for maximum cycle thermal efficiency will be slightly more efficient but would require larger and consequently more expensive hardware. If these units are to be used in combined cycle or cogeneration applications, the design for maximum power output would be better because the air leaves the turbine at a higher temperature. In any case, it is a matter of trade-offs again.

11.17 Brayton Cycle with Regenerative Heating

Improvement in efficiency over the simple gas-turbine cycle can be obtained by heating the air leaving the compressor with the exhaust gas leaving the turbine by means of a heat exchanger known as a *regenerator*. A Brayton cycle with perfect regeneration is shown in Fig. 11.22. In this ideal case, the heat absorbed by the air leaving the compressor is identical to the heat given up by the gas leaving the

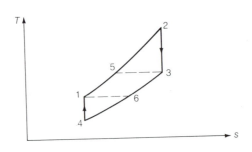

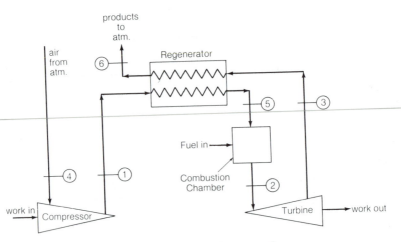

Figure 11.22 Brayton Cycle with Perfect Regeneration

turbine. For this cycle, we have

$$W_{net} = c_p(T_2 - T_3) - c_p(T_1 - T_4)$$

and

$$Q_{in} = Q_{52} = c_p(T_2 - T_5).$$

But $T_5 = T_3$ with perfect regeneration. Thus

$$Q_{in} = c_p(T_2 - T_3).$$

The cycle efficiency is then given by

$$\eta_{th} = \frac{(T_2 - T_3) - (T_1 - T_4)}{T_2 - T_3}$$

$$= 1 - \frac{T_4(T_1/T_4 - 1)}{T_2(1 - T_3/T_2)}$$

$$= 1 - \frac{T_4[(p_1/p_4)^{(k-1)/k} - 1]}{T_2[1 - (p_3/p_2)^{(k-1)/k}]}.$$

But $p_1 = p_2$ and $p_4 = p_3$. Thus

$$\eta_{th} = 1 - \frac{T_4}{T_2}\left(\frac{p_1}{p_4}\right)^{(k-1)/k}, \tag{11.85}$$

where p_1/p_4 is recognized as the pressure ratio of the cycle.

Example 11.14

An air-standard Brayton cycle with perfect regeneration operates according to the following data:

$$\text{Maximum cycle temperature} = 1150 \text{ K}$$

$$\text{Minimum cycle temperature} = 300 \text{ K}$$

$$\text{Maximum cycle pressure} = 505 \text{ kPa}$$

$$\text{Minimum cycle pressure} = 101 \text{ kPa}$$

Determine the cycle thermal efficiency.

Solution (refer to Fig. 11.23)

Using Eq. 11.85, we have

$$\eta_{th} = 1 - \frac{300}{1150} \, (5)^{(1.4-1)/1.4} = 0.587.$$ ∎

Comments

Comparing this with Example 11.10, we note that considerable improvement in cycle efficiency is obtained through the use of a regenerator. In practice, however, the regenerator is costly, heavy, and bulky. These factors must be taken into consideration in our decision making on whether or not it is worthwhile to use a regenerator. We also wish to point out that perfect regeneration is not possible in practice, which means that the temperature of air entering the combustion chamber will be less than the temperature of the gas leaving the turbine (Fig. 11.23). The *regenerator effectiveness* is defined by

$$\eta_{reg} = \frac{h_7 - h_1}{h_3 - h_1},$$ (11.86)

in which the numerator represents the quantity of heat actually transferred while the denominator represents the maximum quantity that can theoretically be transferred. If the specific heat is constant, the regenerator effectiveness may also be given as

$$\eta_{reg} = \frac{T_7 - T_1}{T_3 - T_1}.$$ (11.87)

Equation 11.85 indicates that for a given temperature range, the thermal efficiency of an air-standard Brayton cycle with perfect regeneration increases without limit when the pressure ratio decreases. On the other hand, if we take into consideration the inefficiencies in compressor, turbine, and regenerator, there will be an optimum pressure ratio that will depend on the temperature range of the cycle, compressor efficiency η_c, turbine efficiency η_T, regenerator effectiveness η_{reg} and pressure drops in heat exchangers. The effect on cycle performance due to the use of real hardware will be examined in the next section.

11.18 Regenerative Gas-Turbine Power Plant with Real Compressor, Turbine, Regenerator, and Pressure Drops in Heat Exchangers

The effect on cycle performance of a regenerative gas-turbine power plant due to irreversibilities in compressor, turbine, regener-

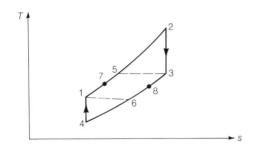

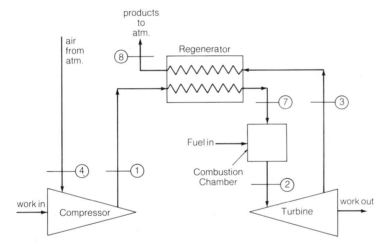

Figure 11.23 Brayton Cycle with Partial Regeneration

ator, and pressure drops in heat exchangers is examined in this section. This cycle is shown in Fig. 11.24.

With reference to Fig. 11.24 and to Fig. 11.23 in which we will consider the combustion chamber as an air heater, and assuming ideal gas with constant specific heats, the compressor work is given by Eq. 11.63:

$$w_C = -\frac{c_p T_4}{\eta_C} [r_p^{(k-1)/k} - 1]. \qquad (11.63)$$

The turbine expansion ratio, p_2/p_3, may be expressed in terms of the compressor compression ratio r_p and pressure drops to be used in each of the heat exchangers. Making use of the concept of relative pressure drop and the quantity pressure drop factor introduced in

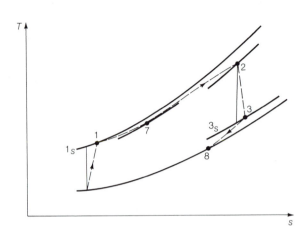

Figure 11.24 Regenerative Gas-Turbine Power Cycle with Irreversible Compressor, Turbine, Regenerator, and Pressure Drops in Heat Exchangers

Section 11.16, we have

$$p_7 = \beta_{17} p_1 \tag{11.88a}$$

$$p_2 = \beta_{72} p_7 \tag{11.88b}$$

$$p_8 = \beta_{38} p_3, \tag{11.88c}$$

where β_{17} is the pressure drop factor in the cold side of the regenerator, β_{38} is the pressure drop factor in the hot side of the regenerator, and β_{72} is the pressure drop factor in the air heater. As pointed out in Section 11.16, each pressure drop factor β is defined as

$$\beta = 1 - \frac{\Delta p}{p},$$

where $(\Delta p)/p$ is known as the relative pressure drop. Then

$$\frac{p_7}{p_1} \times \frac{p_2}{p_7} \times \frac{p_8}{p_3} = \beta_{17} \beta_{72} \beta_{38}$$

or

$$\frac{p_2}{p_3} = (\beta_{17} \beta_{72} \beta_{38}) \frac{p_1}{p_8} = (\beta_{17} \beta_{72} \beta_{38}) r_p, \tag{11.89}$$

where p_2/p_3 is the turbine expansion ratio and r_p is the pressure ratio of the cycle.

Making use of Eq. 11.62, the turbine work may be given as

$$w_T = \eta_T c_p T_2 \left[1 - \frac{1}{(\beta_{17} \beta_{72} \beta_{38} r_p)^{(k-1)/k}} \right]. \qquad (11.90)$$

The cycle net work is then given by

$$w_{net} = \eta_T c_p T_2 \left[1 - \frac{1}{(\beta_{17} \beta_{72} \beta_{38} r_p)^{(k-1)/k}} \right] - \frac{c_p T_4}{\eta_c} [r_p^{(k-1)/k} - 1]. \qquad (11.91)$$

Dividing Eq. 11.91 across by $c_p T_4$, the specific power output of the cycle is given by

$$\frac{w_{net}}{c_p T_4} = \eta_T \left(\frac{T_2}{T_4}\right) \left[1 - \frac{1}{(\beta_{17} \beta_{72} \beta_{38} r_p)^{(k-1)/k}} \right] - \frac{1}{\eta_c} [r_p^{(r-1)/k} - 1] \qquad (11.92)$$

The quantity $w_{net}/(c_p T_4)$ is known as the specific power output of the cycle, which is a dimensionless quantity.

For a given set of values of η_T, η_c, T_2/T_4, c_p, and k, there will be an optimum pressure ratio r_p to give the maximum cycle specific power. This optimum pressure ratio is obtained by differentiating Eq. 11.92 with respect to r_p, equating to zero, and solving for r_p. The result is

$$(r_p)_{opt} = \left[\left(\frac{T_2}{T_4}\right) \frac{\eta_T \eta_c}{\phi} \right]^{k/2(k-1)}, \qquad (11.93)$$

where

$$\phi = (\beta_{17} \beta_{72} \beta_{38})^{(k-1)/k}.$$

It is to be noted that this optimum pressure ratio is independent of the regenerator effectiveness. In fact, Eq. 11.93 is almost identical to Eq. 11.77, which gives the optimum pressure ratio for maximum power for a simple gas-turbine power cycle.

Heat addition for the cycle is given by

$$q_{in} = q_{12} = c_p(T_2 - T_7). \qquad (11.94)$$

Now the regenerator effectiveness is given by Eq. 11.87:

$$\eta_{\text{reg}} = \frac{T_7 - T_1}{T_3 - T_1}. \tag{11.87}$$

Then

$$T_7 = T_1 + \eta_{\text{reg}}(T_3 - T_1). \tag{11.95}$$

From the compressor work, we have

$$T_1 = T_4 + \frac{T_4}{\eta_C}\,[r_p^{(k-1)/k} - 1]. \tag{11.96}$$

From the turbine work, we have

$$T_3 = T_2 - \eta_T\,T_2\left[1 - \frac{1}{(\beta_{17}\,\beta_{72}\,\beta_{38}\,r_p)^{(k-1)/k}}\right]. \tag{11.97}$$

Combining Eqs. 11.94, 11.95, 11.96, and 11.97, we have

$$q_{\text{in}} = c_p\,T_2 - c_p\,T_4\left[1 + \frac{r_p^{(k-1)/k} - 1}{\eta_C}\right]$$

$$- \eta_{\text{reg}}\left\{c_p\,T_2\left[1 - \eta_T\left(1 - \frac{1}{\phi r_p^{(k-1)/k}}\right)\right] - c_p\,T_4\left[1 + \frac{r_p^{(k-1)/k} - 1}{\eta_C}\right]\right\} \tag{11.98}$$

where

$$\phi = (\beta_{17}\,\beta_{72}\,\beta_{38})^{(k-1)/k}.$$

Dividing Eq. 11.98 across by $c_p\,T_4$, the specific heat input for the cycle is given by

$$\frac{q_{\text{in}}}{c_p\,T_4} = \frac{T_2}{T_4} - \left[1 + \frac{r_p^{(k-1)/k} - 1}{\eta_C}\right]$$

$$- \eta_{\text{reg}}\left\{\frac{T_2}{T_4}\left[1 - \eta_T\left(1 - \frac{1}{\phi r_p^{(k-1)/k}}\right)\right] - \left[1 + \frac{r_p^{(k-1)/k} - 1}{\eta_C}\right]\right\}. \tag{11.99}$$

The quantity $q_{in}/(c_p T_4)$ is known as the specific heat input of the cycle, which is a dimensionless quantity. It depends on the regenerator effectiveness as expected. The cycle thermal efficiency is given by the ratio of Eq. 11.91 to Eq. 11.98.

For a given set of values of η_T, η_C, T_2/T_4, c_p, k, η_{reg}, and pressure drop factors, there will be an optimum pressure ratio r_p to give the maximum cycle thermal efficiency. This optimum pressure ratio is obtained by differentiating the expression for cycle thermal efficiency with respect to r_p, equating to zero, and solving for r_p. The result is

$$(r_p)_{opt} = \left[\frac{-b + \sqrt{b^2 - 4ac}}{2a} \right]^{k/(k-1)}, \tag{11.100}$$

where

$$a = \left[\frac{(1 - \eta_{reg})[(T_2/T_4) - 1]}{\eta_C} + \frac{(T_2/T_4)\eta_T (2\eta_{reg} - 1)}{\eta_C} \right] (\beta_{17} \beta_{72} \beta_{38})^{(k-1)/k}$$

$$b = \frac{2(T_2/T_4)\eta_T (1 - 2\eta_{reg})}{\eta_C}$$

$$c = \eta_T \left(\frac{T_2}{T_4} \right)(1 - \eta_{reg})\left(1 - \frac{T_2}{T_4} \right) + \frac{(T_2/T_4)\eta_T(2\eta_{reg} - 1)}{\eta_C}.$$

In Fig. 11.25, the maximum cycle thermal efficiency as a function of pressure ratio, regenerator effectiveness, and ratio of maximum cycle temperature to minimum cycle temperature (with $\eta_T = 0.90$, $\eta_C = 0.90$, $k = 1.4$, and $\beta_{17}\beta_{72}\beta_{38} = 0.90$) is presented graphically. From this figure we can see that a regenerative gas-turbine power cycle can achieve very high cycle thermal efficiencies with a relatively low pressure ratio if we can operate with a large temperature range and a very effective regenerator.

Example 11.15

A regenerative gas-turbine power plant is to be designed according to the following specifications:

Maximum cycle temperature	= 1150 K
Minimum cycle temperature	= 300 K
Adiabatic turbine efficiency	= 85%
Adiabatic compressor efficiency	= 80%

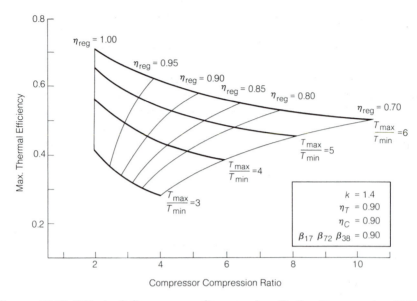

Figure 11.25 Effect of Compressor Compression Ratio, Regenerator Effectiveness, and Ratio of Maximum Cycle Temperature to Minimum Cycle Temperature on Maximum Cycle Thermal Efficiency

Regenerator effectiveness   =   80%

Relative pressure drop in each heat-transfer circuit =   3%

Select pressure ratio for maximum cycle thermal efficiency. Assuming ideal gas with constant specific heats, determine

(a) the cycle net work.
(b) the cycle heat input.
(c) the cycle thermal efficiency.

Solution (refer to Fig. 11.24)

For maximum cycle thermal efficiency, the optimum pressure ratio is given by Eq. 11.100:

$$(r_p)_{opt} = \left(\frac{-b + \sqrt{b^2 - 4ac}}{2a} \right)^{k/(k-1)},$$

where

$$a = \left[\frac{(1 - \eta_{\text{reg}})[(T_2/T_4) - 1]}{\eta_C} + \frac{(T_2/T_4)\eta_T(2\eta_{\text{reg}} - 1)}{\eta_C} \right]$$

$$\times (\beta_{17}\beta_{72}\beta_{38})^{(k-1)/k}$$

$$= \left[\frac{(1 - 0.8)[(1150/300) - 1]}{0.80} + \frac{(1150/300) \times 0.85(2 \times 0.8 - 1)}{0.8} \right]$$

$$\times (0.97 \times 0.97 \times 0.97)^{(1.4-1)/1.4}$$

$$= 3.0708$$

$$b = \frac{2(T_2/T_4)\eta_T(1 - 2\eta_{\text{reg}})}{\eta_C}$$

$$= \frac{2 \times (1150/300) \times 0.85(1 - 2 \times 0.8)}{0.8}$$

$$= -4.8875$$

$$c = \eta_T\left(\frac{T_2}{T_4}\right)(1 - \eta_{\text{reg}})\left(1 - \frac{T_2}{T_4}\right) - \frac{(T_2/T_4)\eta_T(1 - 2\eta_{\text{reg}})}{\eta_C}$$

$$= 0.85\left(\frac{1150}{300}\right)(1 - 0.8)\left(1 - \frac{1150}{300}\right) - \frac{(1150/300) \times 0.85(1 - 2 \times 0.8)}{0.8}$$

$$= 0.5974$$

$$(r_p)_{\text{opt}} = \left[\frac{+4.8875 + \sqrt{(-4.8875)^2 - 4 \times 3.0708 \times 0.5974}}{2 \times 3.0708} \right]^{1.4/(1.4-1)}$$

$$= 3.74.$$

(a) The cycle net work is given by Eq. 11.91:

$$w_{\text{net}} = \eta_T c_p T_2\left[1 - \frac{1}{(\beta_{17}\beta_{72}\beta_{38} r_p)^{(k-1)/k}} \right] - \frac{c_p T_4}{\eta_C}\left[r_p^{(k-1)/k} - 1 \right]$$

$$= 0.85 \times 1.0038 \times 1150\left[1 - \frac{1}{(0.97 \times 0.97 \times 0.97 \times 3.74)^{(1.4-1)/1.4}} \right]$$

$$- \frac{1.0038 \times 300}{0.8}\left[3.74^{(1.4-1)/1.4} - 1 \right]$$

$$= 290.2 - 172.3 = 117.9 \text{ kJ/kg}.$$

(b) The cycle heat input is given by Eq. 11.98:

$$q_{in} = c_p T_2 - c_p T_4 \left[1 + \frac{r_p^{(k-1)/k} - 1}{\eta_C} \right]$$

$$- \eta_{reg} \left\{ c_p T_2 \left[1 - \eta_T \left(1 - \frac{1}{\phi r_p^{(k-1)/k}} \right) \right] - c_p T_4 \left[1 + \frac{r_p^{(k-1)/k} - 1}{\eta_C} \right] \right\}$$

where

$$\phi = (\beta_{17} \beta_{72} \beta_{38})^{(k-1)/k}$$

$$= (0.97 \times 0.97 \times 0.97)^{(1.4-1)/1.4} = 0.9742.$$

Thus

$$q_{in} = 1.0038 \times 1150 - 1.0038 \times 300 \left[1 + \frac{3.74^{(1.4-1)/1.4} - 1}{0.8} \right]$$

$$- 0.8 \left\{ 1.0038 \times 1150 \left[1 - 0.85 \left(1 - \frac{1}{0.9742(3.74)^{(1.4-1)/1.4}} \right) \right] \right.$$

$$\left. - 1.0038 \times 300 \left[1 + \frac{3.74^{(1.4-1)/1.4} - 1}{0.8} \right] \right\}$$

$$= 368.3 \text{ kJ/kg.}$$

(c) The cycle thermal efficiency is

$$\eta_{th} = \frac{w_{net}}{q_{in}}$$

$$= \frac{117.9}{368.3} = 32.0\%. \qquad \blacksquare$$

Comments

Comparing with Example 11.13, we note that considerable improvement in cycle efficiency is obtained through the use of a regenerator. In addition, a pressure ratio of only 3.74 is needed for a regenerative system compared to a pressure ratio of 5.5 used in Example 11.13.

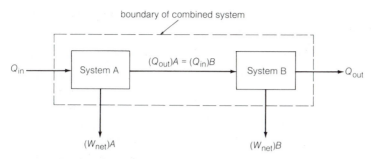

Figure 11.26 Combined Power Plant

We also wish to point out that for a given set of values of η_T, η_C, η_{reg}, c_p, k, and pressure drop factors, the optimum pressure ratio for maximum cycle thermal efficiency is smaller than the optimum pressure ratio for maximum power output. This behavior is exactly the opposite for a simple gas-turbine power cycle, which requires a larger pressure ratio for maximum cycle thermal efficiency than the pressure ratio required for maximum power output for a given set of values of η_T, η_C, c_p, k, and pressure drop factors.

11.19 Thermal Efficiency of a Combined Power Plant

Each type of power plant that we have studied so far employs only one kind of working fluid. We conclude our study of power-producing systems by pointing out the promise of systems in which two kinds of working fluid are used. Figure 11.26 shows a combined system consisting of system A and system B. Let $(\eta_{th})_A$ be the thermal efficiency of system A and $(\eta_{th})_B$ the thermal efficiency of system B. Let all the heat input to the combined system be the heat input to system A and let heat input to system B be the waste heat of system A. Then we have

Net work for system A: $(W_{net})_A = (\eta_{th})_A Q_{in}$

Net work for system B: $(W_{net})_B = (\eta_{th})_B (Q_{out})_A .$

Thermal efficiency for combined system:

$$(\eta_{th})_{comb} = \frac{(W_{net})_A + (W_{net})_B}{Q_{in}}$$

$$= \frac{(\eta_{th})_A Q_{in} + (\eta_{th})_B (Q_{out})_A}{Q_{in}} . \qquad (11.101)$$

But $(Q_{out})_A = Q_{in} - (W_{net})_A$. Thus the thermal efficiency of the combined system may be simplified to

$$(\eta_{th})_{comb} = (\eta_{th})_A + (\eta_{th})_B [1 - (\eta_{th})_A]. \tag{11.102}$$

Equation 11.102 shows that we could obtain a thermal efficiency of 55% if $(\eta_{th})_B$ is 40% and $(\eta_{th})_A$ is 25%. One such combined system that could yield this attractive thermal efficiency is the combined gas–steam power plant.[11] Another combined system considered to have promise is one that hybrids a steam plant with a MHD power generator.[12] A fairly detailed discussion on combined gas and steam plant and binary cycles is given by Haywood.[13]

Figure 11.27 is a schematic diagram of a steam-turbine and gas-turbine combined power cycle that has received a great deal of atten-

[11] H. C. Hottell and J. B. Howard, *New Energy Technology—Some Facts and Assessments*, The MIT Press, Cambridge, Mass., 1971, Chap. 5.

[12] MHD is the abbreviation for magnetohydrodynamics. The working fluid of a MHD power plant is an ionized gas.

[13] R. W. Haywood, *Analysis of Engineering Cycles*, Pergamon Press Ltd., London, 1967.

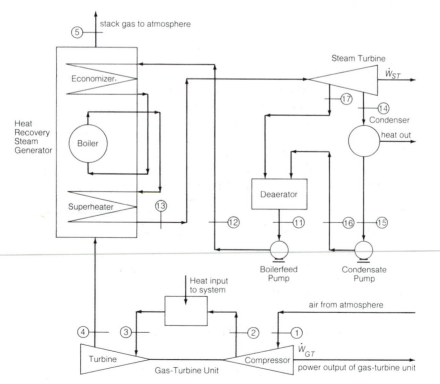

Figure 11.27 Steam-Turbine and Gas-Turbine Combined Cycle Power Plant

tion in recent years. In this system the exhaust from the gas turbine is hot enough to generate steam in a waste-heat boiler (heat recovery steam generator) for the production of more power, thereby increasing the efficiency of the system.

In Fig. 11.27 we have shown the steam-turbine unit having only one feedwater heater (the deaerator). Sometimes it may pay to use more than one feedwater heater. Sometimes it may even pay to have a two-pressure system requiring a low-pressure economizer, a low-pressure boiler, a high-pressure economizer, a high-pressure boiler together with a superheater, thereby resulting in less entropy generation in the heat recovery steam generator. Obviously, it is a matter of trade-offs.

We wish to conclude our study of combined power plants by pointing out that a thermal efficiency of 45% is possible based on state-of-the-art technology. A more complicated topping cycle could increase the combined thermal efficiency considerably. Major effort is being made to develop an advanced gas turbine for such use. This advanced gas turbine under development will have reheat as well as intercooling. It is a high-pressure, high-temperature system. The temperature of hot gas will be 1400°C at the high-pressure turbine inlet. It will be 1200°C at the reheater outlet. The pressure of hot gas will be 55 atm at the high-pressure turbine inlet, making compressor intercooling necessary. Because of reheat, the exhaust gas will leave the topping unit at a fairly high temperature of 610°C, making it possible to have a more efficient bottoming cycle. The thermal efficiency of a combined power plant using the advanced gas turbine as the topping unit is expected to be 55% (based on lower heating value of fuel).[14]

11.20 Cogeneration Systems

Cogeneration is an old engineering concept involving the production of both electricity and useful thermal energy (steam or process heat) in one operation, thereby utilizing fuel more effectively than if the desired products were produced separately. Since the heart of a cogeneration system is a prime mover with waste heat at a usable temperature, it is not surprising that the requirements of cogeneration may be met in many ways, ranging from steam turbine and gas turbine to fuel cell and Stirling engine. But a feasible cogeneration system must be technically sound, economically attractive, and environmentally acceptable. An excellent summary of eight

[14] Masashi Arai, Tetsu Imai, Kiyomi Teshima, and Akinori Koga, "Research and Development on the HPT of the AGTJ-100B", ASME Paper No. 87-GT-263, 1987.

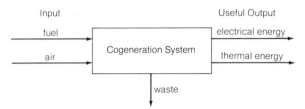

Figure 11.28 General Concept of a Cogeneration System

types of cogeneration system with unique features of each type has been given by Bazques and Strom.[15] The general concept of a cogeneration system is shown in Fig. 11.28.

To assess the performance of a cogeneration system, we may make use of different parameters. A common parameter used to measure the performance of a cogeneration system is the so-called fuel utilization efficiency (η_f), which is the ratio of all the useful energy extracted from the system (electrical energy and process heat) to the energy of fuel input. This parameter is also known as the first-law efficiency since only energy accounting is involved. According to this defini-tion, η_f is then given by

$$\eta_f = \frac{\dot{W}_{el} + \dot{Q}_p}{\dot{E}_f} , \qquad (11.103)$$

where $\dot{W}_{el}$ is the electrical power output of the system, $\dot{Q}_p$ is the amount of process heat production, and $\dot{E}_f$ is the energy of fuel input.

Another common parameter used to assess the performance of a cogeneration system is the electrical-to-thermal energy ratio, which is also known as the power-to-heat ratio. Since power is considered to be the most valuable product of any cogeneration system, the cost-effectiveness of a cogeneration system is directly related to the amount of power it can produce for a given amount of process heat needed. By definition, the power-to-heat ratio (R_{pH}) is given as

$$R_{pH} = \frac{\dot{W}_{el}}{\dot{Q}_p}. \qquad (11.104)$$

In both the fuel utilization efficiency and the electrical-to-thermal energy ratio, power and process heat are treated as equal. This is based on the first law of thermodynamics. But power is much more

[15] E. Bazques and D. Strom, "Cogeneration and Interconnection Technologies for Industrial and Commercial Use," 18th IECEC Conference, Orlando, Florida, August 21–26, 1983.

valuable than process heat, according to the second law of thermodynamics. A process is better if less exergy is consumed. Consequently, the ratio of the amount of exergy in our products to the amount of exergy supplied is a more accurate measure of the thermodynamic performance of a system. By definition, the second-law efficiency of a cogeneration may then be given as

$$\eta_{II} = \frac{\dot{W}_{el} + \dot{B}_p}{\dot{B}_f},$$
(11.105)

where $\dot{W}_{el}$ is all exergy, $\dot{B}_p$ is the exergy content of process heat, and $\dot{B}_f$ is the exergy content of fuel input.

The ratio of $\dot{B}_p$ to $\dot{Q}_p$ may be called the exergy factor of the process heat produced. The ratio of $\dot{B}_f$ to $\dot{E}_f$ may be called the exergy factor of fuel input. Thus we have

$$\varepsilon_p = \frac{\dot{B}_p}{\dot{Q}_p}$$
(11.106)

$$\varepsilon_f = \frac{\dot{B}_f}{\dot{E}_f},$$
(11.107)

where ε_p is the exergy factor of process heat, and ε_f is the exergy factor of fuel input. Equation 11.105 may then be written as

$$\eta_{II} = \frac{\dot{W}_{el} + \varepsilon_p \dot{Q}_p}{\varepsilon_f \dot{E}_f}.$$
(11.108)

Combining Eq. 11.108 with Eq. 11.103, we have

$$\frac{\eta_{II}}{\eta_f} = \left(\frac{\dot{W}_{el}/\dot{Q}_p + \varepsilon_p}{\dot{W}_{el}/\dot{Q}_p + 1}\right) \frac{1}{\varepsilon_f}$$

$$= \left(\frac{R_{pH} + \varepsilon_p}{R_{pH} + 1}\right) \frac{1}{\varepsilon_f}$$
(11.109)

The exergy factor of fuel is close to unity, as the chemical energy in fuel is essentially all exergy.[16] The exergy factor of process steam is always less than unity, but it increases with pressure of the steam

---

[16] M. J. Moran, *Availability Analysis: A Guide to Efficient Energy Use*, Prentice-Hall, Inc., Englewood Cliffs, N.J., 1982.

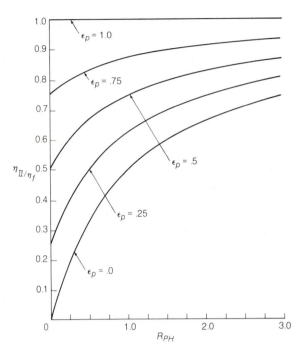

Figure 11.29 Effect of Power-to-Heat Ratio R_{pH} and Exergy Factor of Process Heat ε_p on the Ratio of Second-Law Efficiency η_{II} to Fuel Utilization Efficiency η_f

produced. This is consistent with the second law of thermodynamics, as high-pressure steam is more valuable than low-pressure steam.

If we let $\varepsilon_f = 1.0$, Eq. 11.109 may be written as

$$\frac{\eta_{II}}{\eta_f} = \frac{R_{pH} + \varepsilon_p}{R_{pH} + 1}. \tag{11.110}$$

Since ε_p is always less than unity, Eq. 11.110 indicates that the second-law efficiency of a cogeneration system is always less than the corresponding first-law efficiency. The influence of R_{pH} and ε_p on the ratio of η_{II} to η_f is shown graphically in Fig. 11.29.

11.21 Gas-Turbine Cogeneration System

A simple gas-turbine cogeneration system is shown in Fig. 11.30 with the corresponding enthalpy–entropy diagram shown in Fig. 11.31. Our study will be based on an indirect-fired air turbine system. But the equations developed in this section may be used to obtain

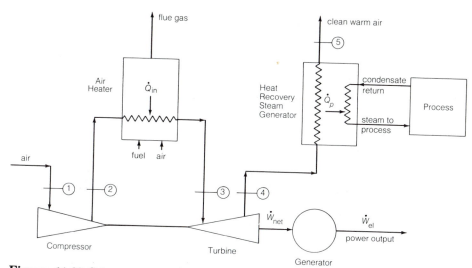

Figure 11.30 Schematic Diagram of an Indirect-Fired Air Turbine Cogeneration System

good approximations of design data for combustion gas-turbine systems.

For a mass flow rate of air of $\dot{m}_a$, the amount of heat addition for the cycle is given by

$$\dot{Q}_{in} = \dot{m}_a(h_3 - h_2), \tag{11.111}$$

the cycle net power output is given by

$$\dot{W}_{net} = \dot{m}_a[(h_3 - h_4) - (h_2 - h_1)], \tag{11.112}$$

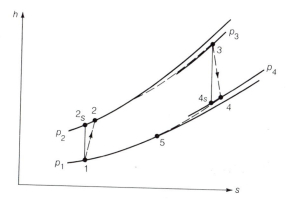

Figure 11.31 Enthalpy–Entropy Diagram of an Air Turbine Cycle

and the amount of process heat production is given by

$$\dot{Q}_p = \dot{m}_a(h_4 - h_5). \tag{11.113}$$

Assuming air to be an ideal gas with constant specific heats, we have

$$\dot{Q}_{in} = \dot{m}_a c_p(T_3 - T_2)$$

$$= \dot{m}_a c_p \left[T_3 - T_1 \frac{1 + (r_{pC}^{(k-1)/k} - 1)}{\eta_C} \right] \tag{11.114}$$

$$\dot{W}_{net} = \dot{m}_a c_p [(T_3 - T_4) - (T_2 - T_1)]$$

$$= \dot{m}_a c_p \left[\eta_T T_3 \left(1 - \frac{1}{r_{pT}^{(k-1)/k}} \right) - \frac{T_1(r_{pC}^{(k-1)/k} - 1)}{\eta_C} \right] \tag{11.115}$$

$$\dot{Q}_p = \dot{m}_a c_p(T_4 - T_5), \tag{11.116}$$

where

$$r_{pC} = \frac{p_2}{p_1}, \text{ the compressor compression ratio}$$

$$r_{pT} = \frac{p_3}{p_4}, \text{ the turbine expansion ratio.}$$

Defining the generator efficiency η_g, and the air heater efficiency η_H as

$$\eta_g = \frac{\dot{W}_{el}}{\dot{W}_{net}} \tag{11.117}$$

and

$$\eta_H = \frac{\dot{Q}_{in}}{\dot{E}_f}, \tag{11.118}$$

we can write the fuel utilization efficiency as

$$\eta_f = \eta_H \left(\eta_g \cdot \eta_{th} + \frac{\dot{Q}_p}{\dot{Q}_{in}} \right). \tag{11.119}$$

Making use of Eqs. 11.114, 11.115, and 11.116, Eq. 11.119 may now be given as

$$\eta_f = \eta_H \left\{ \eta_g \left[\frac{\theta \cdot \eta_T \cdot \psi_T - (\psi_C/\eta_C)}{\theta - 1 - (\psi_C/\eta_C)} \right] + \left[\frac{\theta - \theta \cdot \eta_T \cdot \psi_T - \tau}{\theta - 1 - (\psi_C/\eta_C)} \right] \right\} \quad (11.120)$$

where

$$\theta = \frac{T_3}{T_1} = \frac{T_{max}}{T_{min}} \quad (11.121)$$

$$\psi_C = r_{pC}^{(k-1)/k} - 1 \quad (11.122)$$

$$\psi_T = 1 - 1/r_{pT}^{(k-1)/k} \quad (11.123)$$

$$\tau = \frac{T_5}{T_1} \quad (11.124)$$

$$\eta_{th} = \frac{W_{net}}{Q_{in}}, \text{ the cycle thermal efficiency.}$$

Making use of Eqs. 11.115, 11.116, and 11.117, the power-to-heat ratio R_{pH} may be written as

$$R_{pH} = \frac{\dot{W}_{el}}{\dot{Q}_p}$$

$$= \eta_g \left[\frac{\theta \cdot \eta_T \cdot \psi_T - (\psi_C/\eta_C)}{\theta - \theta \cdot \eta_T \cdot \psi_T - \tau} \right]. \quad (11.125)$$

The exergy factor of process heat depends on the steam generated. If we produce saturated steam, then, with a mass flow rate of steam $\dot{m}_s$,

$$\dot{Q}_p = \dot{m}_s(h_g - h_c) \quad (11.126)$$

$$\dot{B}_p = \dot{m}_s(\text{change of availability per unit mass of steam})$$

$$= \dot{m}_s [(h_g - h_c) - T_0(s_g - s_c)]. \quad (11.127)$$

Thus the exergy factor of process heat is given by

$$\varepsilon_p = 1 - \frac{T_0(s_g - s_c)}{h_g - h_c}, \quad (11.128)$$

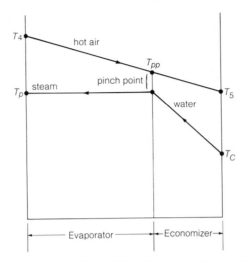

Figure 11.32 Temperature Profile in Heat Recovery Steam Generator

where

T_0 = temperature of the environment

s_g = entropy of saturated vapor at pressure of process steam

s_c = entropy of condensate return

h_g = enthalpy of saturated vapor at pressure of process steam

h_c = enthalpy of condensate return.

The quantity and quality of process steam produced will depend on the temperature of air and temperature of steam in the heat recovery steam generator, as shown in Fig. 11.32.

From energy balance in the evaporator, we have

$$\dot{m}_a c_p(T_4 - T_{pp}) = \dot{m}_s(h_g - h_f). \tag{11.129}$$

From energy balance in the economizer, we have

$$\dot{m}_a c_p(T_{pp} - T_5) = \dot{m}_s(h_f - h_c). \tag{11.130}$$

Combining Eqs. 11.129 and 11.130, we have

$$T_5 = (T_p + \text{pp}) - \frac{[T_4 - (T_p + \text{pp})](h_f - h_c)}{h_g - h_f} \tag{11.131}$$

where

T_p = saturation temperature at pressure of process steam

pp = the closest temperature between the air and the steam

T_{pp} = temperature of air at the pinch point

h_f = enthalpy of saturated liquid at pressure of process steam.

From Eq. 11.131 we note that T_5, the temperature of air leaving the heat recovery steam generator, is directly related to the pinch point. If we use a small pinch point, more heat will be recovered from the hot air, but the steam generator will be more costly.

Knowing ε_p, R_{pH}, and η_f, the second-law efficiency may then be determined using Eq. 11.109:

$$\eta_{II} = \frac{\eta_f}{\varepsilon_f} \left(\frac{R_{pH} + \varepsilon_p}{R_{pH} + 1} \right). \tag{11.109}$$

As we noted before, the exergy factor ε_f for fuel may be taken as unity. Since ε_p is always less than unity, the quantity inside the parentheses in Eq. 11.109 will increase with increase in R_{pH}. This would mean that an increase in R_{pH} will tend to increase the second-law efficiency. However, as R_{pH} increases, η_f will in general decrease. Thus there is a combination of η_f and R_{pH} that will yield a maximum second-law efficiency for a given design.

Example 11.16

An indirect-fired air turbine cogeneration system is to be designed according to the following specifications:

η_g, generator efficiency	=	96%
η_H, air heater efficiency	=	85%
η_C, adiabatic compressor efficiency	=	90%
η_T, adiabatic turbine efficiency	=	90%
r_{pC}, compressor compression ratio	=	6
r_{pT}, turbine expansion ratio	=	$0.94 r_{pC}$
ε_f, exergy factor of fuel input	=	1.0
pp, pinch point	=	20°C
T_3, maximum cycle temperature	=	816°C

$$T_1, \text{ minimum cycle temperature} \quad = \quad 15°C$$

$$T_C, \text{ temperature of condensate return} = 100°C$$

$$T_0, \text{ temperature of the environment} \quad = \quad 15°C$$

$$k, \text{ specific heat ratio} \quad = \quad 1.4$$

$$\text{Pressure of saturated steam produced} = \quad 1.0 \text{ MPa}$$

Determine

(a) the fuel utilization efficiency.
(b) the power-to-heat ratio.
(c) the second-law efficiency.

Solution (refer to Figs. 11.30 and 11.31)

The exergy factor of process heat is given by

$$\varepsilon_p = 1 - \frac{T_0(s_g - s_c)}{h_g - h_c}$$

$$= 1 - \frac{288.15(6.5828 - 1.3069)}{2776.2 - 419.06}$$

$$= 0.355.$$

The temperature of air leaving the system is given by

$$T_5 = (T_p + \text{pp}) - \frac{[T_4 - (T_p + \text{pp})](h_f - h_c)}{h_g - h_f},$$

where

$$T_p = \text{saturation temperature at } 1.0 \text{ MPa}$$

$$= (179.88 + 273.15) \text{ K} = 453.03 \text{ K}$$

and

$$T_4 = T_3 - \eta_T T_3\left[1 - \frac{1}{r_{pT}^{(k-1)/k}}\right]$$

$$= 1089.15 - 0.90 \times 1089.15\left[1 - \frac{1}{(0.94 \times 6)^{(1.4-1)/1.4}}\right]$$

$$= 706.96 \text{ K}.$$

Thus

$$T_5 = (453.03 + 20) - \frac{[706.96 - (453.03 + 20)](762.6 - 419.06)}{2013.6}$$

$$= 433.12 \text{ K}$$

Now

$$\theta = \frac{T_3}{T_1} = \frac{816 + 273.15}{15 + 273.15} = 3.78$$

$$\tau = \frac{T_5}{T_1} = \frac{433.12}{288.15} = 1.5031$$

$$\psi_T = 1 - \frac{1}{r_{pT}^{(k-1)/k}}$$

$$= 1 - \frac{1}{(0.94 \times 6)^{(1.4-1)/1.4}} = 0.3899$$

$$\psi_C = r_{pC}^{(k-1)/k} - 1 = 6^{(1.4-1)/1.4} - 1 = 0.6685.$$

(a) The fuel utilization efficiency is given by Eq. 11.120:

$$\eta_f = \eta_H \left\{ \eta_g \left[\frac{\theta \cdot \eta_T \cdot \psi_T - (\psi_C/\eta_C)}{\theta - 1 - (\psi_C/\eta_C)} \right] + \left[\frac{\theta - \theta \cdot \eta_T \cdot \psi_T - \tau}{\theta - 1 - (\psi_C/\eta_C)} \right] \right\}$$

$$= 0.85 \left\{ 0.96 \left[\frac{3.78 \times 0.90 \times 0.3899 - 0.6685/0.90}{3.78 - 1 - 0.6685/0.90} \right] \right.$$

$$\left. + \left[\frac{3.78 - 3.78 \times 0.90 \times 0.3899 - 1.5031}{3.78 - 1 - 0.6685/0.90} \right] \right\}$$

$$= 0.6304.$$

(b) The power-to-heat ratio is given by Eq. 11.125:

$$R_{pH} = \eta_g \left[\frac{\theta \cdot \eta_T \cdot \psi_T - (\psi_C/\eta_C)}{\theta - \theta \cdot \eta_T \phi_T - \tau} \right]$$

$$= 0.96 \left(\frac{3.78 \times 0.90 \times 0.3899 - 0.6685/0.90}{3.78 - 3.78 - 0.090 \times 0.3899 - 1.5031} \right)$$

$$= 0.5894.$$

(c) The second-law efficiency is given by Eq. 11.109:

$$\eta_{II} = \eta_f \left(\frac{R_{pH} + \varepsilon_p}{R_{pH} + 1} \right)$$

$$= 0.6304 \left(\frac{0.5894 + 0.355}{0.5894 + 1} \right)$$

$$= 0.3746. \qquad \blacksquare$$

Comments

If we repeat the calculations for different compressor compression ratios, we would have the following results:

r_{pC}	η_f	R_{pH}	η_{II}
6	0.6304	0.5894	0.3746
7	0.6141	0.6694	0.3768
8	0.5990	0.7452	0.3776
9	0.5844	0.8173	0.3770

We note that as the compressor compression ratio increases, the fuel utilization efficiency η_f decreases while the power-to-heat ratio R_{pH} increases. This trend has been shown to be true in general for a simple air turbine cogeneration system.[17] The maximum second-law efficiency is obtained with a compressor compression ratio of 8 for our case. But the second-law efficiency does not change very much between the compressor compression ratio of 6 and 9. If we plot both η_{II} and w_{net} against r_{pC} on the same diagram, we would find that these two curves essentially have the same trend, with the optimum pressure ratio for maximum cycle power output very close to the

[17] F. F. Huang and F. Egolfopoulos, *Performance Analysis of an Indirect Fired Air Turbine Cogeneration System*, ASME Paper 85-IGT-3, 1985.

optimum pressure ratio for maximum second-law efficiency. Thus, for a good estimate of the optimum pressure ratio for maximum second-law efficiency, we may make use of Eq. 11.77, which gives the optimum pressure ratio for maximum power output of a simple gas-turbine cycle. With our specifications, the approximate optimum pressure ratio for maximum second-law efficiency would be given as

$$(r_{pC})_{opt} = \left[\frac{T_{max}}{T_{min}} \times \frac{\eta_T \eta_C}{\phi} \right]^{k/2(k-1)},$$

where

$$\phi = (\beta)^{(k-1)/k} = (0.94)^{(1.4-1)/1.4} = 0.9825.$$

Thus

$$(r_{pC})_{opt} = \left(\frac{3.78 \times 0.90 \times 0.90}{0.9825} \right)^{1.4/2(1.4-1)}$$

$$= 7.3,$$

which is indeed very close to the optimum pressure ratio for maximum second-law efficiency.

We wish to conclude by pointing out that in a regenerative gas-turbine cogeneration system as the compressor compression ratio increases, the fuel utilization efficiency η_f will increase while the power-to-heat ratio R_{pH} will decrease.[18]

11.22 The Cheng Cycle Engine and Cogeneration

The Cheng cycle is a new cycle patented only in 1976.[19] It combines the Brayton (gas turbine) and the Rankine (steam turbine) cycles in a unique manner. A schematic diagram of this system is shown in Fig. 11.33.

The first commercial Cheng cycle system, located on the campus of San Jose State University, San Jose, California, has been producing electricity and process steam since 1985. The system essentially consists of a modified gas turbine, a heat recovery steam generator, and water treatment equipment. Superheated steam produced from the

---

[18] F. F. Huang and T. Naumowicz, *Thermodynamic Study of an Indirect Fired Air Turbine Cogeneration System with Regeneration*, ASME Paper 87-GT-34, 1987.

[19] D. Y. Cheng, "Parallel-Compound Dual-Fluid Engine," Patent 3978661, Sept. 7, 1976. See also R. Digumarthi and Chung-Nan Chang, *Cheng Cycle Implementation on a Small Gas Turbine Engine*, ASME Paper 84-GT-150, 1984.

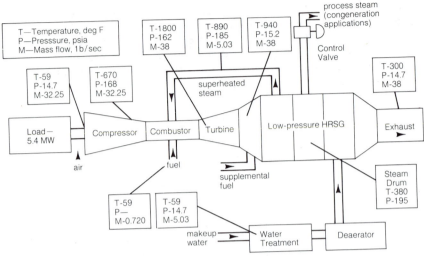

Figure 11.33 Operating Parameters of a Cheng Cycle Engine at Full Rated Output (From *Power*, a McGraw-Hill Publication, February 1983)

gas turbine exhaust is injected back into the combustion region of the turbine. The working fluid in the turbine is thus a combination of combustion gases and superheated steam. The specific heat of the mixture of steam and combustion gases will have a higher value than that of the combustion gases alone. Consequently, for a given mass flow of air through the compressor, the power output of the turbine increases significantly due to steam injection and higher specific heat of the working fluid.

Since the steam injection can vary from zero to 100%, this system is extremely flexible in meeting variable demands of process heat. It is thus quite suitable for cogeneration applications. It also has a thermal efficiency comparable to that of a combined cycle power plant but with less mechanical complexity. In addition, steam injection at the combustion chamber will reduce the generation of NO_x. But the Cheng cycle does have one distinct disadvantage: there is a continuous loss of water.

Problems

11.1 A steam turbine operates on an inlet pressure of 10 MPa and an exhaust pressure of 20 kPa. If the quality is to be 90%, determine the temperature at the inlet for an adiabatic turbine efficiency of 100%.

11.2   A steam turbine operates on an inlet temperature of 500°C and
an exhaust temperature of 60°C. If the quality of exhaust is to
be 90%, determine the pressure at the inlet for an adiabatic
turbine efficiency of 100%.

11.3   A steam turbine operates on an inlet pressure and temperature
of 20 MPa and 550°C. If the quality in the exhaust is not to be
less than 90%, determine the lowest exhaust pressure that can
be used for an adiabatic turbine efficiency of 100%.

11.4   Same as Problem 11.3 except that the adiabatic turbine efficiency is 90%.

11.5   A Carnot vapor cycle uses H_2O as the working fluid. If the
cycle operates within the liquid–vapor region, and upper pressure is 10.0 MPa while lower pressure is 50 kPa, determine the
cycle thermal efficiency.

11.6   A Carnot cycle, using H_2O as the working fluid, operates
between a maximum cycle pressure of 2000 psia and a
minimum cycle pressure of 1.0 psia. All processes are steady-
state steady flow. The fluid states at the beginning and end of
the isothermal heat-addition process are saturated liquid and
dry saturated vapor, respectively. Determine
(a) the cycle thermal efficiency
(b) the ratio of compressor work to turbine work
(c) the steam rate in lbm/kWh.

11.7   An ideal Rankine cycle, using H_2O as the working fluid, operates with a maximum cycle pressure of 2000 psia, a maximum
cycle temperature of 1000°F, and a minimum cycle pressure of
1.0 psia. Determine
(a) the cycle thermal efficiency.
(b) the ratio of pump work to turbine work.

11.8   Steam enters a turbine at 30 MPa and 550°C, and expands
isentropically to a pressure of 5.0 kPa. What is the steam pressure at the section of the turbine where the quality reaches
90%?

11.9   Steam enters a turbine at 5000 psia and 1000°F and expands to
a pressure of 1.0 psia at the outlet. What is the turbine work
for each pound of steam at the inlet
(a) if the expansion is carried out in one turbine?

(b) if the expansion is carried out in two turbines as shown in Fig. P11.9 with 90% quality steam leaving the first turbine and dry saturated steam entering the second turbine? Assume that the expansion is isentropic.

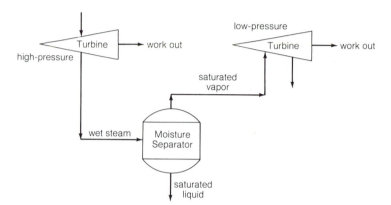

Figure P11.9

11.10 An ideal Rankine cycle, using H_2O as the working fluid, operates between 20.0 MPa and 100 kPa. If the moisture content of the turbine exhaust is not to exceed 10%, determine
(a) the temperature of steam at turbine inlet.
(b) the cycle thermal efficiency.
(c) the steam rate in kg/kWh.

11.11 An ideal Rankine cycle, using H_2O as the working fluid, operates between 550 and 50°C. If the quality of the turbine exhaust is to be 90%, determine
(a) the pressure of steam at turbine inlet.
(b) the cycle thermal efficiency.
(c) the steam rate, in kg/kWh.

11.12 A simple Rankine cycle, using Freon-12 as the working fluid, is to be designed according to the following specifications:
1. Fluid enters turbine as saturated vapor at 100°C.
2. Fluid leaves turbine at 25°C.
3. Adiabatic turbine efficiency is 80%.
4. Adiabatic pump efficiency is 70%.
5. Net cycle power output is 100 kW.
6. Source of energy input is geothermal liquid water.

If geothermal liquid water enters the Freon-12 boiler at 250°C and leaves at 50°C, determine

(a) the cycle thermal efficiency.
(b) the ratio of pump work to turbine work.
(c) the flow rate of Freon-12, in kg/kWh.
(d) the flow rate of geothermal liquid water, in kg/h.

11.13 An ideal Rankine cycle, using Freon-12 as the working fluid, operates with a maximum cycle temperature of 230°F and a minimum cycle temperature of 80°F. The fluid state at the turbine inlet is dry saturated vapor. Determine

(a) the cycle thermal efficiency.
(b) the circulation rate of Freon-12 in lbm/kWh.

11.14 The net output of a modern steam plant is 1 million kW. The plant operates at full capacity for 7500 h/year. Cost of energy input is $3.0/million kJ. Determine the annual energy cost

(a) if the plant thermal efficiency is 40%.
(b) if the plant thermal efficiency is 38%.

11.15 The net output of a modern steam power plant is 1 million kW. The thermal efficiency of the plant is 40%. If 18,000 Btu of heat energy input to the plant requires the burning of 1 lbm of fossil fuel, what is the fuel consumption rate in lbm/h?

11.16 Steam enters the turbine in an ideal Rankine cycle at 20.0 MPa and 550°C. Calculate the cycle thermal efficiency and the moisture content of the steam leaving the turbine for turbine outlet pressures of 5.0, 7.5, 15.0, and 30 kPa. Neglect pump work.

11.17 Steam enters the turbine in an ideal Rankine cycle at 1000 psia and exhausts at 1.0 psia. Calculate the cycle thermal efficiency and the moisture content of the steam leaving the turbine for turbine inlet temperatures of 800, 1000, 1200, and 1400°F. Neglect pump work.

11.18 An ideal Rankine cycle, using H_2O as the working fluid, operates between the pressure limits of 10.0 MPa and 40 kPa. Temperature of steam at turbine inlet is 500°C. On the basis of unit mass, determine

(a) the ratio of pump work to turbine work.
(b) the ratio of heat rejection to heat addition.
(c) the cycle thermal efficiency.
(d) the second-law efficiency of the cycle if heat source is a heat reservoir at 1200 K and the heat sink is the environment at 300 K.

11.19 The operating data of an actual simple steam power cycle are as follows:

Pressure of steam at boiler outlet	= 2100 psia
Temperature of steam at boiler outlet	= 1000°F
Pressure of steam at turbine inlet	= 2000 psia
Temperature of steam at turbine inlet	= 1000°F
Pressure of steam at turbine outlet	= 1 psia
Pump efficiency	= 0.7
Turbine efficiency	= 0.9

Determine the cycle thermal efficiency, the steam rate in lbm/kWh, and the moisture content of the steam leaving the turbine.

11.20 An ideal Rankine cycle with reheat is designed to operate according to the following specifications:

Pressure of steam at high-pressure turbine inlet	= 20.0 MPa
Temperature of steam at high-pressure turbine inlet	= 550°C
Temperature of steam at end of reheat	= 550°C
Pressure of steam at turbine exhaust	= 15 kPa
Quality of steam at turbine exhaust	= 90%

On the basis of unit mass, determine
(a) the pressure of steam in the reheater.
(b) the ratio of pump work to turbine work.
(c) the ratio of heat rejection to heat addition.
(d) the cycle thermal efficiency.
(e) the second-law efficiency if the source of energy is a heat reservoir at 1200 K and the heat sink is the environment at 300 K.

11.21 Steam enters the turbine of an ideal Rankine cycle with reheat at 3000 psia and 1000°F and exhausts at 1.0 psia. The steam is reheated to a temperature of 1000°F. Calculate the cycle thermal efficiency, the steam rate in lbm/kWh, and the moisture content of the steam leaving the turbine for reheat pressures of 1000, 600, and 400 psia. Neglect pump work.

11.22 An ideal Rankine cycle, using H_2O as the working fluid, operates between the temperature limits of 500 and 50°C. If the quality of the turbine exhaust is to be 90%, and neglecting pump work, determine

(a) the average temperature of heat addition.
(b) the average temperature of heat rejection.
(c) the cycle thermal efficiency using the average temperature of heat addition and the average temperature of heat rejection.

11.23 Same as Problem 11.22 except that the cycle operates between the temperature limits of 500 and 30°C.

11.24 Same as Problem 11.22 except that the cycle operates between the temperature limits of 600 and 50°C.

11.25 An ideal Rankine cycle, using H_2O as the working fluid, operates with a maximum cycle pressure of 2000 psia, a maximum cycle temperature of 1000°F, and a minimum cycle pressure of 1.0 psia. Neglecting pump work, determine
(a) the average temperature of heat addition.
(b) the average temperature of heat rejection.
(c) the cycle thermal efficiency using the average temperature of heat addition and the average temperature of heat rejection.

11.26 An ideal Rankine cycle with regenerative feedwater heating operates between the pressure limits of 20.0 MPa and 20 kPa using four open heaters. Determine the operating pressure for each heater on the basis of optimum design.

11.27 In a single-extraction ideal Rankine regenerative cycle, steam is supplied to the turbine at 2000 psia and 1000°F. The condenser pressure is 1.0 psia. Feedwater heating is carried out in an open heater. Neglecting pump work, calculate the cycle thermal efficiency for an extraction pressure of
(a) 30 psia.
(b) 140 psia.
(c) 400 psia.

11.28 An ideal Rankine cycle with regenerative heating operates between the pressure limits of 10.0 MPa and 40 kPa. Tem-

perature of steam at turbine inlet is 500°C. There are two open feedwater heaters. On the basis of optimum design, determine
(a) the ratio of pump work to turbine work.
(b) the ratio of heat rejection to heat addition.
(c) the cycle thermal efficiency.
(d) the second-law efficiency of the cycle if heat source is a heat reservoir at 1200 K and the heat sink is the environment at 300 K.
Make calculations on the basis of 1.0 kg of steam at turbine inlet.

11.29 Same as Problem 11.28 except that there are three open heaters.

11.30 In an ideal Rankine regenerative cycle with reheat, steam is supplied to the high-pressure turbine at 3000 psia and 1000°F and removed from the turbine at 300 psia. Steam is reheated at 300 psia to 1000°F. The condenser pressure is 1.0 psia. Part of the steam leaving the high-pressure turbine is used for feedwater heating in one open feedwater heater. Steam extracted from the low-pressure turbine at 20 psia is used for feedwater heating in another open feedwater heater. Neglecting pump work, determine
(a) the cycle thermal efficiency.
(b) the net work per lbm of steam at the high-pressure turbine inlet.

11.31 Calculate the thermal efficiency of an air-standard Otto cycle for the following compression ratios: 2, 4, 6, 8, 10, 15, and 20. Plot cycle thermal efficiency against compression ratio.

11.32 Sketch on the T–s diagram an air-standard Otto cycle, then superimpose on it a second air-standard Otto cycle having the same inlet conditions and same amount of heat addition but a higher compression ratio. Show by comparison of areas on the T–s diagram that the second cycle has a higher thermal efficiency than the first.

11.33 An air standard Carnot cycle operates between the temperature limits of 500 and 50°C. The maximum cycle pressure is 10 MPa and the minimum cycle pressure is 101 kPa.
(a) Show the cycle on both the T–s and p–v diagrams with the appropriate lines.
(b) Determine the cycle thermal efficiency.
(c) Determine the cycle net work, in kJ/kg.

11.34 An air-standard Otto cycle is to be designed according to the following specifications:

Pressure at start of compression process   = 101 kPa

Temperature at start of compression process = 300 K

Compression ratio   =   8

Maximum pressure of cycle   = 8.0 MPa

(a) Show the cycle on both the T–s and p–v diagrams with the appropriate lines.
(b) Determine the cycle thermal efficiency.
(c) Determine the amount of heat addition, in kJ/kg.
(d) Determine the mean effective pressure.

11.35 Same as Problem 11.34 except that the compression ratio is 10.

11.36 Same as Problem 11.34 except that the maximum cycle pressure is 10 MPa.

11.37 An air-standard Otto cycle operates with a compression ratio of 8. At the beginning of compression, the air is at 14.7 psia and 80°F. The pressure is tripled during the heat-addition process. Determine
(a) the cycle thermal efficiency.
(b) the amount of heat addition, in Btu/lbm.
(c) the mean effective pressure.

11.38 An air-standard Diesel cycle has a compression ratio of 15. Calculate the thermal efficiency of the cycle for the following cutoff ratios: 1.0, 1.5, 2.0, 2.5, and 3.0.

11.39 An air-standard Diesel cycle operates with a compression ratio of 15. At the beginning of compression, the air is at 14.7 psia and 80°F. Determine, for the following quantities of heat added to the air: 500, 750, and 1000 Btu/lbm,
(a) the cycle thermal efficiency using Eq. 11.47.
(b) the mean effective pressure.

11.40   An air-standard Diesel cycle is to be designed according to the following specifications:

Pressure at start of compression process   =   101 kPa

Temperature at start of compression process =   300 K

Compression ratio   =   15

Maximum cycle temperature   = 2800 K

(a) Show the cycle on both the T–s and p–v diagrams with the appropriate lines.
(b) Determine the cycle thermal efficiency.
(c) Determine the amount of heat addition, in kJ/kg.
(d) Determine the mean effective pressure.

11.41   Sketch on the T–s diagram an air-standard Diesel cycle, then superimpose on it a second air-standard Diesel cycle having the same inlet conditions and same amount of heat addition but a higher peak cycle pressure. Show by comparison of areas on the T–s diagram that the second cycle has a higher thermal efficiency than the first.

11.42   Sketch on the T–s diagram an air-standard Otto cycle and an air-standard Diesel cycle having the same compression ratio, the same amount of heat addition, and the same inlet conditions. Show by comparison of areas on the T–s diagram that the Otto cycle has a higher thermal efficiency than the Diesel cycle.

11.43   Sketch on the T–s diagram an air-standard Otto cycle and an air-standard Diesel cycle having the same inlet conditions, the same amount of heat rejection and the same peak cycle pressure. Show by comparison of areas on the T–s diagram that the Diesel cycle has a higher thermal efficiency than the Otto cycle.

11.44   An air-standard Diesel cycle operates with a compression ratio of 15. At the beginning of compression, the air is at 14.7 psia and 80°F. The maximum temperature of the cycle is 3000°F. Determine
(a) the cutoff ratio.
(b) the cycle thermal efficiency.
(c) the mean effective pressure.

11.45   Two Diesel cycles (cycle A and cycle B) are to be designed with the same temperature at start of the compression process, the same pressure at the start of the compression process, and the same compression ratio. But cycle A will have a larger cutoff ratio than cycle B.

(a) Which cycle will have higher thermal efficiency?

(b) Which cycle will have larger mean effective pressure?

Support your conclusion with logical reasoning. Make good use of the T–s and p–v diagrams.

11.46 An ideal Stirling cycle with perfect regeneration, using a monatomic ideal gas as the working fluid, operates between the limits of 500 and 50°C. The maximum cycle pressure is 1.0 MPa, and the minimum cycle pressure is 101 kPa. Determine

(a) the cycle thermal efficiency.

(b) the amount of heat addition, in kJ/kmol.

(c) the amount of cycle net work, in kJ/kmol.

11.47 A gas cycle consisting of two reversible constant-pressure processes and two reversible constant-temperature processes is known as the *Ericsson cycle*.

(a) Sketch this cycle on the T–s diagram.

(b) Show that with perfect regeneration, the thermal efficiency of an air-standard Ericsson cycle operating between the temperatures of T_H and T_L is identical to that of a Carnot cycle operating between the same temperature limits.

11.48 Calculate the thermal efficiency of an air-standard Brayton cycle for the following pressure ratios: 4, 6, 8, 10, 12, and 14. Plot cycle thermal efficiency against pressure ratio. What is the pressure ratio for maximum cycle thermal efficiency?

11.49 An air-standard Brayton cycle operates between the temperatures of 80°F and 1600°F. Calculate the cycle net work in Btu/lbm for the following pressure ratios: 4, 6, 8, 10, 12, and 14. Plot cycle net work against pressure ratio. What is the pressure ratio for maximum cycle net work?

11.50 A closed-cycle, gas-turbine power plant, using air as the working fluid, is to be designed according to the following specifications:

Maximum cycle temperature	$= 850°C$
Minimum cycle temperature	$= 30°C$
Adiabatic turbine efficiency	$= 90\%$
Adiabatic compressor efficiency	$= 90\%$

Neglecting pressure drops, calculate the cycle thermal efficiency for the following pressure ratios: 4, 6, 8, 10, and 12. Plot cycle thermal efficiency against pressure ratio. What is the pressure ratio for maximum cycle thermal efficiency? Assume that air is an ideal gas with constant specific heats of $c_p = 1.045$ kJ/kg · K and $k = 1.39$.

11.51 Same as Problem 11.50 except that the working fluid is argon gas.

11.52 A closed-cycle gas-turbine power plant, using air as the working fluid, is to be designed according to the following specifications:

Maximum cycle temperature	$= 850°C$
Minimum cycle temperature	$= 30°C$
Adiabatic turbine efficiency	$= 90\%$
Adiabatic compressor efficiency	$= 90\%$

Neglecting pressure drops, calculate the specific power output for the following pressure ratios: 4, 6, 8, 10, and 12. Plot specific power output against pressure ratio. What is the pressure ratio for maximum specific power? Assume that air is an ideal gas with constant specific heats of $c_p = 1.045$ kJ/kg · K and $k = 1.39$.

11.53 Same as Problem 11.52 except that the working fluid is argon gas.

11.54 A Brayton cycle operates with a pressure ratio of 5. The pressure and temperature of the working fluid at the compressor inlet are 15 psia and 80°F. The temperature of the fluid at turbine inlet is 1600°F. Assuming that air is the working fluid with constant specific heats of $c_p = 0.240$ Btu/lbm · °R and $c_v = 0.171$ Btu/lbm · °R, determine the cycle thermal efficiency and the ratio of compressor work to turbine work for the following cases:
(a) turbine efficiency is 1.0 and compressor efficiency is 1.0.
(b) turbine efficiency is 0.9 and compressor efficiency is 0.8.
(c) turbine efficiency is 0.8 and compressor efficiency is 0.9.

11.55 A closed-cycle gas turbine power plant, using air as the working fluid, is to be designed according to the following specifications:

Maximum cycle temperature	= 850°C
Minimum cycle temperature	= 30°C
Adiabatic turbine efficiency	= 90%
Adiabatic compressor efficiency	= 90%
Relative pressure drop in each heat exchanger =	3%

On the basis of maximum cycle thermal efficiency, determine
(a) the optimum pressure ratio.
(b) the ratio of compressor work to turbine work.
(c) the cycle thermal efficiency.
(d) the specific power output.
Assume air is an ideal gas with constant specific heat of $c_p = 1.045$ kJ/kg · K and $k = 1.39$.

11.56 Same as Problem 11.55 except that the working fluid is argon gas.

11.57 Same as Problem 11.55 except that the design is on the basis of maximum specific power.

11.58 Same as Problem 11.57 except that the working fluid is argon gas.

11.59 Calculate the second-law efficiency for the plant of Problem 11.55 if the heat source is a heat reservoir at 1200 K and the heat sink is the environment at 300 K.

11.60 Calculate the second-law efficiency for the plant of Problem 11.57 if the heat source is a heat reservoir with 1200 K and the heat sink is the environment at 300 K.

11.61 An air-standard Brayton cycle with perfect regeneration operates between the temperatures of 1600°F and 80°F. Calculate the cycle thermal efficiency for the following pressure ratios: 2, 4, 6, 8, 10, and 12. Plot cycle thermal efficiency against pressure ratio.

11.62   A regenerative gas-turbine power plant is to be designed according to the following specifications:

Maximum cycle temperature	$= 850°C$
Minimum cycle temperature	$= 30°C$
Adiabatic turbine efficiency	$= 90\%$
Adiabatic compressor efficiency	$= 90\%$
Regenerator effectiveness	$= 90\%$
Relative pressure drop in each heat-transfer circuit $=$	3%

Assume that air is an ideal gas with constant specific heats of $c_p = 1.045$ kJ/kg $\cdot$ K and $k = 1.39$. Calculate the cycle thermal efficiency for the pressure ratios of 2, 4, 6, 8, and 10. What is the pressure ratio for maximum cycle thermal efficiency?

11.63   A regenerative gas-turbine power plant is to be designed according to the following specifications:

Maximum cycle temperature	$= 1000°C$
Minimum cycle temperature	$= 25°C$
Adiabatic turbine efficiency	$= 90\%$
Adiabatic compressor efficiency	$= 90\%$
Regenerator effectiveness	$= 90\%$
Relative pressure drop in each heat-transfer circuit $=$	3%

Assume air is an ideal gas with constant specific heats of $c_p = 1.045$ kJ/kg K and $k = 1.39$. On the basis of maximum cycle thermal efficiency, determine
   (a) the cycle thermal efficiency.
   (b) the second-law efficiency of the plant if the heat source is a heat reservoir at 1100°C and the heat sink is the environment at 298 K.

11.64   A regenerative gas-turbine cycle operates between the temperatures of 1600°F and 80°F. The pressure ratio is 4. The turbine efficiency is 0.85 and the compressor efficiency is 0.80. The regenerator effectiveness is 0.80. Assuming that air is the

working fluid and using data from Keenan and Kaye's *Gas Tables*, determine
(a) the cycle thermal efficiency.
(b) the airflow rate for 1.0 kW of net power output.

11.65 Repeat Problem 11.64 for a pressure ratio of 2.

11.66 Repeat Problem 11.64 for a pressure ratio of 6.

11.67 A steam-turbine and gas-turbine combined cycle power plant (see Fig. 11.27) is to be designed according to the following specifications:
Specifications for the gas-turbine unit:

Power output	= 70,000 kW
Maximum cycle temperature	= 1350 K
Minimum cycle temperature	= 289 K
Adiabatic turbine efficiency	= 90%
Adiabatic compressor efficiency	= 90%
Pressure ratio	= 10
Temperature of air leaving steam generator =	150°C

Assume that air is an ideal gas with constant specific heats of $c_p = 1.045$ kJ/kg K and $k = 1.39$.
Specifications for the steam-turbine unit:

Pressure of steam at turbine inlet	= 4.0 MPa
Temperature of steam at turbine inlet	= 400°C
Pressure of steam at turbine outlet	= 7.5 kPa
Adiabatic turbine efficiency	= 75%
Number of feedwater heaters (deaerator) =	1
Pressure in deaerator	= 300 kPa

Neglect pump work.
Determine
(a) the thermal efficiency of the gas-turbine unit.

(b) the thermal efficiency of the steam-turbine unit.

(c) the thermal efficiency of the combined cycle plant.

(d) the second-law efficiency of the combined cycle plant.

11.68 Same as Problem 11.67 except that in the steam-turbine unit there will be two open feedwater heaters operating at the pressures of 100 kPa and 700 kPa.

11.69 Same as Problem 11.67 except that in the gas-turbine unit, the adiabatic turbine efficiency is 85%.

11.70 Same as Problem 11.67 except that in the gas-turbine unit, the adiabatic compressor efficiency is 85%.

11.71 Same as Example 11.16 (an indirect-fired air turbine cogeneration system) except that the saturated steam is produced at a pressure of 2.0 MPa.

11.72 Same as Example 11.16 (an indirect-fired air turbine cogeneration system) except that the saturated steam is produced at a pressure of 700 kPa.

11.73 Same as Example 11.16 (an indirect-fired air turbine cogeneration system) except that the maximum cycle temperature is 900°C.

11.74 Same as Example 11.16 (an indirect-fired air turbine cogeneration system) except that the maximum cycle temperature is 900°C and the compressor compression ratio is 10.

11.75 Write a computer program to study the effect of maximum cycle temperature on the thermal efficiency and the quality of steam at turbine outlet for an ideal Rankine cycle with superheat using H_2O as the working fluid.
Run the program for maximum cycle temperatures of 350, 400, 450, 500, and 550°C with a condenser pressure of 7.5 kPa and a steam generator pressure of 15,000 kPa. Plot the results.

11.76 Modify the computer program in Problem 11.75 to study the effect of maximum cycle pressure on the thermal efficiency and the quality of steam at turbine outlet for an ideal Rankine cycle with superheat using H_2O as the working fluid.
Run the program for maximum cycle pressures of 4000, 5000, 6000, 8000, 10,000, and 15,000 kPa with a condenser pressure of 7.5 kPa and a maximum cycle temperature of 600°C. Plot the results.

11.77 Write a computer program to study the effect of reheat pressure on the thermal efficiency and the quality of steam enter-

ing the condenser for an ideal Rankine cycle with one-stage of reheat. The working fluid is H_2O.
Run the program according to the following specifications:

Pressure in condenser:	7.5 kPa
Pressure at high-pressure turbine inlet:	15,000 kPa
Temperature at high-pressure turbine inlet:	550°C
Temperature at reheater outlet:	550°C
Reheat pressures:	3000, 4000, 5000, 6000, 8000 kPa

Plot the results.

11.78 Write a computer program to study the effect of extraction pressure on the thermal efficiency of an ideal regenerative Rankine cycle with one open feedwater heater. The working fluid is H_2O.
Run the program for operating feedwater heater pressures of 800, 1000, 1200, 1400, 1600, and 1800 kPa with a condenser pressure of 7.5 kPa, a maximum cycle pressure of 15,000 kPa and a maximum cycle temperature of 600°C.
Plot the results and determine the approximate operating pressure in the feedwater heater for maximum cycle thermal efficiency.

11.79 Write a computer program to study the effect of compression ratio on the thermal efficiency, maximum cycle temperature, maximum cycle pressure, and mep for an air-standard Otto cycle.
Run the program for compression ratios of 6, 7, 8, 9, 10, 11, and 12. Pressure and temperature of air at start of compression process are 101.325 kPa and 25°C respectively. The amount of heat addition to the cycle is 2000 kJ/kg of air. Tabulate the results.

11.80 Write a computer program to study the effect of compression ratio on the thermal efficiency, maximum cycle temperature, maximum cycle pressure, and mep for an air-standard Diesel cycle.
Run the program for compression ratios of 10, 11, 12, 13, 14, and 15. Pressure and temperature of air at start of compression process are 101.325 kPa and 25°C, respectively. The

amount of heat addition to the cycle is 2000 kJ/kg of air. Tabulate the results.

11.81 Write a computer program to study the effect of compressor compression ratio on the thermal efficiency of a closed cycle air-standard gas turbine power plant.
Run the program for a plant with the following specifications:

Maximum cycle temperature:	1200 K
Minimum cycle temperature:	300 K
Adiabatic turbine efficiency:	90%
Adiabatic compressor efficiency:	90%
Compressor compression ratios:	4, 5, 6, 7, 8, 9, and 10

Neglect pressure drops.
Plot the results and determine the approximate pressure ratio for maximum cycle thermal efficiency.

11.82 Modify the computer program in Problem 11.81 to study the effect of compressor compression ratio on the specific power output of the plant. Run the program for a plant with the same specifications given in Problem 11.81.
Plot the results and determine the approximate pressure ratio for maximum specific power output.

11.83 Write a computer program to study the effect of compressor compression ratio on the thermal efficiency of a regenerative gas turbine power plant. (Refer to Fig. 11.24.)
Run the program for a plant with the following specifications:

T_{max}/T_{min}:	4.0
Adiabatic turbine efficiency:	90%
Adiabatic compressor efficiency:	90%
Regenerator effectiveness:	90%
Specific heat ratios c_p/c_v:	1.4
Compressor compression ratios:	2, 3, 4, 5, 6, 7, and 8

Neglect pressure drops.

Plot the results and determine the approximate pressure ratio for maximum cycle thermal efficiency.

11.84 Write a computer program to study the effect of compressor compression ratio on

(a) the fuel utilization efficiency

(b) the power-to-heat ratio

(c) the second-law efficiency

for an indirect-fired air turbine cogeneration system. (Refer to Fig. 11.30 and 11.31.)

Using the same specifications given in Example 11.16, except pressure of saturated steam produced will be at 0.7 MPa instead of 1.0 MPa, run the program for compression ratios of 4, 5, 6, 7, 8, 9, and 10. Plot the results and determine the approximate pressure ratio for maximum second-law efficiency.

CHAPTER 12

Refrigeration and Heat Pump Systems

Refrigeration is a term used to denote the maintenance of a body at a temperature lower than that of its surroundings. To maintain or produce the low temperature, it is necessary to transfer heat from the cold body or refrigerated space. A refrigerator is a device that we employ to accomplish this effect by the expenditure of external energy in the form of work or heat or both. For the refrigerator to operate continuously, it must reject heat to an external sink, usually the atmosphere. Thus refrigerators may be considered as heat engines in reverse. There are many ways to produce refrigeration. In this chapter we examine a few refrigeration systems as further illustrations of the use of the first and second laws of thermodynamics and the properties of working fluids.

As mentioned in Chapter 3, the performance of a refrigerator may be measured by the coefficient of performance, β_R, which is defined as

$$\beta_R = \frac{\text{heat removed from cold body}}{\text{work input}}.$$

where heat removed from cold body is known as the refrigeration effect.

The cooling capacity of a refrigeration system is sometimes given in tons of refrigeration. A ton of refrigeration is the removal of heat from the cold body at a rate of 200 Btu/min. Thus

$$1 \text{ ton of refrigeration} = 288,000 \text{ Btu of refrigeration/d}$$

$$= 12,000 \text{ Btu of refrigeration/h}$$

$$= 200 \text{ Btu of refrigeration/min.} \quad (12.1)$$

The term "ton" is derived from the fact that the heat required to melt 1 ton of ice at 32°F in 24 h is about 288,000 Btu. In SI units,

$$1 \text{ ton of refrigeration} = 211 \text{ kJ/min}$$

$$= 3.516 \text{ kW.} \tag{12.1a}$$

The performance of a refrigeration system may also be expressed in terms of the horsepower required to produce 1 ton of refrigeration, which is simply related to the coefficient of performance according to the following equation:

$$\beta_R = \frac{(\text{tons of refrigeration})(12,000 \text{ Btu/h})}{(\text{hp input})(2545 \text{ Btu/h})}$$

$$= \frac{4.715}{\text{hp/ton}}$$

or

$$\text{hp/ton} = \frac{4.715}{\beta_R} \tag{12.2}$$

In SI units, the power required to produce 1 ton of refrigeration is related to the coefficient of performance according to the following expression:

$$\text{kW/ton} = \frac{3.516}{\beta_R}. \tag{12.2a}$$

A machine that operates as a refrigerator also delivers heat to a warm heat reservoir. If the desired effect of the machine is the heat delivered, it is known as a heat pump. Similar to the coefficient of performance of a refrigerator, the coefficient of performance of a heat pump is defined as

$$\beta_{HP} = \frac{\text{desired effect}}{\text{work input}}$$

$$= \frac{\text{heating effect}}{\text{work input}}. \tag{12.3}$$

The basic principles governing the operation and design of refrigerators are the same for heat pumps systems.

12.1 Decision Making in Refrigeration System Design

The applications of refrigeration are many and varied. In addition to the common applications of refrigeration in the manufacture of ice, in the preservation of foods, in the air-conditioning field, and in the creation of skating rinks, the refrigeration effect is also used in the preservation of blood plasma, in the low-temperature treatment of metals, and in the liquefaction of gases. Each application requires a different temperature for the refrigerated space, the determination of which is the first decision the engineer must make in the design of a refrigeration system.

Energy Source

The second law tells us that refrigeration can only be accomplished at the expense of work transfer from a work reservoir or heat transfer from a high-temperature reservoir, or both. Once the decision has been made that there is a need for a refrigeration plant of a certain capacity for a given application, we must next decide on what the energy input should be.

Energy Sink

The second law also tells us that in order for a refrigeration system to operate continuously, it must reject heat to an external sink. If the sink is the atmosphere, which is a natural one, we shall be constrained to operate between the ambient temperature and the temperature of the refrigerated space.

Working Fluid and Cycle Selection

As in the case of power cycles, the decision on what cycle is to be used depends on the decision regarding the working fluid that is to be used. The choice of a working fluid for a given cycle depends upon considerations similar to those which determine the desirability of a working fluid for a power cycle. Consistent with other constraints, we should always search for the combination of working fluid and cycle that will result in the minimum power requirement per unit of refrigeration produced.

Components Selection

Again, as in the case of power cycles, analysis of the cycle will yield the necessary information for the physical design of each component used. One of the major components in a refrigeration cycle is the compressor, of which there are two general types, the reciprocating and the centrifugal. Reciprocating compressors are best adapted to low specific volumes and high pressures; centrifugal compressors are most suitable for low pressures and high specific volumes. We thus see that the selection of a working fluid is influenced by the type of equipment that we plan to use.

12.2 Reversed Carnot Cycle

It has been shown in Chapter 3 that a Carnot refrigeration cycle (Fig. 12.1) is the most efficient theoretical refrigeration cycle. Since all processes are reversible, we have, for this cycle,

$$Q_{in} = Q_{23} = T_2(s_3 - s_2)$$

$$Q_{out} = Q_{41} = T_1(s_1 - s_4)$$

and

$$W_{net} = Q_{net} = -(T_1 - T_2)(s_3 - s_2)$$

since $s_4 = s_3$ and $s_2 = s_1$.

By definition, the coefficient of performance is given by

$$\beta_R = \frac{Q_{in}}{|W_{net}|}.$$

Thus we have, for a Carnot refrigeration cycle,

$$\beta_R = \frac{T_2}{T_1 - T_2}$$

$$= \frac{T_L}{T_H - T_L}. \tag{12.4}$$

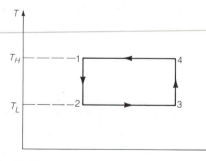

Figure 12.1 Carnot Refrigeration Cycle

Equation 12.4 indicates that β_R increases as the difference between the two temperatures decreases. For a given cold-body temperature T_L, the lower the temperature at which heat is rejected, the greater the coefficient of performance.

In practice, the disadvantage of a Carnot refrigerator is analogous to that of a Carnot heat engine, which requires bulky equipment for a given capacity. Nevertheless, the reversed Carnot cycle is valuable as a standard of comparison, since it represents perfection.

If we operate the reversed Carnot cycle as a heat pump, then

$$\beta_{HP} = \frac{T_H}{T_H - T_L}. \tag{12.5}$$

Comparing Eq. 12.4 to Eq. 12.5, we have, for a reversed Carnot cycle operating between a given temperature range,

$$\beta_{HP} = \beta_R + 1.$$

12.3 Vapor-Compression Refrigeration Cycle

The basic cycle for a mechanical refrigerator is the ideal vapor-compression cycle shown in the T–s diagram of Fig. 12.2a. The equipment arrangement that may be used to carry out the processes for this cycle is shown in Fig. 12.2b. The refrigeration effect is accomplished in the evaporator at constant pressure (process $4 \rightarrow 1$). The isentropic compression process ($1 \rightarrow 2$) is carried out in the compressor. The heat-rejection process ($2 \rightarrow 3$) is carried out at constant pressure in the condenser. The adiabatic pressure-reducing process ($3 \rightarrow 4$) across the throttling valve makes it possible to obtain the low-temperature fluid required for the performance of refrigeration. A reversible expansion engine could have been used in place of the irreversible throttling valve. In practice, however, this is not justified, because the work obtained from expanding a saturated liquid in an engine is much too small.

Example 12.1

An ideal vapor refrigeration cycle uses Freon-12 as the working fluid. The condensing temperature is 35°C and the evaporation temperature is -10°C. Determine

(a) the coefficient of performance for the cycle.
(b) the power required to produce 1 ton of refrigeration.
(c) the circulation rate of fluid for each ton of refrigeration.

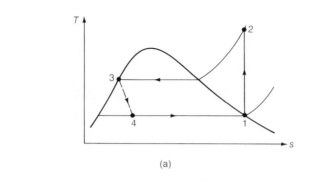

(a)

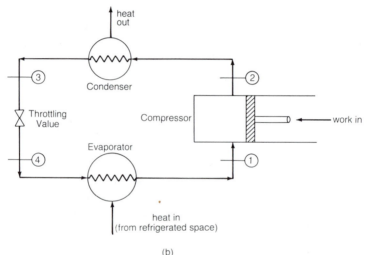

(b)

Figure 12.2 Vapor-Compression Refrigeration Cycle

Solution (refer to Fig. 12.2)

(a) From Freon-12 tables, we have

$$h_1 = h_g \text{ at } -10°C = 183.058 \text{ kJ/kg}$$

$$s_1 = s_g \text{ at } -10°C = 0.7014 \text{ kJ/kg} \cdot \text{K}$$

$$h_3 = h_f \text{ at } 35°C = 69.494 \text{ kJ/kg}$$

$$p_3 = \text{saturation pressure at } 35°C = 0.8477 \text{ MPa} = p_2$$

$$p_1 = \text{saturation pressure at } -10°C = 0.2191 \text{ MPa}.$$

Since $s_2 = s_1$, state 2 is defined by $p_2 = 0.8477$ MPa and $s_2 = 0.7014$ kJ/kg · K. From the superheated tables of Freon-12, we have

$$h_2 = 205.632 \text{ kJ/kg.}$$

For a throttling process, we have $h_3 = h_4 = 69.494$ kJ/kg. Thus the coefficient of performance is given as

$$\beta_R = \frac{h_1 - h_4}{h_2 - h_1}$$

$$= \frac{183.058 - 69.494}{205.632 - 183.058} = 5.03.$$

(b) Using Eq. 12.2a, we have

$$\text{kW/ton} = \frac{3.516}{\beta_R}$$

$$= \frac{3.516}{5.03} = 0.699.$$

(c) $$\dot{m} = \frac{211 \text{ kJ/min}}{(h_1 - h_4) \text{ kJ/kg}}$$

$$= \frac{211}{183.058 - 69.494} \text{ kg/min} = 1.858 \text{ kg/min.} \qquad \blacksquare$$

Comments

The operating temperature range in this example is typical of actual refrigeration systems designed for ordinary food-storage purposes. The power required to produce 1 ton of refrigeration for this type of plant is indeed approximately 0.7 kW.

Example 12.2

An ideal vapor refrigeration cycle uses ammonia as the working fluid. The condensing temperature is 100°F and the evaporation temperature is 20°F. Determine

(a) the coefficient of performance for the cycle.
(b) the horsepower required to produce 1 ton of refrigeration.
(c) the circulation rate of fluid for each ton of refrigeration.

Solution (refer to Fig. 12.2)

(a) From ammonia tables, we have

$$h_1 = h_g \text{ at } 20°F = 617.8 \text{ Btu/lbm}$$

$$s_1 = s_g \text{ at } 20°F = 1.2969 \text{ Btu/lbm} \cdot °R$$

$$h_3 = h_f \text{ at } 100°F = 155.2 \text{ Btu/lbm}$$

$$p_3 = \text{saturation pressure at } 100°F = 211.9 \text{ psia} = p_2$$

$$p_1 = \text{saturation pressure at } 20°F = 48.21 \text{ psia}.$$

Since $s_2 = s_1$, we have, from the superheated tables of ammonia,

$$h_2 = 710.3 \text{ Btu/lbm}.$$

For a throttling process, we have $h_3 = h_4 = 155.2$ Btu/lbm. Thus the coefficient of performance is given as

$$\beta_R = \frac{h_1 - h_4}{h_2 - h_1}$$

$$= \frac{617.8 - 155.2}{710.3 - 617.8} = 5.00.$$

(b) Using Eq. 12.2, we have

$$\text{hp/ton} = \frac{4.715}{\beta_R} = \frac{4.715}{5.00} = 0.943.$$

(c) $$\dot{m} = \frac{200 \text{ Btu/min}}{(617.8 - 155.2) \text{ Btu/lbm}} = 0.432 \text{ lbm/min.} \qquad \blacksquare$$

Comments

Comparing the results of this example with that of Example 12.1, we observe that the performances are quite comparable, except that the mass circulation rate of ammonia is much less than the mass circulation rate of Freon-12. For this reason, ammonia is usually used in industrial refrigeration plants, since it is cheaper than Freon-12.

12.4 Gas Refrigeration Cycle

We can reduce the temperature of a gas considerably by expanding it in a turbine or engine. This process has been exploited in a gas

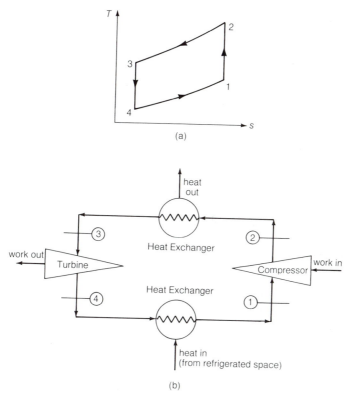

Figure 12.3 Gas Refrigeration Cycle

refrigeration plant. The basic cycle for such a system is the reversed Brayton cycle shown in the T–s diagram of Fig. 12.3a. The gas is compressed isentropically along path $1 \rightarrow 2$. It is then cooled at constant pressure along path $2 \rightarrow 3$. This is followed by an isentropic expansion process along path $3 \rightarrow 4$. Refrigeration is performed along path $4 \rightarrow 1$ at constant pressure. The equipment arrangement that may be used to carry out these processes is shown in Fig. 12.3b. This type of refrigeration system in the form of an open cycle (Fig. 12.4) has been utilized for the cooling of an aircraft.

For this ideal cycle, we have, assuming ideal gas with constant specific heat,

$$W_{in} = h_1 - h_2 = c_p(T_1 - T_2)$$

$$W_{out} = h_3 - h_4 = c_p(T_3 - T_4)$$

$$Q_{in} = h_1 - h_4 = c_p(T_1 - T_4).$$

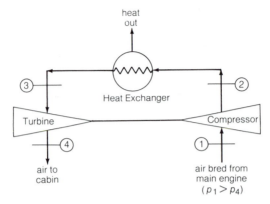

Figure 12.4 Air Refrigeration Cycle for Aircraft Cooling

Thus the coefficient of performance is given as

$$\beta_R = \frac{Q_{in}}{|W_{net}|}$$

$$= \frac{T_1 - T_4}{(T_2 - T_1) - (T_3 - T_4)}$$

$$= \frac{1}{\dfrac{T_2(1 - T_3/T_2)}{T_1(1 - T_4/T_1)} - 1}.$$

For the isentropic processes, we have

$$\frac{T_2}{T_1} = \left(\frac{p_2}{p_1}\right)^{(k-1)/k}$$

and

$$\frac{T_3}{T_4} = \left(\frac{p_3}{p_4}\right)^{(k-1)/k}$$

But $p_2 = p_3$ and $p_4 = p_1$. Thus $T_2/T_1 = T_3/T_4$ and $T_3/T_2 = T_4/T_1$. Consequently, we have

$$\beta_R = \frac{1}{(T_2/T_1) - 1} = \frac{1}{(T_3/T_4) - 1}$$

$$= \frac{1}{r_p^{(k-1)/k} - 1},$$
(12.6)

where $r_p = p_2/p_1 = p_3/p_4$ is the pressure ratio.

Equation 12.6 indicates that the coefficient of performance is increased as the pressure ratio r_p is decreased. Unfortunately, the volume of the gas also increases when the pressure ratio is reduced, thereby making this type of plant very bulky when the pressure ratio is too low.

Example 12.3

Consider the design of an ideal air refrigeration cycle according to the following specifications:

Pressure of air at compressor inlet $= 101 \text{ kPa}$

Pressure of air at turbine inlet $= 404 \text{ kPa}$

Temperature of air at compressor inlet $= -6°C$

Temperature of air at turbine inlet $= 27°C.$

Determine

(a) the coefficient of performance for the cycle.
(b) the power required to produce 1 ton of refrigeration.
(c) the air-circulation rate for each ton of refrigeration.

Solution (refer to Fig. 12.3)

(a) Using Eq. 12.6, we have, with $k = 1.4$,

$$\beta_R = \frac{1}{r_p^{(k-1)/k} - 1}$$

$$= \frac{1}{4^{(1.4-1)/1.4} - 1} = \frac{1}{1.486 - 1}$$

$$= 2.057.$$

(b) Using Eq. 12.2a, we have

$$\text{kW/ton} = \frac{3.516}{\beta_R} = \frac{3.516}{2.057} = 1.709$$

(c)
$$\frac{T_3}{T_4} = \left(\frac{p_3}{p_4}\right)^{(k-1)/k} = 1.486$$

$$T_4 = \frac{300}{1.486} = 201.9 \text{ K}$$

$$\frac{T_2}{T_1} = \left(\frac{p_2}{p_1}\right)^{(k-1)/k} = 1.486$$

$$T_2 = 267 \times 1.486 = 396.8 \text{ K}$$

$$\dot{m} = \frac{211 \text{ kJ/min}}{(h_1 - h_4) \text{ kJ/kg}}$$

$$= \frac{211}{1.0038(267 - 201.9)} = 3.229 \text{ kg/min.} \qquad \blacksquare$$

Comment

This type of refrigeration system requires the circulation of a large volume of gas at fairly low pressures. This is why in practice the compressor is of the centrifugal type.

12.5 Gas Refrigeration Cycle with Real Compressor, Turbine, and Pressure Drops in Heat Exchangers

As in the case of gas-turbine power cycle, the gas refrigeration cycle is also sensitive to inefficiencies in the compressor and turbine. A realistic design must take those irreversibilities into consideration. In this section we examine the performance of a gas refrigeration cycle with real compressor and turbine and pressure drops in the heat exchangers. With reference to Fig. 12.5, we have, assuming ideal gas

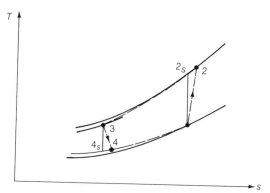

Figure 12.5 Air Refrigeration Cycle with Irreversible Compressor, Turbine, and Pressure Drops in Heat Exchangers

with constant specific heats,

$$w_c = h_1 - h_2$$

$$= c_p(T_1 - T_2)$$

$$= -\frac{c_p T_1}{\eta_C} (r_p^{(k-1)/k} - 1) \qquad (12.7)$$

where $r_p = p_2/p_1$, the compressor compression ratio.

The turbine expansion ratio, p_3/p_4, may be expressed in terms of the compressor compression ratio r_p and pressure drops to be used in each of the two heat exchangers. Making use of the idea of relative pressure drop introduced in Section 11.16, we have

$$p_3 = \beta_{23} p_2 \qquad (12.8)$$

$$p_1 = \beta_{41} p_4, \qquad (12.9)$$

where β_{23} is the pressure drop factor in the high-temperature heat exchanger, and β_{41} is the pressure drop factor in the low-temperature heat exchanger.

Combining Eqs. 12.8 and 12.9 gives us

$$\frac{p_3}{p_2} \times \frac{p_1}{p_4} = \beta_{23} \beta_{41}$$

or

$$\frac{p_3}{p_4} = \beta_{23}\,\beta_{41}\,\frac{p_2}{p_1} = \beta_{23}\,\beta_{41}\,r_p, \tag{12.10}$$

where p_3/p_4 is the turbine expansion ratio.

For each β we have

$$\beta = 1 - \frac{\Delta p}{p},$$

where $(\Delta p)/p$ is known as the relative pressure drop for the heat transfer circuit.

The turbine work may then be given as

$$w_T = h_3 - h_4$$

$$= c_p(T_3 - T_4)$$

$$= \eta_T\,c_p\,T_3\left[1 - \frac{1}{(\beta_{23}\,\beta_{41}\,r_p)^{(k-1)/k}}\right]. \tag{12.11}$$

The cycle net work is then given by

$$w_{\text{net}} = w_T + w_c$$

$$= \eta_T\,c_p\,T_3\left[1 - \frac{1}{(\beta_{23}\,\beta_{41}\,r_p)^{(k-1)/k}}\right] - \frac{c_p\,T_1}{\eta_C}\,(r_p^{(k-1)/k} - 1). \tag{12.12}$$

The refrigeration effect is given by

$$q_L = q_{\text{in}} = h_1 - h_4 = c_p(T_1 - T_4). \tag{12.13}$$

From turbine work,

$$T_4 = T_3 - \eta_T\,T_3\left[1 - \frac{1}{(\beta_{23}\,\beta_{41}\,r_p)^{(k-1)/k}}\right]. \tag{12.14}$$

Thus

$$q_L = c_p\left\{T_1 - T_3\left[1 - \eta_T\left(1 - \frac{1}{(\beta_{23}\,\beta_{41}\,r_p)^{(k-1)/k}}\right)\right]\right\}. \tag{12.15}$$

The coefficient of performance of the cycle may then be given as

$$\beta_R = \frac{|q_L|}{|w_{net}|}$$

$$= \frac{T_1 - T_3\left\{1 - \eta_T\left[1 - \dfrac{1}{(\beta_{23}\beta_{41}r_p)^{(k-1)/k}}\right]\right\}}{\dfrac{T_1}{\eta_C}[r_p^{(k-1)/k} - 1] - \eta_T T_3\left[1 - \dfrac{1}{(\beta_{23}\beta_{41}r_p)^{(k-1)/k}}\right]}, \qquad (12.16)$$

which would reduce to the ideal case given by Eq. 12.6 if $\eta_T = \eta_C = 100\%$ with no pressure drops.

Example 12.4

An air refrigeration system is to be designed according to the following specifications:

Pressure of air at compressor inlet	$= 101$ kPa
Pressure of air at compressor outlet	$= 404$ kPa
Temperature of air at compressor inlet	$= -6°C$
Temperature of air at turbine inlet	$= 27°C$
Adiabatic compressor efficiency (η_C)	$= 85\%$
Adiabatic turbine efficiency (η_T)	$= 85\%$
Relative pressure drop in each heat exchanger $=$	3%

Assuming that air is an ideal gas with constant specific heats of $c_p = 1.0038$ kJ/kg $\cdot$ K and $k = 1.4$, determine

(a) the coefficient of performance for the cycle.
(b) the power required to produce 1 ton of refrigeration.
(c) the air-circulation rate for each ton of refrigeration.

Solution (refer to Fig. 12.5)

(a) The turbine expansion ratio is given as

$$\frac{p_3}{p_4} = \beta_{23}\beta_{41}r_p$$

$$= 0.97 \times 0.97 \times 4 = 3.7636.$$

Using Eq. 12.16, we have

$$\beta_R = \frac{267 - 300\left\{1 - 0.85\left[1 - \dfrac{1}{(3.7636)^{(1.4-1)/1.4}}\right]\right\}}{\dfrac{267}{0.85}\left[4^{(1.4-1)/1.4} - 1\right] - 0.85 \times 300\left[1 - \dfrac{1}{(3.7636)^{(1.4-1)/1.4}}\right]}$$

$$= 0.6554.$$

(b)
$$\text{kW/ton} = \frac{3.516}{\beta_R}$$

$$= \frac{3.516}{0.6554} = 5.365.$$

(c)
$$\dot{m} = \frac{211 \text{ kJ/min}}{(h_1 - h_4) \text{ kJ/kg}}$$

$$= \frac{211 \text{ kJ/min}}{c_p(T_1 - T_4) \text{ kJ/kg}}.$$

Now

$$T_4 = T_3 - \eta_T T_3\left[1 - \frac{1}{(\beta_{23}\beta_{41}r_p)^{(k-1)/k}}\right]$$

$$= 300 - 0.85 \times 300\left[1 - \frac{1}{(3.7636)^{(1.4-1)/1.4}}\right] \text{K} = 219.63 \text{ K}.$$

Thus

$$\dot{m} = \frac{211 \text{ kJ/min}}{1.0038(267 - 219.63) \text{ kJ/kg}} = 4.437 \text{ kg/min}.$$

Comments

Comparing the results of this example with Example 12.3, we note that cycle performance is considerably reduced, from an ideal value of $\beta_R = 2.05$ to an actual value of $\beta_R = 0.6554$. Furthermore, the reader can show that the major reduction in performance is due primarily to irreversibilities in the compressor and turbine, as even with no pressure drops, the coefficient of performance is still only 0.7278.

12.6 Reversed Brayton Cycle with Regenerative Heat Transfer

In the reversed Brayton cycle shown in Fig. 12.3, we observe that we could obtain a colder gas at the turbine outlet if we could cool the gas below T_3 before it enters the turbine. This may be done by the use of a regenerator, as shown in Fig. 12.6.

For this cycle, we have

$$W_{in} = h_1 - h_2 = c_p(T_1 - T_2)$$

$$W_{out} = h_4 - h_5 = c_p(T_4 - T_5)$$

$$Q_{in} = h_6 - h_5 = c_p(T_6 - T_5).$$

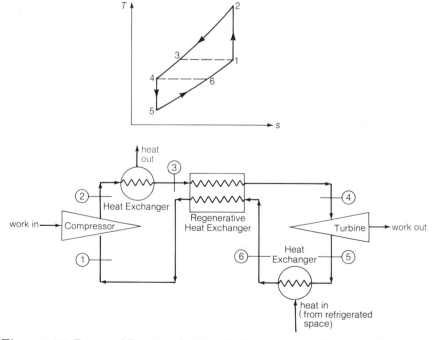

Figure 12.6 Reversed Brayton Cycle with Regenerative Heat Transfer

Thus the coefficient of performance is given as

$$\beta_R = \frac{Q_{in}}{|W_{net}|}$$

$$= \frac{T_6 - T_5}{(T_2 - T_1) - (T_4 - T_5)}.$$

With perfect regeneration, $T_6 = T_4$. Consequently,

$$\beta_R = \frac{T_4 - T_5}{(T_2 - T_1) - (T_4 - T_5)}$$

$$= \frac{T_5(T_4/T_5 - 1)}{T_1(T_2/T_1 - 1) - T_5(T_4/T_5 - 1)}.$$

But

$$\frac{T_2}{T_1} = \left(\frac{p_2}{p_1}\right)^{(k-1)/k}$$

$$\frac{T_4}{T_5} = \left(\frac{p_4}{p_5}\right)^{(k-1)/k}$$

$$p_2 = p_4 \quad \text{and} \quad p_1 = p_5.$$

Making use of these results, we may give the coefficient of performance as

$$\beta_R = \frac{T_5}{T_1 - T_5} = \frac{1}{T_1/T_5 - 1} = \frac{1}{(T_1/T_4)(T_4/T_5) - 1}$$

$$= \frac{1}{\dfrac{T_1}{T_4}\left[\dfrac{p_4}{p_5}\right]^{(k-1)/k} - 1}, \tag{12.17}$$

where p_4/p_5 is the pressure ratio of the cycle.

In Eq. 12.17, T_1 is limited by the ambient temperature and T_4 is governed by the requirement of the refrigerated space. Equation 12.17 indicates that for given values of T_1 and T_4, the coefficient of performance is increased as the pressure ratio of the cycle is decreased.

Example 12.5

Consider the design of an ideal air refrigeration cycle with perfect regeneration according to the following specifications:

Pressure of air at compressor inlet $= 101$ kPa

Pressure of air at turbine inlet $= 404$ kPa

Temperature of air at compressor inlet $= 300$ K

Temperature of air at turbine inlet $= 200$ K

Determine

(a) the coefficient of performance for the cycle.
(b) the power required to produce 1 ton of refrigeration.
(c) the air circulation rate for each ton of refrigeration.

Solution (refer to Fig. 12.6)

(a)
$$\beta_R = \frac{1}{(300/200)4^{(1.4-1)/1.4} - 1}$$

$$= 0.814.$$

(b) Using Eq. 12.2a, we have

$$kW/ton = \frac{3.516}{0.814} = 4.3.$$

(c)
$$\frac{T_4}{T_5} = \left(\frac{p_4}{p_5}\right)^{(k-1)/k} = 1.486$$

$$T_5 = \frac{200}{1.486} = 134.59 \text{ K} = -138.6°C$$

$$\dot{m} = \frac{211 \text{ kJ/min}}{c_p(T_6 - T_5) \text{ kJ/kg}}$$

$$= \frac{211}{1.0038(200 - 134.59)} \text{ kg/min}$$

$$= 3.21 \text{ kg/min.}$$

Comments

This example shows that the turbine must operate at very cold, or *cryogenic*, temperatures. The design of this kind of turbine presents some very challenging problems to engineers. Actual systems operating at these low temperatures are available commercially for use in the freezing of food.

12.7 Regenerative Gas Refrigeration Cycle with Real Compressor, Turbine, and Pressure Drops in Heat Exchangers

A realistic design of any system involving the use of turbomachinery equipment must take into consideration the irreversibilities in these machines. In this section we examine the effects of irreversibilities in compressor, turbine, regenerator, and pressure drops in heat exchangers. With reference to Fig. 12.7, we have, assuming ideal gas with constant specific heats,

$$w_c = h_1 - h_2$$

$$= c_p(T_1 - T_2)$$

$$= -\frac{c_p T_1}{\eta_C}[r_p^{(k-1)/k} - 1],$$

where $r_p = p_2/p_1$, the compressor compression ratio.

The turbine expansion ratio may be expressed in terms of the compressor compression ratio r_p and pressure drops to be used in each of the three heat exchangers. Making use of the idea of relative pressure

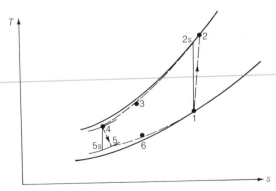

Figure 12.7 Regenerative Gas Refrigeration Cycle with Irreversible Compressor, Turbine, Regenerator, and Pressure Drops in Heat Exchangers

drop introduced in Section 11.16, we have

$$p_3 = \beta_{23} p_2 \tag{12.18}$$

$$p_4 = \beta_{34} p_3 \tag{12.19}$$

$$p_6 = \beta_{56} p_5 \tag{12.20}$$

$$p_1 = \beta_{61} p_6, \tag{12.21}$$

where β_{23} is the pressure drop factor in the high-temperature heat exchanger, β_{34} is the pressure drop factor in the hot side of the regenerator, β_{56} is the pressure drop factor in the low-temperature heat exchanger, and β_{61} is the pressure drop in the cold side of the regenerator.

Combining Eqs. 12.18, 12.19, 12.20, and 12.21 gives us

$$\frac{p_3}{p_2} \times \frac{p_4}{p_3} \times \frac{p_6}{p_5} \times \frac{p_1}{p_6} = \beta_{23} \beta_{34} \beta_{56} \beta_{61}$$

and

$$\frac{p_4}{p_5} = \beta_{23} \beta_{34} \beta_{56} \beta_{61} \frac{p_2}{p_1}$$

$$= \beta_{23} \beta_{34} \beta_{56} \beta_{61} r_p, \tag{12.22}$$

where p_4/p_5 is the turbine expansion ratio.

For each β, we have

$$\beta = 1 - \frac{\Delta p}{p}$$

where $(\Delta p)/p$ is known as the relative pressure drop for the heat-transfer circuit.

The turbine work may now be given as

$$w_T = h_4 - h_5$$

$$= c_p (T_4 - T_5)$$

$$= \eta_T c_p T_4 \left[1 - \frac{1}{(\beta_{23} \beta_{34} \beta_{56} \beta_{61} r_p)^{(k-1)/k}} \right]. \tag{12.23}$$

The cycle net work is then given as

$$w_{net} = w_T + w_c$$

$$= \eta_T c_p T_4 \left[1 - \frac{1}{(\beta_{23} \beta_{34} \beta_{56} \beta_{61} r_p)^{(k-1)/k}} \right] - \frac{c_p T_1}{\eta_c} [r_p^{(k-1)/k} - 1].$$

$$(12.24)$$

The refrigeration effect is given as

$$q_L = h_6 - h_5$$

$$= c_p (T_6 - T_5). \tag{12.25}$$

From turbine work,

$$T_5 = T_4 - \eta_T T_4 \left[1 - \frac{1}{(\beta_{23} \beta_{34} \beta_{56} \beta_{61} r_p)^{(k-1)/k}} \right]. \tag{12.26}$$

Defining the regenerator effectiveness η_{reg} as

$$\eta_{reg} = \frac{T_1 - T_4}{T_1 - T_6} \tag{12.27}$$

$$T_6 = T_1 - \frac{T_1 - T_4}{\eta_{reg}}, \tag{12.28}$$

the refrigeration effect may now be given as

$$q_L = c_p \left\{ T_1 - \frac{T_1 - T_4}{\eta_{reg}} - T_4 + \eta_T T_4 \left[1 - \frac{1}{(\beta_{23} \beta_{34} \beta_{56} \beta_{61} r_p)^{(k-1)/k}} \right] \right\}. $$

$$(12.29)$$

The coefficient of performance for the cycle may now be given as

$$\beta_R = \frac{|q_L|}{|w_{net}|}$$

$$= \frac{(T_1 - T_4)\left(1 - \dfrac{1}{\eta_{reg}}\right) + \eta_T T_4 \left[1 - \dfrac{1}{(\beta_{23} \beta_{34} \beta_{56} \beta_{61} r_p)^{(k-1)/k}} \right]}{\dfrac{T_1}{\eta_c} [r_p^{(k-1)/k} - 1] - \eta_T T_4 \left[1 - \dfrac{1}{(\beta_{23} \beta_{34} \beta_{56} \beta_{61} r_p)^{(k-1)/k}} \right]}$$

$$(12.30)$$

which would be reduced to the ideal case given by Eq. 12.17 if $\eta_T = \eta_C = \eta_{reg} = 100\%$ with no pressure drops in heat exchangers.

If relative pressure drop is small, each pressure-drop factor would be closed to unity. Then the quantity $(\beta_{23}\beta_{34}\beta_{56}\beta_{61})^{(k-1)/k}$ would also be close to unity. Then Eq. 12.30 indicates that the performance of a regenerative gas refrigeration cycle is not very sensitive to small pressure drops in heat exchangers.

Example 12.6

A regenerative air refrigeration cycle is to be designed according to the following specifications:

Pressure of air at compressor inlet	= 101 kPa
Pressure of air at compressor outlet	= 404 kPa
Temperature of air at compressor inlet	= 300 K
Temperature of air at turbine inlet	= 200 K
η_T, adiabatic turbine efficiency	= 85%
η_C, adiabatic compressor efficiency	= 85%
η_{reg}, regenerator effectiveness	= 85%
Relative pressure drop in each heat-transfer circuit =	3%

Assuming that air is an ideal gas with constant specific heats of $c_p = 1.0038$ kJ/kg · K and $k = 1.4$, determine

(a) the coefficient of performance for the cycle.
(b) the power required to produce 1 ton of refrigeration.
(c) the air circulation rate for each ton of refrigeration.

Solution (refer to Fig. 12.7)

The turbine expansion ratio is given as

$$\frac{p_4}{p_5} = \beta_{23}\beta_{34}\beta_{56}\beta_{61}r_p$$

$$= 0.97 \times 0.97 \times 0.97 \times 0.97 \times 4 = 3.5412.$$

(a) Using Eq. 12.30, we have

$$\beta_R = \frac{(300 - 200)\left(1 - \dfrac{1}{0.85}\right) + 0.85 \times 200\left[1 - \dfrac{1}{(3.5412)^{(1.4-1)/1.4}}\right]}{\dfrac{300}{0.85}\left[4^{(1.4-1)/1.4} - 1\right] - 0.85 \times 200\left[-\dfrac{1}{(3.5412)^{(1.4-1)/1.4}}\right]}$$

$$= 0.283.$$

(b)
$$\text{kW/ton} = \frac{3.516}{\beta_R} = \frac{3.516}{0.283} = 12.42.$$

(c)
$$\dot{m} = \frac{211 \text{ kJ/min}}{c_p(T_6 - T_5) \text{ kJ/kg}}.$$

Now

$$T_6 = T_1 - \frac{T_1 - T_4}{\eta_{\text{reg}}}$$

$$= 300 - \frac{300 - 200}{0.85} = 182.35 \text{ K}$$

$$T_5 = T_4 - \eta_T T_4\left[1 - \frac{1}{(\beta_{23}\beta_{34}\beta_{56}\beta_{61}r_p)^{(k-1)/k}}\right]$$

$$= 200 - 0.85 \times 200\left[1 - \frac{1}{(3.5412)^{(1.4-1)/1.4}}\right]$$

$$= 148.46 \text{ K}$$

Thus

$$\dot{m} = \frac{211}{1.0038(182.35 - 148.46)} \text{ kg/min}$$

$$= 6.202 \text{ kg/min.} \qquad \blacksquare$$

Comments

Comparing the result of this example with Example 12.5, we note that the cycle performance is considerably reduced, from an ideal value of $\beta_R = 0.814$ to an actual value of $\beta_R = 0.283$. Further analysis of the problem would yield the following results:

$\beta_R = 0.814$ if $\eta_T = \eta_C = \eta_{reg} = 100\%$ with no pressure drops
(Example 12.5)

$= 0.712$ if $\eta_T = \eta_C = \eta_{reg} = 100\%$ with 3% relative pressure
drops in each heat-transfer circuit

$= 0.594$ if $\eta_T = \eta_C = 100\%$ with no pressure drops but
$\eta_{reg} = 85\%$

$= 0.430$ if $\eta_{reg} = 100\%$ with no pressure drops but
$\eta_T = \eta_C = 85\%$.

These results show that the coefficient of performance of an air refrigeration cycle is relatively small. It is also very sensitive to irreversibilities in compressor, turbine, and regenerator.

12.8 Production of Refrigeration Using Heat Instead of Work

According to the second law of thermodynamics, we can produce refrigeration at the expense of either work or heat, or both. The refrigeration effect of each of the systems we have studied so far is obtained by using work which is "high-grade" energy. It is of interest to explore the idea of producing refrigeration by using heat, since heat is much "cheaper" than work from the entropy-creation point of view. Such a concept is shown in Fig. 12.8.

Taking the refrigerator as the system, we have, according to the first law, for each cycle,

$$Q_H + Q_L - Q_0 = 0$$

or

$$Q_0 = Q_H + Q_L.$$

From the second law, we have

$$(\Delta S)_{system} + (\Delta S)_{surroundings} \geq 0.$$

Since the refrigerator operates in a cycle,

$$(\Delta S)_{system} = 0.$$

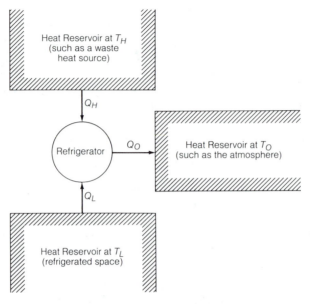

Figure 12.8 Using Heat Instead of Work to Produce Refrigeration

The only surroundings are the three heat reservoirs. Using Eq. 3.11a, we find that

$$(\Delta S)_{\text{surroundings}} = (\Delta S)_{\text{heat reservoir at } T_H} + (\Delta S)_{\text{heat reservoir at } T_L}$$

$$+ (\Delta S)_{\text{heat reservoir at } T_0}$$

$$= -\frac{Q_H}{T_H} - \frac{Q_L}{T_L} + \frac{Q_H + Q_L}{T_0}.$$

Therefore, we get

$$-\frac{Q_H}{T_H} - \frac{Q_L}{T_L} + \frac{Q_H + Q_L}{T_0} \geq 0. \qquad (12.31)$$

Rearranged, Eq. 12.31 becomes

$$\frac{Q_L}{Q_H} \leq \frac{T_L}{T_0 - T_L} \frac{T_H - T_0}{T_H}. \qquad (12.32)$$

The left-hand side of Eq. 12.32 is the ratio of "what you get" (refrigeration effect Q_L) to "what you pay" (heat supplied Q_H), which is simply the definition of the coefficient of performance. Note that T_H must be greater than T_0. For a reversible refrigerator, we have

$$\frac{Q_L}{Q_H} = \frac{T_L}{T_0 - T_L} \frac{T_H - T_0}{T_H}. \tag{12.33}$$

Equation 12.33 may be used as the standard of comparison for this kind of refrigeration system, just as the coefficient of performance of a Carnot refrigerator is used as the standard of comparison for mechanical refrigeration systems. It is of interest to note that the quantity $T_L/(T_0 - T_L)$ is simply the coefficient of performance for a Carnot refrigeration cycle operating between the temperature T_L and T_0, and the quantity $(T_H - T_0)/T_H$ is simply the thermal efficiency of a Carnot heat engine operating between the temperature T_H and T_0.

12.9 Absorption Refrigeration System

The *absorption refrigeration system* shown in Fig. 12.9 is an example of a refrigeration system in which heat instead of work is employed to produce a refrigeration effect. We observe that the compressor is replaced with a vapor generator, an absorber, and a liquid pump which consumes very little work.

The key to success in this system depends on the ability of liquids (absorbents) to absorb certain vapors (refrigerants). In an ammonia absorption system, ammonia is the refrigerant and water is the

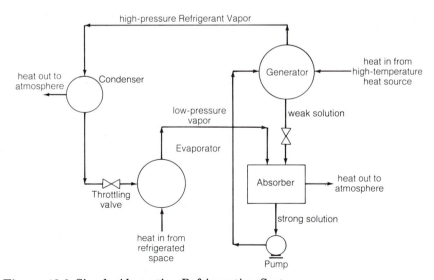

Figure 12.9 Simple Absorption Refrigeration System

absorbent. In a water-vapor absorption system, water is the refrigerant and lithium bromide is a popular absorbent. The search for the best combination of absorbent and refrigerant naturally requires a good understanding of the thermodynamic behavior of solutions, which behavior is different from that of simple compressible substances.

To illustrate the principle of operation of an absorption refrigeration system, let us consider the system shown in Fig. 12.9 to be an ammonia absorption system. The high-pressure refrigerant vapor is condensed to liquid in the condenser. The liquid is throttled to a liquid–vapor mixture. The evaporation of the refrigerant in the evaporator produces the refrigeration effect. The low-pressure vapor is absorbed by a weak solution to form a strong solution in the absorber. The strong solution is then pumped into the generator, which operates at high pressure. Heat is then added to the generator to drive out the refrigerant vapor from the solution. The ammonia absorption process liberates heat which must be removed from the absorber to keep the temperature sufficiently low for absorption to take place.

For applications in which refrigeration temperatures below 32°F are not needed, an absorption refrigeration system using water as the refrigerant and lithium bromide solution as the absorbent has been successfully developed. In recent years, this system has become quite prominent in refrigeration for air conditioning.

For a detailed treatment of different types of refrigeration systems, books on this subject should be consulted.[1]

Example 12.7

A refrigerator operates according to the scheme shown in Fig. 12.8. If $T_H = 245°F$, $T_0 = 70°F$, and $T_L = 43°F$, what is the ideal coefficient of performance?

Solution

Using Eq. 12.33, the ideal coefficient of performance is given as

$$\beta_{ideal} = \frac{T_L}{T_0 - T_L} \frac{T_H - T_0}{T_H}$$

$$= \frac{503}{530 - 503} \frac{705 - 530}{705} = 4.62. \qquad \blacksquare$$

[1] J. L. Threlkeld, *Thermal Environmental Engineering*, Prentice-Hall, Inc., Englewood Cliffs, N.J., 1970; B. H. Jennings, *Environmental Engineering—Analysis and Practice*, International Textbook Company, Scranton, Pa., 1970; W. F. Stoeker, *Refrigeration and Air Conditioning*, McGraw-Hill Book Company, New York, 1958.

Comments

The temperatures used in this example correspond to those used in a practical absorption refrigeration system using water as the refrigerant and lithium bromide solution as the absorbent. In such a system, the heating steam will be condensing at 245°F, corresponding to a steam pressure of about 27 psia. The refrigerant will boil at 43°F. The steam consumption for a 100-ton unit of this type is about 20 lb/h for each ton of refrigeration.

12.10 Production of Low-Temperature Heat Using a Power Generation–Heat Pump System [2]

It will be recalled that a system could be called a refrigerator or a heat pump depending on the purpose of the system. If the purpose of the system is to remove heat from a low-temperature heat reservoir, it is a refrigerator. If the purpose of the system is to deliver heat (such as for space heating) by pumping heat from a low-temperature heat reservoir (such as the atmosphere), it is a heat pump. Consequently, the methodology of analysis for heat-pump systems is identical to that used for the study of refrigeration systems.

Since a very large fraction of fuel consumption in the United States is devoted to providing low-temperature heat for such mundane processes as hot-water heating (50 to 60°C), cooking (100 to 200°C), space heating (20°C), and certain industrial applications, there is considerable interest at the present time in the study of using heat-pump systems for the delivery of such low-temperature heat. Suppose that the temperature of the heat required for a given process is T_p. If a high-temperature heat source at T_H (such as in the burning of a high-quality fuel) is available, the simplest way to deliver the required process heat is to do it by pure conduction. But this is a highly irreversible process. Let us explore the idea of delivering the process heat according to the concept shown in Fig. 12.10.

Taking the heat pump as the system, we have, according to the first law, for each cycle,

$$Q_H + Q_0 - Q_p = 0. \tag{1}$$

From the second law, we have

$$(\Delta S)_{\text{system}} + (\Delta S)_{\text{surroundings}} \geq 0. \tag{2}$$

[2] C. A. Berg, "A Technical Basis for Energy Conservation," *Mechanical Engineering*, May 1974.

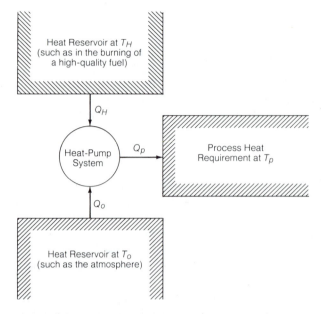

Figure 12.10 Using High-Temperature Heat Instead of Work to Produce Low-Temperature Heat

Since the heat pump operates in a cycle,

$$(\Delta S)_{\text{system}} = 0. \tag{3}$$

The only surroundings are the three heat reservoirs. Thus

$$(\Delta S)_{\text{surroundings}} = (\Delta S)_{\text{heat reservoir at } T_H} + (\Delta S)_{\text{heat reservoir at } T_O}$$

$$+ (\Delta S)_{\text{heat reservoir at } T_p}$$

$$= -\frac{Q_H}{T_H} - \frac{Q_0}{T_0} + \frac{Q_p}{T_p}. \tag{4}$$

Combining Eqs. 1, 2, 3, and 4, we get

$$\frac{Q_p}{Q_H} \leq \frac{T_p}{T_p - T_0} \frac{T_H - T_0}{T_H}. \tag{12.34}$$

The left-hand side of Eq. 12.34 is the ratio of "what you get" to "what you pay," which is simply the definition of the coefficient of performance for a heat pump. The equal sign in Eq. 12.34 is for the case of a reversible system. Thus, for a reversible heat-pump system, we get

$$\frac{Q_p}{Q_H} = \frac{T_p}{T_p - T_0} \frac{T_H - T_0}{T_H}. \qquad (12.35)$$

Equation 12.35 is very similar to Eq. 12.33. The quantity $T_p/(T_p - T_0)$ is simply the coefficient of performance for a Carnot heat-pump cycle operating between the temperature of T_p and T_0, and the quantity $(T_H - T_0)/T_H$ is simply the thermal efficiency of a Carnot heat engine operating between the temperature of T_H and T_0. Thus the heat-pump system scheme shown in Fig. 12.10 may be conceived as a system having a heat pump driven by a heat engine. In other words, the scheme is really a power generation–heat pump system employed for the production of low-temperature heat.

12.11 Heat Pumps

A heat pump is a device that transfers heat from a low-temperature body to a high-temperature body. In principle, it can operate on any of the refrigeration cycles and it can use any of the common refrigerants as the working fluid. In recent years, the heat pump has been increasingly used to provide heating for homes in the winter. It has also been used in industrial applications that require heat sources. This section is intended only as an introduction to the subject. For a detailed treatment of different types of heat pump systems, books on this subject should be consulted.[3]

The heat rejected by a heat pump cycle is the desired effect of the system. Since the heating effect of a heat pump is always greater than the energy input needed, it appears to be an effective heater. But a heating effect can always be provided easily by other means. Thus the application of a heat pump would require a careful analysis of all the

[3] Harry J. Sauer, Jr., and Ronald H. Howell, *Heat Pump Systems*, John Wiley & Sons, Inc., New York, 1983.

factors involved. But it has been shown to be an attractive alternative to the common electric resistance heating in many cases.

The coefficient of performance of a heat pump, thus the work required, is sensitive to the temperature difference between the heat source and the temperature of heat to be delivered. In the case of a vapor compression heat pump, the high temperature will govern the condensing pressure of the working fluid, and the low temperature will determine the evaporating pressure. To minimize the compressor work required, we should design the heat pump to operate with a high pressure in the evaporator. Thus the temperature of the heat source has a pronounced effect on the thermodynamic performance of a heat pump. The usual sources of heat are the atmospheric air, earth, and natural bodies of water. Since well water has a relatively constant temperature, it is an excellent heat source for the heat pump. On the other hand, the use of the atmospheric air as the heat source creates a problem in the operation of a heat pump. During very cold weather, the evaporator will operate at a very low temperature with the consequence that the coefficient of performance will be reduced and the work required will be increased.

Problems

12.1   A reversed Carnot cycle is designed for heating and cooling. The power supplied is 25 kW. The coefficient of performance for cooling is 4.0. Determine
(a) the ratio of T_H/T_L of the two heat reservoirs involved.
(b) the refrigeration effect in tons of refrigeration.
(c) the coefficient of performance for heating.

12.2   A refrigeration system having a cooling capacity of 10 tons requires 12 hp for its operation. What is the coefficient of performance of the system?

12.3   Refrigeration is desired at 10°F. Heat may be rejected at 80°F. Calculate the minimum horsepower required to produce 1 ton of refrigeration.

12.4   A heat pump is to deliver 100,000 Btu/h at 70°F. If the available source of heat for the heat pump is at 20°F, what is the minimum horsepower required?

12.5   A vapor compression refrigeration cycle, using Freon-12 as the working fluid, operates with an evaporator temperature of $-20°C$ and a condenser temperature of 40°C. The fluid enters the compressor as saturated vapor and enters the expansion

valve as saturated liquid. The desired refrigeration effect is 20 tons. If the adiabatic compressor efficiency is 100%, determine
(a) the power required to drive the compressor.
(b) the coefficient of performance.
(c) the circulation rate of Freon-12.

12.6 Same as Problem 12.5 except that the adiabatic compressor efficiency is 85%.

12.7 An ideal refrigeration cycle, using ammonia as the working fluid, operates between a pressure of 180 psia in the condenser and a pressure of 20 psia in the evaporator. The fluid enters the compressor as saturated vapor and leaves the condenser as saturated liquid. Determine the coefficient of performance and refrigeration effect in Btu/lbm for the following cases:
(a) Pressure reduction is carried out in an isentropic expansion engine.
(b) Pressure reduction is accomplished with the use of a throttling valve.

12.8 Repeat Problem 12.7 using Freon-12 as the working fluid.

12.9 Sketch on the T–s diagram an ideal vapor-compression refrigeration cycle in which the fluid enters the compressor as saturated vapor and enters the expansion engine as saturated liquid. Superimpose on this a second ideal cycle that differs from the first one only in having a lower refrigeration temperature. Show by comparison of areas on the T–s diagram that the second cycle has a smaller coefficient of performance than the first.

12.10 Sketch on the T–s diagram an ideal vapor-compression refrigeration cycle in which the fluid enters the compressor as saturated vapor and enters the expansion engine as saturated liquid. Superimpose on this a second ideal cycle that differs from the first one only in having a lower pressure in the condenser. Show by comparison of areas on the T–s diagram that the second cycle has a larger coefficient of performance than the first.

12.11 A vapor compression cycle, using Freon-12 as the working fluid, operates with a pressure of 1.0 MPa in the condenser and a pressure of 0.15 MPa in the evaporator. The fluid enters the compressor as saturated vapor and leaves the condenser as saturated liquid. The desired refrigeration effect is 10 tons. If the adiabatic compressor efficiency is 100%, determine

(a) the power required to drive the compressor.

(b) the coefficient of performance for the cycle.

(c) the circulation rate of Freon-12.

12.12 Same as Problem 12.11 except that the adiabatic compressor efficiency is 80%.

12.13 An ideal refrigeration cycle, using Freon-12 as the working fluid, operates with a condensing temperature of 100°F and an evaporator temperature of −10°F. The fluid enters the compressor as saturated vapor and enters the throttling valve as saturated liquid. Determine

(a) the coefficient of performance.

(b) the refrigeration effect, in Btu/lbm.

12.14 Repeat Problem 12.13 using ammonia as the working fluid.

12.15 A vapor compression refrigeration cycle, using Freon-12 as the working fluid, has a desired refrigeration effect of 150 kJ/s. The refrigerant enters the compressor as saturated vapor at −15°C and leaves the condenser as a saturated liquid. The pressure ratio of the compressor is 6. If the adiabatic compressor efficiency is 100%, determine

(a) the power required to drive the compressor.

(b) the coefficient of performance for the cycle.

(c) the circulation rate of Freon-12.

12.16 Same as Problem 12.15 except that the adiabatic compressor efficiency is 85%.

12.17 A vapor-compression refrigeration cycle, using Freon-12 as the working fluid, operates between a pressure of 175 psia in the condenser and a pressure of 20 psia in the evaporator. The fluid enters the compressor as saturated vapor and enters the expansion valve as saturated liquid. The compressor efficiency is 80%. Determine

(a) the coefficient of performance.

(b) the refrigeration effect, in Btu/lbm.

12.18 Repeat Problem 12.17 using ammonia as the working fluid.

12.19 Air enters a turbine at 960 kPa and 310 K and leaves at 120 kPa. Adiabatic turbine efficiency is 80%. Power produced by the turbine is 50 kW. If the air leaving the turbine is to be used to provide refrigeration for a space to be maintained at 280 K, what is the refrigeration capacity of the air? Assume that air is an ideal gas with constant specific heats of $c_p = 1.0038$ kJ/kg · K and $k = 1.4$.

12.20 An air-conditioning unit for an aircraft receives its air from the jet engine compressor at 100 kPa. The air is cooled at constant pressure to 38°C, and then expanded isentropically to the cabin pressure of 70 kPa. This air is mixed with the cabin air and eventually exhausted to the outside. If 5.0 kJ/s of heat must be removed from the cabin to maintain its temperature at 24°C, determine

(a) the rate of air that must come from the jet engine compressor.

(b) the amount of power produced by the turbine.

12.21 An ideal air refrigeration cycle is to be designed according to the following specifications:

Pressure of air at compressor inlet   = 101 kPa

Pressure of air at compressor outlet   = 505 kPa

Temperature of air at compressor inlet = 10°C

Temperature of air at turbine inlet   = 30°C

Assuming that air is an ideal gas with constant specific heats of $c_p = 1.0038$ kJ/kg · K and $k = 1.4$, determine

(a) the coefficient of performance for the cycle.

(b) the power required to produce 1 ton of refrigeration.

(c) the air circulation rate for each ton of refrigeration.

12.22 Same as Problem 12.21 except that the pressure of air at compressor outlet is 303 kPa.

12.23 Same as Problem 12.21 except that the pressure of air at compressor outlet is 606 kPa.

12.24 An ideal reversed Brayton cycle uses air as the working fluid. At the compressor inlet, the temperature of air is 40°F. At the

turbine inlet, the temperature is 80°F. For a net work input of 2.1 Btu/lbm of air, determine
(a) the coefficient of performance.
(b) the refrigeration effect, in Btu/lbm of air.
Consider air to be an ideal gas with constant specific heats of $c_p = 0.240$ Btu/lbm · °R and $c_v = 0.171$ Btu/lbm · °R.

12.25 In the reversed air-standard Brayton cycle used for the production of refrigeration, could we replace the turbine with an expansion valve? Why?

12.26 An air refrigeration cycle is to be designed according to the following specifications (neglect pressure drops):

Pressure of air at compressor inlet	= 101 kPa
Pressure of air at compressor outlet	= 303 kPa
Temperature of air at compressor inlet =	5°C
Temperature of air at turbine inlet	= 27°C
Adiabatic turbine efficiency	= 85%
Adiabatic compressor efficiency	= 85%

Assume air is an ideal gas with constant specific heats of $c_p = 1.0038$ kJ/kg · K and $k = 1.4$. Determine
(a) the coefficient of performance for the cycle.
(b) the power required for 1 ton of refrigeration.
(c) the air circulation rate for each ton of refrigeration.

12.27 Same as Problem 12.26 except that the pressure of air at compressor outlet is 505 kPa.

12.28 Same as Problem 12.26 except that the pressure of air at compressor outlet is 202 kPa.

12.29 A simple reversed Brayton cycle, using air as the working fluid, operates between the pressures of 30 psia and 15 psia. The temperatures of air at compressor and turbine inlets are 40°F and 80°F, respectively. If the compressor and turbine efficiencies are 80%, determine

(a) the coefficient of performance.

(b) the refrigeration effect, in Btu/lbm of air.

Consider air to be an ideal gas with constant specific heats of $c_p = 0.240$ Btu/lbm · °R and $c_v = 0.171$ Btu/lbm · °R.

12.30 An air refrigeration cycle is to be designed according to the following specifications:

Pressure of air at compressor inlet	= 101 kPa
Pressure of air at compressor outlet	= 303 kPa
Temperature of air at compressor inlet	= 10°C
Temperature of air at turbine inlet	= 27°C
Adiabatic turbine efficiency	= 85%
Adiabatic compressor efficiency	= 85%
Relative pressure drop for each heat exchanger =	3%

Assume that air is an ideal gas with constant specific heats of $c_p = 1.0038$ kJ/kg · K and $k = 1.4$. Determine

(a) the coefficient of performance for the cycle.

(b) the power required for 1 ton of refrigeration.

(c) the air circulation rate for each ton of refrigeration.

12.31 Same as Problem 12.30 except that the relative pressure drop is 2% for each heat exchanger.

12.32 Same as Problem 12.30 except that the relative pressure drop is 4% for each heat exchanger.

12.33 A regenerative air refrigeration cycle is to be designed according to the following specifications:

Pressure of air at compressor inlet	= 101 kPa
Pressure of air at compressor outlet	= 303 kPa
Temperature of air at compressor inlet	= 300 K
Temperature of air at turbine inlet	= 200 K

Adiabatic turbine efficiency	= 90%
Adiabatic compressor efficiency	= 90%
Regenerator effectiveness	= 90%
Relative pressure drop in each heat-transfer circuit =	3%

Assume that air is an ideal gas with constant specific heats of $c_p = 1.0038$ kJ/kg $\cdot$ K and $k = 1.4$. Determine
(a) the coefficient of performance for the cycle.
(b) the power required for 1 ton of refrigeration.
(c) the air circulation rate for each ton of refrigeration.

12.34 Same as Problem 12.33 except that the pressure at compressor outlet is 404 kPa.

12.35 Same as Problem 12.33 except that the pressure at compressor outlet is 505 kPa.

12.36 Same as Problem 12.33 except that the pressure at compressor outlet is 202 kPa.

12.37 Same as Problem 12.33 except that the temperature at turbine inlet is 150 K.

12.38 A simple absorption refrigeration system uses water as the refrigerant and lithium bromide as the absorbent. If the water boils at a temperature of 40°F in the evaporator and condenses at a temperature of 100°F in the condenser, what is the refrigeration effect per pound of water vapor leaving the generator?

12.39 The refrigeration system shown in Fig. 12.8 may be conceived as a system having a refrigerator driven by a heat engine with the refrigerator operating between T_L and T_0 and the heat engine operating between T_H and T_0. Determine Q_L/Q_H for this power generation–refrigeration system for the following design conditions:
(a) $T_H = 245°F$; $T_0 = 70°F$; $T_L = 43°F$.
(b) The thermal efficiency of the heat engine is 60% of that of a Carnot heat engine operating between the same temperature limits.
(c) The coefficient of performance of the refrigerator is 20% of that of a Carnot refrigerator operating between the same temperature limits.

12.40 In a power generation–heat pump system, the heat pump operates between 100 and 20°C, while the heat engine operates between 1000 and 20°C. If the system is reversible, what is the amount of process heat produced for each unit of heat supplied at 1000°C?

12.41 A heat pump is driven by a heat engine operating between 1000 and 20°C. The heat pump operates between 100 and 20°C. If both the heat engine and heat pump are only 60% perfect from the thermodynamic point of view, what is the amount of process heat produced for each unit of heat supplied at 1000°C?

12.42 A heat pump, using Freon-12 as the working fluid, is to be designed according to the following specifications:

Heating effect $\qquad$ = 75,000 kJ/s

Temperature of fluid in condenser =   35°C

Temperature of fluid in evaporator =   15°C

Fluid begins the condensing process as a saturated vapor and ends as a saturated liquid. Assuming the compression process is isentropic, determine
(a) the power required to drive the compressor.
(b) the coefficient of performance for the cycle.
(c) the circulation rate of Freon-12.

12.43 Same as Problem 12.42 except that the temperature of the fluid in the evaporator is 10°C.

12.44 Same as Problem 12.42 except that the temperature of the fluid in the condenser is 30°C.

12.45 A heat pump, using Freon-12 as the working fluid, is to be designed according to the following specifications:

Heating effect $\qquad$ = 75,000 kJ/s

Pressure of fluid in evaporator =   0.22 MPa

Pressure of fluid in condenser =   1.20 MPa

Fluid enters compressor as a saturated vapor and leaves condenser as a saturated liquid. Determine, on the basis of a perfect compressor,
(a) the power required to drive the compressor.
(b) the coefficient of performance for the cycle.
(c) the circulation rate of Freon-12.
(d) the second-law efficiency if temperature of the environment is 300 K.

12.46 Same as Problem 12.45 except that the pressure of fluid in the evaporator is 0.100 MPa.

12.47 Same as Problem 12.45 except that the pressure of fluid in the condenser is 1.60 MPa.

12.48 Write a computer program to study the effect of compressor compression ratio on the performance of an air-standard refrigeration cycle. (Refer to Fig. 12.5.)

Run the program for a cycle with the same specifications given in Example 12.4 but vary the pressure of air at compressor outlet from 202 kPa to 606 kPa. Determine

(a) the coefficient of performance for the cycle.

(b) the power required to produce 1 ton of refrigeration.

(c) the air circulation rate for each ton of refrigeration.

Tabulate the results.

12.49 Write a computer program to study the effect of compressor compression ratio on the performance of a regenerative air-standard refrigeration cycle. (Refer to Fig. 12.7.)

Run the program for a cycle with the same specifications given in Example 12.6 but vary the pressure of air at compressor outlet from 202 kPa to 606 kPa. Determine

(a) the coefficient of performance for the cycle.

(b) the power required to produce 1 ton of refrigeration.

(c) the air circulation rate for each ton of refrigeration.

CHAPTER 13

Thermodynamics of a Simple Compressible Substance

The definition of a simple compressible substance and a description of some of its unique behavior were given in Chapter 4. We wish now to study such a simple but very important substance in more detail. Since the solutions of engineering problems require the evaluation of changes in such properties as internal energy, enthalpy, and entropy, a major task in thermodynamics is to deduce the basic equations that will enable us to determine such changes from measurable properties. Our primary objective in this chapter is to tackle this major task.

13.1 Important Mathematical Relations

A thermodynamic property is a point function whose differential is exact. For a simple compressible substance of fixed mass, one property may be expressed as a function of two independent properties, according to the state postulate given in Chapter 4. Consequently, the differential calculus for the function of three variables will be of great utility to us here.

When a variable z is a continuous function of x and y, we have, from calculus,

$$z = z(x, y)$$

$$dz = M\,dx + N\,dy, \qquad (13.1)$$

where

$$M = \left(\frac{\partial z}{\partial x}\right)_y$$

$$N = \left(\frac{\partial z}{\partial y}\right)_x.$$

Since the order of differentiation is immaterial, we have

$$\left(\frac{\partial M}{\partial y}\right)_x = \left(\frac{\partial N}{\partial x}\right)_y. \tag{13.2}$$

Equation 13.2 is an important mathematical relation that we will use repeatedly in this chapter. Another mathematical relation that is useful to us is

$$\left(\frac{\partial x}{\partial y}\right)_z \left(\frac{\partial y}{\partial z}\right)_x \left(\frac{\partial z}{\partial x}\right)_y = -1, \tag{13.3}$$

We shall refer to Eq. 13.3 as a cyclical relation.

13.2 Maxwell Relations

The four fundamental equations of state for a simple compressible substance in differential form were given in Chapter 4. They are repeated here for convenience.

$$du = T\,ds - p\,dv, \tag{13.4}$$

where

$$T = \left(\frac{\partial u}{\partial s}\right)_v \quad \text{and} \quad p = -\left(\frac{\partial u}{\partial v}\right)_s.$$

$$dh = T\,ds + v\,dp, \tag{13.5}$$

where

$$T = \left(\frac{\partial h}{\partial s}\right)_p \quad \text{and} \quad v = \left(\frac{\partial h}{\partial p}\right)_s.$$

$$da = -s\,dT - p\,dv, \tag{13.6}$$

where

$$s = -\left(\frac{\partial a}{\partial T}\right)_v \quad \text{and} \quad p = -\left(\frac{\partial a}{\partial v}\right)_T.$$

$$dg = -s\,dT + v\,dp, \tag{13.7}$$

where

$$s = -\left(\frac{\partial g}{\partial T}\right)_p \quad \text{and} \quad v = \left(\frac{\partial g}{\partial p}\right)_T.$$

Equations 13.4, 13.5, 13.6, and 13.7 are all of the form

$$dz = M\, dx + N\, dy,$$

in which

$$\left(\frac{\partial M}{\partial y}\right)_x = \left(\frac{\partial N}{\partial x}\right)_y.$$

It follows that we have, from Eq. 13.4,

$$\left(\frac{\partial T}{\partial v}\right)_s = -\left(\frac{\partial p}{\partial s}\right)_v. \tag{13.8}$$

In a similar manner we obtain, from 13.5, 13.6, and 13.7,

$$\left(\frac{\partial T}{\partial p}\right)_s = \left(\frac{\partial v}{\partial s}\right)_p \tag{13.9}$$

$$\left(\frac{\partial s}{\partial v}\right)_T = \left(\frac{\partial p}{\partial T}\right)_v \tag{13.10}$$

$$-\left(\frac{\partial s}{\partial p}\right)_T = \left(\frac{\partial v}{\partial T}\right)_p. \tag{13.11}$$

Equations 13.8, 13.9, 13.10, and 13.11 are known as the *Maxwell relations*. They are important in the correlation of properties of simple compressible substances, because they relate entropy to the directly measurable properties of pressure, specific volume, and temperature.

13.3 Some Quantities Derivable from *p–v–T* Data

As pointed out in Chapter 4, the properties p, v, and T are directly measurable. Specific heat data may also be determined experimentally, although not without difficulty. These two kinds of data are the most commonly available. With such data, it is possible for us to analyze and predict the behavior of many simple compressible substances. In this section we examine some of the information that can be deduced directly from p–v–T measurements.

The partial derivative $(\partial v/\partial T)_p$ is a property that can be deduced from p–v–T data. It is simply the slope of a constant-pressure line at a point in the v–T diagram. If we divide this quantity by v at the point, we arrive at a new property called the *coefficient of expansion*, β. By definition, we have

$$\beta \equiv \frac{1}{v}\left(\frac{\partial v}{\partial T}\right)_p . \tag{13.12}$$

The partial derivative $(\partial v/\partial p)_T$ is another property that can be deduced from p–v–T data. It represents the slope of a constant-temperature line at a point in the v–p diagram. Making use of this quantity, we can form another new property called the *isothermal compressibility*, κ. By definition, we have

$$\kappa \equiv -\frac{1}{v}\left(\frac{\partial v}{\partial p}\right)_T . \tag{13.13}$$

The quantity $(\partial v/\partial p)_T$ is negative for all known simple compressible substances. Consequently, κ is always positive by definition.

Applying the cyclical relation to the p–v–T equation of state, we have

$$\left(\frac{\partial p}{\partial v}\right)_T\left(\frac{\partial v}{\partial T}\right)_p\left(\frac{\partial T}{\partial p}\right)_v = -1. \tag{13.14}$$

Combining Eqs. 13.12, 13.13, and 13.14, we have

$$\left(\frac{\partial p}{\partial T}\right)_v = \frac{\beta}{\kappa}. \tag{13.15}$$

The quantity $(\partial p/\partial T)_v$ is simply the slope of a constant-specific-volume line in the p–T diagram. It can be deduced from p–v–T data directly.

For the p–v–T equation of state, we have

$$dv = \left(\frac{\partial v}{\partial T}\right)_p dT + \left(\frac{\partial v}{\partial p}\right)_T dp, \tag{13.16}$$

which may also be written as

$$dv = v\beta \, dT - v\kappa \, dp. \tag{13.16a}$$

The coefficient of expansion and the isothermal compressibility are physical properties of a simple compressible substance. Values of these properties are found in standard handbooks.

Example 13.1

Show that for an ideal gas

(a) the coefficient of expansion is a function of temperature only.
(b) the isothermal compressibility is a function of pressure only.

Solution

For an ideal gas, we have

$$pv = RT$$

$$p\,dv + v\,dp = R\,dT$$

and

$$dv = \frac{R}{p}\,dT - \frac{v}{p}\,dp. \tag{1}$$

But according to Eq. 13.16a, we have, for any simple compressible substance,

$$dv = v\beta\,dT - v\kappa\,dp. \tag{2}$$

Comparing Eq. 1 with Eq. 2, we have, for an ideal gas,

$$\beta = \frac{R}{pv} = \frac{R}{RT} = \frac{1}{T}.$$

$$\kappa = \frac{1}{p}. \qquad\qquad \blacksquare$$

Example 13.2

One unit mass of liquid undergoes a change of state quasi-statically and isothermally. Derive an expression for the path of this process in the *p–v* diagram.

Solution

Since T is constant, we have, according to Eq. 13.16a,

$$dv = -v\kappa\, dp.$$

For a liquid, κ may be assumed to be a constant. Thus

$$\int_1^2 \frac{dv}{v} = -\kappa \int_1^2 dp$$

and

$$\ln \frac{v_2}{v_1} = -\kappa(p_2 - p_1)$$

or

$$\ln \frac{v}{v_0} = -\kappa(p - p_0),$$

where p_0 and v_0 are the pressure and specific volume at any convenient reference point. ∎

Comments

Students should compare the result in this example with the compression path that we used in Example 2.8. Since κ may also be assumed to be a constant for a solid, the path deduced in this example for a liquid should also be valid for a solid undergoing a change of state quasi-statically and isothermally.

13.4 Some Thermodynamic Relations Involving Specific Heats

We have previously defined the constant-pressure specific heat c_p and the constant-volume specific heat c_v as

$$c_p = \left(\frac{\partial h}{\partial T}\right)_p \tag{13.17}$$

$$c_v = \left(\frac{\partial u}{\partial T}\right)_v . \tag{13.18}$$

We have shown that when a closed system undergoes a quasi-static constant-pressure process, we have

$$dq_p = dh_p.$$

Consequently, if a calorimetric experiment is carried out with a simple compressible substance at constant pressure, the heat required to raise the temperature of a unit mass by ΔT yields the quantity $(\Delta h/\Delta T)_p$, which is the mean constant-pressure specific heat. If the temperature change is made infinitesimally small, such an experiment would provide us with the c_p data.

We have also shown that when a closed system undergoes a quasi-static constant-volume process, we have

$$dq_v = du_v.$$

Analogous to the experimental determination of c_p, a calorimetric experiment carried out at constant volume would yield c_v data. We wish to point out that c_p and c_v data are more difficult to obtain than p–v–T data. For a discussion of experimental methods for the determination of specific heats, books on this subject should be consulted.[1]

To express c_p and c_v in terms of p–v–T data, we shall make use of the equations $s = s(T, v)$ and $s = s(T, p)$. Taking s as a function of T and v, we have

$$ds = \left(\frac{\partial s}{\partial T}\right)_v dT + \left(\frac{\partial s}{\partial v}\right)_T dv$$

and

$$T\,ds = T\left(\frac{\partial s}{\partial T}\right)_v dT + T\left(\frac{\partial s}{\partial v}\right)_T dv. \tag{13.19}$$

But from $T\,ds = du + p\,dv$, we get

$$c_v = T\left(\frac{\partial s}{\partial T}\right)_v,$$

and according to the Maxwell relation of Eq. 13.10, we have

$$\left(\frac{\partial s}{\partial v}\right)_T = \left(\frac{\partial p}{\partial T}\right)_v.$$

[1] D. P. Shoemaker and C. W. Garland, *Experiments in Physical Chemistry*, McGraw-Hill Book Company, New York, 1962.

Thus Eq. 13.19 may be written as

$$T \, ds = c_v \, dT + T\left(\frac{\partial p}{\partial T}\right)_v dv. \tag{13.20}$$

Taking s as a function of T and p, and making use of $T \, ds = dh - v \, dp$ and the Maxwell relation of Eq. 13.11, we would obtain

$$T \, ds = T\left(\frac{\partial s}{\partial T}\right)_p dT + T\left(\frac{\partial s}{\partial p}\right)_T dp \tag{13.21}$$

and

$$T \, ds = c_p \, dT - T\left(\frac{\partial v}{\partial T}\right)_p dp. \tag{13.22}$$

Combining Eqs. 13.20 and 13.22, we have

$$dT = \frac{T(\partial v/\partial T)_p}{c_p - c_v} dp + \frac{T(\partial p/\partial T)_v}{c_p - c_v} dv. \tag{13.23}$$

Equation 13.23 is a general thermodynamic relation. But from $T = T(p, v)$, we also have

$$dT = \left(\frac{\partial T}{\partial p}\right)_v dp + \left(\frac{\partial T}{\partial v}\right)_p dv. \tag{13.24}$$

Comparing Eq. 13.23 with Eq. 13.24, we have

$$\left(\frac{\partial T}{\partial p}\right)_v = \frac{T(\partial v/\partial T)_p}{c_p - c_v}$$

or

$$c_p - c_v = T\left(\frac{\partial v}{\partial T}\right)_p \left(\frac{\partial p}{\partial T}\right)_v. \tag{13.25}$$

From the cyclical relation involving the variables p, v, and T, we have

$$\left(\frac{\partial p}{\partial T}\right)_v = -\left(\frac{\partial v}{\partial T}\right)_p \left(\frac{\partial p}{\partial v}\right)_T.$$

Thus Eq. 13.25 may be written as

$$c_p - c_v = -T\left(\frac{\partial v}{\partial T}\right)_p^2\left(\frac{\partial p}{\partial v}\right)_T. \tag{13.26}$$

Equation 13.26 shows that c_p is always greater than c_v because T and $(\partial v/\partial T)_p^2$ are always positive, while $(\partial p/\partial v)_T$ is negative for all known simple compressible substances.

In terms of the coefficient of expansion and the isothermal compressibility, Eq. 13.26 may be written as

$$c_p - c_v = \frac{Tv\beta^2}{\kappa}. \tag{13.27}$$

Since it is more difficult to measure c_v than c_p, particularly in the case of a liquid or a solid, Eq. 13.26 or Eq. 13.27 makes it possible for us to determine c_v from c_p and p–v–T data.

Example 13.3

Make use of Eq. 13.25 to show that for an ideal gas, we do have $c_p - c_v = R$, as given in Chapter 5.

Solution

For an ideal gas,

$$pv = RT$$

$$p\,dv + v\,dp = R\,dT$$

$$\left(\frac{\partial v}{\partial T}\right)_p = \frac{R}{p}$$

$$\left(\frac{\partial p}{\partial T}\right)_v = \frac{R}{v}.$$

Substituting into Eq. 13.25, we have

$$c_p - c_v = T\left(\frac{R}{p}\right)\left(\frac{R}{v}\right) = R. \qquad \blacksquare$$

Example 13.4

Show that for a simple compressible substance,

$$c_v = -T\left(\frac{\partial^2 a}{\partial T^2}\right)_v.$$

Solution

From $T\,ds = du = p\,dv$, we have

$$c_v = \left(\frac{\partial u}{\partial T}\right)_v = T\left(\frac{\partial s}{\partial T}\right)_v. \tag{1}$$

From $da = -s\,dT - p\,dv$, we have

$$\left(\frac{\partial a}{\partial T}\right)_v = -s.$$

Therefore,

$$\left(\frac{\partial^2 a}{\partial T^2}\right)_v = -\left(\frac{\partial s}{\partial T}\right)_v. \tag{2}$$

Combining Eqs. 1 and 2, we have

$$c_v = -T\left(\frac{\partial^2 a}{\partial T^2}\right)_v. \qquad\blacksquare$$

Comments

The result of this example implies that if we have the Helmholtz function expressed as a function of T and v, we should be able to use it to calculate c_v simply by differentiation. Unfortunately, this is not possible in classical thermodynamics. However, it is possible in statistical thermodynamics, at least for some simple substances. We have seen that in certain applications we need to make use of the specific-heat ratio c_p/c_v. It is therefore of interest to investigate the relation between this ratio and p–v–T data.

Let us consider an isentropic process. Then, for such a process, we have, from Eq. 13.20,

$$c_v\,dT_s = -T\left(\frac{\partial p}{\partial T}\right)_v dv_s, \tag{13.28}$$

and, from Eq. 13.22,

$$c_p\,dT_s = T\left(\frac{\partial v}{\partial T}\right)_p dp_s. \tag{13.29}$$

Combining Eqs. 13.28 and 13.29, we have

$$\frac{c_p}{c_v} = -\left(\frac{\partial v}{\partial T}\right)_p\left(\frac{\partial T}{\partial p}\right)_v\left(\frac{\partial p}{\partial v}\right)_s. \tag{13.30}$$

From the cyclical relation for p, v, and T, we have

$$\left(\frac{\partial v}{\partial T}\right)_p \left(\frac{\partial T}{\partial p}\right)_v = -\left(\frac{\partial v}{\partial p}\right)_T.$$

Substituting this into Eq. 13.30, we have

$$\frac{c_p}{c_v} = \frac{-(\partial v/\partial p)_T}{-(\partial v/\partial p)_s} = \frac{-1/v(\partial v/\partial p)_T}{-1/v(\partial v/\partial p)_s}. \tag{13.31}$$

We recognize the quantity $-1/v(\partial v/\partial p)_T$ to be the isothermal compressibility κ. We may define the quantity $-1/v(\partial v/\partial p)_s$ as the isentropic compressibility α. Thus Eq. 13.31 may be written as

$$\frac{c_p}{c_v} = \frac{\kappa}{\alpha}. \tag{13.22}$$

It is to be noted that c_p is always greater than c_v. Then, according to Eq. 13.32, κ is always greater than α. It was shown in Chapter 9 that α is related to the speed at which sound travels in a substance. Thus Eq. 13.32 makes it possible for us to determine the specific-heat ratio from p–v–T and velocity-of-sound measurements.

13.5 Effect of Pressure and Volume Changes on Specific Heats

From Eq. 13.22, we have

$$ds = \frac{c_p}{T} dT - \left(\frac{\partial v}{\partial T}\right)_p dp.$$

This equation is of the form of

$$dz = M \, dx + N \, dy,$$

in which

$$\left(\frac{\partial M}{\partial y}\right)_x = \left(\frac{\partial N}{\partial x}\right)_y.$$

It follows that we have

$$\left(\frac{\partial c_p}{\partial p}\right)_T = -T\left(\frac{\partial^2 v}{\partial T^2}\right)_p. \tag{13.33}$$

Equation 13.33 indicates that to evaluate changes in c_p due to changes in pressure in an isothermal process, we need only p–v–T data.

In a similar manner, we also obtain

$$\left(\frac{\partial c_v}{\partial v}\right)_T = T\left(\frac{\partial^2 p}{\partial T^2}\right)_v, \tag{13.34}$$

by making use of Eq. 13.20, which is

$$ds = \frac{c_v}{T}\,dT + \left(\frac{\partial p}{\partial T}\right)_v\,dv.$$

Equation 13.34 shows that to evaluate changes in c_v due to changes in v in an isothermal process, we only need p–v–T data.

Example 13.5

The coefficient of expansion and the isothermal compressibility for a solid may be assumed to be constants over a considerable range of temperature and pressure. Show that for a solid that may be modeled as a simple compressible substance,

(a) c_p is a function of temperature only.
(b) c_v is a function of temperature only.

Solution

From Eq. 13.16a, we have, for any simple compressible substance,

$$dv = v\beta\,dT - v\kappa\,dp.$$

If β and κ are constants, we have

$$\ln\frac{v_2}{v_1} = \beta(T_2 - T_1) - \kappa(p_2 - p_1). \tag{1}$$

Let $v_2 = v_1 + \Delta v$. Then

$$\ln\frac{v_2}{v_1} = \ln\left(1 + \frac{\Delta v}{v_1}\right).$$

Since a solid is not very compressible, Δv will be very small. We may therefore obtain the following approximation:

$$\ln\left(1 + \frac{\Delta v}{v_1}\right) \approx \frac{v_2 - v_1}{v_1}.$$

Thus

$$v_2 \approx v_1 + v_1 \beta(T_2 - T_1) - v_1 \kappa(p_2 - p_1) \qquad (13.35)$$

or

$$v \approx v_0 + v_0 \beta(T - T_0) - v_0 \kappa(p - p_0), \qquad (13.35a)$$

where p_0, v_0, and T_0 are the pressure, specific volume, and temperature at any convenient reference point. Equation 13.35a is the approximate p–v–T equation of state for a solid.

(a) According to Eq. 13.33, we have, for any simple compressible substance,

$$\left(\frac{\partial c_p}{\partial p}\right)_T = -T\left(\frac{\partial^2 v}{\partial T^2}\right)_p.$$

Using Eq. 13.35a for a solid,

$$\left(\frac{\partial^2 v}{\partial T^2}\right)_p = 0.$$

Thus

$$\left(\frac{\partial c_p}{\partial p}\right)_T = 0,$$

which means that c_p is a function of temperature only.

(b) According to Eq. 13.34, we have, for any simple compressible substance,

$$\left(\frac{\partial c_v}{\partial v}\right)_T = T\left(\frac{\partial^2 p}{\partial T^2}\right)_v.$$

Using Eq. 13.35a for a solid,

$$\left(\frac{\partial^2 p}{\partial T^2}\right)_v = 0.$$

Thus

$$\left(\frac{\partial c_v}{\partial v}\right)_T = 0,$$

which means that c_v is a function of temperature only. ∎

13.6 Some Thermodynamic Relations for Changes in Entropy, Internal Energy, and Enthalpy

If entropy is expressed as a function of T and v, changes in entropy may be given by Eq. 13.20:

$$ds = \frac{c_v}{T} dT + \left(\frac{\partial p}{\partial T}\right)_v dv. \qquad (13.20)$$

If entropy is expressed as a function of T and p, changes in entropy may be given by Eq. 13.22:

$$ds = \frac{c_p}{T} dT - \left(\frac{\partial v}{\partial T}\right)_p dp. \qquad (13.22)$$

Equations 13.20 and 13.22 indicate that the entropy changes of a simple compressible substance may be evaluated from measurements of p, v, T, and specific heats. Furthermore, at a fixed temperature, the entropy changes of a simple compressible substance depend only on p–v–T data. Students should realize that Eqs. 5.17 and 5.18, developed for an ideal gas, are consistent with Eqs. 13.20 and 13.22.

To obtain an expression for internal energy changes, we may begin with Eq. 13.4:

$$T \, ds = du + p \, dv. \qquad (13.4)$$

Combining Eq. 13.20 with Eq. 13.4, we have

$$du = c_v \, dT + \left[T\left(\frac{\partial p}{\partial T}\right)_v - p \right] dv. \qquad (13.36)$$

Making use of Eq. 13.36, we can express the partial derivatives $(\partial u/\partial v)_T$ and $(\partial u/\partial p)_T$ as

$$\left(\frac{\partial u}{\partial v}\right)_T = T\left(\frac{\partial p}{\partial T}\right)_v - p \qquad (13.37)$$

and

$$\left(\frac{\partial u}{\partial p}\right)_T = \left[T\left(\frac{\partial p}{\partial T}\right)_v - p \right]\left(\frac{\partial v}{\partial p}\right)_T. \qquad (13.38)$$

We can simplify Eq. 13.38 by making use of the cyclical relation for p, v, and T. The result is

$$\left(\frac{\partial u}{\partial p}\right)_T = -T\left(\frac{\partial v}{\partial T}\right)_p - p\left(\frac{\partial v}{\partial p}\right)_T. \qquad (13.39)$$

Equations 13.37 and 13.39 show that at a fixed temperature, the internal energy of a simple compressible substance depends only on the measurements of p, v, and T.

To obtain an expression for enthalpy changes, we may begin with Eq. 13.5:

$$T \, ds = dh - v \, dp. \qquad (13.5)$$

Combining Eq. 13.5 with Eq. 13.22, we have

$$dh = c_p \, dT + \left[v - T\left(\frac{\partial v}{\partial T}\right)_p\right] dp, \qquad (13.40)$$

from which we get

$$\left(\frac{\partial h}{\partial p}\right)_T = v - T\left(\frac{\partial v}{\partial T}\right)_p. \qquad (13.41)$$

Equation 13.41 indicates that at a fixed temperature, the enthalpy of a simple compressible substance also depends only on p–v–T measurements.

Example 13.6

Making use of the equations developed in this section, show that for an ideal gas,

(a) the internal energy is a function of temperature only.
(b) the enthalpy is a function of temperature only.

Solution

For an ideal gas, we have

$$p \, dv + v \, dp = R \, dT$$

$$\left(\frac{\partial v}{\partial T}\right)_p = \frac{R}{p}$$

$$\left(\frac{\partial v}{\partial p}\right)_T = -\frac{v}{p}.$$

Substituting into Eq. 13.39, we have

$$\left(\frac{\partial u}{\partial p}\right)_T = -T\left(\frac{R}{p}\right) - p\left(-\frac{v}{p}\right)$$

$$= -v + v = 0.$$

Substituting into Eq. 13.41, we have

$$\left(\frac{\partial h}{\partial p}\right)_T = v - T\left(\frac{R}{p}\right)$$

$$= v - v = 0. \qquad \blacksquare$$

Comment

The internal energy and the enthalpy of an ideal gas depend indeed only upon temperature, as we have shown in Chapter 5 in a different manner.

13.7 Joule–Thomson Coefficient

From Eq. 13.40, it is readily seen that the partial derivative $(\partial T/\partial p)_h$, known as the Joule–Thomson coefficient μ_{JT}, is given as

$$\left(\frac{\partial T}{\partial p}\right)_h = \frac{T(\partial v/\partial T)_p - v}{c_p}. \qquad (13.42)$$

It may be shown that μ_{JT} is always zero for an ideal gas. However, for a simple compressible substance in general, μ_{JT} will be greater or less than zero as $T(\partial v/\partial T)_p$ is greater or less than v. The importance of this quantity stems partly from the fact that it can be measured in the laboratory, although not without difficulty. To measure the Joule–Thomson coefficient of a gas, we carry out a process identical in principle with the process of adiabatic throttling employed extensively in industrial operation. A schematic diagram of the Joule–Thomson experiment is shown in Fig. 13.1. In this experiment, the pressure drop occurs across the porous plug, which is a device made of porcelain or cotton. The entire apparatus is so constructed that the heat-transfer effect and the changes in kinetic and potential energies are all negligible. Therefore, with steady-state steady-flow conditions, the first law equation becomes simply

$$h_{in} = h_{out}.$$

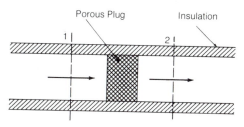

Figure 13.1 Apparatus for the Joule–Thomson Experiment

This result shows that we could in principle obtain the necessary information for the construction of the equation of state involving h, T, and p by carrying out the Joule–Thomson experiment at differing temperatures and pressures. Such data are usually plotted in the T–p diagram, using h as the parameter, as shown in Fig. 13.2. If the temperature is not too great, each of the constant-enthalpy curves will pass through a maximum called the *inversion point*. The locus of the inversion points is the *inversion curve*. The inversion curve gives the cooling or warming characteristics of the fluid in a Joule–Thomson experiment. We note from Fig. 13.2 that at a given inlet pressure, a lower inlet temperature will produce a lower outlet temperature for each of the curves that has an inversion point. Observation of this

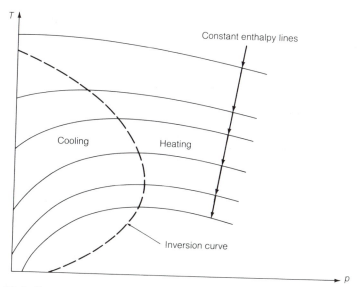

Figure 13.2 Constant-Enthalpy Lines and the Inversion Curve

phenomenon has been exploited in the design of gas-liquefaction systems.

By inserting in the porous plug of the Joule–Thomson experiment an element to effect heating or cooling of the fluid medium, we could carry out the throttling process isothermally instead of adiabatically. This isothermal process would yield information on another partial derivative $(\partial h/\partial p)_T$, known as the *isothermal Joule–Thomson coefficient*.

Now, applying the cyclical relation to the variables h, T, and p, we have

$$\left(\frac{\partial h}{\partial T}\right)_p \left(\frac{\partial T}{\partial p}\right)_h \left(\frac{\partial p}{\partial h}\right)_T = -1$$

or

$$\left(\frac{\partial T}{\partial p}\right)_h = -\frac{(\partial h/\partial p)_T}{(\partial h/\partial T)_p} = -\frac{(\partial h/\partial p)_T}{c_p}. \qquad (13.43)$$

Equation 13.43 is a relation we may use to check the consistency of experimental data on $(\partial T/\partial p)_h$, $(\partial h/\partial p)_T$, and c_p.

Example 13.7

Making use of tabulated data, approximate c_p, $(\partial h/\partial p)_T$, and $(\partial T/\partial p)_h$ for steam at 80 psia and 600°F.

Solution

For steam at 80 psia, we have the following data:

$$T = 500°F \qquad h = 1281.3 \text{ Btu/lbm}$$

$$= 600°F \qquad = 1330.9 \text{ Btu/lbm}$$

$$= 700°F \qquad = 1380.5 \text{ Btu/lbm.}$$

Making use of this information, we have

$$\left(\frac{\partial h}{\partial T}\right)_p \approx \left(\frac{\Delta h}{\Delta T}\right)_p = \frac{1380.5 - 1281.3}{700 - 500} = 0.496 \text{ Btu/lbm} \cdot °F.$$

For steam at 600°F, we have the following data:

$$p = 60 \text{ psia} \qquad h = 1332.3 \text{ Btu/lbm}$$

$$= 80 \text{ psia} \qquad = 1330.9 \text{ Btu/lbm}$$

$$= 100 \text{ psia} \qquad = 1329.6 \text{ Btu/lbm.}$$

Making use of this information, we get

$$\left(\frac{\partial h}{\partial p}\right)_T \approx \left(\frac{\Delta h}{\Delta p}\right)_T = \frac{1329.6 - 1332.3}{100 - 60} = -0.0675 \text{ Btu/lbm} \cdot \text{psi.}$$

Making use of Eq. 13.43, we get

$$\left(\frac{\partial T}{\partial p}\right)_h \approx -\frac{-0.0675}{0.496} = 0.136°\text{F/psi.} \qquad\blacksquare$$

Comment

A positive Joule–Thomson coefficient implies that there will be a temperature drop due to a pressure drop in an adiabatic throttling process.

13.8 Clapeyron Equation

An expression giving the slope of a curve separating two phases in the p–T diagram is known as the *Clapeyron equation*. It can be deduced in several ways.[2] We shall make use of the fact that the chemical potential μ has the same value for two phases of a pure substance in equilibrium, as shown in Chapter 6. For a liquid in equilibrium with its vapor, this means that

$$\mu_f = \mu_g,$$

which is the same as

$$g_f = g_g$$

[2] One way to deduce the Clapeyron equation is to operate a Carnot cycle in the two-phase region. See J. R. Lee and F. W. Sears, *Thermodynamics*, Addison-Wesley Publishing Co., Inc., Reading, Mass., 1955, pp. 134–137.

since the chemical potential is identical to the specific Gibbs function in the case of a single-component simple compressible substance according to Eq. 6.47. It follows that

$$dg_f = dg_g$$

if the temperature of the equilibrium mixture is increased by dT with the corresponding increase of pressure by dp.

Now, according to the fundamental equation of state of $g = g(T, p)$, we have

$$dg = -s\, dT + v\, dp.$$

Thus,

$$-s_f\, dT_f + v_f\, dp_f = -s_g\, dT_g + v_g\, dp_g.$$

But

$$dT_f = dT_g = dT$$

and

$$dp_f = dp_g = dp$$

since the changes in temperature and pressure are the same for both phases. Consequently, we get

$$\frac{dp}{dT} = \frac{s_g - s_f}{v_g - v_f} = \frac{s_{fg}}{v_{fg}}. \tag{13.44}$$

Applying $T\, ds = dh - v\, dp$ to a phase change in which T and p are constant, we get

$$T(s_g - s_f) = h_g - h_f.$$

Substituting into Eq. 13.44, we have

$$\frac{dp}{dT} = \frac{h_g - h_f}{T(v_g - v_f)} = \frac{h_{fg}}{Tv_{fg}}. \tag{13.45}$$

Equation 13.45 is the expression for the slope of the vaporization (vapor-pressure) curve.

In an analogous manner, the slope of the melting (fusion) curve is given as

$$\frac{dp}{dT} = \frac{h_f - h_i}{T(v_f - v_i)} = \frac{h_{if}}{Tv_{if}},$$ (13.46)

and the slope of the sublimation curve is given as

$$\frac{dp}{dT} = \frac{h_g - h_i}{T(v_g - v_i)} = \frac{h_{ig}}{Tv_{ig}}.$$ (13.47)

The enthalpy of vaporization, h_{fg}, is always positive. Since v_g is always greater than v_f, Eq. 13.45 indicates that the slope of the vapor-pressure curve is always positive. Consequently, the vapor pressure of any simple compressible substance increases with temperature. Since h_{ig} and v_{ig} are also always positive, we find that the slope of the sublimation curve is also positive for any pure substance. But the slope of the melting curve could be positive or negative. The enthalpy of melting h_{if} is always positive. If v_{if} is positive, such as in the case of most simple compressible substances, then the slope of the melting curve is positive, indicating that the freeze temperature increases with increasing pressure. However, for a substance that expands on freezing, such as water, v_{if} will be negative, resulting in a negative slope for the melting curve. Hence water will freeze at a lower temperature if the pressure is increased.

For liquid–vapor phase changes at relatively low pressure, v_g is much larger than v_f. We may therefore approximate v_{fg} by v_g in Eq. 13.45. In addition, the vapor may be assumed to behave like an ideal gas at low pressure. Introducing these approximations into Eq. 13.45, we have

$$\frac{dp}{dT} = \frac{h_{fg}}{Tv_g} = \frac{h_{fg}}{TRT/p}$$

or

$$h_{fg} = \frac{RT^2}{p} \frac{dp}{dT}.$$ (13.48)

Equation 13.48 may be used to estimate the enthalpy of vaporization by using vapor-pressure data.

If the T and p dependence of h_{fg} may be neglected, Eq. 13.48 can be integrated to give

$$\ln p = -\frac{h_{fg}}{RT} + \text{constant}$$ (13.49)

or

$$\ln \frac{p_2}{p_1} = -\frac{h_{fg}}{R}\left(\frac{1}{T_2} - \frac{1}{T_1}\right). \tag{13.49a}$$

Equation 13.49 indicates that when h_{fg} may be assumed to be constant and when the vapor may be assumed to behave like an ideal gas, the vapor pressure is a function of temperature only.

In a similar manner, we may also obtain the following expressions in the case of a solid–vapor mixture in equilibrium:

$$h_{ig} = \frac{RT^2}{p}\frac{dp}{dT} \tag{13.50}$$

and

$$\ln \frac{p_2}{p_1} = -\frac{h_{ig}}{R}\left(\frac{1}{T_2} - \frac{1}{T_1}\right). \tag{13.51}$$

Equations 13.48, 13.49, 13.50, and 13.51 are sometimes called the *Clausius–Clapeyron equation.*

Example 13.8

Estimate the melting temperature of ice at a pressure of 3000 psia if we only have the following information:

enthalpy of fusion (h_{if}) for water at $32°F = 143.3$ Btu/lbm

v_f at $32°F = 0.0160$ ft^3/lbm

v_i at $32°F = 0.0175$ ft^3/lbm.

Solution

According to Eq. 13.46, we have

$$\frac{dp}{dT} = \frac{h_{if}}{T(v_f - v_i)}.$$

Assuming h_{if} and v_{if} to be constant gives us

$$\int_{T_1}^{T_2} \frac{dT}{T} \approx \frac{v_f - v_i}{h_{if}} \int_{P_1}^{P_2} dp$$

$$\ln \frac{T_2}{T_1} \approx \frac{(0.0160 - 0.0175)(3000 - 0)(144)}{(143.3)(778)} = -5.81 \times 10^{-3}.$$

Thus

$$T_2 \approx T_1 e^{-5.81 \times 10^{-3}}$$

$$= 492 \times 0.994 = 489°R = 29°F. \qquad ▮$$

Comment

Our result is consistent with the fact that water freezes at a lower temperature as its pressure is increased.

Example 13.9

The enthalpy of vaporization for oxygen at 100 K is 202.613 kJ/kg. The saturated pressure at 100 K is 0.25403 MPa. Making use of Eq. 13.49a, estimate the vapor pressure of oxygen at 105 K.

Solution

From Eq. 13.49a, we have

$$\ln \frac{p_2}{p_1} \approx -\frac{h_{fg}}{R}\left(\frac{1}{T_2} - \frac{1}{T_1}\right)$$

$$= -\frac{202.613}{8.3143/32}\left(\frac{1}{105} - \frac{1}{100}\right)$$

$$= 0.3713.$$

Thus

$$p_2 = p_1 e^{0.3713}$$

$$= 0.25403 e^{0.3713} \text{ MPa} = 0.3682 \text{ MPa}. \qquad ▮$$

Comment

The tabulated vapor pressure at 105 K for oxygen is 0.37859 MPa. Our estimated value is less than 3% off.

13.9 Behavior of a van der Waals Gas

A van der Waals gas is gas obeying the equation of state given by

$$\left(p + \frac{a}{\bar{v}^2}\right)(\bar{v} - b) = \bar{R}T, \qquad (13.52)$$

Table 13.1 Constants for van der Waals Equation of State

Substance	a [kPa (m^3/kg mol)^2]	b (m^3/kg mol)
Air	135.8	0.0366
Ammonia (NH$_3$)	426.5	0.0373
Carbon dioxide (CO$_2$)	364.0	0.0427
Carbon monoxide (CO)	147.7	0.0393
Ethane (C$_2$H$_6$)	556.2	0.0638
Helium (He)	3.46	0.0237
Hydrogen (H$_2$)	24.8	0.0266
Methane (CH$_4$)	228.5	0.0427
Nitrogen (N$_2$)	136.6	0.0386
Oxygen (O$_2$)	137.8	0.0318
Propane (C$_3$H$_8$)	877.9	0.0845
Water (H$_2$O)	553.0	0.0305

which may be rearranged as

$$p = \frac{\bar{R}T}{\bar{v} - b} - \frac{a}{\bar{v}^2}, \tag{13.52a}$$

where a and b are constants. These constants are different for different gases. The van der Waals constants for several gases are given in Table 13.1.

The van der Waals equation of state represents one of the earliest attempts to develop an equation of state that is more accurate than the ideal gas equation. The constant b is intended to account for the volume occupied by the gas molecules. The constant a is intended to account for the intermolecular forces that exist between the molecules of real gases.

Making use of Eq. 13.52a, we may obtain the following expressions for a van der Waals gas:

$$\left(\frac{\partial p}{\partial T}\right)_v = \frac{\bar{R}}{\bar{v} - b} \tag{13.53}$$

$$\left(\frac{\partial p}{\partial v}\right)_T = -\frac{\bar{R}T}{(\bar{v} - b)^2} + \frac{2a}{\bar{v}^3} \tag{13.54}$$

$$\beta = \frac{1}{v}\left(\frac{\partial v}{\partial T}\right)_p = \frac{\bar{R}\bar{v}^2(\bar{v} - b)}{\bar{R}T\bar{v}^3 - 2a(\bar{v} - b)^2} \tag{13.55}$$

$$\kappa = -\frac{1}{v}\left(\frac{\partial v}{\partial p}\right)_T = \frac{\bar{v}^2(\bar{v} - b)^2}{\bar{R}T\bar{v}^3 - 2a(\bar{v} - b)^2}. \tag{13.56}$$

The quantities given by Eqs. 13.53, 13.54, 13.55, and 13.56 would be reduced to those for an ideal gas if the constants a and b are both zeros. This is, of course, as expected.

For any gas in general, we have

$$\left(\frac{\partial u}{\partial v}\right)_T = T\left(\frac{\partial p}{\partial T}\right)_v - p, \tag{13.37}$$

where the property $(\partial u/\partial v)_T$ is sometimes called the internal pressure of the gas due to intermolecular forces.

For a van der Waals gas, we have

$$\left(\frac{\partial u}{\partial v}\right)_T = \frac{\bar{R}T}{\bar{v} - b} - \left(\frac{\bar{R}T}{\bar{v} - b} - \frac{a}{\bar{v}^2}\right)$$

$$= \frac{a}{\bar{v}^2}, \tag{13.57}$$

which means that the internal pressure of a van der Waals gas is exactly the term a/v^2 introduced to account for the intermolecular forces involved.

The entropy change for a real gas is given by

$$ds = c_v\, dT + \left(\frac{\partial p}{\partial T}\right)_v dv.$$

Making use of Eq. 13.53, the entropy change of a van der Waals gas may then be given as

$$ds = c_v\, dT + \left(\frac{\bar{R}}{\bar{v} - b}\right) dv. \tag{13.58}$$

The internal change for a real gas is given by

$$du = c_v\, dT + \left[T\left(\frac{\partial p}{\partial T}\right)_v - p\right] dv.$$

Making use of Eqs. 13.52a and 13.53, the internal energy change of a van der Waals gas may then be given as

$$du = c_v\, dT + \frac{a}{\bar{v}^2}\, dv. \tag{13.59}$$

13.10 Real Gases and the Compressibility Factor

The p–v–T measurements of real gases may be presented in several ways. A simple method involves the use of the compressibility factor z defined by

$$z = \frac{p\bar{v}}{\bar{R}T}. \tag{13.60}$$

For an ideal gas, z is obviously unity under all conditions. For a real gas, the value of z may be less or more than unity. Thus z gives the deviation from ideal-gas behavior. Since z is a thermodynamic property by definition, it may be expressed as a function of two other properties, such as temperature and pressure. That is, there will exist the equation of state

$$z = z(T, p), \tag{13.61}$$

which is the basis of the compressibility-factor chart in which z versus p for various values of T is given. It is evident that a large quantity of data are required to construct a z–p chart.

According to Eq. 13.61, we need a separate chart for each gas. It would be more convenient if one chart could be used for many gases. This turns out to be possible by means of the approximation known as the *principle of corresponding states*, first proposed by van der Waals in 1881. This empirical principle states that all substances obey the same equation of state expressed in terms of the reduced properties. A reduced property is the actual value of the property divided by its value at the critical point. Thus, by definition, the reduced pressure p_R, the reduced temperature T_R, and the reduced volume v_R are given as

$$p_R = \frac{p}{p_C}, \qquad T_R = \frac{T}{T_C}, \qquad \text{and} \qquad v_R = \frac{\bar{v}}{\bar{v}_C}. \tag{13.62}$$

It is found experimentally that for many gases the compressibility factor z may be approximated by the relation

$$z = z(p_R, T_R), \tag{13.63}$$

which is the basis of the generalized compressibility chart in which z versus p_R is given for various values of T_R. Such a chart is shown in Fig. A.4. Equation 13.63 is consistent with the empirical principle of corresponding states.

Let us substitute Eq. 13.62 into Eq. 13.60. Then

$$z = \frac{p_R v_R}{T_R} \frac{p_c \bar{v}_C}{\bar{R} T_C} = \frac{p_R v_R}{T_R} z_C. \tag{13.64}$$

We have in Eq. 13.64 a new variable, z_C, the compressibility factor at the critical point. Unfortunately, z_C is not a constant but varies widely from gas to gas. However, if z is a function of T_R and p_R, the quantity $v_R z_C$, known as the ideal *reduced volume* or *pseudo-reduced volume*,[3] must also be a function of T_R and p_R. By definition, the pseudo-reduced volume $v_{r'}$ is given as

$$v_{r'} \equiv v_R z_C \equiv z \frac{T_R}{p_R} = \frac{\bar{v}}{\bar{R} T_C/p_C}. \tag{13.65}$$

Since $v_{r'}$ is a function of T_R and p_R, it may be used as a parameter in the z–p plot, as shown in Figs. A.4 and A.5. In employing the generalized compressibility chart for engineering calculations, any two of the three properties T, p, and v may be used to find the compressibility factor z. This quantity is then used in the relation $p\bar{v} = z\bar{R}T$ to find the third property of state. To generate p–v–T data for a given gas from the generalized compressibility chart, all we need to know are the critical pressure p_C and the critical temperature T_C of the gas. Consequently, this chart is a very useful tool available to the engineer for predicting the properties of gases for which experimental data are lacking.

We wish to point out that the generalized compressibility chart is constructed by averaging the experimental measurements of several gases. The chart cannot be expected to give accurate information for all gases. This is particularly true in the vicinity of the critical point, since z_C is not the same for all gases. If accuracy is of importance, generalized compressibility chart data should be resorted to only when actual p–v–T data for the particular gas in question are not available.

Example 13.10

Determine the mass of nitrogen contained in a 30 m³ vessel at 20.0 MPa and 200 K by using

(a) the ideal-gas equation of state.

[3] Gouq-Jen Su, "Modified Law of Corresponding States for Real Gases," *Industrial and Engineering Chemistry*, Vol. 38, August 1946, pp. 803–806.

(b) the generalized compressibility chart.
(c) the tabulated p–v–T data given in the Appendix.

Solution

(a)
$$m = \frac{pV}{RT} = \frac{20{,}000 \times 30}{(8.3143/28.013) \times 200} = 10{,}108 \text{ kg.}$$

(b) From Table 4.1, we have, for nitrogen,

$$p_c = 3.40 \text{ MPa} \quad \text{and} \quad T_c = 126.2 \text{ K.}$$

For our problem, we have

$$p_R = \frac{20{,}000}{3400} = 5.88$$

and

$$T_R = \frac{200}{126.2} = 1.585.$$

Using these values of p_R and T_R, we find from the generalized compressibility chart,

$$z = 0.902.$$

Thus

$$v = \frac{zRT}{p} = \frac{0.902 \times (8.3143/28.013) \times 200}{20{,}000} = 0.002677 \text{ m}^3/\text{kg}$$

and

$$m = \frac{V}{v} = \frac{30}{0.002677} = 11{,}210 \text{ kg}$$

(c) From tabulated data, we find

$$v = 0.002687 \text{ m}^3/\text{kg}$$

and

$$m = \frac{30}{0.002687} = 11{,}160 \text{ kg.}$$

Comments

If we use the ideal-gas equation of state for design calculations in this problem, we would be in error by about 10 percent. On the other hand, the result obtained from using the compressibility chart is practically free of error.

13.11 Generalized Enthalpy Chart for Real Gases

According to Eq. 13.40, the enthalpy change of any gas at a fixed temperature is given by

$$dh = \left[v - T\left(\frac{\partial v}{\partial T}\right)_p \right] dp. \tag{13.66}$$

Using $pv = zRT$, we get

$$dh = \left[\frac{zRT}{p} - \frac{zRT}{p} - \frac{RT^2}{p}\left(\frac{\partial z}{\partial T}\right)_p \right] dp$$

$$= -\frac{RT^2}{p}\left(\frac{\partial z}{\partial T}\right)_p dp. \tag{13.67}$$

Integrating Eq. 13.67 from zero pressure to any pressure p, we have, at a fixed temperature,

$$h - h^* = -RT^2 \int_0^p \left(\frac{\partial z}{\partial T}\right)_p \frac{dp}{p}, \tag{13.68}$$

where h^* is the enthalpy of the gas at zero pressure and a given temperature, and h is the enthalpy of the gas at the same temperature and any pressure p. Since all gases behave like an ideal gas at zero pressure, h^* is the ideal-gas value at a given temperature. Thus the right-hand side of Eq. 13.68 represents the enthalpy departure from ideal-gas behavior.

Now, we have, by definition,

$$T = T_c T_R \qquad \text{and} \qquad p = p_c p_R.$$

Hence

$$dT = T_c dT_R \qquad \text{and} \qquad dp = p_c dp_R.$$

Making use of these results, Eq. 13.68 may be expressed in terms of the reduced properties as follows:

$$h - h^* = -RT_C^2 T_R^2 \int_0^{p_R} \frac{1}{T_C} \left(\frac{\partial z}{\partial T_R} \right)_{p_R} \frac{p_C \, dp_R}{p_C p_R}$$

$$= -RT_C T_R^2 \int_0^{p_R} \left(\frac{\partial z}{\partial T_R} \right)_{p_R} \frac{dp_R}{p_R}$$

or

$$\frac{\bar{h}^* - \bar{h}}{T_C} = \bar{R} T_R^2 \int_0^{p_R} \left(\frac{\partial z}{\partial T_R} \right)_{p_R} \frac{dp_R}{p_R}. \tag{13.69}$$

We note that the right-hand side of Eq. 13.69 is a function of T_R and p_R only. Consequently, the integration may be performed by making use of the generalized compressibility-factor data. Equation 13.69 is the basis of the generalized enthalpy correction chart given in Fig. A.6. To use the chart for a particular gas, all we need to know are the critical pressure p_C and the critical temperature T_C of the gas. Making use of this chart, we may give the change of enthalpy between two states as

$$\bar{h}_2 - \bar{h}_1 = T_C \left[\left(\frac{\bar{h}^* - \bar{h}}{T_C} \right)_{T_1, p_1} - \left(\frac{\bar{h}^* - \bar{h}}{T_C} \right)_{T_2, p_2} \right] + (\bar{h}_{T_2, p_2}^* - \bar{h}_{T_1, p_1}^*), \tag{13.70}$$

where

$$\bar{h}_{T_2, p_2}^* - \bar{h}_{T_1, p_1}^* = \int_{T_1}^{T_2} \bar{c}_p^* \, dT$$

is the enthalpy change due to ideal-gas behavior, the evaluation of which requires data giving the idea-gas constant-pressure specific heat $\bar{c}_p^*$ as a function of temperature.

Example 13.11

Making use of the generalized enthalpy correction chart, determine the enthalpy change for nitrogen due to a change of state from 101.325 kPa and 300 K to 20.0 MPa and 200 K.

Solution

At $p_1 = 101.325$ kPa and $T_1 = 300$ K,

$$p_{R1} = \frac{101.325}{3400} = 0.0298$$

$$T_{R1} = \frac{300}{126.2} = 2.377.$$

At $p_2 = 20.0$ MPa and $T_2 = 200$ K,

$$p_{R2} = \frac{20,000}{3400} = 5.88$$

$$T_{R2} = \frac{200}{126.2} = 1.585.$$

From the generalized enthalpy correction chart, we obtain at 101.325 kPa and 300 K,

$$\frac{\bar{h}^* - \bar{h}}{T_C} \approx 0,$$

and at 20.0 MPa and 200 K,

$$\frac{\bar{h}^* - \bar{h}}{T_C} = 4.3 \text{ cal/g mol} \cdot \text{K} = 18.0 \text{ kJ/kg mol} \cdot \text{K}.$$

For ideal-gas behavior for nitrogen in the temperature range of this problem, $\bar{c}_p^* = 29.08$ kJ/kg mol · K. Substituting into Eq. 13.70, we get

$$\bar{h}_2 - \bar{h}_1 = 126.2 \times (0 - 18.0) + 29.08 \times (200 - 300)$$

$$= -2271.6 - 2908 = -5179.6 \text{ kJ/kg mol}$$

or

$$h_2 - h_1 = -\frac{5179.6}{28.013} = -184.9 \text{ kJ/kg}.$$

The enthalpy change would be -181.0 kJ/kg if we have used tabulated data.

∎

13.12 Generalized Entropy Chart for Real Gases

According to Eq. 13.22, the entropy change of any gas at a fixed temperature is given by

$$ds = -\left(\frac{\partial v}{\partial T}\right)_p dp. \qquad (13.71)$$

Integrating Eq. 13.71 from zero pressure to any pressure p for an ideal gas, we have, at a fixed temperature,

$$s_p^* - s_0^* = -\int_0^p \left(\frac{\partial v}{\partial T}\right)_p dp = -\int_0^p \frac{R}{p} dp. \qquad (13.72)$$

In terms of reduced properties, Eq. 13.72 becomes

$$s_p^* - s_0^* = -R \int_0^{p_R} \frac{dp_R}{p_R}. \qquad (13.73)$$

Integrating Eq. 13.71 from zero pressure to any pressure p for a real gas, we have, at a fixed temperature,

$$s_p - s_0^* = -\int_0^p \left(\frac{\partial v}{\partial T}\right)_p dp. \qquad (13.74)$$

Using $pv = zRT$, Eq. 13.74 becomes

$$s_p - s_0^* = -\int_0^p \frac{R}{p}\left[z + T\left(\frac{\partial z}{\partial T}\right)_p\right] dp. \qquad (13.75)$$

In terms of reduced properties, Eq. 13.75 becomes

$$s_p - s_0^* = -R \int_0^{p_R}\left[z + T_R\left(\frac{\partial z}{\partial T_R}\right)_{p_R}\right] \frac{dp_R}{p_R}. \qquad (13.76)$$

Subtracting Eq. 13.73 from Eq. 13.76, we have

$$s_p - s_p^* = -R \int_0^{p_R}\left[(z-1) + T_R\left(\frac{\partial z}{\partial T_R}\right)_{p_R}\right] \frac{dp_R}{p_R}$$

or

$$\bar{s}_p^* - \bar{s}_p = \bar{R} \int_0^{p_R}\left[(z-1) + T_R\left(\frac{\partial z}{\partial T_R}\right)_{p_R}\right] \frac{dp_R}{p_R}. \qquad (13.77)$$

The right-hand side of Eq. 13.77 is a function of T_R and p_R only. The integration may be performed by making use of the generalized compressibility-factor data. Equation 13.77 is the basis of the generalized entropy correction chart given in Fig. A.7.

Making use of the generalized entropy correction chart, we may give the change of entropy between two states as

$$\bar{s}_2 - \bar{s}_1 = (\bar{s}_p^* - \bar{s}_p)_{T_1,p_1} - (\bar{s}_p^* - \bar{s}_p)_{T_2,p_2} + \bar{s}_{T_2,p_2}^* - \bar{s}_{T_1,p_1}^* \quad (13.78)$$

where

$$\bar{s}_{T_2,p_2}^* - \bar{s}_{T_1,p_1}^* = \int_{T_1}^{T_2} \bar{c}_p^* \frac{dT}{T} - \bar{R} \ln \frac{p_2}{p_1}$$

is the entropy change due to ideal-gas behavior.

Example 13.12

Making use of the generalized entropy correction chart, determine the entropy change for nitrogen due to a change of state from 101.325 kPa and 300 K to 20.0 MPa and 200 K.

Solution

From Example 13.11, we have

$$p_{R1} = 0.0298 \qquad p_{R2} = 5.88$$

$$T_{R1} = 2.377 \qquad T_{R2} = 1.585$$

From the generalized entropy chart, we obtain, at 101.325 kPa and 300 K,

$$\bar{s}_p^* - \bar{s}_p \approx 0,$$

and at 20.0 MPa and 200 K,

$$\bar{s}_p^* - \bar{s}_p = 2.2 \text{ cal/g mol} \cdot \text{K} = 9.21 \text{ kJ/kg mol} \cdot \text{K}.$$

We shall also let $\bar{c}_p^* = 29.08$ kJ/kg mol $\cdot$ K, as in Example 13.11. Substituting into Eq. 13.78, we get

$$\bar{s}_2 - \bar{s}_1 = 0 - 9.21 + 29.08 \ln \frac{200}{300} - 8.3143 \ln \frac{20,000}{101.325}$$

$$= -9.21 - 11.79 - 43.94$$

$$= -64.94 \text{ kJ/kg mol} \cdot \text{K}.$$

or

$$s_2 - s_1 = -\frac{64.94}{28.013} = -2.32 \text{ kJ/kg} \cdot \text{K.}$$ ∎

The entropy change would be -2.29 kJ/kg $\cdot$ K if we had used tabulated data.

13.13 Fugacity and the Fugacity Function

Fugacity is a property introduced by G. N. Lewis in 1901 in an attempt to simplify the treatment of cases in which the ideal-gas equation does not apply. Let us start with Eq. 4.15:

$$dg = -s \, dT + v \, dp.$$

At constant temperature, we have

$$dg_T = v \, dp_T. \tag{13.79}$$

Equation 13.79 is valid for real gases as well as ideal gases. In the case of an ideal gas, we have

$$dg_T = \frac{RT}{p} \, dp_T$$

$$= RT \, d(\ln p)_T. \tag{13.80}$$

In order to preserve the form of Eq. 13.80 for real gases, Lewis defines a property by the equation

$$dg_T = RT \, d(\ln f)_T, \tag{13.81}$$

where f is known as the fugacity of a real gas with the requirement that

$$\lim_{p \to 0} \frac{f}{p} = 1. \tag{13.82}$$

Thus as $p \to 0$, $f \to p$. That is, at very low pressure, Eq. 13.81 is reduced to Eq. 13.80. It is to be noted that fugacity must have the units of pressure, by definition.

Combining Eqs. 13.79 and 13.81, we have

$$d(\ln f)_T = \frac{v}{RT} \, dp_T. \tag{13.83}$$

Now the compressibility factor z is defined as

$$z = \frac{pv}{RT}.$$

Thus we may write Eq. 13.83 as

$$d(\ln f)_T = \frac{z}{p} \, dp_T = z \, d(\ln p)_T. \qquad (13.84)$$

Now the reduced pressure p_R is defined as

$$p_R = \frac{p}{p_c}.$$

Then

$$\frac{dp_R}{p_R} = \frac{dp}{p}$$

and

$$d(\ln p_R) = d(\ln p).$$

Thus Eq. 13.84 may be written as

$$d(\ln f)_T = z \, d(\ln p_R)_T$$

and

$$d(\ln f)_T - d(\ln p_R)_T = z \, d(\ln p_R)_T - d(\ln p_R)_T$$

or

$$d\left(\ln \frac{f}{p}\right)_T = (z - 1) \, d(\ln p_R)_T. \qquad (13.85)$$

Integrating at constant temperature from $p = 0$ to any pressure p, we have

$$\int_{f/p=1}^{f/p} d\left(\ln \frac{f}{p}\right) = \int_0^{p_R} (z - 1) \, d(\ln p_R)_T.$$

Since $\ln(f/p) = 0$ for $f/p = 1$,

$$\ln \frac{f}{p} = \int_0^{p_R} (z - 1) \, d(\ln p_R)_T. \tag{13.86}$$

The quantity f/p is known as the fugacity coefficient. Equation 13.86 is the basis of the generalized fugacity coefficient chart given in Fig. A.8.

Example 13.13

Nitrogen is compressed reversibly and isothermally in a steady-state steady-flow compressor from 300 K and 101.325 kPa to 20.0 MPa. Determine the compressor work required.

Solution

System selected for study: compressor
Unique features of process: steady-state steady flow; reversible and isothermal
Assumptions: nitrogen is a simple compressible substance; neglect ΔKE and ΔPE

From the first law we have

$$\bar{d}q = dh + \bar{d}w. \tag{1}$$

From the second law we have

$$\bar{d}q = T \, ds. \tag{2}$$

Combining Eqs. 1 and 2, we have

$$\bar{d}w = T \, ds - dh.$$

Since process is isothermal,

$$w = T(s_2 - s_1) - (h_2 - h_1)$$

$$= -[(h_2 - T_2 s_2) - (h_1 - T_1 s_1)].$$

Since $g = h - Ts$, the work required may also be given as

$$w = -(g_2 - g_1)_T.$$

Making use of Eq. 13.81, we have

$$w = -RT \ln \left(\frac{f_2}{f_1} \right)_T .$$

At $p_1 = 101.325$ kPa and $T_1 = 300$ K,

$$p_{R1} = \frac{101.325}{3400} = 0.0298$$

$$T_{R1} = \frac{300}{126.2} = 2.377.$$

At $p_2 = 20.0$ MPa and $T_2 = T_1 = 300$ K,

$$p_{R2} = \frac{20{,}000}{3400} = 5.88$$

$$T_{R2} = T_{R1} = 2.377.$$

From the generalized fugacity coefficient chart, we obtain at 101.325 kPa and 300 K,

$$\left(\frac{f}{p} \right)_1 \approx 1.0$$

and at 20.0 MPa and 300 K,

$$\left(\frac{f}{p} \right)_2 \approx 0.93.$$

Thus

$$f_1 \approx p_1 = 101.325 \text{ kPa}$$

and

$$f_2 \approx 0.93 \, p_2 = 0.93 \times 20{,}000 \text{ kPa} = 18{,}600 \text{ kPa}$$

$$w = -\frac{8.3143}{28.013} \times 300 \ln \frac{18{,}600}{101.325} \text{ kJ/kg}$$

$$= -464.2 \text{ kJ/kg}.$$

From tabulated data,

$$w = 300(5.1630 - 6.8418) - (279.010 - 311.163) \text{ kJ/kg}$$

$$= -471.5 \text{ kJ/kg.} \qquad\blacksquare$$

Comments

We see that the property fugacity is quite useful for this reversible isothermal process involving a simple compressible substance. However, this property is actually more useful in the treatment of multicomponent systems.

13.14 Virial Equation of State

The p–v–T relation of a real gas may be represented by an infinite series by expanding the compressibility factor z about $p = 0$ or $\rho = 0$ as follows:

$$z = \frac{pv}{RT} = 1 + B'p + C'p^2 + D'p^3 + \cdots \qquad (13.87)$$

$$= 1 + B\rho + C\rho^2 + D\rho^3 + \cdots \qquad (13.88)$$

Equations 13.87 and 13.88 are known as *virial equations of state*. The B', C', and D' are the second, third, and fourth virial coefficients in the pressure series, while B, C, and D are the second, third, and fourth virial coefficients in the density series. All the virial coefficients are functions of temperature only for a pure gas. In principle, the virial coefficients, primarily the second and third, may be obtained by fitting the experimental p, v, and T measurements to either the pressure or the density series. Comparison of the two infinite series shows that the coefficients are related in the following manner:

$$B' = \frac{B}{RT}$$

$$C' = \frac{C - B^2}{(RT)^2}$$

$$D' = \frac{D + 2B - 3BC}{(RT)^3} \quad \text{etc.} \qquad (13.89)$$

The virial equation of state may be derived from statistical thermodynamics. From the microscopic point of view, the second virial coefficient represents the effect of two-body molecular interactions, the third virial coefficient represents the three-body molecular interactions, and so on. Since the probability of three-body interactions will be less than that of two-body interactions, the virial equation of state would yield, at very low pressures, the ideal-gas behavior, as any valid equation of state must. It is obvious that the B term is larger than the C term at low pressures, the C term is important at somewhat higher pressures, and so on. Consequently, it is quite often that the virial equation of state may be truncated as

$$z = 1 + B'p + C'p^2 \tag{13.90}$$

$$= 1 + B\rho + C\rho^2. \tag{13.91}$$

At low pressures, the C term may even be dropped.

It may be shown in statistical thermodynamics that the virial coefficients can be expressed in terms of the intermolecular forces in the gas. Although our present understanding of the intermolecular potential is incomplete, the virial coefficients, primarily the second and the third, can be evaluated through the use of an approximate potential function. For more information on this interesting subject, the excellent summary and discussions given by Hirschfelder, Curtiss, and Bird[4] in their monumental work should be consulted. Because of its exciting promise, the virial equation of state has received considerable attention in recent years.

Problems

13.1 Show that the constant-pressure lines in the two-phase region of an h–s diagram (Mollier diagram) are straight but not parallel.

13.2 Show that the slope of the constant-temperature lines in the h–s diagram (Mollier diagram) become zero in the region where $pv = RT$.

13.3 Derive the relation $c_p = -T(\partial^2 g/\partial T^2)_p$.

[4] J. O. Hirschfelder, C. F. Curtiss, and R. B. Bird, *Molecular Theory of Gases and Lquids*, John Wiley & Sons, Inc., New York, 1964.

13.4   Derive the relation $(\partial T/\partial p)_s = Tv\beta/c_p$.

13.5   Derive the relation $(\partial p/\partial T)_s = c_p/v\beta T$.

13.6   Derive the relation $(\partial v/\partial p)_s = [T(v\beta)^2/c_p] - v\kappa$.

13.7   The velocity of sound in a simple compressible substance is given by

$$c = \sqrt{g_c\left(\frac{\partial p}{\partial \rho}\right)_s}.$$

Make use of this expression to show that the velocity of sound in an ideal gas is given by

$$c = \sqrt{kg_c\, RT}.$$

13.8   Making use of the result in Problem 13.5, derive the following familiar temperature–pressure relation for an ideal gas undergoing a change of state isentropically with constant specific heats:

$$\frac{T_2}{T_1} = \left(\frac{p_2}{p_1}\right)^{(k-1)/k}.$$

13.9   Show that the difference between c_p and c_v is exactly equal to R, the gas constant, for a gas obeying the equation of state $p(v - b) = RT$, where b is a positive constant.

13.10 Show that the slope of a constant-volume line in the superheated-vapor region of an h–s diagram (Mollier diagram) is equal to

$$T + \frac{k-1}{\beta}.$$

13.11 For a gas obeying the equation of state $p(v - b) = RT$, where R is the gas constant and b is a positive constant, show that
(a) the slope of a constant entropy line in the pressure–volume diagram may be given as

$$-\frac{pk}{v-b}.$$

(b) the isentropic path in the pressure–volume diagram may be given by the following expression if specific heats are constant:

$$p(v - b)^k = \text{constant.}$$

13.12 Show that the slope of a constant temperature line in the superheated vapor region of an h–s diagram (Mollier diagram) is equal to

$$T - \frac{1}{\beta}.$$

13.13 One gram of water undergoes a change of state isentropically from 50°C and 1 atm to 1000 atm. Determine the final temperature of the water. For water at 50°C, we may take

$$v = 1.012 \text{ cm}^3/\text{g}$$

$$c_p = 0.998 \text{ cal/g} \cdot \text{K}$$

$$\beta = 465 \times 10^{-6} \, °\text{C}^{-1}.$$

13.14 Derive the relation $(\partial T/\partial v)_s = -T\beta/c_v \kappa$.

13.15 Make use of the relation derived in Problem 13.14 to show that the isentropic path for an ideal gas with constant specific heats may be given by

$$Tv^{k-1} = \text{constant.}$$

13.16 Make use of the relation derived in Problem 13.4 to show that the isentropic path for an ideal gas with constant specific heats may be given by

$$pv^k = \text{constant.}$$

13.17 Show that the internal energy is independent of the volume for a gas which obeys the equation state $p(v - b) = RT$, where R is the gas constant and b is a positive constant.

13.18 For a gas obeying the equation of state $p(v - b) = RT$, where R is the gas constant and b is a positive constant, show that the entropy change for an isothermal process is given by

$$s_2 - s_1 = -R \ln \frac{p_2}{p_1}.$$

13.19 At a constant pressure of 40 psia, the specific volume of ammonia is observed to increase by 0.664 ft³/lbm with a change in temperature from 120°F to 160°F. Making use of this information and the appropriate Maxwell relation, estimate the change in entropy when ammonia undergoes a change of state from 35 psia and 140°F to 45 psia and 140°F.

13.20 For a gas obeying the equation of state $p(v - b) = RT$, in which R is the gas constant and b is a positive constant, can the temperature be reduced through the use of a throttling process?

13.21 The velocity of sound c in a medium is deduced in Chapter 9 as

$$c = \sqrt{g_c \left(\frac{\partial p}{\partial \rho}\right)_s}.$$

Show that the velocity of sound may also be given as

$$c = \sqrt{\frac{g_c vk}{\kappa}}.$$

13.22 Making use of the expression obtained in Problem 13.21, calculate the velocity of sound for the following cases:

(a) water at 75°F, taking

$$v = 0.01606 \text{ ft}^3/\text{lbm}$$

$$k = 1.0$$

$$\kappa = 45 \times 10^{-6} \text{ atm}^{-1}.$$

(b) copper at 32°F, taking

$$v = 1.792 \times 10^{-3} \text{ ft}^3/\text{lbm}$$

$$k = 1.0$$

$$\kappa = 8.09 \times 10^{-10} \text{ atm}^{-1}.$$

13.23 One pound of water at 75°F is compressed reversibly and isothermally inside a cylinder from 1 atm to 1000 atm. Taking $\beta = 140 \times 10^{-6}°R^{-1}$ and $\kappa = 45 \times 10^{-6} \text{ atm}^{-1}$ for water at 75°F, estimate
(a) the amount of heat transferred.
(b) the quantity of work required.

13.24 Demonstrate the validity of the Maxwell relation $(\partial s/\partial p)_T = -(\partial v/\partial T)_p$ by evaluating these two partial derivatives for steam at 700 psia and 800°F.

13.25 Show that the specific-heat ratio k may be given by the expression

$$k = \frac{1}{1 - (Tv\beta^2/c_p\kappa)}.$$

13.26 Making use of tabulated data and the expression given in Problem 13.25, calculate the specific-heat ratio k for steam at 15 psia and 400°F.

13.27 Fifty kilograms of water initially at 75°C are mixed at constant pressure with 50 kg of water initially at 50°C. The specific heat of water may be considered to be a constant of 4.184 kJ/kg · K. Assuming that the mixing process is carried out adiabatically, determine
(a) the final equilibrium temperature of the water.
(b) the thermodynamic lost work due to the mixing process if the temperature of the environment is 300 K.

13.28 A piece of metal initially at 500°C is quenched in a water bath. The initial and final temperatures of the water are 30°C and 80°C, respectively. The specific heat of the metal is 0.450 kJ/kg · K. If the mass of the metal is 100 kg,
(a) determine the mass of the water bath.
(b) determine the thermodynamic lost work (exergy destruction) due to the quenching process if the temperature of the environment is 300 K.

13.29 Make an estimate of the melting temperature of ice at a pressure of 20.0 MPa by making use of the following data for water at 0°C:

s_i, saturated ice $= -1.221$ kJ/kg · K

s_f, saturated water $= 0.$

v_i, specific volume of saturated ice $= 1.0908 \times 10^{-3}$ m³/kg

v_f, specific volume of saturated water $= 1.000 \times 10^{-3}$ m³/kg.

13.30 Making use of tabulated data and the expression given in Problem 13.25, calculate the specific heat ratio k for Freon-12 at 0.25 MPa and 80°C.

13.31 Making use of tabulated data and the expression given in Problem 13.25, calculate the specific heat ratio k for nitrogen at 8.0 MPa and 300 K.

13.32 For water at 75°F and 1.0 atm, we may take

$$c_p = 1.0 \text{ Btu/lbm} \cdot \text{°R}$$

$$\beta = 140 \times 10^{-6} \text{°R}^{-1}$$

$$\kappa = 45 \times 10^{-6} \text{ atm}^{-1}.$$

Calculate the constant-volume specific heat c_v for water at 75°F and 1.0 atm.

13.33 The following boiling temperatures for water are known:

Pressure (kPa)	Boiling temperature (°C)
700	164.96
800	170.41

It is also known that the volume change during vaporization for water at 750 kPa is 0.25431 m³/kg. Make an estimate of the enthalpy of vaporization for water at 750 kPa.

13.34 At 212°F and 1 atm, the enthalpy of vaporization for H_2O is 970.3 Btu/lbm. Estimate the vapor pressure of H_2O at 220°F based on this information only.

13.35 The following boiling temperatures of mercury have been reported for the following pressures:

Pressure (psia)	10	20	40	60	80
Temperature (°F)	637.0	706.0	784.4	835.7	874.8

At 40.0 psia and 784.4°F, the density of liquid mercury is 787.4 lbm/ft³ and the specific volume of the vapor is 1.648 ft³/lbm. Estimate the enthalpy of vaporization of mercury at 40.0 psia.

13.36 A vessel is filled with a liquid–vapor mixture of nitrogen at 115°R. What is the approximate pressure of the mixture?

13.37 Show that the slope of the sublimation curve of a simple compressible substance at the triple point is greater than that of the vaporization curve at the same point.

13.38 Show that the difference between $\bar{c}_p$ and $\bar{c}_v$ for a van der Waals gas may be given by

$$\bar{c}_p - \bar{c}_v = \frac{\bar{R}}{1 - 2a(\bar{v} - b)^2/\bar{R}T\bar{v}^3}.$$

13.39 Derive the relation between T and v for a van der Waals gas for a reversible adiabatic process with constant specific heats. How does it differ from that of an ideal gas?

13.40 Derive the relation between p and v for a van der Waals gas for a reversible adiabatic process with constant specific heats. How does it differ from that of an ideal gas?

13.41 A van der Waals gas is compressed reversibly and isothermally inside a cylinder. Derive an expression for the amount of heat transfer of the process.

13.42 Show that at constant temperature the enthalpy change for a van der Waals gas may be given as

$$\bar{h}_2 - \bar{h}_1 = p_2\bar{v}_2 - p_1\bar{v}_1 - a\left(\frac{1}{\bar{v}_2} - \frac{1}{\bar{v}_1}\right).$$

13.43 A rigid tank having a volume of 2.0 m³ contains oxygen at 20.0 MPa and 40°C. Find the mass of oxygen in the tank using
(a) the ideal-gas equation.
(b) the generalized compressibility chart.
(c) tabulated data.

13.44 Determine the volume occupied by 5 kg of carbon dioxide at 7.0 MPa and 100°C using
(a) the ideal-gas equation.
(b) the generalized compressibility chart.

13.45 Determine the density of nitrogen at 5000 psia and 80°F using
(a) the generalized compressibility chart.
(b) tabulated data.

13.46 Propane gas is compressed reversibly and isothermally in a steady-flow compressor from 14.7 psia to 100 psia. The temperature of the gas is 100°F. Neglecting changes in kinetic energy and potential energy, determine, for each pound of gas compressed, the amount of heat that must be removed from the compressor. Make use of the generalized entropy correction chart and give the answer in Btu/lbm.

13.47 Methane gas is compressed isothermally from 100 psia and 100°F to 2000 psia. Determine the enthalpy change for each pound of methane gas. Make use of the generalized enthalpy correction chart and give the answer in Btu/lbm.

13.48 Methane gas is compressed reversibly and isothermally in a steady-state steady-flow compressor from 300 K and 101.325 kPa to 20.0 MPa. Determine the compressor work required using

(a) the generalized enthalpy and generalized entropy charts.
(b) the generalized fugacity coefficient chart.

13.49 The property fugacity for a real gas at a given temperature and pressure may be calculated from available p–v–T data using the following relation:

$$f = p \, \exp\left[\frac{1}{\bar{R}T} \int_0^p \left(\bar{v} - \frac{\bar{R}T}{p}\right) dp_T\right],$$

where f and p would have the same units of pressure, and temperature is being held constant in carrying out the integration. Show that the expression above may be reduced to the well-known Lewis–Randall rule given by

$$f \approx \frac{p^2}{(\bar{R}T)/\bar{v}}$$

if $pv/RT = 1 + B'p$, where B' is a function of temperature only.

13.50 For a gas obeying the equation of state

$$\frac{pv}{RT} = 1 + B'p,$$

where B' is a function of temperature only, show that

$$\left(\frac{\partial h}{\partial p}\right)_T = -RT\left(T\frac{dB'}{dT}\right).$$

13.51 Determine the fugacity for hydrogen at 0°C and 200 atm ($v = 0.127$ m³/kg) using
(a) the generalized fugacity coefficient chart.
(b) the Lewis–Randall rule given in Problem 13.49.

13.52 One pound of nitrogen gas undergoes a change of state from 20 psia and 540°R to 500 psia and 540°R. Determine the change in enthalphy, in Btu/lbm, using

(a) the generalized enthalpy correction chart.

(b) the expression in Problem 13.50, taking for nitrogen at 540°R,

$$T\frac{dB'}{dT} = 2.5 \times 10^{-3} \text{ atm}^{-1}.$$

(c) tabulated data.

13.53 Expand the van der Waals equation in the form

$$\frac{pv}{RT} = 1 + \frac{B}{\bar{v}} + \frac{C}{\bar{v}^2} + \cdots$$

and show that the second virial coefficient B and the third virial coefficient C are given by

$$B = b - \frac{a}{\bar{R}T}$$

$$C = b^2.$$

13.54 Propane gas is compressed reversibly and isothermally from 300 K and 101.325 kPa to 20.0 MPa. Determine the compressed work required using the generalized fugacity coefficient chart.

13.55 Nitrogen enters a compressor at 300 K and 101.325 kPa and leaves at 300 K and 20.0 MPa. Determine the entropy change due to change of state using

(a) the ideal-gas equation.

(b) the generalized entropy chart.

13.56 Air changes state from 300 K and 200 kPa to 600 K and 1000 kPa. Determine the enthalpy change due to change of state using

(a) the ideal-gas equation.

(b) the generalized enthalpy chart.

13.57 Making use of Newton's method, write a computer program to determine the specific volume of a van der Waals gas with pressure and temperature given.

Run the program to determine the specific volume of nitrogen at the following states:

(a) 300 K and 0.5 MPa.

(b) 300 K and 10.0 MPa.

(c) 300 K and 50.0 MPa.

Compare the calculated results with tabulated data and comment.

13.58 Write a computer program to determine the quantity $(\bar{c}_p - \bar{c}_v)/\bar{R}$ for a van der Waals gas. (See Problems 13.38 and 13.57.)

Run the program to determine this quantity for nitrogen at the following states:

(a) 300 K and 0.5 MPa.

(b) 300 K and 10.0 MPa.

(c) 300 K and 50.0 MPa.

Compare the calculated results with those in the literature which are 1.02 for part (a), 1.45 for part (b), and 1.88 for part (c).

CHAPTER 14

Nonreactive Mixtures:
Gas–Gas and Gas–Vapor

A *nonreactive gas mixture* is a multicomponent system the constituents of which do not react chemically with one another. The properties of such a mixture of fixed proportions can be determined experimentally just as for a simple compressible substance, and they can be correlated, tabulated, and related analytically in the same way. This has been done for common mixtures such as air and for certain combustion products. However, it is evident that an unlimited number of mixtures is possible, and tabulation of the properties of all kinds of mixtures is obviously impractical. Consequently, it is important and necessary to have some method of calculating the properties of any mixture from the properties of its constituents. It is our primary objective in this chapter to explore how this can be done in the case of mixtures of ideal gases, and in the case of mixtures of gases and vapors at low pressures.

14.1 Mole Fraction, Mass Fraction, and Molecular Weight of a Mixture

Let us begin our discussion by considering some basic definitions that are relevant to our study of mixtures. The total mass m of a mixture is the sum of the masses of each component. That is,

$$m = m_1 + m_2 + m_3 + \cdots = \sum_i m_i \tag{14.1}$$

and

$$1 = \frac{m_1}{m} + \frac{m_2}{m} + \frac{m_3}{m} + \cdots = \sum_i \frac{m_i}{m}. \tag{14.2}$$

The ratio m_i/m is defined as the mass fraction of the ith component in the mixture. When an analysis of a gas mixture is made on the basis of mass or weight, it is called a *gravimetric analysis*.

The total number of moles n for a mixture is given by

$$n = n_1 + n_2 + n_3 + \cdots = \sum_i n_i. \tag{14.3}$$

Dividing Eq. 14.3 by n, we have

$$1 = \frac{n_1}{n} + \frac{n_2}{n} + \frac{n_3}{n} + \cdots = \sum_i \frac{n_i}{n} = \sum_i x_i, \tag{14.4}$$

where x_i is called the *mole fraction of the ith component* in the mixture. Thus, by definition, we have

$$x_i \equiv \frac{n_i}{n}. \tag{14.5}$$

When an analysis of a gas mixture is made on the basis of moles, it is called a *molal analysis*.

Now the mass m_i of the ith component may be given as $n_i M_i$, where M_i is the *molal mass* or *molecular weight* of the *ith component*. Therefore, the average molecular weight M of the mixture may be given by

$$M = \frac{\sum n_i M_i}{n} = \sum_i x_i M_i. \tag{14.6}$$

In view of the relation between mass and the number of moles, the mass fraction and mole fraction may readily be shown to be related by

$$\frac{m_i}{m} = \frac{M_i}{M} x_i. \tag{14.7}$$

Example 14.1

Approximating air as a mixture having a mole fraction of 0.79 of nitrogen and 0.21 of oxygen, determine

(a) the average molecular weight of air.
(b) the gravimetric analysis of air.

Solution

Using Eq. 14.6, we have

(a) $M = 0.79 \times 28.02 + 0.21 \times 32$

$= 28.86 \text{ lbm/lb mol} = 28.86 \text{ kg/kg mol}.$

(b) Using Eq. 14.7, we have

$$\text{mass fraction of nitrogen} = \frac{28.02 \times 0.79}{28.86} = 0.767$$

$$\text{mass fraction of oxygen} = \frac{32 \times 0.21}{28.86} = 0.233. \qquad \blacksquare$$

Comment

In many engineering applications, atmospheric air may be assumed to be a mixture of oxygen and nitrogen in the proportions given in this example. Readers should show that the molal analysis and the gravimetric analysis of atmospheric air may then be given alternatively as

4.76 mol of air = 1.0 mol of oxygen + 3.76 mol of nitrogen

4.30 kg of air = 1.0 kg of oxygen + 3.30 kg of nitrogen.

14.2 Dalton's Rule of Partial Pressure

Let us consider a gas mixture of n moles occupying a volume V at the equilibrium temperature and pressure of T and p. If the mixture may be treated as an ideal gas, we have

$$pV = n\bar{R}T, \qquad (14.8)$$

where

$$n = n_1 + n_2 + n_3 + \cdots = \sum_i n_i.$$

We may imagine that such a mixture can be separated into its constituents in such a way that each occupies a volume equal to that of the mixture and each is at the same temperature as the mixture. Let

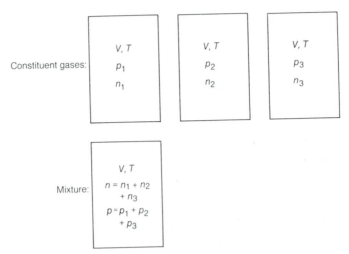

Figure 14.1 Dalton's Rule of Partial Pressure

us assume that each component may also be treated as an ideal gas. Then for each component, we have

$$p_i V = n_i \bar{R} T, \qquad (14.9)$$

where p_i is known as the *partial pressure of the ith component* in the mixture. Substituting Eq. 14.9 into Eq. 14.8, we have

$$p = p_1 + p_2 + p_3 + \cdots = \sum_i p_i. \qquad (14.10)$$

Equation 14.10 states the fact that the total pressure of a mixture of ideal gases is equal to the sum of the partial pressures when the partial pressures are determined at the volume and temperature of the mixture. This is *Dalton's rule of partial pressure*, the physical representation of which is shown in Fig. 14.1. Dalton's rule has been confirmed experimentally for gases at relatively low pressures. It is also obtainable from statistical thermodynamics on the basis that each component is not influenced by the presence of the other components. The microscopic assumption is consistent with the macroscopic assumption that each component occupies the same volume as the mixture and is at the temperature of the mixture.

Dividing Eq. 14.9 by Eq. 14.8, we have

$$\frac{p_i}{p} = \frac{n_i}{n} = x_i$$

or

$$p_i = x_i p. \tag{14.11}$$

Equation 14.11 states that the partial pressure of a component of an ideal-gas mixture is equal to the mole fraction of that component times the total pressure of the mixture.

Now Eq. 14.9 may be written as

$$p_i V = m_i R_i T, \tag{14.12}$$

where R_i is the *specific gas constant of the ith component* in the mixture. Substituting Eq. 14.12 into Eq. 14.10, we have

$$pV = T \sum_i m_i R_i. \tag{14.13}$$

For the ideal-gas mixture itself, we may now write

$$pV = mRT, \tag{14.14}$$

where R is the average specific gas constant of the mixture. Comparing Eq. 14.13 with Eq. 14.14, we have

$$R = \frac{\sum m_i R_i}{m}. \tag{14.15}$$

14.3 Amagat–Leduc Rule of Partial Volume

Let us now imagine that an ideal-gas mixture can be separated into its constituents in such a way that each occupies a volume at the same temperature and pressure as the mixture. The volume occupied by each component is then given by

$$V_i = \frac{n_i \bar{R} T}{p}. \tag{14.16}$$

where V_i is the *partial volume of the ith component* by definition. Dividing V_i by the volume of the mixture given by Eq. 14.8, we have

$$\frac{V_i}{V} = \frac{n_i}{n} = x_i. \tag{14.17}$$

The ratio V_i/V is defined as the *volume fraction*. Equation 14.17 says that for an ideal-gas mixture, the mole fraction is equal to the volume fraction. Consequently, a volumetric analysis is the same as a molal analysis in the case of an ideal-gas mixture. This fact has been exploited for the determination of gas composition experimentally.

Adding up all the partial volumes of an ideal-gas mixture, we have

$$\sum V_i = \sum x_i V$$

$$\sum V_i = V \sum x_i = V \tag{14.18}$$

as $\sum x_i = 1$. Equation 14.18 states the fact that the total volume of a mixture of ideal gases is equal to the sum of the partial volumes of the constituent gases when the partial volumes are determined at the pressure and temperature of the mixture. This is the *Amagat–Leduc rule of partial volume*, the physical representation of which is shown in Fig. 14.2.

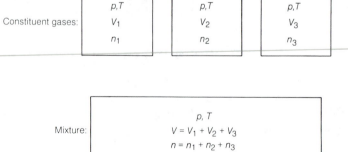

Figure 14.2 Amagat–Leduc Rule of Partial Volume

Example 14.2

The volumetric analysis of a Los Angeles, California, natural gas is as follows:

$$CH_4 \quad 85.8\%$$

$$C_2H_6 \quad 13.2\%$$

$$CO_2 \quad 0.9\%$$

$$N_2 \quad 0.1\%.$$

Determine for this natural gas

(a) the average molecular weight.
(b) the gravimetric analysis.
(c) the average gas constant.

Solution

Assume that the natural gas is an ideal-gas mixture. Then the volumetric analysis is the same as the molal analysis.

(a) Using Eq. 14.6, we have

$$M = 0.858 \times 16 + 0.132 \times 30 + 0.009 \times 44 + 0.001 \times 28$$

$$= 18.11 \text{ lbm/lb mol} = 18.11 \text{ kg/kg mol.}$$

(b) Using Eq. 14.7, we have

$$\frac{m_i}{m} = \frac{M_i}{M} x_i.$$

Therefore,

$$\text{mass fraction of } CH_4 \ = \frac{16}{18.11} \times 0.858 = 0.7580$$

$$\text{mass fraction of } C_2H_6 = \frac{30}{18.11} \times 0.132 = 0.2187$$

$$\text{mass fraction of } CO_2 \ = \frac{44}{18.11} \times 0.009 = 0.0218$$

$$\text{mass fraction of } N_2 \ \ = \frac{28}{18.11} \times 0.001 = 0.0015.$$

(c) We may calculate the average gas constant in two ways. Making use of the average molecular weight, we have

$$R = \frac{\bar{R}}{M}$$

$$= \frac{8.3143}{18.11} = 0.4591 \text{ kJ/kg} \cdot \text{K}.$$

Making use of the mass fractions, we have

$$R = \frac{\sum m_i R_i}{m}$$

$$= 0.7580 \times \frac{8.3143}{16} + 0.2187 \times \frac{8.3143}{30} + 0.0218 \times \frac{8.3143}{44}$$

$$+ 0.0015 \times \frac{8.3143}{28}$$

$$= 0.4591 \text{ kJ/kg} \cdot \text{K.} \qquad \blacksquare$$

14.4 Internal Energy, Enthalpy, and Specific Heats of an Ideal-Gas Mixture

Since each component of an ideal-gas mixture behaves as if it occupies the volume of the mixture alone at the temperature of the mixture, the internal energy of each component may be given as

$$U_i = n_i \bar{u}_i, \qquad (14.19)$$

where $\bar{u}_i$ is the molal internal energy of the ith component and is defined by T and V, or equivalently by T and p_i.

Considering a mixture as a system consisting of several subsystems, we see that the internal energy of a mixture is clearly given by

$$U = U_1 + U_2 + U_3 + \cdots = \sum_i U_i$$

$$= \sum_i n_i \bar{u}_i. \qquad (14.20)$$

Dividing Eq. 14.20 by n, we get the molal internal energy of a mixture:

$$\bar{u} = \sum_i x_i \bar{u}_i . \tag{14.21}$$

We recall that for any ideal gas, the internal energy is a function of temperature only, and the change in its internal energy is related to its constant-volume specific heat by

$$d\bar{u} = \bar{c}_v \, dT . \tag{14.22}$$

Making use of Eqs. 14.22 and 14.21, we have, for a mixture of ideal gases,

$$\bar{c}_v = \sum_i x_i \bar{c}_{vi} . \tag{14.23}$$

From the definition of enthalpy, the enthalpy of each component in a mixture is given by

$$H_i = U_i + p_i V . \tag{14.24}$$

Adding up the enthalpy of all the components in the mixture, we get

$$\sum_i H_i = \sum_i U_i + \sum_i p_i V$$

$$= \sum_i U_i + V \sum_i p_i . \tag{14.25}$$

But

$$\sum_i U_i = U \qquad \text{according to Eq. 14.20}$$

and

$$\sum_i p_i = p \qquad \text{according to Eq. 14.10.}$$

Therefore,

$$\sum_i H_i = U + pV.$$

But, by definition,

$$H = U + pV.$$

Thus we have

$$H = \sum_i H_i. \tag{14.26}$$

Now

$$H_i = n_i \bar{h}_i$$

and

$$H = n\bar{h}.$$

Making use of these expressions, we get

$$H = \sum_i n_i \bar{h}_i$$

and

$$\bar{h} = \sum_i x_i \bar{h}_i. \tag{14.27}$$

For an ideal gas, the enthalpy is also a function of temperature only and the change in enthalpy is related to the constant-pressure specific heat by

$$d\bar{h} = \bar{c}_p \, dT. \tag{14.28}$$

Making use of Eqs. 14.28 and 14.27, we have, for a mixture of ideal gases,

$$\bar{c}_p = \sum_i x_i \bar{c}_{pi}. \tag{14.29}$$

In terms of the mass of each component, the following expressions may readily be shown to be valid for a mixture of ideal gases:

$$U = \sum m_i u_i \tag{14.30}$$

$$u = \frac{\sum m_i u_i}{m} \tag{14.31}$$

$$c_v = \frac{\sum m_i c_{vi}}{m} \tag{14.32}$$

$$H = \sum_i m_i h_i \tag{14.33}$$

$$h = \frac{\sum m_i h_i}{m} \tag{14.34}$$

$$c_p = \frac{\sum m_i c_{pi}}{m}. \tag{14.35}$$

Example 14.3

An ideal-gas mixture of nitrogen and argon flows through a heater at the rate of 50 kg/min. If the mixture enters at 40°C and 1.0 atm, and leaves at 250°C and 1.0 atm, what is the rate of heat addition if the mixture is 40 percent nitrogen by volume?

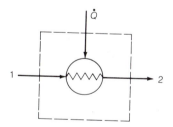

Solution

System selected for study: heater
Unique features of process: steady-state steady flow; no work transfer
Assumptions: neglect ΔKE and ΔPE

From the first law for steady flow, we have

$$\dot{Q} = \dot{m}(h_2 - h_1)$$

$$= \frac{\dot{m}(\bar{h}_2 - \bar{h}_1)}{M}. \tag{1}$$

For an ideal-gas mixture of fixed proportions, we have

$$d\bar{h} = \bar{c}_p \, dT.$$

Assuming constant specific heats, we have

$$\bar{h}_2 - \bar{h}_1 = \bar{c}_p(T_2 - T_1), \tag{2}$$

where

$$\bar{c}_p = \sum_i x_i \bar{c}_{pi}.$$

Using $\bar{c}_{pN} = \frac{7}{2}\bar{R}$ and $\bar{c}_{pA} = \frac{5}{2}\bar{R}$, we have

$$\bar{c}_p = 0.40 \times \tfrac{7}{2}\bar{R} + 0.60 \times \tfrac{5}{2}\bar{R}$$

$$= 2.9\bar{R} = 2.9 \times 8.3143 = 24.111 \text{ kJ/kg mol} \cdot \text{K}. \tag{3}$$

The average molecular weight of the mixture is given by

$$M = \sum_i x_i M_i$$

$$= 0.40 \times 28 + 0.60 \times 39.95$$

$$= 35.17 \text{ kg/kg mol}. \tag{4}$$

Substituting into Eq. 1, we get

$$\dot{Q} = \frac{50 \times 24.111 \times (250 - 40)}{35.17} \text{ kJ/min}$$

$$= 7198 \text{ kJ/min}. \qquad\blacksquare$$

14.5 Entropy of an Ideal-Gas Mixture

Using the same kind of reasoning in arriving at the relation for the internal energy of a mixture of ideal gases, we can express the entropy of the mixture in terms of the entropies of its constituents in the following manner:

$$S = \sum_i S_i = \sum_i n_i \bar{s}_i. \tag{14.36}$$

Equation 14.36 represents the Gibbs rule, which states that the entropy of an ideal-gas mixture is equal to the sum of the entropies that each component of the mixture would have if each alone occupied the volume of the mixture at the temperature of the mixture. Dividing Eq. 14.36 by n, we have

$$\bar{s} = \sum_i x_i \bar{s}_i.$$
(14.37)

The change in entropy of a component in an ideal-gas mixture is obtained by application of Eq. 5.18. For each component, we get

$$d\bar{s}_i = \bar{c}_{pi} \frac{dT_i}{T_i} - \bar{R} \frac{dp_i}{p_i}.$$
(14.38)

Substituting Eq. 14.38 into the differential form of Eq. 14.37, we have, for an ideal-gas mixture of fixed proportions,

$$d\bar{s} = \sum_i x_i \bar{c}_{pi} \frac{dT_i}{T_i} - \sum_i x_i \bar{R} \frac{dp_i}{p_i}.$$
(14.39)

In terms of the mass of each component, we have

$$ds = \frac{1}{m} \sum_i m_i c_{pi} \frac{dT_i}{T_i} - \frac{1}{m} \sum_i m_i R_i \frac{dp_i}{p_i}.$$
(14.40)

It is important to remember that the pressure to be used in Eqs. 14.39 and 14.40 for each component is always the pressure determined by its temperature and its total volume.

Example 14.4

Determine the entropy increase when n_A moles of gas A at a given temperature and pressure mix with n_B moles of gas B at the same temperature and pressure in an adiabatic constant-volume process. Assume ideal-gas behavior.

Solution

System selected for study: n_A moles of gas A and n_B moles of gas B
Unique features of process: adiabatic; constant volume; no work transfer
Assumption: ideal-gas behavior

From the first law, we have

$$U_2 = U_1$$

or

$$n_A(u_{A2} - u_{A1}) + n_B(u_{B2} - u_{B1}) = 0. \tag{1}$$

Since gas A and gas B are ideal gases, Eq. 1 may be written as

$$n_A \int_{T_{A1}}^{T_2} \bar{c}_{vA} \, dT + n_B \int_{T_{B1}}^{T_2} \bar{c}_{vB} \, dT = 0. \tag{2}$$

Since $T_{A1} = T_{B1}$, Eq. 2 can be satisfied only if $T_2 = T_{A1} = T_{B1}$. We conclude that the temperature of the gases will not change.

The initial and final pressures of gas A are given by

$$p_{A1} V_{A1} = n_A \bar{R} T_{A1}$$

$$p_{A2} V_{A2} = n_A \bar{R} T_{A2}.$$

Now $T_{A1} = T_{A2}$, and $V_{A2} = V$, the volume of the mixture. Thus

$$p_{A2} = \frac{V_{A1}}{V} p_{A1}. \tag{3}$$

In a similar manner, we find that

$$p_{B2} = \frac{V_{B1}}{V} p_{B1}. \tag{4}$$

The pressure of the mixture is then given by

$$p_2 = p_{A2} + p_{B2}$$

$$= \frac{1}{V} (V_{A1} + V_{B1}) p_1, \tag{5}$$

where

$$p_1 = p_{A1} = p_{B1}.$$

But $V = V_{A1} + V_{B1}$. Therefore, we have

$$p_2 = p_1.$$

We see that the mixture pressure is equal to the initial pressure.

From the second law, we have

$$S_2 - S_1 = n_A \int_{T_{A1}}^{T_2} \bar{c}_{pA} \frac{dT}{T} + n_B \int_{T_{B1}}^{T_2} \bar{c}_{pB} \frac{dT}{T} - n_A \bar{R} \ln \frac{p_{A2}}{p_{A1}} - n_B \bar{R} \ln \frac{p_{B2}}{p_{B1}}. \tag{6}$$

Since the temperature of the gases will not change, Eq. 6 becomes

$$S_2 - S_1 = -n_A \bar{R} \ln \frac{p_{A2}}{p_{A1}} - n_B \bar{R} \ln \frac{p_{B2}}{p_{B1}} . \tag{7}$$

From Dalton's rule,

$$p_{A2} = x_A p_2$$

$$p_{B2} = x_B p_2 .$$

Now $p_2 = p_1 = p_{A1} = p_{B1}$. Thus Eq. 7 becomes

$$S_2 - S_1 = -\bar{R}(n_A \ln x_A + n_B \ln x_B). \tag{8} \quad \blacksquare$$

Comments

Since x_A and x_B are both fractions, the entropy increase given by Eq. 8 will be positive, which means that the mixing process is an irreversible one. For the general case of mixing any number of ideal gases initially at the same temperature and pressure, the entropy increase is given by

$$S_2 - S_1 = -\bar{R} \sum_i n_i \ln x_i . \tag{9}$$

It is to be noted that Eq. 9 is for mixing different gases. There is no entropy change in mixing samples of the same gas that are initially at the same temperature and pressure.

Example 14.5

A mixture consisting of 7 mol of helium and 3 mol of nitrogen is expanded isentropically in a nozzle from 750 kPa and 60°C to a pressure of 150 kPa. Assuming ideal-gas behavior, determine

(a) the final temperature of the mixture.
(b) the entropy change for each gas.

Solution

The entropy change for 1.0 mol of mixture is given by Eq. 14.39:

$$d\bar{s} = \sum_i x_i \bar{c}_{pi} \frac{dT}{T} - \sum_i x_i \bar{R} \frac{dp_i}{p_i} . \tag{1}$$

Since $x_i = p_i/p$, $dp_i/p_i = dp/p$, and $\sum_i x_i = 1$, Eq. 1 may be written as

$$d\bar{s} = \sum_i x_i \bar{c}_{pi} \frac{dT}{T} - \bar{R} \frac{dp}{p} . \tag{2}$$

For an isentropic process, $d\bar{s} = 0$. Thus

$$\sum x_i \bar{c}_{pi} \frac{dT}{T} - \bar{R} \frac{dp}{p} = 0$$

and

$$(0.7\bar{c}_{pHe} + 0.3\bar{c}_{pN}) \frac{dT}{T} - \bar{R} \frac{dp}{p} = 0. \tag{3}$$

Letting $\bar{c}_p = \frac{7}{2}\bar{R}$ for nitrogen and $\bar{c}_p = \frac{5}{2}\bar{R}$ for helium, we have

$$\left(0.7 \times \frac{5}{2} + 0.3 \times \frac{7}{2}\right) \frac{dT}{T} - \frac{dp}{p} = 0. \tag{4}$$

Integrating Eq. 4, we get

$$2.8 \ln \frac{T_2}{T_1} - \ln \frac{150}{750} = 0.$$

Solving for T_2, we get

(a) $$T_2 = \frac{T_1}{1.777} = \frac{333.15}{1.777} = 187.5 \text{ K}.$$

(b) The entropy change for each component is given by

$$S_{i2} - S_{i1} = n_i \left(\bar{c}_{pi} \ln \frac{T_2}{T_1} - \bar{R} \ln \frac{p_2}{p_1}\right). \tag{5}$$

For the nitrogen gas, we have

$$S_{N2} - S_{N1} = 3\left(\frac{7}{2} \bar{R} \ln \frac{187.5}{333.15} - \bar{R} \ln \frac{150}{750}\right)$$

$$= 3 \times 8.3143\left(3.5 \ln \frac{187.5}{333.15} - \ln \frac{150}{750}\right)$$

$$= -10.0 \text{ kJ/K}.$$

For the helium gas, we have

$$S_{He2} - S_{He1} = 7 \times 8.3143\left(2.5 \ln \frac{187.5}{333.15} - \ln \frac{150}{750}\right)$$

$$= +10.0 \text{ kJ/K.} \qquad \blacksquare$$

Comments

We note that the entropy of the nitrogen gas decreases while the entropy of the helium gas increases in this isentropic expansion process. It can be shown that the reverse will be true in an isentropic compression process involving this ideal-gas mixture of two components. This is because nitrogen has a greater molal specific heat than helium.

14.6 Mixtures of Ideal Gases and a Condensable Vapor

A gas–vapor mixture is an important type of gas mixture from which one or more of the constituent gases can be condensed out. The component that may condense out is called the *vapor of the mixture*. The most important engineering example of mixtures of gases with a condensable vapor is the air–water–vapor mixture such as the atmospheric air. We shall focus our discussion on air–water–vapor mixtures because of their wide range of applications, but the principles involved will apply equally well to any other gas–vapor mixture.

The air–water–vapor mixture encountered in many engineering problems usually exists under such a low pressure that the vapor, as well as the non-condensables generally designated as the dry air, may be treated as an ideal gas. Consequently, the analysis of problems involving such a mixture is, in general, similar to that of ideal-gas mixtures. However, we should keep in mind that there is one important difference between a gas–vapor mixture and a mixture of ideal gases: while the composition of an ideal-gas mixture will remain constant, partial condensation of the vapor in a gas–vapor mixture may result from a decrease of only a few degrees in its temperature.

The properties of an air–water–vapor mixture are important in many industrial processes, such as those in the paper and textile industries that require close control of the vapor content of the atmosphere as well as its temperature. They are also important in the air conditioning of buildings for comfort and in the design of cooling towers for use in heat-removal systems. For these reasons, air–water–vapor mixtures have been studied in great detail, and the name *psychrometric* has been given to this subject. We shall introduce some of the more important terms and definitions from the vocabulary of psychrometrics to facilitate our discussion.

The thermodynamic state of the vapor in an air–water–vapor mixture is fixed by its partial pressure and temperature. Under ordinary conditions, it is a superheated vapor, such as point 1 shown in Fig. 14.3. If the partial pressure of the water vapor corresponds to the saturation pressure of water at the mixture temperature, the mixture

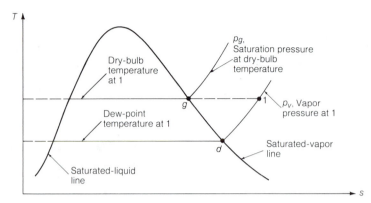

Figure 14.3 Vapor Pressure, Saturation Pressure, Dew-Point Temperature, and Dry-Bulb Temperature

is said to be saturated. *Saturated air* is a mixture of dry air and saturated water vapor. *Unsaturated air* is a mixture of dry air and superheated vapor.

If an unsaturated air is cooled at constant pressure, the mixture will eventually reach the saturation temperature corresponding to the partial pressure of the water vapor. This is called the *dew-point temperature* because at this temperature the vapor starts to condense, resulting in the formation of liquid droplets or dew. The dew-point temperature corresponding to a partial pressure of p_v is shown in Fig. 14.3.

In the study of air–water–vapor mixtures, we shall frequently speak of dry-bulb temperature and wet-bulb temperature in addition to the dew-point temperature. The *dry-bulb temperature* is simply the equilibrium temperature of the mixture indicated by an ordinary thermometer. The *wet-bulb temperature* is the temperature indicated by a wet-bulb thermometer which has its temperature-sensitive element covered with a wicklike material soaked in water (see Fig. 14.4). It will be seen that for an unsaturated air, the wet-bulb temperature is always less than the dry-bulb temperature but greater than the dew-point temperature. For saturated air, the dry-bulb, wet-bulb, and dew-point temperatures are the same.

One way of indicating the composition of an air–water–vapor mixture is through the use of the term *specific humidity*, ω, also known as the *humidity ratio*. It is defined as the ratio of the mass of water vapor to the mass of dry air in a given volume of mixture, or

$$\omega = \frac{m_v}{m_a}, \qquad (14.41)$$

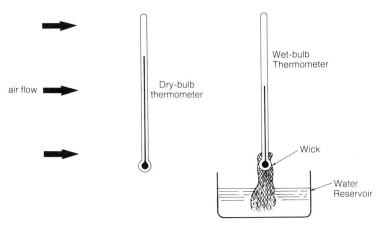

Figure 14.4 Dry-Bulb and Wet-Bulb Temperatures

where m_v is the mass of water vapor in a given volume of mixture and m_a is the mass of dry air in the same volume of mixture.

Assuming ideal-gas behavior for both the dry air and the vapor, we have

$$m_v = \frac{p_v V}{R_v T}$$

and

$$m_a = \frac{p_a V}{R_a T}.$$

Substituting into Eq. 14.42, we get

$$\omega = \frac{R_a}{R_v} \frac{p_v}{p_a}$$

$$= 0.622 \frac{p_v}{p_a}. \tag{14.42}$$

where p_v is the partial pressure of vapor and p_a is the partial pressure of dry air in the same volume of mixture. We have used $R_a = 0.2870$ kJ/kg $\cdot$ K and $R_v = 0.4615$ kJ/kg $\cdot$ K in arriving at the numerical value of 0.622 in Eq. 14.42.

From Dalton's rule of partial pressure, the pressure of the mixture p is given as

$$p = p_a + p_v.$$

Thus Eq. 14.42 may be written as

$$\omega = 0.622 \, \frac{p_v}{p - p_v}$$

or

$$p_v = \frac{\omega p}{\omega + 0.622}. \tag{14.43}$$

Equation 14.43 shows that if the total pressure and the specific humidity are constant, the partial pressure of the vapor must also be constant.

The amount of water vapor diffused through the dry air in an air–water–vapor mixture may vary from practically nothing to that necessary for saturation conditions. As an indication of the degree of saturation of the mixture, the term *relative humidity* is employed. Relative humidity ϕ is defined as the ratio of the partial pressure of water vapor in a mixture to the saturation pressure of water at the dry-bulb temperature, or

$$\phi = \frac{p_v}{p_g}, \tag{14.44}$$

where p_v is the partial pressure of the water vapor in the mixture and p_g is the saturation pressure of water at the temperature of the mixture. On the basis of ideal-gas behavior, the relative humidity may also be given as

$$\phi = \frac{v_g}{v_v} = \frac{\rho_v}{\rho_g}. \tag{14.45}$$

An expression for the relation between the relative humidity ϕ and the specific humidity ω can be obtained by combining Eq. 14.44 with Eq. 14.42. Such an expression for an air–water–vapor mixture is

$$\omega = 0.622\phi \, \frac{p_g}{p_a}. \tag{14.46}$$

Example 14.6

Moist air at 40°C, 101.325 kPa, and a relative humidity of 60% initially is cooled at a constant mixture pressure to 20°C. Determine

(a) the final relative humidity.
(b) the change in specific humidity.

Solution

(a) From saturation tables for steam, we have, at 40°C,

$$p_g = 7.375 \text{ kPa}.$$

Therefore,

$$p_{v1} = 0.60 \times 7.375 = 4.425 \text{ psia}$$

and

$$\omega_1 = 0.622 \, \frac{4.425}{101.325 - 4.425}$$

$$= 0.0284 \text{ kg of water vapor/kg of dry air}.$$

At 20°C, the saturation pressure of water is 2.337 kPa, which is less than the initial partial pressure of water vapor in the mixture. Consequently, we must have saturated air at 20°C. Therefore,

$$p_{v2} = 2.337 \text{ kPa}$$

$$\phi_2 = 100\%.$$

(b)

$$\omega_2 = 0.622 \, \frac{2.337}{101.325 - 2.337}$$

$$= 0.0147 \text{ kg of water vapor/kg of dry air}.$$

Thus

$$\omega_2 - \omega_1 = 0.0147 - 0.0284$$

$$= -0.0137 \text{ kg of water/kg of dry air.} \quad \blacksquare$$

Comment

The change in specific humidity represents the amount of water vapor, per kilogram of dry air, removed by condensation achieved through cooling the

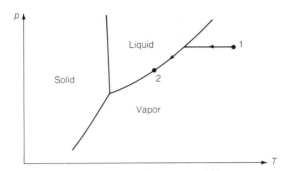

Figure 14.5 Change in Vapor Pressure Due to Cooling Below Dew-Point Temperature

mixture below the dew-point temperature. The initial and final states of the vapor are shown in the phase diagram for water in Fig. 14.5.

14.7 Determination of Vapor Pressure of Water and Dew-Point Temperature

The saturation pressure of water, p_g, in general may be obtained from the steam tables. However, this property of water has been fitted to experimental data over a fairly wide temperature range (0 to 200°C) by the following expression:

$$\ln (p_g) = 70.4346943 - \frac{7362.6981}{T} + 0.006952085\,T - 9.0000 \ln (T),$$

$$(14.47)$$

where p_g is in atmospheres and T is the Kelvin temperature.

For many applications, such as in air-conditioning calculations, an approximation of Eq. 14.47, but a much simpler expression, has been obtained by Liley[1]:

$$\ln (p_g) = 14.43509 - \frac{5333.3}{T},$$
$$(14.48)$$

where p_g is in bar and T is the Kelvin temperature.

Now the dew-point temperature by definition is simply the saturation temperature corresponding to the partial pressure of water in a

[1] P. E. Liley, "Approximations for the Thermodynamic Properties of Air and Steam Useful in Psychrometric Calculations," *Mechanical Engineering News*, Vol. 17, No. 4 (1980).

mixture. In general this temperature can be obtained from the steam tables once the partial pressure of water vapor is known. However, interpolation is usually required. On the other hand, the use of Eq. 14.48 for the determination of the dew-point temperature is a simple matter.

Example 14.7

The pressure and temperature of the air in a room are 101.325 kPa and 25°C. If the relative humidity is 40%, determine

(a) the saturation pressure of water vapor at the dry-bulb temperature using the steam tables, Eq. 14.47, and Eq. 14.48.
(b) the dew-point temperature using the steam tables.
(c) the dew-point temperature using Eq. 14.48.

Solution

(a) From saturation tables for steam, we have at 25°C,

$$p_g = 3.166 \text{ kPa.}$$

Using Eq. 14.47,

$$\ln(p_g) = 70.4346943 - \frac{7362.6981}{298.15} + 0.006952085 \times 298.15$$

$$- 9.000 \ln(298.15).$$

Solving for p_g yields

$$p_g = 0.031256 \text{ atm} = 3.167 \text{ kPa.}$$

Using Eq. 14.48, we have

$$\ln(p_g) = 14.43509 - \frac{5333.3}{298.15}.$$

Solving for p_g gives us

$$p_g = 0.03165 \text{ bar} = 3.165 \text{ kPa.}$$

(b) $$p_v = 0.4 \, p_g$$

$$= 0.4 \times 3.166 \text{ kPa} = 1.2664 \text{ kPa.}$$

The dew-point temperature is the saturation temperature corresponding to 1.2664 kPa. From the steam tables,

$$t_{dp} = \text{saturation temperature at } 1.2664 \text{ kPa} = 10.46°C.$$

(c) Using Eq. 14.48, we have

$$p_v = 0.4 \times 3.165 \text{ kPa}$$

$$= 1.266 \text{ kPa} = 0.03165 \text{ bar}.$$

Then

$$\ln (0.03165) = 14.43509 - \frac{5333.3}{T_{dp}}.$$

Solving for T_{dp}, we obtain

$$T_{dp} = 283.62 \text{ K}$$

$$t_{dp} = 10.47°C. \qquad \blacksquare$$

Comment

We observe that the Liley equation (Eq. 14.48) is very accurate. It makes the use of steam tables not necessary.

14.8 Enthalpy and Entropy of an Air–Water–Vapor Mixture

Since most engineering devices and systems involving air–water–vapor mixtures operate under steady-state steady-flow conditions, the enthalpy is an important property in engineering calculations. Assuming that the Gibbs rule is valid, the enthalpy H of a given mixture is simply the sum of the enthalpies of the dry air and the water vapor. That is,

$$H = H_a + H_v$$

$$= m_a h_a + m_v h_v. \qquad (14.49)$$

It is customary to express the specific enthalpy of a mixture h in Btu/lbm of dry air. Dividing Eq. 14.49 by m_a, we have

$$h = h_a + \frac{m_v}{m_a} h_v$$

or

$$h = h_a + \omega h_v,\qquad(14.50)$$

where

h = specific enthalpy of mixture in kJ/kg of dry air

h_a = specific enthalpy of the dry air in kJ/kg of dry air

h_v = specific enthalpy of water vapor in kJ/kg of water vapor.

The enthalpy of air may be approximated by

$$h_a = c_{pa}\, t,\qquad(14.51)$$

where t is the dry-bulb temperature in °C or °F depending on the units of c_{pa}.

In SI units, c_{pa} may be taken as 1.005 kJ/kg · K. Then

$$h_a = 1.005t \text{ kJ/kg},\qquad(14.52)$$

where t is in °C.

In English units, c_{pa} may be taken as 0.240 Btu/lbm · °R. Then

$$h_a = 0.0240t \text{ Btu/lbm},\qquad(14.53)$$

where t is in °F.

The enthalpy of the water vapor h_v in general depends on the partial pressure p_v and the dry-bulb temperature. However, it happens that for water vapor at the partial pressure existing in atmospheric air, h_v may be approximated by the value h_g (enthalpy of saturation) corresponding to the dry-bulb temperature. But a better approximation is available.

In SI units, the enthalpy of water vapor h_v may be taken as

$$h_v = 2500 + 1.86t \text{ kJ/kg},\qquad(14.54)$$

where t is the dry-bulb temperature in °C, and the datum state is liquid water at 0°C.

In English units, the enthalpy of water vapor h_v may be taken as

$$h_v = 1061 + 0.444t \text{ Btu/lbm},\qquad(14.55)$$

where t is the dry-bulb temperature in °F, and the datum state is liquid water at 32°F.

Making use of Eqs. 14.52 and 14.54, the enthalpy of an air–water–vapor mixture in SI units may now be given by the following approximate relationship:

$$h = 1.005t + \omega(2500 + 1.86t) \text{ kJ/kg of dry air}, \qquad (14.56)$$

where t is in °C.

Making use of Eqs. 14.53 and 14.55, the enthalpy of an air–water–vapor mixture in English units may now be given by the following approximate relationship:

$$h = 0.240t + \omega(1061 + 0.444t) \text{ Btu/lbm of dry air}, \qquad (14.57)$$

where t is in °F.

According to the Gibbs rule, the entropy of the mixture is the sum of the entropies of the dry air and the water vapor. Thus, in an analogous manner, we can obtain the following expression for the specific entropy of the mixture:

$$s = s_a + \omega s_v, \qquad (14.58)$$

where s_a is the specific entropy of dry air and s_v is the specific entropy of water vapor. In using Eq. 14.58, it is important to remember that the entropy values of the components must be evaluated at the dry-bulb temperature and the appropriate partial pressure.

14.9 Adiabatic Saturation Process

We have made use of the concept of specific humidity in our discussions and in our calculations, but we have not mentioned how this quantity may be measured. Actually, there is no convenient way to measure either the specific humidity or the relative humidity directly. In this section, we wish to show how the specific humidity may be calculated.

Let us consider the process shown in Fig. 14.6. An unsaturated air of dry-bulb temperature t_1 enters an insulated chamber containing a body of water that has a very large surface area. It leaves as saturated air with a temperature of t_2. When the unsaturated air passes over the water surface, some of the water is evaporated. The energy for evaporation comes from both the moist air and the water in the chamber. Since the system is insulated, the temperature of the air

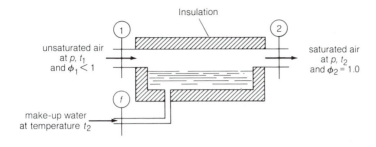

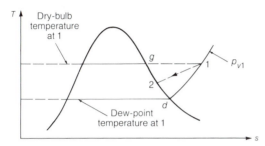

Figure 14.6 Adiabatic Saturation Process

decreases. Therefore, t_2 will be lower than t_1. If the water being added to the air is equal to t_2, the final equilibrium temperature is called the *thermodynamic wet-bulb temperature*, also known as the *adiabatic saturation temperature*.

Let us now apply the first law to the entire apparatus. The process is adiabatic. If the process is also steady-state steady flow, we have, neglecting changes in kinetic and potential energies,

$$m_a h_{a1} + m_{v1} h_{v1} + (m_{v2} - m_{v1})h_{f2} = m_a h_{a2} + m_{v2} h_{v2}. \quad (14.59)$$

Dividing by the mass of dry air m_a, we get

$$h_{a1} + \omega_1 h_{v1} + (\omega_2 - \omega_1)h_{f2} = h_{a2} + \omega_2 h_{v2}. \quad (14.60)$$

Now h_{v1} may be approximated as h_{g1}, the enthalpy of saturation at t_1, and h_{v2} may be approximated as h_{g2}, the enthalpy of saturation at t_2. Thus Eq. 14.60 may be written as

$$h_{a1} + \omega_1 h_{g1} + (\omega_2 - \omega_1)h_{f2} = h_{a2} + \omega_2 h_{g2}$$

and

$$\omega_1 = \frac{(h_{a2} - h_{a1}) + \omega_2(h_{g2} - h_{f2})}{h_{g1} - h_{f2}}$$

$$= \frac{c_{pa}(t_2 - t_1) + \omega_2 h_{fg2}}{h_{g1} - h_{f2}}. \tag{14.61}$$

Now

$$\omega_2 = 0.622 \frac{p_{v2}}{p - p_{v2}}.$$

Since the air is saturated at point 2, the relative humidity at point 2 is unity. Then p_{v2} is simply the saturation pressure of water at t_2. This means that ω_2 may be calculated if we know p and t_2. We thus see that ω_1 may be determined from Eq. 14.61 if we measure t_1, t_2, and the mixture pressure. Knowing ω_1, the relative humidity ϕ_1 and the vapor pressure p_{v1} may also be calculated.

The states of the water vapor in the mixture during the adiabatic process are shown in the T–s diagram in Fig. 14.6. Owing to an increase in the moisture content, the vapor pressure increases during the process as the mixture pressure remains constant. Owing to evaporative cooling, the temperature of the mixture which is also the temperature of the vapor, decreases during the process. The thermodynamic wet-bulb temperature is therefore higher than the dew-point temperature and lower than the dry-bulb temperature.

The exact determination of the thermodynamic wet-bulb temperature is not practical. However, a good approximation may be obtained by means of a wet-bulb thermometer, shown in Fig. 14.4. As air flows around the wet bulb, water is being evaporated from the wick, thereby making the air inside the wick saturated at the temperature of the water in the wick. Because there is continuous heat removal (latent heat of vaporization), the wet-bulb thermometer will register a lower reading of temperature than will the dry-bulb thermometer. It happens that this wet-bulb temperature is very close to the adiabatic saturation temperature for air–water–vapor mixtures. We wish to point out that this equality does not hold for most gas–vapor mixtures. We also wish to point out that a wet-bulb temperature as read from a wet-bulb thermometer is influenced by heat- and mass-transfer rates and is therefore not a sole function of the thermodynamic state of the air to which the thermometer is exposed. The wet-bulb temperature is therefore not a property of the moist air. Thus in psychrometric equations and psychrometric charts where the wet-bulb temperature appears, it is always the thermodynamic wet-bulb temperature that is considered.

It is possible to make use of Eq. 14.61 to determine the vapor pressure of moist air. However, it is more convenient to modify this equation so that calculation of the actual vapor pressure may be made directly from dry-bulb and wet-bulb temperatures. The resulting relation is called *Carrier's equation*[2] and is given as

$$p_v = p_{gw} - \frac{(p - p_{gw})(t_d - t_w)}{2800 - 1.3t_w},$$ (14.62)

where

p_v = actual vapor pressure, psia

p_{gw} = saturation pressure corresponding to wet-bulb temperature, psia

p = total pressure of mixture, psia

t_d = dry-bulb temperature, °F

t_w = wet-bulb temperature, °F.

The Carrier's equation has been obtained by Liley in SI units as follows:

$$p_v = p_{gw} - \frac{(p - p_{gw})(t_d - t_w)}{1532.44 - 1.3t_w},$$ (14.63)

where

p_v = actual vapor pressure, bar

p_{gw} = saturation pressure corresponding to wet-bulb temperature, bar

p = total pressure of mixture, bar

t_d = dry-bulb temperature, °C

t_w = wet-bulb temperature, °C.

In Eq. 14.61, the terms h_{fg2} and $(h_{g1} - h_{f2})$ have been approximated by Liley as $(2501.7 - 2.374t_d)$ and $(2501.7 + 1.82t_d - 4.194t_w)$, respectively. Thus the specific humidity of an air–water–vapor mixture may be given by the following approximate relationship:

$$\omega = \frac{1.005(t_w - t_d) + \omega_w(2501.7 - 2.374t_w)}{2501.7 + 1.82t_d - 4.194t_w}$$ (14.64)

[2] W. H. Carrier, "Rational Psychrometric Formulae," *Transactions of the ASME*, 1911.

where

$$\omega_w = 0.622 \frac{p_{gw}}{p - p_{gw}}$$

p_{gw} = saturation pressure corresponding to wet-bulb temperature, bar

p = total pressure of mixture, bar

t_d = dry-bulb temperature, °C

t_w = wet-bulb temperature, °C.

With the approximate equations given by Liley, we may thus determine the properties of an air–water–vapor mixture without the need of any tables. A simple computer program is all that is needed to make the necessary calculations. These equations are, consequently, very useful in engineering applications.

Example 14.8

Making use of Liley's approximate equations (Eqs. 14.48, 14.63, and 14.64), determine the vapor pressure, dew-point temperature, relative humidity, and specific humidity of an air–water–vapor mixture at a total pressure of 101.325 kPa, a dry-bulb temperature of 40°C, and a wet-bulb temperature of 25°C.

Solution

Making use of Eq. 14.48, we have, at 25°C,

$$\ln (p_{gw}) = 14.43509 - \frac{5333.3}{298.15}.$$

Solving for p_{gw}, we have

$$p_{gw} = 0.03165 \text{ bar} = 3.165 \text{ kPa}.$$

Making use of Eq. 14.63 gives us

$$p_v = 0.03165 - \frac{(1.01325 - 0.03165)(40 - 25)}{1532.44 - 1.3 \times 25} \text{ bar}$$

$$= 0.02183 \text{ bar} = 2.183 \text{ kPa}.$$

Making use of Eq. 14.48 yields

$$\ln (0.02183) = 14.43509 - \frac{5333.3}{T_{dp}}$$

Solving for T_{dp}, we have

$$T_{dp} = 292.08 \text{ K}$$

$$t_{dp} = 18.9°C.$$

Making use of Eq. 14.49, we have, at 40°C,

$$\ln (p_g) = 14.43509 - \frac{5333.3}{313.15}.$$

Solving for p_g, we get

$$p_g = 0.07457 \text{ bar} = 7.457 \text{ kPa}.$$

Thus the relative humidity is given by

$$\phi = \frac{p_v}{p_g} = \frac{2.183}{7.457} = 29.3\%.$$

Now

$$\omega_w = 0.622 \frac{p_{gw}}{p - p_{gw}}$$

$$= 0.622 \frac{3.165}{101.325 - 3.165}$$

$$= 0.020055 \text{ kg of water vapor/kg of dry air.}$$

Making use of Eq. 14.64, the specific humidity is then given by

$$\omega = \frac{1.005(25 - 40) + 0.020055(2501.7 - 2.374 \times 25)}{2501.7 + 1.82 \times 40 - 4.194 \times 25}$$

$$= 0.01373 \text{ kg of water vapor/kg of dry air.}$$

Comment

This example demonstrates that with the approximate equations of Liley we indeed do not need a table to determine the properties of an air–water–vapor mixture at a given state.

14.10 Psychrometric Chart

The various properties of moist air can be calculated through the use of the equations relating them. However, the calculations are tedious and time consuming. It is of considerable advantage to have these properties presented in a chart known as a *psychrometric chart.* Such a chart not only makes it possible to obtain readily the necessary information for engineering calculations but also provides us with a very useful aid in visualizing and understanding various processes involving moist air.

A thermodynamic state of moist air is uniquely fixed if the mixture pressure and two independent properties, such as the dry-bulb temperature and the specific humidity, are known. This means that a psychrometric chart may be constructed for a given mixture pressure, using the dry-bulb temperature and the specific humidity as the coordinates. Such a chart is shown schematically in Fig. 14.7. A complete psychrometric chart is included in the Appendix, Fig. A.9.

At a given mixture pressure, the vapor pressure p_v is a function of the specific humidity only. Consequently, there is only one value of p_v for each value of ω. The lines of constant wet-bulb temperature slope downward to the right because, for unsaturated air, the wet-bulb temperature is higher than the dew-point temperature but lower than the dry-bulb temperature.

From Eq. 14.60 we have, for an adiabatic saturation process,

$$h_{a1} + \omega_1 h_{v1} + (\omega_2 - \omega_1)h_{f2} = h_{a2} + \omega_2 h_{v2}$$

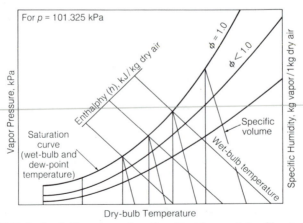

Figure 14.7 Principal Components of a Psychrometric Chart for a Given Mixture Pressure

or

$$h_2 = (\omega_2 - \omega_1)h_{f2} + h_1.$$

The term $(\omega_2 - \omega_1)h_{f2}$ is always very small in comparison with h_1. Thus

$$h_1 \approx h_2.$$

This means that for an adiabatic saturation process, the enthalpy of the mixture remains essentially constant. Since h_2 is a function of the thermodynamic wet-bulb temperature only, lines of constant enthalpy are essentially parallel to lines of constant wet-bulb temperature. For greater accuracy, the correction term $(\omega_2 - \omega_1)h_{f2}$ may be applied.

For a detailed discussion on the method of constructing a psychrometric chart, books on this subject should be consulted.[3]

14.11 Processes Involving Air–Water–Vapor Mixtures

In engineering, some of the common processes involving air–water–vapor mixtures are (1) heating or cooling, (2) humidifying, (3) dehumidifying, and (4) mixing. These generally are steady-state steady-flow processes. Analysis of problems involving these kinds of processes will require the application of

1. The principle of conservation of energy (the energy equation).
2. The principle of conservation of mass (the mass-balance equation).
3. The properties of air–water–vapor mixtures.

In addition, the entropy-creation equation may be used to investigate the extent of irreversibility in such processes and devices.

Heating or Cooling Processes

In a heating or cooling process, there is no change in the moisture content of the working fluid. Consequently, it is a constant-specific-humidity process. Such a process is shown in Fig. 14.8.

[3] William Goodman, *Air Conditioning Analysis*, Macmillan Publishing Co., Inc., New York, 1943, pp. 271–276.

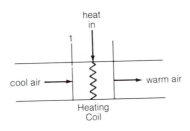

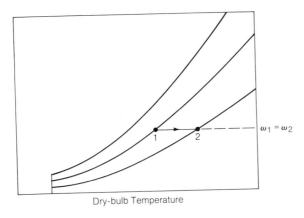

Dry-bulb Temperature

Figure 14.8 Heating Process

On the basis of a flow of unit mass of dry air, the energy equation, neglecting the change of kinetic and potential energies, is given as

$$q + h_{a1} + \omega_1 h_{v1} = h_{a2} + \omega_2 h_{v2}$$

or

$$q = h_2 - h_1, \tag{14.65}$$

in which $\omega_1 = \omega_2$.

Example 14.9

Atmospheric air enters a heater at 40°F and 60% relative humidity, and leaves at a temperature of 70°F. Determine

(a) the heat supplied to the air.
(b) the final relative humidity.

Solution

$$p_{v1} = \phi_1 p_{g1} = 0.60 \times 0.12163 = 0.0730 \text{ psia}$$

$$\omega_1 = 0.622 \frac{p_{v1}}{p - p_{v1}}$$

$$= 0.622 \frac{0.0730}{14.7 - 0.0730} = 0.0031 \text{ lbm of vapor/lbm of dry air.}$$

Using the energy equation and the enthalpy relation given by Eq. 14.57, we have

$$q = 0.240(t_2 - t_1) + \omega \times 0.444(t_2 - t_1)$$

$$= 0.240(70 - 40) + 0.0031 \times 0.444(70 - 40)$$

$$= 7.2 + 0.04 = 7.24 \text{ Btu/lbm of dry air.}$$

Neglecting any pressure drop, we have $p_{v2} = p_{v1} = 0.0730$ psia. Thus

$$\phi_2 = \frac{p_{v2}}{p_{g2}} = \frac{0.0730}{0.3629} = 0.201. \qquad \blacksquare$$

Comment

Note that the heat supplied to the moist air is for all practical purposes the same as the heat supplied to the dry air alone.

Adiabatic Humidification

A humidification process simply means that the specific humidity of the air is increased. If humidification is carried out adiabatically, the energy required for the evaporation of the added moisture must come from the entering air. Consequently, the dry-bulb temperature of the air must decrease. This is why an adiabatic humidification process is also known as an *evaporative cooling process.*

On the basis of a flow of unit mass of dry air, the energy equation, neglecting the change in kinetic and potential energies, is given as

$$h_1 + (\omega_2 - \omega_1)h_f = h_2,$$

where h_f is the enthalpy of the water supply. Neglecting the term $(\omega_2 - \omega_1)h_f$, which is always very small in comparison with the other

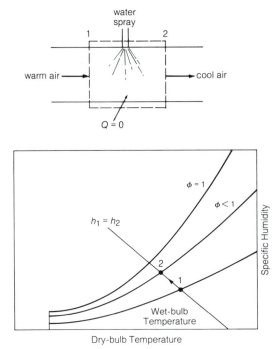

Figure 14.9 Adiabatic Humidification Process

terms, we have

$$h_1 = h_2. \tag{14.66}$$

We thus see that an adiabatic humidification process is essentially a constant-enthalpy process. Since lines of constant enthalpy may be approximated closely as parallel to lines of constant wet-bulb temperature, an adiabatic humidification process is also essentially a constant wet-bulb temperature process. Therefore, the lowest temperature of the air that can be achieved in such a process is the wet-bulb temperature of the air. Such a process is shown in Fig. 14.9. It is clear that not much cooling can be accomplished by this process if the air has a high relative humidity.

Example 14.10

Warm air is to be cooled by an adiabatic humidification process (see Fig. 14.8). At the beginning of the process, the air is at 101.325 kPa, 45°C, and a relative humidity of 30%. If the final temperature is 30°C, determine

(a) the amount of water added to the air.
(b) the final relative humidity.

Solution

From saturation tables for steam, we have, at 45°C,

$$p_{g1} = 9.582 \text{ kPa.}$$

Thus

$$p_{v1} = 0.30 \times 9.582 \text{ kPa} = 2.8746 \text{ kPa}$$

$$\omega_1 = 0.622 \frac{p_{v1}}{p - p_{v1}}$$

$$= 0.622 \frac{2.8746}{101.325 - 2.8746} \text{ kg of vapor/kg of dry air}$$

$$= 0.01816 \text{ kg of vapor/kg of dry air.}$$

Using Eq. 14.56 for the enthalpy of the mixture, we have

$$h_1 = 1.005t_1 + \omega_1(2500 - 1.86t_1)$$

$$= 1.005 \times 45 + 0.01816(2500 - 1.86 \times 45) \text{ kJ/kg of dry air}$$

$$= 89.105 \text{ kJ/kg of dry air.}$$

Now

$$h_2 = 1.005t_2 + \omega_2(2500 - 1.86t_2)$$

$$= 1.005 \times 30 + \omega_2(2500 - 1.86 \times 30)$$

$$= h_1 = 89.105 \text{ kJ/kg of dry air.}$$

Solving for ω_2, we have

$$\omega_2 = 0.02412 \text{ kg of vapor/kg of dry air.}$$

The amount of water added to the air is then given by

$$\omega_2 - \omega_1 = 0.02412 - 0.01816 = 0.00596 \text{ kg of water/kg of dry air.}$$

Now

$$p_{v2} = \frac{\omega_2 p}{\omega_2 + 0.622}$$

$$= \frac{0.02412 \times 101.325}{0.02412 + 0.622} \text{ kPa}$$

$$= 3.783 \text{ kPa.}$$

From saturation tables for steam, we have, at 30°C,

$$p_{g2} = 4.241 \text{ kPa.}$$

Thus

$$\phi_2 = \frac{p_{v2}}{p_{g2}}$$

$$= \frac{3.783}{4.241} = 89.2\%. \qquad \blacksquare$$

Comment

Since the final relative humidity is almost 90%, the final temperature is very close to the lowest temperature we can achieve with this process. From the psychrometric chart, we find that the lowest temperature is about 28°C.

Dehumidification by Cooling

Water vapor may be removed from air by cooling it below its dew-point temperature. As a result of the cooling process, a portion of the vapor in the air is condensed. Such a process is shown in diagram for Example 14.11. Neglecting the change in kinetic and potential energies, the energy equation may be written as

$$Q = m_a(h_2 - h_1) + m_a(\omega_1 - \omega_2)h_f, \qquad (14.67)$$

where h_f is the specific enthalpy of the condensate and $m_a(\omega_1 - \omega_2)$ is the mass of water removed from the air.

Example 14.11

Air is to be conditioned from a dry-bulb temperature of 40°C and a relative humidity of 50% to a final dry-bulb temperature of 20°C and a final relative humidity of 40% by a dehumidification process followed by a reheat process. Assume that the entire process is carried out at a constant pressure of 101.325 kPa. Determine

(a) the amount of water to be removed from the air.
(b) the temperature of the air leaving the dehumidifier.

Solution

The entire process is as shown on sketch.

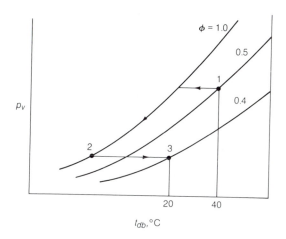

$t_{db}, °C$

From saturation tables for steam, we have, at 40°C,

$$p_{g1} = 7.375 \text{ kPa.}$$

Thus

$$p_{v1} = 0.50 \times 7.375 \text{ kPa} = 3.6875 \text{ kPa}$$

and

$$\omega_1 = 0.622 \frac{3.6875}{101.325 - 3.6875}$$

$$= 0.02349 \text{ kg of vapor/kg of dry air.}$$

From the saturation tables for steam, we have, at 20°C,

$$p_{g3} = 2.337 \text{ kPa.}$$

Thus

$$p_{v3} = 0.40 \times 2.337 \text{ kPa} = 0.9348 \text{ kPa}$$

and

$$\omega_3 = 0.622 \frac{0.9348}{101.325 - 0.9348}$$

$$= 0.00579 \text{ kg of vapor/kg of dry air.}$$

The amount of moisture removed from the air is given by

$$\omega_1 - \omega_3 = 0.02349 - 0.00579$$

$$= 0.0177 \text{ kg of water/kg of dry air.}$$

(b) The air leaves the dehumidifier as saturated air at a vapor pressure of 0.9348 kPa or 0.009348 bar. Using Eq. 14.48, we have

$$\ln (0.009348) = 14.43509 - \frac{5333.3}{T_2}.$$

Solving for T_2, we have

$$T_2 = \frac{5333.3}{14.43509 + 4.67259} \text{ K} = 279.12 \text{ K}$$

$$t_2 = 5.97°\text{C.} \qquad \blacksquare$$

Comment

This means that the refrigerant in the cooling coil in the dehumidifier must operate at a temperature below 5.97°C.

Mixing Process

In the design of an air-conditioning system, it is almost always necessary to mix two or more streams of air to produce a stream with the desirable state of temperature and relative humidity. Fig. 14.10a shows schematically the mixing of two streams.

If mixing is carried out adiabatically, and if changes in kinetic and potential energies are negligible, we have, from the first law for steady-state steady flow,

$$m_{a1}(h_{a1} + \omega_1 h_{v1}) + m_{a2}(h_{a2} + \omega_2 h_{v2}) = m_{a3}(h_{a3} + \omega_3 h_{v3})$$

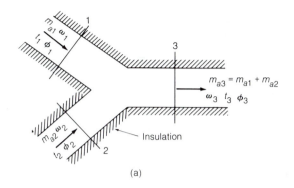

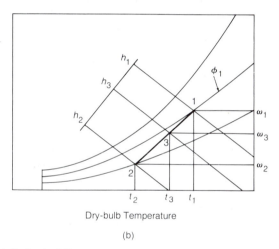

(b)

Figure 14.10 Adiabatic Mixing of Two Air Streams

or

$$m_{a1} h_1 + m_{a2} h_2 = m_{a3} h_3. \qquad (14.68)$$

From mass balance, we have

$$m_{a1} + m_{a2} = m_{a3} \qquad (14.69)$$

$$m_{a1} \omega_1 + m_{a2} \omega_2 = m_{a3} \omega_3. \qquad (14.70)$$

Equations 14.68, 14.69, and 14.70 are the fundamental equations that we need to solve algebraically the problem of mixing two streams. Combining Eq. 14.68 with Eq. 14.69, we have

$$\frac{m_{a1}}{m_{a2}} = \frac{h_3 - h_2}{h_1 - h_3}. \qquad (14.71)$$

Combining Eq. 14.69 with Eq. 14.70, we have

$$\frac{m_{a1}}{m_{a2}} = \frac{\omega_3 - \omega_2}{\omega_1 - \omega_3}. \tag{14.72}$$

Therefore, we have

$$\frac{m_{a1}}{m_{a2}} = \frac{h_3 - h_2}{h_1 - h_3} = \frac{\omega_3 - \omega_2}{\omega_1 - \omega_3}. \tag{14.73}$$

Now suppose that the states of the two incoming streams are as shown in Fig. 14.10b. Then Eq. 14.73 tells us that the resulting state 3 must lie on a straight line connecting states 1 and 2. Thus the final state may be found by dividing the line 1–2 into segments proportional to the relative masses of dry air before mixing takes place. Consequently, the problem of mixing two streams can be solved graphically if the appropriate psychrometric chart is available.

Example 14.12

Two kilograms of air at 200 kPa, 5°20C, and 80% relative humidity is mixed with 4 kg of air at 200 kPa, 40°C, and 20% relative humidity in a steady-state steady-flow device. Neglecting pressure drop, change in kinetic and potential energies, and assuming that the mixing is carried out adiabatically, determine the specific humidity, temperature, and relative humidity of the stream leaving the device.

Solution

From saturation tables for steam, we have, at 5°,

$$p_{g1} = 0.8718 \text{ kPa}.$$

Thus

$$p_{v1} = 0.80 \times 0.8718 \text{ kPa} = 0.6974 \text{ kPa}$$

$$\omega_1 = 0.622 \frac{0.6974}{200 - 0.6974} = 0.002177 \text{ kg of vapor/kg of dry air}.$$

From saturation tables for steam, we have, at 40°C,

$$p_{g2} = 7.375 \text{ kPa}.$$

Thus

$$p_{v2} = 0.20 \times 7.375 \text{ kPa} = 1.475 \text{ kPa}$$

$$\omega_2 = 0.622 \frac{1.475}{200 - 1.475} = 0.004621 \text{ kg of vapor/kg of dry air.}$$

Now the dry-air portion in a mixture is related to the moist air in the following manner:

$$m_a = \frac{m_{\text{moist air}}}{1 + \omega}.$$

Thus

$$m_{a1} = \frac{2}{1 + 0.002177} \text{ kg} = 1.9957 \text{ kg}$$

$$m_{a2} = \frac{4}{1 + 0.004621} \text{ kg} = 3.9816 \text{ kg.}$$

From mass balance of dry air, we have

$$m_{a3} = m_{a1} + m_{a2}$$

$$= 1.9957 + 3.9816 = 5.9773 \text{ kg.}$$

From mass balance of water vapor, we have

$$m_{a1}\,\omega_1 + m_{a2}\,\omega_2 = m_{a3}\,\omega_3$$

$$1.9957 \times 0.002177 + 3.9816 \times 0.004621 = 5.9773 \times \omega_3.$$

Solving for ω_3, we have

$$\omega_3 = 0.003805 \text{ kg of vapor/kg of dry air.}$$

From energy balance, we have

$$m_{a1} h_1 + m_{a2} h_2 = m_{a3} h_3.$$

Making use of Eq. 14.56 for the enthalpy of an air–water–vapor mixture, we have

$$1.9957[1.005 \times 5 + 0.002177(2500 + 1.86 \times 5)]$$

$$+ 3.9816[1.005 \times 40 + 0.004621(2500 + 1.86 \times 40)]$$

$$= 5.9773[1.005 \times t_3 + 0.003805(2500 + 1.86 \times t_3)].$$

Solving for t_3, we have

$$t_3 = 28.35°C.$$

Making use of Eq. 14.48, we have

$$\ln (p_{g3}) = 14.43509 - \frac{5333.3}{T_3}$$

$$= 14.43509 - \frac{5333.3}{273.15 + 28.35}$$

$$p_{g3} = 0.03861 \text{ bar} = 3.861 \text{ kPa}.$$

Now

$$p_{v3} = \frac{\omega_3 \, p_3}{\omega_3 + 0.622}$$

$$= \frac{0.003805 \times 200}{0.003805 + 0.622} \text{ kPa}$$

$$= 1.216 \text{ kPa}.$$

Thus

$$\phi_3 = \frac{p_{v3}}{p_{g3}}$$

$$= \frac{1.216}{3.861}$$

$$= 32.15\%.$$

Comment

Since the moisture content is small, we could have assumed this problem is just like mixing two streams of dry air. Then we would have, from the energy balance,

$$2 \times 1.005 \times 5 + 4 \times 1.005 \times 40 = 6 \times 1.005 \times t_3,$$

yielding $t_3 = 28.33°C$, which is almost the same as the correct answer.

Cooling Towers

We may cool water by allowing it to exchange heat with air through a dividing wall such as in a simple heat exchanger. With this method of heat exchange, the water temperature can only be made to approach the dry-bulb temperature of the air. On the other hand, if the water is in direct contact with the air, such as in a cooling tower, the water temperature may be theoretically cooled to the wet-bulb temperature of the air. Since the cooling tower utilizes the phenomenon of evaporative cooling, much more heat can be transferred to the air with a cooling tower than with a simple heat exchanger. Consequently, a cooling tower is smaller than a simple heat exchanger for a given heat-transfer requirement. A disadvantage of the cooling tower is that there is a loss of the evaporated water.

Figure 14.11 shows a schematic arrangement of a cooling tower. The warm water is sprayed into the tower near the top and allowed to fall through a network of wood arranged to facilitate evaporation by breaking up the water stream into small droplets. Atmospheric air passes upward in the tower, where it picks up vapor and is also heated. It then leaves at the top, very nearly saturated.

Neglecting heat transfer between the cooling tower and the surroundings, and neglecting the changes in kinetic and potential energies, we have, from the first law for steady-state steady flow,

$$m_{w1} h_{f1} + m_a(h_{a3} + \omega_3 h_{v3}) = m_{w2} h_{f2} + m_a(h_{a4} + \omega_4 h_{v4}). \quad (14.74)$$

From mass balance on water, we have

$$m_{w1} + m_a \omega_3 = m_{w2} + m_a \omega_4$$

or

$$m_{w1} - m_{w2} = m_a(\omega_4 - \omega_3). \quad (14.75)$$

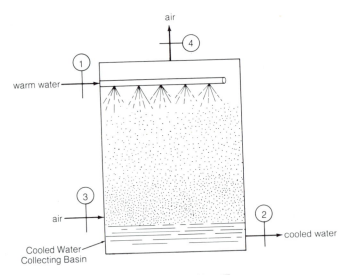

air

warm water

air

Cooled Water
Collecting Basin

cooled water

Figure 14.11 Schematic Arrangement of Cooling Tower

Combining Eq. 14.74 with Eq. 14.75, we have

$$m_a = \frac{m_{w1}(h_{f1} - h_{f2})}{h_{a4} - h_{a3} + \omega_4 h_{v4} - \omega_3 h_{v3} - (\omega_4 - \omega_3)h_{f2}}$$

or

$$m_a = \frac{m_{w1}(h_{f1} - h_{f2})}{h_4 - h_3 - (\omega_4 - \omega_3)h_{f2}}, \qquad (14.76)$$

where h_4 and h_3 may be evaluated by using Eq. 14.57.

Making use of Eq. 14.76, we can calculate the air requirement for the cooling of a given amount of warm water. Equation 14.75 gives the makeup-water requirement for the collecting basin.

Example 14.13

Water enters a cooling tower at 45°C at the rate of 100 kg/s. Air enters the cooling tower at a dry-bulb temperature of 30°C and a wet-bulb temperature of 20°C at the rate of 75 m³/s. If the air leaves saturated at 40°C, determine the temperature of water leaving the cooling tower. The barometric pressure is 101.325 kPa.

Solution

From the psychrometric chart, the relative humidity of air at the inlet of cooling tower is 40%. From the saturation tables for steam, we have, at 30°C,

$$P_{g1} = 4.241 \text{ kPa.}$$

Thus

$$P_{v1} = 0.40 \times 4.241 \text{ kPa} = 1.6964 \text{ kPa}$$

and

$$\omega_1 = 0.622 \frac{P_{v1}}{p - P_{v1}}$$

$$= 0.622 \frac{1.6964}{101.325 - 1.6964} = 0.01059 \text{ kg of vapor/kg of dry air.}$$

From the steam tables, we have, at 40°C,

$$P_{g2} = 7.375 \text{ kPa.}$$

Since we have saturated air at outlet of tower,

$$\omega_2 = 100\%.$$

Thus

$$P_{v2} = 1.0 \times 7.375 \text{ kPa} = 7.375 \text{ kPa}$$

$$\omega_2 = 0.622 \frac{7.375}{101.325 - 7.375} = 0.04883 \text{ kg of vapor/kg of dry air.}$$

The moisture gained by air is given by

$$\omega_2 - \omega_1 = 0.04883 - 0.01059 = 0.03824 \text{ kg of water/kg of dry air.}$$

The mass flow rate of dry air is given by

$$\dot{m}_a = \frac{P_a \dot{V}_1}{R_a T_1}$$

$$= \frac{(101.325 - 1.6964) \times 10^3 \times 75}{(8314.3/28.97)(30 + 273.15)} \text{ kg/s} = 85.884 \text{ kg/s.}$$

Then the mass flow rate of water leaving the tower is given by

$$(\dot{m}_w)_{out} = (\dot{m}_w)_{in} - \dot{m}_a(\omega_2 - \omega_1)$$

$$= 100 - 85.884 \times 0.03824 = 96.716 \text{ kg/s}.$$

Making the energy balance around the cooling tower, we have

$$(\dot{m}_w)_{in}(h_f)_{in} + \dot{m}_a h_1 = (\dot{m}_w)_{out}(h_f)_{out} + \dot{m}_a h_2. \tag{1}$$

Making use of Eq. 14.56 for the enthalpy of an air–water–vapor mixture, and assuming the enthalpy of liquid water to be $4.186t$, Eq. 1 may be given as

$$100 \times 4.186 \times 45 + 85.884[1.005 \times 30 + 0.01059(2500 + 1.86 \times 30)]$$

$$= 96.716 \times 4.186 \times (t_w)_{out}$$

$$+ 85.884[1.005 \times 40 + 0.04883(2500 + 1.86 \times 40)].$$

Solving for $(t_w)_{out}$, we have

$$(t_w)_{out} = 23.5°\text{C}. \qquad \blacksquare$$

Comments

The actual cooling of the water is from 45°C to 23.5°C. The ideal cooling of the water is from 45°C to the wet-bulb temperature of the entering moist air, which is 20°C. The ratio of actual cooling to ideal cooling is sometimes called the cooling efficiency of the cooling tower. For our problem, we have

$$\text{cooling efficiency} = \frac{45 - 23.5}{45 - 20}$$

$$= 86.0\%.$$

14.12 Real-Gas Mixtures

When the constituents of a gas mixture deviate from ideal-gas behavior, we have a real-gas mixture. The simple relations we have developed for ideal-gas mixtures do not hold. In principle, a real-gas mixture of fixed compositions can be studied experimentally just like a simple compressible substance. This has been done for some common mixtures such as air. But there are an unlimited number of possible mixtures. Consequently, it is desirable to have a method whereby the properties of a real-gas mixture may be determined in terms of properties of its constituents. Unfortunately, no general theory has yet been developed. We must resort to approximation methods at the present time. An excellent source of approximation methods is the work of Reid and Sherwood.[4] In this section we give a brief presentation of the Kay's rule.

Kay's Rule

A very simple method suggested by W. B. Kay is that the critical temperature T_C and the critical pressure P_C of a real-gas mixture may be defined by the following relations:

$$T_C = \sum x_i T_{ci} \tag{14.77}$$

$$P_C = \sum x_i P_{ci}, \tag{14.78}$$

where x_i, T_{ci}, and P_{ci} are the mole fraction, critical temperature, and critical pressure of the ith species in the mixture.

The critical parameters, defined by Eqs. 14.77 and 14.78, are known as the pseudocritical properties of the mixture. We may make use of these properties together with generalized data to study the behavior of a real-gas mixture as if it is a single-component simple compressible substance.

For example, if we have a mixture consisting of 21% by mole of oxygen and 79% by mole of nitrogen, then the critical temperature and critical pressure of this mixture would be given by

$$T_C = x_{O_2}(T_C)_{O_2} + x_{N_2}(T_C)_{N_2}$$

$$= 0.21 \times 154.6 + 0.79 \times 126.2 \text{ K} = 132.2 \text{ K}$$

$$P_C = x_{O_2}(P_C)_{O_2} + x_{N_2}(P_C)_{N_2}$$

$$= 0.21 \times 5.04 + 0.79 \times 3.4 \text{ MPa} = 3.74 \text{ MPa}.$$

It is interesting to note that the composition of our mixture is essentially that of dry air. The critical temperature and critical pressure for air are 132.4 K and 3.77 MPa, respectively. We thus see that the Kay's rule appears to be exact for our composition. But not enough experimental data for real-gas mixtures are available, so that we may draw conclusions relative to the accuracy of any approximation method.

---

[4] Robert C. Reid and Thomas K. Sherwood, *The Properties of Gases and Liquids—Their Estimation and Correlation*, 2nd ed., McGraw-Hill Book Company, New York, 1966.

Problems

When specific heats for gases at low pressure at 300 K are needed refer to Table 5.2.

14.1 Determine the volume of 1 kg mol of any ideal gas at 101.325 kPa and 0°C.

14.2 An ideal-gas mixture consists of 2 kg of methane (CH_4) and 2 kg of nitrogen (N_2). Determine
(a) the volumetric analysis of the mixture.
(b) the gas constant of the mixture in SI units.

14.3 An ideal-gas mixture consists of 0.5 kg of CO_2, 0.2 kg of CH_4, and 0.5 kg of O_2. Determine
(a) the volumetric analysis of the mixture.
(b) the gas constant of the mixture in SI units.

14.4 The volumetric analysis of an ideal-gas mixture is as follows: CO_2, 40%; N_2, 40%; CO, 10%; O_2, 10%. Determine the average molecular weight, the gas constant, and the gravimetric analysis of the mixture.

14.5 The gravimetric analysis of an ideal gas mixture is as follows: N_2, 85%; CO_2, 13%; CO, 2%. Determine the average molecular weight, the gas constant, and the volumetric analysis of the mixture.

14.6 The volumetric analysis of a flue gas mixture is as follows: N_2, 81%; CO_2, 14%; O_2, 4.2%; CO, 0.8%. Determine
(a) the gravimetric analysis of the mixture.
(b) the gas constant of the mixture in SI units.

14.7 A rigid tank, having a volume of 0.15 m³, contains 1.0 kg of O_2 and 1.5 kg of CO_2 at a temperature of 20°C. Determine
(a) the total pressure of the mixture.
(b) the partial pressure of each constituent in the mixture.

14.8 A rigid tank contains an ideal-gas mixture consisting of 25 kg of O_2, 10 kg of N_2, and 5 kg of H_2. The temperature and pressure of the mixture are 50°C and 120 kPa, respectively. Determine
(a) the volume of the tank.
(b) the partial pressure of each constituent in the mixture.

14.9 Four kilograms of CO_2 is mixed with 10 kg of an unknown gas. If the mixture occupies a volume of 4.0 m³ when the mixture is at 350 kPa and 100°C, determine
(a) the molecular weight of the unknown gas.
(b) the partial pressure of each constituent in the mixture.

14.10 It is desired to prepare an ideal-gas mixture having the following volumetric analysis: CO_2, 60%; O_2, 40%. If this mixture contains 5 kg of CO_2, what is the mass of O_2 in the mixture?

14.11 A mixture, containing 30 lbm of nitrogen and the remainder carbon dioxide, occupies a volume of 100 ft³ at a pressure of 80 psia and temperature cf 80°F. Determine the average molecular weight of the mixture and the partial pressure of the nitrogen.

14.12 Determine the specific heats c_p and c_v at 300 K for the ideal-gas mixture given in Problem 14.6 in SI units.

14.13 Determine the specific heats c_p and c_v at 80°F for the mixture given in Problem 14.4. Give the answers in Btu/lbm · °R.

14.14 Determine the specific heats c_p and c_v at 80°F for the mixture given in Problem 14.5. Give the answers in Btu/lbm · °R.

14.15 Four kilograms of nitrogen at 1.0 MPa and 25°C and 3 kg of oxygen at 500 kPa and 30°C is mixed adiabatically with no change in the total volume. Determine
(a) the temperature and pressure of the resulting mixture.
(b) the entropy generation due to the mixing process.

14.16 A divided chamber has one compartment filled with 5 lbm of nitrogen at 25 psia and 80°F, and the second compartment filled with 3 lbm of carbon dioxide at 75 psia and 200°F. The partition is removed and the gases mix. Determine
(a) the temperature and pressure of the resulting mixture.
(b) the entropy creation in the universe due to the mixing process.
Consider the mixing to be adiabatic.

14.17 An ideal-gas mixture is produced by mixing 0.30 m³ of nitrogen at 250 kPa and 30°C with 0.15 m³ of oxygen at 125 kPa and 20°C. The mixing process is carried out adiabatically at constant volume. Determine
(a) the final temperature and pressure of the mixture.
(b) the entropy generation due to the mixing process.

14.18 A rigid vessel, having a volume of 15 ft³, contains a mixture of O_2 and CO_2 at 50 psia and 80°F. If adding 15 Btu of heat to the mixture raises its temperature to 100°F, how many pounds of oxygen and how many pounds of carbon dioxide does the vessel contain?

14.19 An ideal-gas mixture consisting of 2 kg of H_2 and 3 kg of O_2 is cooled from 50°C to 30°C at a constant pressure of 200 kPa.

Determine the change in internal energy, enthalpy, and entropy per kilogram of the mixture.

14.20 The density of a mixture of CO_2 and N_2 is 0.0830 lbm/ft^3 at 14.7 psia and 70°F. How many pounds of carbon dioxide are present in 1 lbm of the mixture?

14.21 An ideal-gas mixture consisting of 2 kg of CO, 6 kg of CO_2, and 12 kg of N_2 is to be heated at constant volume from 25°C to 50°C. Determine the amount of heat that must be supplied.

14.22 If the ideal-gas mixture given in Problem 14.21 is to be heated at a constant pressure from 25 to 50°C, determine the amount of heat that must be supplied.

14.23 A rigid vessel, having a volume of 10 ft^3, contains nitrogen at 20 psia and 70°F. If 0.50 lbm of 70°F oxygen is introduced into the vessel adiabatically, what are the temperature and pressure of the resulting mixture?

14.24 An ideal-gas mixture inside a rigid tank has a volumetric analysis as follows: H_2, 80%; N_2, 20%. If we want to alter the composition to 50% by volume of H_2 and 50% of N_2 by removing some of the mixture and adding some N_2, calculate per mole of mixture the mass of mixture to be removed, and the mass of N_2 to be added. The pressure and temperature of the mixture in the tank remain constant during the process.

14.25 The pressure and temperature of a mixture of N_2 and CO_2 contained in a rigid vessel of 10 ft^3 capacity are to be maintained at 100 psia and 100°F. When some of the mixture at a volumetric composition of 30% N_2 and 70% CO_2 is removed and nitrogen is introduced into the vessel until a new volumetric composition of 50% N_2 and 50% CO_2 is produced, how many pounds of the mixture are removed and how many pounds of nitrogen are added?

14.26 An ideal-gas mixture consisting of 1.0 kg of O_2 and 3.30 kg of N_2 is to be heated at constant pressure from 300 K to 1000 K. Determine the amount of heat addition needed
(a) assuming constant specific heats at 300 K.
(b) when the variations of $\bar{c}_p$ with temperature are as follows:

For O_2: $\bar{c}_p = 25.48 + 1.520 \times 10^{-2}T - 0.7155 \times 10^{-5}T^2$

$+ 1.312 \times 10^{-9}T^3$ kJ/kg mol · K

For N_2: $\bar{c}_p = 28.90 - 0.1571 \times 10^{-2}T + 0.8081 \times 10^{-5}T^2$

$$- 2.873 \times 10^{-9}T^3 \text{ kJ/kg mol} \cdot \text{K}$$

where T is in kelvin.

14.27 An ideal-gas mixture consisting of 15 kg of air and 1.0 kg of CH_4 is to be heated at constant volume from 300 K to 500 K. Determine the amount of heat addition required
(a) assuming constant specific heats at 300 K.
(b) when the variations of $\bar{c}_p$ with temperature are as follows:

For air: $\bar{c}_p = 28.11 + 0.1967 \times 10^{-2}T + 0.4802 \times 10^{-5}T^2$

$$- 1.966 \times 10^{-9}T^3 \text{ kJ/kg mol} \cdot \text{K}$$

For CH_4: $\bar{c}_p = 19.89 + 5.024 \times 10^{-2}T + 1.269 \times 10^{-5}T^2$

$$- 11.01 \times 10^{-9}T^3 \text{ kJ/kg mol} \cdot \text{K},$$

where T is in kelvin.

14.28 The gas in an engine is an ideal-gas mixture of 15 kg of dry air and 1.0 kg of CH_4. At the beginning of the compression stroke, the pressure and temperature of the mixture are 101.325 kPa and 25°C. If the mixture is compressed quasi-statically with a compression ratio of 8 according to the path $pv^{1.3} = $ constant, determine
(a) the work of compression.
(b) the amount of heat transfer involved.
The variations of $\bar{c}_p$ with temperature for air and CH_4 are the same as those given in Problem 14.27.

14.29 Calculate the change of entropy due to the compression process for the ideal-gas mixture in Problem 14.28.

14.30 A gaseous mixture consisting of 5 lbm of nitrogen and 2 lbm of helium is compressed from 20 psia and 80°F to 80 psia and 240°F. Is the compression process reversible adiabatic?

14.31 A gaseous mixture consisting of 5 lbm of argon and 5 lbm of helium is compressed from 80°F and 1 atm to 5 atm isentropically in a steady-flow machine. Determine the compressor work.

14.32 An ideal-gas mixture consists of 0.012 kg of water vapor and 1.0 kg of dry air. If the pressure and temperature of the mixture are

101.325 kPa and 50°C, determine for the mixture
(a) the dew-point temperature.
(b) the relative humidity.

14.33 When a mixture of ideal gases undergoes a reversible adiabatic process, show that in order for the individual component entropies to remain constant, the molal specific heats for the mixture and the components must be identical.

14.34 A mixture consisting of 0.2 lbm of dry saturated water vapor and 0.3 lbm of nitrogen is contained in a rigid vessel at a temperature of 200°F. Calculate the total pressure in the vessel.

14.35 A rigid vessel contains an ideal-gas mixture consisting of 1.0 kg of nitrogen and 0.01 kg of water vapor. If the pressure and temperature of the mixture are 200 kPa and 100°C, determine
(a) the volume of the vessel.
(b) the dew-point temperature of the mixture.

14.36 One pound of nitrogen and 0.2 lbm of water vapor occupy a volume of 5.0 ft³. Calculate the total pressure for the mixture if the temperature is 450°F.

14.37 An ideal-gas mixture consists of dry saturated water vapor and nitrogen at 500 kPa and 25°C. Determine
(a) the partial pressure of water vapor in the mixture.
(b) the specific humidity of the mixture.

14.38 A mixture of air and water vapor at 14.5 psia and 80°F has a dew point of 65°F. Determine the relative humidity and specific humidity.

14.39 A rigid tank contains 1.5 lbm of dry air at 14.5 psia and 70°F initially. Water vapor is then introduced into the tank until a saturated mixture at 70°F is reached. Determine
(a) the amount of water vapor added.
(b) the pressure of the mixture.

14.40 A room having the dimensions of 6 m × 6 m × 3 m contains an air–water vapor mixture at a total pressure of 101.325 kPa and 25°C. If the partial pressure of the water vapor is 1.5 kPa, determine
(a) the relative humidity.
(b) the specific humidity.
(c) the total mass of water vapor in the room.

14.41 One method of removing moisture from atmospheric air is to cool the air until the moisture condenses out. If a specific humidity of 0.01 kg/kg of dry air is desired, to what temperature must the air be cooled at a pressure of 101.325 kPa?

14.42 The pressure and temperature of the air in a greenhouse are 14.7 psia and 90°F, respectively. The relative humidity is 85%. Determine
(a) the partial pressure of the water vapor.
(b) the dew point.
(c) the specific humidity.

14.43 Air enters a heater at 15°C, 101.325 kPa, and 80% relative humidity and leaves at 25°C. Neglecting pressure drop, determine
(a) the relative humidity of air leaving the heater.
(b) the amount of heat added in kilojoules per kilogram of dry air.

14.44 Air enters a dehumidifier at 35°C, 101.325 kPa, and 60% relative humidity. After being dehumidified the air is heated to 20°C with a relative humidity of 50%. Determine
(a) the temperature of air leaving the dehumidifier.
(b) the amount of heat removed from the air in the dehumidifier in kilojoules per kilogram of dry air.

14.45 Determine the amount of heat required for the heating process in Problem 14.44 in kilojoules per kilogram of dry air.

14.46 Air enters a heater at 10°C, 101.325 kPa, and 80% relative humidity at the rate of 200 m³/min of dry air. If heat added to the air is at the rate of 14,000 kJ/min, determine
(a) the temperature of air leaving the heater.
(b) the relative humidity of the air leaving the heater.

14.47 Air enters a compressor at 25°C, 101.0 kPa, and 50% relative humidity and leaves with a pressure of 404.0 kPa. It is then cooled in a heat exchanger. If condensation of water vapor is to be prevented, determine the lowest temperature to which the air can be cooled in the heat exchanger.

14.48 Air enters a nozzle at 40°C, 300 kPa, and a relative humidity of 50%. If the expansion process is carried out reversibly and adiabatically, what is the lowest pressure at which air may leave the nozzle if no condensation is allowed?

14.49 Air enters a compressor at 75°F, 14.0 psia, and 75% relative humidity, and leaves the compressor at 70 psia and 280°F. What is the relative humidity at the compressor discharge?

14.50 A mixture of nitrogen and water vapor at 200°F has the following volumetric analysis: 85% of nitrogen and 15% of water vapor. Determine the temperature at which the water vapor will start to condense if the mixture is cooled at a constant pressure of 14.8 psia.

14.51 A rigid tank having a volume of 0.5 m³ contains air at 101.325 kPa, 30°C, and 60% relative humidity. If the air is cooled to 15°C, determine
   (a) the temperature at which condensation will begin.
   (b) the amount of water that will condense out from the mixture.

14.52 The air in a room has a dry-bulb temperature of 20°C. The wall temperature is to be 15°C. What is the maximum relative humidity at which condensation on the wall can be avoided?

14.53 The inside surface of a wall in a home is found to be 60°F. If the air within the room is at 70°F, what is the maximum relative humidity at which condensation on the wall can be avoided?

14.54 A mixture of air and water vapor has a relative humidity of 50% at a total pressure of 14.5 psia and a temperature of 80°F. This mixture is compressed isothermally. At what total pressure will condensation of the water vapor begin?

14.55 A mixture of air and water vapor has a relative humidity of 60% at a total pressure of 14.7 psia and a temperature of 80°F. When this mixture is cooled to 40°F at a constant total pressure, how many pounds of moisture will be removed?

14.56 Air at 10°C, 101.325 kPa, and 50% relative humidity enters a heater at the rate of 20 m³/s. If the air leaves the heater at 30°C, determine
   (a) the relative humidity of the air leaving the heater.
   (b) the rate of heat input needed for the process.

14.57 Air enters a drier at 14.7 psia, 120°F, and a relative humidity of 10%. It leaves the drier at 14.7 psia and a relative humidity of 90%. For an airflow rate of 5000 lbm/h, determine
   (a) the amount of moisture picked up by the air in lbm/hr.
   (b) the dry-bulb temperature of the air at drier outlet.

14.58 One stream of air enters a mixing chamber at 20°C, 101.325 kPa, and 40% relative humidity at the rate of 40 kg/s. Another stream enters at 10°C, 101.325 kPa, and 80% relative humidity at the rate of 20 kg/s. Determine the specific humidity, the relative humidity, and the temperature of the air leaving the mixing chamber. Assume that the mixing process is carried out adiabatically.

14.59 Air enters an evaporative cooler at a pressure of 14.7 psia, a dry-bulb temperature of 100°F, and a wet-bulb temperature of 70°F. The air leaves the cooler at 14.7 psia and a dry-bulb temperature of 75°F. Assuming the process to be adiabatic, determine
(a) the relative humidity of the cool air.
(b) the quantity of water that must be supplied to the cooler for each pound of dry air.

14.60 Air enters an evaporative cooler at a pressure of 101.325 kPa, a dry-bulb temperature of 35°C, and a relative humidity of 20%. If we want the air to leave the cooler at 25°C, determine
(a) the relative humidity of the cool air.
(b) the quantity of water that must be added to the air per kilogram of dry air.

14.61 Air flowing 5000 ft³/min at 45°F dry-bulb and 40°F wet-bulb is to be pre-heated, passed through water sprays until the relative humidity is 100%, and then reheated to a final condition of 80°F dry-bulb and 30% relative humidity. Assume that the total pressure is 14.7 psia throughout.
(a) Sketch the complete process on a psychrometric chart.
(b) What is the heating load of the preheater?
(c) What is the heating load of the reheater?

14.62 Air at 35°C and 60% relative humidity is cooled in a dehumidifier until the air is saturated. It is then heated in a heater. The desired condition of the air leaving the heater is 20°C and 60% relative humidity. Assume that the total pressure of the air is 101.325 kPa throughout. The flow rate of air is 2.5 m³/s at the inlet of the dehumidifier.
(a) Sketch the complete process on a psychrometric chart.
(b) Determine the cooling load of the dehumidifier.
(c) Determine the heating load of the heater.

14.63 Water is cooled in a cooling tower from 50°C to 25°C. Air enters the tower at 25°C, 101.325 kPa, and a relative humidity of 40%, and leaves the tower at 35°C and a relative humidity of 95%.

Determine
(a) the mass-flow rate of water that can be cooled for a flow rate of 1.0 kg/s of dry air.
(b) the makeup water required for a flow rate of 1.0 kg/s of dry air.
Assume that the atmospheric pressure is 101.325 kPa.

14.64 Water enters a cooling tower at 120°F. Air enters the cooling tower at a dry-bulb temperature of 80°F and a relative humidity of 30%, and leaves as saturated air at 100°F. For water entering the tower at the rate of 500,000 lbm/h, calculate the mass-flow rate of dry air in lbm/h that must be supplied to the cooling tower
(a) if the water leaves the tower at 80°F.
(b) if the water leaves the tower at the lowest possible temperature.
Atmospheric pressure is 14.7 psia.

14.65 Making use of the approximations of Liley, write a computer program to determine the psychrometric properties of atmospheric air given dry bulb temperature and relative humidity. Run the program to determine the vapor pressure, the specific humidity, the enthalpy, the specific volume, and the dew point temperature of atmospheric air at
(a) dry bulb temperature of 25°C and relative humidity of 40%.
(b) dry bulb temperature of 35°C and relative humidity of 60%.
Compare the results with those obtained from a psychrometric chart.

14.66 Modify the computer program in Problem 14.65 to determine the psychrometric properties of atmospheric air given dry bulb and wet bulb temperatures.
Run the program to determine the vapor pressure, the specific humidity, the relative humidity, the specific volume, and the dew point temperature of atmospheric air at
(a) dry bulb and wet bulb temperatures of 25°C and 20°C, respectively.
(b) dry bulb and wet bulb temperatures of 35°C and 25°C, respectively.
Compare the results with those obtained from a psychrometric chart.

14.67 Modify the computer program in Problem 14.65 to study the effect of temperature and relative humidity of incoming air on the performance of a natural draught cooling tower.

Run the program to determine the volumetric flow rate of air and the mass flow rate of make-up water according to the following specifications:

1. Mass flow rate of water entering cooling tower = 100 kg/s
2. Temperature of water entering cooling tower = 45°C
3. Temperature of water leaving cooling tower = 25°C
4. Temperature of air leaving cooling tower = 35°C
5. Relative humidity of air leaving cooling tower = 1.0
6. Vary the atmospheric conditions to allow for inlet states of dry bulb temperature of 5, 10, 15, and 25°C with relative humidity of 0.25, 0.50, 0.75, and 1.0.

Plot the results.

CHAPTER 15

Chemical Reactions and Reactive Mixtures

A reactive mixture is a multicomponent system the constituents of which may react chemically with one another. The study of systems involving chemical reactions is an important topic in chemical thermodynamics. It has three main aspects, corresponding to the three laws of nature: the principle of conservation of matter, the first law of thermodynamics, and the second law of thermodynamics.

The first aspect is known as *stoichiometry*. It is concerned with the relations between the composition of the reactants and the composition of the products. Since a chemical reaction can be defined as the rearrangement of atoms due to a redistribution of electrons, it is necessary for us to account for the masses as well as for each of the chemical elements involved in a given reaction.

The second aspect involves the application of the energy-conservation principle. But we are no longer dealing with simple compressible substances. In our energy accounting, when chemical reactions are involved, we must recognize that the chemical properties of the system will be altered.

The third aspect involves the application of the second law of thermodynamics, which is useful to us in two important respects. First, by distinguishing between reversible and irreversible processes, the second law tells us whether a given reaction will go forward or backward. Second, the second law dictates that the equilibrium composition of a reactive mixture depends on temperature and pressure.

Although the basic principles in this chapter apply to any chemical reaction, we shall focus our attention on an important type of chemical reaction: combustion. Combustion is a process involving the reaction of a fuel and oxidizer in which the stored chemical energy in the fuel is released. Since the primary source of energy that supports our industries, transportation, and comfort heating systems is obtained through the burning of fossil fuels, combustion is of great importance in engineering.

We wish to point out that there is a fourth aspect of chemical reactions—the rate at which reactions proceed. This aspect involves principles covered in the discipline of chemical kinetics and is beyond the scope of this book. However, we should be aware of the fact that reaction rates play a very important role in the determination of the physical size of the hardware.

15.1 Stoichiometry and the Chemical Equation

Associated with every chemical reaction is a chemical equation. For example, the reaction of carbon monoxide and oxygen may be represented by the following equation:

$$2CO + 1O_2 \rightleftharpoons 2CO_2 \tag{15.1}$$

The symbols in this chemical equation mean that 2 mol of CO may react with 1 mol of O_2 to form 2 mol of CO_2. The two arrows in the chemical equation indicate that the reaction may also go in the opposite direction; that is, 2 mol of CO_2 may decompose into 2 mol of CO and 1 mol of O_2. Which way the process will actually go cannot be determined from Eq. 15.1 alone.

The coefficients in the chemical equation are called *stoichiometric coefficients*. In Eq. 15.1, these coefficients are 2, 1, and 2. Note that the number of carbon atoms and the number of oxygen atoms are conserved. The mass of all the substances on one side of the equation must be equal to the mass of all the substances on the other side of the equation. In short, a chemical equation expresses the principle of the conservation of mass in terms of the conservation of atoms.

15.2 Basic Combustion Equations for Fossil Fuels

The combustible constituents of fuels are carbon, hydrogen, and sulfur and their compounds. If fuel is burned completely, that is, if complete release of the chemical energy in the fuel is realized, all carbon is burned to carbon dioxide, all hydrogen is burned to water vapor, and all sulfur is burned to sulfur dioxide. The basic equations for the complete combustion of carbon, hydrogen, and sulfur can be written as

$$C + O_2 \rightarrow CO_2 \qquad\qquad (15.2)$$

$$H_2 + \tfrac{1}{2}O_2 \rightarrow H_2O \qquad\qquad (15.3)$$

$$S + O_2 \rightarrow SO_2. \qquad\qquad (15.4)$$

The complete combustion of hydrocarbon compounds result in the formation of carbon dioxide and water. For example, we have, in the case of burning methane (CH_4),

$$CH_4 + 2O_2 \rightarrow CO_2 + 2H_2O. \qquad\qquad (15.5)$$

If the burning of carbon is incomplete, carbon monoxide will be formed according to the equation

$$C + \tfrac{1}{2}O_2 \rightarrow CO. \qquad\qquad (15.6)$$

We shall see that only 9210 kJ (out of a possible 32,800 kJ) of the chemical energy in the fuel will be released in burning 1 kg of carbon to carbon monoxide. The presence of carbon monoxide in the products of combustion is a direct indication of inefficient energy utilization. In addition, carbon monoxide pollutes the quality of the air that we breathe.

15.3 Theoretical Air, Excess Air, and Air–Fuel Ratio

The oxygen required for most combustion is obtained from air, which for nearly all combustion calculations may be considered as a mixture of 21% of oxygen and 79% of nitrogen by volume. Thus the composition of dry air may be given in the following manner:

$$1 \text{ mol of } O_2 + 3.76 \text{ mol of } N_2 = 4.76 \text{ mol of air} \qquad (15.7)$$

$$1 \text{ kg of } O_2 + 3.30 \text{ kg of } N_2 = 4.30 \text{ kg of air}. \qquad (15.8)$$

It is of interest to calculate the minimum amount of dry air that would supply sufficient oxygen for complete combustion. This is know as the *theoretical air* or *stoichiometric air*. For the complete burning of methane with dry air, we would have

$$CH_4 + 2O_2 + (2)(3.76)N_2 \rightarrow CO_2 + 2H_2O + (2)(3.76)N_2. \quad (15.9)$$

This means that we need (2)(4.76) mol of air to burn 1 mol of methane completely.

In the great majority of practical combustion problems, the difficulties involved in realizing satisfactory mixing between the fuel and the air supply are so great that it becomes necessary to provide excess air to assure complete combustion. Excess air is air that is supplied for combustion in excess of the theoretical air requirements of the fuel. It is usually expressed as a percentage of the theoretical air. Thus "110% theoretical air" is equivalent to 10% excess air. If methane is burned with 10% excess air, the complete combustion equation is then given as

$$CH_4 + (1.1)(2)O_2 + (1.1)(2)(3.76)N_2$$

$$\rightarrow CO_2 + 2H_2O + 0.2O_2 + (1.1)(2)(3.76)N_2 . \quad (15.10)$$

Another important parameter applied to combustion processes is the air–fuel ratio (AF). For the complete burning of methane with theoretical air, we have, on a molal basis,

$$AF = \frac{n_{air}}{n_{CH_4}} = \frac{2 + 7.52}{1} = 9.52 \text{ mol of air/mol of fuel.}$$

Taking the molecular weight of air to be 28.9 kg/kg mol, the air–fuel ratio may also be expressed on a mass basis:

$$AF = \frac{9.52 \times 28.9}{1 \times 16} = 17.2 \text{ kg of air/kg of fuel.}$$

Example 15.1

Calculate the air–fuel ratio for the burning of propane (C_3H_8) with 120% theoretical air.

Solution

The complete combustion equation for C_3H_8 is

$$C_3H_8 + 5O_2 \rightarrow 3CO_2 + 4H_2O.$$

With 120% theoretical air, the combustion equation is

$$C_3H_8 + (1.2)(5)O_2 + (1.2)(5)(3.76)N_2$$

$$\rightarrow 3CO_2 + 4H_2O + 1O_2 + (1.2)(5)(3.76)N_2 .$$

On a molal basis, we have

$$AF = \frac{(1.2)(5)(4.76)}{1} = 28.56 \text{ mol of air/mol of fuel.}$$

On a mass basis, we have

$$AF = \frac{28.56 \times 28.9}{1 \times 44} = 18.8 \text{ kg of air/kg of fuel.}$$

Example 15.2

Octane (C_8H_{18}) is burned with 95% of the theoretical amount of dry air. Assuming that the products of combustion are a mixture of CO_2, CO, H_2O, and N_2, determine the combustion equation.

Solution

The complete combustion equation for C_8H_{18} is

$$C_8H_{18} + 12.5O_2 \rightarrow 8CO_2 + 9H_2O.$$

The combustion equation for C_8H_{18} with 95% theoretical air is

$$C_8H_{18} + (0.95)(12.5)O_2 + (0.95)(12.5)(3.76)N_2$$

$$\rightarrow aCO_2 + bCO + 9H_2O + (0.95)(12.5)(3.76)N_2.$$

From carbon balance, we have

$$8 = a + b. \tag{1}$$

From oxygen balance, we have

$$(0.95)(12.5)(2) = 2a + b + 9. \tag{2}$$

Solving Eqs. 1 and 2 simultaneously, we have

$$a = 6.75$$

$$b = 1.25.$$

Thus the combustion equation is

$$C_8H_{18} + (0.95)(12.5)O_2 + (0.95)(12.5)(3.76)N_2$$

$$\rightarrow 6.75CO_2 + 1.25CO + 9H_2O + (0.95)(12.5)(3.76)N_2. \quad \blacksquare$$

Comments

In the burning of a hydrocarbon fuel with insufficient air, it is reasonable to assume that all the hydrogen will burn to water due to its affinity for oxygen, with the result that there will be inadequate oxygen to burn all the carbon to carbon dioxide. We should keep in mind that if the amount of air supplied is considerably less than the theoretical air, there may even be some hydrocarbons in the products of combustion.

15.4 Analysis of the Products of Combustion

To determine the completeness of actual combustion processes and the amount of air being supplied, it is necessary to acquire information on the composition of the products of combustion. One of the most commonly used methods of obtaining this kind of information is by means of an *Orsat analyzer*, which selectively achieves absorption in various chemical solutions of the several constituent gases.

In using an Orsat analyzer, a representative sample of the gas to be analyzed is first obtained. This sample, of known volume and saturated with water vapor, is brought into the analyzer at a known temperature and atmospheric pressure. It is then passed sequentially into vessels containing reagents that absorb the constituent gases. Since the absorption of each constituent is carried out at constant temperature and pressure, the change in volume of the gas sample during each separate absorption process is a measure of the volume fraction of that particular constituent in the sample. On the basis of ideal-gas behavior, the volume fraction is the same as the mole fraction. Thus the Orsat analyzer will yield a molal analysis of the gas sample.

Since the gas sample is assumed to be saturated with water vapor, and since the temperature is constant during the entire operation, the partial pressure of the saturated water vapor remains constant. Since the total pressure is also constant, an appropriate amount of water condenses out with each absorption. Thus an Orsat analysis may be shown[1] to give results which are the same as those if the sample had been composed of dry gases. This means that an Orsat apparatus measures the percentage of each constituent in the dry mixture and not the actual percent in the real mixture.

---

[1] E. F. Obert, *Concepts of Thermodynamics*, McGraw-Hill Book Company, New York, 1960, p. 450.

Example 15.3

The Orsat analysis of the products obtained from combustion of a hydrocarbon fuel of unknown composition is as follows:

$$CO_2 = 12.5\%$$

$$CO = 0.5\%$$

$$O_2 = 3.0\%$$

$$N_2 = 84.0\%.$$

Determine the actual air–fuel ratio on a mass basis.

Solution

On the basis of 100 mol of dry products, the combustion equation is

$$C_x H_y + aO_2 + 3.76aN_2 \rightarrow 12.5CO_2 + 0.5CO + 3.00_2 + 84.0N_2 + bH_2O.$$

From nitrogen balance, we have

$$3.76a = 84.0$$

and

$$a = \frac{84.0}{3.76} = 22.34.$$

From oxygen balance, we have

$$2a = (2)(12.5) + 0.5 + (2)(3.0) + b$$

and

$$b = 2 \times 22.34 - 2 \times 12.5 - 0.5 - 2 \times 3 = 13.18.$$

From hydrogen balance, we have

$$y = 2b = 2 \times 13.18 = 26.36.$$

From carbon balance, we have

$$x = 12.5 + 0.5 = 13.0.$$

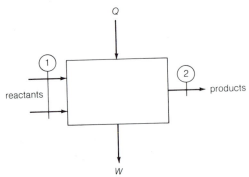

Figure 15.1 Generalized Steady-Flow System with Chemical Reactions

On a mass basis, we have

$$AF = \frac{22.34 \times 4.76 \times 28.9}{13 \times 12 + 26.36 \times 1} = 16.9 \text{ kg of air/kg of fuel.} \qquad \blacksquare$$

15.5 Energy Balance for Steady-State Steady-Flow Systems with Chemical Reactions

The first law of thermodynamics may be used for any system. Up to now, we have used it only for systems involving no chemical reactions. Let us now apply it to a generalized steady-flow system with chemical reactions, as shown in Fig. 15.1. Neglecting the change in kinetic and potential energies, we have, from the first law for steady-state steady flow,

$$Q = H_{p2} - H_{R1} + W, \qquad (15.11)$$

where

$$H_{p2} = \text{enthalpy of the products at state 2}$$

$$H_{R1} = \text{enthalpy of the reactants at state 1.}$$

Since there is a change in the composition of the system, there are changes in the chemical properties of the substances involved. The enthalpy data needed for the evaluation of $(H_{p2} - H_{R1})$ must be, in principle, absolute values. This means that the various tables of thermodynamic properties for each substance that we have employed for engineering calculations up to now cannot be used for the determination of $(H_{p2} - H_{R1})$. To circumvent this difficulty, let us rewrite

$(H_{p2} - H_{R1})$ in the following manner:

$$H_{p2} - H_{R1} = (H_{p2} - H_{p0}) - (H_{R1} - H_{R0}) + (H_{p0} - H_{R0}), \quad (15.12)$$

where

H_{p0} = enthalpy of the products at a standard state of T_0 (usually 25°C or 77°F) and p_0 (usually 1.0 atm)

H_{R0} = enthalpy of the reactants at the same standard state as H_{p0}.

In terms of the various species involved, the first two terms in parentheses on the right-hand side of Eq. 15.12 may be written as

$$H_{p2} - H_{p0} = \sum_p n_i(\bar{h}_{p2} - \bar{h}_{p0})_i \qquad (15.13)$$

$$H_{R1} - H_{R0} = \sum_R n_i(\bar{h}_{R1} - \bar{h}_{R0})_i \qquad (15.14)$$

It is clear that the evaluation of Eqs. 15.13 and 15.14 may be carried out by making use of the tables of thermodynamic properties for various simple compressible substances, since we are only dealing with enthalpy changes for the same substance between two states. So, the problem of how to determine $(H_{p2} - H_{R1})$ boils down to the problem of how to determine $(H_{p0} - H_{R0})$. It turns out that the quantity $(H_{p0} - H_{R0})$ may be measured in the laboratory. This quantity is commonly known as the *standard enthalpy of reaction* and is the topic of discussion in the next section.

In terms of the standard enthalpy of reaction, Eq. 15.11 may now be written as

$$Q = \sum_p n_i(\bar{h}_{p2} - \bar{h}_{p0})_i - \sum_R n_i(\bar{h}_{R1} - \bar{h}_{R0})_i + \Delta H^0 + W \quad (15.15)$$

where $\Delta H^0 = H_{p0} - H_{R0}$ = standard enthalpy of reaction.

Equation 15.15 is the working form of the first law of thermodynamics for the study of steady-state steady-flow systems with chemical reactions. In its development we have neglected the changes of kinetic and potential energies, but these quantities may be brought back in very readily if they are significant.

15.6 Enthalpy of Reaction and Heating Values of Fuels

Let us carry out a chemical reaction in a steady-state steady-flow device at constant pressure with no work transfer, as shown in

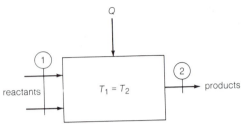

Figure 15.2 Schematic Arrangement for the Determination of Enthalpy of Reaction

Fig. 15.2. Neglecting the changes in kinetic and potential energies, we have, from the first law for steady-state steady flow,

$$Q = H_{p2} - H_{R1} = \Delta H. \tag{15.16}$$

If the reactants and products are both at the same temperature, the quantity ΔH is called the *enthalpy of reaction*. It is simply given by the heat-transfer quantity, and thus it is often called the *heat of reaction*. If the chemical reaction involved is a combustion process, the enthalpy of reaction is also known as the *enthalpy of combustion*. Since the reaction may be carried out at any temperature, there is one value of the heat of reaction corresponding to a given reaction temperature. The enthalpy of reaction that is of particular importance to us is the one corresponding to a reaction carried out at the reference temperature T_0, usually at 25°C or 77°F and a pressure of 1 atm. This is the standard enthalpy of reaction mentioned in the previous section. Some values of the standard enthalpy of combustion are given in Table A.9 in the Appendix.

Chemical reactions which liberate energy in the form of heat are known as *exothermic*, and those which absorb energy are called *endothermic*. In order to be consistent with the convention of heat adopted in this book, we shall consider ΔH as negative when the reaction is exothermic and positive when the reaction is endothermic. We wish to point out that although this convention on the enthalpy of reaction is used in a majority of textbooks and data references, it is not universal.

The enthalpy of reaction for any chemical reaction depends on the phases in which the reactants and products appear. The phase of each constituent is indicated by an *s*, *l*, or *g* (for solid, liquid, and gas). We should always use this notation whenever there is any doubt.

Let us consider the reaction of hydrogen with oxygen to form water. Hydrogen and oxygen generally will exist in the gaseous phase, but water could exist as liquid or vapor. Consequently, we can write the chemical equation for this reaction in the following manner:

$$H_2(g) + \tfrac{1}{2}O_2(g) \rightarrow H_2O(l) \qquad\qquad (15.17)$$

or

$$H_2(g) + \tfrac{1}{2}O_2(g) \rightarrow H_2O(g). \qquad\qquad (15.18)$$

For the reaction given by Eq. 15.17, the product is in the liquid phase and the standard enthalpy of reaction is $-286,030$ kJ/kg mol at $25°C$ of H_2. That is

$$\Delta H^0 = -286,030 \text{ kJ/kg mol of } H_2.$$

For the reaction given by Eq. 15.18, the product is in the gas phase and the standard enthalpy of reaction is $-242,000$ kJ/kg mol at $25°C$ of H_2. That is,

$$\Delta H^0 = -242,000 \text{ kJ/kg mol of } H_2.$$

This means that 286,030 kJ of energy in the form of heat is released when 1 kg mol of hydrogen reacts with $\tfrac{1}{2}$ kg mol of oxygen to form 1 kg mol of liquid water at 298 K. On the other hand, only 242,000 kJ of energy in the form of heat is released when 1 kg mol of hydrogen reacts with $\tfrac{1}{2}$ kg mol of oxygen to form 1 kg mol of water vapor at 298 K.

In engineering practice, the heat obtainable from the burning of a fuel when the reactants and products are both at the same temperature is known as the *heating value* of the fuel. By definition, we have

$$\text{heating value} \equiv H_R - H_p$$

$$= -\Delta H. \qquad\qquad (15.19)$$

We thus see that the heating value of a given fuel is numerically equal to its enthalpy of combustion but is opposite in sign. Most common fuels contain combustible hydrogen, which, upon burning, forms H_2O. The H_2O in the products may appear in either the liquid or vapor phase. When the H_2O appears in the liquid phase, the heating value is called the higher heating value (HHV) of the fuel. When the H_2O appears in the gas phase, the heating value is called the lower heating value (LHV) of the fuel. The difference between the HHV and the LHV is simply the energy associated with the vaporization of the H_2O formed in the burning of the fuel. Thus we have

$$\text{HHV} = \text{LHV} + m_{H_2O} h_{fg}, \qquad (15.20)$$

where, in SI units,

HHV = higher heating value, kJ/kg of fuel

LHV = lower heating value, kJ/kg of fuel

m_{H_2O} = amount of H_2O formed, kg of H_2O/kg of fuel

h_{fg} = enthalpy of vaporization of H_2O, kJ/kg of H_2O.

Example 15.4

Determine the lower heating value (LHV) of methane gas at 101.325 kPa and 25°C.

Solution

The combustion equation is

$$CH_4 + 2O_2 \rightarrow CO_2 + 2H_2O.$$

Thus the amount of H_2O formed is given by

$$m_{H_2O} = \frac{2 \times 18.016}{16.043} = 2.246 \text{ kg/kg of fuel.}$$

From Table A.9 we have

$$\Delta H^\circ = -890{,}900 \text{ kJ/kg mol}$$

$$= \frac{-890{,}900}{16.043} \text{ kJ/kg of fuel} = -55{,}532.0 \text{ kJ/kg of fuel.}$$

At 25°C, the enthalpy of vaporization of water is 2442.5 kJ/kg. Making use of Eq. 15.20, we have

$$\text{LHV} = \text{HHV} - m_{H_2O} h_{fg}$$

$$= 55{,}532.0 - 2.246 \times 2442.5 \text{ kJ/kg of fuel}$$

$$= 50{,}046 \text{ kJ/kg of fuel.} \qquad \blacksquare$$

Comments

We see that the LHV of methane is about 10% less than its HHV. Consequently, if a power plant uses methane gas as its fuel, a 40% thermal efficiency based on LHV is only 36% based on HHV of the fuel. To avoid misunderstanding, we should always state which heating value is used for thermal efficiency calculations. In general, thermal efficiency of a steam power plant is based on the higher heating value of the fuel. On the other hand, the thermal efficiency of a gas-turbine power plant is based on the lower heating value of the fuel.

Example 15.5

Calculate the enthalpy of combustion for gaseous propane at 500°F and 1 atm. Assume that the water in the products is in the vapor phase.

Solution

The combustion equation is

$$C_3H_8(g) + 5O_2(g) \rightarrow 3CO_2(g) + 4H_2O(g).$$

Making use of Eq. 15.15 and the definition of enthalpy of combustion, we have

$$\Delta H_T = \sum_p n_i(\bar{h}_{pT} - \bar{h}_{p0})_i - \sum_R n_i(\bar{h}_{RT} - \bar{h}_{R0})_i + \Delta H^0,$$

where ΔH_T is the enthalpy of combustion at any temperature T and ΔH^0 is the enthalpy of combustion at the reference temperature of 25°C or 77°F. For the given reaction, we have

$$\Delta H_{500} = 3(\bar{h}_{500} - \bar{h}_{77})_{CO_2} + 4(\bar{h}_{500} - \bar{h}_{77})_{H_2O}$$

$$- 1(\bar{h}_{500} - \bar{h}_{77})_{C_3H_8} - 5(\bar{h}_{500} - \bar{h}_{77})_{O_2} + \Delta H_{77}. \quad (1)$$

From Table A.9, we have

$$\Delta H^0 = -879,380 \text{ Btu/lb mol of } C_3H_8.$$

Making use of data from Keenan and Kaye's *Gas Tables*, we find that

$$(\bar{h}_{500} - \bar{h}_{77})_{CO_2} = 8243.8 - 4030.0 = 4213.8 \text{ Btu/lb mol}$$

$$(\bar{h}_{500} - \bar{h}_{77})_{H_2O} = 7738.0 - 4258.3 = 3479.7 \text{ Btu/lb mol}$$

$$(\bar{h}_{500} - \bar{h}_{77})_{O_2} = 6786.0 - 3725.1 = 3060.9 \text{ Btu/lb mol.}$$

Assuming constant specific heat for C_3H_8 and using the value of $\bar{c}_p = 17.876$ Btu/lb mol $\cdot$ °R, we have

$$(\bar{h}_{500} - \bar{h}_{77})_{C_3H_8} = \bar{c}_p(T_{500} - T_{77})_{C_3H_8}$$

$$= 17.876(500 - 77) = 7561.5 \text{ Btu/lb mol.}$$

Substituting into Eq. 1, we have

$$\Delta H_{500} = 3 \times 4213.8 + 4 \times 3479.7 - 1 \times 7561.5 - 5 \times 3060.9 - 879{,}380$$

$$= 3694.2 - 879{,}380 = -875{,}690 \text{ Btu/lb mol of } C_3H_8. \qquad \blacksquare$$

Comments

This example shows that for C_3H_8 the enthalpy of combustion at 500°F does not differ very much from that at 77°F. This is true for fuels in general: Although the enthalpy of combustion is a function of temperature, it is not very sensitive to temperature change.

15.7 Enthalpy of Formation

We have seen the usefulness of standard enthalpy of reaction data. In principle, the standard enthalpy of reaction for any chemical reaction may be determined experimentally. But since the number of possible chemical reactions is so enormous, it would be advantageous to have a method whereby the standard enthalpies of reactions could be obtained from a minimum amount of data. Such a method has been developed through the use of an experimentally determined quantity designated as the *enthalpy formation.*

Let us consider the reaction of burning carbon with oxygen to yield carbon dioxide as the product. This reaction may also be viewed as forming carbon dioxide from its elements of carbon and oxygen. The chemical equation is

$$C(s) + O_2(g) \rightarrow CO_2(g).$$

If the reaction is carried out at 1 atm and if the reactants and product are both at the same reference temperature of 25°C, then the heat transferred from the system is the standard enthalpy of reaction given by

$$Q = \Delta H^0 = H_{p0} - H_{R0} = \bar{h}_{CO_2}^0 - \bar{h}_C^0 - \bar{h}_{O_2}^0, \qquad (15.21)$$

where $\bar{h}^0_{CO_2}$, $\bar{h}^0_C$, and $\bar{h}^0_{O_2}$ must be the molal enthalpies of CO_2, C, and O_2, respectively, measured with respect to the same reference datum.

Let us now assign a value of zero to the enthalpy of all stable elements at 1 atm and 25°C. Since solid carbon and gaseous diatomic oxygen are the stable forms of these elements, both $\bar{h}^0_C$ and $\bar{h}^0_{O_2}$ in Eq. 15.21 will be zero according to this convention.

We may now consider the heat-transfer quantity as the heat involved in the forming of 1 mol of CO_2 gas. The enthalpy of reaction for this process is found to be $-393{,}790$ kJ/kg mol. Thus the enthalpy of formation of carbon dioxide is $-393{,}790$ kJ/kg mol. That is,

$$(\bar{h}^0_f)_{CO_2} = -393{,}790 \text{ kJ/kg mol.}$$

This means that the absolute enthalpy of carbon dioxide may be given as

$$\bar{h}^0_{CO_2} = (\bar{h}^0_f)_{CO_2} + \bar{h}^0_C + \bar{h}^0_{O_2}.$$

Thus, by definition, the standard enthalpy of formation of a compound is simply the enthalpy of reaction for the formation of the compound from its elements at the stable state at 25°C and 1 atm. The enthalpy of formation of various substances at the standard state of 25°C (77°F) at 1 atm is given in Table A.10 in the Appendix.

To illustrate the calculation of the standard enthalpy of reaction from standard enthalpy of formation data, let us consider the reaction

$$C_3H_8(g) + 5O_2(g) \rightarrow 3CO_2(g) + 4H_2O(g).$$

For this reaction, the standard enthalpy of reaction is given by

$$\Delta H^0 = H_{p0} - H_{R0} = 3\bar{h}^0_{CO_2} + 4\bar{h}^0_{H_2O} - \bar{h}^0_{C_3H_8} - 5\bar{h}^0_{O_2}. \qquad (15.22)$$

Now suppose that we have the standard enthalpy of formation data for C_3H_8, CO_2, and H_2O. Therefore, from consideration of the reactions

$$3C(s) + 4H_2(g) \rightarrow C_3H_8(g)$$

$$C(s) + O_2(g) \rightarrow CO_2(g)$$

$$H_2(g) + \tfrac{1}{2}O_2(g) \rightarrow H_2O(g),$$

and making use of the concept of the standard enthalpy of formation, we have

$$\bar{h}_{C_3H_8} = (\bar{h}_f^0)_{C_3H_8} + 3\bar{h}_C^0 + 4\bar{h}_{H_2}^0$$

$$\bar{h}_{CO_2}^0 = (\bar{h}_f^0)_{CO_2} + \bar{h}_C^0 + \bar{h}_{O_2}^0$$

$$\bar{h}_{H_2O}^0 = (\bar{h}_f^0)_{H_2O} + \bar{h}_{H_2}^0 + \tfrac{1}{2}\bar{h}_{O_2}^0 .$$

Substituting into Eq. 15.22, we have

$$\Delta H^0 = 3[(\bar{h}_f^0)_{CO_2} + \bar{h}_C^0 + \bar{h}_{O_2}^0] + 4[(\bar{h}_f^0)_{H_2O} + \bar{h}_{H_2}^0 + \tfrac{1}{2}\bar{h}_{O_2}^0]$$

$$- [(\bar{h}_f^0)_{C_3H_8} + 3\bar{h}_C^0 + 4\bar{h}_{H_2}^0] - 5\bar{h}_{O_2}^0 .$$

Collecting terms, we get

$$\Delta H^0 = 3(\bar{h}_f^0)_{CO_2} + 4(\bar{h}_f^0)_{H_2O} - (\bar{h}_f^0)_{C_3H_8} .$$

Thus we have obtained the standard enthalpy of reaction solely in terms of the standard enthalpy of formation for the various constituents involved in the reaction. In general, the standard enthalpy of reaction for any reaction is given by

$$\Delta H^0 = \sum_{prod} n_i(\bar{h}_f^0)_i - \sum_{reac} n_i(\bar{h}_f^0)_i . \tag{15.23}$$

Equation 15.23 is a widely accepted relationship of thermochemistry. It applies to the reaction involving a stoichiometric mixture as well as to arbitrary mixtures of reactants and to incomplete reactions. In using Eq. 15.23, it is important to remember that by definition, the enthalpy of formation of an element is zero.

Making use of Eq. 15.23, Eq. 15.15, the first law of thermodynamics for steady-state steady flow neglecting change in kinetic and potential energies, may now be written in terms of enthalpy of formation as

$$Q = \sum_{prod} n_i(\bar{h}_{p2} - \bar{h}_{p0})_i - \sum_{reac} n_i(\bar{h}_{R1} - \bar{h}_{R0})_i$$

$$+ \sum_{prod} n_i(\bar{h}_f^0)_i - \sum_{reac} n_i(\bar{h}_f^0)_i + W, \tag{15.15a}$$

which may also be given as

$$Q = \sum_{prod} n_i \bar{h}_i - \sum_{reac} n_i \bar{h}_i + W, \tag{15.15b}$$

where $\bar{h}_i$ is the molal enthalpy of the ith species involved.

In terms of enthalpy of formation, it is seen that the total molal enthalpy $\bar{h}_i$ is given by

$$\bar{h}_i = (\bar{h}_f^0)_i + (\bar{h}_{T,p} - \bar{h}_{298,\,1\,\text{atm}})_i, \qquad (15.24)$$

where the term $(\bar{h}_{T,p} - \bar{h}_{298,\,1\,\text{atm}})$ is the difference in enthalpy between any given state and the enthalpy at the reference state of 298 K and 1 atm.

Example 15.6

Making use of standard enthalpy of formation data, determine the standard enthalpy of reaction for the following:

$$C_8H_{18}(l) + 1.2 \times 12.5O_2(g) \rightarrow 8CO_2(g) + 9H_2O(g) + 0.2 \times 12.5O_2(g).$$

Solution

Making use of Eq. 15.23, we have

$$\Delta H^0 = 8(\bar{h}_f^0)_{CO_2(g)} + 9(\bar{h}_f^0)_{H_2O(g)} + 0.2 \times 12.5(\bar{h}_f^0)_{O_2(g)}$$

$$- (\bar{h}_f^0)_{C_8H_{18}(l)} - 1.2 \times 12.5(\bar{h}_f^0)_{O_2(g)}.$$

Using standard enthalpy of formation data from Table A.10, we get

$$\Delta H^0 = 8(-393,790) + 9(-242,000) + 0.2 \times 12.5(0)$$

$$- (-250,100) - 1.2 \times 12.5(0)$$

$$= -5,078,200 \text{ kJ/kg mol of } C_8H_{18}. \qquad \blacksquare$$

Comments

For this problem, the standard enthalpy of reaction is the same as the enthalpy of combustion at 25°C and 1 atm. With H_2O appearing as vapor in the products of combustion, the enthalpy of combustion at 25°C and 1 atm for liquid C_8H_{18} is given as $-5,078,200$ kJ/kg mol in Table A.9.

15.8 Applications of the First Law to Steady-State Steady-Flow Systems with Chemical Reactions

We now have the necessary tools to analyze steady-state steady-flow systems with chemical reactions from energy consideration. The basic equation we need is Eq. 15.15. We shall illustrate with examples.

Example 15.7

Consider the burning of methane gas in a steady-state steady-flow device with theoretical air involving no work transfer. Neglecting heat transfer and the changes in kinetic and potential energies, determine the temperature of the products of combustion if combustion is complete. Assume that both air and fuel enter the combustion chamber at 1 atm and 77°F, and the products leave at 1 atm.

Solution

We shall make calculations on the basis of 1 mol of methane gas. The combustion equation is

$$CH_4(g) + 2O_2(g) + 2 \times 3.76N_2(g) \rightarrow CO_2(g) + 2H_2O(g) + 2 \times 3.76N_2(g).$$

Referring to Fig. 15.1, and making use of Eq. 15.15 (the steady-state steady-flow energy equation), we have

$$0 = \sum_P n_i(\bar{h}_{p2} - \bar{h}_{p0})_i - \sum_R n_i(\bar{h}_{R1} - \bar{h}_{R0})_i + \Delta H^0. \tag{1}$$

Since both air and fuel enter at 77°F,

$$\sum_R n_i(\bar{h}_{R1} - \bar{h}_{R0})_i = 0.$$

Thus we have

$$\sum_P n_i(\bar{h}_{p2} - \bar{h}_{p0})_i = -\Delta H^0. \tag{2}$$

Now

$$\Delta H^0 = \sum_{prod} n_i(\bar{h}_f^0)_i - \sum_{reac} n_i(\bar{h}_f^0)_i.$$

Therefore, we have

$$\Delta H^0 = (\bar{h}_f^0)_{CO_2} + 2(\bar{h}_f^0)_{H_2O} + 7.52(\bar{h}_f^0)_{N_2} - (\bar{h}_f^0)_{CH_4} - 2(\bar{h}_f^0)_{O_2} - 7.52(\bar{h}_f^0)_{N_2}.$$

Making use of standard enthalpy of formation data, we get

$$\Delta H^0 = -169,300 + 2(-104,040) + 7.52(0)$$

$$- (-32,211) - 2(0) - 7.52(0)$$

$$= -345,170 \text{ Btu.} \tag{3}$$

Now

$$\sum_p n_i(\bar{h}_{p2} - \bar{h}_{p0})_i = (\bar{h}_{T_2} - \bar{h}_{77})_{CO_2} + 2(\bar{h}_{T_2} - \bar{h}_{77})_{H_2O}$$

$$+ 7.52(\bar{h}_{T_2} - \bar{h}_{77})_{N_2}, \tag{4}$$

where T_2 is the temperature of the products. Combining Eq. 3 with Eq. 4, we get

$$(\bar{h}_{T_2} - \bar{h}_{77})_{CO_2} + 2(\bar{h}_{T_2} - \bar{h}_{77})_{H_2O} + 7.52(\bar{h}_{T_2} - \bar{h}_{77})_{N_2} = 345,170, \tag{5}$$

in which the enthalpy terms must be in Btu/lb mol.

On the basis of ideal-gas behavior, enthalpy is a function of temperature only. This means that there is only one temperature of the products that will satisfy Eq. 5. This temperature is determined by means of a trial-and-error solution. We shall make use of data from Keenan and Kaye's *Gas Tables*.

Let us investigate what products temperature we should use in our first trial. Since we have $(1 + 2 + 7.52)$ or 10.52 mol of products, we may write the left-hand side of Eq. 2 as

$$\sum_p n_i(\bar{h}_{p2} - \bar{h}_{p0})_i = 10.52(\bar{h}_{T_2} - \bar{h}_{77})$$

$$\approx 10.52(\bar{c}_p)_{av}(T_2 - T_0),$$

where $(\bar{c}_p)_{av}$ is the average molal $\bar{c}_p$ of the products.

Now, $\bar{c}_p$ for nitrogen varies from about $\frac{7}{2}\bar{R}$ at room temperature to about $\frac{9}{2}\bar{R}$ at very high temperature, $\bar{c}_p$ for carbon dioxide varies from about $\frac{9}{2}\bar{R}$ at room temperature to about $\frac{15}{2}\bar{R}$ at very high temperature, and $\bar{c}_p$ for water vapor varies from about $\frac{8}{2}\bar{R}$ at room temperature to about $\frac{13}{2}\bar{R}$ at very high temperature. Since our products are dominated by the presence of nitrogen, let us use $(\bar{c}_p)_{av} = \frac{9}{2}\bar{R}$ to arrive at a products temperature for our first trial. Thus we have

$$(10.52)(\tfrac{9}{2} \times 1.986)(T_2 - T_0) = 345,170$$

or

$$T_2 - T_0 = 3670°R$$

$$T_2 = 3670 + 537 = 4207°R.$$

Let us use $T_2 = 4200°R$ as the products temperature for our first trial. Using data from the gas tables, we have

$$\sum_p n_i(\bar{h}_{p2} - \bar{h}_{p0}) = (52,162.0 - 4030.1) + 2(43,008.4 - 4258.3)$$

$$+ 7.52(33,068.1 - 3730)$$

$$= 346,255 \text{ Btu}$$

which is sufficiently close to 345,170 Btu. Consequently, the products temperature is about 4200°R. Another trial would yield the correct temperature of 4190°R.

Comments

The temperature of the products obtained in this example is called the *adiabatic flame temperature* for an adiabatic steady-state steady-flow combustion process with no work transfer. Since all the released chemical energy is used toward increasing the enthalpy of the products, the adiabatic flame temperature is the maximum temperature that can be achieved for this process. It is therefore of interest to the engineer engaged in the design of the combustion chamber.

Example 15.8

Figure 15.3 shows the arrangement of a simple gas-turbine power plant. Theoretically, we should take full advantage of the high-temperature products

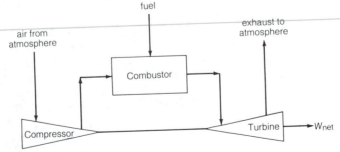

Figure 15.3 Simple Gas-Turbine Power Plant

such as those obtained in Example 15.7. Because of metallurgical consider-ations, the gas turbine must operate with an inlet temperature considerably lower than the adiabatic flame temperature. Suppose that we want the pro-ducts to leave the combustor at 1600°F. Suppose that the air enters the com-bustor at 400°F and the fuel enters at 77°F. Determine the air–fuel ratio if the fuel is methane gas.

Solution

We shall make calculations on the basis of 1 mol of fuel. Assuming complete combustion, the combustion equation is

$$CH_4(g) + aO_2(g) + 3.76aN_2(g) \rightarrow CO_2(g) + 2H_2O(g)$$

$$+ bO_2(g) + 3.76aN_2(g). \quad (1)$$

From oxygen balance, we have

$$2a = 2 + 2 + 2b$$

or

$$b = a - 2.$$

Thus Eq. 1 may be written as

$$CH_4 + aO_2 + 3.76aN_2 \rightarrow CO_2 + 2H_2O + (a - 2)O_2 + 3.76aN_2. \quad (2)$$

Taking the combustor as our control volume, and neglecting heat transfer and changes in kinetic and potential energies, we have, from the steady-state steady-flow energy equation,

$$0 = \sum_p n_i(\bar{h}_{p2} - \bar{h}_{p0})_i - \sum_R n_i(\bar{h}_{R1} - \bar{h}_{R0})_i + \Delta H^0. \quad (3)$$

Making use of standard enthalpy of formation data, we have

$$\Delta H^0 = (\bar{h}_f^0)_{CO_2} + 2(\bar{h}_f^0)_{H_2O(g)} + (a - 2)(\bar{h}_f^0)_{O_2} + 3.76a(\bar{h}_f^0)_{N_2}$$

$$- (\bar{h}_f^0)_{CH_4} - a(\bar{h}_f^0)_{O_2} - 3.76a(\bar{h}_f^0)_{N_2}$$

$$= -169{,}300 + 2(-104{,}040) + (a - 2)(0) + 3.76a(0)$$

$$- (-32{,}211) - a(0) - 3.76a(0)$$

$$= -345{,}170 \text{ Btu.} \quad (4)$$

Making use of data from the gas tables, we have

$$\sum_p n_i(\bar{h}_{p2} - \bar{h}_{po})_i = (\bar{h}_{1600} - \bar{h}_{77})_{CO_2} + 2(\bar{h}_{1600} - \bar{h}_{77})_{H_2O}$$

$$+ (a - 2)(\bar{h}_{1600} - \bar{h}_{77})_{O_2}$$

$$+ 3.76a(\bar{h}_{1600} - \bar{h}_{77})_{N_2}$$

$$= (21{,}817.6 - 4030.1) + 2(18{,}053.6 - 4258.3)$$

$$+ (a - 2)(15{,}671.5 - 3725.1)$$

$$+ 3.76a(15{,}013.3 - 3730.0)$$

$$= 21{,}485.3 + 54{,}371.6a \tag{5}$$

$$\sum_R n_i(\bar{h}_{R1} - \bar{h}_{RO}) = (\bar{h}_{77} - \bar{h}_{77})_{CH_4} + a(\bar{h}_{400} - \bar{h}_{77})_{O_2}$$

$$+ 3.76a(\bar{h}_{400} - \bar{h}_{77})_{N_2}$$

$$= 0 + a(6041.9 - 3725.1) + 3.76a(5985.9 - 3730.1)$$

$$= 10{,}799.0a. \tag{6}$$

Substituting into Eq. 3, we have

$$0 = 21{,}485.3 + 54{,}371.6a - 10{,}799.0a - 345{,}170.$$

Solving for a, we get

$$a = 7.428.$$

On a molal basis, we have

$$AF = 7.428 \times 4.76 = 35.36 \text{ lb mol of air/lb mol of fuel.}$$

On a mass basis, we have

$$AF = \frac{7.428 \times 4.76 \times 28.9}{1 \times 16} = 63.9 \text{ lbm of air/lbm of fuel.} \qquad \blacksquare$$

Comments

As found in Section 15.3, the theoretical air required to burn 1 lb of methane gas is 17.2 lbm. Thus we see that the combustor in this example operates with 63.9/17.2, or 372% theoretical air. It is indeed quite common for a simple gas-turbine power plant to be designed with 400% theoretical air, using a hydrocarbon fuel.

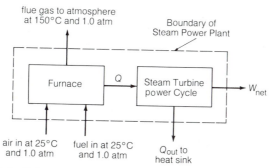

Figure 15.4 Main Subsystems of a Steam Power Plant

Example 15.9

Let us divide a steam power plant into two parts, the furnace and the steam-turbine power cycle, as shown in Fig. 15.4. Determine the fuel consumption of such a power plant designed according to the following specifications:

Net power output	$= 1.0 \times 10^6$ kW
Thermal efficiency of steam-turbine power cycle	$= 40\%$
Temperature of air supply	$= 25°C$
Pressure of air supply	$= 1.0$ atm
Temperature of fuel (methane gas) supply	$= 25°C$
Pressure of fuel supply	$= 1.0$ atm
Excess air used for combustion	$= 10\%$
Temperature of flue gas	$= 150°C$
Pressure of flue gas	$= 1.0$ atm

Solution

Let us first make calculations for the amount of heat transferred out of the furnace on the basis of 1.0 mol of methane gas. Assuming complete combustion, the combustion equation with 110% theoretical air is

$$CH_4 + 1.1 \times 2O_2 + 1.1 \times 2 \times 3.76N_2$$

$$\rightarrow CO_2 + 2H_2O(g) + 0.2O_2 + 1.1 \times 2 \times 3.76N_2. \quad (1)$$

Taking the furnace as our control volume, and neglecting changes in kinetic and potential energies, we have, from the steady-state steady-flow energy equation,

$$Q = \sum_p n_i (\bar{h}_{p2} - \bar{h}_{p0})_i - \sum_R n_i (\bar{h}_{R1} - \bar{h}_{R0})_i + \Delta H^0. \tag{2}$$

Now

$$\Delta H^0 = -802{,}800 \text{ kJ per kg mol of } CH_4.$$

Since both air and fuel enter at 25°C,

$$\sum_R n_i (\bar{h}_{R1} - \bar{h}_{R0}) = 0. \tag{3}$$

We shall assume that the flue gas is an ideal-gas mixture with constant specific heats. Then

$$\sum_p n_i (\bar{h}_{p2} - \bar{h}_{p0})_i = n_p (\bar{c}_p)_p (T_2 - T_0),$$

where

$$n_p = \text{total number of moles of flue gas}$$

$$(\bar{c}_p)_p = \sum x_i (\bar{c}_p)_i \text{ is the weighted specific heat of the flue gas,}$$

with x_i being the mole fraction of the ith species in the flue gas.
For our problem, on the basis of 1 mol of fuel burned,

$$n_p = 1 + 2 + 0.2 + 1.1 \times 2 \times 3.76 = 11.472 \text{ mol}$$

$$(\bar{c}_p)_p = x_{CO_2} (\bar{c}_p)_{CO_2} + x_{H_2O} (\bar{c}_p)_{H_2O} + x_{O_2} (\bar{c}_p)_{O_2} + x_{N_2} (\bar{c}_p)_{N_2}$$

$$= \left(\frac{1}{11.472} \right)(37.2539) + \left(\frac{2}{11.472} \right)(33.6693) + \left(\frac{0.2}{11.472} \right)(29.4315)$$

$$+ \left(\frac{8.272}{11.472} \right)(29.0784)$$

$$= 30.6 \text{ kJ/kg mol} \cdot \text{K}.$$

Then

$$\sum_p n_i (\bar{h}_{p2} - \bar{h}_{p0})_i = 11.472 \times 30.6 \times (150 - 25)$$

$$= 43{,}900 \text{ kJ}. \tag{4}$$

Substituting into Eq. 2, we get

$$Q = 43,900 - 0 - 802,800$$

$$= -758,900 \text{ kJ}$$

for each kg mol of fuel burned.

Let us now turn our attention to the steam-turbine power cycle. From the definition of thermal efficiency, we have

$$\dot{Q}_{in} = \frac{\dot{W}_{net}}{\eta_{th}}$$

$$= \frac{1.0 \times 10^6 \text{ kW}}{0.4} = 2.5 \times 10^6 \text{ kW} = 9.0 \times 10^9 \text{ kJ/h}.$$

Thus the rate of fuel supply is given as

$$m_{fuel} = \frac{9.0 \times 10^9 \text{ kJ/h}}{758,900 \text{ kJ/kg mol of fuel burned}}$$

$$= 1.1859 \times 10^4 \text{ kg mol/h}$$

$$= 1.1859 \times 10^4 \times 16.043 \text{ kg/h} = 1.903 \times 10^5 \text{ kg/h.} \quad \blacksquare$$

Comments

The flue gas leaving the furnace must be kept at a high enough temperature (around 150°C) to prevent the water vapor in it from condensing. A wet flue gas would be corrosive because of the fact that a hydrocarbon fuel usually has a small amount of sulfur, which burns to form sulfur dioxide. Condensed water vapor would combine with sulfur dioxide to form dilute sulfuric acid, which is very corrosive.

Example 15.10

Determine the efficiency of the furnace in Example 15.9.

Solution

The furnace efficiency, also known as the steam-generator efficiency in the case of a steam power plant, is defined as

$$\eta_{sg} = \frac{\dot{Q}}{\dot{m}_f(\text{HHV})}, \tag{15.25}$$

where

$\dot{Q}$ = amount of heat transferred to the working fluid of the steam power cycle, kJ/h

$\dot{m}_f$ = amount of fuel burned, kg/h

HHV = higher heating value of the fuel burned, kJ/kg.

For Example 15.9,

$$\dot{Q} = 9.0 \times 10^9 \text{ kJ/h}$$

$$\dot{m}_f = 1.903 \times 10^5 \text{ kg/h}$$

$$\text{HHV} = \frac{890,900}{16.043} = 55,532 \text{ kJ/kg}.$$

Substituting into Eq. 15.25, we have

$$\eta_{sg} = \frac{9.0 \times 10^9}{1.903 \times 10^5 \times 55,532} = 85.1\%.$$ ∎

15.9 Absolute Entropy and the Third Law of Thermodynamics

As we pointed out in Chapter 6, the third law of thermodynamics states that at the absolute zero of temperature, the entropy of a perfect crystal is zero. This law leads to a number of consequences, all of which have been verified experimentally. Some of these are that the specific heats $\bar{c}_p$ and $\bar{c}_v$ must vanish as the absolute zero of temperature is approached and that thermal expansion must approach zero. As far as chemical reactions are concerned, the important result of the third law is that it permits the determination of absolute entropies from thermal data. The absolute entropy of some substances, each at a standard state of 1 atm and 25°C, is given in Table A.10 in the Appendix.

Knowing the absolute entropy at a reference state, the absolute entropy of a substance in any other state is given by

$$\bar{s}_{T,p} = \bar{s}^0 + \int_{T_0, p_0}^{T, p} d\bar{s}, \qquad (15.26)$$

where the integral is evaluated by the methods already discussed for simple compressible substances. For ideal gases, the entropy change between two state points is given by Eq. 5.18.

$$d\bar{s} = \bar{c}_p \frac{dT}{T} - \bar{R} \frac{dp}{p}.$$

Thus the absolute entropy of an ideal gas at a given state may be given by

$$\bar{s}_{T,p} = \bar{s}^0 + \int_{T_0}^{T} \bar{c}_p \frac{dT}{T} - \bar{R} \int_{p_0}^{p} \frac{dp}{p}$$

$$= \bar{s}^0 + \int_{T_0}^{T} \bar{c}_p \frac{dT}{T} - \bar{R} \ln \frac{p}{p_0}. \tag{15.27}$$

Equation 15.27 shows that the determination of the absolute entropy of an ideal gas at a given state is a relatively simple matter if we have $\bar{c}_p$ data and the value of the absolute entropy of the gas at the reference state.

The first two terms on the right-hand side of Eq. 15.27 may be combined to form a new function, which is a function of T only. Denoting this new function by ϕ, we may write Eq. 15.27 as

$$\bar{s}_{T,p} = \phi_T - \bar{R} \ln \frac{p}{p_0}. \tag{15.28}$$

The ϕ values for N_2, O_2, H_2O vapor, CO_2, H_2, CO, and monatomic gases at various temperatures are given by Keenan and Kaye's *Gas Tables*, where $p_0 = 1$ atm. Thus for ideal gases at 1 atm, the absolute entropy values are the same as the ϕ values. (*Caution*: The ϕ values for air given by Keenan and Kaye are not the ϕ values to be used in Eq. 15.28.[2])

15.10 Applications of the Second Law to Steady-State Steady-Flow Systems with Chemical Reactions

Any working system must satisfy the first law as well as the second law. Up to now, in our study of systems involving chemical reactions,

---

[2] For ϕ values for air to be used in Eq. 15.28 see Table B.13 in W. R. Reynolds and H. C. Perkins, *Engineering Thermodynamics*, McGraw-Hill Book Company, New York, 1970, p. 548.

we have examined them based on the first law only. Some of the processes that we assumed to have occurred may or may not satisfy the second law. We shall illustrate the usefulness of the second law with examples.

The second law for any steady-state steady-flow process is given by Eq. 3.30:

$$S_{gen} = \sum_{out} \dot{m}s - \sum_{in} \dot{m}s - \sum \frac{\dot{Q}_k}{T_k} \geq 0,$$

which, in the case of a reactive process, may be written as

$$S_{gen} = S_p - S_R - \sum \frac{\dot{Q}_k}{T_k} \geq 0, \qquad (15.29)$$

where

S_p = entropy of the products

S_R = entropy of the reactants

$\sum \frac{\dot{Q}_k}{T_k}$ = entropy transfer into control volume due to heat transfer.

If the products and reactants are ideal gas mixtures, S_p and S_R may be written as

$$S_p = \sum_{prod} n_i \bar{s}_i \qquad (15.30)$$

and

$$S_R = \sum_{reac} n_i \bar{s}_i. \qquad (15.31)$$

The molal entropy for each species to be used in Eqs. 15.30 and 15.31 must be the absolute molal entropy.

Example 15.11

In Example 15.7, we studied the adiabatic burning of methane gas with theoretical air. The results, based only on the use of the first law and the principle of conservation of matter, are shown in Fig. 15.5. Determine whether these results are permitted by the second law of thermodynamics.

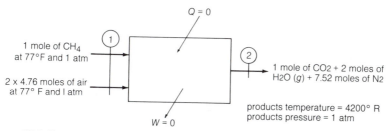

Figure 15.5 Results of Burning Methane Gas with Theoretical Air Based on First-Law Analysis Only

Solution

Taking the combustion chamber as our control volume, we have, from the second law for steady-state steady flow,

$$\Delta S_{creation} = \sum_{prod} n_i \bar{s}_i - \sum_{reac} n_i \bar{s}_i, \tag{1}$$

where $\sum_{prod} n_i \bar{s}_i = 1.0\bar{s}_{CO_2} + 2.0\bar{s}_{H_2O} + 7.52\bar{s}_{N_2}$, evaluated at the product temperature of 4200°R and pressure of 1 atm; and $\sum_{reac} n_i \bar{s}_i = 1.0\bar{s}_{CH_4} + 2.0\bar{s}_{O_2} + 7.52\bar{s}_{N_2}$ evaluated at the inlet conditions.

To determine the molal entropy of CO_2, H_2O, N_2, and O_2, we need to know the partial pressures of these species. For the oxygen and nitrogen in the reactants, we have

$$(p_1)_{O_2} = 0.21 \times 1 = 0.21 \text{ atm}$$

$$(p_1)_{N_2} = 0.79 \times 1 = 0.79 \text{ atm.}$$

For the species in the products, we have

$$(p_2)_{CO_2} = \frac{1}{10.52} \times 1 = \frac{1}{10.52} \text{ atm}$$

$$(p_2)_{H_2O} = \frac{2}{10.52} \times 1 = \frac{2}{10.52} \text{ atm}$$

$$(p_2)_{N_2} = \frac{7.52}{10.52} \times 1 = \frac{7.52}{10.52} \text{ atm.}$$

Making use of Eq. 15.27 and absolute entropy data given in Table A.10, we have

$$(\bar{s}_1)_{CH_4} = 44.490 + 0 - 1.986 \ln \frac{1}{1} = 44.49 \text{ Btu/lb mol} \cdot {}^{\circ}R$$

$$(\bar{s}_1)_{O_2} = 49.004 + 0 - 1.986 \ln \frac{0.21}{1} = 52.103 \text{ Btu/lb mol} \cdot {}^{\circ}R$$

$$(\bar{s}_1)_{N_2} = 45.770 + 0 - 1.986 \ln \frac{0.79}{1} = 46.238 \text{ Btu/lb mol} \cdot {}^{\circ}R.$$

Making use of Eq. 15.28 and the ϕ values in the gas tables,

$$(\bar{s}_2)_{CO_2} = 76.119 - 1.986 \ln \frac{1/10.52}{1} = 80.793 \text{ Btu/lb mol} \cdot {}^{\circ}R$$

$$(\bar{s}_2)_{H_2O} = 65.114 - 1.986 \ln \frac{2/10.52}{1} = 68.411 \text{ Btu/lb mol} \cdot {}^{\circ}R$$

$$(\bar{s}_2)_{N_2} = 61.520 - 1.986 \ln \frac{7.52/10.52}{1} = 62.187 \text{ Btu/lb mol} \cdot {}^{\circ}R.$$

Substituting into Eq. 1, we have

$$(\Delta S)_{creation} = 1.0 \times 80.793 + 2.0 \times 68.411 + 7.52 \times 62.187$$

$$- 1.0 \times 44.49 - 2.0 \times 52.103 - 7.52 \times 46.238$$

$$= 685.26 - 496.40$$

$$= +188.9 \text{ Btu/}{}^{\circ}R.$$

Since the entropy creation is positive, the assumed process appears to have satisfied the second law. However, the results have been based on the assumption that the reaction could proceed to completion. If complete combustion is possible, the entropy of the ideal products must be greater than the entropy of the products resulting from incomplete combustion for this adiabatic combustion process. Let us assume that the adiabatic burning of methane gas with theoretical air may proceed and that it will yield the products given in the following chemical equation:

$$CH_4 + 2O_2 + 2 \times 3.76N_2$$

$$\rightarrow 0.9CO_2 + 0.1CO + 0.05O_2 + 2H_2O(g) + 7.52N_2.$$

Let us designate the state of these products as state 3. Then, applying the first law as in Example 15.7, we find the temperature of these products to be about 4080°R. Assuming that the pressure of these products is also at 1 atm, the molal entropy of each species is then given as

$$(\bar{s}_3)_{CO_2} = 75.693 - 1.986 \ln \frac{0.9/10.57}{1} = 80.585 \text{ Btu/lb mol} \cdot {}^\circ R$$

$$(\bar{s}_3)_{CO} = 62.945 - 1.986 \ln \frac{0.1/10.57}{1} = 72.201 \text{ Btu/lb mol} \cdot {}^\circ R$$

$$(\bar{s}_3)_{O_2} = 65.305 - 1.986 \ln \frac{0.05/10.57}{1} = 75.938 \text{ Btu/lb mol} \cdot {}^\circ R$$

$$(\bar{s}_3)_{H_2O} = 64.777 - 1.986 \ln \frac{2.0/10.57}{1} = 68.083 \text{ Btu/lb mol} \cdot {}^\circ R$$

$$(\bar{s}_3)_{N_2} = 61.268 - 1.986 \ln \frac{7.52/10.57}{1} = 61.944 \text{ Btu/lb mol} \cdot {}^\circ R.$$

The entropy content of the products at state 3 is then given by

$$(S_3)_p = 0.9 \times 80.585 + 0.1 \times 72.201 + 0.05 \times 75.938$$

$$+ 2 \times 68.083 + 7.52 \times 61.944$$

$$= 685.53 \text{ Btu/}^\circ R.$$

The entropy content of the products at state 2 was found to be

$$(S_2)_p = 685.26 \text{ Btu/}^\circ R.$$

We thus see that the entropy creation in going from state 3 to state 2 is given by

$$S_{gen} = 685.26 - 685.53 = -0.27 \text{ Btu/}^\circ R.$$

Since the entropy generation in this case is negative, we may now conclude that it is not possible to have the products as shown in Fig. 15.5.   ∎

Comments

Our analysis shows that the adiabatic burning of methane gas with theoretical air may not proceed to completion. But we should not conclude that the products at state 3 are possible. To determine the extent to which this adiabatic combustion process will proceed, we could make repeated calculations like those in this example. We shall see that there is a more direct method for such a determination.

Example 15.12

The flow of fluid between the furnace and the steam turbine power cycle for Example 15.9 (Fig. 15.4) is as shown. The rate of heat input to the steam turbine power cycle is that provided by the furnace in the figure. Determine

(a) the second-law efficiency of the furnace.
(b) the second-law efficiency of the steam power cycle.

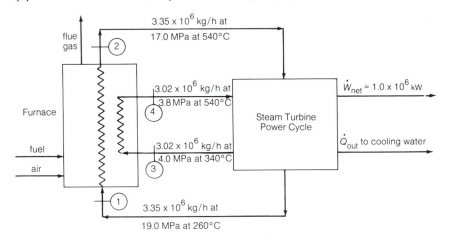

Solution

(a) The second-law efficiency of the furnace may be given as

$$\eta_{\mathrm{II}} = \frac{\text{exergy input to steam turbine power cycle}}{\text{exergy content of fuel input}}. \tag{1}$$

From Example 15.9,

$$\dot{m}_{\mathrm{fuel}} = 1.903 \times 10^5 \text{ kg/h}.$$

The higher heating value of the fuel is 55,532 kJ/kg.
    The energy content of fuel input is then given by

$$\dot{E}_f = 1.903 \times 10^5 \times 55{,}532 \text{ kJ/h} = 1.0568 \times 10^{10} \text{ kJ/h}.$$

Assume that the exergy factor of fuel input is 1.0. Then the exergy content of fuel input is

$$\dot{B}_f \approx \dot{E}_f = 1.0568 \times 10^{10} \text{ kJ/h}. \tag{2}$$

The exergy input to the steam turbine power cycle, $\dot{B}_{in}$, is the exergy gained by the working fluid of the cycle:

$$\dot{B}_{in} = \dot{m}_1[(h_2 - h_1) - T_0(s_2 - s_1)] + \dot{m}_3[(h_4 - h_3) - T_0(s_4 - s_3)]. \quad (3)$$

From the steam tables, we have

$$h_1 = 1134.0 \text{ kJ/kg} \qquad h_3 = 3069.8 \text{ kJ/kg}$$

$$s_1 = 2.8419 \text{ kJ/kg} \cdot \text{K} \qquad s_3 = 6.5461 \text{ kJ/kg} \cdot \text{K}$$

$$h_2 = 3398.7 \text{ kJ/kg} \qquad h_4 = 3537.8 \text{ kJ/kg}$$

$$s_2 = 6.4137 \text{ kJ/kg} \cdot \text{K} \qquad s_4 = 7.2339 \text{ kJ/kg} \cdot \text{K}.$$

Let T_0, the temperature of the environment, be 25°C. Then

$$\dot{B}_{in} = 3.35 \times 10^6[(3398.7 - 1134.0 - 298.15(6.4137 - 2.8419)]$$

$$+ \, 3.02 \times 10^6[(3537.8 - 3069.8) - 298.15(7.2339 - 6.5461)]$$

$$= 4.8134 \times 10^9 \text{ kJ/h}.$$

The second-law efficiency of the furnace is then given by

$$\eta_{II} = \frac{\dot{B}_{in}}{\dot{B}_f}$$

$$= \frac{4.8134 \times 10^9}{1.0568 \times 10^{10}}$$

$$= 45.5\%.$$

(b) The second-law efficiency of the steam turbine power cycle is given as

$$\eta_{II} = \frac{\dot{W}_{net}}{\dot{B}_{in}}$$

$$= \frac{1.0 \times 10^6 \times 3600 \text{ kJ/h}}{4.8134 \times 10^9 \text{ kJ/h}}$$

$$= 74.8\%. \qquad \blacksquare$$

Comments

Although the first-law efficiency of the furnace (the steam-generator efficiency) is quite high (about 85%), its second-law efficiency is only 45.5%. This means that over 50% of the exergy was destroyed in the furnace. This is because the conversion of chemical energy (organized energy) into thermal energy (disorganized energy) is a highly irreversible process.

The thermal efficiency of the steam power cycle (first-law efficiency) appears to be low, only 40%, yet its second-law efficiency is quite good, 74.8%. This means that the engineers have done a pretty good job in the design of the modern steam power cycle.

15.11 Standard Gibbs Function of Formation

Let us consider the formation of liquid water from hydrogen and oxygen according to the following reaction:

$$H_2(g) + \tfrac{1}{2}O_2(g) \rightarrow H_2O(l).$$

If the reaction is carried out at 1 atm and if the reactants and product are both at 25°C, the difference between the Gibbs function of the product and the Gibbs function of the reactants, known as the *standard Gibbs function change* of the reaction, is given as

$$G_{p0} - G_{R0} = \bar{g}^0_{H_2O(l)} - \bar{g}^0_{H_2} - \tfrac{1}{2}\bar{g}^0_{O_2},$$

where $\bar{g}^0_{H_2O(l)}$, $\bar{g}^0_{H_2}$, and $\bar{g}^0_{O_2}$ are the molal Gibbs function of $H_2O(l)$, H_2, and O_2, respectively, measured with respect to the same reference datum.

We may now introduce the concept of the standard Gibbs function of formation in a manner analogous to the standard enthalpy of formation. Assigning a value of zero to the molal Gibbs function of all stable elements at 1 atm and 25°C, we see that the absolute molal Gibbs function of formation of a compound is simply the molal Gibbs function of the compound formed in a constant-pressure isothermal reaction in which the compound is the only product and the reactants are the stable chemical elements at the standard state of 1 atm and 25°C. With this definition of the standard Gibbs function of formation, $\bar{g}^0_{H_2O(l)} = (\bar{g}^0_f)_{H_2O(l)}$. The Gibbs function of formation of various substances at the standard state of 25°C (77°F) and 1 atm is given in Table A.10 in the Appendix.

Analogous to the use of standard enthalpy of formation data in the calculation of the standard enthalpy of reaction for any reaction, we may also make use of the standard Gibbs function of formation data in calculating the standard Gibbs function change of any reaction. In general, the standard Gibbs function change for any reaction is given by

$$G_{p0} - G_{R0} = \sum_{\text{prod}} n_i(\bar{g}_f^0)_i - \sum_{\text{reac}} n_i(\bar{g}_f^0)_i. \tag{15.32}$$

Example 15.13

Making use of enthalpy of formation and absolute entropy data at 1 atm and 25°C, determine the standard Gibbs function of formation for liquid water.

Solution

The chemical equation is

$$H_2(g) + \tfrac{1}{2}O_2(g) \rightarrow H_2O(l).$$

The Gibbs function change for this reaction at 1 atm and 25°C is given as

$$G_{p0} - G_{R0} = H_{p0} - H_{R0} - T_0(S_{p0} - S_{R0})$$

$$= (\bar{h}_f^0)_{H_2O(l)} - T_0[(\bar{s}^0)_{H_2O(l)} - (\bar{s}^0)_{H_2} - \tfrac{1}{2}(\bar{s}^0)_{O_2}]$$

$$= -286,030 - 298.15(69.99 - 130.66 - \tfrac{1}{2} \times 205.17)$$

$$= -237,350 \text{ kJ/kg mol.}$$

From the definition of the standard Gibbs function of formation, we have

$$G_{p0} - G_{R0} = (\bar{g}_f^0)_{H_2O(l)}.$$

Thus

$$(\bar{g}_f^0)_{H_2O(l)} = -237,350 \text{ kJ/kg mol,}$$

which is the value given in Table A.10 in the Appendix.   ∎

Example 15.14

A reversible isothermal reaction may be approached in a steady-state steady-flow device known as the *fuel cell* (see Fig. 15.6). Determine the reversible isothermal work obtainable from a fuel cell when the reaction is

$$H_2(g) + \tfrac{1}{2}O_2(g) \rightarrow H_2O(l).$$

Assume that the reaction is carried out at 25°C and 1 atm.

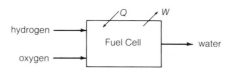

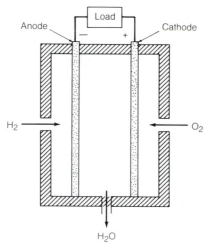

Figure 15.6 Fuel Cell

Solution

Taking the fuel cell as our system, and neglecting changes in kinetic and potential energies, the reversible isothermal work is, according to the first and second laws of thermodynamics,

$$W = -(G_{p0} - G_{R0}). \qquad (1)$$

Making use of Eq. 15.32 and the standard Gibbs function of formation data from Table A.10, we have, on the basis of 1 kg mol of hydrogen gas,

$$W = -[1.0(g_f^0)_{H_2O(l)} - 1.0(g_f^0)_{H_2} - \tfrac{1}{2}(g_f^0)_{O_2}]$$

$$= -[1.0(-237,350) - 1.0(0) - \tfrac{1}{2}(0)]$$

$$= +237,350 \text{ kJ},$$

which is practically equal to the heating value of hydrogen. ∎

Comments

A fuel cell is a direct-energy-conversion device, in which the chemical energy of the fuel is harnessed directly to produce electrical power. Thus a fuel cell is not a heat engine and its performance is not limited by the Carnot efficiency. Fuel cells employing hydrogen and oxygen as the reactants have been developed for use in space vehicles, but these fuel cells are much too expensive for ordinary commercial applications. Some effort is being made toward the development of a fuel cell that utilizes common hydrocarbons as the source of energy and air as the source of oxygen. If this effort is fruitful, fuel cells may become the power plants of our cars of the future, and the seriousness of our pollution problems will be reduced significantly. For more information on fuel cells, books on this topic should be consulted.[3]

15.12 Chemical Equilibrium

It was shown in Chapter 3 that for a system of constant mass the direction of change at constant temperature and pressure is governed by Eq. 3.37:

$$dG_{T,p} \leq 0. \tag{3.37}$$

This important result is simply a more convenient way of expressing the second law of thermodynamics for systems undergoing a change of state at constant temperature and pressure. Applying it to a reactive system, this means that a chemical reaction carried out at constant temperature and pressure can proceed only if the Gibbs function of the system will continually decrease, and the reaction will stop when the Gibbs function of the system has reached a minimum. This is illustrated in Fig. 15.7, where the Gibbs function of the system is plotted against its composition. We thus see that the equilibrium composition of any reactive system of known temperature and pressure is governed by

$$dG_{T,p} = 0. \tag{15.33}$$

Let us now make use of Eq. 15.33 to study the equilibrium composition of a simple homogeneous reactive mixture in which the following chemical reaction occurs:

$$v_A A + v_B B \rightleftharpoons v_C C + v_D D, \tag{15.34}$$

where A, B, C, and D are the chemical constituents, and v_A, v_B, v_C, and v_D are the stoichiometric coefficients.

The state of such a system may be fixed by specifying T, p, and the mole numbers of the species involved. Thus, in accordance with

[3] S. L. Soo, *Direct Energy Conversion*, Prentice-Hall, Inc., Englewood Cliffs, N.J., 1968; S. W. Angrist, *Direct Energy Conversion*, Allyn and Bacon, Inc., Boston, 1965.

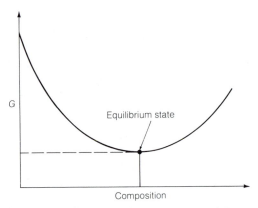

Figure 15.7 Equilibrium at Constant Temperature and Pressure

Eq. 15.34, the differential of the total Gibbs function of the system is given by

$$dG = -S\,dT + V\,dp + \mu_A\,dn_A + \mu_B\,dn_B + \mu_C\,dn_C + \mu_D\,dn_D. \quad (15.35)$$

The changes in the amounts of the species involved are related to one another in the following manner:

$$-\frac{dn_A}{\nu_A} = -\frac{dn_B}{\nu_B} = \frac{dn_C}{\nu_C} = \frac{dn_D}{\nu_D} \qquad (15.36)$$

since the disappearance of the reactants and the appearance of the products are constrained by Eq. 15.34.

By introducing the proportionality factor, $d\varepsilon$, Eq. 15.36 may be replaced by the following equations:

$$dn_A = -\nu_A\,d\varepsilon \qquad\qquad (15.37a)$$

$$dn_B = -\nu_B\,d\varepsilon \qquad\qquad (15.37b)$$

$$dn_C = +\nu_C\,d\varepsilon \qquad\qquad (15.37c)$$

$$dn_D = +\nu_D\,d\varepsilon. \qquad\qquad (15.37d)$$

The ε is known as the *reaction variable* or *reaction coordinate*. It gives the extent to which a reaction has occurred; so $d\varepsilon$ will measure an increment of reaction.

Combining Eq. 15.35 with Eq. 15.37, we have

$$dG = -S\,dT + V\,dp + (v_C\,\mu_C + v_D\,\mu_D - v_A\,\mu_A - v_B\,\mu_B)\,d\varepsilon. \quad (15.38)$$

When T and p are constant, Eq. 15.38 becomes

$$dG_{T,p} = (v_C\,\mu_C + v_D\,\mu_D - v_A\,\mu_A - v_B\,\mu_B)\,d\varepsilon. \quad (15.39)$$

Equation 15.39 gives the change in G at constant T and p when ε is changed by $d\varepsilon$. For equilibrium at constant T and p, $dG_{T,p}$ must be zero, according to Eq. 15.33, which requires that

$$(v_C\,\mu_C + v_D\,\mu_D - v_A\,\mu_A - v_B\,\mu_B) = 0 \quad (15.40)$$

since $d\varepsilon$ is, by definition, not zero. Equation 15.40 is the condition of chemical equilibrium for a simple reactive mixture. It must be kept in mind that when we apply this equation in our study of reactive systems, we are applying the second law of thermodynamics.

The conditions for equilibrium when two or more chemical reactions occur simultaneously are obtained in a similar manner. To illustrate, let us consider a simple reactive mixture having the following pair of reactions occurring in it:

$$\text{(I)} \qquad CO + \tfrac{1}{2}O_2 \rightleftharpoons CO_2 \qquad (15.41)$$

$$\text{(II)} \qquad CO_2 + H_2 \rightleftharpoons CO + H_2O \qquad (15.42)$$

If we denote $d\varepsilon_I$ and $d\varepsilon_{II}$ as the differential extents of the reactions, the changes in the mole numbers are then given by

$$dn_{CO} = -d\varepsilon_I + d\varepsilon_{II} \qquad (15.43a)$$

$$dn_{O_2} = -\tfrac{1}{2}d\varepsilon_I \qquad (15.43b)$$

$$dn_{CO_2} = d\varepsilon_I - d\varepsilon_{II} \qquad (15.43c)$$

$$dn_{H_2} = -d\varepsilon_{II} \qquad (15.43d)$$

$$dn_{H_2O} = d\varepsilon_{II}. \qquad (15.43e)$$

Substituting Eqs. 15.43 into the general expression for dG, we have

$$dG = -S\,dT + V\,dp + (\mu_{CO_2} - \mu_{CO} - \tfrac{1}{2}\mu_{O_2})\,d\varepsilon_I$$

$$+ (\mu_{CO} + \mu_{H_2O} - \mu_{CO_2} - \mu_{H_2})\,d\varepsilon_{II}. \qquad (15.44)$$

The conditions for equilibrium at constant T and p are then given by

$$\mu_{CO_2} - \mu_{CO} - \tfrac{1}{2}\mu_{O_2} = 0 \qquad (15.45a)$$

$$\mu_{CO} + \mu_{H_2O} - \mu_{CO_2} - \mu_{H_2} = 0 \qquad (15.45b)$$

since $d\varepsilon_I$ and $d\varepsilon_{II}$ are not zero.

To make use of the equilibrium conditions for the determination of the equilibrium composition of a reactive mixture, we must know how to evaluate the chemical potential of each species involved.

15.13 Equilibrium Constant of a Reactive Mixture of Ideal Gases

We wish now to apply the condition of equilibrium to a reactive mixture of ideal gases. To perform this task, we must first develop an expression for the chemical potential of each constituent in such a mixture. It will be recalled that, according to the Gibbs–Dalton rule, an ideal gas in a mixture of gases behaves like a simple compressible substance existing at the temperature of the mixture and its partial pressure. Now for a simple compressible substance, we have, according to Eq. 4.15,

$$d\bar{g} = -\bar{s}\, dT + \bar{v}\, dp.$$

Applying this equation to an ideal gas at constant T, we have

$$d\bar{g}_T = \bar{R}T \frac{dp}{p}. \qquad (15.46)$$

Integrating Eq. 15.46 at constant temperature from $\bar{g}^0$ (the molal Gibbs function in the standard state of $p = 1$ atm) to $\bar{g}$ (the molal Gibbs function at any pressure p and T), we have

$$\bar{g} = \bar{g}^0 + \bar{R}T \ln p, \qquad (15.47)$$

where p must be in atmospheres. Note that $\bar{g}^0$ is a function of temperature only.

We have shown in Chapter 6 (Eq. 6.47) that the chemical potential of a pure substance is simply its molal Gibbs function. Thus the chemical potential of the ith species in a reactive mixture of ideal gas is given by

$$\mu_i = \bar{g}_i^0 + \bar{R}T \ln p_i, \qquad (15.48)$$

where p_i is the partial pressure of the ith species in atmospheres.

Making use of Eq. 15.48, the condition of equilibrium given by Eq. 15.40 becomes

$$v_C(\bar{g}_C^0 + \bar{R}T \ln p_C) + v_D(\bar{g}_D^0 + \bar{R}T \ln p_D)$$

$$- v_A(\bar{g}_A^0 + \bar{R}T \ln p_A) - v_B(\bar{g}_B^0 + \bar{R}T \ln p_B) = 0. \quad (15.49)$$

Collecting terms and rearranging, we have

$$v_C \bar{g}_C^0 + v_D \bar{g}_D^0 - v_A \bar{g}_A^0 - v_B \bar{g}_B^0 = -\bar{R}T \ln \frac{(p_C)^{v_C}(p_D)^{v_D}}{(p_A)^{v_A}(p_B)^{v_B}}. \quad (15.50)$$

We shall designate the left-hand side of Eq. 15.50 by ΔG_r^0, which is known as the standard-state Gibbs function change of the reaction. That is,

$$\Delta G_r^0 = v_C \bar{g}_C^0 + v_D \bar{g}_D^0 - v_A \bar{g}_A^0 - v_B \bar{g}_B^0. \quad (15.51)$$

For convenience, we define an equilibrium constant K_p for ideal-gas reactions by the relation

$$K_p = \frac{(p_C)^{v_C}(p_D)^{v_D}}{(p_A)^{v_A}(p_B)^{v_B}}. \quad (15.52)$$

Thus we may now write Eq. 15.50 as

$$\ln K_p = -\frac{\Delta G_r^0}{\bar{R}T}. \quad (15.53)$$

Note that ΔG_r^0 is a function of temperature only. Therefore, K_p is a function of temperature only. The equilibrium constant can be calculated if ΔG_r^0 is known. Values of K_p for several reactions are given in Table A.11 in the Appendix.

The partial pressures of the constituents may be expressed in terms of the mole fractions of these constituents and the mixture pressure. Thus

$$p_A = x_A p$$

$$p_B = x_B p$$

$$p_C = x_C p$$

$$p_D = x_D p.$$

In terms of these mole fractions and the mixture pressure, K_p may be given as

$$K_p = \frac{x_C^{\nu_C} x_D^{\nu_D}}{x_A^{\nu_A} x_B^{\nu_B}} (p)^{\nu_C + \nu_D - \nu_A - \nu_B}, \qquad (15.54)$$

where p is the mixture pressure in atmospheres.

Equation 15.54 is often called the *law of mass action*. It is really the second law of thermodynamics in disguise for a reactive mixture of ideal gases. It clearly shows the dependence of composition (the mole fractions) on temperature (K_p) and the mixture pressure.

For any reaction carried out at constant T and p, we have, from the definition of the Gibbs function,

$$\Delta G = \Delta H - T \, \Delta S.$$

For a reaction at the standard-state pressure of 1 atm, we have

$$\Delta G_r^0 = \Delta H_r^0 - T\Delta S_r^0. \qquad (15.55)$$

For the reaction of

$$\nu_A A + \nu_B B \rightleftharpoons \nu_C C + \nu_D D$$

we have

$$\Delta H_r^0 = \nu_C \bar{h}_C^0 + \nu_D \bar{h}_D^0 - \nu_A \bar{h}_A^0 - \nu_B \bar{h}_B^0 \qquad (15.56)$$

$$\Delta S_r^0 = \nu_C \bar{s}_C^0 + \nu_D \bar{s}_D^0 - \nu_A \bar{s}_A^0 - \nu_B \bar{s}_B^0. \qquad (15.57)$$

From Eqs. 15.56 and 15.57, we thus see that we can calculate ΔG_r^0 for any reaction if we have the enthalpy and entropy data for the species involved.

Example 15.15

Determine the equilibrium constant K_p, expressed as $\log_{10} K_p$, for the reaction

$$H_2O(g) \rightleftharpoons H_2 + \tfrac{1}{2}O_2$$

at 298 K and 2000 K.

Solution

From Eq. 15.53, we have

$$\ln K_p = -\frac{\Delta G_r^0}{\bar{R}T}.$$

Since $\ln K_p = 2.3026 \log_{10} K_p$, we get

$$\log_{10} K_p = -\frac{\Delta G_r^0}{2.3026 \bar{R} T}. \qquad (1)$$

To make use of Eq. 1, we must now determine ΔG_r^0. We shall do this by making use of Eq. 15.50:

$$\Delta G_r^0 = \Delta H_r^0 - T\Delta S_r^0.$$

For the given reaction, we have

$$\Delta H_r^0 = \bar{h}_{H_2}^0 + \tfrac{1}{2}\bar{h}_{O_2}^0 - \bar{h}_{H_2O}$$

$$= (\bar{h}_T - \bar{h}_{77})_{H_2} + \tfrac{1}{2}(\bar{h}_T - \bar{h}_{77})_{O_2} - (\bar{h}_T - \bar{h}_{77})_{H_2O} - (\bar{h}_f^0)_{H_2O(g)}$$

$$\Delta S_r^0 = \bar{s}_{H_2}^0 + \tfrac{1}{2}\bar{s}_{O_2}^0 - \bar{s}_{H_2O(g)}^0.$$

Note that we are dealing with properties at the standard-state pressure of 1 atm. Consequently, the absolute entropy values for H_2, O_2, and H_2O vapor are the same as the tabulated ϕ values in Keenan and Kaye's *Gas Tables*.

Making use of available data, we have, at 298 K:

$$\Delta H_r^0 = -(-242{,}000) = +242{,}000 \text{ kJ}$$

$$\Delta S_r^0 = 130.66 + \tfrac{1}{2} \times 205.17 - 188.85 = +44.40 \text{ kJ/K}$$

$$\Delta G_r^0 = 242{,}000 - 298 \times 44.40 = +228{,}769 \text{ kJ}$$

$$\log_{10} K_p = -\frac{228{,}769}{2.3026 \times 8.3143 \times 298} = -40.099,$$

compared to a value of -40.048 given in Table A.11 in the Appendix. At 2000 K, we have

$$\Delta H_r^0 = (61{,}419 - 8463.1) + \tfrac{1}{2}(67{,}846.8 - 8678.9)$$

$$- (82{,}693.7 - 9899.0) - (-242{,}000)$$

$$= +251{,}746 \text{ kJ}$$

$$\Delta S_r^0 = 188.313 + \tfrac{1}{2} \times 268.586 - 264.652 = 57.954 \text{ kJ/K}$$

$$\Delta G_r^0 = 251{,}745 - 2000 \times 57.954 = +135{,}837 \text{ kJ}$$

$$\log_{10} K_p = -\frac{135{,}837}{2.3026 \times 8.3143 \times 2000} = -3.548,$$

compared to a value of -3.540 given in Table A.11.   ∎

Example 15.16

Carbon dioxide gas enters a steady-state steady-flow hot-gas generator at 298 K and 6 atm. Determine the amount of heat addition required if the desired outlet temperature and pressure are 2800 K at 5 atm.

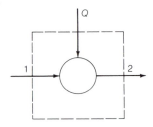

Solution

System selected for study: hot-gas generator
Unique features of process: steady-state steady flow; no work transfer
Assumptions: neglect ΔKE and ΔPE; assume ideal-gas behavior; assume outgoing stream is an equilibrium mixture of CO_2, CO, and O_2

We shall make calculations on the basis of 1 mol of CO_2 at the hot-gas-generator inlet. The chemical equation is

$$CO_2 \rightarrow (1 - \varepsilon)CO_2 + \varepsilon CO + \frac{\varepsilon}{2} O_2. \tag{1}$$

In Eq. 1, ε may be interpreted as the extent to which CO_2 will be dissociated into CO and O_2. The equilibrium composition at the outlet is made up of $(1 - \varepsilon)$ mol of CO_2, ε mol of CO, and $\varepsilon/2$ mol of O_2.

From the first law for a steady-state steady-flow process involving reactive mixtures, we have

$$Q = H_2 - H_1$$

$$= (1 - \varepsilon)(\bar{h}_{2800} - \bar{h}_{298})_{CO_2} + \varepsilon(\bar{h}_{2800} - \bar{h}_{298})_{CO}$$

$$+ \frac{\varepsilon}{2}(\bar{h}_{2800} - \bar{h}_{298})_{O_2} - 1.0(\bar{h}_{298} - \bar{h}_{298})_{CO_2}$$

$$+ (1 - \varepsilon)(\bar{h}_f^0)_{CO_2} + \varepsilon(\bar{h}_f^0)_{CO} - 1.0(\bar{h}_f^0)_{CO_2}$$

$$= (1 - \varepsilon)(149{,}796.8 - 9359.7) + \varepsilon(94{,}744.3 - 8666.7)$$

$$+ \frac{\varepsilon}{2}(98{,}691.6 - 8678.9) - 1.0(0) + (1 - \varepsilon)(-393{,}790)$$

$$+ \varepsilon(-110{,}600) - 1.0(-393{,}790)$$

$$= 140{,}437 + (273{,}837)\varepsilon \text{ kJ.} \tag{2}$$

Equation 2 contains two unknowns, Q and ε. We must therefore look for another equation.

In the equilibrium mixture consisting of CO_2, CO, and O_2, we have the reaction

$$CO_2 \rightleftharpoons CO + \tfrac{1}{2}O_2 ,$$

the equilibrium constant K_p of which is given by

$$K_p = \frac{x_{CO}\, x_{O_2}^{1/2}}{x_{CO_2}}\, p^{1+1/2-1} = \frac{x_{CO}\, x_{O_2}^{1/2}}{x_{CO_2}}\, p^{1/2}, \qquad (3)$$

where x_{CO}, x_{O_2}, and x_{CO_2} are the mole fractions of CO, O_2, and CO_2, respectively, in the equilibrium mixture, and p is the mixture pressure.

The total number of moles in the equilibrium mixture is

$$n_{\text{total}} = (1-\varepsilon) + \varepsilon + \frac{\varepsilon}{2} = 1 + \frac{\varepsilon}{2}.$$

Thus the mole fractions are

$$x_{CO} = \frac{\varepsilon}{1 + \varepsilon/2}$$

$$x_{O_2} = \frac{\varepsilon/2}{1 + \varepsilon/2}$$

$$x_{CO_2} = \frac{1-\varepsilon}{1 + \varepsilon/2}.$$

Substituting into Eq. 3, we have

$$K_p = \frac{\left(\dfrac{\varepsilon}{1+\varepsilon/2}\right)\left(\dfrac{\varepsilon/2}{1+\varepsilon/2}\right)^{1/2}}{\dfrac{1-\varepsilon}{1+\varepsilon/2}}\, p^{1/2},$$

from which we get

$$\frac{\varepsilon^3}{(1-\varepsilon)^2(2+\varepsilon)} = \frac{K_p^2}{p}. \qquad (4)$$

The only unknown in Eq. 4 is ε, as p and T are known. But we must first find the value of K_p for the given reaction at $2800°K$.

From Table A.11, we have, for the reaction of $CO_2 \rightleftharpoons CO + \frac{1}{2}O_2$ at 2800 K,

$$\log_{10} K_p = -0.825$$

and

$$K_p = 0.150.$$

Substituting into Eq. 4 we get

$$\frac{\varepsilon^3}{(1 - \varepsilon)^2(2 + \varepsilon)} = \frac{(0.150)^2}{5}$$

$$= 0.0045. \tag{5}$$

Equation 5 may be solved by the iterative method to give

$$\varepsilon = 0.1867.$$

Making use of this value of ε, we have, from Eq. 2,

$$Q = 140{,}437 + 273{,}837 \times 0.1867 \text{ kJ}$$

$$= 191{,}562 \text{ kJ}. \qquad\blacksquare$$

Comment

If we had ignored the dissociation of CO_2 (that is, if we had assumed that $\varepsilon = 0$), Q would have been 140,437 kJ, which is only about 74 percent of the heat addition required. The hot-gas generator would have been under-designed if we did not take dissociation into consideration.

Example 15.17

According to the following reaction, the equilibrium constant K_p for the ionization reaction of cesium is 15.63 at 2000 K.

$$Cs \rightleftharpoons Cs^+ + e^-.$$

In this reaction, Cs is the neutral cesium atom, Cs^+ is the singly ionized cesium ion, and e^- is the electron gas.

(a) Determine the extent to which cesium will be ionized at 2000 K and 1 atm.

(b) Determine the ionization pressure if we desire to have the ionization process 99% complete at 2000 K.

Solution

The equilibrium constant of the given reaction is

$$K_p = \frac{x_{Cs^+} x_{e^-}}{x_{Cs}} p. \tag{1}$$

Let ε be the extent of the ionization reaction. Then the chemical equation is

$$Cs \rightarrow (1 - \varepsilon)Cs + \varepsilon Cs^+ + \varepsilon e^-$$

and the total number of moles in the equilibrium mixture is

$$n_{total} = (1 - \varepsilon) + \varepsilon + \varepsilon = 1 + \varepsilon.$$

The mole fractions are

$$x_{Cs} = \frac{1 - \varepsilon}{1 + \varepsilon}$$

$$x_{Cs^+} = \frac{\varepsilon}{1 + \varepsilon}$$

$$x_{e^-} = \frac{\varepsilon}{1 + \varepsilon}.$$

Substituting into Eq. 1, we have

$$K_p = \frac{\left(\dfrac{\varepsilon}{1 + \varepsilon}\right)\left(\dfrac{\varepsilon}{1 + \varepsilon}\right)}{\dfrac{1 - \varepsilon}{1 + \varepsilon}} p.$$

Simplifying, we have

$$K_p = \frac{\varepsilon^2}{1 - \varepsilon^2} p$$

or

$$\varepsilon = \sqrt{\frac{K_p}{p + K_p}}, \tag{2}$$

where p is the pressure of the equilibrium mixture in atmospheres.

Making use of Eq. 2, we have

(a)
$$\varepsilon = \sqrt{\frac{15.63}{1 + 15.63}} = 0.9695.$$

(b) To get $\varepsilon = 0.99$, we have

$$p = \frac{1 - \varepsilon^2}{\varepsilon^2} K_p$$

$$= \frac{1 - (0.99)^2}{(0.99)^2} \times 15.63$$

$$= 0.3174 \text{ atm.}$$

Comment

We note from Eq. 2 that if we want to approach 100% ionization, we must carry out the ionization process in a very high vacuum.

Example 15.18

Let us now return to the problem of the adiabatic burning of methane gas with theoretical air that we studied in Examples 15.7 and 15.11. Let us determine the temperature of the products of combustion on the basis that the equilibrium mixture will consist of only CO_2, CO, O_2, H_2O, and N_2.

Solution (refer to Fig. 15.8)

The chemical equation is

$$CH_4 + 2O_2 + 2 \times 3.76N_2 \rightarrow (1 - \varepsilon)CO_2 + \varepsilon CO + \frac{\varepsilon}{2} O_2$$

$$+ 2H_2O(g) + 2 \times 3.76N_2.$$

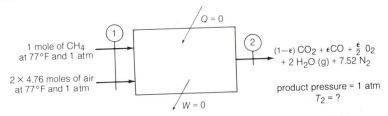

Figure 15.8 Adiabatic Burning of Methane Gas with Theoretical Air

Neglecting change in kinetic and potential energies, we have, from the first law,

$$0 = H_2 - H_1$$

$$= (1 - \varepsilon)(\bar{h}_T - \bar{h}_{77})_{CO_2} + \varepsilon(\bar{h}_T - \bar{h}_{77})_{CO} + \frac{\varepsilon}{2}(\bar{h}_T - \bar{h}_{77})_{O_2}$$

$$+ 2(\bar{h}_T - \bar{h}_{77})_{H_2O} + 2 \times 3.76(\bar{h}_T - \bar{h}_{77})_{N_2} + (1 - \varepsilon)(\bar{h}_f^0)_{CO_2}$$

$$+ \varepsilon(\bar{h}_f^0)_{CO} + 2(\bar{h}_f^0)_{H_2O(g)} - (\bar{h}_f^0)_{CH_4}. \tag{1}$$

Equation 1 contains two unknowns: ε and the temperature of the equilibrium mixture at the outlet.

Since we have CO_2, CO, and O_2 in the equilibrium products of combustion, we have the reaction

$$CO_2 \rightleftharpoons CO + \tfrac{1}{2}O_2,$$

the equilibrium constant of which is

$$K_p = \frac{x_{CO}\, x_{O_2}^{1/2}}{x_{CO_2}}\, p^{1/2}. \tag{2}$$

The total number of moles in the products is

$$n_{total} = (1 - \varepsilon) + \varepsilon + \frac{\varepsilon}{2} + 2 + 7.52$$

$$= 10.52 + \frac{\varepsilon}{2}.$$

The mole fractions of CO_2, CO, and O_2 are thus given by

$$x_{CO_2} = \frac{1 - \varepsilon}{10.52 + \varepsilon/2}$$

$$x_{CO} = \frac{\varepsilon}{10.52 + \varepsilon/2}$$

$$x_{O_2} = \frac{\varepsilon/2}{10.52 + \varepsilon/2}.$$

Substituting into Eq. 2 and simplifying, we have

$$\frac{\varepsilon^3}{(1-\varepsilon)^2(21.04+\varepsilon)} = \frac{K_p^2}{p}. \tag{3}$$

Since the pressure of the equilibrium products is known and since K_p is a function of temperature only, Eq. 3 is also an equation containing two unknowns, ε and T.

To determine the temperature of the products, we must solve Eqs. 1 and 3 simultaneously. One method is as follows:

1. Assume a temperature and calculate the corresponding value of ε from Eq. 3.
2. Make use of this assumed temperature and the corresponding value of ε to see if they will satisfy Eq. 1.

It will be recalled that in Example 15.7 the adiabatic flame temperature was found to be about 4200°R. Then it was found in Example 15.11 that, according to the second law, 4200°R was not possible. For our trial, let us assume that $T = 4100°R$.

For the reaction $CO_2 \rightleftharpoons CO + \frac{1}{2}O_2$ at 4100°R, we have, from Table A.11,

$$\log_{10} K_p = -2.013$$

and

$$K_p = 0.009711.$$

Then, we have, from Eq. 3,

$$\frac{\varepsilon^3}{(1-\varepsilon)^2(21.04+\varepsilon)} = \frac{(0.009711)^2}{1} = 0.0000943,$$

which may be solved by the iterative method to give

$$\varepsilon = 0.116.$$

Using $T = 4100°R$ and $\varepsilon = 0.116$, the right-hand side of Eq. 1 becomes

$$(1-0.116)(50,695.1 - 4030) + 0.116(32,479.1 - 3730)$$

$$+ \frac{0.116}{2}(33,721.6 - 3725) + 2(41,745.4 - 4258)$$

$$+ 7.52(32,198.0 - 3730) + (1 - 0.116)(-169,300)$$

$$+ 0.116 \times (-47,551) + 2(-104,040) - (-32,211)$$

$$= 335,381 - 331,046 = +4335 \text{ Btu} \neq 0.$$

Although Eq. 1 was not satisfied with $T = 4100°R$, our calculations showed that the correct temperature would be very close to $4100°R$. The correct temperature turns out to be about $4060°R$ with $\varepsilon = 0.1065$. ∎

Example 15.19

Water vapor may react with carbon monoxide gas to form hydrogen and carbon dioxide according to the reaction

$$H_2O(g) + CO \rightleftharpoons H_2 + CO_2. \tag{I}$$

Equation (I) is known as the *water–gas reaction equation*. Determine the equilibrium constant K_p, expressed as $\log_{10} K_p$, for this reaction at 2000 K.

Solution

We first make the observation that the water–gas reaction may be looked upon as the overall reaction of the following two reactions:

$$H_2O(g) \rightleftharpoons H_2 + \tfrac{1}{2}O_2 \tag{II}$$

$$CO + \tfrac{1}{2}O_2 \rightleftharpoons CO_2. \tag{III}$$

From the definition of the equilibrium constant K_p, we have, in terms of partial pressures,

$$(K_p)_I = \frac{p_{H_2} p_{CO_2}}{p_{CO} p_{H_2O}}$$

$$(K_p)_{II} = \frac{p_{H_2} p_{O_2}^{1/2}}{p_{H_2O}}$$

$$(K_p)_{III} = \frac{p_{CO_2}}{p_{CO} p_{O_2}^{1/2}},$$

from which we see that $(K_p)_I = (K_p)_{II}(K_p)_{III}$. Thus we have

$$\log_{10} (K_p)_I = \log_{10} (K_p)_{II} + \log_{10} (K_p)_{III}. \tag{1}$$

From Table A.11, we have, for the reaction $H_2O(g) \rightleftharpoons H_2 = \tfrac{1}{2}O_2$ at 2000 K,

$$\log_{10} (K_p)_{II} = -3.540.$$

From Table A.11, we have, for the reaction $CO_2 \rightleftharpoons CO + \frac{1}{2}O_2$ at 2000 K,

$$\log_{10} K_p = -2.884.$$

It is clear from the definition of the equilibrium constant that K_p for the reaction $CO + \frac{1}{2}O_2 \rightleftharpoons CO_2$ is simply the negative of the K_p for the reaction $CO_2 \rightleftharpoons CO + \frac{1}{2}O_2$. Thus we have

$$\log_{10} (K_p)_{III} = +2.884.$$

Substituting into Eq. 1, we have, for the water–gas reaction,

$$\log_{10} (K_p) = -3.540 + 2.884 = -0.656. \qquad \blacksquare$$

Comments

For any reaction carried out at given T and p, the change in the Gibbs function of the reaction is independent of the path of the process. Thus

$$(\Delta G_r^0)_I = (\Delta G_r^0)_{II} + (\Delta G_r^0)_{III},$$

from which we also get

$$-\frac{(\Delta G_r^0)_I}{\bar{R}T} = -\frac{(\Delta G_r^0)_{II}}{\bar{R}T} - \frac{(\Delta G_r^0)_{III}}{\bar{R}T}.$$

Now, according to Eq. 15.53, we have, for any reaction involving ideal gases,

$$\ln K_p = -\frac{\Delta G_r^0}{\bar{R}T}.$$

Thus we must have

$$\ln (K_p)_I = \ln (K_p)_{II} + \ln (K_p)_{III}$$

and

$$\log_{10} (K_p)_I = \log_{10} (K_p)_{II} + \log_{10} (K_p)_{III}.$$

Note that the changes in enthalpy and entropy of any reaction are also independent of the path of the process. Consequently, we must have

$$(\Delta H_r^0)_I = (\Delta H_r^0)_{II} + (\Delta H_r^0)_{III}$$

and

$$(\Delta S_r^0)_I = (\Delta S_r^0)_{II} + (\Delta S_r^0)_{III}.$$

This example shows that thermochemical data for complex reactions may be obtained from thermochemical data for simple reactions.

15.14 Temperature Dependence of the Equilibrium Constant

For a simple reactive mixture having the reaction

$$v_A A + v_B B \rightleftharpoons v_C C + v_D D,$$

the expression for dG of the mixture is given by Eq. 15.38:

$$dG = -S\,dT + V\,dp + (v_C \mu_C + v_D \mu_D - v_A \mu_A - v_B \mu_B)\,d\varepsilon,$$

which may be written as

$$dG = -S\,dT + V\,dp + \Delta G_r\,d\varepsilon, \tag{15.58}$$

where

$$\Delta G_r = v_C \mu_C + v_D \mu_D - v_A \mu_A - v_B \mu_B. \tag{15.59}$$

We have seen that at constant T and p, the quantity ΔG_r becomes zero when equilibrium is reached. This is why ΔG_r is known as the *reaction potential.*

From Eq. 15.58, it is evident that the independent variables of G can be taken as T, p, and ε. Thus we have

$$G = G(T, p, \varepsilon)$$

and

$$dG = \left(\frac{\partial G}{\partial T}\right)_{p,\varepsilon} dT + \left(\frac{\partial G}{\partial p}\right)_{T,\varepsilon} dp + \left(\frac{\partial G}{\partial \varepsilon}\right)_{T,p} d\varepsilon. \tag{15.60}$$

Comparing Eq. 15.60 with Eq. 15.58, we have

$$\left(\frac{\partial G}{\partial T}\right)_{p,\varepsilon} = -S \tag{15.61}$$

$$\left(\frac{\partial G}{\partial p}\right)_{T,\varepsilon} = V \tag{15.62}$$

$$\left(\frac{\partial G}{\partial \varepsilon}\right)_{T,p} = \Delta G_r. \tag{15.63}$$

From Eq. 15.61, we have

$$\left(\frac{\partial^2 G}{\partial \varepsilon \, \partial T}\right)_p = -\left(\frac{\partial S}{\partial \varepsilon}\right)_{T,p}. \tag{15.64}$$

From Eq. 15.63, we have

$$\left(\frac{\partial^2 G}{\partial T \, \partial \varepsilon}\right)_p = \left(\frac{\partial \Delta G_r}{\partial T}\right)_{p,\varepsilon}. \tag{15.65}$$

Since the order of differentiation is immaterial, we have

$$\left(\frac{\partial \Delta G_r}{\partial T}\right)_{p,\varepsilon} = -\left(\frac{\partial S}{\partial \varepsilon}\right)_{T,p}. \tag{15.66}$$

Since

$$\left(\frac{\partial G}{\partial \varepsilon}\right)_{T,p} = \Delta G_r,$$

we have

$$\left(\frac{\partial S}{\partial \varepsilon}\right)_{T,p} = \Delta S_r, \tag{15.67}$$

where

$$\Delta S_r = \nu_C \bar{s}_C + \nu_D \bar{s}_D - \nu_A \bar{s}_A - \nu_B \bar{s}_B. \tag{15.68}$$

Thus we have

$$\left(\frac{\partial \Delta G_r}{\partial T}\right)_{p,\varepsilon} = -\Delta S_r. \tag{15.69}$$

Now, $\Delta G_r = \Delta H_r - T\Delta S_r$. Consequently, Eq. 15.69 may be written as

$$\left(\frac{\partial \Delta G_r}{\partial T}\right)_{p,\varepsilon} = \frac{\Delta G_r - \Delta H_r}{T}. \tag{15.70}$$

Equation 15.70 is the famous *Gibbs–Helmholtz equation*. It gives the temperature dependence of ΔG_r along a constant-pressure path.

Applying Eq. 15.70 to a reaction at the standard state of 1 atm, we have

$$\left(\frac{\partial \Delta G_r^0}{\partial T}\right)_{p,\varepsilon} = \frac{\Delta G_r^0 - \Delta H_r^0}{T}. \tag{15.71}$$

Since

$$\left[\frac{\partial (\Delta G_r^0)/T}{\partial T}\right]_{p,\varepsilon} = \frac{1}{T}\left(\frac{\partial \Delta G_r^0}{\partial T}\right)_{p,\varepsilon} - \frac{\Delta G_r^0}{T^2}.$$

Equation 15.71 may be written as

$$\left[\frac{\partial (\Delta G_r^0)/T}{\partial T}\right]_{p,\varepsilon} = -\frac{\Delta H_r^0}{T^2}. \tag{15.72}$$

For ideal-gas reactions, we have

$$\ln K_p = -\frac{\Delta G_r^0}{RT}.$$

which is a function of temperature only. Thus, applying Eq. 15.72 to ideal-gas reactions, we have

$$\frac{d(\ln K_p)}{dT} = \frac{\Delta H_r^0}{\bar{R}T^2}. \tag{15.73}$$

Equation 15.73 is known as the *van't Hoff equation*, which is one of the most important equations in chemical thermodynamics. It is a valuable equation because, if the temperature variation of K_p is known, ΔH_r^0 can be calculated. This equation also shows that for an exothermic reaction (ΔH_r^0 will be negative according to our convention), K_p will decrease with increasing temperature, while for an endothermic reaction (ΔH_r^0 will be positive), K_p will increase with increasing temperature. Equation 15.73 may be written as

$$\frac{d(\ln K_p)}{d(1/T)} = -\frac{\Delta H_r^0}{\bar{R}}. \tag{15.74}$$

This form of the van't Hoff equation shows that if ΔH_r^0 varies only slightly with temperature, $\ln K_p$ will be nearly a linear function of $1/T$. This is indeed true for many reactions, at least in a narrow range of temperature.

If we assume ΔH_r^0 to be a constant, we can integrate Eq. 15.74 to yield the expression

$$\ln \frac{(K_p)_2}{(K_p)_1} = -\frac{\Delta H_r^0}{\bar{R}}\left(\frac{1}{T_2} - \frac{1}{T_1}\right) \tag{15.75}$$

or

$$\log_{10} \frac{(K_p)_2}{(K_p)_1} = -\frac{\Delta H_r^0}{2.3026\bar{R}}\left(\frac{1}{T_2} - \frac{1}{T_1}\right). \tag{15.75a}$$

Equation 15.75 makes it possible for us to determine an average value for ΔH_r^0 by making measurements of K_p at only two temperatures.

Example 15.20

Determine the enthalpy of reaction at the standard-state pressure of 1.0 atm and 2000 K for the water–gas reaction

$$H_2O + CO \rightleftharpoons H_2 + CO_2.$$

Solution

(We shall make use of Eq. 15.75a.)

$$\Delta H_r^0 = 2.3026\bar{R}[\log_{10} (K_p)_2 - \log_{10} (K_p)_1]\bigg/\left(\frac{1}{T_1} - \frac{1}{T_2}\right). \tag{1}$$

To obtain $\log_{10} K_p$ for the water–gas reaction, we shall make use of $\log_{10} K_p$ for the reactions $H_2O \rightleftharpoons H_2 + \frac{1}{2}O_2$ and $CO + \frac{1}{2}O_2 \rightleftharpoons CO_2$, as in Example 15.19. Let $T_1 = 1800$ K and $T_2 = 2200$ K. Making use of data from Table A.11, we have

At 1800 K: $\log_{10} (K_p)_1 = -4.270 + 3.693 = -0.577$

At 2200 K: $\log_{10} (K_p)_2 = -2.942 + 2.226 = -0.716.$

Substituting into Eq. 1, we have, at 2000 K,

$$\Delta H_r^0 = 2.3026 \times 8.3143(-0.716 + 0.577)\bigg/\left(\frac{1}{1800} - \frac{1}{2200}\right)$$

$$= -26,345 \text{ kJ}.$$

The minus sign indicates that the water–gas reaction is an exothermic one. ∎

Example 15.21

A fuel cell (see Fig. 15.6) operates according to the following reaction:

$$H_2(g) + \tfrac{1}{2}O_2(g) \rightarrow H_2O(g).$$

Determine the maximum conversion efficiency of the fuel cell if the reaction is carried out at

(a) 298 K and 1.0 atm.
(b) 500 K and 1.0 atm.
(c) 1000 K and 1.0 atm.

Solution

The conversion efficiency of a fuel cell is defined as

$$\eta_C = \frac{W}{-\Delta H_r^0}, \tag{1}$$

where W is work output of the fuel cell, and ΔH_r^0 is the enthalpy of reaction of the fuel cell.

To have maximum conversion efficiency, the fuel cell must operate reversibly. Then, according to the first and second laws of thermodynamics,

$$W_{\text{max}} = -\Delta G_r^0,$$

where ΔG_r^0 is the Gibbs function change of the reaction. Thus the maximum conversion efficiency is given by

$$(\eta_C)_{\text{max}} = \frac{-\Delta G_r^0}{-\Delta H_r^0}. \tag{2}$$

(a) At 298 K and 1.0 atm:

$$\Delta H_r^0 = 1.0(\bar{h}_f^0)_{H_2O(g)} = -242,000 \text{ kJ}.$$

$$\Delta G_r^0 = -2.3026RT \log_{10} K_p, \text{ according to Eq. 15.53.}$$

From Table A.11, we have, at 298 K,

$$\log_{10} K_p = 40.048.$$

Thus

$$\Delta G_r^0 = -2.3026 \times 8.3143 \times 298 \times 40.048 \text{ kJ}$$

$$= -228,476 \text{ kJ}$$

and

$$(\eta_C)_{max} = \frac{228,476}{242,000} = 94.4\%.$$

(b) At 500 K and 1.0 atm: From Table A.11, we have, at 500 K,

$$\log_{10} K_p = 22.886.$$

Thus

$$\Delta G_r^0 = -2.3026 \times 8.3143 \times 500 \times 22.886 \text{ kJ}$$

$$= -219,070 \text{ kJ}.$$

To obtain ΔH_r^0 at 500 K, we make use of Eq. 15.75a:

$$\log_{10} \frac{(K_p)_2}{(K_p)_1} = -\frac{\Delta H_r^0}{2.3026\bar{R}} \left(\frac{1}{T_2} - \frac{1}{T_1} \right).$$

Let $T_1 = 298$ K. Then, we have, at 500 K,

$$\Delta H_r^0 = -\frac{2.3026\bar{R}[\log_{10} (K_p)_{500} - \log_{10} (K_p)_{298}]}{1/500 - 1/298}$$

$$= -\frac{2.3026 \times 8.3143 \times (22.886 - 40.048)}{1/500 - 1/298} \text{ kJ} = -242,350 \text{ kJ}.$$

Then

$$(\eta_C)_{max} = \frac{219,070}{242,350} = 90.4\%.$$

(c) At 1000 K and 1.0 atm: From Table A.11, we have, at 1000 K,

$$\log_{10} K_p = 10.062.$$

Thus

$$\Delta G_r^0 = -2.3026 \times 8.3143 \times 1000 \times 10.062 \text{ kJ}$$

$$= -192,630 \text{ kJ}.$$

Making use of Eq. 15.75a, we have

$$\Delta H_r^0 = -\frac{2.3026 \bar{R}[\log_{10} (K_p)_2 - \log_{10} (K_p)_1]}{1/T_2 - 1/T_1}.$$

Let $T_1 = 500$ K. Then we have, at 1000 K,

$$\Delta H_r^0 = -\frac{2.3026 \times 8.3143 \times (10.062 - 22.886)}{1/1000 - 1/500} \text{ kJ}$$

$$= -245,510 \text{ kJ}$$

and

$$(\eta_C)_{max} = \frac{192,630}{245,510} = 78.5\%.$$

Comments

We note that for a given reaction the enthalpy of reaction increases with increase in the reaction temperature, while the Gibbs function change of the reaction decreases with increase in the reaction temperature. Thus the maximum conversion efficiency of the fuel cell decreases with an increase in the reaction temperature. However, at higher reaction temperatures we get better reaction rates, thereby requiring a smaller fuel cell for a given amount of power output. In addition, the heat that must be removed from a high-temperature fuel cell may be used as the heat source for more power generation with a conventional system and for cogeneration applications.

15.15 Concluding Remarks on Reactive Systems

Although we have concerned ourselves only with ideal-gas reactions, we wish to point out once again that the basic principles in this chapter apply to any chemical reaction. In particular, the condition of equilibrium that we have found to be so useful in the study of ideal-gas reactions is just as useful in the study of reactive mixtures of real gases and homogeneous systems of liquids. However, the chemical potential of a real gas or liquid is more complicated than that of an ideal gas. But this simply makes the problem more interesting and challenging. It is hoped that our readers will have the opportunity to explore the fascinating world of chemical thermodynamics beyond the scope of this text.

Problems

15.1 One kilogram of methane gas (CH_4) is burned with 25% excess air. Determine
(a) the mass of air supplied.
(b) the dew point of the products at a total pressure of 101.325 kPa. Assume complete combustion.

15.2 Ethane gas (C_2H_6) is burned with 150% theoretical air. Determine
(a) the mass of air supplied per kilogram of fuel burned.
(b) the dew point of the products at a total pressure of 101.325 kPa. Assume complete combustion.

15.3 The volumetric analysis of a natural gas is as follows:

$$CH_4: \quad 77.5\%$$

$$C_2H_6: \quad 16.0\%$$

$$CO_2: \quad 6.5\%$$

Determine the theoretical air required in kilograms per kilogram of fuel.

15.4 One pound of propane (C_3H_8) is burned with 20 lbm of air. Calculate the percent of excess air supplied.

15.5 One pound of octane (C_8H_{18}) is burned with 100% excess air. Determine the dew point of the products at a total pressure of 14.7 psia. Assume complete combustion.

15.6 One pound of ethane (C_2H_6) is burned with 80% of theoretical air. Determine the volumetric analysis of the dry products of combustion. Assume complete combustion for the hydrogen in the fuel.

15.7 The natural gas in Problem 15.3 is burned with 10% excess air. Assuming that combustion is complete, determine
(a) the volumetric analysis of the total products.
(b) the volumetric analysis of the dry products.

15.8 One pound of methane (CH_4) is burned with 120% of theoretical air. What is the volume of combustion products per lbm of fuel at 15 psia and 1000°F? Assume complete combustion.

15.9 Five kilograms of carbon monoxide is burned with 100% theoretical air. Assuming combustion is complete, determine
(a) the mass of air supplied.
(b) the volume of the products at 25°C and 101.325 kPa.

15.10 Same as Problem 15.9 except that the carbon monoxide gas is burned with 30% excess air.

15.11 The volumetric analysis of a natural gas is 79% CH_4 and 21% C_2H_6. If this fuel is burned with air, what is the theoretical air–fuel ratio on both a molal and mass basis?

15.12 Determine the enthalpy of combustion of CH_4 at 1000 K and 1.0 atm. Make use of data in Keenan and Kaye's *Gas Tables*.

15.13 Determine the enthalpy of combustion of CH_4 at 1000 K and 1.0 atm without making use of data from *Gas Tables* but making use of reasonable assumptions.

15.14 The dry exhaust from an internal combustion engine has the following volumetric analysis: 9.9% CO_2, 7.2% CO, 3.3% H_2, 0.3% CH_4, and 79.3% N_2. If the fuel is octane (C_8H_{18}), determine the air–fuel ratio (on a mass basis)
(a) by a carbon balance.
(b) by a hydrogen–oxygen balance.

15.15 Determine the enthalpy of combustion at 1000°R and 1 atm for gaseous propane. Give the answer in Btu/lbm of fuel.

15.16 Determine the adiabatic flame temperature for burning methane gas with 25% excess air in a steady-state steady-flow device. Assume complete combustion with air and fuel supplied at 25°C and 1.0 atm.

15.17 Determine the adiabatic flame temperature for burning propane with 200% theoretical air in a steady-state steady-flow process. Assume complete combustion with air and fuel supplied at 77°F and 1 atm.

15.18 Propane at 25°C is burned with air that enters the combustion chamber at 200°C. If the products of combustion leave at 827°C, determine the air–fuel ratio in kg air/kg fuel. Assume that combustion is complete and neglect heat loss.

15.19 Liquid octane at 77°F is burned with 400% of theoretical air that enters the combustion chamber at 500°F. Determine the temperature of the products at the outlet of the chamber. Assume that combustion is complete and neglect heat loss.

15.20 Propane gas at 25°C is burned with 300% theoretical air that enters the combustion chamber at 200°C. Determine the temperature of the products of combustion. Assume that combustion is complete and neglect heat loss.

15.21 Liquid octane at 77°F is burned with air that enters the combustion chamber at 400°F. If the products of combustion leave at 2000°R, what is the excess air supplied? Assume that combustion is complete and neglect heat loss.

15.22 A natural gas consisting of 77.5% by volume of CH_4, 16.0% of C_2H_6, and 6.5% of CO_2, is burned in a furnace with 15% excess air. The flue gas leaves at a temperature of 150°C. Make a good estimate of the furnace efficiency. The higher heating value (HHV) of the natural gas is 38.86 kJ/liter at 1.0 atm and 20°C.

15.23 Same as Problem 15.22 except that the fuel is burned with 50% excess air.

15.24 One pound of methane gas is burned with 120% of theoretical air in the furnace of a steam power plant. The products of combustion leave the stack at 300°F. How much heat is transferred to the steam? Assume that combustion is complete with air and fuel supplied at 77°F and 1 atm.

15.25 The *enthalpy of combustion*, also known as the *heat of reaction at constant pressure*, is, by definition, the heat released in a steady-flow process in which the reactants and the products are both at the same temperature. In a similar manner, we can define the *internal energy of combustion*, also known as the *heat of reaction at constant volume*, as the heat released in a constant-volume process in which the reactants and the products are both at the same temperature. Show that in a chemical reaction involving ideal gases, we have

$$H_p - H_R = U_p - U_R + \bar{R}T(N_p - N_R),$$

where

$$H_p - H_R = \text{enthalpy of combustion}$$

$$U_p - U_R = \text{internal energy of combustion}$$

$$N_p = \text{number of moles of products}$$

$$N_R = \text{number of moles of reactants.}$$

15.26 Determine the internal energy of combustion (see Problem 15.25) at 25°C and 1.0 atm for propane gas assuming liquid water in the products.

15.27 Determine the internal energy of combustion (see Problem 15.25) at 77°F and 1 atm for methane gas assuming liquid water in the products.

15.28 For the water–gas reaction

$$CO(g) + H_2O(g) \rightarrow CO_2(g) + H_2(g).$$

(a) Determine the enthalpy of reaction at 77°F and 1 atm making use of enthalpy-of-formation data.
(b) Show that the enthalpy of reaction is the sum of the enthalpy of reaction for $H_2O(g) \rightarrow H_2(g) + \frac{1}{2}O_2(g)$ and the enthalpy of reaction for $CO(g) + \frac{1}{2}O_2(g) \rightarrow CO_2(g)$. Is this a coincidence or is it a consequence of the laws of thermodynamics?

15.29 The higher heating value of ethyl alcohol (C_2H_5OH) at 77°F and 1 atm is 588,000 Btu/lb mol.
(a) What is the lower heating value of ethyl alcohol at 77°F and 1 atm? Give the answer in Btu/lbm of fuel.
(b) What is the enthalpy of formation of ethyl alcohol at 77°F and 1 atm? Give the answer in Btu/lb mol of fuel.

15.30 What fraction of CO_2 becomes dissociated at
(a) 1000 K and 1 atm?
(b) 1000 K and 10 atm?

15.31 What fraction of CO_2 becomes dissociated at
(a) 3000 K and 1 atm?
(b) 3000 K and 10 atm?

15.32 What fraction of H_2O becomes dissociated to H_2 and O_2 at
(a) 1000 K and 1 atm?
(b) 1000 K and 10 atm?

15.33 What fraction of H_2O becomes dissociated to H_2 and O_2 at
(a) 3000 K and 1 atm?
(b) 3000 K and 10 atm?

15.34 Compute the equilibrium constant for the reaction

$$CH_4(g) + 2O_2(g) \rightleftharpoons CO_2(g) + 2H_2O(g)$$

at 25°C and 1 atm.

15.35 Compute the equilibrium constant for the reaction

$$C_3H_8(g) + 5O_2(g) \rightleftharpoons 3CO_2(g) + 4H_2O(g)$$

at 25°C and 1 atm.

15.36 Compute the change in the Gibbs function for the reaction

$$CH_4(g) + 2O_2(g) \rightleftharpoons CO_2(g) + 2H_2O(g)$$

at 25°C and 1 atm using
(a) data on the Gibbs function of formation.
(b) data on enthalpy of formation and absolute entropy.

15.37 Hydrogen peroxide may be used as the oxidizer in rocket engines. Determine the enthalpy of combustion at 25°C and 1 atm per pound of reactants for the following combustion process:

$$H_2O_2(l) + H_2(g) \rightarrow 2H_2O(l).$$

The enthalpy of formation for $H_2O_2(l)$ at 25°C and 1 atm is $-187,700$ kJ/kg mol.

15.38 For an initial mixture of 1 mol of carbon monoxide gas and 1 mol of water vapor, determine the composition of the equilibrium mixture at
(a) 1800 K and 1 atm.
(b) 1800 K and 10 atm.

15.39 The volumetric analysis of an equilibrium mixture consisting of CO_2, CO, and O_2 at 4320°R and 1 atm is as follows: 86.53% CO_2, 8.98% CO and 4.49% O_2. Making use of this information, determine the equilibrium constant for the reaction $CO_2 \rightleftharpoons CO + \frac{1}{2}O_2$ at 4320°R.

15.40 Water vapor is to be dissociated into H_2 and O_2 in a steady-flow device at a constant pressure of 1 atm. Water vapor enters the device at 680°R and the products leave at 4500°R. How much heat must be added for each mole of water vapor?

15.41 For the reaction $SO_3 \rightleftharpoons SO_2 + \frac{1}{2}O_2$, the following data are available:

T (K)	Equilibrium Constant, K_p
900	0.146
1000	0.518
1100	1.45

Estimate the enthalpy of reaction at 1000 K. Give the answer in SI units.

15.42 Propane gas is burned with 90% theoretical air in a steady-state steady-flow process at 101.325 kPa. Both the fuel and the air enter the combustion chamber at 25°C. The equilibrium products may be assumed to consist of only CO_2, CO, H_2O, O_2, and N_2. If the products leave the combustion chamber at 1500 K and 101.325 kPa, determine
(a) the volumetric analysis of the total products.
(b) the amount of heat transfer per kilogram of fuel burned.

15.43 Same as Problem 15.42 except that the fuel is burned with 100% theoretical air.

15.44 Same as Problem 15.42 except that the fuel is burned with 110% theoretical air.

15.45 Same as Problem 15.42 except that the products of combustion leave at 1000 K.

15.46 Same as Problem 15.42 except that the fuel, instead of propane gas, is a natural gas consisting of 75% by volume of methane (CH_4) and 25% by volume of ethane (C_2H_6).

15.47 Hydrogen gas is burned with 100% of theoretical air at a constant pressure of 1 atm. If there is a 5% dissociation of H_2O to H_2 and O_2 in the products, determine the equilibrium temperature of the products if both the hydrogen gas and air enter the combustion chamber at 77°F and 1 atm.

15.48 Gaseous propane is burned with 100% of theoretical air in a steady-state steady-flow system at a pressure of 1 atm. If both the fuel and air enter at 77°F, and the products leave at 2000 K, determine the amount of heat transferred from the system.

Assume that the products contain no constituents other than CO, CO_2, O_2, N_2, and H_2O.

15.49 The efficiency of a fuel cell may be defined as the ratio of its work output to the enthalpy of reaction involved. For a fuel cell involving the reaction

$$CH_4(g) + 2O_2 \rightleftharpoons CO_2(g) + 2H_2O(l)$$

at 25°C and 1 atm, what is the maximum efficiency?

15.50 For a fuel cell involving the reaction

$$CO + \tfrac{1}{2}O_2 \rightleftharpoons CO_2,$$

determine the maximum conversion efficiency of this fuel cell
(a) if the reaction is carried out at 298 K and 1.0 atm.
(b) if the reaction is carried out at 500 K and 1.0 atm.
(c) if the reaction is carried out at 1000 K and 1.0 atm.

15.51 (a) For the reaction

$$CH_4(g) + 2O_2(g) \rightleftharpoons CO_2(g) + 2H_2O(g)$$

develop an expression giving the enthalpy of reaction as a function of temperature by making use of the following specific heat data:

$$(\bar{c}_p)_{CH_4} = 19.89 + 5.024 \times 10^{-2}T + 1.269 \times 10^{-5}T^2$$

$$- 11.01 \times 10^{-9}T^3 \text{ kJ/kg mol} \cdot \text{K}$$

$$(\bar{c}_p)_{O_2} = 25.48 + 1.520 \times 10^{-2}T - 0.7155 \times 10^{-5}T^2$$

$$+ 1.312 \times 10^{-9}T^3 \text{ kJ/kg mol} \cdot \text{K}$$

$$(\bar{c}_p)_{CO_2} = 22.26 + 5.981 \times 10^{-2}T - 3.501 \times 10^{-5}T^2$$

$$+ 7.469 \times 10^{-9}T^3 \text{ kJ/kg mol} \cdot \text{K}$$

$$(\bar{c}_p)_{H_2O} = 32.24 + 0.1923 \times 10^{-2}T + 1.055 \times 10^{-5}T^2$$

$$- 3.595 \times 10^{-9}T^3 \text{ kJ/kg mol} \cdot \text{K}$$

where T is in kelvin.
(b) Calculate the enthalpy of reaction at 500 K.
(c) Calculate the enthalpy of reaction at 1000 K.

15.52 Show that at the standard state of 1.0 atm,

$$\left(\frac{\Delta G_r^0}{T}\right)_T - \left(\frac{\Delta G_r^0}{T}\right)_{T_0} = -\int_{T_0}^T (\Delta H_r^0) \frac{dT}{T^2},$$

where

ΔH_r^0 = enthalpy of reaction at 1.0 atm and temperature T

ΔG_r^0 = Gibbs function change at 1.0 atm and temperature T.

15.53 For the reaction

$$CH_4(g) + 2O_2(g) \rightleftharpoons CO_2(g) + 2H_2O(g)$$

calculate the equilibrium constant
(a) at 500 K.
(b) at 1000 K.
Use the expression in Problem 15.52 and the enthalpy of reaction data in Problem 15.51.

15.54 For a fuel cell involving the reaction

$$CH_4(g) + 2O_2 \rightleftharpoons CO_2 + 2H_2O(g),$$

determine the maximum conversion efficiency of the fuel cell
(a) if the reaction is carried out at 1.0 atm and 500 K.
(b) if the reaction is carried out at 1.0 atm and 1000 K.
Make use of the results from Problems 15.51 and 15.53.

15.55 Methane gas at 25°C and 7 atm is burned in a steady-state steady-flow device such as the combustor of a gas-turbine power plant with air entering at 200°C and 7 atm. Assume that combustion is complete and neglect heat loss and pressure drop. If the products leave the combustion chamber at 827°C, determine, on the basis of 1 kg of fuel burned,
(a) the thermodynamic lost work of the combustion process.
(b) the second-law efficiency of the combustion process.
(c) the availability of the products leaving the combustion chamber.

15.56 Same as Problem 15.55 except that the products leave the combustion chamber at 1127°C.

15.57 Same as Problem 15.55 except that the fuel is burned with 400% theoretical air.

15.58 Same as Problem 15.55 except that the fuel is burned with 300% theoretical air.

15.59 Write a computer program to determine the actual air-fuel ratio and the percentage of excess air used in the burning of a hydrocarbon fuel, and the mass analysis of fuel burned given the Orsat analysis of the dry flue gas.
Run the program for the following results of Orsat analysis:

	CO_2 (%)	CO (%)	O_2 (%)	N_2 (%)
(a)	10.0	1.0	8.0	81.0
(b)	12.0	0	3.0	85.0
(c)	13.0	0	1.0	86.0

15.60 Write a computer program to study the effect of percent theoretical air on the adiabatic flame temperature for a hydrocarbon fuel C_aH_b, assuming that both the fuel and air enter the combustion chamber at 25°C, and also assuming complete combustion. Make use of the expressions given in Table A.8 for the specific heat of the various species in the products of combustion.
Run the program for the burning of methane gas for percent theoretical air of 100, 150, 200, 300, and 400. Plot the results.

15.61 Write a computer program to study the effect of percent theoretical air on the adiabatic flame temperature and equilibrium composition of the products for a hydrocarbon fuel C_aH_b assuming there will be only dissociation of CO_2 into CO and O_2. Assume that both the fuel and air enter the combustion chamber at 25°C and 5.0 atm, and neglect pressure drop. Make use of the expressions given in Table A.8 for the specific heat of the various species in the products of combustion.
Run the program for the burning of methane gas for percent theoretical air of 100, 150, 200, 300, and 400. Plot the results.

15.62 Equation 15.72 implies that the equilibrium constant K_p (thus $\log_{10} K_p$) of a reaction may be determined at a given temperature if K_p of the reaction is known for one temperature (such as at 25°C) and if ΔH_r (the enthalpy of reaction) of the reaction may be expressed as a function of temperature.
Write a computer program to determine the effect of temperature on the equilibrium constant K_p for the reaction of $H_2O \rightleftharpoons H_2 + \frac{1}{2}O_2$. Make use of the expressions given in Table A.8 for the specific heat of species involved.
Run the program to determine $\log_{10} K_p$ for the temperatures of 500, 1000, 1500, and 2000 K. Compare the results with those given in Table A.11.

CHAPTER 16

System Design Involving Heat Reservoirs, Work Reservoirs, and Matter Reservoirs

In the study of systems for the production of power and refrigeration, we make use of the changes of state of different kinds of substances only as a means to an end. In this chapter, we shall examine systems the objectives of which are changes in matter. In a gas-liquefaction system, our objective is to change a substance from its gaseous state to a liquid state. In a gas-separation system, our objective is to produce pure gases from a gaseous mixture. In a saline-water conversion system, our objective is to separate the fresh water from the salt solution. It will be seen that the knowledge required for the design and analysis of these and similar systems is the same: the first and second laws of thermodynamics and an understanding of the thermodynamic behavior of the substances involved.

GAS-LIQUEFACTION SYSTEMS

16.1 Work Requirement for an Ideal Liquefaction System

The major elements of a scheme to liquefy a gas are shown in Fig. 16.1. Before going into the study of specific liquefaction systems, we wish to determine the minimum work required to liquefy a unit mass of gas so that we can have a basis for comparison of actual systems.

Let us take the entire gas-liquefaction system in Fig. 16.1 as our control volume. Neglecting changes in kinetic and potential energies,

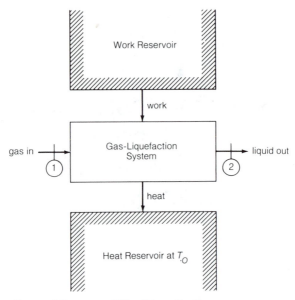

Figure 16.1 General Process of Gas Liquefaction

we have, from the first law for steady-state steady flow,

$$Q = m(h_2 - h_1) + W_{net}. \qquad (16.1)$$

We are interested in the minimum work required; consequently, we want to consider the case of a reversible process. Then, from the second law for steady-state steady flow, we have, in the case of reversible process,

$$m(s_2 - s_1) - \frac{Q}{T_0} = 0. \qquad (16.2)$$

Combining Eq. 16.1 with Eq. 16.2, we have

$$\frac{W_{net}}{m} = T_0(s_2 - s_1) - (h_2 - h_1). \qquad (16.3)$$

Equation 16.3 gives the minimum work required to produce 1.0 kg of liquefied gas according to the scheme shown in Fig. 16.1. This equation also indicates that we should use as low a value as possible for T_0, which in practice is usually the ambient temperature. Conceptually, a system that will meet the minimum work requirement is one consisting of an infinite number of Carnot refrigerators, as shown in Fig. 16.2, if the gas enters the system at a temperature equal to or less than that of the surroundings. If the gas enters the system at a tem-

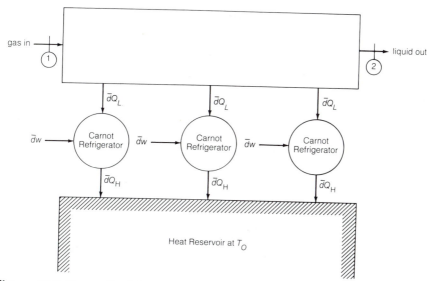

Figure 16.2 Generalized Gas-Liquefaction System

perature greater than that of the surroundings, the system would consist of Carnot heat engines as well as Carnot refrigerators.

Example 16.1

Determine the minimum work required to liquefy 1.0 kg of nitrogen from 1.0 atm and 300 K to saturated liquid at 1.0 atm. Assume the surroundings to be at 300 K.

Solution (refer to Fig. 16.1)

From the nitrogen tables, we have

$$h_1 = 311.163 \text{ kJ/kg}$$

$$s_1 = 6.8418 \text{ kJ/kg} \cdot \text{K}$$

$$h_2 = -122.150 \text{ kJ/kg}$$

$$s_2 = 2.8326 \text{ kJ/kg} \cdot \text{K.}$$

Substituting these values into Eq. 16.3, we have

$$\frac{W_{net}}{m} = 300(2.8326 - 6.8418) - (-122.150 - 311.163)$$

$$= -769.4 \text{ kJ/kg.}$$

16.2 Reversible Gas-Liquefaction System

In order to liquefy a gas with the minimum work requirement, we need a system in which all processes are reversible. Let us consider the case of changing a gas from conditions p_1 and T_1 (point 1 in the T–s diagram of Fig. 16.3a) to a saturated liquid at the same pressure (point f in the same diagram). A reversible system may simply consist of two reversible processes: a reversible isothermal compression (process $1 \rightarrow 2$) followed by an isentropic expansion (process $2 \rightarrow f$). The reversible isothermal process may be carried out in an isothermal compressor while the isentropic process may be carried out in an isentropic turbine, as shown in Fig. 16.3b.

To determine the work requirement for this system, we apply the laws of thermodynamics. We shall neglect change in kinetic and potential energies. For the compressor, we have, from the first law for

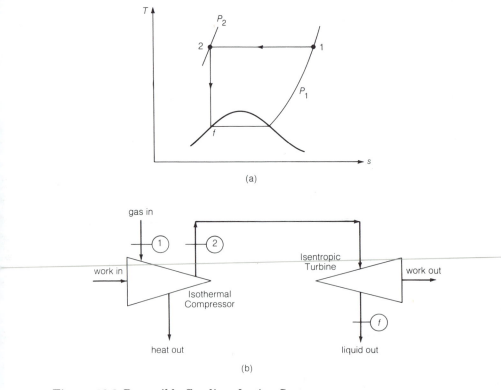

(a)

(b)

Figure 16.3 Reversible Gas-liquefaction System

steady-state steady flow,

$$Q = m(h_2 - h_1) + W_C,$$ (16.4)

and from the second law,

$$Q = mT_1(s_2 - s_1).$$ (16.5)

For the turbine, we have, from the first law,

$$W_T = m(h_2 - h_f).$$ (16.6)

Combining Eqs. 16.4, 16.5, and 16.6, we have

$$\frac{W_{net}}{m} = T_1(s_2 - s_1) - (h_f - h_1)$$

or

$$= T_1(s_f - s_1) - (h_f - h_1)$$ (16.7)

as $s_2 = s_f$.

Note that we need a heat reservoir at T_0 equal to T_1 in order to accomplish the reversible isothermal process. We thus see that Eq. 16.7 is consistent with Eq. 16.3. Equation 16.7 shows that the minimum work required is only a function of the initial conditions of the gas. However, the pressure attained at the end of the isothermal compression is impractically high. Nevertheless, this reversible system is useful as it serves as the basis of modified systems, one of which will be given in the next section.

16.3 Simple Linde–Hampson Gas-Liquefaction System

A gas may be liquefied by passing it through a throttling valve if it enters the valve at a sufficiently low temperature. The mechanism that makes this possible is known as the *Joule–Thomson effect,* and the throttling valve is commonly called the *Joule–Thomson valve.* The simple *Linde–Hampson system,* shown Fig. 16.4, is a system that makes use of such an effect to produce liquefied gas.

To analyze the Linde–Hampson system, we shall assume reversible processes throughout with the exception of the throttling process, which is of course irreversible. Taking the heat exchanger, expansion

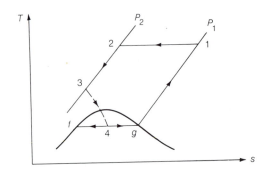

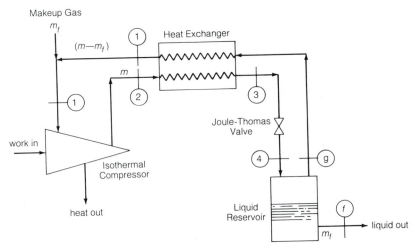

Figure 16.4 Simple Linde–Hampson System

valve, liquid reservoir, and the piping connecting these components as our control volume, we have, from the first law for steady-state steady flow, and neglecting change in kinetic and potential energies,

$$(m - m_f)h_1 + m_f h_f - mh_2 = 0, \tag{16.8}$$

from which we get the ratio of the mass of liquid produced to the mass of gas compressed as

$$Y = \frac{m_f}{m} = \frac{h_1 - h_2}{h_1 - h_f}. \tag{16.9}$$

This ratio is known as the *liquid yield of the system* and is an important performance parameter of the system. Equation 16.9 indicates that for a given initial state of the gas, the yield is only a function of h_2, which is fixed by p_2 and $T_2 = T_1$. To maximize the yield, we must minimize h_2. For air at $p_1 = 1.0$ atm and $T_1 = 20°C$, p_2 must be approximately 400 atm to get maximum yield. In practice, p_2 is about 200 atm. It is to be noted here that to obtain a minimum h_2, we must have $(\partial h/\partial p)_{T=T_1} = 0$. The quantity $(\partial h/\partial p)_T$ is known as the *isothermal* Joule–Thomson coefficient of the gas. As shown in Chapter 13, it is related to c_p and $(\partial T/\partial p)_h$, the more familiar Joule–Thomson coefficient.

To determine the work required for this system, we apply the first and second laws to the compressor. From the first law, we get

$$Q = m(h_2 - h_1) + W. \qquad (16.10)$$

From the second law, we get

$$Q = mT_1(s_2 - s_1). \qquad (16.11)$$

Combining Eqs. 16.10 and 16.11, we have

$$\frac{W}{m} = T_1(s_2 - s_1) - (h_2 - h_1). \qquad (16.12)$$

Equation 16.12 gives the work required for each unit mass of gas compressed. To get an expression for the work required for each unit mass of liquid produced, we combine Eq. 16.12 with Eq. 16.9 to yield

$$\frac{W}{m_f} = \frac{1}{Y}\frac{W}{m} = \frac{h_1 - h_f}{h_1 - h_2}[T_1(s_2 - s_1) - (h_2 - h_1)]. \qquad (16.13)$$

Equation 16.13 indicates that for a given initial state of the gas, the work required to produce a unit mass of liquid depends only on the state of the gas at the compressor outlet.

We conclude our study of gas-liquefaction systems by pointing out that there are many modifications that we can make to improve the performance of the simple Linde–Hampson system. For a more detailed treatment of gas liquefaction, books on this subject should be consulted.[1]

---

[1] Randall Barron, *Cryogenic Systems*, McGraw-Hill Book Company, New York, 1966, Chap. 3.

Example 16.2

A simple Linde–Hampson system is used to liquefy nitrogen gas according to the following specifications:

Temperature of nitrogen gas at compressor inlet = 300 K

Pressure of nitrogen gas at compressor inlet   = 101.325 kPa

Pressure of nitrogen gas at compressor outlet   = 20 MPa

Temperature of the surroundings   = 300 K

Determine

(a) the liquid yield.
(b) the work required to produce 1.0 kg of liquid nitrogen.

Solution (refer to Fig. 16.4)

From nitrogen tables, we have

$$h_1 = 311.163 \text{ kJ/kg}$$

$$s_1 = 6.8418 \text{ kJ/kg} \cdot \text{K}$$

$$h_f = -122.150 \text{ kJ/kg}$$

$$h_2 = 279.010 \text{ kJ/kg}$$

$$s_2 = 5.1630 \text{ kJ/kg} \cdot \text{K}.$$

(a) Substituting the data above into Eq. 16.9, we have

$$Y = \frac{h_1 - h_2}{h_1 - h_f} = \frac{311.163 - 279.010}{311.163 - (-122.150)} = 0.0742.$$

(b) Substituting into Eq. 16.13, we have

$$\frac{W}{m_f} = \frac{1}{0.0742} \left[300(5.1630 - 6.8418) - (279.010 - 311.163) \right]$$

$$= -6354.3 \text{ kJ/kg liquefied.} \qquad \blacksquare$$

Comments

One other performance parameter used in the study of gas liquefaction is the *figure of merit*, which is defined as the theoretical minimum work requirement divided by the actual work requirement for the system. For the simple Linde–Hampson system of this example, the figure of merit is only 769.4/6354.3, or 0.121, which is far from being perfect, and yet this system uses a reversible compressor and a reversible heat exchanger. Obviously there is a lot of room for improvement.

GAS-SEPARATION SYSTEMS

16.4 Work Requirement for an Ideal Gas-Separation System

The industrial pure gases we need usually exist in a mixture of which they are the constituents. For example, atmospheric air is the major source of nitrogen and oxygen, while natural gas is the major source of helium. The development of efficient gas-separation systems is an important and challenging endeavor for engineers and scientists.

The major elements of a scheme to separate the constituents from a gas mixture are shown in Fig. 16.5. Let us consider the performance of an ideal process for separating the gases so that we can have a

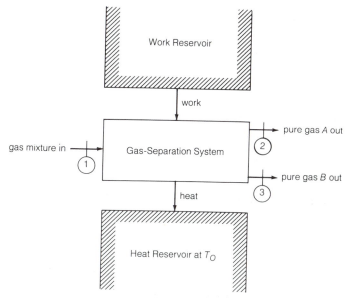

Figure 16.5 General Process of Gas Separation

basis for comparison of actual systems. In our analysis we shall neglect changes in kinetic and potential energies. Taking the gas-separation system in Fig. 16.5 as our control volume, we have, from the first law for steady-state steady flow,

$$Q = W + H_2 + H_3 - H_1, \tag{16.14}$$

and from the second law, if the process is reversible,

$$S_2 + S_3 - S_1 - \frac{Q}{T_0} = 0. \tag{16.15}$$

Combining Eq. 16.14 with Eq. 16.15, we have

$$W = T_0(S_2 + S_3 - S_1) - (H_2 + H_3 - H_1). \tag{16.16}$$

Since entropy and enthalpy are extensive properties, we may express Eq. 16.16 as

$$W = T_0 N_A(\bar{s}_{A2} - \bar{s}_{A1}) + T_0 N_B(\bar{s}_{B3} - \bar{s}_{B1})$$
$$- N_A(\bar{h}_{A2} - \bar{h}_{A1}) - N_B(\bar{h}_{B3} - \bar{h}_{B1}) \tag{16.17}$$

or

$$\frac{W}{N_m} = T_0 X_A(\bar{s}_{A2} - \bar{s}_{A1}) + T_0 X_B(\bar{s}_{B3} - \bar{s}_{B1})$$
$$- X_A(\bar{h}_{A2} - \bar{h}_{A1}) - X_B(\bar{h}_{B3} - \bar{h}_{B1}), \tag{16.18}$$

where

N_m = total number of moles of mixture entering the system

$X_A = \dfrac{N_A}{N_m}$ is the mole fraction of species A in the mixture

$X_B = \dfrac{N_B}{N_m}$ is the mole fraction of species B in the mixture.

If the gas mixture and the pure species all behave like ideal gases, and if we assume constant specific heats, then from property relations for ideal gases, we have

$$\bar{s}_{A2} - \bar{s}_{A1} = \bar{c}_{pA} \ln \frac{T_{A2}}{T_{A1}} - \bar{R} \ln \frac{p_{A2}}{p_{A1}}$$

$$\bar{s}_{B3} - \bar{s}_{B1} = \bar{c}_{pB} \ln \frac{T_{B3}}{T_{B1}} - \bar{R} \ln \frac{p_{B3}}{p_{B1}}$$

$$\bar{h}_{A2} - \bar{h}_{A1} = \bar{c}_{pA}(T_{A2} - T_{A1})$$

and

$$\bar{h}_{B3} - \bar{h}_{B1} = \bar{c}_{pB}(T_{B3} - T_{B1}).$$

Substituting into Eq. 16.18, we have

$$\frac{W}{N_m} = T_0 X_A \left(\bar{c}_{pA} \ln \frac{T_{A2}}{T_{A1}} - \bar{R} \ln \frac{p_{A2}}{p_{A1}} \right)$$

$$+ T_0 X_B \left(\bar{c}_{pB} \ln \frac{T_{B3}}{T_{B1}} - \bar{R} \ln \frac{p_{B3}}{p_{B1}} \right)$$

$$- X_A \bar{c}_{pA}(T_{A2} - T_{A1}) - X_B \bar{c}_{pB}(T_{B3} - T_{B1}). \qquad (16.19)$$

If the process is isothermal as well as reversible, Eq. 16.19 becomes

$$\frac{W}{N_m} = - T_0 \bar{R} \left(X_A \ln \frac{p_{A2}}{p_{A1}} + X_B \ln \frac{p_{B3}}{p_{B1}} \right). \qquad (16.20)$$

Now p_{A1}, the pressure exerted by species A in the mixture, is known as the *partial pressure of species A*. From elementary chemistry (Dalton's law of partial pressures), this partial pressure is given as

$$p_{A1} = X_A p_{m1}.$$

Similarly,

$$p_{B1} = X_B p_{m1}.$$

Making use of these relationships, we have

$$\frac{W}{N_m} = T_0 \bar{R} \left(X_A \ln \frac{X_A p_{m1}}{p_{A2}} + X_B \ln \frac{X_B p_{m1}}{p_{B3}} \right). \qquad (16.21)$$

If $p_{A2} = p_{B3} = p_{m1}$, then Eq. 16.21 becomes

$$\frac{W}{N_m} = T_0 \bar{R}(X_A \ln X_A + X_B \ln X_B). \tag{16.22}$$

We may extend Eq. 16.22 to the case of several gases:

$$\frac{W}{N_m} = T_0 \bar{R} \sum_i X_i \ln X_i. \tag{16.23}$$

Equation 16.23 gives the minimum work requirement for the separation of an ideal-gas mixture with each pure species leaving the system at the same temperature and pressure as the mixture. Note that this equation does not contain the temperature of the gas.

Example 16.3

Determine the minimum work requirement for the separation of an ideal-gas mixture consisting of 79% nitrogen by mole and 21% oxygen. The mixture is at a temperature of 25°C and a pressure of 1.0 atm. The pure gases are also at 25°C and 1.0 atm. The surroundings are at 25°C.

Solution (refer to Fig. 16.5)

Using Eq. 16.23, we have

$$\frac{W}{N_m} = (298)(8314.3)(0.79 \ln 0.79 + 0.21 \ln 0.21)$$

$$= -1.273 \times 10^6 \text{ J/kg mol of mixture.}$$

In terms of the constituents,

$$\frac{W}{N_{N_2}} = \frac{-1.273 \times 10^6}{0.79} = -1.611 \times 10^6 \text{ J/kg mol of nitrogen}$$

$$\frac{W}{N_{O_2}} = \frac{-1.273 \times 10^6}{0.21} = -6.062 \times 10^6 \text{ J/kg mol of oxygen.} \qquad \blacksquare$$

Example 16.4

Determine the minimum work required to produce 1.0 mol of pure gas A from a mixture consisting of 0.9% of species A by volume and 99.1% of species B. The mixture is at a temperature of 25°C and a pressure of 1.0 atm. The pure gases are also at 25°C and 1.0 atm. The surroundings are at 25°C. Express the result in SI units.

Solution (refer to Fig. 16.5)

Assuming ideal-gas behavior, we have, using Eq. 16.23,

$$\frac{W}{N_m} = (298)(8314.3)(0.009 \ln 0.009 + 0.991 \ln 0.991)$$

$$= -1.272 \times 10^5 \text{ J/kg mol of mixture}$$

and

$$\frac{W}{N_A} = -\frac{1.272 \times 10^5}{0.009} = -1.413 \times 10^7 \text{ J/kg mol of pure gas } A. \quad \blacksquare$$

Comments

The results of Examples 16.3 and 16.4 show that as the percentage of a component becomes smaller, more work is required to produce 1 mol of that component in the pure state. The amount of argon gas in the atmosphere is about 0.9% by volume. If we want to obtain argon from the atmosphere, the minimum work requirement is given by the result of Example 16.4.

16.5 Reversible Gas-Separation System

The gas-separation system shown in Fig. 16.5 is simply a "black box." There are many physical arrangements, some still unimagined, that could be used in its place to accomplish the given task. A classical theoretical scheme, shown in Fig. 16.6, depends on the availability of semipermeable membranes, that is, membranes that allow the passage of one gas but block the passage of all other gases.

Let us place a mixture of species A and B in the cylinder. Let us fit the cylinder with two frictionless pistons, with one permeable to species A only and the other permeable to species B only. Now, if we move the pistons very slowly by the application of external work, all of the species A will eventually occupy the left-hand side of the cylinder and all of the species B will occupy the right-hand side of the cylinder. The pressures of the mixture and the pure gases are all the same, since the process is not only quasi-static but also isothermal.

Focusing our attention on only the species A at all times, we have a closed system undergoing a change of state quasi-statically and isothermally. Then, from the first and second laws of thermodynamics,

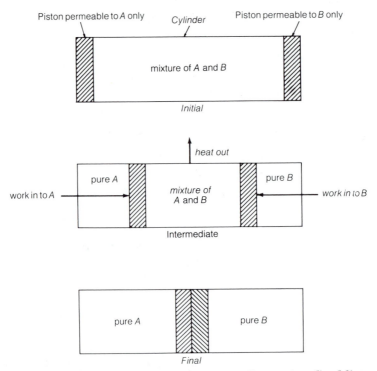

Figure 16.6 Scheme for the Separation of a Two-Component Gas Mixture

the work required to separate species A is given by

$$W_A = \int_{V_{A1}}^{V_{A2}} p_A \, dV_A.$$ (16.24)

Assuming ideal-gas behavior, Eq. 16.24 may be transformed into the following, since the temperature is constant:

$$W_A = -\int_{p_{A1}}^{p_{A2}} V_A \, dp_A$$

$$= -N_A \bar{R} T \int_{p_{A1}}^{p_{A2}} \frac{dp_A}{p_A} = -N_A \bar{R} T \ln \frac{p_{A2}}{p_{A1}}.$$ (16.25)

In a similar manner, the work required to separate species B is given by

$$W_B = -N_B \bar{R} T \ln \frac{p_{B2}}{p_{B1}}.$$ (16.26)

The total work required for the separation of N_m mole of mixture is then given by

$$W = W_A + W_B$$

$$= -\bar{R}T\left(N_A \ln \frac{p_{A2}}{p_{A1}} + N_B \ln \frac{p_{B2}}{p_{B1}}\right), \qquad (16.27)$$

where p_{A1} and p_{B1} are the partial pressures of A and B in the mixture. Thus, making use of Dalton's law of partial pressures, and since $p_{A2} = p_{B2} = p_{m1}$, we may express Eq. 16.27 as

$$\frac{W}{N_m} = \bar{R}T(X_A \ln X_A + X_B \ln X_B). \qquad (16.28)$$

In Eq. 16.28, the temperature T is the temperature of the mixture as well as the temperature of the pure gases. In order to have a reversible isothermal process, the surroundings must also be at the same temperature. We thus see that Eq. 16.28 is the same as Eq. 16.22.

16.6 Phase Diagram of a Two-Component Fluid

Multicomponent gases such as air can be separated into their pure gases by absorption and other chemical processes. However, the most important commercial systems make use of low-temperature distillation methods. These methods exploit the fact that at a given pressure, the equilibrium composition of a liquid–vapor mixture of a multicomponent fluid varies with temperature. For example, a typical temperature–composition diagram for a two-component system such as nitrogen–oxygen is shown in Fig. 16.7.

In Fig. 16.7, the liquid–vapor two-phase region is enclosed by the dew line and the bubble line. Above the dew line, we have a homogeneous vapor. Below the bubble line, we have a homogeneous liquid. When we cool a homogeneous vapor from point 1 to point 2, the mixture begins to condense. This is why the dew line is also known as the *condensation curve*. When we heat a homogeneous liquid from point 5 to point 4, the mixture begins to vaporize, and consequently the bubble line is also known as the *vaporization curve*. For a given temperature of equilibrium between liquid and vapor, the corresponding point on the upper curve gives the composition of the vapor and the point at the same ordinate on the lower curve gives the composition of the liquid. At the equilibrium point 3 in Fig. 16.7, the mole fraction of nitrogen in the liquid phase is given by point 3f. It is evident that a simple evaporation process will effect a partial separa-

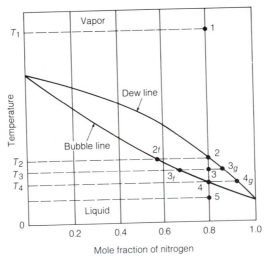

Figure 16.7 Temperature–Composition Diagram for Oxygen–Nitrogen Mixtures at a Given Pressure

tion; the vapor will be richer in the lower-boiling component, which is nitrogen, while the liquid will become richer in the higher-boiling component, which is oxygen. For a complete separation, we need a large number of evaporation processes. A good understanding of the thermodynamic behavior of multicomponent fluids is therefore essential to the study of gas-separation systems.

16.7 Simple Linde Single-Column Air-Separation System

The simplest air-separation system is the Linde single-column system shown in Fig. 16.8. This system produces nitrogen and oxygen by the method of distillation at low temperature. There are three basic steps involved in this system: purification, refrigeration, and separation.

The *purification* step consists of the removal of impurities, particularly of water and carbon dioxide, so that they will not block the flow path. *Refrigeration* is necessary to liquefy the air. The compressed air, after being cooled in the heat exchanger, is further cooled and partially condensed in the coil located at the bottom of the rectification column. The cold air is then expanded through a Joule–Thomson valve to about atmospheric pressure. The mixture of liquid and vapor leaving the expansion valve is then passed to the top of the column. The *separation* process is carried out in the rectification column,

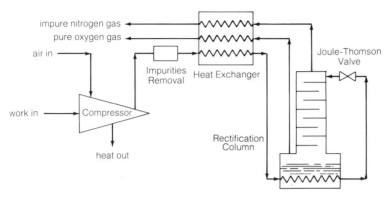

Figure 16.8 Simple Linde Single-Column Air-Separation System

which may be called the air separator. The rectification column is simply a device providing the effects of a large number of evaporations that are necessary for more complete separation. A typical rectification column consists of a boiler (heating coil) at the bottom and a number of trays or plates through which the ascending vapor is caused to percolate. As the liquid flows down the column, it is enriched in oxygen by exchange with the upward-flowing vapor. As the vapor approaches the top of the column, it becomes very rich in the lower-boiling constituent, nitrogen. From the phase equilibrium data of an oxygen–nitrogen mixture at 1.0 atm, it may be shown that for a liquid mixture of 79 percent nitrogen and 21 percent oxygen, the composition of the equilibrium vapor will be about 7 percent oxygen. Therefore, with the Linde single-column system, there will be some oxygen in the nitrogen product. Modifications must be made to this simple system in order to produce very pure nitrogen. For detailed treatment of gas-separation systems, books on this subject should be consulted.[2]

Example 16.5

A simple Linde single-column air-separation system is to be designed according to the following specifications:

Temperature of air at the compressor inlet $= 25°C$

Pressure of air at the compressor inlet $\quad = 1.0$ atm

Pressure of air at the compressor outlet $\quad = 50$ atm

[2] M. Ruhemann, *The Separation of Gases*, 2nd ed., Oxford University Press, New York, 1949; also Randall Barron, *Cryogenic Systems*, McGraw-Hill Book Company, New York, 1966.

If the compression process is reversible and isothermal, determine the work required

(a) for the separation of 1.0 mol of air.
(b) for the production of 1.0 mol of pure oxygen gas.

Solution

We shall neglect changes in kinetic and potential energies. Taking the compressor as our control volume, we have, from the first and second laws for steady flow,

$$W = -\int_{p_1}^{p_2} V \, dp.$$

Assuming ideal-gas behavior, we may express the compressor work as

$$W = -N_{air} \bar{R} T \int_{p_1}^{p_2} \frac{dp}{p} = -N_{air} \bar{R} T \ln \frac{p_2}{p_1}$$

and

(a)
$$\frac{W}{N_{air}} = -\bar{R} T \ln \frac{p_2}{p_1} = -(8314.3)(298) \ln \frac{50}{1}$$

$$= -9.693 \times 10^6 \text{ J/kg mol of air.}$$

Note that the theoretical minimum work requirement is obtained in Example 16.3 as -1.273×10^6 J/kg mol of air.
(b) Assuming the incoming air consists of 21% oxygen by volume, we have

$$\frac{W}{N_{O_2}} = -\frac{9.693 \times 10^6}{0.21} = -4.616 \times 10^7 \text{ J/kg mol of oxygen.} \quad \blacksquare$$

Comment

With the simple Linde single-column system, pressures of 30 to 60 atm are needed if gaseous oxygen is desired. If liquid oxygen is desired (to be withdrawn from the bottom of the rectification column), the air must be compressed to about 200 atm since more refrigeration is needed.

WATER-DESALINATION SYSTEMS
16.8 Minimum Energy Requirement for Desalination

The United States is currently consuming fresh water at a rate of approximately 360 billion gallons per day. Furthermore, freshwater consumption in the United States is increasing at a rate of 25,000 gal/min.[3] The quest for potable water is of course by no means confined to the United States. Water shortages exist in many parts of the world. The water problem is a worldwide one and is of challenging proportion. The need for the development of new sources of fresh water is evident. Because three fourths of the earth's surface is covered by salt water, it is obvious that we should devote our effort toward the development of effective means of obtaining fresh water economically from oceans and other bodies of saline water.

Saline-water conversion can be accomplished in many ways. Some methods have been known in concept for centuries, while others have been developed only in recent years. An excellent survey of the most promising desalination processes and an examination of the critical fluid engineering problems associated with those processes have been given by Probstein.[4]

Let us refer to the general desalination process shown in Fig. 16.9. In order that we may establish a basis for comparison of actual systems, we shall deduce the minimum energy requirement for a desalination process regardless of the mechanism involved. Taking the entire saline-water conversion system shown in Fig. 16.9 as our control volume, and neglecting changes in kinetic and potential energies, we have, from the first law for steady-state steady flow,

$$Q = W + H_2 + H_3 - H_1, \qquad (16.29)$$

and from the second law, if the process is reversible,

$$S_2 + S_3 - S_1 - \frac{Q}{T_0} = 0. \qquad (16.30)$$

Combining Eq. 16.29 with Eq. 16.30, we have

$$W = T_0(S_2 + S_3 - S_1) - (H_2 + H_3 - H_1). \qquad (16.31)$$

---

[3] *The A-B-Seas of Desalting*, Office of Saline Water, U.S. Department of the Interior, U.S. Government Printing Office, Washington, D.C., 1968.
[4] R. F. Probstein, "Desalination: Some Fluid Mechanical Problems," *Journal of Basic Engineering*, June 1972, pp. 286–313.

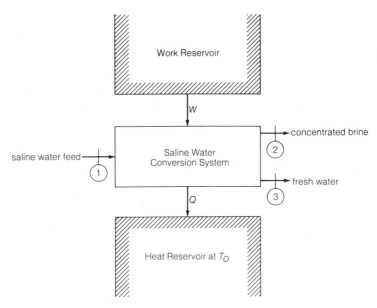

Figure 16.9 General Process of Water Desalination

Equation 16.31 gives the minimum work requirement for the separation of fresh water from the saline solution. It applies to any reversible process.

If the process is isothermal as well as reversible, Eq. 16.31 becomes

$$W = T(S_2 + S_3 - S_1) - (H_2 + H_3 - H_1). \qquad (16.32)$$

Making use of the definition of the Gibbs function, $G = H - TS$, Eq. 16.32 may be written as

$$-W = G_2 + G_3 - G_1. \qquad (16.33)$$

It will be recalled that the Gibbs function is related to the chemical potential by the following general relation (Eq. 6.46),

$$G = \sum N_i \mu_i.$$

Thus, in terms of the chemical potentials, Eq. 16.33 may be written as

$$-W = N_2 \mu_{B2} + N_3 \mu_3 - N_1 \mu_{B1}, \qquad (16.34)$$

where μ_{B2} is the chemical potential of the concentrated brine at point 2, μ_3 is the chemical potential of the fresh water at point 3, and μ_{B1} is the chemical potential of the saline water at point 1.

To make use of either Eq. 16.31 or Eq. 16.34 to obtain the minimum work required for water desalination, we must have the necessary data for the solutions involved. For a solution consisting of 3.5% by weight of NaCl, Dodge and Eshaya[5] shows that at 25°C, the minimum work is approximately 3 kWh for the production of 1000 gal of fresh water. Since normal seawater has a total solid concentration of about 35,000 ppm, it is generally considered that the minimum work required to produce 1000 gal of fresh water from seawater is about 3 kWh. It is to be noted that this minimum-energy requirement is only valid for the limiting case in which the mass of the recovered fresh water is negligibly small compared to that of the feed. For 50% recovery, Dodge and Eshaya gives the minimum work required as 5.8 kW/1000 gal.

If heat instead of work is used as the energy source for water desalination, we may establish the minimum heat requirement through the Carnot cycle efficiency as

$$Q_{min} = W_{min} \frac{T}{T - T_0},$$

where T is the temperature of the heat source and T_0 is the temperature of the heat sink.

For example, if steam at 100°C and 1.0 atm is available as the energy source, and the surroundings is at 20°C, then

$$Q_{min} = (3.0) \frac{373}{373 - 293} \text{ kWh/1000 gal of fresh water}$$

$$= 14.0 \text{ kWh/1000 gal of fresh water.}$$

This is equivalent to the minimum requirement of approximately 22 kg of 100°C steam to produce 1000 gal of fresh water, since the amount of heat that may be extracted from each pound of steam is about 2257 kJ.

16.9 Reverse-Osmosis Process

When a semipermeable membrane is placed between saline water and pure water that are both at the same pressure, diffusion of fresh water into the saline water will occur because the concentration of

[5] B. F. Dodge and A. M. Eshaya, *Thermodynamics of Some Desalting Processes* (Advances in Chemistry Series, No. 27), American Chemical Society, Washington, D.C., 1960, pp. 8–11.

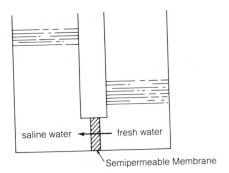

Figure 16.10 Osmosis Process

fresh water in the saline water is less. This phenomenon is known as *osmosis* (see Fig. 16.10). by applying sufficient pressure to the concentrated solution, the osmotic process can be reversed. The pure water then diffuses through the membrane from the concentrated solution to the pure-water side. Hence the term *reverse osmosis* (see Fig. 16.11). A reverse-osmosis desalination system basically requires only the membranes and a pump to bring the pressure of the feed up to the operating conditions (see Fig. 16.12). It is inherently very simple. Furthermore, the work required to pump a liquid is relatively small. Because of important advances in membrane technology in recent years, the reverse-osmosis process is the most heralded desalination process today.

The amount by which the pressure on the saline water must be increased over the pressure on the pure water in order just to prevent the osmosis process from occurring is called the *osmotic pressure* of the saline water. Denoting the osmotic pressure by the symbol π, the saline water will be in equilibrium with the fresh water if we apply to

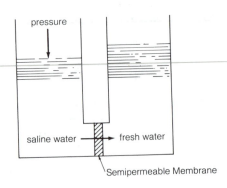

Figure 16.11 Reverse-Osmosis Process

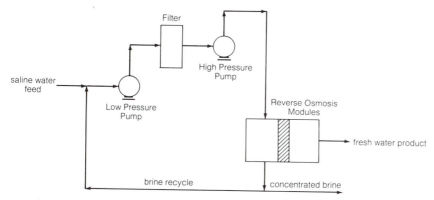

Figure 16.12 Reverse-Osmosis Desalination System

the saline water a pressure equal to $p + \pi$, in which p is the pressure of the pure water. To separate the fresh water from the saline water reversibly, we must apply a pressure to the saline water that will exceed $p + \pi$ by only an infinitesimal amount. When 1 mol of pure water passes through the membrane, the volume of the solution is decreased by v at a pressure difference of π. Thus the minimum work requirement is simply πv. For normal seawater, the osmotic pressure is about 24.8 atm. Taking v as 0.001 m³/kg, the minimum work amounts to about 2.63 kWh/1000 gal of fresh water. For a rigorous derivation of the minimum work requirement, books of this subject should be consulted.[6]

We wish to point out that the minimum work requirement given here is deduced on the basis that the osmotic pressure will remain constant. Actually the osmotic pressure increases with increased concentration. In addition, we wish to point out that in order to get decent permeability from the present types of membranes, the operating pressure of a practical system must be considerably higher than the osmotic pressure.

Example 16.6

A reverse-osmosis desalination system operates with a high pressure of 5515 kPa. If 1 kg of fresh water is produced from 3 kg of seawater, what is the minimum work required to produce 1000 gal of fresh water?

---

[6] K. S. Spiegler, *Salt-Water Purification*, John Wiley & Sons, Inc., New York, 1962, pp. 139–142.

Solution

System selected for study: seawater pump
Assumptions: process is steady-state steady flow and adiabatic; neglect changes in kinetic energy and potential energy

According to the combined first and second laws, the minimum pump work may be given as

$$w_{\min} = -\int_{p_{\text{in}}}^{p_{\text{out}}} v \, dp. \tag{1}$$

Neglecting the compressibility effect and taking v as 0.001 m³/kg, we have, for each kilogram of seawater flowing through the pump,

$$w_{\min} = -0.001(5515 - 101.325) = -5.414 \text{ kJ/kg}.$$

Now we must pump 3000 gal of seawater to produce 1000 gal of fresh water. Therefore, the minimum work required for the system may then be given as

$$W_{\min} = -5.414 \text{ kJ/kg} \times \frac{3000 \text{ gal} \times 3.7854 \times 10^{-3} \text{ m}^3/\text{gal}}{0.001 \text{ m}^3/\text{kg}}$$

$$= -61{,}480 \text{ kJ} = -17.1 \text{ kWh}. \qquad \blacksquare$$

Comments

If the adiabatic pump efficiency is 75%, the actual pump work required would be 22.8 kWh per 1000 gal of fresh water, which is considerably higher than the theoretical minimum of 2.63 kWh. Nevertheless, this amount of energy required is still quite favorable compared to that required by other methods of desalination. But we wish to point out that based on current technology, the energy cost represents only about 20% of the total cost of freshwater production from seawater. For a plant that produces 1 million gallons of fresh water per day, the total cost is about $5.00 per 1000 gal, which is not cost-effective in many instances. However, desalination of seawater by reverse osmosis is important for the future. It is also a fascinating technology. A book that contains many U.S. patents on reverse osmosis and related topics has been published by Noyes Data Corporation.[7] It should be a useful reference for persons having an interest in this topic.

An interesting process contained in the Noyes Data book is one that utilizes osmotic pressure as a driving force to operate a reverse osmosis cycle

[7] Jeanette Scott, ed., *Desalination of Seawater by Reverse Osmosis*, Noyes Data Corporation, Park Ridge, N.J., 1981.

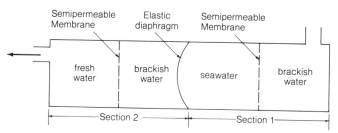

Figure 16.13 Osmotic Pressure Used to Operate Reverse Osmosis Cycle

(U.S. Patent 3,423,310, January 1969). As described by K. Popper, the process encompasses a system where osmotic pressure is generated in one cell (one involving direct osmosis), and this pressure is then utilized to drive a second osmotic cell (one involving reverse osmosis). The necessary requirement of this process is essentially the availability of a pair of liquids having different amounts of dissolved solids. For example, a system could operate with seawater that has 3.5% dissolved solids and brackish water that has 0.5% dissolved solids. The conceptual design is as shown in Fig. 16.13. A vessel is divided into two sections. In section 1, brackish water is separated from the seawater by a semipermeable membrane. The osmotic pressure generated in this section is the driving force needed to operate the reverse osmosis process in section 2. Theoretically, the energy needed to operate this cycle is simply the pump work required to fill the vessel.

16.10 Multieffect Distillation System

Distillation is one of the oldest methods known of separating fresh water from saline water. Basically, the process is one of vaporizing the saline water leaving the dissolved salts in a concentrated solution, and then condensing the pure water vapor. A simple multieffect distillation system is shown in Fig. 16.14. The series of condenser–evaporators or "effects" represents a regenerative heating scheme capitalizing on the fact that water boils at lower temperatures as it is subjected to lower pressures. It can be seen that the heat given up by 1 lb of condensing steam in each effect is reused for the vaporization of more saline water. Since the heat of vaporization for pure water does not differ too much from that of saline water, ideally, in a system of n effects we would get n pounds of fresh water per pound of steam supplied. However, the heating steam requirement represents only the energy cost. More effects would mean more capital investment. As in many other industrial systems, the optimization of a desalination system is based on operating expenses and fixed charges.

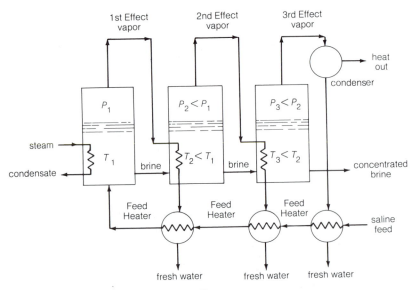

Figure 16.14 Multieffect Distillation System

Following Spiegler,[8] we may give the cost C per unit of fresh water produced by the following approximate equation:

$$C = An + \frac{S}{n},\qquad(16.35)$$

where A is the fixed charges for each effect per unit of fresh water produced, S is the cost of steam per unit of water produced in a single-effect evaporator, and n is the number of effects. By differentiating C with respect to n and setting the derivative equal to zero, we would find that the optimum number of effects is given by

$$n_{\mathrm{opt}} = \sqrt{\frac{S}{A}}.\qquad(16.36)$$

A detailed analysis of the various parameters involved in the optimization of a water-desalination system has been given by Sherwood.[9]

[8] K. S. Spiegler, *Salt-Water Purification*, John Wiley & Sons, Inc., New York, 1962, pp. 49–50.

[9] T. K. Sherwood, *A Course in Process Design*, The MIT Press, Cambridge, Mass., 1963, Chap. 7.

16.11 Vapor-Compression Distillation

In a multieffect distillation system, the heat source is an external supply of heating steam. In a vapor-compression distillation system, such as that shown in Fig. 16.15, the heat source is literally its own steam after it has been compressed.

As shown in Fig. 16.15, the saline feed, preheated in a heat exchanger by the outgoing streams of concentrated brine and fresh water, boils in the still. The steam flows from the still to the compressor, which raises its pressure and temperature. The temperature differential between the steam and brine causes the vapor to condense as fresh water, thus providing the heat necessary for the boiling process. If the still operates at 1 atm with doubly concentrated seawater, the steam will leave the still at a temperature of about 101°C. This means that the steam enters the compressor as a slightly superheated vapor.

It is seen that in this process the operating cost is the compressor work. To minimize operating cost, it is desirable to operate with a small pressure differential across the compressor. But a small pressure differential across the compressor would mean a small temperature differential between the brine and condensing steam, resulting in a very large heat-transfer area and a very large still. Obviously, it is necessary to weigh the operating cost against capital cost and thus determine the optimum design conditions.

Let us investigate the compressor work required for the production of a unit of fresh water. We shall assume that the system operates under steady-state steady-flow conditions. If the steam is compressed reversibly and adiabatically, the minimum compressor work may be

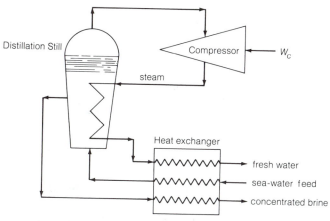

Figure 16.15 Simple Vapor-Compression Distillation System

given by

$$W_{min} = - \int_{P_{in}}^{P_{out}} V \, dp. \tag{16.37}$$

If we assume the low-pressure steam to behave like an ideal gas with constant specific heats, the relation between V and p for in isentropic process is given by

$$pV^k = \text{constant}, \tag{16.38}$$

where k is the specific-heat ratio c_p/c_v. For our application, k may be taken to be equal to 1.32. Substituting V from Eq. 16.38 into Eq. 16.37, we get

$$W_{min} = - \frac{k}{k-1} (pV)_{in} [r_p^{(k-1)/k} - 1], \tag{16.39}$$

where $r_p = p_{out}/p_{in}$ is the compression ratio. For an ideal gas, we have

$$pV = mRT.$$

Thus Eq. 16.39 may be written as

$$W_{min} = - \frac{k}{k-1} mRT_{in} [r_p^{(k-1)/k} - 1]. \tag{16.39a}$$

Example 16.7

A vapor-compression distillation system is to be designed to operate according to the following conditions:

Temperature of steam at compressor inlet	101°C
Compression ratio	1.5
Adiabatic compressor efficiency η_c	75%

Determine the compressor work required to produce 1000 gal of fresh water.

Solution

The actual compressor work is given as

$$W_{act} = \frac{W_{min}}{\eta_c}, \tag{1}$$

where W_{min} may be evaluated through the use of Eq. 16.39a.

Assuming that fresh water leaves the system at 20°C, and taking the density of fresh water at 20°C to be 1000 kg/m³, the total mass m for 1000 gal of fresh water is given by

$$m = 1000 \text{ gal} \times 3.7854 \times 10^{-3} \text{ m}^3/\text{gal} \times 1000 \text{ kg/m}^3$$

$$= 3785.4 \text{ kg}.$$

Substituting into Eq. 16.39a, we get, for the production of 1000 gal of fresh water,

$$W_{min} = -\frac{1.32}{1.32 - 1} \times 3785.4 \times \frac{8.3143}{18} \times 374 \times [(1.5)^{(1.32-1)/1.32} - 1] \text{ kJ}$$

$$= -278,600 \text{ kJ}$$

$$= -77.4 \text{ kWh}.$$

Therefore,

$$W_{net} = -\frac{77.4}{0.75} = -103 \text{ kWh}.$$

If electric power costs $0.03/kWh, the power cost for the production of 1000 gal of fresh water would be $3.09. ∎

Comments

The compressor work can be reduced by reducing the compression ratio. For example, if the compression ratio is only 1.2, the minimum compressor work required would only be about 34 kWh. The reduction in compression ratio would also result in a smaller compressor. However, the distillation still would be much larger, owing to the fact that the temperature differential between the brine and the condensing steam will be reduced. Once again we are facing a real engineering problem in which we must make the proper trade-offs between operating cost and capital cost.

16.12 Chemical Exergy of Hydrocarbon Fuels

"ENERGY TURNS THE WORLD" ... This was the theme of the 1982 World's Fair held at Knoxville, Tennessee.

"ENERGY ... THE SPARK AND LIFELINE OF CIVILIZATION" ... This was proclaimed at the 1982 Intersociety Energy Conversion and Engineering Conference.

In view of the second law of thermodynamics, it may be more appropriate to say that it is "EXERGY" or "AVAILABILITY" that makes the world go around. Energy is definitely important to the survival of the kind of society and of "good life" that we know now; and we need a great amount of it. At the end of the 1970s, the world annual energy consumption was equivalent to that of 50 billion (50×10^9) barrels of oil, about 25% of which was consumed by the United States. Moreover, most of our energy needs must come from fossil fuels (coal, natural gas, and petroleum). In the United States, about 96% of our energy consumption comes from fossil fuels. This percentage has been gradually declining, but it was still about 90% in 1985. It is thus appropriate to conclude our study of engineering thermodynamics by examining the exergy content in the chemical energy of our hydrocarbon fuels.

Let us begin with the following chemical reaction involving the release of the chemical energy in a fuel:

$$A_F + \sum_R v_i B_i \rightarrow \sum_p v_j C_j, \qquad (16.40)$$

where

A_F is the chemical symbol of the fuel

B_i are the reactants, v_i their stoichiometric coefficients

C_j are the products, v_j their stoichiometric coefficients.

Let us now consider the release of the chemical energy in the fuel under the following conditions:

1. The fuel or fuel mixture is at thermal and mechanical equilibrium with the environment. That is, it is at T_0 and p_0, the temperature and pressure of the environment, respectively.
2. The needed oxygen will come from the environment. That is, the oxygen will exist at its state in the environment.
3. The products will be part of the environment. That is, each of the species involved will exist at its state in the environment.
4. Heat interaction is with the environment only.
5. The reaction will be carried out reversibly and isothermally under steady-state steady-flow conditions, neglecting changes in kinetic and potential energies.

The conceptual system for this reaction is shown in Fig. 16.16 The work obtained from this system, W_F, is the maximum useful work pro-

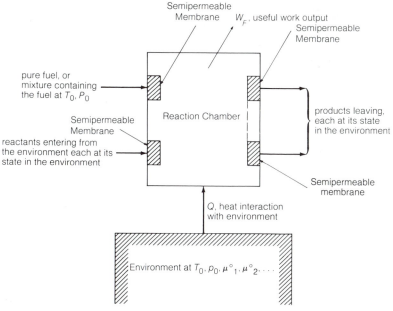

Figure 16.16 Conceptual System for the Determination of Exergy Content in the Chemical Energy of a Fuel

duced per mole of fuel. By definition, this is the chemical exergy of the fuel at T_0 and p_0. Let us now apply the first law of thermodynamics, the second law of thermodynamics, the appropriate property relations, and the conservation of species to deduce an expression for this quantity.

Taking the reaction chamber as our control volume, we have, from the first law of thermodynamics,

$$W_F = Q + H_F^\circ + H_R^\circ - H_p^\circ. \tag{16.41}$$

From the second law of thermodynamics, we have

$$Q = T_0(S_p^\circ - S_R^\circ - S_F^\circ). \tag{16.42}$$

Combining Eqs. 16.41 and 16.42, we have

$$W_F = (H_F^\circ - T_0 S_F^\circ) + (H_R^\circ - T_0 S_R^\circ) - (H_p^\circ - T_0 S_p^\circ)$$

$$= G_F^\circ + G_R^\circ - G_p^\circ \tag{16.43}$$

From Eq. 6.46, we have

$$G_F^{\circ} = \mu_F^{\circ}$$

$$G_R^{\circ} = \sum v_i \mu_i$$

$$G_P^{\circ} = \sum v_j \mu_j.$$

Thus Eq. 16.43 may be written as

$$W_F = \mu_F^{\circ} - \left[\sum_P v_j \mu_j^{\circ} - \sum_R v_i \mu_i^{\circ} \right]. \qquad (16.44)$$

Equation 16.44 is the equation for the exergy of a fuel at T_0 and p_0. We will denote this quantity by the symbol ψ_F^{Ch}. Thus

$$\psi_F^{Ch} = \mu_F^{\circ} - \left[\sum_P v_j \mu_j^{\circ} - \sum_R v_i \mu_i^{\circ} \right]. \qquad (16.44a)$$

If the fuel is a pure substance, then

$$\mu_F^{\circ} = \bar{g}_F(T_0, p_0)$$

and

$$\psi_F^{Ch}(T_0, p_0) = \bar{g}_F(T_0, p_0) - \left[\sum_P v_j \mu_j^{\circ} - \sum_R v_i \mu_i^{\circ} \right]. \qquad (16.45)$$

If the fuel is a hydrocarbon fuel having the formula $C_a H_b$, the reaction equation is given by

$$C_a H_b + \left(a + \frac{b}{4} \right) O_2 \rightarrow a CO_2 + \frac{b}{2} H_2 O(v) \qquad (16.46)$$

and Eq. 16.45 becomes

$$\psi_F^{Ch}(T_0, p_0) = \bar{g}_F(T_0, p_0) + \left(a + \frac{b}{4} \right) \mu_{O_2}^{\circ} - a \mu_{CO_2}^{\circ} - \frac{b}{2} \mu_{H_2 O}^{\circ}. \qquad (16.47)$$

In the environment, the oxygen, carbon dioxide, and water vapor may all be treated as components of an ideal-gas mixture. Making use of Eq. 16.47, we have

$$\mu_{O_2}^\circ = \bar{g}_{O_2}(T_0, p_0) + \bar{R}T \ln (x_{O_2}^\circ) \tag{16.48}$$

$$\mu_{CO_2}^\circ = \bar{g}_{CO_2}(T_0, p_0) + \bar{R}T \ln (x_{CO_2}^\circ) \tag{16.49}$$

$$\mu_{H_2O}^\circ = \bar{g}_{H_2O}(T_0, p_0) + \bar{R}T \ln (x_{H_2O}^\circ), \tag{16.50}$$

where $x_{O_2}^\circ$, $x_{CO_2}^\circ$, and $x_{H_2O}^\circ$ are the mole fractions of oxygen, carbon dioxide, and water vapor within the environment, respectively.

Combining Eqs. 16.47, 16.48, 16.49, and 16.50, the chemical exergy of a pure hydrocarbon fuel having the formula C_aH_b is then given by

$$\psi_F^{Ch}(T_0, p_0) = \left[\bar{g}_F + \left(a + \frac{b}{4} \right) \bar{g}_{O_2} - \left(a\bar{g}_{CO_2} + \frac{b}{2} \bar{g}_{H_2O} \right) \right]$$

$$+ \bar{R}T_0 \ln \frac{(x_{O_2}^\circ)^{(a+b/4)}}{(x_{CO_2}^\circ)^a (x_{H_2O}^\circ)^{b/2}}. \tag{16.51}$$

Now

$$\left[\bar{g}_F + \left(a + \frac{b}{4} \right) \bar{g}_{O_2} - \left(a\bar{g}_{CO_2} + \frac{b}{2} \bar{g}_{H_2O} \right) \right] = G_R - G_p$$

$$= -\Delta G(T_0, p_0), \tag{16.52}$$

where ΔG is simply the Gibbs function change of a reaction at T_0, p_0 which may be evaluated by making use of Gibbs function of formation data. Thus the chemical exergy of a pure fuel having the formula C_aH_b may now be given as

$$\psi_F^{Ch}(T_0, p_0) = -\Delta G(T_0, p_0) + \bar{R}T \ln \frac{(x_{O_2}^\circ)^{(a+b/4)}}{(x_{CO_2}^\circ)^a (x_{H_2O}^\circ)^{b/2}}. \tag{16.51a}$$

To make use of Eq. 16.51a, we must specify the temperature, the pressure, and the composition of the environment. Following Moran,[10] we will make use of the information given in Table 16.1.

---

[10] Michael J. Moran, *Availability Analysis: A Guide to Efficient Energy Use*, Prentice-Hall, Inc., Englewood Cliffs, N.J., 1982, p. 154.

Table 16.1 Environment for Calculating the Chemical Exergy of Hydrocarbon Fuels

Temperature, T_0	Pressure, p_0	Composition	
		Substance (All in Gas Phase)	Mole Fraction, x_i
298.15 K	101.325 kPa	N_2	0.7567
		O_2	0.2035
		H_2O	0.0303
		CO_2	0.0003
		Other	0.0092

Example 16.8

Calculate the chemical exergy in methane gas at 25°C and 101.325 kPa.

Solution

The reaction equation is

$$CH_4 + 2O_2 \rightarrow CO_2 + 2H_2O(v).$$

Making use of Eq. 16.51, we have

$$\psi_{CH_4}^{Ch} = -[(\bar{g}_{CO_2} + 2\bar{g}_{H_2O(v)}) - (\bar{g}_{CH_4} + 2\bar{g}_{O_2})] + \bar{R}T_0 \ln \frac{(x_{O_2}^\circ)^2}{(x_{CO_2}^\circ)(x_{H_2O}^\circ)^2}. \quad (1)$$

From Table 16.1,

$$x_{O_2}^\circ = 0.2035$$

$$x_{CO_2}^\circ = 0.0003$$

$$x_{H_2O}^\circ = 0.0303.$$

From Table A.10, we have at 25°C and 101.325 kPa,

$$\bar{g}_{CO_2} = -394{,}630 \text{ kJ/kg mol}$$

$$\bar{g}_{H_2O} = -228{,}730 \text{ kJ/kg mol}$$

$$\bar{g}_{CH_4} = -50{,}844 \text{ kJ/kg mol}$$

$$\bar{g}_{O_2} = 0.$$

Substituting into Eq. 1, we have

$$\psi_{CH_4}^{Ch} = -[(-394,630) + 2(-228,730) - (-50,844 + 0)]$$

$$+ 8.3143 \times 298.15 \ln \frac{(0.2035)^2}{(0.0003)(0.0303)^2}$$

$$= 801,250 + 29,550 \text{ kJ/kg mol} = 830,800 \text{ kJ/kg mol}. \qquad \blacksquare$$

Comments

Our result is very close to the lower heating value of the fuel, which is 802,840 kJ/kg mol. Our result is also very close to ΔG at 25°C and 101.325 kPa. This is not surprising, as the Gibbs function is a so-called free-energy function the change of which in an ideal reaction is a measure of its work-producing potential.

Example 16.9

An ideal-gas mixture consists of 75% by mole of CH_4 and 25% of nitrogen. Calculate the exergy content of the mixture in kJ/kg mol of mixture at 25°C and 101.325 kPa.

Solution

The reaction equation is the same as for Example 16.8:

$$CH_4 + 2O_2 \rightarrow CO_2 + 2H_2O.$$

This problem is similar to Example 16.8 except that the fuel now is not a pure substance. Its chemical potential must be evaluated at its temperature of 25°C and its partial pressure in the mixture. That is, we have

$$\mu_{CH_4} = \bar{g}_{CH_4}(T_0, p_0) + \bar{R}T_0 \ln (x_{CH_4}), \qquad (1)$$

where x_{CH_4} is the mole fraction of methane in the mixture.

Making use of Eq. 1 and Eqs. 16.48, 16.49, and 16.50, the chemical exergy of each mole of methane is given by Eq. 16.44:

$$\psi_{CH_4}^{Ch} = [(\bar{g}_{CH_4} + 2\bar{g}_{O_2}) - (\bar{g}_{CO_2} + 2\bar{g}_{H_2O})]$$

$$+ \bar{R}T_0 \ln \frac{(x_{O_2}^\circ)^2}{(x_{CO_2}^\circ)(x_{H_2O}^\circ)^2} + \bar{R}T \ln (x_{CH_4}). \qquad (2)$$

We observe that Eq. 2 differs from the result of Example 16.8 just by the last term $\bar{R}T \ln (x_{CH_4})$. Making use of the result in Example 16.8, we have

$$\psi_{CH_4}^{Ch} = 830{,}800 + 8.3143 \times 298.15 \ln (0.75) \text{ kJ/kg mol of } CH_4$$

$$= 830{,}800 - 713 = 830{,}090 \text{ kJ/kg mol of } CH_4.$$

Since the nitrogen has no chemical energy for this reaction, the chemical exergy of the mixture is due only to the methane content. Thus we have

$$\psi_{mix}^{Ch} = 830{,}090 \text{ kJ/kg mol of } CH_4$$

$$\times 0.75 \text{ kg mol of } CH_4/\text{kg mol of mixture}$$

$$= 622{,}570 \text{ kJ/kg mol of mixture.} \qquad \blacksquare$$

Comments

We observe from Eq. 2 that the exergy content will be decreased from that for a pure fuel due to the partial pressure in the mixture. The reduction is not significant if the partial pressure is not too small. However, if the mole fraction is small, it may be desirable to separate some of the nitrogen out so that the exergy content per mole of mixture will be increased.

Problems

16.1   Air is to be liquefied in a reversible system from 101.325 kPa and 300 K to saturated liquid at 101.325 kPa. Determine
   (a) the minimum work required, in kJ/kg.
   (b) the amount of heat rejected in the ideal isothermal compressor, in kJ/kg.
   The following data on air are taken from *Thermodynamic Functions of Gases*, Vol. 2, by F. Din:[11]

   At 101.325 kPa and 300 K:
     $\bar{h} = 12{,}388$ kJ/kg mol; $\bar{s} = 112.62$ kJ/kg mol · K
   At 101.325 kPa: $\bar{h}_f = 0$; $\bar{s}_f = 0$.

16.2   Argon is to be liquefied in a reversible system from 101.325 kPa and 300 K to saturated liquid at 101.325 kPa. Determine
   (a) the minimum work required, in kJ/kg.

---

[11] F. Din, editor, *Thermodynamic Functions of Gases*, Butterworth, London. Vol. 1 & 2 (1956, reprinted 1962), Vol. 3 (1961).

(b) the amount of heat rejected in the ideal isothermal compressor, in kJ/kg.

The following data on argon are taken from *Thermodynamic Functions of Gases*, Vol. 1, by F. Din:[11]

At 101.325 kPa and 300 K:
  $\bar{h}$ = 13,952 kJ/kg mol; $\bar{s}$ = 154.56 kJ/kg mol · K
  At 101.325 kPa: $\bar{h}_f$ = 2972.0 kJ/kg; $\bar{s}_f$ = 0.

16.3 Oxygen is to be liquefied in a reversible system from 1 atm and 80°F to saturated liquid at 1 atm. Determine
(a) the minimum work required, in Btu/lbm and in SI units.
(b) the amount of heat rejected in the ideal isothermal compressor, in Btu/lbm and in in SI units.

16.4 Ammonia is to be liquefied in a reversible system from 1 atm and 80°F to saturated liquid at 1 atm. Determine
(a) the minimum work required, in Btu/lbm and in SI units.
(b) the amount of heat rejected in the ideal isothermal compressor, in Btu/lbm and SI units.

16.5 Oxygen is liquefied from 1 atm and 80°F to saturated liquid at 1 atm in a system that has a figure of merit of 0.20. What is the actual work required to produce 1 lbm of liquid oxygen?

16.6 Air is liquefied from 101.325 kPa and 300 K to saturated liquid at 101.325 kPa. If the actual work required to produce 1 kg of liquid air is 3000 kJ, what is the figure of merit of the system?

16.7 Methane is liquefied from 101.325 kPa and 300 K to saturated liquid at 101.325 kPa. If the actual work required to produce 1 kg of liquid methane is 3000 kJ, what is the figure of the system?

16.8 In an ideal liquefaction system, nitrogen is compressed isothermally at 300 K from 101.325 kPa to a very high pressure. Estimate the pressure of the gas at the end of the isothermal compression process using the van der Waals equation of state:

$$p = \frac{\bar{R}T}{\bar{v} - b} + \frac{a}{\bar{v}^2},$$

where the constants a and b may be found in Table 13.1 in Chapter 13. [*Hint:* For a van der Waals gas, the entropy change

[11] F. Din, editor, *Thermodynamic Functions of Gases*, Butterworth, London. Vol. 1 & 2 (1956, reprinted 1962), Vol. 3 (1961).

due to an isothermal change of state is given by

$$(\bar{s}_2 - \bar{s}_1)_T = \bar{R} \ln \frac{\bar{v}_2 - b}{\bar{v}_1 - b}.]$$

16.9 Same as Problem 16.8 except that the gas is oxygen.

16.10 Same as Problem 16.8 except that the gas is methane. The following data on methane are taken from *Thermodynamic Functions of Gases*, Vol. 3, by F. Din:[11]

At 101.325 kPa and 300 K:
$\bar{v}$ = 24,575 cc/g mol; $\bar{s}$ = 44.54 cal/g mol · K
At 101.325 kPa: $\bar{s}_f$ = 18.916 cal/g mol · K.

16.11 Same as Problem 16.8 except that the gas is carbon monoxide and that the gas is initially at 25°C instead of 300 K. The following data on carbon monoxide gas are taken from *Thermodynamic Functions of Gases*, Vol. 1, by F. Din:[11]

At 101.325 kPa and 25°C:
$\bar{v}$ = 24,457 cc/g mol; $\bar{s}$ = 47.266 cal/g mol · K
At 101.325 kPa: $\bar{s}_f$ = 20.124 cal/g mol · K.

16.12 If a source of high-pressure nitrogen gas is available, liquid nitrogen may be produced using the system shown in Fig. P16.12. Determine the fraction of the high-pressure gas that is liquefied if the temperature of the low-pressure gas leaving the system is at 300 K. Assume that all equipment are well insulated.

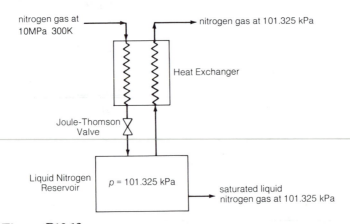

nitrogen gas at 10MPa 300K

nitrogen gas at 101.325 kPa

Heat Exchanger

Joule-Thomson Valve

Liquid Nitrogen Reservoir

p = 101.325 kPa

saturated liquid nitrogen gas at 101.325 kPa

Figure P16.12

[11] F. Din, editor, *Thermodynamic Functions of Gases*, Butterworth, London. Vol. 1 & 2 (1956, reprinted 1962), Vol. 3 (1961).

16.13 Same as Problem 16.12 except that the high-pressure nitrogen gas enters the heat exchanger at 20.0 MPa.

16.14 Same as Problem 16.12 except that the low-pressure nitrogen gas leaves the heat exchanger at 290 K.

16.15 Same as Problem 16.12 except that the gas is oxygen.

16.16 Same as Problem 16.12 except that the gas is oxygen and that the high-pressure oxygen enters the heat exchanger at 20.0 MPa.

16.17 In Problem 16.12, determine
    (a) the temperature of high-pressure fluid leaving the heat exchanger.
    (b) the thermodynamic lost work in the heat exchanger, in kilojoules per kilogram of gas entering the heat exchanger.
    (c) the thermodynamic lost work in the Joule–Thomson valve, in kilojoules per kilogram of gas entering the heat exchanger.
Neglect pressure drops. Let the temperature of the environment be 300 K.

16.18 In Problem 16.13, determine
    (a) the temperature of high-pressure fluid leaving the heat exchanger.
    (b) the amount of exergy destruction in the heat exchanger, in kilojoules per kilogram of gas entering the heat exchanger.
    (c) the amount of exergy destruction in the Joule–Thomson valve, in kilojoules per kilogram of gas entering the heat exchanger.
Neglect pressure drops. Let the temperature of the environment be 300 K.

16.19 In Problem 16.14, determine
    (a) the temperature of high-pressure fluid leaving the heat exchanger.
    (b) the amount of irreversibility in the heat exchanger, in kilojoules per kilogram of gas entering the heat exchanger.
    (c) the amount of irreversibility in the Joule–Thomson valve, in kilojoules per kilogram of gas entering the heat exchanger.
Neglect pressure drops. Let the temperature of the environment be 300 K.

[11] F. Din, editor, *Thermodynamic Functions of Gases*, Butterworth, London. Vol. 1 & 2 (1956, reprinted 1962), Vol. 3 (1961).

16.20 In Problem 16.15, determine
  (a) the temperature of high-pressure fluid leaving the heat exchanger.
  (b) the thermodynamic lost work in the heat exchanger, in kilojoules per kilogram of gas entering the heat exchanger.
  (c) the thermodynamic lost work in the Joule–Thomson valve, in kilojoules per kilogram of gas entering the heat exchanger.
  Neglect pressure drops. Let the temperature of the environment be 300 K.

16.21 In Problem 16.16, determine
  (a) the temperature of high-pressure fluid leaving the heat exchanger.
  (b) the thermodynamic lost work in the heat exchanger, in kilojoules per kilogram of gas entering the heat exchanger.
  (c) the thermodynamic lost work in the Joule–Thomson valve, in kilojoules per kilogram of gas entering the heat exchanger.

16.22 A simple Linde–Hampson gas-liquefaction system (Fig. 16.4) is modified by replacing the Joule–Thomson valve with an adiabatic expander. Show that the liquid yield is given by the following expression:

$$Y = \frac{m_f}{m} = \frac{h_1 - h_2}{h_1 - h_f} + \frac{h_3 - h_4}{h_1 - h_f},$$

where h_3 is the enthalpy of the high-pressure fluid entering expander and h_4 is the enthalpy of fluid leaving the expander.

16.23 Nitrogen gas at 80°F and 1 atm is to be liquefied to saturated liquid at 1 atm using an ideal Linde–Hampson system. Determine the liquid yield for a compressor discharge pressure of 2000, 3000, 4000, 5000, 6000, 7000, and 8000 psia. What conclusion can one make from these results?

16.24 Oxygen gas at 80°F and 1 atm is to be liquefied using an ideal simple Linde–Hampson system. Determine the liquid yield for a compressor discharge pressure of 1000, 2000, 3000, 4000, and 5000 psia. What conclusion can one make from these results?

16.25 Plot enthalpy against pressure for the isotherm of 80°F for nitrogen gas in the pressure range 1000 to 8000 psia. Making use of this chart, determine the compressor discharge pressure of an ideal simple Linde–Hampson liquefaction system for maximum liquid yield with nitrogen coming in at 1 atm and 80°F.

16.26 Plot enthalpy against pressure for the isotherm of 80°F for oxygen gas in the pressure range 1000 to 5000 psia. Making use of this chart, can we determine the discharge pressure of an ideal simple Linde–Hampson liquefaction system for maximum liquid yield with oxygen coming in at 1 atm and 80°F?

16.27 The effect of heat leaking into any liquefaction system is to reduce the liquid yield. Show that the liquid yield for a simple Linde–Hampson liquefaction system including heat leak is given by

$$Y = \frac{h_1 - h_2}{h_1 - h_f} - \frac{q}{h_1 - h_f},$$

where q is the quantity of heat leaking into the system per unit mass of gas compressed.

16.28 Nitrogen gas at 300 K and 101.325 kPa is to be liquefied to saturated liquid at 101.325 kPa using an ideal Linde–Hampson system. Determine the liquid yield for a compressor discharge pressure of 10.0, 15.0, 20.0, 30.0, 40.0, 50.0, and 60.0 MPa. What conclusions can one draw from these results?

16.29 Plot enthalpy against pressure for the isotherm of 300 K for nitrogen gas in the pressure range 10.0 to 60.0 MPa. Making use of this chart, determine the compressor discharge pressure of an ideal simple Linde–Hampson liquefaction system for maximum liquid yield with nitrogen gas coming in at 101.325 kPa and 300 K.

16.30 Oxygen gas at 101.325 kPa and 300 K is to be liquefied to saturated liquid at 101.325 kPa using an ideal Linde–Hampson system. Determine the liquid yield for a compressor discharge pressure of 10.0, 15.0, 20.0, 30.0, 40.0, 50.0, and 60.0 MPa. What conclusions can one draw from these results?

16.31 Plot enthalpy against pressure for the isotherm of 300 K for oxygen gas in the pressure range 10.0 to 60.0 MPa. Making use of this chart, determine the compressor discharge pressure of an ideal simple Linde–Hampson liquefaction system for maximum liquid yield with oxygen coming in at 101.325 kPa and 300 K.

16.32 Nitrogen gas at 101.325 kPa and 300 K is to be liquefied to saturated liquid at 101.325 kPa using an ideal Linde–Hampson system with a compressor discharge pressure of 40.0 MPa. Determine the second-law efficiency of the process.

16.33 Same as Problem 16.32 except that the compressor discharge pressure is 50.0 MPa.

16.34 Same as Problem 16.32 except that the compressor discharge pressure is 30.0 MPa.

16.35 Oxygen gas at 101.325 kPa and 300 K is to be liquefied to saturated liquid at 101.325 kPa using an ideal Linde–Hampson system with a compressor discharge pressure of 40.0 MPa. Determine the second-law efficiency of the process.

16.36 Same as Problem 16.35 except that the compressor discharge pressure is 50.0 MPa.

16.37 Same as Problem 16.35 except that the compressor discharge pressure is 30.0 MPa.

16.38 An ideal-gas mixture consists of 45% hydrogen, 30% methane, 12% nitrogen, 6% carbon monoxide, 3% carbon dioxide, and 4% oxygen by mole at 20°C and 1 atm. For complete separation with an isothermal process, what is the minimum work required for 100 g mol of mixture? Give the answer in SI units.

16.39 An ideal-gas mixture consists of 60% methane gas and 40% nitrogen by mole at 101.325 kPa and 300 K. It is desired to separate this stream into two streams, with one stream having all nitrogen and another stream having a composition of 90% by mole of methane and 10% by mole of nitrogen. For an isothermal process, what is the minimum work required, in kilojoules, to separate 1.0 kg mol of mixture?

16.40 Fresh water is to be obtained from seawater using a reverse-osmosis system with an adiabatic pump efficiency of 70%. If we must pump 2000 gal to produce 1000 gal of fresh water, determine
   (a) the energy needed, in kWh, to produce 1000 gal of fresh water.
   (b) the energy cost to produce 1000 gal of fresh water if the electricity cost is 3 cents per kWh.

16.41 In a simple vapor-compression distillation system, the compression ratio r_p is closed to unity in practice. Making use of this fact, show that Eq. 16.39a may be simplified to yield the following approximate relation:

$$W_{\min} \approx -mRT_{\text{in}}(r_p - 1),$$

indicating that the minimum compressor work increases linearly with the compression ratio for given inlet conditions. [*Hint:* Expand the quantity $r_p^{(k-1)/k}$ in Eq. 16.39a using the binomial law.]

16.42 A simple vapor-compression distillation system is to be designed to operate according to the following conditions:

Pressure of steam at compressor inlet   = 1 atm

Temperature of steam at compressor inlet = 101°C

Compression ratio   = 1.2

Determine the minimum compressor work required, in Btu and in SI units, to produce 1000 gal of fresh water using
(a) Eq. 16.39a.
(b) the approximate relation given in Problem 16.41.
(c) real-gas data from superheated steam tables.

16.43 Determine the chemical exergy of hydrogen at 25°C and 101.325 kPa. Compare your results with the lower heating value and the higher heating value of this fuel.

16.44 A natural gas consists of 75% by mole of methane (CH_4) and 25% by mole of ethane (C_2H_6). Calculate the chemical exergy of the mixture at 25°C and 101.325 kPa, in kJ/kg mol of mixture.

16.45 An ideal-gas mixture consists of 75% by mole of ethane (C_2H_6) and 25% of carbon dioxide. Calculate the chemical exergy of the mixture at 25°C and 101.325 kPa, in kJ/kg mol of mixture.

16.46 Write a computer program to study the effect of compressor compression ratio on the work required to produce 1000 gals of fresh water at 20°C for a vapor-compression distillation system, assuming steam is an ideal gas with constant specific heats. (See Example 16.7.)
Run the program according to the following specifications:
1. Temperature of steam at compressor inlet = 101°C
2. Adiabatic compressor efficiency = 80%
3. Compressor compression ratio: 1.2, 1.4, 1.6, 1.8, 2.0.
Plot the results.

16.47 Write a computer program to determine the chemical exergy of a hydrocarbon fuel C_aH_b. (See Example 16.8.)

Run the program to determine the chemical exergy of the following gaseous fuels:
(a) C_2H_2.
(b) C_2H_4.
(c) C_2H_6.
(d) C_3H_8.
(e) C_8H_{18}.
Tabulate the results and compare them with the higher and lower heating values of the fuels.

Bibliography

A. General References on Thermodynamics

Bacon, D. H. *Engineering Thermodynamics*. Butterworths, London, 1972.

Balzhiser, R. E., and Samuels, M. R. *Engineering Thermodynamics*. Prentice-Hall, Englewood Cliffs, New Jersey, 1977.

Black, W. Z. and Hartley, J. G. *Thermodynamics*. Harper and Row, New York, 1985.

Benson, R. S. *Advanced Engineering Thermodynamics*. 2nd ed. Pergamon Press, Oxford, 1977.

Burghardt, M. D. *Engineering Thermodynamics with Applications*. 3rd ed. Harper and Row, New York, 1986.

Callen, H. B. *Thermodynamics*. John Wiley & Sons, New York, 1960.

Cravalho, E. G. and Smith, J. L., Jr. *Engineering Thermodynamics*. Pitman, Boston, 1981.

Faires, V. M. and Simmang, C. M. *Thermodynamics*. 6th ed. Macmillan, New York, 1978.

Guggenheim, E. A. *Thermodynamics*. 4th ed. North-Holland Publishing Company, Amsterdam, 1959.

Gibbs, J. W. *Collected Works*. Vol. 1. Yale University Press, New Haven, Conn., 1948.

Haberman, W. L. and John, J. E. A. *Engineering Thermodynamics*. Allyn and Bacon, Boston, 1980.

Hall, N. A. and Ibele, W. E. *Engineering Thermodynamics*. Prentice-Hall, Englewood Cliffs, New Jersey, 1960.

Hatsopoulos, G. N. and Keenan, J. H. *Principles of General Thermodynamics*. John Wiley & Sons, New York, 1965.

Haywood, R. W. *Equilibrium Thermodynamics for Engineers and Scientists*. John Wiley & Sons, New York, 1980.

Holman, J. P. *Thermodynamics*. 3rd ed. McGraw-Hill, New York, 1980.

Howell, J. R. and Buckius, R. O. *Fundamentals of Engineering Thermodynamics*. SI Version. McGraw-Hill, New York, 1987.

Hsieh, J. S. *Principles of Thermodynamics*. McGraw-Hill, New York, 1975.

Jones, J. B. and Hawkins, G. A. *Engineering Thermodynamics*. 2nd ed. John Wiley & Sons, New York, 1986.

Karlekar, B. V. *Thermodynamics for Engineers*. Prentice-Hall, Englewood Cliffs, New Jersey, 1983.

Keenan, J. H. *Thermodynamics*. MIT Press, Cambridge, Mass., 1970.

Kestin, J. *A Course in Thermodynamics*. Blaisdell, Waltham, Mass., 1966.

Klotz, I. and Rosenberg, R. M. *Chemical Thermodynamics*. 4th ed. Benjamin/Cummings, Menlo Park, Calif. 1986.

Lay, J. E. *Thermodynamics*. Merrill, Columbus, Ohio, 1963.

Lee, J. F. and Sears, F. W. *Thermodynamics*. 2nd ed. Addison Wesley, Reading, Mass., 1963.

Lewis, G. N. and Randall, M. *Thermodynamics*. 2nd ed. (revised by K. S. Pitzer and L. Brewer), McGraw-Hill, New York, 1961.

Look, D. C., Jr. and Sauer, H. S., Jr. *Thermodynamics*. Brooks/Cole, Monterey, Calif. 1982.

Obert, E. F. *Concepts of Thermodynamics*. McGraw-Hill, New York, 1960.

Obert, E. F. and Gaggioli, R. A. *Thermodynamics*. McGraw-Hill, New York, 1963.

Rolle, K. C. *Introduction to Thermodynamics*. 2nd ed. Merrill, Columbus, Ohio, 1980.

Silver, H. F. and Nydahl, J. E. *Introduction to Engineering Thermodynamics*. West, St. Paul, Minn., 1977.

Smith, J. M. and Van Ness, H. C. *Introduction to Chemical Engineering Thermodynamics*. 4th ed. McGraw-Hill, New York, 1987.

Soo, S. L. *Thermodynamics of Engineering Science*. Prentice-Hall, Englewood Cliffs, New Jersey, 1958.

Tribus, M. *Thermostatics and Thermodynamics*. Van Nostrand, New York, 1961.

Van Wylen, G. J. and Sonntag, R. E. *Fundamentals of Classical Thermodynamics*. 3rd ed. (SI Version), John Wiley & Sons, New York, 1985.

Wark, K. *Thermodynamics*. 4th ed. McGraw-Hill, New York, 1983.

Wood, B. D. *Applications of Thermodynamics*. 2nd ed. Addison-Wesley, Reading, Mass., 1982.

Zemansky, M. W., Abbott, M. M., and Van Ness, H. C. *Basic Engineering Thermodynamics*. 2nd ed. McGraw-Hill, New York, 1975.

B. References on Exergy and Second-Law Analysis

Ahern, J. E. *The Exergy Method of Energy Systems Analysis*. John Wiley & Sons, New York, 1980.

Bejan, Adrian. *Entropy Generation through Heat and Fluid Flow*. John Wiley & Sons, New York, 1982.

Bejan, Adrian, editor. *Second Law Aspects of Thermal Design*. ASME, New York, 1984.

Bruges, E. A. *Available Energy and the Second Law Analysis*. Academic Press, New York, 1959.

de Nevers, N. "Two Fundamental Approaches to Second-Law Analysis," in *Foundations of Computer-Aided Chemical Process Design*. Vol. II (R. Mah and W. Seider, eds.). AIChE, New York, pp. 501–536, 1981.

de Nevers, N. and Seader, J. D. "Lost Work: A Measure of Thermodynamic Efficiency," *ENERGY*, Vol. 5, pp. 757–769, 1980.

Didion, D., Garvin, D. and Snell, J. *A Report on the Relevance of the Second Law of Thermodynamics to Energy Conservation*. NBS Technical Note 1115, 1980.

Evans, R. B. *A Proof that Essergy is the only consistent measure of Potential Work*. PhD thesis, Dartmouth, 1969.

Gaggioli, R. A., ed. *Thermodynamics: Second Law Analysis*. ACS Symposium Series 122, American Chemical Society, Washington, D.C. 1980.

Gaggioli, R. A., ed. *Efficiency and Costing: Second Law Analysis of Processes*. ACS Symposium Series 235. American Chemical Society, Washington, D.C. 1983.

Kotas, T. J. *The Exergy Method of Thermal Plant Analysis*. Butterworths, London, 1985.

London, A. L. "Economics and the Second Law: An Engineering View and Methodology." *International Journal of Heat and Mass Transfer*, Vol. 25, pp. 743–751, 1982.

London, A. L. and Shah, R. "Cost of Irreversibilities in Heat Exchange Design," *Heat Transfer Engineering*, Vol. 4, No. 2, pp. 59–73, 1983.

Moran, M. J. *Availability Analysis: A Guide to Efficient Energy Use*. Prentice-Hall, Englewood Cliffs, New Jersey, 1982.

Reisted, G. M. *Availability: Concepts and Applications*. PhD thesis, University of Wisconsin-Madison, 1970.

Sussman, M. V. *Availability (Exergy) Analysis ... A Self Instruction Manual*. Mulliken House, Lexington, Mass., 1980.

Szargut, J. "International Progress in Second Law Analysis." *ENERGY*, Vol. 5, pp. 709–718, 1980.

Wepfer, W. J. *Application of the Second Law to the Analysis and Design of Energy Systems*. PhD thesis, University of Wisconsin-Madison, 1979. (This thesis has a comprehensive list of references compiled from literature throughout the world, including very early work as well as current work.)

APPENDIX

859

Table A.1.1 (SI) Properties of Saturated Water and Saturated Steam (Temperature)*

Temp. °C	Press. kPa	Volume, m³/kg			Enthalpy, kJ/kg			Entropy, kJ/kg·K		
		Water v_f	Evap. v_{fg}	Steam v_g	Water h_f	Evap. h_{fg}	Steam h_g	Water s_f	Evap. s_{fg}	Steam s_g
0.01	0.6112	0.0010002	206.16	206.16	0.00	2501.6	2501.6	0.0000	9.1575	9.1575
5	0.8718	0.0010000	147.16	147.16	21.01	2489.7	2510.7	0.0762	8.9507	9.0269
10	1.2270	0.0010003	106.43	106.43	41.99	2477.9	2519.9	0.1510	8.7510	8.9020
15	1.7039	0.0010008	77.98	77.98	62.94	2466.1	2529.0	0.2244	8.5583	8.7827
20	2.337	0.0010017	57.84	57.84	83.86	2454.3	2538.2	0.2963	8.3721	8.6684
25	3.166	0.0010029	43.40	43.40	104.77	2442.5	2547.2	0.3672	8.1923	8.5591
30	4.241	0.0010043	32.93	32.93	125.66	2430.7	2556.4	0.4365	8.0181	8.4546
35	5.622	0.0010060	25.25	25.25	146.56	2418.8	2565.3	0.5049	7.8494	8.3543
40	7.375	0.0010078	19.545	19.546	167.45	2406.9	2574.4	0.5721	7.6862	8.2583
45	9.582	0.0010099	15.280	15.281	188.35	2394.9	2583.3	0.6383	7.5278	8.1661
50	12.335	0.0010121	12.045	12.046	209.26	2382.9	2592.2	0.7035	7.3741	8.0776
55	15.741	0.0010145	9.578	9.579	230.17	2370.8	2601.0	0.7677	7.2249	7.9926
60	19.920	0.0010171	7.678	7.679	251.09	2358.6	2609.7	0.8310	7.0798	7.9108
65	25.009	0.0010199	6.201	6.202	272.02	2346.3	2618.4	0.8933	6.9389	7.8322
70	31.16	0.0010228	5.045	5.046	292.97	2334.0	2626.9	0.9548	6.8017	7.7565
75	38.549	0.0010259	4.133	4.134	313.94	2321.5	2635.4	1.0154	6.6681	7.6835
80	47.36	0.0010292	3.408	3.409	334.92	2308.8	2643.8	1.0753	6.5379	7.6132
85	57.81	0.0010326	2.828	2.829	355.92	2296.1	2652.0	1.1343	6.4111	7.5454
90	70.11	0.0010361	2.3603	2.3613	376.94	2283.2	2660.1	1.1925	6.2874	7.4799
95	84.52	0.0010399	1.9810	1.9820	397.99	2270.1	2668.1	1.2501	6.1665	7.4166
100	101.33	0.0010437	1.6720	1.6730	419.06	2256.9	2676.0	1.3069	6.0485	7.3554
105	120.80	0.0010477	1.4182	1.4193	440.17	2243.6	2683.7	1.3630	5.9332	7.2962
110	143.27	0.0010519	1.2089	1.2099	461.32	2230.0	2691.3	1.4185	5.8203	7.2388
115	169.06	0.0010562	1.0352	1.0363	482.50	2216.2	2698.7	1.4733	5.7099	7.1832
120	198.54	0.0010606	0.8905	0.8915	503.72	2202.3	2706.0	1.5276	5.6017	7.1293
125	232.1	0.0010652	0.7692	0.7702	524.99	2188.0	2713.0	1.5813	5.4956	7.0769
130	270.1	0.0010700	0.6671	0.6681	546.31	2173.6	2719.9	1.6344	5.3917	7.0261
135	313.1	0.0010750	0.5807	0.5818	567.68	2158.9	2726.6	1.6869	5.2897	6.9766
140	361.4	0.0010801	0.5074	0.5085	589.10	2144.0	2733.1	1.7390	5.1894	6.9284
145	415.5	0.0010853	0.4449	0.4460	610.59	2128.7	2739.3	1.7906	5.0909	6.8815
150	476.0	0.0010908	0.3914	0.3924	632.15	2113.3	2745.4	1.8416	4.9942	6.8358

155	543.3	0.0010964	0.3453	0.3464	653.77	2097.4	2751.2	1.8923	4.8988	6.7911
160	618.1	0.0011022	0.3057	0.3068	675.47	2081.2	2756.7	1.9425	4.8050	6.7475
165	700.8	0.0011082	0.2713	0.2724	697.25	2064.8	2762.0	1.9923	4.7125	6.7048
170	792.0	0.0011145	0.2414	0.2426	719.12	2048.0	2767.1	2.0416	4.6214	6.6630
175	892.4	0.0011209	0.21542	0.21654	741.07	2030.7	2771.8	2.0906	4.5315	6.6221
180	1002.7	0.0011275	0.19267	0.19380	763.12	2013.2	2776.3	2.1393	4.4426	6.5819
185	1123.3	0.0011344	0.17272	0.17386	785.26	1995.2	2780.4	2.1876	4.3548	6.5424
190	1255.1	0.0011415	0.15517	0.15632	807.52	1976.7	2784.3	2.2356	4.2680	6.5036
195	1398.7	0.0011489	0.13969	0.14084	829.88	1957.9	2787.8	2.2833	4.1821	6.4654
200	1554.9	0.0011565	0.12600	0.12716	852.37	1938.5	2790.9	2.3307	4.0971	6.4278
205	1724.3	0.0011644	0.11386	0.11503	874.99	1918.8	2793.8	2.3778	4.0128	6.3906
210	1907.7	0.0011726	0.10307	0.10424	897.73	1898.5	2796.2	2.4247	3.9292	6.3539
215	2106.0	0.0011811	0.09344	0.09463	920.63	1877.6	2798.3	2.4713	3.8463	6.3176
220	2319.8	0.0011900	0.08485	0.08604	943.67	1856.2	2799.9	2.5178	3.7639	6.2817
225	2550	0.0011992	0.07715	0.07835	966.88	1834.3	2801.2	2.5641	3.6820	6.2461
230	2798	0.0012087	0.07024	0.07145	990.27	1811.7	2802.0	2.6102	3.6005	6.2107
235	3063	0.0012187	0.06403	0.06525	1013.83	1788.5	2802.3	2.6561	3.5195	6.1756
240	3348	0.0012291	0.05843	0.05965	1037.60	1764.6	2802.2	2.7020	3.4386	6.1406
245	3652	0.0012399	0.05337	0.05461	1061.58	1740.0	2801.6	2.7478	3.3579	6.1057
250	3978	0.0012513	0.04879	0.05004	1085.78	1714.6	2800.4	2.7935	3.2773	6.0708
255	4325	0.0012632	0.04463	0.04590	1110.23	1688.5	2798.7	2.8392	3.1967	6.0359
260	4694	0.0012756	0.04086	0.04213	1134.94	1661.5	2796.4	2.8848	3.1162	6.0010
270	5506	0.0013025	0.03429	0.03559	1185.23	1604.6	2789.9	2.9763	2.9541	5.9304
280	6420	0.0013324	0.02879	0.03013	1236.84	1543.6	2780.4	3.0683	2.7903	5.8586
290	7446	0.0013659	0.02417	0.02554	1290.01	1477.6	2767.6	3.1611	2.6237	5.7848
300	8593	0.0014041	0.020245	0.021649	1345.05	1406.0	2751.0	3.2552	2.4529	5.7081
310	9870	0.0014480	0.016886	0.018334	1402.39	1327.6	2730.0	3.3512	2.2766	5.6278
320	11289	0.0014995	0.013980	0.015480	1462.60	1241.1	2703.7	3.4500	2.0923	5.5423
330	12863	0.0015615	0.011428	0.012989	1526.52	1143.7	2670.2	3.5528	1.8962	5.4490
340	14605	0.0016387	0.009142	0.010780	1595.47	1030.7	2626.2	3.6616	1.6811	5.3427
350	16535	0.0017411	0.007058	0.008799	1671.94	895.8	2567.7	3.7800	1.4377	5.2177
360	18675	0.0018959	0.005044	0.006940	1764.17	721.2	2485.4	3.9210	1.1390	5.0600
370	21054	0.0022136	0.002759	0.004973	1890.21	452.6	2342.8	4.1108	0.7036	4.8144
374.15	22120	0.00317	0.0	0.00317	2107.37	0.0	2107.37	4.4429	0.0	4.4429

* Abstracted with permission from *ASME Steam Tables in SI (Metric) Units* (Copyright 1977, The American Society of Mechanical Engineers.)

Table A.1.2 (SI) Properties of Saturated Water and Saturated Steam (Pressure)*

Press. kPa	Temp. °C	Volume, m³/kg			Enthalpy, kJ/kg			Entropy, kJ/kg · K			Energy, kJ/kg	
		Water v_f	Evap. v_{fg}	Steam v_g	Water h_f	Evap. h_{fg}	Steam h_g	Water s_f	Evap. s_{fg}	Steam s_g	Water u_f	Steam u_g
0.6112	0.01	0.0010002	206.16	206.16	0.00	2501.6	2501.6	0.0000	9.1575	9.1575	0	2375.6
1.0	6.983	0.0010001	129.21	129.21	29.34	2485.0	2514.4	0.1060	8.8706	8.9767	29.33	2385.2
1.5	13.036	0.0010006	87.98	87.98	54.71	2470.7	2525.5	0.1957	8.6332	8.8288	54.71	2393.5
2.0	17.513	0.0010012	67.01	67.01	73.46	2460.2	2533.6	0.2607	8.4639	8.7246	73.46	2399.6
2.5	21.10	0.0010020	54.30	54.30	88.45	2451.7	2540.2	0.3119	8.3321	8.6440	88.45	2404.5
3.0	24.100	0.0010027	45.67	45.67	101.00	2444.6	2545.6	0.3544	8.2241	8.5785	101.00	2408.6
4.0	28.983	0.0010040	34.80	34.80	121.41	2433.1	2554.5	0.4225	8.0530	8.4755	121.41	2415.3
5.0	32.898	0.0010052	28.19	28.19	137.77	2423.8	2561.6	0.4763	7.9197	8.3960	137.77	2420.6
7.5	40.316	0.0010079	19.238	19.239	168.77	2406.2	2574.9	0.5763	7.6760	8.2523	168.76	2430.6
10	45.833	0.0010102	14.674	14.675	191.83	2392.9	2584.8	0.6493	7.5018	8.1511	191.82	2438.0
15	53.997	0.0010140	10.022	10.023	225.97	2373.2	2599.2	0.7549	7.2544	8.0093	225.96	2448.9
20	60.086	0.0010172	7.649	7.650	251.45	2358.4	2609.9	0.8321	7.0774	7.9094	251.43	2456.9
25	64.99	0.0010199	6.203	6.204	271.99	2346.4	2618.3	0.8932	6.9391	7.8323	271.96	2463.2
30	69.124	0.0010223	5.228	5.229	289.30	2336.1	2625.4	0.9441	6.8254	7.7695	289.27	2468.6
40	75.886	0.0010265	3.992	3.993	317.65	2319.2	2636.9	1.0261	6.6448	7.6709	317.61	2477.1
50	81.345	0.0010301	3.239	3.240	340.56	2305.4	2646.0	1.0912	6.5035	7.5947	340.51	2484.0
75	91.785	0.0010375	2.2158	2.2169	384.45	2278.6	2663.0	1.2131	6.2439	7.4570	384.37	2496.7
100	99.632	0.0010434	1.6927	1.6937	417.51	2257.9	2675.4	1.3027	6.0571	7.3598	417.41	2506.1
150	111.37	0.0010530	1.1580	1.1590	467.13	2226.2	2693.4	1.4336	5.7898	7.2234	466.97	2519.5
200	120.23	0.0010608	0.8844	0.8854	504.70	2201.6	2706.3	1.5301	5.5967	7.1268	504.49	2529.2
250	127.44	0.0010670	0.7176	0.7187	535.25	2181.3	2716.5	1.6072	5.4452	7.0524	534.98	2536.8
300	133.54	0.0010735	0.6045	0.6056	561.4	2163.2	2724.7	1.6716	5.3193	6.9909	561.11	2543.0
350	138.87	0.0010789	0.5229	0.5240	584.3	2147.4	2731.6	1.7273	5.2119	6.9392	583.89	2548.2
400	143.62	0.0010839	0.4611	0.4622	604.7	2133.0	2737.6	1.7764	5.1179	6.8943	604.24	2552.7
450	147.92	0.0010885	0.4127	0.4138	623.2	2119.7	2742.9	1.8204	5.0343	6.8547	622.67	2556.7
500	151.84	0.0010928	0.3736	0.3747	640.1	2107.4	2747.5	1.8604	4.9588	6.8192	639.57	2560.2
550	155.47	0.0010969	0.3414	0.3425	655.8	2095.9	2751.7	1.8970	4.8900	6.7870	655.20	2563.3
600	158.84	0.0011009	0.3144	0.3155	670.4	2085.0	2755.5	1.9308	4.8267	6.7575	669.76	2566.2
650	161.99	0.0011046	0.29138	0.29249	684.1	2074.7	2758.9	1.9623	4.7681	6.7304	683.42	2568.7
700	164.96	0.0011082	0.27157	0.27268	697.1	2064.9	2762.0	1.9918	4.7134	6.7052	696.29	2571.1
750	167.76	0.0011116	0.25431	0.25543	709.3	2055.5	2764.8	2.0195	4.6621	6.6817	708.47	2573.3
800	170.41	0.0011150	0.23914	0.24026	720.9	2046.5	2767.5	2.0457	4.6139	6.6596	720.04	2575.3

900	175.36	0.0011213	0.21369	0.21481	742.6	2029.5	2772.1	2.0941	4.5250	6.6192	741.63	2578.8
1000	179.88	0.0011274	0.19317	0.19429	762.6	2013.6	2776.2	2.1382	4.4446	6.5828	761.48	2581.9
1100	184.07	0.0011331	0.17625	0.17738	781.1	1998.5	2779.7	2.1786	4.3711	6.5497	779.88	2584.5
1200	187.96	0.0011386	0.16206	0.16320	798.4	1984.3	2782.7	2.2161	4.3033	6.5194	797.06	2586.9
1300	191.61	0.0011438	0.14998	0.15113	814.7	1970.7	2785.4	2.2510	4.2403	6.4913	813.21	2589.0
1400	195.04	0.0011489	0.13957	0.14072	830.1	1957.7	2787.8	2.2837	4.1814	6.4651	828.47	2590.8
1500	198.29	0.0011539	0.13050	0.13166	844.7	1945.2	2789.9	2.3145	4.1261	6.4406	842.93	2592.4
1600	201.37	0.0011586	0.12253	0.12369	858.6	1933.2	2791.7	2.3436	4.0739	6.4175	856.71	2593.8
1800	207.11	0.0011678	0.10915	0.11032	884.6	1910.3	2794.8	2.3976	3.9775	6.3751	882.47	2596.3
2000	212.37	0.0011766	0.09836	0.09954	908.6	1888.6	2797.2	2.4469	3.8898	6.3367	906.24	2598.2
2200	217.24	0.0011850	0.08947	0.09065	931.0	1868.1	2799.1	2.4922	3.8093	6.3015	928.35	2599.6
2400	221.78	0.0011932	0.08201	0.08320	951.9	1848.5	2800.4	2.5343	3.7347	6.2690	949.07	2600.7
2600	226.04	0.0012011	0.07565	0.07686	971.7	1829.6	2801.4	2.5736	3.6651	6.2387	968.60	2601.5
2800	230.05	0.0012088	0.07018	0.07139	990.5	1811.5	2802.0	2.6106	3.5998	6.2104	987.10	2602.1
3000	233.84	0.0012163	0.06541	0.06663	1008.4	1793.9	2802.3	2.6455	3.5382	6.1837	1004.70	2602.4
3500	242.54	0.0012345	0.05579	0.05703	1049.8	1752.2	2802.0	2.7253	3.3976	6.1228	1045.44	2602.4
4000	250.33	0.0012521	0.04850	0.04975	1087.4	1712.9	2800.3	2.7965	3.2720	6.0685	1082.4	2601.3
5000	263.91	0.0012858	0.03814	0.03943	1154.5	1639.7	2794.2	2.9206	3.0529	5.9735	1148.0	2597.0
6000	275.55	0.0013187	0.03112	0.03244	1213.7	1571.3	2785.0	3.0273	2.8635	5.8908	1205.8	2590.4
7000	285.79	0.0013513	0.026022	0.027373	1267.4	1506.0	2773.5	3.1219	2.6943	5.8162	1258.0	2581.8
8000	294.97	0.0013842	0.022141	0.023525	1317.1	1442.8	2759.9	3.2076	2.5395	5.7471	1306.0	2571.7
9000	303.31	0.0014179	0.019078	0.020495	1363.7	1380.9	2744.6	3.2867	2.3953	5.6820	1351.0	2560.1
10000	310.96	0.0014526	0.016589	0.018041	1408.0	1319.7	2727.7	3.3605	2.2593	5.6198	1393.5	2547.3
11000	318.05	0.0014887	0.014517	0.016006	1450.6	1258.7	2709.3	3.4304	2.1291	5.5595	1434.2	2533.2
12000	324.65	0.0015268	0.012756	0.014283	1491.8	1197.4	2689.2	3.4972	2.0030	5.5002	1473.4	2517.8
13000	330.83	0.0015672	0.011230	0.012797	1532.0	1135.0	2667.0	3.5616	1.8792	5.4408	1511.6	2500.6
14000	336.64	0.0016106	0.009884	0.011495	1571.6	1070.7	2642.4	3.6242	1.7560	5.3803	1549.1	2481.4
15000	342.13	0.0016579	0.008682	0.010340	1611.0	1004.0	2615.0	3.6859	1.6320	5.3178	1586.1	2459.9
16000	347.33	0.0017103	0.007597	0.009308	1650.5	934.3	2584.9	3.7471	1.5060	5.2531	1623.2	2436.0
17000	352.26	0.0017696	0.006601	0.008371	1691.7	859.9	2551.6	3.8107	1.3748	5.1855	1661.6	2409.3
18000	356.96	0.0018399	0.005658	0.007498	1734.8	779.1	2513.9	3.8765	1.2362	5.1128	1701.7	2378.9
19000	361.43	0.0019260	0.004751	0.006678	1778.7	692.0	2470.6	3.9429	1.0903	5.0332	1742.1	2343.8
20000	365.70	0.0020370	0.003840	0.005877	1826.5	591.9	2418.4	4.0149	0.9263	4.9412	1785.7	2300.8
21000	369.78	0.0022015	0.002822	0.005023	1886.3	461.3	2347.6	4.1048	0.7175	4.8223	1840.0	2242.1
22000	373.69	0.0026714	0.001056	0.003728	2011.1	184.5	2195.6	4.2947	0.2852	4.5799	1952.4	2113.6
22120	374.15	0.00317	0.0	0.00317	2107.4	0.0	2107.4	4.4429	0.0	4.4429	2037.3	2037.3

Table A.1.3 (SI) Properties of Superheated Steam* v in m³/kg, h in kJ/kg, s in kJ/kg · K

Abs. Press. kPa (Sat. Temp.)°C		50	100	150	200	250	300	350	400	450	500	550	600	650	700	750	800
5.0 (32.90)	v	29.782	34.417	39.042	43.661	48.280	52.897	57.513	62.129	66.745	71.360	75.976	80.592	85.207	89.822	94.438	99.053
	h	2593.7	2688.1	2784.3	2879.9	2977.6	3076.7	3177.4	3279.7	3383.6	3489.2	3596.5	3705.6	3816.3	3928.8	4043.0	4158.7
	s	8.4981	8.7698	9.0094	9.2248	9.4211	9.6021	9.7705	9.9283	10.0772	10.2184	10.3529	10.4815	10.6049	10.7235	10.8379	10.9483
10.0 (45.83)	v	14.869	17.195	19.512	21.825	24.136	26.445	28.754	31.062	33.370	35.679	37.987	40.295	42.603	44.910	47.218	49.526
	h	2592.7	2687.5	2783.1	2879.6	2977.5	3076.6	3177.3	3279.6	3383.5	3489.1	3596.5	3705.5	3816.3	3928.8	4042.9	4158.7
	s	8.1757	8.4486	8.6889	8.9045	9.1010	9.2820	9.4505	9.6083	9.7572	9.8984	10.0329	10.1616	10.2849	10.4036	10.5180	10.6284
20.0 (60.09)	v		8.5847	9.748	10.907	12.064	13.210	14.374	15.529	16.684	17.838	18.992	20.146	21.300	22.455	23.609	24.762
	h		2686.3	2782.4	2879.2	2977.2	3076.4	3177.2	3279.4	3383.4	3489.0	3596.4	3705.4	3816.2	3928.7	4042.9	4158.7
	s		8.1261	8.3676	8.5839	8.7806	8.9618	9.1304	9.2882	9.4372	9.5784	9.7130	9.8416	9.9650	10.0836	10.1980	10.3085
40.0 (75.89)	v		4.2792	4.8657	5.4478	6.0277	6.6065	7.1846	7.7625	8.3401	8.9176	9.4950	10.072	10.649	11.227	11.804	12.381
	h		2683.8	2780.9	2878.2	2976.5	3075.9	3176.8	3279.1	3383.1	3488.8	3596.2	3705.3	3816.1	3928.6	4042.8	4158.6
	s		7.8009	8.0449	8.2625	8.4598	8.6413	8.8101	8.9680	9.1171	9.2583	9.3929	9.5216	9.6450	9.7636	9.8780	9.9885
60.0 (85.95)	v		2.8440	3.2382	3.6281	4.0157	4.4022	4.7881	5.1736	5.5590	5.9441	6.3292	6.7141	7.0991	7.4839	7.8687	8.2535
	h		2681.3	2779.4	2877.3	2975.8	3075.4	3176.4	3278.8	3382.8	3488.6	3596.0	3705.1	3816.0	3928.5	4042.7	4158.5
	s		7.6085	7.8551	8.0738	8.2718	8.4536	8.6225	8.7806	8.9297	9.0710	9.2056	9.3343	9.4577	9.5764	9.6908	9.8013
80 (93.51)	v		2.1262	2.4245	2.7183	3.0097	3.3000	3.5898	3.8792	4.1683	4.4574	4.7463	5.0351	5.3239	5.6126	5.9013	6.1899
	h		2678.8	2777.8	2876.3	2975.2	3075.0	3176.0	3278.5	3382.6	3488.4	3595.8	3705.0	3815.8	3928.4	4042.6	4158.4
	s		7.4703	7.7195	7.9396	8.1381	8.3202	8.4893	8.6475	8.7967	8.9380	9.0727	9.2014	9.3248	9.4436	9.5580	9.6685
100.0 (99.63)	v		1.6955	1.9362	2.1723	2.4061	2.6387	2.8708	3.1025	3.3340	3.5653	3.7966	4.0277	4.2588	4.4898	4.7208	4.9517
	h		2676.3	2776.3	2875.4	2974.5	3074.5	3175.6	3278.2	3382.4	3488.1	3595.6	3704.8	3815.7	3928.2	4042.5	4158.3
	s		7.3618	7.6137	7.8349	8.0342	8.2166	8.3859	8.5442	8.6935	8.8347	8.9696	9.0982	9.2217	9.3405	9.4549	9.5654
150.0 (111.4)	v			1.2851	1.4444	1.6013	1.7570	1.9122	2.0669	2.2215	2.3759	2.5303	2.6845	2.8386	2.9927	3.1468	3.3008
	h			2772.5	2872.9	2972.9	3073.3	3174.7	3277.5	3381.8	3487.6	3595.2	3704.4	3815.3	3927.9	4042.2	4158.0
	s			7.4194	7.6439	7.8447	8.0280	8.1976	8.3562	8.5057	8.6472	8.7820	8.9108	9.0343	9.1531	9.2676	9.3781
200.0 (120.2)	v			0.9595	1.0804	1.1989	1.3162	1.4328	1.5492	1.6653	1.7812	1.8971	2.0129	2.1286	2.2442	2.3598	2.4754
	h			2768.5	2870.5	2971.2	3072.1	3173.8	3276.7	3381.1	3487.0	3594.7	3704.0	3815.0	3927.6	4041.9	4157.8
	s			7.2794	7.5072	7.7096	7.8937	8.0638	8.2226	8.3722	8.5139	8.6488	8.7776	8.9012	9.0201	9.1346	9.2452
300.0 (133.5)	v			0.6337	0.7164	0.7964	0.8753	0.9535	1.0314	1.1090	1.1865	1.2639	1.3412	1.4185	1.4957	1.5728	1.6499
	h			2760.3	2865.5	2967.9	3069.7	3171.9	3275.2	3379.8	3486.0	3593.7	3703.2	3814.2	3927.0	4041.4	4157.3
	s			7.0771	7.3119	7.5176	7.7034	7.8745	8.0338	8.1838	8.3257	8.4608	8.5898	8.7135	8.8325	8.9471	9.0577
400.0 (143.6)	v			0.4707	0.5343	0.5962	0.6549	0.7139	0.7725	0.8309	0.8892	0.9474	1.0054	1.0634	1.1214	1.1793	1.2372
	h			2752.0	2860.4	2964.5	3067.2	3170.0	3273.6	3378.5	3484.9	3592.8	3702.3	3813.5	3926.4	4040.8	4156.9
	s			6.9285	7.1708	7.3800	7.5675	7.7395	7.8994	8.0497	8.1919	8.3271	8.4563	8.5802	8.6992	8.8139	8.9246
500.0 (151.8)	v				0.4250	0.4744	0.5226	0.5701	0.6172	0.6640	0.7108	0.7574	0.8039	0.8504	0.8968	0.9432	0.9896
	h				2855.1	2961.1	3064.8	3168.1	3272.1	3377.2	3483.8	3591.8	3701.5	3812.8	3925.8	4040.3	4156.4
	s				7.0592	7.2721	7.4614	7.6343	7.7948	7.9454	8.0879	8.2233	8.3526	8.4766	8.5957	8.7105	8.8213
600.0 (158.8)	v				0.3520	0.3939	0.4344	0.4742	0.5136	0.5528	0.5918	0.6308	0.6696	0.7084	0.7471	0.7858	0.8245
	h				2849.7	2957.6	3062.3	3166.2	3270.6	3376.0	3482.7	3590.9	3700.7	3812.1	3925.1	4039.8	4155.9
	s				6.9662	7.1829	7.3740	7.5479	7.7090	7.8600	8.0027	8.1383	8.2678	8.3919	8.5111	8.6259	8.7368
800 (170.4)	v				0.2608	0.2932	0.3241	0.3543	0.3842	0.4137	0.4432	0.4725	0.5017	0.5309	0.5600	0.5891	0.6181
	h				2838.6	2950.4	3057.3	3162.4	3267.5	3373.4	3480.5	3589.0	3699.1	3810.7	3923.9	4038.7	4155.0
	s				6.8148	7.0397	7.2348	7.4107	7.5729	7.7246	7.8678	8.0038	8.1336	8.2579	8.3773	8.4923	8.6033

p (kPa) (Tsat, °C)															
1000.0 (179.9)	v	0.2059	0.2327	0.2580	0.2824	0.3065	0.3303	0.3540	0.3775	0.4010	0.4244	0.4477	0.4710	0.4943
	h	2826.8	2942.9	3052.1	3158.5	3264.4	3370.8	3478.3	3587.1	3697.4	3809.3	3922.7	4037.6	4154.1
	s	6.6922	6.9259	7.1251	7.3031	7.4665	7.6190	7.7627	7.8991	8.0292	8.1537	8.2734	8.3885	8.4997
1500 (198.3)	v	0.1324	0.1520	0.1697	0.1865	0.2029	0.2191	0.2350	0.2509	0.2667	0.2824	0.2980	0.3136	0.3292
	h	2794.7	2923.3	3038.9	3148.7	3256.6	3364.4	3472.8	3582.4	3693.3	3805.7	3919.6	4034.9	4151.7
	s	6.4508	6.7099	6.9207	7.1044	7.2709	7.4253	7.5703	7.7077	7.8385	7.9636	8.0838	8.1993	8.3108
2000 (212.4)	v		0.1114	0.1255	0.1386	0.1511	0.1634	0.1756	0.1876	0.1995	0.2114	0.2232	0.2349	0.2467
	h		2902.4	3025.0	3138.6	3248.7	3357.8	3467.3	3577.7	3689.2	3802.1	3916.5	4032.2	4149.4
	s		6.5454	6.7696	6.9596	7.1296	7.2859	7.4323	7.5706	7.7022	7.8279	7.9485	8.0645	8.1763
3000 (233.8)	v		0.07050	0.08116	0.09063	0.09931	0.1078	0.1161	0.1243	0.1323	0.1404	0.1483	0.1562	0.1641
	h		2854.0	2995.1	3117.5	3232.5	3344.6	3456.2	3568.1	3681.0	3795.0	3910.3	4026.8	4144.7
	s		6.2857	6.5422	6.7471	6.9246	7.0854	7.2345	7.3748	7.5079	7.6349	7.7564	7.8733	7.9857
4000.0 (250.3)	v			0.05883	0.06643	0.07338	0.07996	0.08634	0.09260	0.09876	0.1049	0.1109	0.1169	0.1229
	h			2962.0	3095.1	3215.7	3331.2	3445.0	3558.6	3672.8	3787.9	3904.1	4021.4	4140.0
	s			6.3642	6.5870	6.7733	6.9388	7.0909	7.2333	7.3680	7.4961	7.6187	7.7363	7.8495
5000.0 (263.9)	v			0.04530	0.05193	0.05779	0.06325	0.06849	0.07360	0.07862	0.08356	0.08845	0.09329	0.09809
	h			2925.5	3071.2	3198.3	3317.5	3433.7	3549.0	3664.5	3780.7	3897.9	4016.1	4135.3
	s			6.2105	6.4545	6.6508	6.8217	6.9770	7.1215	7.2578	7.3872	7.5108	7.6292	7.7431
6000.0 (275.5)	v			0.03614	0.04221	0.04738	0.05210	0.05659	0.06094	0.06518	0.06936	0.07348	0.07755	0.08159
	h			2885.0	3045.8	3180.1	3303.5	3422.2	3539.3	3656.2	3773.5	3891.7	4010.7	4130.7
	s			6.0692	6.3386	6.5462	6.7230	6.8818	7.0285	7.1664	7.2971	7.4217	7.5409	7.6554
8000.0 (295.0)	v			0.02426	0.02995	0.03431	0.03814	0.04170	0.04510	0.04839	0.05161	0.05407	0.05788	0.06096
	h			2786.8	2989.9	3141.6	3274.3	3398.8	3519.7	3639.5	3759.2	3879.2	3999.9	4121.3
	s			5.7942	6.1349	6.3694	6.5597	6.7262	6.8778	7.0191	7.1523	7.2790	7.3999	7.5158
10000.0 (311.0)	v				0.02242	0.02641	0.02974	0.03276	0.03560	0.03832	0.04096	0.04355	0.04608	0.04858
	h				2925.8	3099.9	3243.6	3374.6	3499.8	3622.7	3744.7	3866.8	3989.1	4112.0
	s				5.9489	6.2182	6.4243	6.5994	6.7564	6.9013	7.0373	7.1660	7.2886	7.4058
15000.0 (342.1)	v				0.01147	0.01566	0.01845	0.02080	0.02291	0.02488	0.02677	0.02859	0.03036	0.03209
	h				2696.0	2979.1	3159.8	3310.6	3450.6	3579.8	3708.3	3835.4	3962.1	4088.6
	s				5.4486	5.8876	6.1469	6.3487	6.5243	6.6764	6.8195	6.9536	7.0806	7.2013
20000.0 (365.7)	v					0.009947	0.01271	0.01477	0.01655	0.01816	0.01967	0.02111	0.02250	0.02385
	h					2820.5	3064.3	3241.1	3394.1	3535.5	3671.1	3803.8	3935.0	4065.3
	s					5.5585	5.9089	6.1456	6.3374	6.5043	6.6554	6.7953	6.9267	7.0511
30000.0	v					0.002831	0.006735	0.008681	0.01017	0.01144	0.01258	0.01365	0.01465	0.01562
	h					2161.8	2825.6	3085.0	3277.4	3443.0	3595.0	3739.7	3880.3	4018.5
	s					4.4896	5.4495	5.7972	6.0386	6.2340	6.4033	6.5560	6.6970	6.8288
40000.0	v					0.001909	0.003675	0.005616	0.006982	0.008088	0.009053	0.009930	0.01075	0.01152
	h					1934.1	2515.6	2906.8	3151.6	3346.4	3517.0	3674.8	3825.5	3971.7
	s					4.1190	4.9611	5.4762	5.7835	6.0135	6.2035	6.3701	6.5210	6.6606
50000.0	v					0.001729	0.002492	0.003882	0.005113	0.006111	0.006960	0.007720	0.008420	0.009076
	h					1877.7	2293.2	2723.0	3021.1	3248.3	3438.9	3610.2	3770.9	3925.3
	s					4.0083	4.6026	5.1782	5.5525	5.8207	6.0331	6.2138	6.3749	6.5222

* Abstracted with permission from *ASME Steam Tables in SI (Metric) Units* (Copyright 1977, The American Society of Mechanical Engineers.)

Table A.1.4 (SI) Properties of Compressed Water* v in m³/kg, h in kJ/kg, s in kJ/kg · K

Press. MPa (Sat. Temp.) °C		20	40	80	120	160	200	240	280	320	360
						Temperature, °C					
1 (179.9)	v	0.0010013	0.0010074	0.0010287	0.0010602	0.0011019					
	h	84.8	168.3	335.7	504.3	675.7					
	s	0.2961	0.5717	1.0746	1.5269	1.9420					
2 (212.4)	v	0.0010008	0.0010069	0.0010282	0.0010596	0.0011012					
	h	85.7	169.2	336.5	505.0	676.3					
	s	0.2959	0.5713	1.0740	1.5260	1.9408					
3 (233.8)	v	0.0010004	0.0010065	0.0010278	0.0010590	0.0011005	0.0011550				
	h	86.7	170.1	337.3	505.7	676.9	853.0				
	s	0.2957	0.5709	1.0733	1.5251	1.9396	2.3284				
4 (250.3)	v	0.0009999	0.0010060	0.0010273	0.0010584	0.0010997	0.0011540	0.0012280			
	h	87.6	171.0	338.1	506.4	677.5	853.4	1037.7			
	s	0.2955	0.5706	1.0726	1.5242	1.9385	2.3268	2.7006			
5 (263.9)	v	0.0009995	0.0010056	0.0010268	0.0010579	0.0010990	0.0011530	0.0012264			
	h	88.6	171.9	338.8	507.1	678.1	853.8	1037.8			
	s	0.2952	0.5702	1.0720	1.5233	1.9373	2.3253	2.6984			

10 (311.0)	v	0.0009972	0.0010034	0.0010245	0.0010551	0.0010954	0.0011480	0.0012188	0.0013221		
	h	93.2	176.3	342.8	510.6	681.0	855.9	1038.4	1235.0		
	s	0.2942	0.5682	1.0687	1.5188	1.9315	2.3176	2.6877	3.0563		
15 (342.1)	v	0.0009950	0.0010013	0.0010221	0.0010523	0.0010919	0.0011433	0.0012115	0.0013090	0.0014736	
	h	97.9	180.7	346.8	514.2	684.0	858.1	1039.2	1232.9	1454.3	
	s	0.2931	0.5663	1.0655	1.5144	1.9258	2.3102	2.6775	3.0407	3.4267	
20 (365.7)	v	0.0009929	0.0009992	0.0010199	0.0010497	0.0010886	0.0011387	0.0012047	0.0012971	0.0014451	0.0018269
	h	102.5	185.1	350.8	517.7	687.1	860.4	1040.3	1231.4	1445.6	1742.9
	s	0.2919	0.5643	1.0623	1.5101	1.9203	2.3030	2.6677	3.0262	3.3998	3.8835
30	v	0.0009886	0.0009951	0.0010155	0.0010445	0.0010821	0.0011301	0.0011922	0.0012763	0.0014012	0.0016285
	h	111.7	193.8	358.7	524.9	693.3	865.2	1042.8	1229.7	1433.6	1678.0
	s	0.2895	0.5604	1.0560	1.5017	1.9095	2.2891	2.6492	2.9998	3.3556	3.7541
40	v	0.0009845	0.0009910	0.0010112	0.0010395	0.0010760	0.0011220	0.0011808	0.0012583	0.0013677	0.0015425
	h	120.8	202.5	366.7	532.1	699.6	870.2	1045.8	1229.2	1425.9	1650.5
	s	0.2870	0.5565	1.0498	1.4935	1.8991	2.2758	2.6320	2.9761	3.3193	3.6856
50	v	0.0009804	0.0009872	0.0010071	0.0010347	0.0010701	0.0011144	0.0011703	0.0012426	0.0013406	0.0014862
	h	129.9	211.2	374.7	539.4	705.9	875.4	1049.2	1229.8	1421.0	1633.9
	s	0.2843	0.5525	1.0438	1.4856	1.8890	2.2632	2.6158	2.9545	3.2882	3.6355

* Abstracted with permission from *ASME Steam Tables in SI (Metric) Units* (Copyright 1977, The American Society of Mechanical Engineers.)

Table A.1.1 (E) Properties of Saturated Water and Saturated Steam (Temperature)*

Temp., °F	Press., psia	Volume, ft³/lbm			Enthalpy, Btu/lbm			Entropy, Btu/lbm-°R			Temp., °F
		Water v_f	Evap. v_{fg}	Steam v_g	Water h_f	Evap. h_{fg}	Steam h_g	Water s_f	Evap. s_{fg}	Steam s_g	
32	0.08859	0.01602	3305	3305	−0.02	1075.5	1075.5	0.0000	2.1873	2.1873	32
35	0.09991	0.01602	2948	2948	3.00	1073.8	1076.8	0.0061	2.1706	2.1767	35
40	0.12163	0.01602	2446	2446	8.03	1071.0	1079.0	0.0162	2.1432	2.1594	40
45	0.14744	0.01602	2037.7	2037.8	13.04	1068.1	1081.2	0.0262	2.1164	2.1426	45
50	0.17796	0.01602	1704.8	1704.8	18.05	1065.3	1083.4	0.0361	2.0901	2.1262	50
60	0.2561	0.01603	1207.6	1207.6	28.06	1059.7	1087.7	0.0555	2.0391	2.0946	60
70	0.3629	0.01605	868.3	868.4	38.05	1054.0	1092.1	0.0745	1.9900	2.0645	70
80	0.5068	0.01607	633.3	633.3	48.04	1048.4	1096.4	0.0932	1.9426	2.0359	80
90	0.6981	0.01610	468.1	468.1	58.02	1042.7	1100.8	0.1115	1.8970	2.0086	90
100	0.9492	0.01613	350.4	350.4	68.00	1037.1	1105.1	0.1295	1.8530	1.9825	100
110	1.2750	0.01617	265.4	265.4	77.98	1031.4	1109.3	0.1472	1.8105	1.9577	110
120	1.6927	0.01620	203.25	203.26	87.97	1025.6	1113.6	0.1646	1.7693	1.9339	120
130	2.2230	0.01625	157.32	157.33	97.96	1019.8	1117.8	0.1817	1.7295	1.9112	130
140	2.8892	0.01629	122.98	123.00	107.95	1014.0	1122.0	0.1985	1.6910	1.8895	140
150	3.718	0.01634	97.05	97.07	117.95	1008.2	1126.1	0.2150	1.6536	1.8686	150
160	4.741	0.01640	77.27	77.29	127.96	1002.2	1130.2	0.2313	1.6174	1.8487	160
170	5.993	0.01645	62.04	62.06	137.97	996.2	1134.2	0.2473	1.5822	1.8295	170
180	7.511	0.01651	50.21	50.22	148.00	990.2	1138.2	0.2631	1.5480	1.8111	180
190	9.340	0.01657	40.94	40.96	158.04	984.1	1142.1	0.2787	1.5148	1.7934	190
200	11.526	0.01664	33.62	33.64	168.09	977.9	1146.0	0.2940	1.4824	1.7764	200
210	14.123	0.01671	27.80	27.82	178.15	971.6	1149.7	0.3091	1.4509	1.7600	210
212	14.696	0.01672	26.78	26.80	180.17	970.3	1150.5	0.3121	1.4447	1.7568	212
220	17.186	0.01678	23.13	23.15	188.23	965.2	1153.4	0.3241	1.4201	1.7442	220
230	20.779	0.01685	19.364	19.381	198.33	958.7	1157.1	0.3388	1.3902	1.7290	230
240	24.968	0.01693	16.304	16.321	208.45	952.1	1160.6	0.3533	1.3609	1.7142	240
250	29.825	0.01701	13.802	13.819	218.59	945.4	1164.0	0.3677	1.3323	1.7000	250

Temp (°F)	Press (psia)	v_f	v_{fg}	v_g	h_f	h_{fg}	h_g	s_f	s_{fg}	s_g	Temp (°F)
260	35.427	0.01709	11.745	11.762	228.76	938.6	1167.4	0.3819	1.3043	1.6862	260
270	41.856	0.01718	10.042	10.060	238.95	931.7	1170.6	0.3960	1.2769	1.6729	270
280	49.200	0.01726	8.627	8.644	249.17	924.6	1173.8	0.4098	1.2501	1.6599	280
290	57.550	0.01736	7.443	7.460	259.4	917.4	1176.8	0.4236	1.2238	1.6473	290
300	67.005	0.01745	6.448	6.466	269.7	910.0	1179.7	0.4372	1.1979	1.6351	300
310	77.67	0.01755	5.609	5.626	280.0	902.5	1182.5	0.4506	1.1726	1.6232	310
320	89.64	0.01766	4.896	4.914	290.4	894.8	1185.2	0.4640	1.1477	1.6116	320
340	117.99	0.01787	3.770	3.788	311.3	878.8	1190.1	0.4902	1.0990	1.5892	340
360	153.01	0.01811	2.939	2.957	332.3	862.1	1194.4	0.5161	1.0517	1.5678	360
380	195.73	0.01836	2.317	2.335	353.6	844.5	1198.0	0.5416	1.0057	1.5473	380
400	247.26	0.01864	1.8444	1.8630	375.1	825.9	1201.0	0.5667	0.9607	1.5274	400
420	308.78	0.01894	1.4808	1.4997	396.9	806.2	1203.1	0.5915	0.9165	1.5080	420
440	381.54	0.01926	1.1976	1.2169	419.0	785.4	1204.4	0.6161	0.8729	1.4890	440
460	466.9	0.0196	0.9746	0.9942	441.5	763.2	1204.8	0.6405	0.8299	1.4704	460
480	566.2	0.0200	0.7972	0.8172	464.5	739.6	1204.1	0.6648	0.7871	1.4518	480
500	680.9	0.0204	0.6545	0.6749	487.9	714.3	1202.2	0.6890	0.7443	1.4333	500
520	812.5	0.0209	0.5386	0.5596	512.0	687.0	1199.0	0.7133	0.7013	1.4146	520
540	962.8	0.0215	0.4437	0.4651	536.8	657.5	1194.3	0.7378	0.6577	1.3954	540
560	1133.4	0.0221	0.3651	0.3871	562.4	625.3	1187.7	0.7625	0.6132	1.3757	560
580	1326.2	0.0228	0.2994	0.3222	589.1	589.9	1179.0	0.7876	0.5673	1.3550	580
600	1543.2	0.0236	0.2438	0.2675	617.1	550.6	1167.7	0.8134	0.5196	1.3330	600
620	1786.9	0.0247	0.1962	0.2208	646.9	506.3	1153.2	0.8403	0.4689	1.3092	620
640	2059.9	0.0260	0.1543	0.1802	679.1	454.6	1133.7	0.8686	0.4134	1.2821	640
660	2365.7	0.0277	0.1166	0.1443	714.9	392.1	1107.0	0.8995	0.3502	1.2498	660
680	2708.6	0.0304	0.0808	0.1112	758.5	310.1	1068.5	0.9365	0.2720	1.2086	680
700	3094.3	0.0366	0.0386	0.0752	822.4	172.7	995.2	0.9901	0.1490	1.1390	700
705.5	3208.2	0.0508	0	0.0508	906.0	0	906.0	1.0612	0	1.0612	705.5

*Abstracted with permission from *Thermodynamics and Transport Properties of Steam*. (Copyright, 1967, by The American Society of Mechanical Engineers.)

Table A.1.2 (E) Properties of Saturated Water and Saturated Steam (Pressure)*

Press., psia	Temp., °F	Volume, ft³/lbm Water v_f	Evap. v_{fg}	Steam v_g	Enthalpy, Btu/lbm Water h_f	Evap. h_{fg}	Steam h_g	Entropy, Btu/lbm-°R Water s_f	Evap. s_{fg}	Steam s_g	Energy, Btu/lbm Water u_f	Steam u_g	Press., psia
0.0886	32.018	0.01602	3302.4	3302.4	0.00	1075.5	1075.5	0	2.1872	2.1872	0	1021.3	0.0886
0.10	35.023	0.01602	2945.5	2945.5	3.03	1073.8	1076.8	0.0061	2.1705	2.1766	3.03	1022.3	0.10
0.15	45.453	0.01602	2004.7	2004.7	13.50	1067.9	1081.4	0.0271	2.1140	2.1411	13.50	1025.7	0.15
0.20	53.160	0.01603	1526.3	1526.3	21.22	1063.5	1084.7	0.0422	2.0738	2.1160	21.22	1028.3	0.20
0.30	64.484	0.01604	1039.7	1039.7	32.54	1057.1	1089.7	0.0641	2.0168	2.0809	32.54	1032.0	0.30
0.40	72.869	0.01606	792.0	792.1	40.92	1052.4	1093.3	0.0799	1.9762	2.0562	40.92	1034.7	0.40
0.5	79.586	0.01607	641.5	641.5	47.62	1048.6	1096.3	0.0925	1.9446	2.0370	47.62	1036.9	0.5
0.6	85.218	0.01609	540.0	540.1	53.25	1045.5	1098.7	0.1028	1.9186	2.0215	53.24	1038.7	0.6
0.7	90.09	0.01610	466.93	466.94	58.10	1042.7	1100.8	0.1117	1.8966	2.0083	58.10	1040.3	0.7
0.8	94.38	0.01611	411.67	411.69	62.39	1040.3	1102.6	0.1195	1.8775	1.9970	62.39	1041.7	0.8
0.9	98.24	0.01612	368.41	368.43	66.24	1038.1	1104.3	0.1264	1.8606	1.9870	66.24	1042.9	0.9
1.0	101.74	0.01614	333.59	333.60	69.73	1036.1	1105.8	0.1326	1.8455	1.9781	69.73	1044.1	1.0
2.0	126.07	0.01623	173.74	173.76	94.03	1022.1	1116.2	0.1750	1.7450	1.9200	94.03	1051.8	2.0
3.0	141.47	0.01630	118.71	118.73	109.42	1013.2	1122.6	0.2009	1.6854	1.8864	109.41	1056.7	3.0
4.0	152.96	0.01636	90.63	90.64	120.92	1006.4	1127.3	0.2199	1.6428	1.8626	120.90	1060.2	4.0
5.0	162.24	0.01641	73.515	73.53	130.20	1000.9	1131.1	0.2349	1.6094	1.8443	130.18	1063.1	5.0
6.0	170.05	0.01645	61.967	61.98	138.03	996.2	1134.2	0.2474	1.5820	1.8294	138.01	1065.4	6.0
7.0	176.84	0.01649	53.634	53.65	144.83	992.1	1136.9	0.2581	1.5587	1.8168	144.81	1067.4	7.0
8.0	182.86	0.01653	47.328	47.35	150.87	988.5	1139.3	0.2676	1.5384	1.8060	150.84	1069.2	8.0
9.0	188.27	0.01656	42.385	42.40	156.30	985.1	1141.4	0.2760	1.5204	1.7964	156.28	1070.8	9.0
10	193.21	0.01659	38.404	38.42	161.26	982.1	1143.3	0.2836	1.5043	1.7879	161.23	1072.3	10
14.696	212.00	0.01672	26.782	26.80	180.17	970.3	1150.5	0.3121	1.4447	1.7568	180.12	1077.6	14.696
15	213.03	0.01673	26.274	26.29	181.21	969.7	1150.9	0.3137	1.4415	1.7552	181.16	1077.9	15
20	227.96	0.01683	20.070	20.087	196.27	960.1	1156.3	0.3358	1.3962	1.7320	196.21	1082.0	20
30	250.34	0.01701	13.7266	13.744	218.9	945.2	1164.1	0.3682	1.3313	1.6995	218.8	1087.9	30
40	267.25	0.01715	10.4794	10.497	236.1	933.6	1169.8	0.3921	1.2844	1.6765	236.0	1092.1	40
50	281.02	0.01727	8.4967	8.514	250.2	923.9	1174.1	0.4112	1.2474	1.6586	250.1	1095.3	50

60	292.71	0.01738	7.1562	7.174	262.2	915.4	1177.6	0.4273	1.2167	1.6440	262.0	1098.0	60
70	302.93	0.01748	6.1875	6.205	272.7	907.8	1180.6	0.4411	1.1905	1.6316	272.5	1100.2	70
80	312.04	0.01757	5.4536	5.471	282.1	900.9	1183.1	0.4534	1.1675	1.6208	281.9	1102.1	80
90	320.28	0.01766	4.8777	4.895	290.7	894.6	1185.3	0.4643	1.1470	1.6113	290.4	1103.7	90
100	327.82	0.01774	4.4133	4.431	298.5	888.6	1187.2	0.4743	1.1284	1.6027	298.2	1105.2	100
120	341.27	0.01789	3.7097	3.728	312.6	877.8	1190.4	0.4919	1.0960	1.5879	312.2	1107.6	120
140	353.04	0.01803	3.2010	3.219	325.0	868.0	1193.0	0.5071	1.0681	1.5752	324.5	1109.6	140
160	363.55	0.01815	2.8155	2.834	336.1	859.0	1195.1	0.5206	1.0435	1.5641	335.5	1111.2	160
180	373.08	0.01827	2.5129	2.531	346.2	850.7	1196.9	0.5328	1.0215	1.5543	345.6	1112.5	180
200	381.80	0.01839	2.2689	2.287	355.5	842.8	1198.3	0.5438	1.0016	1.5454	354.8	1113.7	200
250	400.97	0.01865	1.8245	1.8432	376.1	825.0	1201.1	0.5679	0.9585	1.5264	375.3	1115.8	250
300	417.35	0.01889	1.5238	1.5427	394.0	808.9	1202.9	0.5882	0.9223	1.5105	392.9	1117.2	300
350	431.73	0.01913	1.3064	1.3255	409.8	794.2	1204.0	0.6059	0.8909	1.4968	408.6	1118.1	350
400	444.60	0.0193	1.14162	1.1610	424.2	780.4	1204.6	0.6217	0.8630	1.4847	422.7	1118.7	400
450	456.28	0.0195	1.01224	1.0318	437.3	767.5	1204.8	0.6360	0.8378	1.4738	435.7	1118.9	450
500	467.01	0.0198	0.90787	0.9276	449.5	755.1	1204.7	0.6490	0.8148	1.4639	447.7	1118.8	500
550	476.94	0.0199	0.82183	0.8418	460.9	743.3	1204.3	0.6611	0.7936	1.4547	458.9	1118.6	550
600	486.20	0.0201	0.74962	0.7698	471.7	732.0	1203.7	0.6723	0.7738	1.4461	469.5	1118.2	600
700	503.08	0.0205	0.63505	0.6556	491.6	710.2	1201.8	0.6928	0.7377	1.4304	488.9	1116.9	700
800	518.21	0.0209	0.54809	0.5690	509.8	689.6	1199.4	0.7111	0.7051	1.4163	506.7	1115.2	800
900	531.95	0.0212	0.47968	0.5009	526.7	669.7	1196.4	0.7279	0.6753	1.4032	523.2	1113.0	900
1000	544.58	0.0216	0.42436	0.4460	542.6	650.4	1192.9	0.7434	0.6476	1.3910	538.6	1110.4	1000
1100	556.28	0.0220	0.37863	0.4006	557.5	631.5	1189.1	0.7578	0.6216	1.3794	553.1	1107.5	1100
1200	567.19	0.0223	0.34013	0.3625	571.9	613.0	1184.8	0.7714	0.5969	1.3683	566.9	1104.3	1200
1300	577.42	0.0227	0.30722	0.3299	585.6	594.6	1180.2	0.7843	0.5733	1.3577	580.1	1100.9	1300
1400	587.07	0.0231	0.27871	0.3018	598.8	576.5	1175.3	0.7966	0.5507	1.3474	592.9	1097.1	1400
1500	596.20	0.0235	0.25372	0.2772	611.7	558.4	1170.1	0.8085	0.5288	1.3373	605.2	1093.1	1500
2000	635.80	0.0257	0.16266	0.1883	672.1	466.2	1138.3	0.8625	0.4256	1.2881	662.6	1068.6	2000
2500	668.11	0.0286	0.10209	0.1307	731.7	361.6	1093.3	0.9139	0.3206	1.2345	718.5	1032.9	2500
3000	695.33	0.0343	0.05073	0.0850	801.8	218.4	1020.3	0.9728	0.1891	1.1619	782.8	973.1	3000
3208.2	705.47	0.0508	0	0.0508	906.0	0	906.0	1.0612	0	1.0612	875.9	875.9	3208.2

Table A.1.3 (E) Properties of Superheated Steam*

Abs. press., psia (sat. temp.)		100	200	300	400	500	600	700	800	900	1000	1100	1200	1300	1400	1500
							Temperature, °F									
1 (101.74)	v	...	392.5	452.3	511.9	571.5	631.1	690.7								
	h	...	1150.2	1195.7	1241.8	1288.6	1336.1	1384.5								
	s	...	2.0509	2.1152	2.1722	2.2237	2.2708	2.3144								
5 (162.24)	v	...	78.14	90.24	102.24	114.21	126.15	138.08	150.01	161.94	173.86	185.78	197.70	209.62	221.53	233.45
	h	...	1148.6	1194.8	1241.3	1288.2	1335.9	1384.3	1433.6	1483.7	1534.7	1586.7	1639.6	1693.3	1748.0	1803.5
	s	...	1.8716	1.9369	1.9943	2.0460	2.0932	2.1369	2.1776	2.2159	2.2521	2.2866	2.3194	2.3509	2.3811	2.4101
10 (193.21)	v	...	38.84	44.98	51.03	57.04	63.03	69.00	74.98	80.94	86.91	92.87	98.84	104.80	110.76	116.72
	h	...	1146.6	1193.7	1240.6	1287.8	1335.5	1384.0	1433.4	1483.5	1534.6	1586.6	1639.5	1693.3	1747.9	1803.4
	s	...	1.7928	1.8593	1.9173	1.9692	2.0166	2.0603	2.1011	2.1394	2.1757	2.2101	2.2430	2.2744	2.3046	2.3337
15 (213.03)	v	...	...	29.899	33.963	37.985	41.986	45.978	49.964	53.946	57.926	61.905	65.882	69.858	73.833	77.807
	h	...	...	1192.5	1239.9	1287.3	1335.2	1383.8	1433.2	1483.4	1534.5	1586.5	1639.4	1693.2	1747.8	1803.4
	s	...	...	1.8134	1.8720	1.9242	1.9717	2.0155	2.0563	2.0946	2.1309	2.1653	2.1982	2.2297	2.2599	2.2890
20 (227.96)	v	...	...	22.356	25.428	28.457	31.466	34.465	37.458	40.447	43.435	46.420	49.405	52.388	55.370	58.352
	h	...	...	1191.4	1239.2	1286.9	1334.9	1383.5	1432.9	1483.2	1534.3	1586.3	1639.3	1693.1	1747.8	1803.3
	s	...	...	1.7805	1.8397	1.8921	1.9397	1.9836	2.0244	2.0628	2.0991	2.1336	2.1665	2.1979	2.2282	2.2572
40 (267.25)	v	...	...	11.036	12.624	14.165	15.685	17.195	18.699	20.199	21.697	23.194	24.689	26.183	27.676	29.168
	h	...	...	1186.6	1236.4	1285.0	1333.6	1382.5	1432.1	1482.5	1533.7	1585.8	1638.8	1692.7	1747.5	1803.0
	s	...	...	1.6992	1.7608	1.8143	1.8624	1.9065	1.9476	1.9860	2.0224	2.0569	2.0899	2.1224	2.1516	2.1807
60 (292.71)	v	...	...	7.257	8.354	9.400	10.425	11.438	12.446	13.450	14.452	15.452	16.450	17.448	18.445	19.441
	h	...	...	1181.6	1233.5	1283.2	1332.3	1381.5	1431.3	1481.8	1533.2	1585.3	1638.4	1692.4	1747.1	1802.8
	s	...	...	1.6492	1.7134	1.7681	1.8168	1.8612	1.9024	1.9410	1.9774	2.0120	2.0450	2.0765	2.1068	2.1359
80 (312.04)	v	...	...	...	6.218	7.018	7.794	8.560	9.319	10.075	10.829	11.581	12.331	13.081	13.829	14.577
	h	...	...	...	1230.5	1281.3	1330.5	1380.5	1430.5	1481.1	1532.6	1584.9	1638.0	1692.0	1746.8	1802.5
	s	...	...	...	1.6790	1.7349	1.7842	1.8289	1.8702	1.9089	1.9454	1.9800	2.0131	2.0446	2.0750	2.1041

P (psia) (T sat, °F)																	
100 (327.82)	v	...	...	...	...	4.935	5.588	6.216	6.833	7.443	8.050	8.655	9.258	9.860	10.460	11.060	11.659
	h	...	...	...	...	1227.4	1279.3	1329.6	1379.5	1429.7	1480.4	1532.0	1584.4	1637.6	1691.6	1746.5	1802.2
	s	...	...	...	...	1.6516	1.7088	1.7586	1.8036	1.8451	1.8839	1.9205	1.9552	1.9883	2.0199	2.0502	2.0794
120 (341.27)	v	...	...	...	...	4.0786	4.6341	5.1637	5.6831	6.1928	6.7006	7.2060	7.7096	8.2119	8.7130	9.2134	9.7130
	h	...	...	...	...	1224.1	1277.4	1328.2	1378.4	1428.8	1479.8	1531.4	1583.9	1637.1	1691.3	1746.2	1802.0
	s	...	...	...	...	1.6286	1.6872	1.7376	1.7829	1.8246	1.8635	1.9001	1.9349	1.9680	1.9996	2.0300	2.0592
140 (353.04)	v	...	...	...	...	3.4661	3.9526	4.4119	4.8585	5.2995	5.7364	6.1709	6.6036	7.0349	7.4652	7.8946	8.3233
	h	...	...	...	...	1220.8	1275.3	1326.8	1377.4	1428.0	1479.1	1530.8	1583.4	1636.7	1690.9	1745.9	1801.7
	s	...	...	...	...	1.6085	1.6686	1.7196	1.7652	1.8071	1.8461	1.8828	1.9176	1.9508	1.9825	2.0129	2.0421
160 (363.55)	v	...	...	...	...	3.0060	3.4413	3.8480	4.2420	4.6295	5.0132	5.3945	5.7741	6.1522	6.5293	6.9055	7.2811
	h	...	...	...	...	1217.4	1273.3	1325.4	1376.4	1427.2	1478.4	1530.3	1582.9	1636.3	1690.5	1745.6	1801.4
	s	...	...	...	...	1.5906	1.6522	1.7039	1.7499	1.7919	1.8310	1.8678	1.9027	1.9359	1.9676	1.9980	2.0273
180 (373.08)	v	...	...	...	...	2.6474	3.0433	3.4093	3.7621	4.1084	4.4508	4.7907	5.1289	5.4657	5.8014	6.1363	6.4704
	h	...	...	...	...	1213.8	1271.2	1324.0	1375.3	1426.3	1477.7	1529.7	1582.4	1635.9	1690.2	1745.3	1801.2
	s	...	...	...	...	1.5743	1.6376	1.6900	1.7362	1.7784	1.8176	1.8545	1.8894	1.9227	1.9545	1.9849	2.0142
200 (381.80)	v	...	...	...	...	2.3598	2.7247	3.0583	3.3783	3.6915	4.0008	4.3077	4.6128	4.9165	5.2191	5.5209	5.8219
	h	...	...	...	...	1210.1	1269.0	1322.6	1374.3	1425.5	1477.0	1529.1	1581.9	1635.4	1689.8	1745.0	1800.9
	s	...	...	...	...	1.5593	1.6242	1.6773	1.7239	1.7663	1.8057	1.8426	1.8776	1.9109	1.9427	1.9732	2.0025
250 (400.97)	v	...	...	...	...	...	2.1504	2.4662	2.6872	2.9410	3.1909	3.4382	3.6837	3.9278	4.1709	4.4131	4.6546
	h	...	...	...	...	...	1263.5	1319.0	1371.6	1423.4	1475.3	1527.6	1580.6	1634.4	1688.9	1744.2	1800.2
	s	...	...	...	...	...	1.5951	1.6502	1.6976	1.7405	1.7801	1.8173	1.8524	1.8858	1.9177	1.9482	1.9776
300 (417.35)	v	...	...	...	...	...	1.7665	2.0044	2.2263	2.4407	2.6509	2.8585	3.0643	3.2688	3.4721	3.6746	3.8764
	h	...	...	...	...	...	1257.7	1315.2	1368.9	1421.3	1473.6	1526.2	1579.4	1633.3	1688.0	1743.4	1799.6
	s	...	...	...	...	...	1.5703	1.6274	1.6758	1.7192	1.7591	1.7964	1.8317	1.8652	1.8972	1.9278	1.9572
350 (431.73)	v	...	...	...	...	...	1.4913	1.7028	1.8970	2.0832	2.2652	2.4445	2.6219	2.7980	2.9730	3.1471	3.3205
	h	...	...	...	...	...	1251.5	1311.4	1366.2	1419.2	1471.8	1524.7	1578.2	1632.3	1687.1	1742.6	1798.9
	s	...	...	...	...	...	1.5483	1.6077	1.6571	1.7009	1.7411	1.7787	1.8141	1.8477	1.8798	1.9105	1.9400

*Abstracted with permission from *Thermodynamic and Transport Properties of Steam.* (Copyright, 1967, by The American Society of Mechanical Engineers.)

(continued)

Table A.1.3 (E) (continued)

Abs. press., psia (sat. temp.)		100	200	300	400	500	600	700	800	900	1000	1100	1200	1300	1400	1500
								Temperature, °F								
400 (444.60)	v	...	...	...	...	1.2841	1.4763	1.6499	1.8151	1.9759	2.1339	2.2901	2.4450	2.5987	2.7515	2.9037
	h	...	...	...	...	1245.1	1307.4	1363.4	1417.0	1470.1	1523.3	1576.9	1631.2	1686.2	1741.9	1798.2
	s	...	...	...	...	1.5282	1.5901	1.6406	1.6850	1.7255	1.7632	1.7988	1.8325	1.8647	1.8955	1.9250
500 (467.01)	v	...	...	...	...	0.9919	1.1584	1.3037	1.4397	1.5708	1.6992	1.8256	1.9507	2.0746	2.1977	2.3200
	h	...	...	...	...	1231.2	1299.1	1357.7	1412.7	1466.6	1520.3	1574.4	1629.1	1684.4	1740.3	1796.9
	s	...	...	...	...	1.4921	1.5595	1.6123	1.6578	1.6990	1.7371	1.7730	1.8069	1.8393	1.8702	1.8998
600 (486.20)	v	...	...	...	...	0.7944	0.9456	1.0726	1.1892	1.3008	1.4093	1.5160	1.6211	1.7252	1.8284	1.9309
	h	...	...	...	...	1215.9	1290.3	1351.8	1408.3	1463.0	1517.4	1571.9	1627.0	1682.6	1738.8	1795.6
	s	...	...	...	...	1.4590	1.5329	1.5844	1.6351	1.6769	1.7155	1.7517	1.7859	1.8184	1.8494	1.8792
700 (503.08)	v	...	...	...	...	...	0.7928	0.9072	1.0102	1.1078	1.2023	1.2948	1.3858	1.4757	1.5647	1.6530
	h	...	...	...	...	...	1281.0	1345.6	1403.7	1459.4	1514.4	1569.4	1624.8	1680.7	1737.2	1794.3
	s	...	...	...	...	...	1.5090	1.5673	1.6154	1.6580	1.6970	1.7335	1.7679	1.8006	1.8318	1.8617
800 (518.21)	v	...	...	...	...	...	0.6774	0.7828	0.8759	0.9631	1.0470	1.1289	1.2093	1.2885	1.3669	1.4446
	h	...	...	...	...	...	1271.1	1339.3	1399.1	1455.8	1511.4	1566.9	1622.7	1678.9	1735.7	1792.9
	s	...	...	...	...	...	1.4869	1.5484	1.5980	1.6413	1.6807	1.7175	1.7522	1.7851	1.8164	1.8464
900 (531.95)	v	...	...	...	...	...	0.5869	0.6858	0.7713	0.8504	0.9262	0.9998	1.0720	1.1430	1.2131	1.2825
	h	...	...	...	...	...	1260.6	1332.7	1394.4	1452.2	1508.5	1564.4	1620.6	1677.1	1734.1	1791.6
	s	...	...	...	...	...	1.4659	1.5311	1.5822	1.6263	1.6662	1.7033	1.7382	1.7713	1.8028	1.8329
1000 (544.58)	v	...	...	...	...	...	0.5137	0.6080	0.6875	0.7603	0.8295	0.8966	0.9622	1.0266	1.0901	1.1529
	h	...	...	...	...	...	1249.3	1325.9	1389.6	1448.5	1504.4	1561.9	1618.4	1675.3	1732.5	1790.3
	s	...	...	...	...	...	1.4457	1.5149	1.5677	1.6126	1.6530	1.6905	1.7256	1.7589	1.7905	1.8207
1100 (556.28)	v	...	...	...	...	...	0.4531	0.5440	0.6188	0.6865	0.7505	0.8121	0.8723	0.9313	0.9894	1.0468
	h	...	...	...	...	...	1237.3	1318.8	1384.7	1444.7	1502.4	1559.4	1616.3	1673.5	1731.0	1789.0
	s	...	...	...	...	...	1.4259	1.4996	1.5542	1.6000	1.6410	1.6787	1.7141	1.7475	1.7793	1.8097

(continued)

1200	v	0.4016	0.4905	0.5615	0.6250	0.6845	0.7418	0.7974	0.8519	0.9055	0.9584
(567.19)	h	1224.2	1311.5	1379.7	1440.9	1499.4	1556.9	1614.2	1671.6	1729.4	1787.6
	s	1.4061	1.4851	1.5415	1.5883	1.6298	1.6679	1.7035	1.7371	1.7691	1.7996
1400	v	0.3176	0.4059	0.4712	0.5282	0.5809	0.6311	0.6798	0.7272	0.7737	0.8195
(587.07)	h	1194.1	1296.1	1369.3	1433.2	1493.2	1551.8	1609.9	1668.0	1726.3	1785.0
	s	1.3652	1.4575	1.5182	1.5670	1.6096	1.6484	1.6845	1.7185	1.7508	1.7815
1600	v	…	0.3415	0.4032	0.4555	0.5031	0.5482	0.5915	0.6336	0.6748	0.7153
(604.87)	h	…	1279.4	1358.5	1425.2	1486.9	1546.6	1605.6	1664.3	1723.2	1782.3
	s	…	1.4312	1.4968	1.5478	1.5916	1.6312	1.6678	1.7022	1.7347	1.7657
1800	v	…	0.2906	0.3500	0.3988	0.4426	0.4836	0.5229	0.5609	0.5980	0.6343
(621.02)	h	…	1261.1	1347.2	1417.1	1480.6	1541.1	1601.2	1660.7	1720.1	1779.7
	s	…	1.4054	1.4768	1.5302	1.5753	1.6156	1.6528	1.6876	1.7204	1.7516
2000	v	…	0.2488	0.3072	0.3534	0.3942	0.4320	0.4680	0.5027	0.5365	0.5695
(635.80)	h	…	1240.9	1335.4	1408.7	1474.1	1536.2	1596.9	1657.0	1717.0	1777.1
	s	…	1.3794	1.4578	1.5138	1.5603	1.6014	1.6391	1.6743	1.7075	1.7389
2500	v	…	0.1681	0.2293	0.2712	0.3068	0.3390	0.3692	0.3980	0.4259	0.4529
(668.11)	h	…	1176.7	1303.4	1386.7	1457.5	1522.9	1585.9	1647.8	1709.2	1770.4
	s	…	1.3076	1.4129	1.4766	1.5269	1.5703	1.6094	1.6456	1.6796	1.7116
3000	v	…	0.0982	0.1759	0.2161	0.2484	0.2770	0.3033	0.3282	0.3522	0.3753
(695.33)	h	…	1060.5	1267.0	1363.2	1440.2	1509.4	1574.8	1638.5	1701.4	1763.8
	s	…	1.1966	1.3692	1.4429	1.4976	1.5434	1.5841	1.6214	1.6561	1.6888
3200	v	…	…	0.1588	0.1987	0.2301	0.2576	0.2827	0.3065	0.3291	0.3510
(705.08)	h	…	…	1250.9	1353.4	1433.1	1503.8	1570.3	1634.8	1698.3	1761.2
	s	…	…	1.3515	1.4300	1.4866	1.5335	1.5749	1.6126	1.6477	1.6806
3500	v	…	…	0.1364	0.1764	0.2066	0.2326	0.2563	0.2784	0.2995	0.3198
	h	…	…	1224.6	1338.2	1422.2	1495.5	1563.3	1629.2	1693.6	1757.2
	s	…	…	1.3242	1.4112	1.4709	1.5194	1.5618	1.6002	1.6358	1.6691

Table A.1.3 (E) (continued)

Abs. press., psia (sat. temp.)		100	200	300	400	500	600	700	Temperature, °F 800	900	1000	1100	1200	1300	1400	1500
4000	v	⋯	⋯	⋯	⋯	⋯	⋯	⋯	0.1052	0.1463	0.1752	0.1994	0.2210	0.2411	0.2601	0.2783
	h	⋯	⋯	⋯	⋯	⋯	⋯	⋯	1174.3	1311.6	1403.6	1481.3	1552.2	1619.8	1685.7	1750.6
	s	⋯	⋯	⋯	⋯	⋯	⋯	⋯	1.2754	1.3807	1.4461	1.4976	1.5417	1.5812	1.6177	1.6516
5000	v	⋯	⋯	⋯	⋯	⋯	⋯	⋯	0.0591	0.1038	0.1312	0.1529	0.1718	0.1890	0.2050	0.2203
	h	⋯	⋯	⋯	⋯	⋯	⋯	⋯	1042.9	1252.9	1364.6	1452.1	1529.1	1600.9	1670.0	1737.4
	s	⋯	⋯	⋯	⋯	⋯	⋯	⋯	1.1593	1.3207	1.4001	1.4582	1.5061	1.5481	1.5863	1.6216
6000	v	⋯	⋯	⋯	⋯	⋯	⋯	⋯	0.0397	0.0757	0.1020	0.1221	0.1391	0.1544	0.1684	0.1817
	h	⋯	⋯	⋯	⋯	⋯	⋯	⋯	945.1	1188.8	1323.6	1422.3	1505.9	1582.0	1654.2	1724.2
	s	⋯	⋯	⋯	⋯	⋯	⋯	⋯	1.0746	1.2615	1.3574	1.4229	1.4748	1.5194	1.5593	1.5960
7000	v	⋯	⋯	⋯	⋯	⋯	⋯	⋯	0.0334	0.0573	0.0816	0.1004	0.1160	0.1298	0.1424	0.1542
	h	⋯	⋯	⋯	⋯	⋯	⋯	⋯	901.8	1124.9	1281.7	1392.2	1482.6	1563.1	1638.6	1711.1
	s	⋯	⋯	⋯	⋯	⋯	⋯	⋯	1.0350	1.2055	1.3171	1.3904	1.4466	1.4938	1.5355	1.5735

Table A.1.4 (E) Properties of Compressed Water*

Abs. press., psia (sat. temp.)		70	100	200	300	Temperature, °F 400	500	600	700
250 (400.97)	v	0.01604	0.01612	0.01662	0.01744	0.01864	⋯	⋯	⋯
	h	38.74	68.66	168.63	270.05	375.10	⋯	⋯	⋯
	s	0.0745	0.1294	0.2937	0.4368	0.5667	⋯	⋯	⋯
500 (467.01)	v	0.01602	0.01611	0.01661	0.01742	0.01861	⋯	⋯	⋯
	h	39.44	69.32	169.19	270.51	375.38	⋯	⋯	⋯
	s	0.0744	0.1292	0.2934	0.4364	0.5660	⋯	⋯	⋯
1000 (544.58)	v	0.01600	0.01608	0.01658	0.01738	0.01855	0.02036	⋯	⋯
	h	40.82	70.63	170.33	271.44	375.96	487.79	⋯	⋯
	s	0.0742	0.1289	0.2928	0.4355	0.5647	0.6876	⋯	⋯

Press (T sat)									
1500 (596.20)	v	0.01598	0.01606	0.01655	0.01735	0.01849	0.02025	⋯	⋯
	h	42.20	71.95	171.47	272.38	376.56	487.63	⋯	⋯
	s	0.0740	0.1286	0.2922	0.4346	0.5634	0.6855	⋯	⋯
2000 (635.80)	v	0.01595	0.01603	0.01653	0.01731	0.01844	0.02014	0.02332	⋯
	h	43.58	73.26	172.60	273.32	377.19	487.53	614.48	⋯
	s	0.0738	0.1283	0.2916	0.4337	0.5621	0.6834	0.8091	⋯
2500 (668.11)	v	0.01593	0.01601	0.01650	0.01728	0.01838	0.02004	0.02302	⋯
	h	44.95	74.57	173.74	274.27	377.82	487.50	612.08	⋯
	s	0.0736	0.1280	0.2910	0.4329	0.5609	0.6815	0.8048	⋯
3000 (695.33)	v	0.01590	0.01599	0.01648	0.01724	0.01833	0.01995	0.02276	⋯
	h	46.31	75.88	174.88	275.22	378.47	487.52	610.08	⋯
	s	0.0734	0.1277	0.2904	0.4320	0.5597	0.6796	0.8009	⋯
3200 (705.08)	v	0.0159	0.0160	0.0165	0.0172	0.0183	0.0199	0.0227	0.0335
	h	46.9	76.4	175.3	275.6	378.7	487.5	609.4	800.8
	s	0.0733	0.1276	0.2902	0.4317	0.5592	0.6788	0.7994	0.9708
4000	v	0.0159	0.0159	0.0164	0.0172	0.0182	0.0198	0.0223	0.0287
	h	49.0	78.5	177.2	277.1	379.8	487.7	606.9	763.0
	s	0.0730	0.1271	0.2893	0.4304	0.5573	0.6760	0.7940	0.9343
5000	v	0.0158	0.0159	0.0164	0.0171	0.0181	0.0196	0.0219	0.0268
	h	51.7	81.1	179.5	279.1	381.2	488.1	604.6	746.0
	s	0.0726	0.1265	0.2881	0.4287	0.5550	0.6726	0.7880	0.9153
6000	v	0.0158	0.0159	0.0163	0.0170	0.0180	0.0195	0.0216	0.0256
	h	54.4	83.7	181.7	281.0	382.7	488.6	602.9	736.1
	s	0.0721	0.1258	0.2870	0.4271	0.5528	0.6693	0.7826	0.9026
7000	v	0.0157	0.0158	0.0163	0.0170	0.0180	0.0193	0.0213	0.0248
	h	57.1	86.2	184.0	283.0	384.2	489.3	601.7	729.3
	s	0.0717	0.1252	0.2859	0.4256	0.5507	0.6663	0.7777	0.8926

*Abstracted with permission from *Thermodynamic and Transport Properties of Steam*. (Copyright, 1967, by The American Society of Mechanical Engineers.)

Table A.2.1 (SI) Properties of Saturated Freon-12*

Temp. °C	Abs. Press. MPa P	Specific Volume m³/kg			Enthalpy kJ/kg			Entropy kJ/kg·K		
		Sat. Liquid v_f	Evap. v_{fg}	Sat. Vapor v_g	Sat. Liquid h_f	Evap. h_{fg}	Sat. Vapor h_g	Sat. Liquid s_f	Evap. s_{fg}	Sat. Vapor s_g
−90	0.0028	0.000608	4.414937	4.415545	−43.243	189.618	146.375	−0.2084	1.0352	0.8268
−85	0.0042	0.000612	3.036704	3.037316	−38.968	187.608	148.640	−0.1854	0.9970	0.8116
−80	0.0062	0.000617	2.137728	2.138345	−34.688	185.612	150.924	−0.1630	0.9609	0.7979
−75	0.0088	0.000622	1.537030	1.537651	−30.401	183.625	153.224	−0.1411	0.9266	0.7855
−70	0.0123	0.000627	1.126654	1.127280	−26.103	181.640	155.536	−0.1197	0.8940	0.7744
−65	0.0168	0.000632	0.840534	0.841166	−21.793	179.651	157.857	−0.0987	0.8630	0.7643
−60	0.0226	0.000637	0.637274	0.637910	−17.469	177.653	160.184	−0.0782	0.8334	0.7552
−55	0.0300	0.000642	0.490358	0.491000	−13.129	175.641	162.512	−0.0581	0.8051	0.7470
−50	0.0391	0.000648	0.382457	0.383105	−8.772	173.611	164.840	−0.0384	0.7779	0.7396
−45	0.0504	0.000654	0.302029	0.302682	−4.396	171.558	167.163	−0.0190	0.7519	0.7329
−40	0.0642	0.000659	0.241251	0.241910	−0.000	169.479	169.479	−0.0000	0.7269	0.7269
−35	0.0807	0.000666	0.194732	0.195398	4.416	167.368	171.784	0.0187	0.7027	0.7214
−30	0.1004	0.000672	0.158703	0.159375	8.854	165.222	174.076	0.0371	0.6795	0.7165
−25	0.1237	0.000679	0.130487	0.131166	13.315	163.037	176.352	0.0552	0.6570	0.7121
−20	0.1509	0.000685	0.108162	0.108847	17.800	160.810	178.610	0.0730	0.6352	0.7082
−15	0.1826	0.000693	0.090326	0.091018	22.312	158.534	180.846	0.0906	0.6141	0.7046
−10	0.2191	0.000700	0.075946	0.076646	26.851	156.207	183.058	0.1079	0.5936	0.7014
−5	0.2610	0.000708	0.064255	0.064963	31.420	153.823	185.243	0.1250	0.5736	0.6986

0	0.3086	0.000716	0.054673	0.055389	36.022	151.376	187.397	0.1418	0.5542	0.6960
5	0.3626	0.000724	0.046761	0.047485	40.659	148.859	189.518	0.1585	0.5351	0.6937
10	0.4233	0.000733	0.040180	0.040914	45.337	146.265	191.602	0.1750	0.5165	0.6916
15	0.4914	0.000743	0.034671	0.035413	50.058	143.586	193.644	0.1914	0.4983	0.6897
20	0.5673	0.000752	0.030028	0.030780	54.828	140.812	195.641	0.2076	0.4803	0.6879
25	0.6516	0.000763	0.026091	0.026854	59.653	137.933	197.586	0.2237	0.4626	0.6863
30	0.7449	0.000774	0.022734	0.023508	64.539	134.936	199.475	0.2397	0.4451	0.6848
35	0.8477	0.000786	0.019855	0.020641	69.494	131.805	201.299	0.2557	0.4277	0.6834
40	0.9607	0.000798	0.017373	0.018171	74.527	128.525	203.051	0.2716	0.4104	0.6820
45	1.0843	0.000811	0.015220	0.016032	79.647	125.074	204.722	0.2875	0.3931	0.6806
50	1.2193	0.000826	0.013344	0.014170	84.868	121.430	206.298	0.3034	0.3758	0.6792
55	1.3663	0.000841	0.011701	0.012542	90.201	117.565	207.766	0.3194	0.3582	0.6777
60	1.5259	0.000858	0.010253	0.011111	95.665	113.443	209.109	0.3355	0.3405	0.6760
65	1.6988	0.000877	0.008971	0.009847	101.279	109.024	210.303	0.3518	0.3224	0.6742
70	1.8858	0.000897	0.007828	0.008725	107.067	104.255	211.321	0.3683	0.3038	0.6721
75	2.0874	0.000920	0.006802	0.007723	113.058	99.068	212.126	0.3851	0.2845	0.6697
80	2.3046	0.000946	0.005875	0.006821	119.291	93.373	212.665	0.4023	0.2644	0.6667
85	2.5380	0.000976	0.005029	0.006005	125.818	87.047	212.865	0.4201	0.2430	0.6631
90	2.7885	0.001012	0.004246	0.005258	132.708	79.907	212.614	0.4385	0.2200	0.6585
95	3.0569	0.001056	0.003508	0.004563	140.068	71.658	211.726	0.4579	0.1946	0.6526
100	3.3440	0.001113	0.002790	0.003903	148.076	61.768	209.843	0.4788	0.1655	0.6444
105	3.6509	0.001197	0.002045	0.003242	157.085	49.014	206.099	0.5023	0.1296	0.6319
110	3.9784	0.001364	0.001098	0.002462	168.059	28.425	196.484	0.5322	0.0742	0.6064
112	4.1155	0.001792	0.000005	0.001797	174.920	0.151	175.071	0.5651	0.0004	0.5655

* SOURCE: Gordon J. Van Wylen and Richard E. Sonntag, *Fundamentals of Classical Thermodynamics*, SI Version 3rd ed., John Wiley & Sons, Inc., 1985, pp. 636–637, Table A.3.1. (Copyright 1955 and 1956, E. I. du Pont de Nemours & Company, Inc.)

Table A.2.2 (SI) Properties of Superheated Freon-12*

Temp. °C	0.05 MPa			0.10 MPa			0.15 MPa		
	v m³/kg	h kJ/kg	s kJ/kg·K	v m³/kg	h kJ/kg	s kJ/kg·K	v m³/kg	h kJ/kg	s kJ/kg·K
−20.0	0.341857	181.042	0.7912	0.167701	179.861	0.7401			
−10.0	0.356227	186.757	0.8133	0.175222	185.707	0.7628	0.114716	184.619	0.7318
0.0	0.370508	192.567	0.8350	0.182647	191.628	0.7849	0.119866	190.660	0.7543
10.0	0.384716	198.471	0.8562	0.189994	197.628	0.8064	0.124932	196.762	0.7763
20.0	0.398863	204.469	0.8770	0.197277	203.707	0.8275	0.129930	202.927	0.7977
30.0	0.412959	210.557	0.8974	0.204506	209.866	0.8482	0.134873	209.160	0.8186
40.0	0.427012	216.733	0.9175	0.211691	216.104	0.8684	0.139768	215.463	0.8390
50.0	0.441030	222.997	0.9372	0.218839	222.421	0.8883	0.144625	221.835	0.8591
60.0	0.455017	229.344	0.9565	0.225955	228.815	0.9078	0.149450	228.277	0.8787
70.0	0.468978	235.774	0.9755	0.233044	235.285	0.9269	0.154247	234.789	0.8980
80.0	0.482917	242.282	0.9942	0.240111	241.829	0.9457	0.159020	241.371	0.9169
90.0	0.496838	248.868	1.0126	0.247159	248.446	0.9642	0.163774	248.020	0.9354

Temp. °C	0.20 MPa			0.25 MPa			0.30 MPa		
	v m³/kg	h kJ/kg	s kJ/kg·K	v m³/kg	h kJ/kg	s kJ/kg·K	v m³/kg	h kJ/kg	s kJ/kg·K
0.0	0.088608	189.669	0.7320	0.069752	188.644	0.7139	0.057150	187.583	0.6984
10.0	0.092550	195.878	0.7543	0.073024	194.969	0.7366	0.059984	194.034	0.7216
20.0	0.096418	202.135	0.7760	0.076218	201.322	0.7587	0.062734	200.490	0.7440
30.0	0.100228	208.446	0.7972	0.079350	207.715	0.7801	0.065418	206.969	0.7658
40.0	0.103989	214.814	0.8178	0.082431	214.153	0.8010	0.068049	213.480	0.7869
50.0	0.107710	221.243	0.8381	0.085470	220.642	0.8214	0.070635	220.030	0.8075
60.0	0.111397	227.735	0.8578	0.088474	227.185	0.8413	0.073185	226.627	0.8276
70.0	0.115055	234.291	0.8772	0.091449	233.785	0.8608	0.075705	233.273	0.8473
80.0	0.118690	240.910	0.8962	0.094398	240.443	0.8800	0.078200	239.971	0.8665
90.0	0.122304	247.593	0.9149	0.097327	247.160	0.8987	0.080673	246.723	0.8853
100.0	0.125901	254.339	0.9332	0.100238	253.936	0.9171	0.083127	253.530	0.9038
110.0	0.129483	261.147	0.9512	0.103134	260.770	0.9352	0.085566	260.391	0.9220

	0.40 MPa			0.50 MPa			0.60 MPa		
20.0	0.045836	198.762	0.7199	0.035646	196.935	0.6999			
30.0	0.047971	205.428	0.7423	0.037464	203.814	0.7230	0.030422	202.116	0.7063
40.0	0.050046	212.095	0.7639	0.039214	210.656	0.7452	0.031966	209.154	0.7291
50.0	0.052072	218.779	0.7849	0.040911	217.484	0.7667	0.033450	216.141	0.7511
60.0	0.054059	225.488	0.8054	0.042565	224.315	0.7875	0.034887	223.104	0.7723
70.0	0.056014	232.230	0.8253	0.044184	231.161	0.8077	0.036285	230.062	0.7929
80.0	0.057941	239.012	0.8448	0.045774	238.031	0.8275	0.037653	237.027	0.8129
90.0	0.059846	245.837	0.8638	0.047340	244.932	0.8467	0.038995	244.009	0.8324
100.0	0.061731	252.707	0.8825	0.048886	251.869	0.8656	0.040316	251.016	0.8514
110.0	0.063600	259.624	0.9008	0.050415	258.845	0.8840	0.041619	258.053	0.8700
120.0	0.065455	266.590	0.9187	0.051929	265.862	0.9021	0.042907	265.124	0.8882
130.0	0.067298	273.605	0.9364	0.053430	272.923	0.9198	0.044181	272.231	0.9061

	0.70 MPa			0.80 MPa			0.90 MPa		
40.0	0.026761	207.580	0.7148	0.022830	205.924	0.7016	0.019744	204.170	0.6982
50.0	0.028100	214.745	0.7373	0.024068	213.290	0.7284	0.020912	211.765	0.7131
60.0	0.029387	221.854	0.7590	0.025247	220.558	0.7469	0.022012	219.212	0.7358
70.0	0.030632	228.931	0.7799	0.026380	227.766	0.7682	0.023062	226.564	0.7575
80.0	0.031843	235.997	0.8002	0.027477	234.941	0.7888	0.024072	233.856	0.7785
90.0	0.033027	243.066	0.8199	0.028545	242.101	0.8088	0.025051	241.113	0.7987
100.0	0.034189	250.146	0.8392	0.029588	249.260	0.8283	0.026005	248.355	0.8184
110.0	0.035332	257.247	0.8579	0.030612	256.428	0.8472	0.026937	255.593	0.8376
120.0	0.036458	264.374	0.8763	0.031619	263.613	0.8657	0.027851	262.839	0.8562
130.0	0.037572	271.531	0.8943	0.032612	270.820	0.8838	0.028751	270.100	0.8745
140.0	0.038673	278.720	0.9119	0.033592	278.055	0.9016	0.029639	277.381	0.8923
150.0	0.039764	285.946	0.9292	0.034563	285.320	0.9189	0.030515	284.687	0.9098

(*continued*)

Table A.2.2 (SI) (*continued*)

Temp. °C	v m³/kg	h kJ/kg	s kJ/kg·K	v m³/kg	h kJ/kg	s kJ/kg·K	v m³/kg	h kJ/kg	s kJ/kg·K
		1.00 MPa			1.20 MPa			1.40 MPa	
50.0	0.018366	210.162	0.7021	0.014483	206.661	0.6812	0.012579	211.457	0.6876
60.0	0.019410	217.810	0.7254	0.015463	214.805	0.7060	0.013448	219.822	0.7123
70.0	0.020397	225.319	0.7476	0.016368	222.687	0.7293	0.014247	227.891	0.7355
80.0	0.021341	232.739	0.7689	0.017221	230.398	0.7514	0.014997	235.766	0.7575
90.0	0.022251	240.101	0.7895	0.018032	237.995	0.7727	0.015710	243.512	0.7785
100.0	0.023133	247.430	0.8094	0.018812	245.518	0.7931	0.016393	251.170	0.7988
110.0	0.023993	254.743	0.8287	0.019567	252.993	0.8129	0.017053	258.770	0.8183
120.0	0.024835	262.053	0.8475	0.020301	260.441	0.8320	0.017695	266.334	0.8373
130.0	0.025661	269.369	0.8659	0.021018	267.875	0.8507	0.018321	273.877	0.8558
140.0	0.026474	276.699	0.8839	0.021721	275.307	0.8689	0.018934	281.411	0.8738
150.0	0.027275	284.047	0.9015	0.022412	282.745	0.8867	0.019535	288.946	0.8914
160.0	0.028068	291.419	0.9187	0.023093	290.195	0.9041			
		1.60 MPa			1.80 MPa			2.00 MPa	
70.0	0.011208	216.650	0.6959	0.009406	213.049	0.6794	0.008704	218.859	0.6909
80.0	0.011984	225.177	0.7204	0.010187	222.198	0.7057	0.009406	228.056	0.7166
90.0	0.012698	233.390	0.7433	0.010884	230.835	0.7298	0.010035	236.760	0.7402
100.0	0.013366	241.397	0.7651	0.011526	239.155	0.7524	0.010615	245.154	0.7624
110.0	0.014000	249.264	0.7859	0.012126	247.264	0.7739	0.011159	253.341	0.7835
120.0	0.014608	257.035	0.8059	0.012697	255.228	0.7944	0.011676	261.384	0.8037
130.0	0.015195	264.742	0.8253	0.013244	263.094	0.8141	0.012172	269.327	0.8232
140.0	0.015765	272.406	0.8440	0.013772	270.891	0.8332	0.012651	277.201	0.8420
150.0	0.016320	280.044	0.8623	0.014284	278.642	0.8518	0.013116	285.027	0.8603
160.0	0.016864	287.669	0.8801	0.014784	286.364	0.8698	0.013570	292.822	0.8781
170.0	0.017398	295.290	0.8975	0.015272	294.069	0.8874	0.014013	300.598	0.8955
180.0	0.017923	302.914	0.9145	0.015752	301.767	0.9046			

Temp	2.50 MPa			3.00 MPa			3.50 MPa		
90.0	0.006595	219.562	0.6823						
100.0	0.007264	229.852	0.7103	0.005231	220.529	0.6770			
110.0	0.007837	239.271	0.7352	0.005886	232.068	0.7075	0.004324	222.121	0.6750
120.0	0.008351	248.192	0.7582	0.006419	242.208	0.7336	0.004959	234.875	0.7078
130.0	0.008827	256.794	0.7798	0.006887	251.632	0.7573	0.005456	245.661	0.7349
140.0	0.009273	265.180	0.8003	0.007313	260.620	0.7793	0.005884	255.524	0.7591
150.0	0.009697	273.414	0.8200	0.007709	269.319	0.8001	0.006270	264.846	0.7814
160.0	0.010104	281.540	0.8390	0.008083	277.817	0.8200	0.006626	273.817	0.8023
170.0	0.010497	289.589	0.8574	0.008439	286.171	0.8391	0.006961	282.545	0.8222
180.0	0.010879	297.583	0.8752	0.008782	294.422	0.8575	0.007279	291.100	0.8413
190.0	0.011250	305.540	0.8926	0.009114	302.597	0.8753	0.007584	299.528	0.8597
200.0	0.011614	313.472	0.9095	0.009436	310.718	0.8927	0.007878	307.864	0.8775

Temp	4.00 MPa		
120.0	0.003736	224.863	0.6771
130.0	0.004325	238.443	0.7111
140.0	0.004781	249.703	0.7386
150.0	0.005172	259.904	0.7630
160.0	0.005522	269.492	0.7854
170.0	0.005845	278.684	0.8063
180.0	0.006147	287.602	0.8262
190.0	0.006434	296.326	0.8453
200.0	0.006708	304.906	0.8636
210.0	0.006972	313.380	0.8813
220.0	0.007228	321.774	0.8985
230.0	0.007477	330.108	0.9152

* SOURCE: Gordon J. Van Wylen and Richard E. Sonntag, *Fundamentals of Classical Thermodynamics*, SI Version 3rd ed., John Wiley & Sons, Inc., 1985, pp. 638–641, Table A.3.2. (Copyright 1955 and 1956, E. I. du Pont de Nemours & Company, Inc.)

Table A.2.1 (E) Properties of Saturated Freon-12 (Dichlorodifluoromethane)*

Temp., °F	Press., psia	Volume, ft³/lbm			Enthalpy, Btu/lbm			Entropy, Btu/lbm-°R		
		Sat. Liquid v_f	Evap. v_{fg}	Sat. Vapor v_g	Sat. Liquid h_f	Evap. h_{fg}	Sat. Vapor h_g	Sat. Liquid s_f	Evap. s_{fg}	Sat. Vapor s_g
−130	0.41224	0.009736	70.7203	70.730	−18.609	81.577	62.968	−0.04983	0.24743	0.19760
−120	0.64190	0.009816	46.7312	46.741	−16.565	80.617	64.052	−0.04372	0.23731	0.19359
−110	0.97034	0.009899	31.7671	31.777	−14.518	79.663	65.145	−0.03779	0.22780	0.19002
−100	1.4280	0.009985	21.1541	22.164	−12.466	78.714	66.248	−0.03200	0.21883	0.18683
−90	2.0509	0.010073	15.8109	15.821	−10.409	77.764	67.355	−0.02637	0.21034	0.18398
−80	2.8807	0.010164	11.5228	11.533	−8.3451	76.812	68.467	−0.02086	0.20229	0.18143
−70	3.9651	0.010259	8.5584	8.5687	−6.2730	75.853	69.580	−0.01548	0.19464	0.17916
−60	5.3575	0.010357	6.4670	6.4774	−4.1919	74.885	70.693	−0.01021	0.18716	0.17714
−50	7.1168	0.010459	4.9637	4.9742	−2.1011	73.906	71.805	−0.00506	0.18038	0.17533
−40	9.3076	0.010564	3.8644	3.8750	0	72.913	72.913	0	0.17373	0.17373
−30	11.999	0.010674	3.0478	3.0585	2.1120	71.903	74.015	0.00496	0.16733	0.17229
−20	15.267	0.010788	2.4321	2.4429	4.2357	70.874	75.110	0.00983	0.16119	0.17102
−10	19.189	0.010906	1.9628	1.9727	6.3716	69.824	76.196	0.01462	0.15527	0.16989
0	23.849	0.011030	1.5979	1.6089	8.5207	68.750	77.271	0.01932	0.14956	0.16888
10	29.335	0.011160	1.3129	1.3241	10.684	67.651	78.335	0.02395	0.14403	0.16798
20	35.736	0.011296	1.0875	1.0988	12.863	66.522	79.385	0.02852	0.13867	0.16719
30	43.148	0.011438	0.90736	0.91880	15.058	65.361	80.419	0.03301	0.13347	0.16648

40	51.667	0.011588	0.76198	0.77357	17.273	64.163	81.436	0.03745	0.12841	0.16586
50	61.394	0.011746	0.64362	0.65537	19.507	62.926	82.433	0.04184	0.12346	0.16530
60	72.433	0.011913	0.54648	0.55839	21.766	61.643	83.409	0.04618	0.11861	0.16479
70	84.888	0.012089	0.46609	0.47818	24.050	60.309	84.359	0.05048	0.11386	0.16434
80	98.870	0.012277	0.39907	0.41135	26.365	58.917	85.282	0.05475	0.10917	0.16392
90	114.49	0.012478	0.34281	0.35529	28.713	57.461	86.174	0.05900	0.10453	0.16353
100	131.86	0.012693	0.29525	0.30794	31.100	55.929	87.029	0.06323	0.09992	0.16315
110	151.11	0.012924	0.25577	0.26769	33.531	54.313	87.844	0.06745	0.09534	0.16279
120	172.35	0.013174	0.22019	0.23326	36.013	52.597	88.610	0.07168	0.09073	0.16241
130	195.71	0.013447	0.19019	0.20364	38.553	50.768	89.321	0.07593	0.08609	0.16202
140	221.32	0.013746	0.16424	0.17799	41.162	48.805	89.967	0.08021	0.08138	0.16159
150	249.31	0.014078	0.14156	0.15564	43.850	46.684	90.534	0.08453	0.07657	0.16110
160	279.82	0.014449	0.12159	0.13604	46.633	44.373	91.006	0.08893	0.07260	0.16053
170	313.00	0.014871	0.10386	0.11873	49.529	41.830	91.359	0.09342	0.06643	0.15985
180	349.00	0.015360	0.08794	0.10330	52.562	38.999	91.561	0.09804	0.06096	0.15900
190	387.98	0.015942	0.073476	0.089418	55.769	35.792	91.561	0.10284	0.05511	0.15793
200	430.09	0.016659	0.060069	0.076728	59.203	32.075	91.278	0.10789	0.04862	0.15651
210	475.52	0.017601	0.047242	0.064843	62.959	27.599	90.558	0.11332	0.03921	0.15453
220	524.43	0.018986	0.035154	0.053140	67.246	21.790	89.036	0.11943	0.03206	0.15149
230	577.03	0.021854	0.017581	0.039435	72.893	12.229	85.122	0.12739	0.01773	0.14512
233.6	596.9	0.02870	0	0.02870	78.86	0	78.86	0.1359	0	0.1359

Table A.2.2 (E) Properties of Superheated Freon-12*

Temp., °F	v	h	s	v	h	s	v	h	s
	5 lbf/in²			10 lbf/in²			15 lbf/in²		
0	8.0611	78.582	0.19663	3.9809	78.246	0.18471	2.6201	77.902	0.17751
20	8.4265	81.309	0.20244	4.1691	81.014	0.19061	2.7494	80.712	0.18349
40	8.7903	84.090	0.20812	4.3556	83.828	0.19635	2.8770	83.561	0.18931
60	9.1528	86.922	0.21367	4.5408	86.689	0.20197	3.0031	86.451	0.19498
80	9.5142	89.806	0.21912	4.7248	89.596	0.20746	3.1281	89.383	0.20051
100	9.8747	92.738	0.22445	4.9079	92.548	0.21283	3.2521	92.357	0.20593
120	10.234	95.717	0.22968	5.0903	95.546	0.21809	3.3754	95.373	0.21122
140	10.594	98.743	0.23481	5.2720	98.586	0.22325	3.4981	98.429	0.21640
160	10.952	101.812	0.23985	5.4533	101.669	0.22830	3.6202	101.525	0.22148
180	11.311	104.925	0.24479	5.6341	104.793	0.23326	3.7419	104.661	0.22646
200	11.668	108.079	0.24964	5.8145	107.957	0.23813	3.8632	107.835	0.23135
220	12.026	111.272	0.25441	5.9946	111.159	0.24291	3.9841	111.046	0.23614
	20 lbf/in²			25 lbf/in²			30 lbf/in²		
20	2.0391	80.403	0.17829	1.6125	80.088	0.17414	1.3278	79.765	0.17065
40	2.1373	83.289	0.18419	1.6932	83.012	0.18012	1.3969	82.730	0.17671
60	2.2340	86.210	0.18992	1.7723	85.965	0.18591	1.4644	85.716	0.18257
80	2.3295	89.168	0.19550	1.8502	88.950	0.19155	1.5306	88.729	0.18826
100	2.4241	92.164	0.20095	1.9271	91.968	0.19704	1.5957	91.770	0.19379
120	2.5179	95.198	0.20628	2.0032	95.021	0.20240	1.6600	94.843	0.19918
140	2.6110	98.270	0.21149	2.0786	98.110	0.20763	1.7237	97.948	0.20445
160	2.7036	101.380	0.21659	2.1535	101.234	0.21276	1.7868	101.086	0.20960
180	2.7957	104.528	0.22159	2.2279	104.393	0.21778	1.8494	104.258	0.21463
200	2.8874	107.712	0.22649	2.3019	107.588	0.22269	1.9116	107.464	0.21957
220	2.9789	110.932	0.23130	2.3756	110.817	0.22752	1.9735	110.702	0.22440
240	3.0700	114.186	0.23602	2.4491	114.080	0.23225	2.0351	113.973	0.22915

	35 lbf/in²			40 lbf/in²			50 lbf/in²		
40	1.1850	82.442	0.17375	1.0258	82.148	0.17112	0.80248	81.540	0.16655
60	1.2442	85.463	0.17968	1.0789	85.206	0.17712	0.84713	84.676	0.17271
80	1.3021	88.504	0.18542	1.1306	88.277	0.18292	0.89025	87.811	0.17862
100	1.3589	91.570	0.19100	1.1812	91.367	0.18854	0.93216	90.953	0.18434
120	1.4148	94.663	0.19643	1.2309	94.480	0.19401	0.97313	94.110	0.18988
140	1.4701	97.785	0.20172	1.2798	97.620	0.19933	1.0133	97.286	0.19527
160	1.5248	100.938	0.20689	1.3282	100.788	0.20453	1.0529	100.485	0.20051
180	1.5789	104.122	0.21195	1.3761	103.985	0.20961	1.0920	103.708	0.20563
200	1.6327	107.338	0.21690	1.4236	107.212	0.21457	1.1307	106.958	0.21064
220	1.6862	110.586	0.22175	1.4707	110.469	0.21944	1.1690	110.235	0.21553
240	1.7394	113.865	0.22651	1.5176	113.757	0.22420	1.2070	113.539	0.22032
260	1.7923	117.175	0.23117	1.5642	117.074	0.22888	1.2447	116.871	0.22502

	60 lbf/in²			70 lbf/in²			80 lbf/in²		
60	0.69210	84.126	0.16892	0.58088	83.552	0.16556	...	...	...
80	0.72964	87.330	0.17497	0.61458	86.832	0.17175	0.52795	86.316	0.16885
100	0.76588	90.528	0.18079	0.64685	90.091	0.17768	0.55734	89.640	0.17489
120	0.80110	93.731	0.18641	0.67803	93.343	0.18339	0.58556	92.945	0.18070
140	0.83551	96.945	0.19186	0.70836	96.597	0.18891	0.61286	96.242	0.18629
160	0.86928	100.176	0.19716	0.73800	99.862	0.19427	0.63943	99.542	0.19170
180	0.90252	103.427	0.20233	0.76708	103.141	0.19948	0.66543	102.851	0.19696
200	0.93531	106.700	0.20736	0.79571	106.439	0.20455	0.69095	106.174	0.20207
220	0.96775	109.997	0.21229	0.82397	109.756	0.20951	0.71609	109.513	0.20706
240	0.99988	113.319	0.21710	0.85191	113.096	0.21435	0.74090	112.872	0.21193
260	1.0318	116.666	0.22182	0.87959	116.459	0.21909	0.76544	116.251	0.21669
280	1.0634	120.039	0.22644	0.90705	119.846	0.22373	0.78975	119.652	0.22135

(continued)

Table A.2.2 (E) (continued)

Temp., °F	90 lbf/in² v	h	s	100 lbf/in² v	h	s	125 lbf/in² v	h	s
100	0.48749	89.175	0.17234	0.43138	88.694	0.16996	0.32943	87.407	0.16455
120	0.51346	92.536	0.17824	0.45562	92.116	0.17597	0.35086	91.008	0.17087
140	0.53845	95.879	0.18391	0.47881	95.507	0.18172	0.37098	94.537	0.17686
160	0.56268	99.216	0.18938	0.50118	98.884	0.18726	0.39015	98.023	0.18258
180	0.58629	102.557	0.19469	0.52291	102.257	0.19262	0.40857	101.484	0.18807
200	0.60941	105.905	0.19984	0.54413	105.633	0.19782	0.42642	104.934	0.19338
220	0.63213	109.267	0.20486	0.56492	109.018	0.20287	0.44380	108.380	0.19853
240	0.65451	112.644	0.20976	0.58538	112.415	0.20780	0.46081	111.829	0.20353
260	0.67662	116.040	0.21455	0.60554	115.828	0.21261	0.47750	115.287	0.20840
280	0.69849	119.456	0.21923	0.62546	119.258	0.21731	0.49394	118.756	0.21316
300	0.72016	122.892	0.22381	0.64518	122.707	0.22191	0.51016	122.238	0.21780
320	0.74166	126.349	0.22830	0.66472	126.176	0.22641	0.52619	125.737	0.22235

Temp., °F	150 lbf/in² v	h	s	175 lbf/in² v	h	s	200 lbf/in² v	h	s
120	0.28007	89.800	0.16629	....	....	....	....	....	....
140	0.29845	93.498	0.17256	0.24595	92.373	0.16859	0.20579	91.137	0.16480
160	0.31566	97.112	0.17849	0.26198	96.142	0.17478	0.22121	95.100	0.17130
180	0.33200	100.675	0.18415	0.27697	99.823	0.18062	0.23535	98.921	0.17737
200	0.34769	104.206	0.18958	0.29120	103.447	0.18620	0.24860	102.652	0.18311
220	0.36285	107.720	0.19483	0.30485	107.036	0.19156	0.26117	106.325	0.18860
240	0.37761	111.226	0.19992	0.31804	110.605	0.19674	0.27323	109.962	0.19387
260	0.39203	114.732	0.20485	0.33087	114.162	0.20175	0.28489	113.576	0.19896
280	0.40617	118.242	0.20967	0.34339	117.717	0.20662	0.29623	117.178	0.20390
300	0.42008	121.761	0.21436	0.35567	121.273	0.21137	0.30730	120.775	0.20870
320	0.43379	125.290	0.21894	0.36773	124.835	0.21599	0.31815	124.373	0.21337
340	0.44733	128.833	0.22343	0.37963	128.407	0.22052	0.32881	127.974	0.21793

250 lbf/in²

240	0.21014	108.607	0.18877
260	0.22027	112.351	0.19404
280	0.23001	116.060	0.19913
300	0.23944	119.747	0.20405
320	0.24862	123.420	0.20882
340	0.25759	127.088	0.21346
360	0.26639	130.754	0.21799
380	0.27504	134.423	0.22241

300 lbf/in²

240	0.16761	107.140	0.18419
260	0.17685	111.043	0.18969
280	0.18562	114.879	0.19495
300	0.19402	118.670	0.20000
320	0.20214	122.430	0.20489
340	0.21002	126.171	0.20963
360	0.21770	129.900	0.21423
380	0.22522	133.624	0.21872

400 lbf/in²

240	0.11300	103.735	0.17568
260	0.12163	108.105	0.18183
280	0.12949	112.286	0.18756
300	0.13680	116.343	0.19298
320	0.14372	120.318	0.19814
340	0.15032	124.235	0.20310
360	0.15668	128.112	0.20789
380	0.16285	131.961	0.21253

500 lbf/in²

220	0.064207	92.397	0.15683
240	0.077620	99.218	0.16672
260	0.087054	104.526	0.17421
280	0.094923	109.277	0.18072
300	0.10190	113.729	0.18666
320	0.10829	117.997	0.19221
340	0.11426	122.143	0.19746
360	0.11992	126.205	0.20247
380	0.12533	130.207	0.20730
400	0.13054	134.166	0.21196
420	0.13559	138.096	0.21648
440	0.14051	142.004	0.22087

600 lbf/in²

220	. . .	. . .	. . .
240	0.047488	91.024	0.15335
260	0.061922	99.741	0.16566
280	0.070859	105.637	0.17374
300	0.078059	110.729	0.18053
320	0.084333	115.420	0.18663
340	0.090017	119.871	0.19227
360	0.095289	124.167	0.19757
380	0.10025	128.355	0.20262
400	0.10498	132.466	0.20746
420	0.10952	136.523	0.21213
440	0.11391	140.539	0.21664

Table A.3.1 (SI) Properties of Saturated Ammonia*

Temp. °C	Press. kPa	Volume, m³/kg Sat. Liquid v_f	Evap. v_{fg}	Sat. Vapor v_g	Enthalpy, kJ/kg Sat. Liquid h_f	Evap. h_{fg}	Sat. Vapor h_g	Entropy, kJ/kg · K Sat. Liquid s_f	Evap. s_{fg}	Sat. Vapor s_g
−50	40.9	0.001424	2.62558	2.62700	−986.87	1416.30	429.30	4.7933	6.3468	11.1401
−48	45.9	0.001429	2.35320	2.35463	−978.15	1410.91	432.76	4.8321	6.2665	11.0986
−46	51.5	0.001434	2.11373	2.11516	−969.41	1405.46	436.05	4.8707	6.1874	11.0581
−44	57.7	0.001439	1.90265	1.90409	−960.66	1399.95	439.29	4.9090	6.1094	11.0184
−42	64.4	0.001444	1.71620	1.71764	−951.89	1394.39	442.50	4.9471	6.0324	10.9795
−40	71.7	0.001450	1.55111	1.55256	−943.11	1388.78	445.67	4.9849	5.9566	10.9415
−38	79.7	0.001455	1.40463	1.40608	−934.30	1383.09	448.80	5.0225	5.8817	10.9042
−36	88.5	0.001460	1.27436	1.27582	−925.47	1377.34	451.88	5.0598	5.8079	10.8677
−34	98.0	0.001465	1.15829	1.15975	−916.62	1371.53	454.92	5.0969	5.7350	10.8319
−32	108.3	0.001471	1.05464	1.05611	−907.74	1365.66	457.91	5.1338	5.6631	10.7969
−30	119.5	0.001476	0.96191	0.96339	−898.85	1359.71	460.86	5.1705	5.5920	10.7625
−28	131.6	0.001482	0.87880	0.88028	−889.94	1353.70	463.76	5.2069	5.5219	10.7288
−26	144.6	0.001487	0.80416	0.80565	−881.00	1347.62	466.62	5.2431	5.4526	10.6957
−24	158.7	0.001493	0.73701	0.73850	−872.04	1341.47	469.43	5.2791	5.3842	10.6633
−22	173.9	0.001498	0.67648	0.67798	−863.06	1335.24	472.18	5.3149	5.3166	10.6315
−20	190.2	0.001504	0.62184	0.62334	−854.06	1328.95	474.89	5.3506	5.2496	10.6002
−18	207.7	0.001510	0.57242	0.57393	−845.03	1322.58	477.55	5.3860	5.1835	10.5695
−16	226.4	0.001516	0.52765	0.52917	−835.98	1316.14	480.16	5.4212	5.1182	10.5394
−14	246.5	0.001522	0.48705	0.48857	−826.91	1309.62	482.71	5.4562	5.0535	10.5097
−12	267.9	0.001528	0.45013	0.45166	−817.23	1303.03	485.21	5.4910	4.9896	10.4806
−10	290.8	0.001534	0.41655	0.41808	−808.70	1296.36	487.66	5.5257	4.9263	10.4520
−8	315.3	0.001540	0.38593	0.38747	−799.56	1289.61	490.05	5.5601	4.8638	10.4239
−6	341.3	0.001546	0.35798	0.35953	−790.39	1282.78	492.39	5.5944	4.8018	10.3962

−4	368.9	0.001553	0.33245	0.33400	−781.20	1275.87	494.67	5.6285	4.7404	10.3689
−2	398.3	0.001559	0.30906	0.31062	−771.99	1268.87	496.89	5.6625	4.6796	10.3421
0	429.5	0.001566	0.28763	0.28920	−762.75	1261.79	499.05	5.6963	4.6194	10.3157
2	462.6	0.001573	0.26797	0.26954	−753.49	1254.63	501.14	5.7299	4.5598	10.2897
4	497.6	0.001579	0.24989	0.25147	−744.19	1247.37	503.18	5.7633	4.5007	10.2640
6	534.7	0.001586	0.23326	0.23485	−734.86	1240.01	505.15	5.7966	4.4421	10.2387
8	573.8	0.001593	0.21794	0.21953	−725.52	1232.57	507.05	5.8298	4.3840	10.2138
10	615.2	0.001600	0.20381	0.20541	−716.14	1225.03	508.89	5.8628	4.3264	10.1892
12	658.8	0.001608	0.19076	0.19237	−706.73	1217.39	510.66	5.8956	4.2693	10.1649
14	704.7	0.001615	0.17870	0.18031	−697.30	1209.66	512.36	5.9283	4.2127	10.1410
16	753.1	0.001623	0.16753	0.16915	−687.84	1201.82	513.99	5.9609	4.1564	10.1173
18	804.0	0.001630	0.15718	0.15881	−678.34	1193.88	515.54	5.9934	4.1005	10.0939
20	857.5	0.001638	0.14758	0.14922	−668.82	1185.83	517.01	6.0257	4.0451	10.0708
22	913.6	0.001646	0.13867	0.14032	−659.26	1177.67	518.41	6.0578	3.9901	10.0479
24	972.6	0.001654	0.13039	0.13204	−649.67	1169.40	519.73	6.0899	3.9354	10.0253
26	1034.4	0.001663	0.12269	0.12435	−640.05	1161.01	520.96	6.1218	3.8810	10.0028
28	1099.1	0.001671	0.11551	0.11718	−630.39	1152.50	522.11	6.1536	3.8270	9.9806
30	1166.9	0.001680	0.10881	0.11049	−620.70	1143.87	523.17	6.1853	3.7733	9.9586
32	1237.9	0.001688	0.10257	0.10426	−610.97	1135.11	524.14	6.2169	3.7199	9.9368
34	1312.0	0.001697	0.09674	0.09844	−601.21	1126.23	525.02	6.2484	3.6667	9.9151
36	1389.5	0.001707	0.09128	0.09299	−591.41	1117.21	525.80	6.2798	3.6138	9.8936
38	1470.4	0.001716	0.08618	0.08790	−581.57	1108.05	526.49	6.3111	3.5611	9.8722
40	1554.8	0.001725	0.08141	0.08313	−571.69	1098.76	527.07	6.3422	3.5088	9.8510
42	1642.8	0.001735	0.07693	0.07866	−561.77	1089.32	527.55	6.3733	3.4565	9.8298
44	1734.6	0.001745	0.07273	0.07447	−551.80	1079.72	527.92	6.4043	3.4045	9.8088
46	1830.1	0.001755	0.06879	0.07054	−541.80	1069.98	528.18	6.4353	3.3526	9.7879
48	1929.6	0.001766	0.06507	0.06684	−531.74	1060.07	528.33	6.4661	3.3009	9.7670
50	2033.0	0.001777	0.06159	0.06337	−521.64	1050.00	528.36	6.4969	3.2493	9.7462

* Abstracted with permission from "Thermodynamic Properties of Ammonia" by Lester Haar and John S. Gallagher, *Journal of Physical and Chemical Reference Data*, Vol. 7, No. 3, pp. 635–792, 1978.

Table A.3.2 (SI) Properties of Superheated Ammonia* v in m^3/kg, h in kJ/kg, s in kJ/kg · K

Press. kPa (Sat. Temp.) °C		250	260	270	280	290	300	310	320	340	360	380	400
							Temperature, K						
40 (−50.36)	v	3.02458	3.14996	3.27475	3.39908	3.52303	3.64668	3.77009	3.89330	4.13928	4.38483	4.63009	4.87515
	h	485.95	506.94	527.95	549.00	570.12	591.32	612.62	634.04	677.28	721.12	765.63	810.85
	s	11.3895	11.4718	11.5511	11.6276	11.7017	11.7736	11.8435	11.9115	12.0425	12.1678	12.2881	12.4041
60 (−43.28)	v	2.00739	2.09212	2.17626	2.25993	2.34321	2.42619	2.50892	2.59146	2.75608	2.92027	3.08417	3.24787
	h	483.57	504.89	526.18	547.47	568.78	590.14	611.57	633.10	676.51	720.47	765.07	810.37
	s	11.1841	11.2678	11.3481	11.4255	11.5003	11.5727	11.6430	11.7113	11.8429	11.9686	12.0891	12.2053
80 (−37.94)	v	1.49871	1.56315	1.62698	1.69032	1.75328	1.81593	1.87833	1.94053	2.06447	2.18799	2.31121	2.43422
	h	481.16	502.83	524.40	545.92	567.43	588.95	610.52	632.16	675.74	719.83	764.52	809.88
	s	11.0362	11.1212	11.2026	11.2809	11.3563	11.4293	11.5000	11.5687	11.7008	11.8268	11.9476	12.0639
100 (−33.60)	v	1.19345	1.24572	1.29738	1.34854	1.39930	1.44976	1.49996	1.54996	1.64951	1.74862	1.84743	1.94604
	h	478.73	500.74	522.61	544.37	566.07	587.76	609.46	631.21	674.97	719.18	763.96	809.39
	s	10.9197	11.0060	11.0885	11.1677	11.2438	11.3174	11.3885	11.4576	11.5902	11.7166	11.8376	11.9541
120 (−29.91)	v	0.98988	1.03407	1.07761	1.12066	1.16330	1.20563	1.24771	1.28958	1.37286	1.45570	1.53825	1.62058
	h	476.26	498.64	520.80	542.80	564.71	586.56	608.40	630.27	674.20	718.53	763.41	809.91
	s	10.8230	10.9107	10.9944	11.0744	11.1513	11.2253	11.2970	11.3664	11.4995	11.6262	11.7475	11.8642

140	v	0.84443	0.88285	0.92062	0.95787	0.99472	1.03125	1.06752	1.10358	1.17525	1.24647	1.31740	1.38811
(−26.69)	h	473.77	496.51	518.97	541.22	563.33	585.35	607.33	629.32	673.42	717.88	762.85	808.42
	s	10.7399	10.8291	10.9139	10.9948	11.0724	11.1471	11.2191	11.2889	11.4226	11.5497	11.6712	11.7881
160	v	0.73530	0.76941	0.80285	0.83577	0.86827	0.90045	0.93237	0.96408	1.02704	1.08955	1.15176	1.21375
(−23.83)	h	471.24	494.36	517.13	539.64	561.95	584.14	606.26	628.36	672.64	717.23	762.29	807.93
	s	10.6668	10.7575	10.8434	10.9253	11.0036	11.0788	11.1513	11.2215	11.3557	11.4832	11.6050	11.7220
180	v		0.68115	0.71123	0.74078	0.76991	0.79871	0.82725	0.85558	0.91176	0.96750	1.02293	1.07814
(−21.23)	h		492.19	515.27	538.04	560.56	582.92	605.19	627.40	671.86	716.58	761.73	807.45
	s		10.6935	10.7806	10.8634	10.9424	11.0182	11.0912	11.1618	11.2965	11.4243	11.5464	11.6636
200	v		0.61052	0.63792	0.66478	0.69122	0.71732	0.74315	0.76877	0.81954	0.86985	0.91986	0.96965
(−18.86)	h		489.99	513.40	536.42	559.16	581.70	604.11	626.44	671.08	715.92	761.17	806.96
	s		10.6354	10.7238	10.8075	10.8873	10.9637	11.0372	11.1081	11.2434	11.3716	11.4939	11.6113
250	v		0.48328	0.50590	0.52794	0.54953	0.57078	0.59175	0.61250	0.65353	0.69409	0.73434	0.77437
(−13.66)	h		484.40	508.64	532.34	555.63	578.61	601.38	624.02	669.12	714.28	759.77	805.73
	s		10.5097	10.6012	10.6874	10.7691	10.8470	10.9217	10.9935	11.1302	11.2593	11.3823	11.5001
300	v			0.41780	0.43665	0.45503	0.47305	0.49079	0.50831	0.54284	0.57691	0.61066	0.64418
(−9.24)	h			503.78	528.18	552.04	575.48	598.63	621.57	667.14	712.63	758.36	804.51
	s			10.4981	10.5869	10.6706	10.7501	10.8260	10.8988	11.0369	11.1670	11.2906	11.4089
350	v			0.35480	0.37139	0.38749	0.40322	0.41866	0.43387	0.46377	0.49320	0.52231	0.55119
(−5.36)	h			498.79	523.94	548.39	572.31	595.84	619.10	665.15	710.97	756.94	803.27
	s			10.4084	10.4999	10.5857	10.6668	10.7439	10.8178	10.9574	11.0883	11.2126	11.3314

(continued)

Table A.3.2 (SI) (continued)

Temperature, K

Press. kPa (Sat. Temp.) °C		250	260	270	280	290	300	310	320	340	360	380	400
400	v				0.322403	0.33681	0.35083	0.36454	0.37802	0.40446	0.43042	0.45604	0.48144
(−1.89)	h				519.61	544.67	569.09	593.02	616.61	663.14	709.31	755.52	802.04
	s				10.4227	10.5106	10.5934	10.6719	10.7467	10.8878	11.0198	11.1447	11.2640
450	v				0.284258	0.297358	0.310054	0.322438	0.33458	0.35833	0.38158	0.40450	0.42719
(1.25)	h				515.19	540.90	565.83	590.17	614.09	661.12	707.63	754.10	800.80
	s				10.3528	10.4431	10.5276	10.6074	10.6833	10.8259	10.9588	11.0844	11.2042
500	v				0.253701	0.265768	0.277414	0.288738	0.299812	0.321412	0.34251	0.36326	0.38379
(4.13)	h				510.68	537.06	562.51	587.28	611.54	659.08	705.95	752.66	799.56
	s				10.2888	10.3814	10.4677	10.5489	10.6259	10.7700	10.9040	11.0303	11.1505
600	v					0.218307	0.228401	0.238148	0.247631	0.266022	0.283386	0.301404	0.318683
(9.28)	h					529.16	555.74	581.40	606.37	654.96	702.55	749.78	797.07
	s					10.2712	10.3613	10.4454	10.5247	10.6720	10.8080	10.9357	11.0570
700	v					0.184313	0.193326	0.201967	0.210326	0.226439	0.242004	0.257211	0.272174
(13.8)	h					520.97	548.76	575.36	601.09	650.78	699.12	746.88	794.56
	s					10.1738	10.2680	10.3553	10.4369	10.5876	10.7258	10.8549	10.9772
800	v						0.166959	0.174787	0.182315	0.196734	0.210582	0.224060	0.237289
(17.85)	h						541.55	569.17	595.69	646.54	695.65	743.96	792.04
	s						10.1841	10.2747	10.3589	10.5131	10.6534	10.7840	10.9074

P (Tsat)		(1)	(2)	(3)	(4)	(5)	(6)	(7)
900 (21.52)	v	0.146392	0.153606	0.160499	0.173614	0.186133	0.198271	0.210152
	h	534.09	562.80	590.17	642.22	692.14	741.01	789.50
	s	10.1072	10.2013	10.2882	10.4461	10.5887	10.7208	10.8452
1000 (24.9)	v	0.129880	0.136622	0.143019	0.155103	0.166565	0.177634	0.188439
	h	526.36	556.24	584.52	637.84	688.59	738.03	786.95
	s	10.0354	10.1335	10.2232	10.3849	10.5300	10.6637	10.7891
1200 (30.94)	v		0.111037	0.116723	0.127297	0.137191	0.146665	0.155861
	h		542.53	572.79	628.85	681.36	732.01	781.79
	s		10.0098	10.1059	10.2759	10.4260	10.5630	10.6906
1400 (36.26)	v		0.092619	0.097844	0.107388	0.116183	0.124528	0.132582
	h		527.89	560.44	619.55	673.96	725.89	776.58
	s		9.8972	10.0005	10.1798	10.3354	10.4758	10.6058
1600 (41.04)	v			0.083590	0.092409	0.100402	0.107912	0.115113
	h			547.37	609.90	666.37	719.65	771.30
	s			9.9032	10.0929	10.2543	10.3984	10.5308
1800 (45.38)	v			0.072404	0.080714	0.088104	0.094974	0.101518
	h			533.48	599.88	658.58	713.31	765.95
	s			9.8112	10.0126	10.1805	10.3284	10.4634
2000 (49.37)	v				0.071313	0.078243	0.084611	0.090635
	h				589.43	650.59	706.84	760.53
	s				9.9373	10.1121	10.2642	10.4019
2200 (53.07)	v				0.063577	0.070153	0.076120	0.081723
	h				578.53	642.36	700.25	755.03
	s				9.8656	10.0481	10.2047	10.3452

* Abstracted with permission from "Thermodynamic Properties of Ammonia" by Lester Haar and John S. Gallagher, *Journal of Physical and Chemical Reference Data*, Vol. 7, No. 3; pp. 635–792, 1978.

Table A.3.1 (E) Properties of Saturated Ammonia*

Temp., °F	Press., psia	Specific Volume, ft³/lbm			Enthalpy, Btu/lbm			Entropy, Btu/lbm-°R		
		Sat. Liquid v_f	Evap. v_{fg}	Sat. Vapor v_g	Sat. Liquid h_f	Evap. h_{fg}	Sat. Vapor h_g	Sat. Liquid s_f	Evap. s_{fg}	Sat. Vapor s_g
−60	5.55	0.0228	44.707	44.73	−21.2	610.8	589.6	−0.0517	1.5286	1.4769
−55	6.54	0.0229	38.357	38.38	−15.9	607.5	591.6	−0.0386	1.5017	1.4631
−50	7.67	0.0230	33.057	33.08	−10.6	604.3	593.7	−0.0256	1.4753	1.4497
−45	8.95	0.0231	28.597	28.62	−5.3	600.9	595.6	−0.0127	1.4495	1.4368
−40	10.41	0.02322	24.837	24.86	0	597.6	597.6	0.000	1.4242	1.4242
−35	12.05	0.02333	21.657	21.68	5.3	594.2	599.5	0.0126	1.3994	1.4120
−30	13.90	0.0235	18.947	18.97	10.7	590.7	601.4	0.0250	1.3751	1.4001
−25	15.98	0.0236	16.636	16.66	16.0	587.2	603.2	0.0374	1.3512	1.3886
−20	18.30	0.0237	14.656	14.68	21.4	583.6	605.0	0.0497	1.3277	1.3774
−15	20.88	0.02381	12.946	12.97	26.7	580.0	606.7	0.0618	1.3044	1.3664
−10	23.74	0.02393	11.476	11.50	32.1	576.4	608.5	0.0738	1.2820	1.3558
−5	26.92	0.02406	10.206	10.23	37.5	572.6	610.1	0.0857	1.2597	1.3454
0	30.42	0.02419	9.092	9.116	42.9	568.9	611.8	0.0975	1.2377	1.3352
5	34.27	0.02432	8.1257	8.150	48.3	565.0	613.3	0.1092	1.2161	1.3253
10	38.51	0.02446	7.2795	7.304	53.8	561.1	614.9	0.1208	1.1949	1.3157
15	43.14	0.02460	6.5374	6.562	59.2	557.1	616.3	0.1323	1.1739	1.3062

20	48.21	0.02474	5.8853	5.910	64.7	553.1	617.8	0.1437	1.1532	1.2969
25	53.73	0.02488	5.3091	5.334	70.2	548.9	619.1	0.1551	1.1328	1.2879
30	59.74	0.02503	4.8000	4.825	75.7	544.8	620.5	0.1663	1.1127	1.2790
35	66.26	0.02518	4.3478	4.373	81.2	540.5	621.7	0.1775	1.0929	1.2704
40	73.32	0.02533	3.9457	3.971	86.8	536.2	623.0	0.1885	1.0733	1.2618
45	80.96	0.02548	3.5885	3.614	92.3	531.8	624.1	0.1996	1.0539	1.2535
50	89.19	0.02564	3.2684	3.294	97.9	527.3	625.2	0.2105	1.0348	1.2453
55	98.06	0.02581	2.9822	3.008	103.5	522.8	626.3	0.2214	1.0159	1.2373
60	107.6	0.02597	2.7250	2.751	109.2	518.1	627.3	0.2322	0.9972	1.2294
65	117.8	0.02614	2.4939	2.520	114.8	513.4	628.2	0.2430	0.9786	1.2216
70	128.8	0.02632	2.2857	2.312	120.5	508.6	629.1	0.2537	0.9603	1.2140
75	140.5	0.02650	2.0985	2.125	126.2	503.7	629.9	0.2643	0.9422	1.2065
80	153.0	0.02668	1.9283	1.955	132.0	498.7	630.7	0.2749	0.9242	1.1991
85	166.4	0.02687	1.7741	1.801	137.8	493.6	631.4	0.2854	0.9064	1.1918
90	180.6	0.02707	1.6339	1.661	143.5	488.5	632.0	0.2958	0.8888	1.1846
95	195.8	0.02727	1.5067	1.534	149.4	483.2	632.6	0.3062	0.8713	1.1775
100	211.9	0.02747	1.3915	1.419	155.2	477.8	633.0	0.3166	0.8539	1.1705
105	228.9	0.02769	1.2853	1.313	161.1	472.3	633.4	0.3269	0.8366	1.1635
110	247.0	0.02790	1.1891	1.217	167.0	466.7	633.7	0.3372	0.8194	1.1566
115	266.2	0.02813	1.0999	1.128	173.0	460.9	633.9	0.3474	0.8023	1.1497
120	286.4	0.02836	1.0186	1.047	179.0	455.0	634.0	0.3576	0.7851	1.1427
125	307.8	0.02860	0.9444	0.973	185.1	448.9	634.0	0.3679	0.7679	1.1358

*Reprinted from National Bureau of Standards Circular No. 142, *Tables of Thermodynamic Properties of Ammonia.*

Table A.3.2 (E) Properties of Superheated Ammonia*

Abs. press., psia (sat. temp.)		Temperature, °F									
		0	20	40	60	80	100	150	200	250	300
5 (−63.11)	v	57.55	60.12	62.69	65.24	67.79	70.33	76.68	...	...	...
	h	620.4	630.4	640.4	650.5	660.6	670.7	696.4	...	...	...
	s	1.5608	1.5821	1.6026	1.6223	1.6413	1.6598	1.7038	...	...	...
10 (−41.34)	v	28.58	29.90	31.20	32.49	33.78	35.07	38.26	41.45	...	...
	h	618.9	629.1	639.3	649.5	659.7	670.0	695.8	722.2	...	...
	s	1.4773	1.4992	1.5200	1.5400	1.5593	1.5779	1.6222	1.6637	...	...
15 (−27.29)	v	18.92	19.82	20.70	21.58	22.44	23.31	25.46	27.59	...	...
	h	617.2	627.8	638.2	648.5	658.9	669.2	695.3	721.7	...	...
	s	1.4272	1.4497	1.4709	1.4912	1.5108	1.5296	1.5742	1.6158	...	...
20 (−16.64)	v	14.09	14.78	15.45	16.12	16.78	17.43	19.05	20.66	...	...
	h	615.5	626.4	637.0	647.5	658.0	668.5	694.7	721.2	...	...
	s	1.3907	1.4138	1.4356	1.4562	1.4760	1.4950	1.5399	1.5817	...	...
25 (−7.96)	v	11.19	11.75	12.30	12.84	13.37	13.90	15.21	16.50	17.79	...
	h	613.8	625.0	635.8	646.5	657.1	667.7	694.1	720.8	748.0	...
	s	1.3616	1.3855	1.4077	1.4287	1.4487	1.4679	1.5131	1.5552	1.5948	...
30 (−0.57)	v	...	9.731	10.20	10.65	11.10	11.55	12.65	13.73	14.81	...
	h	...	623.5	634.6	645.5	656.2	666.9	693.5	720.3	747.5	...
	s	...	1.3618	1.3845	1.4059	1.4261	1.4456	1.4911	1.5334	1.5733	...
35 (5.89)	v	...	8.287	8.695	9.093	9.484	9.869	10.82	11.75	12.68	...
	h	...	622.0	633.4	644.4	655.3	666.1	692.9	719.9	747.2	...
	s	...	1.3413	1.3646	1.3863	1.4069	1.4265	1.4724	1.5148	1.5547	...
40 (11.66)	v	...	7.203	7.568	7.922	8.268	8.609	9.444	10.27	11.08	11.88
	h	...	620.4	632.1	643.4	654.4	665.3	692.3	719.4	746.8	774.6
	s	...	1.3231	1.3470	1.3692	1.3900	1.4098	1.4561	1.4987	1.5387	1.5766

Press. (Sat. Temp.)											
50 (21.67)	v	...	...	5.988	6.280	6.564	6.843	7.521	8.185	8.840	9.489
	h	...	...	629.5	641.2	652.6	663.7	691.1	718.5	746.1	774.0
	s	...	...	1.3169	1.3399	1.3613	1.3816	1.4286	1.4716	1.5219	1.5500
60 (30.21)	v	...	...	4.933	5.184	5.428	5.665	6.239	6.798	7.348	7.892
	h	...	...	626.8	639.0	650.7	662.1	689.9	717.5	745.3	773.3
	s	...	...	1.2913	1.3152	1.3373	1.3581	1.4058	1.4493	1.4898	1.5281
70 (37.70)	v	...	...	4.177	4.401	4.615	4.822	5.323	5.807	6.282	6.750
	h	...	...	623.9	636.6	648.7	660.4	688.7	716.6	744.5	772.7
	s	...	...	1.2688	1.2937	1.3166	1.3378	1.3863	1.4302	1.4710	1.5095
80 (44.40)	v	...	...	...	3.812	4.005	4.190	4.635	5.063	5.482	5.894
	h	...	...	...	634.3	646.7	658.7	687.5	715.6	743.8	772.1
	s	...	...	...	1.2745	1.2981	1.3199	1.3692	1.4136	1.4547	1.4933
90 (50.47)	v	...	...	...	3.353	3.529	3.698	4.100	4.484	4.859	5.228
	h	...	...	...	631.8	644.7	657.0	686.3	714.7	743.0	771.5
	s	...	...	...	1.2571	1.2814	1.3038	1.3539	1.3988	1.4401	1.4789
100 (56.05)	v	...	...	...	2.985	3.149	3.304	3.672	4.021	4.361	4.695
	h	...	...	...	629.3	642.6	655.2	685.0	713.7	742.2	770.8
	s	...	...	...	1.2409	1.2661	1.2891	1.3401	1.3854	1.4271	1.4660
120 (66.02)	v	...	...	...	...	2.576	2.712	3.029	3.326	3.614	3.895
	h	...	...	...	...	638.3	651.6	682.5	711.8	740.7	769.6
	s	...	...	...	...	1.2386	1.2628	1.3157	1.3620	1.4042	1.4435
140 (74.79)	v	...	...	...	...	2.166	2.288	2.569	2.830	3.080	3.323
	h	...	...	...	...	633.8	647.8	679.9	709.9	739.2	768.3
	s	...	...	...	...	1.2140	1.2396	1.2945	1.3418	1.3846	1.4243
160 (82.64)	v	...	...	...	...	...	1.969	2.224	2.457	2.679	2.895
	h	...	...	...	...	...	643.9	677.2	707.9	737.6	767.1
	s	...	...	...	...	...	1.2186	1.2757	1.3240	1.3675	1.4076

(continued)

*Reprinted from National Bureau of Standards Circular No. 142, Tables of Thermodynamic Properties of Ammonia.

Table A.3.2 (E) Properties of Superheated Ammonia (*continued*)

Abs. press., psia (sat. temp.)		Temperature, °F									
		0	20	40	60	80	100	150	200	250	300
180 (89.78)	v	...	...	...	...	...	1.720	1.955	2.167	2.367	2.561
	h	...	...	...	...	...	639.9	674.6	705.9	736.1	765.8
	s	...	...	...	...	...	1.1992	1.2586	1.3081	1.3521	1.3926
200 (96.34)	v	...	...	...	...	...	...	1.740	1.935	2.118	2.295
	h	...	...	...	...	...	...	671.8	703.9	734.5	764.5
	s	...	...	...	...	...	...	1.2429	1.2935	1.3382	1.3791
220 (102.42)	v	...	...	...	...	...	...	1.564	1.745	1.914	2.076
	h	...	...	...	...	...	...	669.0	701.9	732.9	763.2
	s	...	...	...	...	...	...	1.2281	1.2801	1.3255	1.3668
240 (108.09)	v	...	...	...	...	...	...	1.416	1.587	1.745	1.895
	h	...	...	...	...	...	...	666.1	699.8	731.3	762.0
	s	...	...	...	...	...	...	1.2145	1.2677	1.3137	1.3554
260 (113.42)	v	...	...	...	...	...	...	1.292	1.453	1.601	1.741
	h	...	...	...	...	...	...	663.1	697.7	729.7	760.7
	s	...	...	...	...	...	...	1.2014	1.2560	1.3027	1.3449
280 (118.45)	v	...	...	...	...	...	...	1.184	1.339	1.478	1.610
	h	...	...	...	...	...	...	660.1	695.6	728.1	759.4
	s	...	...	...	...	...	...	1.1888	1.2449	1.2924	1.3350
300 (123.21)	v	...	...	...	...	...	...	1.091	1.239	1.372	1.496
	h	...	...	...	...	...	...	656.9	693.5	726.5	758.1
	s	...	...	...	...	...	...	1.1767	1.2344	1.2827	1.3257

Table A.4.1 (SI) Properties of Saturated Nitrogen*

Temp. K	Press. MPa	Volume, m³/kg — Sat. Liquid v_f	Volume — Evap. v_{fg}	Volume — Sat. Vapor v_g	Enthalpy, kJ/kg — Sat. Liquid h_f	Enthalpy — Evap. h_{fg}	Enthalpy — Sat. Vapor h_g	Entropy, kJ/kg·K — Sat. Liquid s_f	Entropy — Evap. s_{fg}	Entropy — Sat. Vapor s_g
63.15	0.01252	0.001150	1.480698	1.481848	−150.912	215.396	64.484	2.4235	3.4109	5.8344
65	0.01741	0.001161	1.092184	1.093345	−147.175	213.387	66.212	2.4817	3.2828	5.7645
70	0.03857	0.001191	0.525169	0.526360	−137.090	207.793	70.703	2.6309	2.9683	5.5992
75	0.07610	0.001223	0.280527	0.281750	−126.952	201.821	74.869	2.7701	2.6909	5.4610
77.35	0.101325	0.001240	0.215149	0.216389	−122.150	198.839	76.689	2.8326	2.5706	5.4032
80	0.13698	0.001259	0.162499	0.163758	−116.692	195.323	78.631	2.9015	2.4414	5.3429
85	0.22908	0.001299	0.100184	0.101483	−106.254	188.152	81.898	3.0265	2.2136	5.2401
90	0.36083	0.001343	0.064769	0.066112	−95.577	180.134	84.557	3.1468	2.0012	5.1480
95	0.54112	0.001393	0.043370	0.044763	−84.593	171.067	86.474	3.2628	1.8006	5.0634
100	0.77917	0.001452	0.029747	0.031199	−73.198	160.682	87.484	3.3763	1.6067	4.9830
105	1.0846	0.001522	0.020663	0.022185	−61.239	148.588	87.349	3.4884	1.4150	4.9034
110	1.4676	0.001610	0.014336	0.015946	−48.445	134.152	85.707	3.6019	1.2194	4.8213
115	1.9393	0.001729	0.009710	0.011439	−34.310	116.193	81.883	3.7204	1.0103	4.7307
120	2.5130	0.001915	0.006079	0.007994	−17.609	91.910	74.301	3.8536	0.7660	4.6196
125	3.2080	0.002355	0.002541	0.004896	+6.725	48.878	55.603	4.0399	0.3913	4.4312
126.19	3.3978	0.003194	0.000000	0.003194	+29.792	0.000	29.792	4.2195	0.0000	4.2195

* Abstracted with permission from "Thermodynamic Properties of Nitrogen from the Freezing Line to 2000 K at Pressures to 1000 MPa" by R. T. Jacobsen, R. B. Stewart, and M. Jahangiri, *Journal of Physical and Chemical Reference Data*, Vol. 15 (2), 1986.

Table A.4.2 (SI) Properties of Superheated Nitrogen*

Temp. K	v m³/kg	h kJ/kg	s kJ/kg·K	v m³/kg	h kJ/kg	s kJ/kg·K	v m³/kg	h kJ/kg	s kJ/kg·K
	0.101325 MPa (sat. temp. 77.35 K)			0.2 MPa (sat. temp. 83.62 K)			0.5 MPa (sat. temp. 93.99 K)		
80	0.225222	80.027	5.4457						
100	0.287144	101.917	5.6906	0.142523	100.239	5.4775	0.053063	94.460	5.1662
120	0.347423	123.136	5.8837	0.173974	121.936	5.6756	0.067012	118.127	5.3821
140	0.407090	144.194	6.0461	0.204759	143.280	5.8398	0.080067	140.442	5.5542
160	0.466454	165.163	6.1861	0.235194	164.442	5.9812	0.092721	162.221	5.6995
180	0.525585	186.081	6.3092	0.265430	185.496	6.1054	0.105154	183.700	5.8262
200	0.584538	206.965	6.4195	0.295511	206.479	6.2157	0.117442	205.001	5.9383
220	0.643434	227.826	6.5188	0.325531	227.419	6.3156	0.129640	226.181	6.0393
240	0.702296	248.670	6.6094	0.355449	248.324	6.4067	0.141770	247.278	6.1311
260	0.761145	269.507	6.6930	0.385338	269.211	6.4902	0.153856	268.315	6.2153
280	0.819883	290.337	6.7701	0.415235	290.080	6.5673	0.165905	289.312	6.2931
300	0.878604	311.163	6.8418	0.445053	310.941	6.6394	0.177919	310.281	6.3653
320	0.937194	331.989	6.9093	0.474830	331.800	6.7069	0.189922	331.228	6.4331

Temp. K	v m³/kg	h kJ/kg	s kJ/kg·K	v m³/kg	h kJ/kg	s kJ/kg·K	v m³/kg	h kJ/kg	s kJ/kg·K
	1.0 MPa (sat. temp. 103.73 K)			2.0 MPa (sat. temp. 115.58 K)			4.0 MPa		
120	0.031169	111.081	5.1358	0.012598	92.100	4.8117			
140	0.038450	135.473	5.3240	0.017516	124.403	5.0619	0.006643	94.131	4.6896
160	0.045219	158.424	5.4771	0.021444	150.437	5.2358	0.009513	132.581	4.9477
180	0.051732	180.677	5.6085	0.025030	174.480	5.3775	0.011702	161.521	5.1183
200	0.058094	202.524	5.7234	0.028438	197.530	5.4989	0.013650	187.420	5.2551
220	0.064357	224.117	5.8262	0.031737	219.994	5.6060	0.015470	221.802	5.3711
240	0.070553	245.536	5.9194	0.034963	242.080	5.7024	0.017213	235.291	5.4735
260	0.076700	266.833	6.0047	0.038142	263.902	5.7895	0.018903	258.194	5.5649
280	0.082812	288.041	6.0832	0.041283	285.539	5.8698	0.020555	280.695	5.6484
300	0.088893	309.185	6.1564	0.044396	307.040	5.9440	0.022179	302.902	5.7252
320	0.094956	330.286	6.2243	0.047487	328.437	6.0129	0.023783	324.892	5.7959

(continued)

* Abstracted with permission from "Thermodynamic Properties of Nitrogen from the Freezing Line to 2000 K at Pressures to 1000 MPa" by R. T. Jacobsen, R. B. Stewart, and M. Jahangiri. *Journal of Physical and Chemical Reference Data*, Vol. 15 (2), 1986.

Table A.4.2 (SI) (continued)

Temp. K	v m³/kg	h kJ/kg	s kJ/kg·K	v m³/kg	h kJ/kg	s kJ/kg·K	v m³/kg	h kJ/kg	s kJ/kg·K
	6.0 MPa			8.0 MPa			4.0 MPa		
140	0.002941	47.442	4.2926	0.002224	27.784	4.1167	0.002003	20.867	4.0374
160	0.005556	112.159	4.7292	0.003748	91.800	4.5454	0.002908	76.518	4.4090
180	0.007309	148.021	4.9413	0.005193	134.687	4.7988	0.004021	122.650	4.6814
200	0.008772	177.296	5.0955	0.006387	167.472	4.9720	0.005014	158.355	4.8699
220	0.010095	203.777	5.2219	0.007449	196.073	5.1083	0.005902	188.884	5.0155
240	0.011337	228.730	5.3304	0.008434	222.479	5.2233	0.006722	216.639	5.1362
260	0.012526	252.729	5.4264	0.009367	247.553	5.3236	0.007495	242.719	5.2408
280	0.013678	276.090	5.5132	0.010264	271.749	5.4132	0.008236	267.697	5.3332
300	0.014803	298.993	5.5920	0.011136	295.324	5.4946	0.008952	291.907	5.4168
320	0.015907	321.561	5.6649	0.011987	318.445	5.5692	0.009650	315.553	5.4932

Temp. K	v m³/kg	15.0 MPa h kJ/kg	s kJ/kg·K	v m³/kg	20.0 MPa h kJ/kg	s kJ/kg·K	v m³/kg	30.0 MPa h kJ/kg	s kJ/kg·K
140	0.001770	14.808	3.9275	0.001655	13.747	3.8589	0.001526	16.391	3.7647
160	0.002183	59.140	4.2234	0.001929	53.625	4.1249	0.001697	52.522	4.0056
180	0.002749	102.345	4.4779	0.002281	93.021	4.3569	0.001899	87.998	4.2148
200	0.003365	140.603	4.6796	0.002687	130.172	4.5529	0.002127	122.407	4.3962
220	0.003964	174.098	4.8395	0.003108	164.259	4.7157	0.002373	155.381	4.5532
240	0.004531	204.330	4.9709	0.003525	195.598	4.8520	0.002628	186.770	4.6900
260	0.005071	232.410	5.0834	0.003930	224.828	4.9688	0.002885	216.649	4.8096
280	0.005589	259.012	5.1822	0.004323	252.501	5.0716	0.003140	245.218	4.9156
300	0.006088	284.568	5.2704	0.004704	279.010	5.1630	0.003392	272.702	5.0102
320	0.006573	309.353	5.3504	0.005075	304.644	5.2458	0.003640	299.304	5.0962

(*continued*)

Table A.4.2 (SI) (*continued*)

Temp. K	v m³/kg	h kJ/kg	s kJ/kg·K	v m³/kg	h kJ/kg	s kJ/kg·K	v m³/kg	h kJ/kg	s kJ/kg·K
		40.0 MPa			50.0 MPa			60.0 MPa	
140	0.001447	21.627	3.6958	0.001391	28.049	3.6405	0.001347	35.137	3.5933
160	0.001577	56.113	3.9264	0.001497	61.618	3.8646	0.001438	68.129	3.8136
180	0.001722	89.794	4.1245	0.001612	94.317	4.0574	0.001535	100.221	4.0028
200	0.001881	122.579	4.2973	0.001736	126.149	4.2252	0.001638	131.460	4.1674
220	0.002051	154.364	4.4490	0.001867	157.120	4.3726	0.001745	161.889	4.3126
240	0.002229	185.086	4.5825	0.002003	187.238	4.5036	0.001855	191.561	4.4415
260	0.002411	214.750	4.7014	0.002143	216.528	4.6211	0.001969	220.519	4.5575
280	0.002595	243.412	4.8074	0.002285	245.026	4.7267	0.002084	248.802	4.6621
300	0.002779	271.81	4.9034	0.002429	272.788	4.8224	0.002201	276.457	4.7578
320	0.002962	298.169	4.9905	0.002572	299.886	4.9099	0.002318	303.541	4.8453

Table A.4.1 (E) Properties of Saturated Nitrogen*

Temp., °R	Press., psia	Volume, ft³/lbm Sat. Liquid v_f	Sat. Vapor v_g	Internal Energy, Btu/lbm Sat. Liquid u_f	Sat. Vapor u_g	Enthalpy, Btu/lbm Sat. Liquid h_f	Sat. Vapor h_g	Entropy, Btu/lbm-°R Sat. Liquid s_f	Sat. Vapor s_g
113.666	1.818	0.01846	23.730	−64.690	19.908	−64.684	27.895	0.58022	1.39546
120	3.341	0.01875	13.556	−61.683	20.933	−61.671	29.320	0.60596	1.36499
130	7.665	0.01927	6.3211	−56.821	22.468	−56.794	31.440	0.64487	1.32426
139.224	14.696	0.01981	3.4734	−52.297	23.767	−52.243	33.218	0.67850	1.29290
140	15.457	0.01986	3.3146	−51.916	23.871	−51.859	33.359	0.68123	1.29048
150	28.194	0.02053	1.8993	−46.987	25.107	−46.879	35.023	0.71526	1.26171
160	47.520	0.02130	1.1636	−42.007	26.136	−41.819	36.376	0.74743	1.23646
170	75.182	0.02220	0.7496	−36.932	26.910	−36.623	37.349	0.77825	1.21358
180	112.990	0.02326	0.5015	−31.708	27.364	−31.222	37.857	0.80821	1.19208
190	162.849	0.02455	0.3439	−26.264	27.400	−25.524	37.770	0.83781	1.17096
200	226.851	0.02620	0.2385	−20.491	26.842	−19.390	36.863	0.86772	1.14895
210	307.316	0.02849	0.1642	−14.150	25.333	−12.529	34.676	0.89920	1.12390
220	406.863	0.03246	0.1070	−6.432	21.800	−3.986	29.862	0.93629	1.09010
226	477.892	0.03942	0.0707	1.460	15.661	4.949	21.919	0.97421	1.04930
227.160	493.123	0.0510	0.0510	8.589	8.589	13.247	13.247	1.01026	1.01026

*Abstracted with permission from *Table of Thermodynamic Properties of Nitrogen*, by R. T. Jacobsen and R. B. Stewart, Department of Mechanical Engineering, University of Idaho, 1972.

Table A.4.2 (E) Properties of Superheated Nitrogen*

Temp., °R	14.7 psia v ft³/lbm	14.7 psia h Btu/lbm	14.7 psia s Btu/lbm-°R	20 psia v ft³/lbm	20 psia h Btu/lbm	20 psia s Btu/lbm-°R	50 psia v ft³/lbm	50 psia h Btu/lbm	50 psia s Btu/lbm-°R
	(sat. temp. 139.224 °R)			(sat. temp. 144.108 °R)			(sat. temp. 161.049 °R)		
140	3.4949	33.426	1.29439						
180	4.5988	43.858	1.35999	3.3539	43.585	1.33713	1.2815	41.955	1.26602
220	5.6718	54.017	1.41097	4.1508	53.832	1.38855	1.6215	52.760	1.32028
260	6.7326	64.076	1.45299	4.9353	63.940	1.43077	1.9470	63.157	1.36372
300	7.7876	74.089	1.48881	5.7136	73.982	1.46670	2.2658	73.378	1.40029
340	8.8386	84.076	1.52006	6.4885	83.990	1.49802	2.5810	83.507	1.43199
380	9.8873	94.047	1.54778	7.2611	93.977	1.52579	2.8939	93.582	1.46000
420	10.935	104.008	1.57271	8.0321	103.950	1.55074	3.2054	103.622	1.48512
460	11.982	113.963	1.59535	8.8020	113.914	1.57340	3.5159	113.639	1.50791
500	13.028	123.914	1.61609	9.5712	123.873	1.59416	3.8256	123.640	1.52875
540	14.073	133.862	1.63523	10.340	133.827	1.61331	4.1348	133.630	1.54797
580	15.117	143.812	1.65301	11.109	143.782	1.63110	4.4435	143.614	1.56581
600	15.640	148.787	1.66144	11.493	148.760	1.63953	4.5977	148.605	1.57427

Temp., °R	100 psia v ft³/lbm	100 psia h Btu/lbm	100 psia s Btu/lbm-°R	200 psia v ft³/lbm	200 psia h Btu/lbm	200 psia s Btu/lbm-°R	500 psia v ft³/lbm	500 psia h Btu/lbm	500 psia s Btu/lbm-°R
	(sat. temp. 176.882 °R)			(sat. temp. 196.090 °R)					
180	0.5836	38.804	1.20457						
220	0.7767	50.863	1.26522	0.3503	46.508	1.20202			
260	0.9504	61.814	1.31100	0.4512	58.966	1.25416	0.1477	48.524	1.15946
300	1.1165	72.355	1.34872	0.5417	70.254	1.29457	0.1965	63.449	1.21302
340	1.2786	82.696	1.38108	0.6275	81.054	1.32838	0.2374	75.990	1.25231
380	1.4384	92.922	1.40952	0.7108	91.597	1.35770	0.2749	87.610	1.28464
420	1.5967	103.076	1.43492	0.7925	101.985	1.38369	0.3107	98.750	1.31252

T									
460	1.7539	113.182	1.45791	0.8731	112.272	1.40709	0.3452	109.598	1.33719
500	1.9103	123.254	1.47891	0.9529	122.489	1.42839	0.3790	120.254	1.35941
540	2.0662	133.303	1.49824	1.0322	132.656	1.44795	0.4122	130.776	1.37965
580	2.2217	143.337	1.51617	1.1110	142.788	1.46605	0.4450	141.201	1.39828
600	2.2993	148.349	1.52466	1.1503	147.844	1.47462	0.4613	146.385	1.40707

T	1000 psia			2000 psia			3000 psia		
260	0.0469	23.113	1.03206	0.0303	11.002	0.96039	0.0271	9.677	0.93498
300	0.0828	50.590	1.13118	0.0399	33.170	1.03970	0.0322	28.600	1.00268
340	0.1091	67.336	1.18374	0.0520	53.287	1.10276	0.0385	46.958	1.06016
380	0.1313	81.095	1.22204	0.0641	70.205	1.14987	0.0456	63.913	1.10735
420	0.1515	93.588	1.25332	0.0753	84.946	1.18679	0.0528	79.371	1.14605
460	0.1706	105.394	1.28018	0.0859	98.383	1.21737	0.0598	93.627	1.17849
500	0.1888	116.775	1.30391	0.0959	111.004	1.24369	0.0666	107.001	1.20638
540	0.2065	127.871	1.32526	0.1055	123.081	1.26693	0.0732	119.735	1.23089
580	0.2238	138.765	1.34472	0.1148	134.777	1.28783	0.0796	132.000	1.25280
600	0.2324	144.153	1.35386	0.1193	140.516	1.29756	0.0827	137.993	1.26296

T	4000 psia			5000 psia			6000 psia		
260	0.0254	10.126	0.91804	0.0243	11.307	0.90490	0.0234	12.887	0.89399
300	0.0290	27.693	0.98089	0.0271	28.135	0.96511	0.0258	29.248	0.95254
340	0.0332	44.821	1.03451	0.0303	44.529	1.01643	0.0284	45.173	1.00238
380	0.0380	61.142	1.07991	0.0338	60.313	1.06033	0.0312	60.567	1.04520
420	0.0429	76.496	1.11835	0.0376	75.393	1.09808	0.0342	75.388	1.08230
460	0.0480	90.941	1.15122	0.0414	89.777	1.13081	0.0373	89.642	1.11473
500	0.0530	104.629	1.17976	0.0453	103.537	1.15950	0.0404	103.374	1.14336
540	0.0579	117.713	1.20494	0.0492	116.769	1.18496	0.0436	116.647	1.16890
580	0.0627	130.323	1.22747	0.0530	129.563	1.20782	0.0467	129.525	1.19191
600	0.0651	136.483	1.23791	0.0549	135.821	1.21843	0.0483	135.836	1.20261

*Abstracted with permission from Table of Thermodynamic Properties of Nitrogen, by R. T. Jacobsen and R. B. Stewart, Department of Mechanical Engineering, University of Idaho, 1972.

(continued)

Table A.4.2 (E) (continued)

Temp., °R	7000 psia			8000 psia			9000 psia		
	v ft³/lbm	h Btu/lbm	s Btu/lbm-°R	v ft³/lbm	h Btu/lbm	s Btu/lbm-°R	v ft³/lbm	h Btu/lbm	s Btu/lbm-°R
260	0.0228	14.716	0.88458	0.0222	16.712	0.87625	0.0217	18.828	0.86874
300	0.0248	30.753	0.94197	0.0240	32.514	0.93280	0.0233	34.451	0.92465
340	0.0270	46.358	0.99082	0.0259	47.887	0.98092	0.0251	49.651	0.97223
380	0.0294	61.471	1.03285	0.0280	62.794	1.02238	0.0269	64.402	1.01326
420	0.0319	76.080	1.06942	0.0301	77.237	1.05853	0.0288	78.715	1.04908
460	0.0344	90.201	1.10154	0.0323	91.240	1.09038	0.0307	92.620	1.08071
500	0.0371	103.870	1.13004	0.0346	104.839	1.11874	0.0327	106.153	1.10892
540	0.0397	117.134	1.15557	0.0369	118.073	1.14420	0.0347	119.350	1.13432
580	0.0423	130.043	1.17863	0.0391	130.984	1.16727	0.0367	132.250	1.15737
600	0.0437	136.380	1.18937	0.0403	137.331	1.17803	0.0377	138.599	1.16813

Table A.5.1 (SI) Properties of Saturated Oxygen*

Temp. K	Press. MPa	Volume, m³/kg			Enthalpy, kJ/kg			Entropy, kJ/kg · K		
		Sat. Liquid v_f	Evap. v_{fg}	Sat. Vapor v_g	Sat. Liquid h_f	Evap. h_{fg}	Sat. Vapor h_g	Sat. Liquid s_f	Evap. s_{fg}	Sat. Vapor s_g
54.36	0.00015	0.000765	97.655485	97.656250	−193.828	242.909	49.081	2.0869	4.4684	6.5553
55	0.00018	0.000767	80.127438	80.128205	−192.700	242.356	49.656	2.1075	4.4066	6.5141
60	0.00073	0.000780	21.403330	21.404110	−184.106	238.275	54.169	2.2572	3.9712	6.2284
65	0.00234	0.000794	7.216296	7.217090	−175.722	234.422	58.700	2.3913	3.6065	5.9978
70	0.00626	0.000809	2.892710	2.893519	−167.381	230.590	63.209	2.5147	3.2944	5.8091
75	0.01454	0.000824	1.330096	1.330920	−159.028	226.659	67.631	2.6300	3.0222	5.6522
80	0.03012	0.000840	0.680879	0.681719	−150.634	222.534	71.900	2.7381	2.7819	5.5200
85	0.05683	0.000857	0.379962	0.380819	−142.188	218.138	75.950	2.8403	2.5663	5.4066
90	0.09935	0.000875	0.227194	0.228069	−133.669	213.397	79.728	2.9372	2.3712	5.3084
90.19	0.101325	0.000876	0.223106	0.223982	−133.347	213.210	79.863	2.9409	2.3641	5.3050
95	0.16309	0.000895	0.143640	0.144535	−125.066	208.247	83.181	3.0297	2.1922	5.2219
100	0.25403	0.000917	0.094992	0.095909	−116.350	202.613	86.263	3.1181	2.0263	5.1444
105	0.37859	0.000940	0.065142	0.066082	−107.500	196.425	88.925	3.2034	1.8707	5.0741
110	0.54350	0.000966	0.045985	0.046951	−98.475	189.588	91.113	3.2859	1.7235	5.0094
115	0.75575	0.000994	0.033206	0.034200	−89.228	181.994	92.766	3.3663	1.5825	4.9488
120	1.0225	0.001027	0.024384	0.025411	−79.700	173.503	93.803	3.4450	1.4459	4.8909
125	1.3512	0.001064	0.018107	0.019171	−69.806	163.925	94.119	3.5231	1.3113	4.8344
130	1.7494	0.001108	0.013508	0.014616	−59.441	153.004	93.563	3.6009	1.1769	4.7778
135	2.2252	0.001162	0.010042	0.011204	−48.425	140.322	91.897	3.6800	1.0394	4.7194
140	2.7875	0.001230	0.007345	0.008575	−36.481	125.203	88.722	3.7619	0.8944	4.6563
145	3.4466	0.001324	0.005149	0.006473	−23.040	106.256	83.216	3.8503	0.7328	4.5831
150	4.2166	0.001480	0.003193	0.004673	−6.555	79.708	73.153	3.9547	0.5316	4.4863
154.58	5.0430	0.002293	0.000000	0.002293	+33.013	0.000	33.013	4.2041	0.0000	4.2041

* Abstracted with permission from *Thermodynamic Properties of Oxygen from the Triple Point to 350 K with Pressures to 80 MPa* by R. B. Stewart, R. T. Jacobsen, and W. Wagner. Center for Applied Thermodynamic Studies, Report 87-1, University of Idaho, Moscow, Idaho, June 1987.

Table A.5.2 (SI) Properties of Superheated Oxygen*

Temp. K	0.101325 MPa (sat. temp. 90.19 K)			0.2 MPa (sat. temp. 97.24 K)			0.5 MPa (sat. temp. 108.81 K)		
	v m³/kg	h kJ/kg	s kJ/kg·K	v m³/kg	h kJ/kg	s kJ/kg·K	v m³/kg	h kJ/kg	s kJ/kg·K
100	0.250441	89.113	5.4022	0.123694	87.350	5.2141			
120	0.303398	107.694	5.5716	0.151515	106.475	5.3884	0.057780	102.466	5.1281
140	0.355760	126.156	5.7141	0.178602	125.222	5.5328	0.069411	122.281	5.2809
160	0.407751	144.541	5.8369	0.205335	143.800	5.6569	0.080597	141.500	5.4094
180	0.459559	162.875	5.9447	0.231859	162.272	5.7656	0.091548	160.413	5.5206
200	0.511289	181.181	6.0413	0.258243	180.678	5.8628	0.102352	179.138	5.6194
220	0.562860	199.472	6.1284	0.284557	199.044	5.9503	0.113061	197.747	5.7081
240	0.614432	217.759	6.2078	0.310791	217.394	6.0300	0.123708	216.284	5.7888
260	0.665885	236.063	6.2813	0.336964	235.747	6.1034	0.134305	234.788	5.8628
280	0.717401	254.391	6.3491	0.363119	254.116	6.1716	0.144864	253.281	5.9313
300	0.768758	272.763	6.4125	0.389214	272.522	6.2350	0.155395	271.791	5.9950
320	0.820210	291.191	6.4719	0.415338	290.978	6.2947	0.165915	290.334	6.0550

* Abstracted with permission from *Thermodynamic Properties of Oxygen from the Triple Point to 350 K with Pressure to 80 MPa* by R. B. Stewart, R. T. Jacobsen, and W. Wagner. Center for Applied Thermodynamic Studies, Report 87-1, University of Idaho, Moscow, Idaho, June 1987.

Temp. K	1.0 MPa (sat. temp. 119.62 K)			2.0 MPa (sat. temp. 132.74 K)			4.0 MPa (sat. temp. 148.66 K)		
	v m³/kg	h kJ/kg	s kJ/kg·K	v m³/kg	h kJ/kg	s kJ/kg·K	v m³/kg	h kJ/kg	s kJ/kg·K
120	0.026127	94.244	4.8994						
140	0.032898	116.975	5.0750	0.014315	104.003	4.8284			
160	0.038972	137.500	5.2122	0.018063	128.731	4.9941	0.007276	105.819	4.7069
180	0.044759	157.231	5.3284	0.021331	150.541	5.1225	0.009534	135.456	4.8819
200	0.050382	176.528	5.4300	0.024387	171.156	5.2313	0.011374	159.716	5.0100
220	0.055902	195.559	5.5209	0.027324	191.116	5.3263	0.013041	181.931	5.1159
240	0.061354	214.425	5.6028	0.030182	210.675	5.4113	0.014614	203.059	5.2078
260	0.066755	233.188	5.6781	0.032989	229.975	5.4888	0.016126	223.531	5.2897
280	0.072119	251.891	5.7472	0.035757	249.113	5.5594	0.017596	243.581	5.3641
300	0.077457	270.572	5.8119	0.038496	268.150	5.6253	0.019035	263.356	5.4322
320	0.082773	289.263	5.8722	0.041213	287.138	5.6866	0.020453	282.950	5.4953

(continued)

Table A.5.2 (SI) Properties of Superheated Oxygen (*continued*)

Temp. K	v m³/kg	h kJ/kg	s kJ/kg·K	v m³/kg	h kJ/kg	s kJ/kg·K	v m³/kg	h kJ/kg	s kJ/kg·K
	6.0 MPa			8.0 MPa			10.0 MPa		
160	0.002564	49.225	4.2925	0.001526	10.790	4.0300	0.001396	4.180	3.9706
180	0.005511	117.078	4.6991	0.003458	94.234	4.5234	0.002362	70.416	4.3597
200	0.007027	147.228	4.8584	0.004862	133.713	4.7325	0.003602	119.684	4.6203
220	0.008293	172.363	4.9781	0.005937	162.503	4.8697	0.004549	152.563	4.7775
240	0.009443	195.328	5.0781	0.006876	187.553	4.9788	0.005357	179.844	4.8963
260	0.010524	217.094	5.1653	0.007742	210.709	5.0716	0.006090	204.447	4.9947
280	0.011561	238.116	5.2431	0.008561	232.744	5.1531	0.006776	227.513	5.0803
300	0.012567	258.650	5.3141	0.009348	254.059	5.2269	0.007430	249.609	5.1566
320	0.013549	278.866	5.3794	0.010111	274.897	5.2941	0.008060	271.066	5.2256

Temp. K	15.0 MPa			20.0 MPa			30.0 MPa		
	v m³/kg	h kJ/kg	s kJ/kg·K	v m³/kg	h kJ/kg	s kJ/kg·K	v m³/kg	h kJ/kg	s kJ/kg·K
160	0.001261	−1.889	3.8916	0.001192	−3.728	3.8419	0.001111	−3.298	3.7728
180	0.001600	+42.866	4.1547	0.001405	+34.741	4.0681	0.001242	+30.396	3.9713
200	0.002183	90.303	4.4047	0.001726	75.188	4.2813	0.001409	64.628	4.1516
220	0.002828	129.788	4.5931	0.002130	113.619	4.4644	0.001614	98.550	4.3131
240	0.003415	161.897	4.7331	0.002550	147.588	4.6125	0.001844	131.100	4.4547
260	0.003949	189.838	4.8450	0.002954	177.622	4.7328	0.002086	161.697	4.5772
280	0.004445	215.328	4.9394	0.003336	204.916	4.8338	0.002330	190.344	4.6834
300	0.004913	239.275	5.0222	0.003700	230.347	4.9216	0.002570	217.322	4.7766
320	0.005360	262.191	5.0963	0.004048	254.491	4.9997	0.002804	242.969	4.8594

(continued)

Table A.5.2 (SI) Properties of Superheated Oxygen (continued)

Temp. K	40.0 MPa			50.0 MPa			60.0 MPa		
	v m³/kg	h kJ/kg	s kJ/kg·K	v m³/kg	h kJ/kg	s kJ/kg·K	v m³/kg	h kJ/kg	s kJ/kg·K
160	0.001060	−0.532	3.7222	0.001023	3.353	3.6818	0.000994	7.886	3.6469
180	0.001158	+31.074	3.9084	0.001103	33.759	3.8606	0.001063	37.509	3.8213
200	0.001275	62.725	4.0753	0.001195	63.994	4.0200	0.001139	66.844	3.9759
220	0.001411	94.100	4.2247	0.001298	93.903	4.1625	0.001223	95.819	4.1141
240	0.001563	124.772	4.3581	0.001411	123.291	4.2903	0.001313	124.331	4.2381
260	0.001725	154.394	4.4769	0.001532	151.969	4.4050	0.001410	152.284	4.3500
280	0.001894	182.806	4.5822	0.001658	179.816	4.5084	0.001510	179.600	4.4513
300	0.002064	210.025	4.6759	0.001787	206.806	4.6016	0.001614	206.256	4.5431
320	0.002235	236.178	4.7603	0.001918	232.972	4.6859	0.001719	232.263	4.6272

Table A.5.1 (E) Properties of Saturated Oxygen*

| Temp., °R | Press., psia | Volume, ft³/lbm | | Internal Energy, Btu/lbm | | Enthalpy, Btu/lbm | | Entropy, Btu/lbm-°R | |
		Sat. Liquid v_f	Sat. Vapor v_g	Sat. Liquid u_f	Sat. Vapor u_g	Sat. Liquid h_f	Sat. Vapor h_g	Sat. Liquid s_f	Sat. Vapor s_g
97.832	0.0212	0.01225	1550.6	−83.867	15.079	−83.867	21.153	0.49593	1.56736
100	0.0307	0.01230	1093.1	−82.982	15.415	−82.982	21.623	0.50488	1.54904
110	0.1388	0.01254	265.53	−78.879	16.957	−78.879	23.781	0.54399	1.47592
120	0.4784	0.01280	83.893	−74.795	18.483	−74.794	25.916	0.57952	1.41778
130	1.340	0.01306	32.342	−70.742	19.981	−70.739	28.008	0.61197	1.37080
140	3.198	0.01334	14.522	−66.704	21.437	−66.696	30.037	0.64189	1.33226
150	6.727	0.01364	7.3368	−62.664	22.835	−62.647	31.977	0.66976	1.30013
160	12.798	0.01396	4.0683	−58.608	24.159	−58.575	33.801	0.69595	1.27292
162.343	14.696	0.01404	3.5842	−57.654	24.456	−57.616	34.209	0.70187	1.26713
170	22.446	0.01431	2.4278	−54.523	25.390	−54.463	35.482	0.72072	1.24949
180	36.843	0.01469	1.5361	−50.396	26.508	−50.296	36.988	0.74432	1.22895
190	57.255	0.01510	1.0182	−46.215	27.490	−46.055	38.285	0.76694	1.21060
200	85.019	0.01557	0.7003	−41.961	28.309	−41.716	39.337	0.78879	1.19385
210	121.529	0.01609	0.4960	−37.612	28.932	−37.249	40.097	0.81006	1.17820
220	168.233	0.01670	0.3593	−33.136	29.313	−32.616	40.506	0.83095	1.16318
230	226.649	0.01742	0.2643	−28.491	29.392	−27.760	40.484	0.85171	1.14831
240	298.369	0.01830	0.1962	−23.610	29.074	−22.599	39.911	0.87266	1.13302
250	385.083	0.01943	0.1457	−18.384	28.203	−16.999	38.591	0.89428	1.11657
260	488.651	0.02101	0.1069	−12.595	26.475	−10.694	36.147	0.91748	1.09760
270	611.527	0.02368	0.07484	−5.584	23.030	−2.903	31.505	0.94495	1.07238
278.246	731.417	0.03673	0.03673	9.359	9.359	14.334	14.334	1.00530	1.00530

*Abstracted with permission from *Table of Thermodynamic Properties of Oxygen*, by R. B. Stewart and R. T. Jacobsen, Department of Mechanical Engineering, University of Idaho, 1972.

Table A.5.2 (E) Properties of Superheated Oxygen*

Temp., °R	14.7 psia (sat. temp. 162.343 °R)			20 psia (sat. temp. 167.838 °R)			50 psia (sat. temp. 186.805 °R)		
	v ft³/lbm	h Btu/lbm	s Btu/lbm-°R	v ft³/lbm	h Btu/lbm	s Btu/lbm-°R	v ft³/lbm	h Btu/lbm	s Btu/lbm-°R
180	4.0098	38.255	1.29079	2.9195	37.964	1.27057			
220	4.9539	47.211	1.33574	3.6223	47.022	1.31604	1.4070	45.910	1.25579
260	5.8848	56.041	1.37262	4.3111	55.903	1.35313	1.6946	55.103	1.29420
300	6.8092	64.821	1.40403	4.9938	64.713	1.38465	1.9752	64.092	1.32636
340	7.7304	73.576	1.43143	5.6728	73.488	1.41211	2.2520	72.987	1.35420
380	8.6498	82.320	1.45574	6.3500	82.247	1.43647	2.5268	81.831	1.37879
420	9.5675	91.061	1.47761	7.0254	91.000	1.45837	2.8000	90.650	1.40086
460	10.484	99.808	1.49750	7.7000	99.755	1.47828	3.0722	99.458	1.42089
500	11.400	108.568	1.51576	8.3738	108.523	1.49656	3.3437	108.268	1.43925
540	12.315	117.349	1.53266	9.0473	117.311	1.51346	3.6147	117.091	1.45623
580	13.231	126.161	1.54840	9.7201	126.128	1.52922	3.8852	125.939	1.47204
600	13.687	130.581	1.55589	10.056	130.550	1.53671	4.0203	130.374	1.47955

Temp., °R	100 psia (sat. temp. 204.422 °R)			200 psia (sat. temp. 225.701 °R)			500 psia (sat. temp. 260.998 °R)		
	v ft³/lbm	h Btu/lbm	s Btu/lbm-°R	v ft³/lbm	h Btu/lbm	s Btu/lbm-°R	v ft³/lbm	h Btu/lbm	s Btu/lbm-°R
220	0.6654	43.864	1.20641						
260	0.8214	53.705	1.24755	0.3825	50.607	1.19629			
300	0.9686	63.031	1.28093	0.4645	60.797	1.23278	0.1589	52.870	1.15692
340	1.1116	72.138	1.30943	0.5411	70.392	1.26282	0.1979	64.714	1.19405
380	1.2523	81.132	1.33444	0.6150	79.711	1.28874	0.2324	75.262	1.22340
420	1.3915	90.063	1.35679	0.6872	88.879	1.31168	0.2646	85.246	1.24839
460	1.5296	98.960	1.37702	0.7583	97.959	1.33233	0.2956	94.925	1.27040

T									
500	1.6669	107.842	1.39554	0.8285	106.989	1.35115	0.3257	104.420	1.29020
540	1.8037	116.726	1.41263	0.8983	115.994	1.36848	0.3552	113.802	1.30825
580	1.9400	125.624	1.42853	0.9675	124.995	1.38456	0.3842	123.115	1.32489
600	2.0080	130.082	1.43609	1.0020	129.498	1.39219	0.3986	127.757	1.33276

T	1000 psia			2000 psia			3000 psia		
300	0.0458	28.227	1.04640	0.0222	5.916	0.95532	0.0199	3.663	0.93492
340	0.0826	53.444	1.12625	0.0316	29.924	1.03035	0.0242	22.290	0.99317
380	0.1051	67.311	1.16490	0.0449	51.193	1.08966	0.0304	41.090	1.04547
420	0.1243	79.034	1.19426	0.0567	67.090	1.12951	0.0375	58.045	1.08794
460	0.1420	89.852	1.21888	0.0671	80.364	1.15973	0.0445	72.807	1.12155
500	0.1586	100.181	1.24041	0.0767	92.360	1.18475	0.0511	86.028	1.14912
540	0.1746	110.214	1.25972	0.0857	103.639	1.20646	0.0574	98.271	1.17269
580	0.1902	120.055	1.27730	0.0942	114.469	1.22581	0.0634	109.885	1.19344
600	0.1978	124.926	1.28556	0.0984	119.766	1.23479	0.0663	115.522	1.20300

T	4000 psia			5000 psia		
300	0.0187	3.294	0.92179	0.0179	3.681	0.91177
340	0.0216	19.985	0.97400	0.0201	19.359	0.96082
380	0.0254	36.775	1.02070	0.0229	35.011	1.00435
420	0.0299	52.870	1.06099	0.0261	50.262	1.04252
460	0.0347	67.757	1.09487	0.0296	64.796	1.07559
500	0.0395	81.467	1.12346	0.0332	78.520	1.10421
540	0.0442	94.261	1.14809	0.0368	91.513	1.12922
580	0.0487	106.391	1.16976	0.0404	103.914	1.15138
600	0.0509	112.268	1.17973	0.0422	109.936	1.16158

*Abstracted with permission from *Table of Thermodynamic Properties of Oxygen*, by R. B. Stewart and R. T. Jacobsen, Department of Mechanical Engineering, University of Idaho, 1972.

Table A.6 (SI) Properties of Air at Low Pressure* h and u in kJ/kg, ϕ in kJ/kg · K

T, K	h	p_r	u	v_r	ϕ	T, K	h	p_r	u	v_r	ϕ
200	200.13	0.33468	142.72	171.52	5.2950	700	713.51	28.679	512.59	7.0058	6.5725
210	210.15	0.39684	149.88	151.89	5.3439	710	724.27	30.245	520.47	6.7380	6.5877
220	220.18	0.46684	157.03	135.26	5.3905	720	735.05	31.876	528.39	6.4832	6.6028
230	230.20	0.54524	164.18	121.08	5.4350	730	745.85	33.575	536.32	6.2407	6.6177
240	240.22	0.63263	171.34	108.89	5.4777	740	756.68	35.344	544.28	6.0097	6.6324
250	250.25	0.7296	178.49	98.353	5.5186	750	767.53	37.184	552.26	5.7894	6.6470
260	260.28	0.8368	185.65	89.188	5.5580	760	778.41	39.098	560.27	5.5795	6.6614
270	270.31	0.9547	192.81	81.173	5.5958	770	789.31	41.088	568.30	5.3791	6.6757
280	280.35	1.0842	199.98	74.129	5.6323	780	800.23	43.156	576.35	5.1878	6.6898
290	290.39	1.2258	207.15	67.909	5.6676	790	811.18	45.304	584.43	5.0052	6.7037
300	300.43	1.3801	214.32	62.393	5.7016	800	822.15	47.535	592.53	4.8306	6.7175
310	310.48	1.5480	221.50	57.481	5.7346	810	833.15	49.852	600.65	4.6637	6.7312
320	320.53	1.7301	228.68	53.091	5.7665	820	844.16	52.256	608.80	4.5041	6.7447
330	330.59	1.9271	235.87	49.152	5.7974	830	855.20	54.749	616.97	4.3514	6.7581
340	340.66	2.1398	243.07	45.608	5.8275	840	866.26	57.336	625.16	4.2052	6.7713
350	350.73	2.3689	250.27	42.407	5.8567	850	877.35	60.017	633.37	4.0652	6.7844
360	360.81	2.6154	257.48	39.509	5.8851	860	888.45	62.795	641.61	3.9310	6.7974
370	370.91	2.8799	264.71	36.877	5.9127	870	899.58	65.674	649.87	3.8024	6.8103
380	381.01	3.1633	271.94	34.481	5.9397	880	910.73	68.655	658.15	3.6791	6.8230
390	391.12	3.4664	279.18	32.293	5.9660	890	921.90	71.742	666.45	3.5608	6.8356
400	401.25	3.7902	286.43	30.292	5.9916	900	933.10	74.937	674.77	3.4473	6.8482
410	411.38	4.1356	293.70	28.456	6.0166	910	944.31	78.243	683.11	3.3383	6.8605
420	421.54	4.5035	300.98	26.769	6.0411	920	955.55	81.663	691.48	3.2336	6.8728
430	431.70	4.8949	308.28	25.215	6.0650	930	966.80	85.200	699.87	3.1331	6.8850
440	441.88	5.3106	315.59	23.781	6.0884	940	978.08	88.856	708.27	3.0365	6.8971

450	452.07	5.7519	322.91	22.456	6.1113	950	989.38	92.63	716.70	2.9436	6.9090
460	462.28	6.2197	330.25	21.228	6.1338	960	1000.69	96.54	725.14	2.8543	6.9209
470	472.51	6.7150	337.61	20.090	6.1557	970	1012.03	100.57	733.61	2.7683	6.9326
480	482.76	7.2391	344.98	19.032	6.1773	980	1023.39	104.74	742.10	2.6857	6.9443
490	493.02	7.7930	352.38	18.048	6.1985	990	1034.76	109.04	750.60	2.6061	6.9558
500	503.30	8.378	359.79	17.130	6.2193	1000	1046.16	113.48	759.13	2.5294	6.9673
510	513.60	8.995	367.22	16.274	6.2396	1010	1057.57	118.06	767.67	2.4556	6.9786
520	523.93	9.645	374.67	15.474	6.2597	1020	1069.01	122.78	776.23	2.3845	6.9899
530	534.27	10.331	338.14	14.726	6.2794	1030	1080.46	127.65	784.81	2.3160	7.0010
540	544.63	11.052	389.63	14.025	6.2988	1040	1091.93	132.68	793.41	2.2499	7.0121
550	555.01	11.810	397.15	13.367	6.3178	1050	1103.41	137.86	802.03	2.1862	7.0231
560	565.42	12.608	404.68	12.749	6.3366	1060	1114.92	143.20	810.66	2.1247	7.0340
570	575.84	13.445	412.24	12.169	6.3550	1070	1126.44	148.70	819.32	2.0654	7.0448
580	586.29	14.324	419.82	11.623	6.3732	1080	1137.98	154.37	827.99	2.0082	7.0556
590	596.77	15.245	427.42	11.108	6.3911	1090	1149.54	160.20	836.67	1.9529	7.0662
600	607.26	16.212	435.04	10.623	6.4087	1100	1161.11	166.21	845.38	1.8996	7.0768
610	617.78	17.224	442.69	10.165	6.4261	1110	1172.70	172.40	854.10	1.8481	7.0873
620	628.32	18.284	450.36	9.733	6.4433	1120	1184.31	178.76	862.83	1.7983	7.0977
630	638.89	19.393	458.06	9.324	6.4602	1130	1195.93	185.32	871.59	1.7502	7.1080
640	649.47	20.553	465.77	8.938	6.4768	1140	1207.57	192.06	880.35	1.7038	7.1183
650	660.09	21.766	473.52	8.5718	6.4933	1150	1219.23	198.99	889.14	1.6588	7.1285
660	670.72	23.033	481.28	8.2249	6.5095	1160	1230.90	206.12	897.94	1.6154	7.1386
670	681.38	24.356	489.07	7.8960	6.5256	1170	1242.58	213.45	906.76	1.5733	7.1486
680	692.07	25.736	496.89	7.5839	6.5414	1180	1254.28	220.99	915.59	1.5327	7.1586
690	702.78	27.177	504.73	7.2875	6.5570	1190	1266.00	228.73	924.43	1.4933	7.1684

(continued)

Table A.6 (SI) (continued)

T, K	h	p_r	u	v_r	ϕ	T, K	h	p_r	u	v_r	ϕ
1200	1277.73	236.69	933.29	1.4552	7.1783	1700	1879.58	1017.9	1391.63	0.47937	7.5970
1210	1289.48	244.86	942.17	1.4184	7.1880	1725	1910.31	1083.6	1415.18	0.45694	7.6149
1220	1301.24	253.26	951.06	1.3827	7.1977	1750	1941.09	1152.6	1438.78	0.43582	7.6326
1230	1313.01	261.89	959.96	1.3481	7.2073	1775	1971.91	1225.0	1462.43	0.41592	7.6501
1240	1324.80	270.74	968.88	1.3146	7.2168	1800	2002.78	1300.9	1486.12	0.39714	7.6674
1250	1336.60	279.83	977.81	1.2822	7.2263	1825	2033.69	1380.6	1509.86	0.37943	7.6844
1260	1348.42	289.16	986.76	1.2507	7.2357	1850	2064.65	1464.0	1533.65	0.36270	7.7013
1270	1360.25	298.74	995.72	1.2202	7.2451	1875	2095.66	1551.5	1557.47	0.34689	7.7179
1280	1372.09	308.57	1004.69	1.1907	7.2544	1900	2126.70	1643.0	1581.34	0.33194	7.7344
1290	1383.95	318.65	1013.68	1.1620	7.2636	1925	2157.79	1738.7	1605.25	0.31779	7.7506
1300	1395.81	328.98	1022.67	1.1342	7.2728	1950	2188.91	1838.8	1629.20	0.30439	7.7667
1310	1407.70	339.59	1031.69	1.1073	7.2819	1975	2220.08	1943.4	1653.19	0.29170	7.7826
1320	1419.59	350.46	1040.71	1.0811	7.2909	2000	2251.28	2052.6	1677.22	0.27967	7.7983
1330	1431.50	361.61	1049.75	1.0557	7.2999	2025	2282.52	2166.7	1701.28	0.26826	7.8138
1340	1443.42	373.03	1058.79	1.0311	7.3088	2050	2313.80	2285.7	1725.38	0.25743	7.8292
1350	1455.35	384.74	1067.86	1.00716	7.3177	2075	2345.11	2409.9	1749.52	0.24714	7.8443
1360	1467.29	396.74	1076.93	0.98393	7.3265	2100	2376.46	2539.3	1773.70	0.23737	7.8594
1370	1479.25	409.03	1086.01	0.96138	7.3353	2125	2407.85	2674.2	1797.90	0.22808	7.8742
1380	1491.21	421.62	1095.11	0.93947	7.3440	2150	2439.26	2814.7	1822.15	0.21924	7.8889
1390	1503.19	434.52	1104.22	0.91819	7.3526	2175	2470.71	2961.0	1846.42	0.21084	7.9035
1400	1515.18	447.73	1113.34	0.89752	7.3612	2200	2502.20	3113.3	1870.73	0.20283	7.9179
1410	1527.18	461.25	1122.47	0.87742	7.3698	2225	2533.71	3271.7	1895.07	0.19520	7.9321
1420	1539.20	475.10	1131.61	0.85789	7.3783	2250	2565.26	3436.4	1919.44	0.18793	7.9462
1430	1551.22	489.27	1140.77	0.83891	7.3867	2275	2596.84	3607.7	1943.84	0.18100	7.9602
1440	1563.26	503.78	1149.93	0.82045	7.3951	2300	2628.45	3785.6	1968.27	0.17439	7.9740

T						T					
1450	1575.30	518.63	0.80250	1159.11	7.4034	2325	2660.08	3970.4	1992.74	0.16808	7.9877
1460	1587.36	533.81	0.78504	1168.29	7.4117	2350	2691.75	4162.3	2017.23	0.16206	8.0012
1470	1599.42	549.35	0.76806	1177.49	7.4199	2375	2723.45	4361.4	2041.75	0.15630	8.0146
1480	1611.50	565.25	0.75154	1186.69	7.4281	2400	2755.17	4568.1	2066.29	0.15080	8.0279
1490	1623.59	581.51	0.73546	1195.91	7.4363	2425	2786.92	4782.4	2090.87	0.14554	8.0411
1500	1635.68	598.14	0.71982	1205.14	7.4444	2450	2818.70	5004.7	2115.47	0.14051	8.0541
1510	1647.79	615.14	0.70459	1214.37	7.4524	2475	2850.50	5235.0	2140.10	0.13570	8.0670
1520	1659.91	632.52	0.68976	1223.62	7.4604	2500	2882.34	5473.7	2164.76	0.13110	8.0798
1530	1672.03	650.29	0.67533	1232.88	7.4684	2525	2914.19	5720.9	2189.44	0.12669	8.0925
1540	1684.17	668.45	0.66128	1242.14	7.4763	2550	2946.07	5976.9	2214.15	0.12246	8.1051
1550	1696.32	687.01	0.64759	1251.42	7.4841	2575	2977.98	6241.9	2238.88	0.11841	8.1175
1560	1708.47	705.98	0.63425	1260.70	7.4919	2600	3009.91	6516.1	2263.63	0.11453	8.1299
1570	1720.64	725.36	0.62127	1270.00	7.4997	2625	3041.87	6799.8	2288.41	0.11081	8.1421
1580	1732.81	745.16	0.60861	1279.30	7.5074	2650	3073.85	7093.2	2313.22	0.10723	8.1542
1590	1744.99	765.38	0.59628	1288.61	7.5151	2675	3105.85	7396.5	2338.05	0.10381	8.1662
1600	1757.19	786.04	0.58426	1297.94	7.5228	2700	3137.88	7710.1	2362.90	0.10052	8.1781
1610	1769.39	807.13	0.57255	1307.27	7.5304	2725	3169.93	8034.1	2387.77	0.09736	8.1900
1620	1781.60	828.67	0.56113	1316.61	7.5379	2750	3202.00	8368.8	2412.67	0.09432	8.2017
1630	1793.82	850.67	0.54999	1325.95	7.5455	2775	3234.09	8714.5	2437.58	0.09140	8.2133
1640	1806.04	873.12	0.53913	1335.31	7.5529	2800	3266.21	9071.4	2462.52	0.08860	8.2248
1650	1818.28	896.05	0.52855	1344.68	7.5604	2825	3298.35	9439.8	2487.48	0.08590	8.2362
1660	1830.52	919.44	0.51822	1354.05	7.5678	2850	3330.50	9820	2512.46	0.08330	8.2476
1670	1842.78	943.32	0.50814	1363.43	7.5751	2875	3362.68	10212	2537.47	0.08081	8.2588
1680	1855.04	967.69	0.49832	1372.82	7.5825	2900	3394.88	10617	2562.49	0.07840	8.2700
1690	1867.30	992.55	0.48873	1382.22	7.5897	2925	3427.10	11034	2587.53	0.07609	8.2810

* Abridged from Table 1 in *Gas Tables*, by Joseph H. Keenan, Jing Chao, and Joseph Kaye, 2nd ed., SI, John Wiley & Sons, New York, 1983.

Table A.6 (E) Properties of Air at Low Pressure*

T, °R	h, Btu/lbm	P_r	u, Btu/lbm	v_r	ϕ, Btu/lbm-°R
200	47.67	0.04320	33.96	1714.9	0.36303
220	52.46	0.06026	37.38	1352.5	0.38584
240	57.25	0.08165	40.80	1088.8	0.40666
260	62.03	0.10797	44.21	892.0	0.42582
280	66.82	0.13986	47.63	741.6	0.44356
300	71.61	0.17795	51.04	624.5	0.46007
320	76.40	0.22290	54.46	531.8	0.47550
340	81.18	0.27545	57.87	457.2	0.49002
360	85.97	0.3363	61.29	396.6	0.50369
380	90.75	0.4061	64.70	346.6	0.51663
400	95.53	0.4858	68.11	305.0	0.52890
420	100.32	0.5760	71.52	270.1	0.54058
440	105.11	0.6776	74.93	240.6	0.55172
460	109.90	0.7913	78.36	215.33	0.56235
480	114.69	0.9182	81.77	193.65	0.57255
500	119.48	1.0590	85.20	174.90	0.58233
520	124.27	1.2147	88.62	158.58	0.59173
540	129.16	1.3801	92.04	144.32	0.60078
560	133.86	1.5742	95.47	131.78	0.60950
580	138.66	1.7800	98.90	120.70	0.61793
600	143.47	2.005	102.34	110.88	0.62607
620	148.28	2.249	105.78	102.12	0.63395
640	153.09	2.514	109.21	94.30	0.64159
660	157.92	2.801	112.67	87.27	0.64902
680	162.73	3.111	116.12	80.96	0.65621
700	167.56	3.446	119.58	75.25	0.66321
720	172.39	3.806	123.04	70.07	0.67002
740	177.23	4.193	126.51	65.38	0.67665
760	182.08	4.607	129.99	61.10	0.68312
780	186.94	5.051	133.47	57.20	0.68942
800	191.81	5.526	136.97	53.63	0.69558
820	196.69	6.033	140.47	50.35	0.70160
840	201.56	6.573	143.98	47.34	0.70747
860	206.46	7.149	147.50	44.57	0.71323
880	211.35	7.761	151.02	42.01	0.71886
900	216.26	8.411	154.57	39.64	0.72438
920	221.18	9.102	158.12	37.44	0.72979
940	226.11	9.834	161.68	35.41	0.73509
960	231.06	10.610	165.26	33.52	0.74030
980	236.02	11.430	168.83	31.76	0.74540

*Abridged from Table 1 in *Gas Tables*, by Joseph H. Keenan and Joseph Kaye. Copyright 1948, by Joseph H. Keenan and Joseph Kaye. Published by John Wiley & Sons, Inc., New York.

Table A.6 (E) Properties of Air at Low Pressure (*continued*)

T, °R	h, Btu/lbm	P_r	u, Btu/lbm	v_r	ϕ, Btu/lbm-°R
1000	240.98	12.298	172.43	30.12	0.75042
1020	245.97	13.215	176.04	28.59	0.75536
1040	250.95	14.182	179.66	27.17	0.76019
1060	255.96	15.203	183.29	25.82	0.76496
1080	260.97	16.278	186.93	24.58	0.76964
1100	265.99	17.413	190.58	23.40	0.77426
1120	271.03	18.604	194.25	22.30	0.77880
1140	276.08	19.858	197.94	21.27	0.78326
1160	281.14	21.18	201.63	20.293	0.78767
1180	286.21	22.56	205.33	19.377	0.79201
1200	291.30	24.01	209.05	18.514	0.79628
1220	296.41	25.53	212.78	17.700	0.80050
1240	301.52	27.13	216.53	16.932	0.80466
1260	306.65	28.80	220.28	16.205	0.80876
1280	311.79	30.55	224.05	15.518	0.81280
1300	316.94	32.39	227.83	14.868	0.81680
1320	322.11	34.31	231.63	14.253	0.82075
1340	327.29	36.31	235.43	13.670	0.82464
1360	332.48	38.41	239.25	13.118	0.82848
1380	337.68	40.59	243.08	12.593	0.83229
1400	342.90	42.88	246.93	12.095	0.83604
1420	348.14	45.26	250.79	11.622	0.83975
1440	353.37	47.75	254.66	11.172	0.84341
1460	358.63	50.34	258.54	10.743	0.84704
1480	363.89	53.04	262.44	10.336	0.85062
1500	369.17	55.86	266.34	9.948	0.85416
1520	374.47	58.78	270.26	9.578	0.85767
1540	379.77	61.83	274.20	9.226	0.86113
1560	385.08	65.00	278.13	8.890	0.86456
1580	390.40	68.30	282.09	8.569	0.86794
1600	395.74	71.73	286.06	8.263	0.87130
1620	401.09	75.29	290.04	7.971	0.87462
1640	406.45	78.99	294.03	7.691	0.87791
1660	411.82	82.83	298.02	7.424	0.88116
1680	417.20	86.82	302.04	7.168	0.88439
1700	422.59	90.95	306.06	6.924	0.88758
1720	428.00	95.24	310.09	6.690	0.89074
1740	433.41	99.69	314.13	6.465	0.89387
1760	438.83	104.30	318.18	6.251	0.89697
1780	444.26	109.08	322.24	6.045	0.90003
1800	449.71	114.03	326.32	5.847	0.90308
1820	455.17	119.16	330.40	5.658	0.90609
1840	460.63	124.47	334.50	5.476	0.90908
1860	466.12	129.95	338.61	5.302	0.91203

(*continued*)

Table A.6 (E) Properties of Air at Low Pressure (*continued*)

T, °R	h, Btu/lbm	P_r	u, Btu/lbm	v_r	ϕ Btu/lbm-°R
1880	471.60	135.64	342.73	5.134	0.91497
1900	477.09	141.51	346.85	4.974	0.91788
1920	482.60	147.59	350.98	4.819	0.92076
1940	488.12	153.87	355.12	4.670	0.92362
1960	493.64	160.37	359.28	4.527	0.92645
1980	499.17	167.07	363.43	4.390	0.92926
2000	504.71	174.00	367.61	4.258	0.93205
2020	510.26	181.16	371.79	4.130	0.93481
2040	515.82	188.54	375.98	4.008	0.93756
2060	521.39	196.16	380.18	3.890	0.94026
2080	526.97	204.02	384.39	3.777	0.94296
2100	532.55	212.1	388.60	3.667	0.94564
2120	538.15	220.5	392.83	3.561	0.94829
2140	543.74	229.1	397.05	3.460	0.95092
2160	549.35	238.0	401.29	3.362	0.95352
2180	554.97	247.2	405.53	3.267	0.95611
2200	560.59	256.6	409.78	3.176	0.95868
2220	566.23	266.3	414.05	3.088	0.96123
2240	571.86	276.3	418.31	3.003	0.96376
2260	577.51	286.6	422.59	2.921	0.96626
2280	583.16	297.2	426.87	2.841	0.96876
2300	588.82	308.1	431.16	2.765	0.97123
2320	594.49	319.4	435.46	2.691	0.97369
2340	600.16	330.9	439.76	2.619	0.97611
2360	605.84	342.8	444.07	2.550	0.97853
2380	611.53	355.0	448.38	2.483	0.98092
2400	617.22	367.6	452.70	2.419	0.98331

Table A.7 (SI) Properties of Ideal Gases* $\bar{h}$ in kJ/kg mol, $\bar{\phi}$ in kJ/kg mol · K

T, K	N₂ $\bar{h}$	N₂ $\bar{\phi}$	O₂ $\bar{h}$	O₂ $\bar{\phi}$	H₂O $\bar{h}$	H₂O $\bar{\phi}$	CO₂ $\bar{h}$	CO₂ $\bar{\phi}$	H₂ $\bar{h}$	H₂ $\bar{\phi}$	CO $\bar{h}$	CO $\bar{\phi}$
200	5812.8	179.842	5815.7	193.330	6621.8	175.369	5952.3	199.859	5697.2	119.328	5813.2	185.891
250	7268.2	186.337	7273.5	199.836	8291.1	182.819	7631.8	207.344	7089.9	125.539	7268.7	192.387
300	8724.1	191.646	8737.6	205.174	9966.1	188.925	9433.7	213.908	8520.7	130.752	8725.0	197.697
350	10181.2	196.138	10214.1	209.726	11652.7	194.124	11349.7	219.810	9969.7	135.225	10183.7	202.194
400	11641.3	200.037	11708.8	213.717	13355.6	198.672	13368.5	225.199	11426.7	139.117	11647.1	206.102
450	13107.0	203.490	13225.9	217.290	15080.1	202.733	15478.9	230.168	12887.0	142.550	13118.7	209.568
500	14581.0	206.596	14767.7	220.538	16828.4	206.417	17671.7	234.788	14349.2	145.635	14601.7	212.693
550	16065.9	209.426	16335.0	223.526	18603.1	209.799	19938.8	239.108	15813.0	148.429	16098.8	215.547
600	17564.1	212.033	17927.3	226.296	20405.3	212.935	22273.3	243.170	17278.3	150.973	17612.1	218.180
650	19077.4	214.455	19543.5	228.883	22235.8	215.866	24669.1	247.005	18745.5	153.326	19143.0	220.631
700	20606.8	216.722	21182.0	231.312	24095.7	218.622	27120.9	250.639	20216.1	155.504	20692.3	222.927
750	22153.0	218.856	22841.0	233.601	25985.7	221.229	29623.9	254.091	21690.3	157.533	22260.0	225.090
800	23716.2	220.873	24518.8	235.766	27906.2	223.708	32173.5	257.382	23168.1	159.445	23845.9	227.137
850	25296.2	222.789	26213.7	237.821	29858.1	226.074	34765.8	260.525	24653.5	161.241	25449.5	229.081
900	26892.6	224.614	27924.2	239.776	31841.2	228.342	37397.4	263.533	26143.3	162.946	27069.9	230.933
950	28504.9	226.357	29648.8	241.641	33856.7	230.521	40064.7	266.417	27642.3	164.567	28706.4	232.703
1000	30132.4	228.027	31386.2	243.423	35904.1	232.621	42764.6	269.186	29148.1	166.105	30357.9	234.397
1050	31774.2	229.629	33135.3	245.130	37983.4	234.650	45493.7	271.848	30663.0	167.585	32023.7	236.022
1100	33429.7	231.169	34895.2	246.768	40094.4	236.614	48250.8	274.417	32186.9	169.007	33702.6	237.584
1150	35098.0	232.652	36664.9	248.341	42236.8	238.518	51033.0	276.886	33719.9	170.370	35393.9	239.088
1200	36778.3	234.082	38443.8	249.855	44410.1	240.368	53840.2	279.272	35264.3	171.684	37096.7	240.537
1250	38469.9	235.463	40231.2	251.314	46613.3	242.167	56667.8	281.592	36817.8	172.948	38810.2	241.936
1300	40172.0	236.798	42026.6	252.726	48846.1	243.918	59515.8	283.820	38384.9	174.178	40533.7	243.288
1350	41884.0	238.091	43829.5	254.083	51107.5	245.625	62379.7	285.991	39961.1	175.376	42266.3	244.596
1400	43605.3	239.343	45639.6	255.400	53396.6	247.290	65261.8	288.086	41548.6	176.531	44007.6	245.862
1450	45335.2	240.557	47456.6	256.675	55712.5	248.915	68159.9	290.114	43147.5	177.646	45756.9	247.090
1500	47073.2	241.735	49280.1	257.912	58054.5	250.503	71071.5	292.093	44757.8	178.743	47513.7	248.281
1550	48818.8	242.880	51110.1	259.112	60421.3	252.055	73996.7	294.014	46379.4	179.807	49277.4	249.438
1600	50571.4	243.993	52946.2	260.278	62811.9	253.573	76935.6	295.868	48010.1	180.838	51047.6	250.562
1650	52330.7	245.075	54788.5	261.411	65226.0	255.059	79885.9	297.681	49652.2	181.844	52823.9	251.655

(*continued*)

Table A.7 (SI) (continued)

T, K	N₂ $\bar{h}$	N₂ $\bar{\phi}$	O₂ $\bar{h}$	O₂ $\bar{\phi}$	H₂O $\bar{h}$	H₂O $\bar{\phi}$	CO₂ $\bar{h}$	CO₂ $\bar{\phi}$	H₂ $\bar{h}$	H₂ $\bar{\phi}$	CO $\bar{h}$	CO $\bar{\phi}$
1700	54096.3	246.129	56636.7	262.515	67662.0	256.513	82847.5	299.452	51303.3	182.834	54605.8	252.719
1750	55867.7	247.156	58490.7	263.590	70119.2	257.938	85818.1	301.173	52963.5	183.798	56393.1	253.755
1800	57644.5	248.158	60350.6	264.638	72596.8	259.334	88797.9	302.852	54635.1	184.738	58185.4	254.765
1850	59426.6	249.134	62216.2	265.660	75093.7	260.702	91786.8	304.490	56318.0	185.661	59982.3	255.750
1900	61213.5	250.087	64087.5	266.658	77609.5	262.044	94784.7	306.095	58007.8	186.559	61783.7	256.710
1950	63005.0	251.018	65964.4	267.633	80143.0	263.360	97791.7	307.658	59708.9	187.448	63589.2	257.648
2000	64800.9	251.927	67846.8	268.586	82693.7	264.652	100805.5	309.179	61419.0	188.313	65398.6	258.565
2050	66600.8	252.816	69734.9	269.519	85260.8	265.919	103826.2	310.668	63138.3	189.161	67211.7	259.460
2100	68404.6	253.685	71628.4	270.341	87843.6	267.164	106853.6	312.131	64864.4	189.992	69028.3	260.336
2150	70212.1	254.536	73527.4	271.325	90441.5	268.387	109887.9	313.561	66601.8	190.807	70848.3	261.192
2200	72023.1	255.369	75431.8	272.201	93053.8	269.588	112929.0	314.958	68343.8	191.614	72671.3	262.030
2250	73837.3	256.184	77341.7	273.059	95680.0	270.768	115974.6	316.330	70097.2	192.395	74497.4	262.851
2300	75654.7	256.983	79256.9	273.901	98319.5	271.928	119024.8	317.668	71855.0	193.169	76326.2	263.655
2350	77475.1	257.766	81177.4	274.727	100971.8	273.069	122081.8	318.982	73622.0	193.933	78157.8	264.443
2400	79298.3	258.534	83103.1	275.538	103636.5	274.191	125143.3	320.271	75395.8	194.682	79991.9	265.215
2450	81124.2	259.287	85034.1	276.334	106313.3	275.295	128211.7	321.535	77178.6	195.413	81828.4	265.972
2500	82952.7	260.025	86970.2	277.116	109001.3	276.381	131282.3	322.782	78966.0	196.137	83667.3	266.715
2550	84783.7	260.751	88911.4	277.885	111700.3	277.450	134357.4	323.996	80760.3	196.852	85508.4	267.444
2600	86617.0	261.463	90857.6	278.641	114410.3	278.502	137437.1	325.193	82563.6	197.550	87351.7	268.160
2650	88452.5	262.162	92808.8	279.384	117130.7	279.539	140521.4	326.365	84371.4	198.240	89197.0	268.863
2700	90290.3	262.849	94764.9	280.116	119860.8	280.559	143610.1	327.521	86186.1	198.914	91044.2	269.554
2750	92130.1	263.524	96725.9	280.835	122600.5	181.565	146701.2	328.660	88007.5	199.587	92893.4	270.232
2800	93971.8	264.188	98691.6	281.544	125349.8	282.556	149796.8	329.774	89833.6	200.244	94744.3	270.900
2850	95815.5	264.840	100662.0	282.241	128108.1	283.532	152896.9	330.872	91666.4	200.893	96597.0	271.555
2900	97661.0	265.482	102637.0	282.928	130875.5	284.495	155999.3	331.953	93506.0	201.533	98451.3	272.200

* Abridged from Table 11 (N₂), Table 13 (O₂), Table 15 (H₂O), Table 17 (CO₂), Table 19 (H₂), and Table 21 (CO) in *Gas Tables* by Joseph H. Keenan, Jing Chao, and Joseph Kaye, 2nd ed., SI, John Wiley & Sons, New York, 1983.

Table A.7 (E) Properties of Ideal Gases $\bar{h}$ in Btu/lb mol, $\bar{\phi}$ in Btu/lb mol · °R

T, °R	N_2 $\bar{h}$	N_2 $\bar{\phi}$	O_2 $\bar{h}$	O_2 $\bar{\phi}$	H_2O $\bar{h}$	H_2O $\bar{\phi}$	CO_2 $\bar{h}$	CO_2 $\bar{\phi}$	H_2 $\bar{h}$	H_2 $\bar{\phi}$	CO $\bar{h}$	CO $\bar{\phi}$
300	2081.9	41.687	2083.0	44.908	2369.1	40.434	2106.3	46.361	2066.2	27.337	2082.1	43.132
400	2777.1	43.687	2778.6	46.909	3165.7	42.726	2873.3	48.563	2712.6	29.194	2777.3	45.132
500	3472.5	45.238	3476.2	48.466	3964.2	44.507	3705.1	50.417	3388.5	30.701	3472.7	46.684
540	3750.7	45.774	3756.5	49.005	4284.6	45.124	4055.8	51.091	3663.2	31.230	3751.1	47.219
600	4168.2	46.507	4178.9	49.747	4767.4	45.971	4599.8	52.046	4078.0	31.958	4169.0	47.953
700	4865.2	47.581	4890.3	50.843	5578.5	47.222	5551.0	53.511	4773.1	33.031	4867.3	49.029
800	5564.8	48.516	5613.1	51.808	6400.4	48.319	6552.1	54.847	5470.6	33.960	5569.5	49.967
900	6268.7	49.344	6349.0	52.675	7234.9	49.302	7597.5	56.078	6169.0	34.784	6277.6	50.801
1000	6978.4	50.092	7098.3	53.464	8083.4	50.196	8682.3	57.221	6868.4	35.521	6993.2	51.555
1100	7695.2	50.775	7860.9	54.191	8946.5	51.018	9802.5	58.288	7568.5	36.188	7717.4	52.245
1200	8420.2	51.406	8636.0	54.865	9824.9	51.782	10954.6	59.290	8269.7	36.798	8451.2	52.883
1300	9153.9	51.993	9422.6	55.495	10718.8	52.498	12135.6	60.236	8972.7	37.360	9194.7	53.478
1400	9896.5	52.544	10219.6	56.085	11628.9	53.172	13342.5	61.130	9678.0	37.884	9947.9	54.037
1500	10648.2	53.062	11026.2	56.642	12555.3	53.811	14573.2	61.979	10386.2	38.371	10710.7	54.563
1600	11408.6	53.553	11841.3	57.168	13498.7	54.420	15825.2	62.787	11097.1	38.830	11482.5	55.061
1700	12177.6	54.019	12664.0	57.666	14458.8	55.002	17096.6	63.557	11812.8	39.264	12263.0	55.534
1800	12954.6	54.463	13493.6	58.141	15436.0	55.561	18385.1	64.294	12531.4	39.674	13051.6	55.985
1900	13739.3	54.887	14329.5	58.593	16430.1	56.098	19689.6	64.999	13254.9	40.065	13847.6	56.415
2000	14531.1	55.294	15170.9	59.024	17441.0	56.616	21008.7	65.676	13984.3	40.440	14650.7	56.827
2100	15329.6	55.683	16017.6	59.437	18468.5	57.118	22341.5	66.326	14717.6	40.798	15460.1	57.222
2200	16134.4	56.058	16868.9	59.833	19512.3	57.603	23686.1	66.951	15457.8	41.141	16275.6	57.601
2300	16945.1	56.418	17724.7	60.214	20571.9	58.074	25042.3	67.555	16202.8	41.473	17096.5	57.966
2400	17761.1	56.765	18584.6	60.580	21646.8	58.532	26407.4	68.137	16953.6	41.794	17922.5	58.318
2500	18582.1	57.100	19448.3	60.932	22736.7	58.977	27781.2	68.699	17710.4	42.104	18753.2	58.657

(continued)

Table A.7 (E) (continued)

T, °R	N₂ $\bar{h}$	N₂ $\bar{\phi}$	O₂ $\bar{h}$	O₂ $\bar{\phi}$	H₂O $\bar{h}$	H₂O $\bar{\phi}$	CO₂ $\bar{h}$	CO₂ $\bar{\phi}$	H₂ $\bar{h}$	H₂ $\bar{\phi}$	CO $\bar{h}$	CO $\bar{\phi}$
2600	19407.8	57.424	20315.7	61.272	23840.8	59.410	29164.8	69.239	18473.9	42.400	19588.2	58.984
2700	20237.8	57.737	21186.6	61.601	24958.9	59.832	30555.2	69.765	19242.4	42.692	20427.2	59.301
2800	21071.9	58.041	22060.9	61.919	26090.1	60.243	31953.5	70.272	20017.7	42.974	21269.9	59.607
2900	21909.6	58.335	22938.6	62.227	27234.1	60.644	33357.6	70.764	20796.8	43.246	22116.0	59.904
3000	22750.9	58.620	23819.4	62.526	28390.1	61.036	34768.5	71.243	21582.9	43.512	22965.3	60.192
3100	23595.4	58.897	24703.3	62.815	29557.8	61.419	36185.3	71.707	22372.8	43.772	23817.5	60.472
3200	24442.9	59.166	25590.4	63.097	30736.4	61.793	37606.0	72.158	23168.6	44.025	24672.5	60.743
3300	25293.2	59.428	26480.5	63.371	31925.7	62.159	39032.5	72.597	23971.2	44.271	25530.0	61.007
3400	26146.2	59.682	27373.6	63.638	33125.0	62.517	40463.0	73.026	24776.7	44.511	26389.9	61.264
3500	27001.6	59.930	28269.8	63.897	34333.9	62.868	41899.3	73.441	25589.1	44.747	27252.1	61.514
3600	27859.3	60.172	29168.9	64.151	35551.9	63.211	43338.6	73.846	26405.4	44.978	28116.3	61.757
3700	28719.3	60.407	30070.9	64.398	36778.6	63.547	44781.7	74.241	27226.6	45.202	28982.5	61.994
3800	29581.2	60.637	30975.9	64.639	38013.5	63.876	46228.8	74.629	28052.6	45.423	29850.6	62.226
3900	30445.1	60.862	31883.7	64.875	39256.5	64.199	47678.8	75.004	28882.6	45.639	30720.4	62.452
4000	31310.8	61.081	32794.5	65.106	40507.0	64.516	49132.7	75.373	29716.5	45.850	31591.8	62.672
4100	32178.2	61.295	33708.1	65.331	41764.7	64.826	50588.5	75.733	30556.2	46.056	32464.7	62.888
4200	33047.3	61.504	34624.5	65.552	43029.3	65.131	52047.3	76.084	31397.9	46.261	33339.2	63.099
4300	33917.8	61.709	35543.7	65.768	44300.6	65.430	53509.0	76.428	32244.4	46.459	34215.0	63.305
4400	34789.8	61.910	36465.7	65.980	45578.4	65.724	54973.6	76.765	33095.4	46.654	35092.1	63.506
4500	35663.2	62.106	37390.4	66.188	46862.1	66.012	56441.2	77.095	33949.3	46.846	35970.5	63.704
4600	36537.9	62.298	38317.9	66.392	48151.7	66.296	57909.7	77.417	34806.6	47.035	36850.0	63.897
4700	37413.9	62.487	39248.0	66.592	49447.1	66.574	59382.2	77.735	35668.8	47.222	37730.7	64.087
4800	38291.0	62.671	40180.7	66.788	50748.0	66.848	60855.6	78.044	36532.9	47.402	38612.4	64.272
4900	39169.2	62.852	41116.0	66.981	52053.9	67.118	62331.9	78.348	37400.9	47.581	39495.2	64.454
5000	40048.6	63.030	42053.9	67.171	53364.9	67.382	63809.3	78.648	38271.9	47.758	40378.9	64.633

* Abridged from Table 11 (N₂), Table 13 (O₂), Table 15 (H₂O), Table 17 (CO₂), Table 19 (H₂), and Table 21 (CO) in *Gas Tables* by Joseph H. Keenan, Jing Chao, and Joseph Kaye, 2nd ed. (English Units), John Wiley & Sons, New York, 1980.

Table A.8 (SI) Molal Heat Capacities of Gases at Zero Pressure*

$$\overline{c_p^*} = a + bT + cT^2 + dT^3$$
$$(T \text{ in K}, \overline{c_p^*} \text{ in kJ/kg mol} \cdot \text{K})$$

Substance	Formula	a	b	c	d	Temperature Range, K	Error, % Max.	Error, % Avg.
Nitrogen	N_2	28.90	-0.1571×10^{-2}	0.8081×10^{-5}	-2.873×10^{-9}	273–1800	0.59	0.34
Oxygen	O_2	25.48	1.520×10^{-2}	-0.7155×10^{-5}	1.312×10^{-9}	273–1800	1.19	0.28
Air		28.11	0.1967×10^{-2}	0.4802×10^{-5}	-1.966×10^{-9}	273–1800	0.72	0.33
Hydrogen	H_2	29.11	-0.1916×10^{-2}	0.4003×10^{-5}	-0.8704×10^{-9}	273–1800	1.01	0.26
Carbon monoxide	CO	28.16	0.1675×10^{-2}	0.5372×10^{-5}	-2.222×10^{-9}	273–1800	0.89	0.37
Carbon dioxide	CO_2	22.26	5.981×10^{-2}	-3.501×10^{-5}	7.469×10^{-9}	273–1800	0.67	0.22
Water vapor	H_2O	32.24	0.1923×10^{-2}	1.055×10^{-5}	-3.595×10^{-9}	273–1800	0.53	0.24
Nitric oxide	NO	29.34	-0.09395×10^{-2}	0.9747×10^{-5}	-4.187×10^{-9}	273–1500	0.97	0.36
Nitrous oxide	N_2O	24.11	5.8632×10^{-2}	-3.562×10^{-5}	10.58×10^{-9}	273–1500	0.59	0.26
Nitrogen dioxide	NO_2	22.9	5.715×10^{-2}	-3.52×10^{-5}	7.87×10^{-9}	273–1500	0.46	0.18
Ammonia	NH_3	27.568	2.5646×10^{-2}	0.99072×10^{-5}	-6.6909×10^{-9}	273–1800	0.91	0.36
Sulfur dioxide	SO_2	25.78	5.795×10^{-2}	-3.811×10^{-5}	8.612×10^{-9}	273–1800	0.45	0.24
Acetylene	C_2H_2	21.8	9.2143×10^{-2}	-6.527×10^{-5}	18.21×10^{-9}	273–1500	1.46	0.59
Benzene	C_6H_6	-36.22	48.475×10^{-2}	-31.57×10^{-5}	77.62×10^{-9}	273–1500	0.34	0.20
Methanol	CH_4O	19.0	9.152×10^{-2}	-1.22×10^{-5}	-8.039×10^{-9}	273–1000	0.18	0.08
Ethanol	C_2H_6O	19.9	20.96×10^{-2}	-10.38×10^{-5}	20.05×10^{-9}	273–1500	0.40	0.22
Methane	CH_4	19.89	5.024×10^{-2}	1.269×10^{-5}	-11.01×10^{-9}	273–1500	1.33	0.57
Ethane	C_2H_6	6.900	17.27×10^{-2}	-6.406×10^{-5}	7.285×10^{-9}	273–1500	0.83	0.28
Propane	C_3H_8	-4.04	30.48×10^{-2}	-15.72×10^{-5}	31.74×10^{-9}	273–1500	0.40	0.12
n-Butane	C_4H_{10}	3.96	37.15×10^{-2}	-18.34×10^{-5}	35.00×10^{-9}	273–1500	0.54	0.24
i-Butane	C_4H_{10}	-7.913	41.60×10^{-2}	-23.01×10^{-5}	49.91×10^{-9}	273–1500	0.25	0.13
n-Pentane	C_5H_{12}	6.774	45.43×10^{-2}	-22.46×10^{-5}	42.29×10^{-9}	273–1500	0.56	0.21
n-Hexane	C_6H_{14}	6.938	55.22×10^{-2}	-28.65×10^{-5}	57.69×10^{-9}	273–1500	0.72	0.20
Ethylene	C_2H_4	3.95	15.64×10^{-2}	-8.344×10^{-5}	17.67×10^{-9}	273–1500	0.54	0.13
Propylene	C_3H_6	3.15	23.83×10^{-2}	-12.18×10^{-5}	24.62×10^{-9}	273–1500	0.73	0.17

* Source: B. G. Kyle, *Chemical and Process Thermodynamics*, Copyright 1984, pp. 495–496. Adapted by permission of Prentice-Hall, Inc., Englewood Cliffs, New Jersey.

Table A.8 (E) Molal Heat Capacities of Gases at Zero Pressure*

$$\overline{c_p^*} = a + bT + cT^2 + dT^3$$
(T in °R, c_p^* in Btu/lb mol · °R)

Substance	Formula	a	b	c	d	Temperature Range, °R	Error, % Max.	Error, % Avg.
Nitrogen	N_2	6.903	-0.02085×10^{-2}	0.05957×10^{-5}	-0.1176×10^{-9}	491–3240	0.59	0.34
Oxygen	O_2	6.085	0.2017×10^{-2}	-0.05275×10^{-5}	0.05372×10^{-9}	491–3240	1.19	0.28
Air		6.713	0.02609×10^{-2}	0.03540×10^{-5}	-0.08052×10^{-9}	491–3240	0.72	0.33
Hydrogen	H_2	6.952	-0.02542×10^{-2}	0.02952×10^{-5}	-0.03565×10^{-9}	491–3240	1.01	0.26
Carbon monoxide	CO	6.726	0.02222×10^{-2}	0.03960×10^{-5}	-0.09100×10^{-9}	491–3240	0.89	0.37
Carbon dioxide	CO_2	5.316	0.79361×10^{-2}	-0.2581×10^{-5}	0.3059×10^{-9}	491–3240	0.67	0.22
Water vapor	H_2O	7.700	0.02552×10^{-2}	0.07781×10^{-5}	-0.1472×10^{-9}	491–3240	0.53	0.24
Nitric oxide	NO	7.008	-0.01247×10^{-2}	0.07185×10^{-5}	-0.1715×10^{-9}	491–2700	0.97	0.36
Nitrous oxide	N_2O	5.758	0.7780×10^{-2}	-0.2626×10^{-5}	0.4331×10^{-9}	491–2700	0.59	0.26
Nitrogen dioxide	NO_2	5.48	0.7583×10^{-2}	-0.260×10^{-5}	0.322×10^{-9}	491–2700	0.46	0.18
Ammonia	NH_3	6.5846	0.34028×10^{-2}	0.073034×10^{-5}	-0.27402×10^{-9}	491–2700	0.91	0.36
Sulfur dioxide	SO_2	6.157	0.7689×10^{-2}	-0.2810×10^{-5}	0.3527×10^{-9}	491–3240	0.45	0.24
Acetylene	C_2H_2	5.21	1.2227×10^{-2}	-0.4812×10^{-5}	0.7457×10^{-9}	491–2700	1.46	0.59
Benzene	C_6H_6	-8.650	6.4322×10^{-2}	-2.327×10^{-5}	3.179×10^{-9}	491–2700	0.34	0.20
Methanol	CH_4O	4.55	1.214×10^{-2}	-0.0898×10^{-5}	-0.329×10^{-9}	491–1800	0.18	0.08
Ethanol	C_2H_6O	4.75	2.781×10^{-2}	-0.7651×10^{-5}	0.821×10^{-9}	491–2700	0.40	0.22
Methane	CH_4	4.750	0.6667×10^{-2}	0.09352×10^{-5}	-0.4510×10^{-9}	491–2740	1.33	0.57
Ethane	C_2H_6	1.648	2.291×10^{-2}	-0.4722×10^{-5}	0.2984×10^{-9}	491–2740	0.83	0.28
Propane	C_3H_8	-0.966	4.044×10^{-2}	-1.159×10^{-5}	1.300×10^{-9}	491–2740	0.40	0.12
n-Butane	C_4H_{10}	0.945	4.929×10^{-2}	-1.352×10^{-5}	1.433×10^{-9}	491–2740	0.54	0.24
i-Butane	C_4H_{10}	-1.890	5.520×10^{-2}	-1.696×10^{-5}	2.044×10^{-9}	491–2740	0.25	0.13
n-Pentane	C_5H_{12}	1.618	6.028×10^{-2}	-1.656×10^{-5}	1.732×10^{-9}	491–2740	0.56	0.21
n-Hexane	C_6H_{14}	1.657	7.328×10^{-2}	-2.112×10^{-5}	2.363×10^{-9}	491–2740	0.72	0.20
Ethylene	C_2H_4	0.944	2.075×10^{-2}	-0.6151×10^{-5}	0.7326×10^{-9}	491–2740	0.54	0.13
Propylene	C_3H_6	0.753	3.162×10^{-2}	-0.8981×10^{-5}	1.008×10^{-9}	491–2740	0.73	0.17

* Source: B. G. Kyle, *Chemical and Process Thermodynamics*, Copyright 1984, pp. 495–496. Adapted by permission of Prentice-Hall, Inc., Englewood Cliffs, New Jersey.

Table A.9 Enthalpy of Combustion of Substances at 25°C (77°F) and 1 Atm

Substance	Formula	Molecular Weight	H_2O Appears as Liquid in Products of Combustion		H_2O Appears as Vapor in Products of Combustion	
			kJ/kg mol	Btu/lb mol	kJ/kg mol	Btu/lb mol
Hydrogen	H_2(g)	2.016	−286030	−122970	−242000	−104040
Carbon (graphite)	C(s)	12.011	−393790	−169300	−393790	−169300
Carbon monoxide	CO(g)	28.011	−283190	−121750	−283190	−121750
Methane	CH_4(g)	16.043	−890900	−383030	−802800	−345160
Acetylene	C_2H_2(g)	26.038	−1300500	−559110	−1256400	−540170
Ethylene	C_2H_4(g)	28.054	−1411900	−607010	−1323800	−569150
Ethane	C_2H_6(g)	30.070	−1560900	−671080	−1428800	−614280
Propane	C_3H_8(g)	44.097	−2221500	−955090	−2045400	−879380
Benzene	C_6H_6(g)	78.114	−3303800	−1420400	−3171700	−1363600
Octane	C_8H_{18}(g)	114.23	−5515900	−2371400	−5119500	−2201000
Octane	C_8H_{18}(l)	114.23	−5474500	−2353600	−5078200	−2183200

Sources:
1. *JANAF Thermochemical Tables*, Second Edition, NSRDS-NBS-37 (Catalog No. C13.48:37), 1971.
2. *Circular No. 500*, National Bureau of Standards, 1952.
3. *API Research Project No. 40*, National Bureau of Standards, 1952.

Table A.10 Enthalpy of Formation, Gibbs Function of Formation, and Absolute Entropy at 25°C (77°F) and 1 Atm

Substance	Formula	Molecular Weight	$\bar{h}_f^\circ$ kJ/kg mol	$\bar{h}_f^\circ$ Btu/lb mol	$\bar{g}_f^\circ$ kJ/kg mol	$\bar{g}_f^\circ$ Btu/lb mol	$\bar{s}^\circ$ kJ/kg mol·K	$\bar{s}^\circ$ Btu/lb mol·°R
Carbon monoxide	CO(g)	28.011	−110600	−47551	−137250	−59009	197.68	47.214
Carbon dioxide	CO_2(g)	44.010	−393790	−169300	−394630	−169680	213.83	51.072
Water	H_2O(g)	18.016	−242000	−104040	−228730	−98345	188.85	45.106
Water	H_2O(l)	18.016	−286030	−122970	−237350	−102040	69.99	16.716
Methane	CH_4(g)	16.043	−74920	−32211	−50840	−21860	186.27	44.490
Acetylene	C_2H_2(g)	26.038	+226870	+97542	+209300	+89987	200.98	48.004
Ethylene	C_2H_4(g)	28.054	+52260	+22470	+68150	+29300	219.56	52.442
Ethane	C_2H_6(g)	30.070	−84720	−36425	−32900	−14148	229.60	54.85
Propane	C_3H_8(g)	44.097	−103910	−44676	−23502	−10105	270.09	64.51
Benzene	C_6H_6(g)	78.114	+82976	+35676	+129730	+55780	269.38	64.34
Octane	C_8H_{18}(g)	114.23	−208580	−89676	+16540	+7110	467.04	111.55
Octane	C_8H_{18}(l)	114.23	−250100	−107530	+6614	+2844	361.03	86.23
Hydrogen	H_2(g)	2.016	0	0	0	0	130.66	31.208
Oxygen	O_2(g)	32.000	0	0	0	0	205.17	49.004
Nitrogen	N_2(g)	28.013	0	0	0	0	191.61	45.770
Carbon (graphite)	C(s)	12.011	0	0	0	0	5.69	1.359

Sources:
1. *JANAF Thermochemical Tables*, Second Edition, NSRDS-NBS-37 (Catalog No. C13.48:37), 1971.
2. *Circular No. 500*, National Bureau of Standards, 1952.
3. *API Research Project No. 44*, National Bureau of Standards, 1952.

Table A.11 Logarithms to the Base 10 of the Equilibrium Constant K_p

$$K_p = \frac{p_C^{\nu_C} p_D^{\nu_D}}{p_A^{\nu_A} p_B^{\nu_B}} \text{ for the Reaction } \nu_A A + \nu_B B \rightleftharpoons \nu_C C + \nu_D D$$

T, K	$H_2 \rightleftharpoons 2H$	$O_2 \rightleftharpoons 2O$	$N_2 \rightleftharpoons 2N$	$H_2O(g) \rightleftharpoons$ $H_2 + \frac{1}{2}O_2$	$H_2O(g) \rightleftharpoons$ $OH + \frac{1}{2}H_2$	$CO_2 \rightleftharpoons$ $CO + \frac{1}{2}O_2$	$\frac{1}{2}O_2 + \frac{1}{2}N_2 \rightleftharpoons$ NO	T, K
298	−71.224	−81.208	−159.600	−40.048	−46.181	−45.066	−15.171	298
500	−40.316	−45.880	−92.672	−22.886	−26.208	−25.025	−8.783	500
1000	−17.292	−19.614	−43.056	−10.062	−11.322	−10.221	−4.062	1000
1500	−9.512	−10.790	−26.434	−5.725	−6.314	−5.316	−2.487	1500
1800	−6.896	−7.836	−20.874	−4.270	−4.638	−3.693	−1.962	1800
2000	−5.580	−6.356	−18.092	−3.540	−3.799	−2.884	−1.699	2000
2200	−4.502	−5.142	−15.810	−2.942	−3.113	−2.226	−1.484	2200
2400	−3.600	−4.130	−13.908	−2.443	−2.541	−1.679	−1.305	2400
2500	−3.202	−3.684	−13.070	−2.224	−2.158	−1.440	−1.227	2500
2600	−2.834	−3.272	−12.298	−2.021	−2.057	−1.219	−1.154	2600
2800	−2.178	−2.536	−10.914	−1.658	−1.642	−0.825	−1.025	2800
3000	−1.606	−1.898	−9.716	−1.343	−1.282	−0.485	−0.913	3000
3200	−1.106	−1.340	−8.664	−1.067	−0.967	−0.189	−0.815	3200
3500	−0.462	−0.620	−7.312	−0.712	−0.563	+0.190	−0.690	3500
4000	+0.402	+0.340	−5.504	−0.238	−0.025	+0.692	−0.524	4000
4500	+1.074	+1.086	−4.094	+0.133	+0.394	+1.079	−0.397	4500
5000	+1.612	+1.686	−2.962	+0.430	+0.728	+1.386	−0.296	5000

SOURCE: *JANAF Thermochemical Tables*, Second Edition, NSRDS-NBS-37 (Catalog No. C13.48:37), 1971.

935

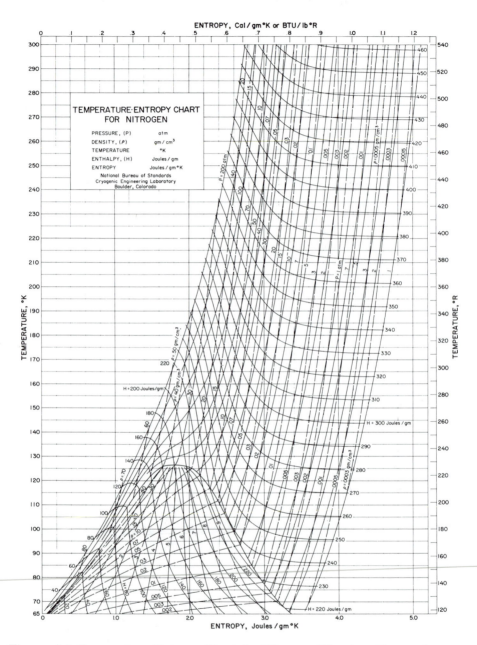

Figure A.1 Temperature–Entropy Chart for Nitrogen [Courtesy Cryogenics Division, National Bureau of Standards.]

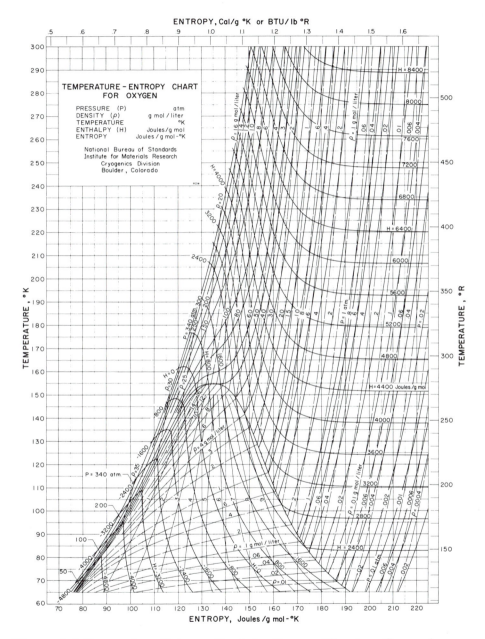

Figure A.2 Temperature–Entropy Chart for Oxygen [Courtesy Cryogenics Division, National Bureau of Standards.]

Figure A.3 (SI) Pressure–Enthalpy Diagram for Refrigerant 12 (Dichlorodifluoromethane) From 1985 FUNDAMENTALS SI ed., ASHRE HANDBOOK. Reproduced with Permission of The American Society of Heating, Refrigerating and Air-Conditioning Engineers, Inc.

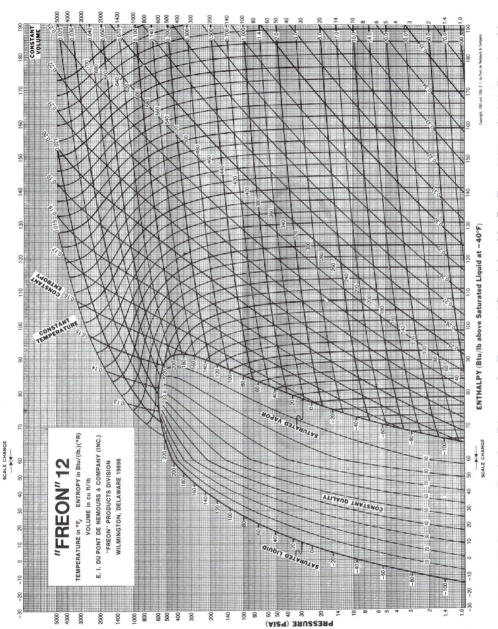

Figure A.3 (E) Pressure–Enthalpy Diagram for Freon-12 [Copyrighted by Du Pont 1955 and 1956. Reprinted by Permission.]

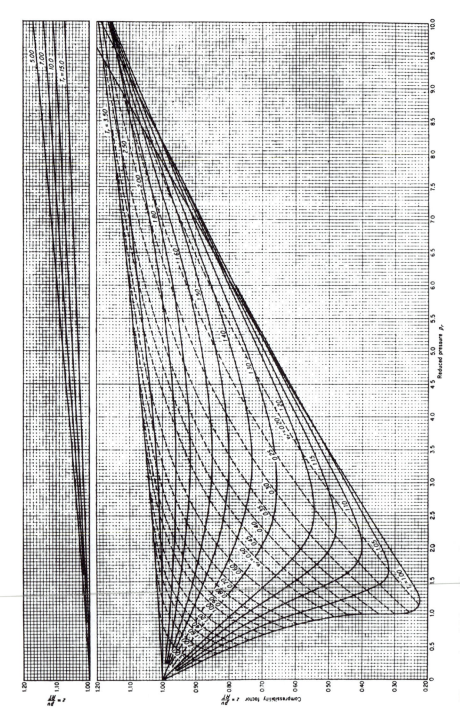

Figure A.4 Nelson-Obert Generalized Compressibility Chart (Medium-Pressure Region) [From *Concepts of The Thermodynamics* by Edward T. Obert. Copyright 1960 by McGraw-Hill Book Company. Used with Permission of McGraw-Hill Book Company.]

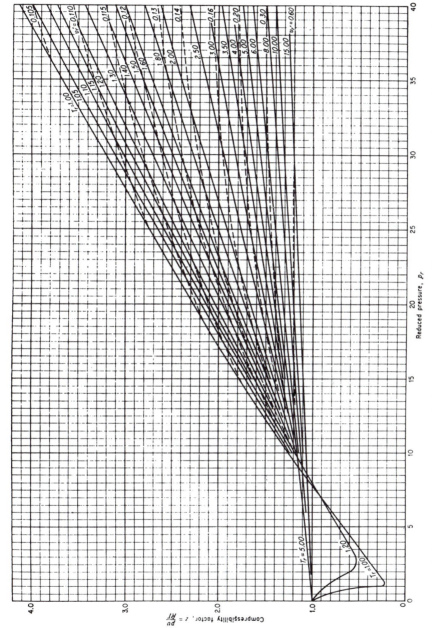

Figure A.5 Nelson–Obert Generalized Compressibility Chart (High-Pressure Region) [From *Concepts of Thermodynamics* by Edward T. Obert. Copyright 1960 by McGraw-Hill Book Company. Used with Permission of McGraw-Hill Book Company.]

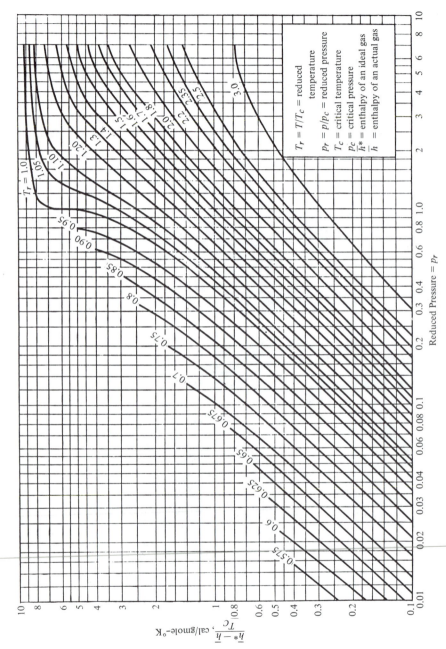

Figure A.6 Generalized Enthalpy Correction Chart [From *Chemical Process Principles* by O. A. Hougen and K. M. Watson, John Wiley & Sons, Inc., New York, 1947. Used with Permission of John Wiley & Sons, Inc.]

Reduced Pressure = p_r

$\dfrac{\bar{h}^* - \bar{h}}{T_c}$, cal/gmole-°K

$T_r = T/T_c$ = reduced temperature
$p_r = p/p_c$ = reduced pressure
T_c = critical temperature
p_c = critical pressure
$\bar{h}^*$ = enthalpy of an ideal gas
$\bar{h}$ = enthalpy of an actual gas

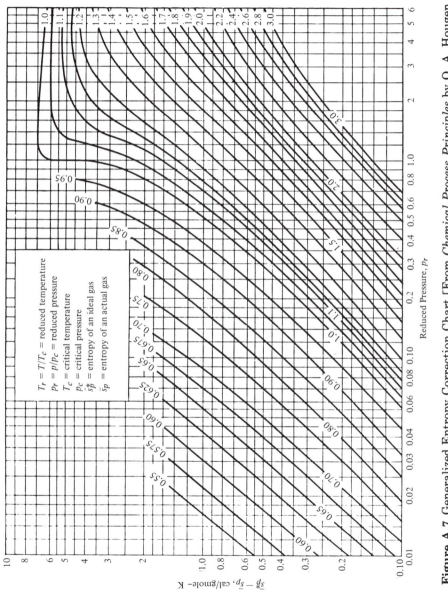

Figure A.7 Generalized Entropy Correction Chart [From *Chemical Process Principles* by O. A. Hougen and K. M. Watson, John Wiley & Sons, Inc., New York, 1947. Used with Permission of John Wiley & Sons, Inc.]

$T_r = T/T_c$ = reduced temperature
$p_r = p/p_c$ = reduced pressure
T_c = critical temperature
p_c = critical pressure
$\bar{s}_p^*$ = entropy of an ideal gas
$\bar{s}_p$ = entropy of an actual gas

Reduced Pressure, p_r

$\bar{s}_p^* - \bar{s}_p$, cal/gmole-K

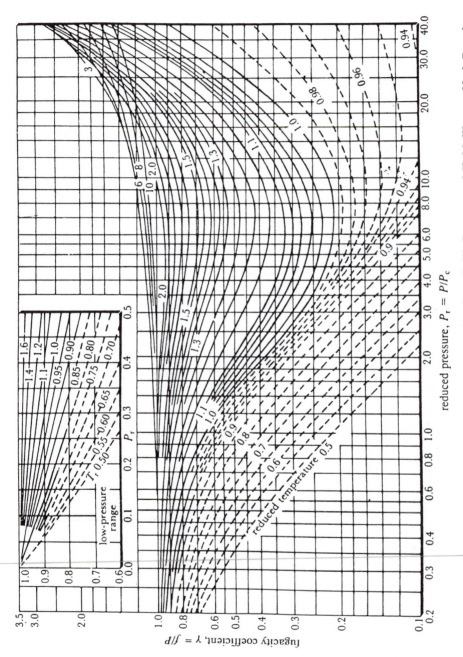

Figure A.8 *Fugacity coefficient of gases. Based on data taken from B. W. Gamson and K. M. Watson, Natl. Petrol. News. Tech. Sec. 36, R623 (Sept. 6, 1944).* (Reproduced with Permission from CHEMICAL THERMODYNAMICS 4th ed., copyright 1986 by Irving M. Klotz and Robert M. Rosenberg, Benjamin/Cummings Publishing Company, Menlo Park, California.)

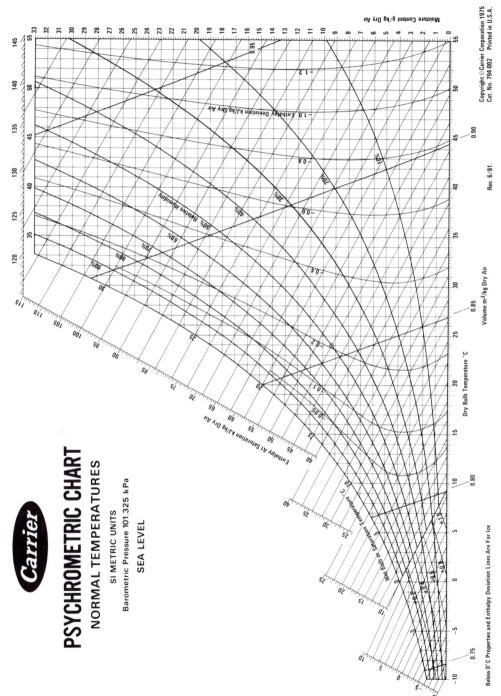

Figure A.9 (SI) Psychrometric Chart (Reproduced Courtesy of Carrier Corporation)

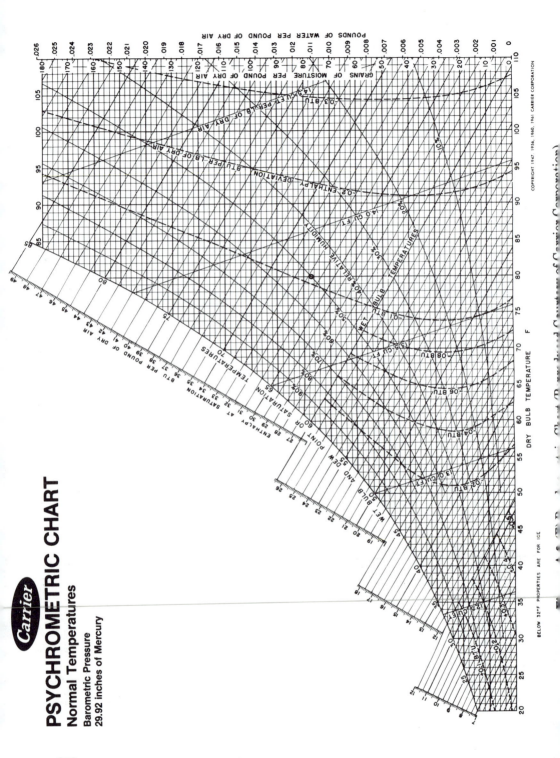

PSYCHROMETRIC CHART

Normal Temperatures

Barometric Pressure
29.92 inches of Mercury

946

Answers to Selected Problems

1.2 86.68 kg
1.4 0.102 kg
1.6 62.14 miles/h
1.9 19629 dynes
1.12 199.43 lbf
1.15 (a) 1000 kg/m^3;
    (b) 62.4 lbm/ft^3
1.16 988.8 Btu/lbm
1.17 713.56 kg/m^3
1.19 2.12 m^3/kg
1.20 53.1 × 10^9
1.23 1636 kg/m^3
1.24 1000 kg/m^3
1.27 37°C, 310.15 K, 558.27°R
1.28 −136,58°C
1.33 134700 Pa
1.35 64.7 psia; 446.1 kPa
1.38 1.3332 × 10^{-6} N/m^2
1.40 1795 m
1.41 34 in
1.45 16670 Pa
1.47 401325 Pa
1.49 53.1 kg

2.2 1.472 × 10^7 J
2.4 51000 kg/s
2.6 597300 J
2.8 −12.5 J
2.10 −954 kJ
2.12 25 kJ
2.13 −55 kJ
2.14 $W_{41} = -2$ kJ;
    $E_1 - E_4 = +2$ kJ
2.15 2680 kJ/kg
2.18 (a) 0; (b) −75 kJ
2.21 230.26 kJ
2.25 279.6 kJ
2.27 197 kJ
2.28 −15 J
2.31 5.8905 kg/s
2.34 −80000 kg
2.37 −27.8 kJ/kg
2.39 1.30 kW
2.41 2553 ft/s

2.44 328.3 K

3.2 $\dot{Q}_{add} = 31250$ kJ/s
3.3 $\dot{W} = 52.78$ kW
3.6 13.89 kW
3.8 $\Delta S^{HR}_{1800} = -2.778$ kj/K;
    $\Delta S^{HR}_{320} = +2.778$ kJ/K
3.11 (a) irreversible;
    (b) reversible;
    (c) reversible
3.14 2.347 kW
3.15 188.7 hp
3.22 3333.3 K
3.24 268.9 K
3.27 4.175
3.29 (b) 50.0 kJ
3.33 positive
3.36 no
3.38 0.036 Btu/lbm·°R
3.40 59.93 kJ/kg
3.43 positive

4.1 (a) subcooled liquid;
    (b) superheated vapor
4.2 (a) superheated vapor;
    (b) liq-vap mixture;
    (c) superheated vapor
4.3 (a) subcooled liquid;
    (b) subcooled liquid;
    (c) superheated vapor
4.4 (a) subcooled liquid;
    (b) liq-vap mixture;
    (c) compressed liquid
4.5 254.03 kPa
4.7 338.0 kg
4.8 0.9771 m^3
4.10 2.026 lbm
4.11 (a) 0.0001655;
    (b) 584.9491 lbm
4.13 (a) 7000 psia;
    (b) 3520 psia
4.14 (a) $p = 567.3$ kPa;
    $x = 67.59\%$;
    $v = 0.021048$ m^3/kg;

$u = 138.\overset{\cdot}{0}5$ kJ/kg
4.15 0.03704 m^3
4.17 4000 kPa
4.20 (a) −18.752 Btu/lbm;
    (b) −0.35792 Btu/lbm·°R
4.22 (a) 65.957 kJ/kg;
    (b) 1.8109 kJ/kg·K
4.24 $x_1 = 0.0186$
4.26 2568.38 kJ/kg
4.29 (a) 948°F;
    (b) 0.0386 Btu/lbm·°F
4.31 (a) 40°C; (b) 40.3°C;
    (c) 41.6°C
4.33 (a) 101.1°F;
    (b) 8.98 Btu/lbm
4.35 (a) 349.3°C;
    (b) 134.827 kJ/kg
4.37 (a) −1074.9 kJ/kg;
    (b) 1.2666 kJ/kg·K
4.39 (a) 77.35 K; (b) 90.19 K;
    (c) 372.782 K
4.41 (b) zero
4.44 (a) 1.045 kJ/kg·K;
    (b) 1.041 kJ/kg·K;
    (c) 2.046 kJ/kg·K;
    (d) 1.196 kJ/kg·K
4.45 (a) 0.916 kJ/kg·K;
    (b) 0.920 kJ/kg·K;
    (c) 2.888 kJ/kg·K;
    (d) 1.089 kJ/kg·K
4.47 (a) 1.050 kJ/kg·K;
    (b) 1.044 kJ/kg·K;
    (c) 2.074 kJ/kg·K;
    (d) 1.200 kJ/kg·K
4.48 (a) 0.920kJ/kg·K;
    (b) 0.921 kJ/kg·K;
    (c) 2.924 kJ/kg·K;
    (d) 1.090 kJ/kg·K

5.2 A: $v = 0.7168$ m^3/kg;
    $R = 0.2867$ kJ/kg·K
5.3 $A: R = 0.2771$ kJ/kg·K;
    $c_p = 1.2008$ kJ/kg·K;
    $c_v = 0.9237$ kJ/kg·K

5.5 38.06 kPa
5.6 2.1763 m^3
5.8 $u_2 - u_1 = 64.96$ kJ/kg; $h_2 - h_1 = 90.94$ kJ/kg
5.10 $u_2 - u_1 = 8.948$ Btu/lbm; $h_2 - h_1 = 14.914$ Btu/lbm
5.12 0.1497 kJ/kg·K
5.14 (a) -0.3813 kJ/kg·K; (b) 0.6830
5.16 $c_p = 0.8647$ kJ/kg·K; $c_v = 0.5780$ kJ/kg·K
5.17 (a) 662.5°R; (b) 0.408 lbm
5.19 326.2 Btu/lbm
5.20 zero
5.21 115.1 lbm/s
5.25 (b) -0.3261 kJ/kg·K
5.29 (a) 475.1 K; (b) 473.5 K
5.31 (a) -397.3 kJ/kg; (b) -406.8 kJ/kg
5.32 (a) -0.1595 kJ/kg·K

6.1 6.0485 kJ/K
6.2 0.134 J/K
6.3 150 J
6.4 zero
6.9 zero
6.11 7.0072 kJ/kg·K
6.13 (a) 274.589 kJ/kg; (b) 251.931 kJ/kg
6.14 558.0 kJ
6.17 120.6 m^3
6.19 (a) 34.7 kJ/kg; (b) 34.7 kJ/kg
6.26 (a) 32.87 kJ/kg; (b) 475.3 kJ/kg
6.28 25890 kW
6.30 0.82 kJ/kg
6.32 (a) -4.425×10^6 Btu/h; (b) 2.605×10^6 Btu/h

7.1 193 h
7.3 (b) 466.7 kPa
7.5 (a) $p_2 = 233.3$ kPa; $T_2 = 328.51$ K
7.7 -1071 Btu
7.7 (a) 84.0 Btu/lbm; (b) 0.0491 Btu/lbm·°R
7.10 (a) 56.31 kJ/kg; (b) 39.84 kJ/kg
7.12 (b) -30.0 kJ/kg
7.15 (a) -21.11 kJ; (b) -146.4 kJ
7.17 (a) 386.8 K; (b) 785.8 K; (c) 0.9068 kJ/K
7.20 12500 Btu
7.22 (a) 542.8 K
7.24 (a) 313.531 kPa;

(b) 1.4424 m^3; (c) 8.797 kJ; (d) 779.9 kJ
7.26 (b) 200 kPa; (c) 0.0174 kJ/K
7.28 (a) -627.8 kJ; (b) 215.1 kJ
7.30 (a) -22.55 Btu/lbm; (b) 0.0536 Btu/lbm·°R
7.32 (a) 80.167 kJ; (b) 44.012 kJ
7.34 0.2527 Btu/lbm·°R; 0.1786 Btu/lbm·°R
7.36 (a) -167.2 kJ/kg; (b) -41.8 kJ/kg
7.37 (a) 1.674 MPa; (b) 754.92 kJ
7.39 (a) 78.78 kJ/kg; (b) -171.69 kJ/kg
7.41 (a) -50907.8 kJ; (b) 36385 kJ; (c) 14522.8 kJ

8.1 (a) -1.8×10^4 kW; (b) -1.97×10^4 kW
8.3 -254.9 hp
8.5 $W_{ideal} = -140$ kJ/kg
8.7 (a) -175.7 kJ/kg; (b) -175.9 kJ/kg
8.8 $W_{ideal} = 734.9$ kJ/kg
8.9 (a) 434.8 kJ/kg; (b) 437.8 kJ/kg
8.10 (a) -402.1 kJ/kg; (b) -409.7 kJ/kg
8.13 166.7 K
8.15 (a) 207.2 K; (b) 15.4 MPa
8.17 0.94
8.19 (a) 34.13 kg/h; (b) 0.759 kJ/K·h
8.21 353.5 lbm/h
8.23 (a) -0.4929 kW; (b) 35176 kJ/h; (c) 128.0 kJ/K·h
8.25 (a) 0.046 lbm; (b) 0.0108 Btu/°R
8.27 447.2 kPa
8.30 3.18 Btu
8.32 433.4 K
8.33 (a) 452.4 kJ/kg; (b) -452.4 kJ/kg
8.35 (a) 5.6423×10^7 kJ/h; (b) 8.0604×10^7 kJ/h; (c) 2.418×10^7 kJ/h; (d) 2.9243×10^7 kJ/h; (e) 1.10575×10^7 kJ/h
8.36 (a) 128.52 kJ/kg; (b) 471.487 kJ/kg
8.38 $W_{ideal} = -82.0$ Btu/lbm
8.39 184370 kJ

8.41 (a) -9767.0 kJ; (b) 5790 kJ
8.43 -10100 Btu
8.44 (a) 715°R; (b) 0.902 lbm

9.2 1398 kPa
9.5 1070 m/s
9.7 999.5 kPa
9.9 32.05 psia
9.13 1741 kPa
9.15 (a) 346 m/s; (b) 352 m/s; (c) 322 m/s
9.18 237°R
9.21 0.44; 397.2 K; 292.3 kPa; 32.26 kg/s
9.23 50.07°C
9.25 (a) 0.145 m^2; (b) 0.685
9.26 (a) 2.08; (b) 168.8 kPa
9.28 1127 m/s
9.29 4.53 kg/s
9.32 29.3 kg/s
9.34 0.664 in^2
9.36 (a) 7.42 cm^2; (b) 454.8 m/s
9.38 43.3 cm^2
9.41 greater
9.43 1790 kPa; 1687.4 m/s
9.45 (a) 2.51 lbm/s; (b) 1.655 in^2

10.1 167.64 kJ/kg
10.2 4626.3 kJ
10.5 396.48 kJ
10.7 (a) 248.7 kJ/kg; (b) 109.1 kJ/kg
10.8 (a) -98.7 kJ/kg; (b) 128.3 kJ/kg
10.10 (a) -49.36 Btu/lbm; (b) -33.6 Btu/lbm; (c) 15.76 Btu/lbm
10.11 (a) 92.905 kJ/kg; (b) 171.689 kJ/kg; (c) 78.784 kJ/kg
10.14 52.3% or 73.4%
10.16 76.4% or 62.8%
10.18 21.6%
10.20 19.8%
10.24 53.1%
10.29 277.14 kW
10.30 (a) 82.6%; (b) 84.6%
10.32 (a) 1.4611; (b) 179.88°C; (c) 11.6 kJ
10.34 (a) $\eta_I = 24.9\%$; (b) $(\eta_{II})^{cycle} = 86.3\%$ if we take credit for energy in turbine exhaust
10.35 (a) 381.8 kJ; (b) 242.4 kJ

11.1 714.6°C
11.3 474.8 kPa
11.5 39.3%
11.6 (a) 48.8%; (b) 45.9%;
(c) 15.0 lbm/kWh
11.7 (a) 42.7%; (b) 0.99%
11.8 1444 kPa
11.12 (a) 11.5%; (b) 14.8%;
(c) 213.4 kg/kWh;
(d) 3585 kg/h
11.14 (a) 2.025×10^8
11.16 For $p_{out} = 5.0$ kPa;
44.9% and 0.2599
11.17 For $T_{max} = 800°F$: 39.0%
and 0.2224
11.20 (a) 3642 kPa; (b) 0.0121;
(c) 0.5642; (d) 43.58%;
(e) 58.1%
11.23 (a) 464.2 K; (b) 303.2 K;
(c) 34.7%
11.26 192 kPa; 953 kPa; 3153
kPa; 8054 kPa
11.28 (a) 0.0116; (b) 0.602;
(c) 39.8%; (d) 53.0%
11.31 24.2%; 42.6%; 51.2%;
56.5%; 60.2%; 66.1%;
69.8%
11.33 (b) 58.2%;
(c) 199.4 kJ/kg
11.34 (b) 56.5%;
(c) 1634 kJ/kg;
(d) 1238 kPa
11.40 (b) 55.2%;
(c) 1921.3 kJ/kg;
(d) 1333 kPa
11.44 (a) 2.171; (b) 59.5%;
(c) 113 psi
11.45 (a) cycle B; (b) cycle A
11.49 98.7, 111.6, 116.3, 117.7,
117.2, 115.9 Btu/lbm
11.51 31.6%; 35.2%; 34.9%;
25.7%
11.55 (a) 15; (b) 0.723;
(c) 33.6%; (d) 0.484
11.57 (a) 7.3; (b) 0.596;
(c) 30.0%; (d) 0.563
11.59 44.8%
11.61 68.0, 61.0, 56.3, 52.5, 49.4,
46.7%
11.63 (a) 48.4%; (b) 61.8%
11.67 (a) 37.3; (b) 29.0%;
(c) 50.4%
11.71 (a) 60.1%; (b) 0.6364;
(c) 37.7%
11.73 (a) 66.8%; (b) 0.5619;
(c) 39.2%

12.1 (a) 1.25; (b) 28.44; (c) 5.0

12.5 (a) 21.9 kW; (b) 3.211;
(c) 40.54 kg/min
12.11 (a) 11.8 kW; (b) 2.976;
(c) 20.6 kg/min
12.13 (a) 3.13;
(b) 45.1 Btu/lbm
12.15 (a) 49.65 kW; (b) 3.021;
(c) 89.32 kg/min
12.17 (a) 1.92;
(b) 40.082 Btu/lbm
12.19 36.5 kJ/s
12.22 (a) 2.712; (b) 1.2964 kW;
(c) 3.41kg/min
12.24 (a) 5.27;
(b) 11.06Btu/lbm
12.26 (a) 0.900; (b) 3.907 kW;
(c) 4.50 kg/min
12.28 (a) 0.923; (b) 3.809 kW;
(c) 8.821 kg/min
12.30 (a) 0.845; (b) 4.161 kW;
(c) 4.338 kg/min
12.32 (a) 0.810; (b) 4.342 kW;
(c) 4.439 kg/min
12.34 (a) 0.405; (b) 8.688 kW;
(c) 4.836 kg/min
12.37 (a) 0.180; (b) 19.544 kW;
(c) 12.975 kg/min
12.42 (a) 9385.2 kW;
(b) 13.927; (c) 569.02 kg/s
12.45 (a) 17712 kW; (b) 4.234;
(c) 587.34 kg/s; (d) 29.7%

13.13 326.7 K
13.19 −0.0307 Btu/lbm·°R
13.22 (a) 4930 ft/s;
(b) 3.88×10^5 ft/s
13.23 (a) −3.27 Btu/lbm;
(b) −0.98 Btu/lbm
13.27 (a) 62.5°C; (b) 90.99 kJ
13.28 (a) 90.34 kg; (b) 67250 kJ
13.29 −1.5°C
13.31 1.484
13.33 2057.1 kJ/kg
13.36 2.099 psia
13.43 (a) 491.6 kg;
(b) 506.8 kg; (c) 509.0 kg
13.48 (a) 779.1 kJ/kg;
(b) 781.3 kJ/kg
13.51 (a) 24.3 MPa;
(b) 23.0 MPa
13.55 (a) −43.942 kJ/kg
mol·K; (b) −47.082
kJ/kg mol·K

14.1 22.41 m³
14.3 (a) 0.2878 CO_2,
0.3165 CH_4, 0.3957 O_2;
(b) 0.2736 kJ/kg·K

14.6 (a) 0.7458 N_2, 0.2026 CO_2,
0.0442 O_2; 0.0074 CO;
(b) 0.2734 kJ/kg·K
14.8 (a) 81.463 m³;
(b) 25.77, 11.78, 82.46
kPa
14.10 2.4242 kg
14.12 0.9944, 0.7210 kJ/kg·K
14.15 (a) 300.15 K; 714.2 kPa;
(b) 1.4431 kJ/K
14.19 −89.50 kJ/kg;
−125.61 kJ/kg;
−0.4012 kJ/kg·K
14.21 358.1 kJ
14.24 2.7 and 10.5 kg
14.27 (a) 2493.1 kJ;
(b) 2611.5 kJ
14.29 −2.2224 kJ/K
14.30 no
14.32 (a) 16.8°C; (b) 15.55%
14.34 22.64 psia
14.37 (a) 3.166 kPa; (b) 0.0041
kg/kg of nitrogen
14.39 (a) 0.02333 lbm;
(b) 14.9 psia
14.40 (a) 47.38%; (b) 0.009346
kg/kg of dry air; (c)
1.1763 kg
14.43 (a) 43.05%;
(b) 10.21 kJ/kg of dry air
14.46 (a) 65.2°C; (b) 3.89%
14.49 3.32%
14.52 72.91%
14.56 (a) 14.47%;
(b) 501.7 kJ/s
14.58 0.005890 kg/kg of dry
air; 49.7%; 16.66°C
14.60 (a) 55.62%;
(b) 0.004 kg/kg of dry air

15.1 (a) 21.5 kg; (b) 54.9°C
15.3 14.47
15.5 104°F
15.7 (a) 0.09565 CO_2, 0.16738
H_2O, 0.01740 O_2, 0.71957
N_2;
(b) 0.11487 CO_2, 0.02090
O_2, 0.86423 N_2
15.9 (a) 12.285 kg;
(b) 12.582 m³
15.11 11.02; 16.8
15.12 −49123 kJ/kg
15.16 2017 K
15.18 63.8
15.19 2090°R
15.21 302%
15.24 20300 Btu/lbm of fuel
15.28 (a) −17709 Btu/lbmol

15.30 (a) nil; (b) nil

15.34 e^{323}

15.38 (a) 0.66 CO, 0.66 H_2O, 0.34 CO_2, 0.34 H_2

15.41 9.448×10^7 J/kg mol

15.43 (a) 0.11619 CO_2, 0.00009 CO, 0.00004 O_2, 0.15503 H_2O, 0.72865 N_2; (b) -21478 kJ

15.47 2460 K

15.49 91.9%

16.1 (a) 737.9 kJ/kg; (b) 1165.0 kJ/kg

16.4 (a) 3.607×10^5 J/kg;
(b) 1.864×10^6 J/kg

16.6 0.246

16.8 7.42×10^7 kPa

16.12 0.044

16.14 0.021

16.16 0.104

16.18 (a) 164.2 K; (b) 102.81; (c) 311.61

16.20 (a) 179.2 K; (b) 69.48; (c) 247.92

16.24 0.0408, 0.0784, 0.1090, 0.1320, 0.1477

16.26 no

16.28 0.0444, 0.06138, 0.07420, 0.08876, 0.9082, 0.08856, 0.08009

16.30 0.0570, 0.0825, 0.1044, 0.1365, 0.1545, 0.1624, 0.1638

16.32 12.9%

16.34 13.4%

16.36 21.6%

16.38 3.356×10^7 J

16.40 (a) 16.3 kWh; (b) $0.49

16.42 (a) 1.214×10^8 J;
(b) 1.307×10^8 J;
(c) 1.196×10^8 J

16.45 1120700 kJ/kg mol of fuel

Index